$Ax = b$

(213) 386

$$\frac{x}{A} = B$$

= cancel

$A + b = C$

ELEMENTARY ALGEBRA

ELEMENTARY ALGEBRA

SECOND EDITION

Jack Barker James Rogers James Van Dyke

Portland Community College
Portland, Oregon

SAUNDERS COLLEGE PUBLISHING
A Harcourt Brace Jovanovich College Publisher

FORT WORTH PHILADELPHIA SAN DIEGO
NEW YORK ORLANDO AUSTIN SAN ANTONIO
TORONTO MONTREAL LONDON SYDNEY TOKYO

Text Typeface • *Times Roman*
Compositor • *York Graphic Services, Inc.*
Acquisitions Editor • *Robert B. Stern*
Developmental Editor • *Ellen Newman*
Managing Editor • *Carol Field*
Copy Editor • *York Production Services*
Manager of Art and Design • *Carol Bleistine*
Art Creation • *York Production Services*
Cover Designer • *Lawrence R. Didona*
Director of EDP • *Tim Frelick*
Production Manager • *Bob Butler*
Marketing Manager • *Monica Wilson*
Cover Credit • *Circle, Isoceles Triangle, Rectangle and Square,*
© 1991 Bishop/PHOTOTAKE, NYC

Printed in the United States of America

ELEMENTARY ALGEBRA, second edition

ISBN 0-03-072881-9

Library of Congress Catalog Card Number: 91-46521

234 039 987654321

To our children:

Ken, Norm, and Linda Barker
John, Heather, and Eddie Fincher, Paul, Patty, and Pamela Hurst, Terry, Becky, and Perry Washington, Jim, Michelle, and Tyler Raible
Dan, Claudia, Avalon, Tom, Larry, Karla, Greg, and Ann Van Dyke

PREFACE

Elementary Algebra is a text in beginning algebra that serves as an introduction to algebra or as a review course. It is suitable for students at any level who need a solid foundation in the basic skills of algebra in order to fulfill competency requirements, achieve adequate scores on placement exams, prepare for technical mathematics courses, prepare for problem-solving in applied courses, or complete prerequisites for intermediate or college algebra.

Changes in the Second Edition The second edition contains the following new features:

- Each chapter begins with a Preview to introduce the student to the material discussed in the chapter.
- The text of each section has been rewritten so that real-life applications and examples with solutions are integrated into the discussion.
- The step-by-step strategies printed in blue to the right of the example have been expanded and clarified to give students more help in solving problems.
- A four-color system is used to highlight the various pedagogical features of the text, such as definitions, cautions, procedures, properties, and formulas. Color is also used to distinguish between two lines that are graphed simultaneously. The complete color system is described in more detail on p. xxiii.
- Exercises have been completely revised, and all problems have been checked by two accuracy reviewers. Every section of the text now contains State Your Understanding and Challenge Exercises. Maintain Your Skills Exercises are now keyed to sections of the text.
- Each chapter concludes with a Chapter Summary, Chapter Review Exercises (keyed to section number and objective), a True–False Concept Review, and a Chapter Test. In addition, there are Cumulative Reviews following Chapters 4 and 8.
- Particular attention has been paid to the testing requirements for various states (e.g., ELM, TASP, CLAST, etc.). Please see pages xv–xviii for additional information on how the text meets the requirements for these states.

A complete pedagogical system that is designed to motivate the student and make mathematics more accessible includes the following features:

Chapter Preview Each chapter begins with a Preview to introduce the student to the material and show how it is integrated with the study of algebra.

Objectives Objectives are identified at the beginning of each section to help students focus on skills to be learned.

Examples with Solutions Examples are used to illustrate the concepts explained within the section. As each example is worked out, there is a step-by-step explanation that expands on the procedures and the thinking necessary to work the problem. The examples also illustrate shortcuts. Where applicable, caution comments about common errors are included, highlighted in red for the student.

Calculator Examples Some examples contain problems solved by using a calculator. These examples, set off with a calculator symbol, demonstrate how to use a calculator and signal to the student that these problems are suited for calculator practice. However, the use of the calculator is left to the discretion of the instructor and/or the student. Nowhere is the use of the calculator required, and all sections of the text can be studied without a calculator.

Pedagogical Use of Color This text uses color to highlight and distinguish definitions, caution comments, procedures, properties, and formulas. Color is also used to distinguish between two lines that are graphed simultaneously. The complete color system is described in more detail on page xxiii.

EXERCISES

The second edition of *Elementary Algebra* continues to organize the exercises in terms of difficulty. As before, "A" indicates that the problems are relatively easy; the "A" problems have been increased in number and can be used as class or "oral" exercises. The "B" problems may require paper and pencil. The "C" problems are more difficult and in some instances offer a challenge for the advanced students. At the discretion of the instructor, the "C" problems may provide calculator drill for the students.

Applications Whenever appropriate, the "D" problems (word problems) have been included so that the student constantly practices translating word phrases and statements into mathematical equations. The word problems are realistic and are taken from geometry, science, business, and economics. The structure of problem solving is emphasized throughout the text, beginning with the simple conversion of word phrases into mathematical expressions.

State Your Understanding This set of exercises requires the students to explain their answer in words, thus addressing the trend of writing across the curriculum.

Challenge Exercises These exercises appear near the end of each section and provide problems for the more capable students to solve.

Maintain Your Skills The purpose of this portion of the exercises is to review material previously covered. The exercises in each Maintain Your Skills section have been referenced to the section or sections reviewed.

In all exercise sets, an effort has been made to pair the problems so that the set of odd-numbered problems is equivalent in kind and difficulty to the set of even-numbered problems. Answers to the odd-numbered exercises are provided in the back of the text.

Problem Solving One of the most important features of the new edition is the emphasis on problem solving because students need to learn how to read a word problem and then think about how to obtain the solution. We have therefore included several learning aids to help the student with the problem-solving process. In each example section a strategy solution column contains a step-by-step explanation to provide a good model for working the many different exercises in the text. The procedure box, another new pedagogical device, provides instructions on how to solve a specific kind of problem, for example, how to factor a perfect square trinomial. In addition, we have also included exercises at the end of each section entitled "State Your Understanding" where the student is asked to explain how the problem is solved rather than give a numerical answer. Many educators now agree that if students can write and think about a problem clearly, they most likely understand the concepts involved.

Within the text itself we have devoted an entire section to solving word problems. This material is found in Chapter 3 on p. 192. A flow chart shows how to translate words into mathematical equations, along with a complete explanation of how the process works. The strategy solution column following each example is broken down into more manageable pieces for solution.

Chapter Summary Each chapter concludes with a comprehensive review, which includes definitions and procedures learned in the chapter. Each item is keyed to the page number where it is presented in the text for easy reference.

Chapter Review Exercises These exercises are included at the end of the chapter and review the objectives to ensure that the student has attained a desired level of proficiency and is comfortable proceeding to the next chapter. All exercises are keyed to sections and objectives so the student can refer to the text for assistance. Answers to the odd-numbered exercises can be found in the back of the book.

True–False Concept Review Serves as a check on the students' understanding of the new concepts presented in the chapter. Answers to all of the questions appear in the back of the book.

Chapter Test Each chapter concludes with a Chapter Test as a final review of student comprehension. The test contains representative problems from the entire chapter; answers to all of the test items appear in the back of the book.

Cumulative Review Exercises After Chapters 4 and 8 students can work the Cumulative Review Exercises to help reinforce concepts and check retention. Answers to all test exercises appear in the back of the book.

Timetable The text is intended to be used in a one-quarter or one-semester course. In a Math Lab or Learning Center the time lines are variable. Other variations are possible, but preparation for more difficult algebra courses is the intent.

State Requirements Particular attention has been paid to the testing requirements for various states (e.g., TASP, ELM, CLAST, etc.). Please see pages xv–xviii for additional information on how this text meets requirements for your state.

CONTENT OVERVIEW

In **Chapter 1** the operations of addition, subtraction, multiplication, and division of positive rational numbers are reviewed. The basic properties of operations that are used in both arithmetic and algebra—commutative properties, associative properties, distributive property, multiplication properties of one and zero, and the reciprocal property are reviewed.

The study of algebra begins with the introduction of exponents and algebraic logic through order of operations. This leads to the uses of variables and expressions containing variables. This "language of algebra" is reinforced by translating from English to algebra and vice versa. Elementary equations are introduced. Solutions of these equations and other equations play a major role throughout the text. The study of numbers includes the set of real numbers. Absolute value leads to the rules for operations on signed numbers. The chapter concludes by checking equations with positive and negative numbers and by reviewing the order of operations.

Chapter 2 introduces the solution of equations. A great many of the practical applications of algebra involve solving equations and inequalities. In Sections 2.1 and 2.2, we examine the properties of equality that are used to solve linear equations. Throughout the text, you will find a variety of equations and their applications. Section 2.3 exhibits equations that require more than one property to solve. Section 2.4 shows how to simplify expressions. Combining terms in equations occurs in Section 2.5. Percents in Section 2.6 model one of the most common uses of equations in business. Finally, in Section 2.7, inequalities and ways to show their solution, including graphing are studied.

Chapter 3 is devoted to the algebra of polynomials. Operations of addition, subtraction, multiplication, and division are covered. Solutions of equations containing polynomials are covered at each appropriate opportunity. The FOIL method of multiplying binomials is stressed and special products are identified. The associated patterns are used to "shortcut" the usual multiplication process.

Since polynomials play such a major role in the study of algebra, mastery of the skills presented in this chapter is necessary to be successful in the following chapters.

Chapter 4 leads off with the zero-product property to allow the solution of equations solved by factoring. A thorough discussion of factoring follows. At the completion of the chapter, you will be able to factor a trinomial with integer coefficients or know that it is prime. Also included are special polynomials: differences of squares, perfect square trinomial, and the sum and differences of two cubes. The final section mixes up all the types of factoring to provide practice in selecting the correct method.

Factoring is used in Chapter 5 to simplify rational expressions and in Chapter 8 to solve other quadratic equations.

In **Chapter 5** algebraic expressions that contain fractions are covered, for the most part, using properties parallel to those for rational numbers. In Sections 5.1 and 5.2, we examine the properties that are used to reduce and multiply rational expressions. Section 5.3 shows how to find the least common multiple of two or more rational expressions. A comparison shows that the method used here is a generalization of the method used for finding common multiples of sets of whole numbers. Section 5.4 establishes that rational expressions are combined in a manner akin to that for rational numbers. Ratios and proportions in Section 5.6 model one of the most common uses of equations that contain rational expressions. Lastly, in Section

5.7, variation is investigated as an important application of equations incorporating products and quotients.

In **Chapter 6** the solution of equations in two variables with solutions written as ordered pairs are introduced. The rectangular coordinate system is presented so that these equations can be graphed. Properties of slope and intercept are discussed. The slope and the y-intercept are used to draw graphs of linear equations. Linear inequalities follow naturally and utilize the concept of half-planes. We then solve systems of linear equations by graphing, substitution, and linear combinations. Less emphasis is placed on solutions by graphing since nonintegral solutions can only be estimated by this method.

In **Chapter 7** it is noted that the inverse of squaring a number is the operation of finding the square root of a number. In Section 7.1, a radical sign is used to designate the positive square root of a positive number. This leads to a consideration of numbers that are not rational and to the mention of the existence of numbers that are not real numbers. Section 7.2 illustrates the procedure for simplifying radical expressions, and this leads to the operations on radical expressions in Sections 7.3, 7.4, and 7.5. Section 7.5 concludes the chapter with an explanation of the use of the squaring property of equality to solve radical equations and the necessity of checking for extraneous roots.

Chapter 8 continues the solution of quadratic equations. Progression is made from taking the square root of both sides of an equation to completing the square to the quadratic formula. The section on special cases includes the Pythagorean theorem for solving right triangles. Complex numbers are introduced to make it possible to solve every quadratic equation. Also introduced are graphs of quadratic equations, parabolas, to prepare for the study of functions in later courses.

A new **Appendix** on **Sets** is included for those instructors who wish to cover sets because of interest or state requirements. In that appendix, you will find complete coverage of sets, including the complement, union, and intersection of the set, along with examples for the student to study and problems for the student to work.

ANCILLARY ITEMS

The following supplements are available to accompany this text:

Instructor's Resource Manual This manual contains complete, worked-out solutions to all of the problems in the text to complement this Annotated Instructor's Edition, which contains answers to all exercises in red next to the problem and teaching tips in the margin.

Instructor's Testing Manual This manual contains six written tests for each chapter, a final examination, and a Diagnostic Test. Half of the tests have multiple-choice questions and half use open-ended questions to provide flexibility in testing. Also included in this manual is a Printed Test Bank containing multiple-choice tests generated from the Computerized Test Bank. These questions are keyed to the section number and objective of the text, and are graded in difficulty.

ExaMaster™ Computerized Test Bank Available for Apple II, IBM, and Macintosh computers, this test bank contains over 2,000 questions, representing every section of the text. Keyed to section number, objective, and level of difficulty, these questions can be combined to create customized tests. The instructor can add and edit questions, and separate grading keys are provided.

MathCue Interactive Software This program disk contains practice problems from each section of the textbook. Using an interactive approach, the software provides students with an alternate way to learn the material and, at the same time, provides the student with more individualized attention. The program will automatically advance to the next level of difficulty once the student has successfully solved a few problems; the student may also ask to see the solution to check his or her understanding of the process used. The software is keyed to the textbook and will refer the student to the appropriate section of the text if an incorrect answer is input. A useful tool to check skills and to identify and correct any difficulties in finding solutions, this software is available for the Apple II, Macintosh and IBM PC microcomputers.

MathCue Solution Finder Software Available for IBM and Macintosh, this software allows students to input their own questions through the use of an expert system, a branch of artificial intelligence. Students may check their answers or receive help as if they were working with a tutor. The software will refer the student to the appropriate section of the text and will record the number of problems entered and number of correct answers given. Featuring a function grapher, the students can zoom in and out, evaluate a function at a point, graph up to four functions simultaneously, and save and retrieve function setups via disk files.

Videotapes A complete set of videotapes (19 hours) is available to give added assistance or to serve as a quick review of the book. The tapes review problem-solving methods and guide students through practice problems; students can stop the tape to work the problems and begin it again to check their solutions. Keyed to the text and providing coverage of every section of the book, these tapes offer another approach to mastery of the given topic.

Student Solutions Manual This manual contains worked-out solutions to one quarter of the problems in the exercise sets (every other odd-numbered problem) to help the student learn and practice the techniques used in solving problems.

ACKNOWLEDGMENTS

The authors appreciate the unfailing patience and continuing support of their wives, Mary Barker, Elinore Rogers, and Carol Van Dyke, who made the completion of this work possible. Thanks go to Bob Malena of Community College of Allegheny County for his help in revising the exercises. We also thank our colleagues for their help and suggestions for the improved second edition.

We are grateful to Bob Stern and Ellen Newman of Saunders College Publishing for their suggestions during the preparation of the text and to Kirsten Kauffman of York Production Services. We would also like to express our gratitude to the following reviewers for their many excellent contributions to the development of the text:

Helen Burrier, Kirkwood Community College

Arthur Dull, Diablo Valley College

Robert B. Eicken, Illinois Central College

Catherine Hess, Anne Arundel Community College

Linda B. Holden, Indiana University

Mary T. Long, Erie Community College

Carol Metz, Westchester Community College

Barbara Jane Sparks, Camden County College

Special thanks to Kathy Butts-Bernunzio of Portland Community College and to John R. Martin of Tarrant County Junior College for their accuracy reviews of all the problems and exercises in the text.

We would also like to thank the following people for their excellent work on the various ancillary items that accompany Elementary Algebra:

John R. Martin, Tarrant County Junior College (Instructor's Resource Manual and Student Solutions Manual)

Mark Serebransky, Camden County College and John Garlow, Tarrant County Junior College (Instructor's Testing Manual)

George W. Bergeman, Northern Virginia Community College (MathCue Interactive Software and MathCue Solution Finder Software)

Bob Finnell and Hollis Adams, Portland Community College (Videotapes)

John Garlow, Tarrant County Junior College (Computerized Test Bank)

Jack Barker
Jim Rogers
Jim Van Dyke

ELM MATHEMATICAL SKILLS

The following table lists the California ELM MATHEMATICAL SKILLS and where coverage of these skills can be found in the text. Skills not covered in this text can be found in Fundamentals of Mathematics, 5th Edition, or Intermediate Algebra, 2nd Edition. Location of skills are indicated by text section or chapter.

Skill	Location in Text
Whole numbers and their operations	1.1
Fractions and their operations	1.1
Decimals and their operations	1.1
Exponentiation and square roots	1.2, 7.1
Applications (averages, percents, word problems)	2.6, In applications throughout the text
Simplifying polynomials by grouping (1 and 2 variables)	3.5
Evaluating polynomials (1 and 2 variables)	1.2
Addition and subtraction of polynomials	3.5
Multiplication of a polynomial by a monomial	3.7
Multiplying two binomials	3.7
Squaring a binomial	3.8
Divide a polynomial by a monomial (no remainder)	3.9
Divide a polynomial by a linear binomial	3.9
Factor out the GCF from a polynomial	4.2
Factor a trinomial	4.3, 4.4, 4.5, 4.6, 4.7
Factor the difference of squares	4.5
Simplify a rational expression by cancelling common factors	5.1
Evaluate a rational expression (1 or 2 variables)	5.3, 5.4
Addition and subtraction of rational expressions	5.4
Multiplication and division of rational expressions	5.1, 5.2
Simplification of complex fractions	5.5
Positive exponents	1.5
Laws of exponents (positive)	3.1
Simplifying an expression (+ exponents)	3.1
Integer exponents	3.2
Laws of exponents (integers)	3.2
Simplifying an expression (integer exponents)	3.2
Scientific notation	3.3
Radical sign (square roots)	7.1
Simplify products under a radical	7.2
Addition and subtraction of radicals	7.5
Multiplication of radicals	7.3
Solving radical equations	7.6
Solving linear equations, one variable, numerical coefficients	2.1, 2.2, 2.3, 2.6
Solving linear equations, one variable, literal coefficients	2.1, 2.2, 2.3, 2.6
Ratio, proportion and variance	5.6, 5.7
Solving equations reducible to linear	5.3, 5.4, 5.5, 7.6
Solving linear inequalities: 1 variable numerical coefficients	2.7
Solving 2 equations, 2 unknowns, numerical coefficients by substitution	6.6

Skill	Location in Text
Solving 2 equations, 2 unknowns, numerical coefficients by elimination	6.7
Solving quadratic equations from factors	4.1
Solving quadratic equations by factoring	4.2, 4.3, 4.4, 4.5, 4.6, 4.7, 8.1
Graphing points on number lines	2.7
Graphing linear inequalities (one unknown)	2.7
Graphing points in the coordinate plane	6.2
Graphing linear equations: $y = mx$, $y = b$, $x = b$	6.2, 6.3

TASP MATHEMATICS SKILLS

The following table lists the Texas TASP MATHEMATICS SKILLS and where coverage of these skills can be found in the text. Skills not covered in this text can be found in Fundamentals of Mathematics, 5th Edition, or Intermediate Algebra, 2nd Edition. Location of skills are indicated by text section or chapter.

Skill	Location in Text
Use number concepts and computation skills.	Ch. 1, 2.7, 3.2, 3.3
Solve word problems involving integers, fractions, or decimals (including percents, ratios, and proportions).	2.3, 5.6, 5.7
Graph numbers or number relationships.	Ch. 6
Solve one- and two-variable equations.	2.1, 2.2, 2.3, 2.4, 6.1, 6.5, 6.6, 6.7
Solve word problems involving one and two variables.	Ch. 2, 3.4, Ch. 6, In applications throughout the text
Understand operations with algebraic expressions.	Ch. 3, Ch. 4, Ch. 5
Solve problems involving quadratic equations.	Ch. 4, Ch. 8
Solve problems involving geometric figures.	In applications throughout the text

CLAST MATHEMATICAL SKILLS

The following table lists the Florida CLAST MATHEMATICAL SKILLS and where coverage of these skills can be found in the text. Skills not covered in this text can be found in Fundamentals of Mathematics, 5th Edition, or Intermediate Algebra, 2nd Edition. Location of skills are indicated by text section or chapter.

Skill	Location in Text
1A1a—Adds and subtracts rational numbers	1.1
1A1b—Multiplies and divides rational numbers	1.1
1A2a—Adds and subtracts rational numbers in decimal form	1.1
1A2b—Multiplies and divides rational numbers in decimal form	1.1
1A3 —Calculates percent increase and percent decrease	2.6
2A1 —Recognizes the meaning of exponents	1.2, 3.1, 3.2, 3.3
2A4 —Determines the order-relation between magnitudes	2.7
4A3 —Solves real-world problems which do not require the use of variables and which do require the use of percent	2.3
4B1 —Solves real-world problems involving perimeters, areas, and volumes of geometric figures	In applications throughout the text
4B2 —Solves real-world problems involving the Pythagorean property	8.1
1C1a—Adds and subtracts real numbers	1.5, 1.6
1C1b—Multiplies and divides real numbers	1.7, 1.8
1C2 —Applies the order-of-operations agreement to computations involving numbers and variables	1.9
1C3 —Uses scientific notation in calculations involving very large or very small numbers	3.3
1C4 —Solves linear equations and inequalities	2.1, 2.2, 2.3, 2.4, 2.7
1C5 —Uses given formulas to compute results when geometric measurements are not involved	In applications throughout the text
1C7 —Factors a quadratic expression	4.2, 4.3, 4.4, 4.5, 4.6, 4.7
1C8 —Finds the roots of a quadratic equation	4.1, 4.2, 4.3, 4.4, 4.5, 4.6, 4.7, Ch. 8
2C1 —Recognizes and uses properties of operations	1.1
2C2 —Determines whether a particular number is among the solutions of a given equation or inequality	Ch. 1
2C3 —Recognizes statements and conditions of proportionality and variation	5.6, 5.7
2C4 —Recognizes regions of the coordinate plane which correspond to specific conditions	6.2, 6.3, 6.4, 6.5
3C1 —Infers simple relations among variables	1.3, 5.7, 6.3
3C2 —Selects applicable properties for solving equations and inequalities	Throughout the text
4C1 —Solves real-world problems inviting the use of variables, aside from commonly used geometric formulas	In applications throughout the text

CONTENTS

CHAPTER 1 BASIC CONCEPTS AND PROPERTIES OF ALGEBRA 1

CHAPTER 2 LINEAR EQUATIONS AND INEQUALITIES 91

CHAPTER 3 ALGEBRA OF POLYNOMIALS AND RELATED EQUATIONS 149

CHAPTER 4 FACTORING POLYNOMIALS AND RELATED EQUATIONS 234

CHAPTER 5 RATIONAL EXPRESSIONS AND RELATED EQUATIONS 296

CHAPTER 6 LINEAR EQUATIONS, SYSTEMS AND GRAPHS 372

CHAPTER 7 ROOTS, RADICALS, AND RELATED EQUATIONS 470

CHAPTER 8 QUADRATIC EQUATIONS 521

PEDAGOGICAL USE OF COLOR

The various colors in the text figures are used to improve clarity and understanding. Any figures with three-dimensional representations are shown in various colors to make them as realistic as possible. Color is used in those graphs where different lines are being plotted simulanteously and need to be distinguished.

In addition to the use of color in the figures, the pedagogical system in the text has been enhanced with color as well. We have used the following colors to distinguish the various pedagogical features:

■ PROPERTY

■ DEFINITION

■ PROCEDURE

■ CAUTION

RULE

FORMULA

Table Title

Table Head

INTRODUCING THE BOOK

Previews introduce the student to the material and show its relevance to the study of algebra.

Objectives are identified at the beginning of each section to help students focus on skills to be learned.

Definitions are highlighted with blue-green.

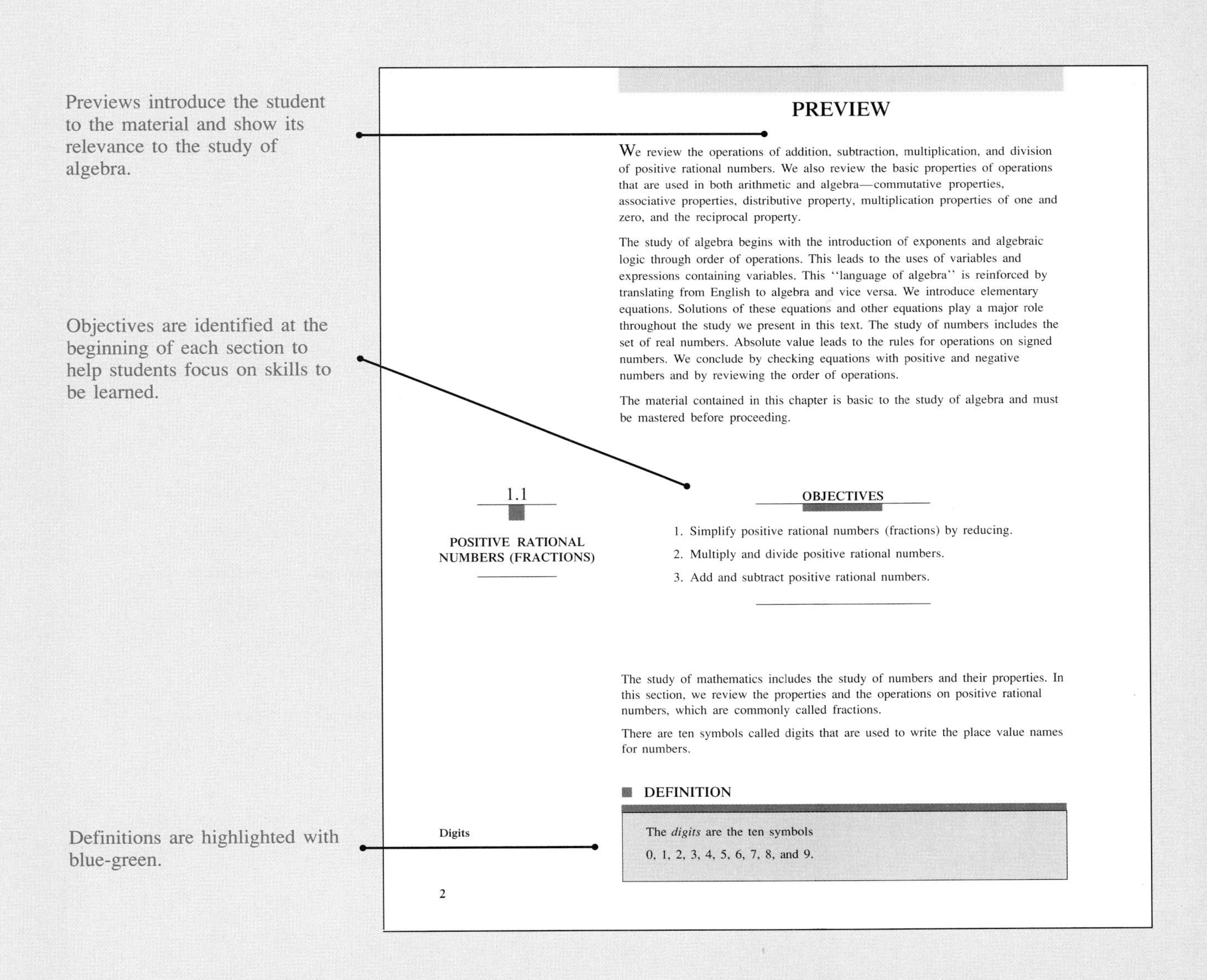

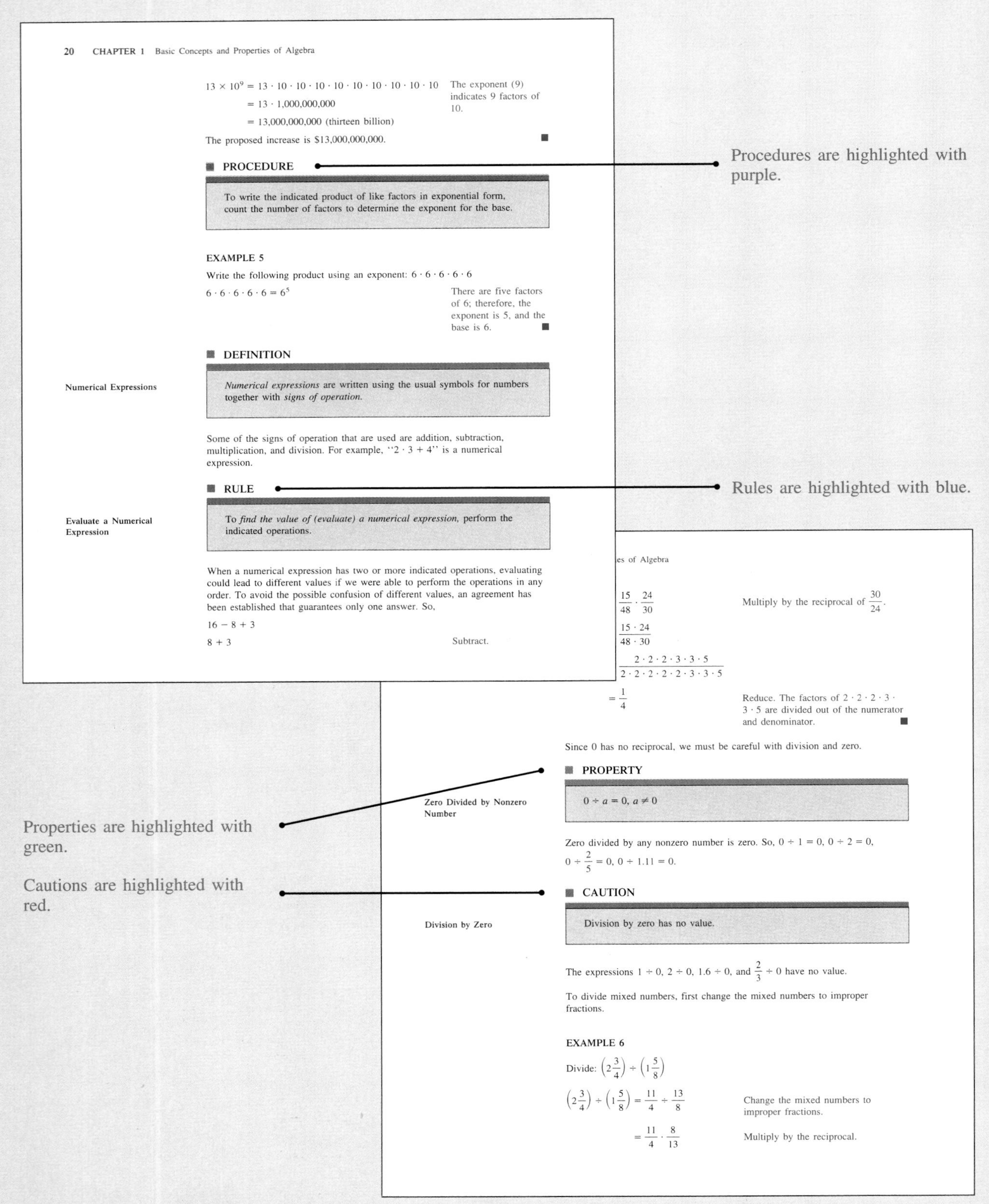

20 CHAPTER 1 Basic Concepts and Properties of Algebra

$13 \times 10^9 = 13 \cdot 10 \cdot 10 \cdot 10 \cdot 10 \cdot 10 \cdot 10 \cdot 10 \cdot 10 \cdot 10$ The exponent (9) indicates 9 factors of 10.

$= 13 \cdot 1{,}000{,}000{,}000$

$= 13{,}000{,}000{,}000$ (thirteen billion)

The proposed increase is $13,000,000,000. ■

■ PROCEDURE

To write the indicated product of like factors in exponential form, count the number of factors to determine the exponent for the base.

EXAMPLE 5

Write the following product using an exponent: $6 \cdot 6 \cdot 6 \cdot 6 \cdot 6$

$6 \cdot 6 \cdot 6 \cdot 6 \cdot 6 = 6^5$ There are five factors of 6; therefore, the exponent is 5, and the base is 6. ■

■ DEFINITION

Numerical Expressions

Numerical expressions are written using the usual symbols for numbers together with *signs of operation.*

Some of the signs of operation that are used are addition, subtraction, multiplication, and division. For example, "$2 \cdot 3 + 4$" is a numerical expression.

■ RULE

Evaluate a Numerical Expression

To *find the value of (evaluate) a numerical expression,* perform the indicated operations.

When a numerical expression has two or more indicated operations, evaluating could lead to different values if we were able to perform the operations in any order. To avoid the possible confusion of different values, an agreement has been established that guarantees only one answer. So,

$16 - 8 + 3$

$8 + 3$ Subtract.

es of Algebra

$\frac{15}{48} \cdot \frac{24}{30}$ Multiply by the reciprocal of $\frac{30}{24}$.

$\frac{15 \cdot 24}{48 \cdot 30}$

$\frac{2 \cdot 2 \cdot 2 \cdot 3 \cdot 3 \cdot 5}{2 \cdot 2 \cdot 2 \cdot 2 \cdot 2 \cdot 3 \cdot 3 \cdot 5}$

$= \frac{1}{4}$ Reduce. The factors of $2 \cdot 2 \cdot 2 \cdot 3 \cdot 3 \cdot 5$ are divided out of the numerator and denominator. ■

Since 0 has no reciprocal, we must be careful with division and zero.

■ PROPERTY

Zero Divided by Nonzero Number

$0 \div a = 0, a \neq 0$

Zero divided by any nonzero number is zero. So, $0 \div 1 = 0$, $0 \div 2 = 0$, $0 \div \frac{2}{5} = 0$, $0 \div 1.11 = 0$.

■ CAUTION

Division by Zero

Division by zero has no value.

The expressions $1 \div 0$, $2 \div 0$, $1.6 \div 0$, and $\frac{2}{3} \div 0$ have no value.

To divide mixed numbers, first change the mixed numbers to improper fractions.

EXAMPLE 6

Divide: $\left(2\frac{3}{4}\right) \div \left(1\frac{5}{8}\right)$

$\left(2\frac{3}{4}\right) \div \left(1\frac{5}{8}\right) = \frac{11}{4} \div \frac{13}{8}$ Change the mixed numbers to improper fractions.

$= \frac{11}{4} \cdot \frac{8}{13}$ Multiply by the reciprocal.

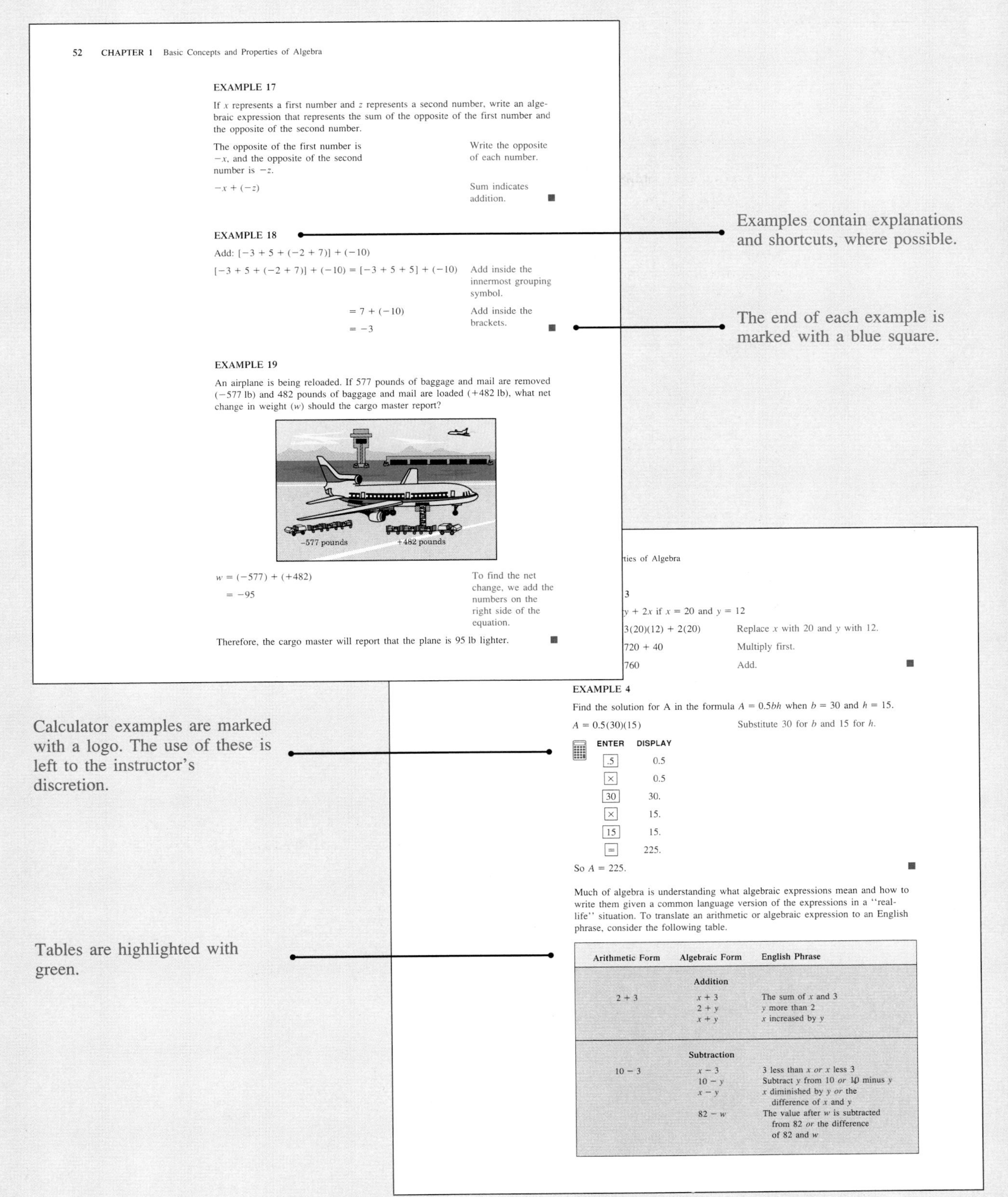

52 CHAPTER 1 Basic Concepts and Properties of Algebra

EXAMPLE 17

If x represents a first number and z represents a second number, write an algebraic expression that represents the sum of the opposite of the first number and the opposite of the second number.

The opposite of the first number is $-x$, and the opposite of the second number is $-z$.	Write the opposite of each number.
$-x + (-z)$	Sum indicates addition. ■

EXAMPLE 18

Add: $[-3 + 5 + (-2 + 7)] + (-10)$

$[-3 + 5 + (-2 + 7)] + (-10) = [-3 + 5 + 5] + (-10)$	Add inside the innermost grouping symbol.
$= 7 + (-10)$	Add inside the brackets.
$= -3$	■

EXAMPLE 19

An airplane is being reloaded. If 577 pounds of baggage and mail are removed (-577 lb) and 482 pounds of baggage and mail are loaded ($+482$ lb), what net change in weight (w) should the cargo master report?

$w = (-577) + (+482)$	To find the net change, we add the numbers on the right side of the equation.
$= -95$	

Therefore, the cargo master will report that the plane is 95 lb lighter. ■

ties of Algebra

3

$y + 2x$ if $x = 20$ and $y = 12$

$3(20)(12) + 2(20)$	Replace x with 20 and y with 12.
$720 + 40$	Multiply first.
760	Add. ■

EXAMPLE 4

Find the solution for A in the formula $A = 0.5bh$ when $b = 30$ and $h = 15$.

$A = 0.5(30)(15)$ Substitute 30 for b and 15 for h.

ENTER	DISPLAY
.5	0.5
×	0.5
30	30.
×	15.
15	15.
=	225.

So $A = 225$. ■

Much of algebra is understanding what algebraic expressions mean and how to write them given a common language version of the expressions in a "real-life" situation. To translate an arithmetic or algebraic expression to an English phrase, consider the following table.

Arithmetic Form	Algebraic Form	English Phrase
	Addition	
2 + 3	$x + 3$	The sum of x and 3
	$2 + y$	y more than 2
	$x + y$	x increased by y
	Subtraction	
10 − 3	$x - 3$	3 less than x *or* x less 3
	$10 - y$	Subtract y from 10 *or* 10 minus y
	$x - y$	x diminished by y *or* the difference of x and y
	$82 - w$	The value after w is subtracted from 82 *or* the difference of 82 and w

5.7

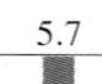

VARIATION

OBJECTIVES

1. Solve problems involving direct variation.
2. Solve problems involving inverse variation.

In our physical world, most things are in a state of change or variation. Measurements of changes in temperature, rainfall, light, heat, and fuel supplies are recorded. Changes in height, weight, and blood pressure can be important to a person's health. Many relationships between such measurements can be expressed in formulas. Here we are concerned about only two types of variation: *direct* and *inverse*.

Direct Variation

DEFINITION

y varies directly as x when the quotient of y and x is a constant.

FORMULA

Formulas are highlighted with orange.

$\frac{y}{x} = k$

$y = kx$

Direct variation:

$\frac{y}{x} = k, k \neq 0$ or $y = kx, k \neq 0$

Several traditional formulas are examples of direct variation.

Distance (D) traveled at a constant speed (r) varies directly as time (t).

$D = rt$ or $\frac{D}{t} = r$

The weight (w) of a gold ingot varies directly as the volume (v).

$w = kv$ or $\frac{w}{v} = k$

EXAMPLE 1

Write an equation expressing the following relationship: Weight varies directly as volume. Use k as the constant of variation.

$\frac{w}{v} = k$ In direct variation, the quotient is constant. Let w represent weight and v represent the volume. ■

2.3 Solving Linear Equations 115

EXERCISE 2.3

A

Exercises are graded in difficulty; A Exercises are relatively easy and may be done mentally.

Solve:

1. $7x + 5 = 40$ **2.** $8y + 3 = 51$ **3.** $3w - 4 = -16$

4. $2x - 5 = -17$ **5.** $2y + 17 = -8$ **6.** $4x + 21 = -11$

7. $13 = 18w - 23$ **8.** $-14 = 37y - 88$ **9.** $0 = -3x + 15$

10. $19 = -8w + 3$ **11.** $\frac{x}{3} + 6 = 16$ **12.** $\frac{y}{5} - 4 = -16$

13. Solve $ax + b = c$ for x if $a = 3$, $b = 4$, and $c = 1$.

14. Solve $ax + b = c$ for x if $a = 6$, $b = 14$, and $c = 2$.

15. Solve for x: $tx + c = d$ **16.** Solve for y: $my + r = w$

Solve:

17. $22 = 4 + \frac{x}{7}$ **18.** $19 = 9 + \frac{y}{10}$ **19.** $15 = \frac{x}{6} - 5$ **20.** $3 = \frac{y}{11} - 8$

B

The "B" exercises may require paper and pencil.

Solve:

21. $29y + 2 = 205$ **22.** $74w + 3 = 225$ **23.** $-2x + 5 = -17$

24. $9z - 15 = -87$ **25.** $67w + 34 = 369$ **26.** $50w - 16 = 384$

27. $-18y - 10 = -1$ **28.** $-12x - 10 = -2$ **29.** $4 = 22x - 150$

30. $2 = -3x - 61$ **31.** $\frac{w}{9} - 8 = -2$ **32.** $\frac{x}{7} + 6 = -5$

33. $\frac{p}{-2} - 14 = 6$ **34.** $\frac{t}{-3} + 23 = 17$

36. $\frac{y}{6} - 8 = \frac{5}{2}$

37. Solve $ax - b = c$ for x if $a = 12$, $b = 6$, and $c = -14$.

38. Solve $ax - b = c$ for x if $a = 3$, $b = 5$, and $c = -50$.

39. Solve for x: $gx - t = k$

116 CHAPTER 2 Linear Equations and Inequalities

C

These problems are more challenging. At the discretion of the instructor, "C" problems may sometimes be solved with a calculator.

Solve:

41. $2x - \frac{1}{3} = \frac{1}{2}$ **42.** $-6x + \frac{1}{5} = \frac{3}{10}$ **43.** $-2p + \frac{1}{6} = \frac{2}{3}$

44. $3y - \frac{1}{4} = \frac{3}{8}$ **45.** $\frac{x}{4} + \frac{1}{2} = \frac{1}{3}$ **46.** $\frac{y}{3} + \frac{1}{4} = \frac{1}{2}$

47. $22.8x + 56.6 = 170.6$ **48.** $44.5y + 9.4 = 320.9$ **49.** $-13y + 20 = 35.6$

50. $-5x + 17 = 30.5$ **51.** $-18x + 3.8 = 65$ **52.** $-14y + 3.2 = 90$

53. $2.3p - 19.3 = -83.7$ **54.** $9.45p + 55.5 = -246.9$

55. Solve $ax + b = c$ for x if $a = 7$, $b = 1.3$, and $c = -2.9$.

56. Solve $ax + b = c$ for x if $a = 3$, $b = \frac{4}{7}$, and $c = -\frac{3}{2}$.

57. Solve for w: $2\ell + 2w = P$ **58.** Solve for ℓ: $2\ell + 2w = P$

59. Solve for p: $c = 18p + 20$ **60.** Solve for t: $s = v + gt$

D

"D" problems provide application of the concepts presented in this section.

61. Find the time (t) it takes a free-falling skydiver to reach a falling speed of 173 feet per second if $v = -3$ and $g = 32$. The formula is $s = v + gt$.

62. Find the time (t) it takes a free-falling skydiver to reach a falling speed of 77 feet per second if $v = -3$ and $g = 32$. The formula is $s = v + gt$.

63. A formula for distance traveled is $2d = t^2a + 2tv$, where d represents distance, v represents initial velocity, t represents time, and a represents acceleration. Find a if $d = 240$, $v = 20$, and $t = 3$.

64. Use the formula in Exercise 63 to find a if $d = 240$, $v = 20$, and $t = 5$.

65. If 98 is added to six times some number, the sum is 266. What is the number?

66. If 73 is added to eleven times a number, the sum is 117. What is the number?

67. The difference of 15 times a number and 87 is -39. What is the number?

68. The difference of 24 times a number and 78 is 78. What is the number?

69. Using the formula $D = B - NP$ (see Example 8), find the number of payments made (N) if the balance (D) is \$205, the monthly payment (P) is \$35, and the amount borrowed (B) is \$450.

70. Use the formula in Exercise 69 to find the number of payments made if the balance is \$459, the monthly payment is \$57, and the amount borrowed is \$1200.

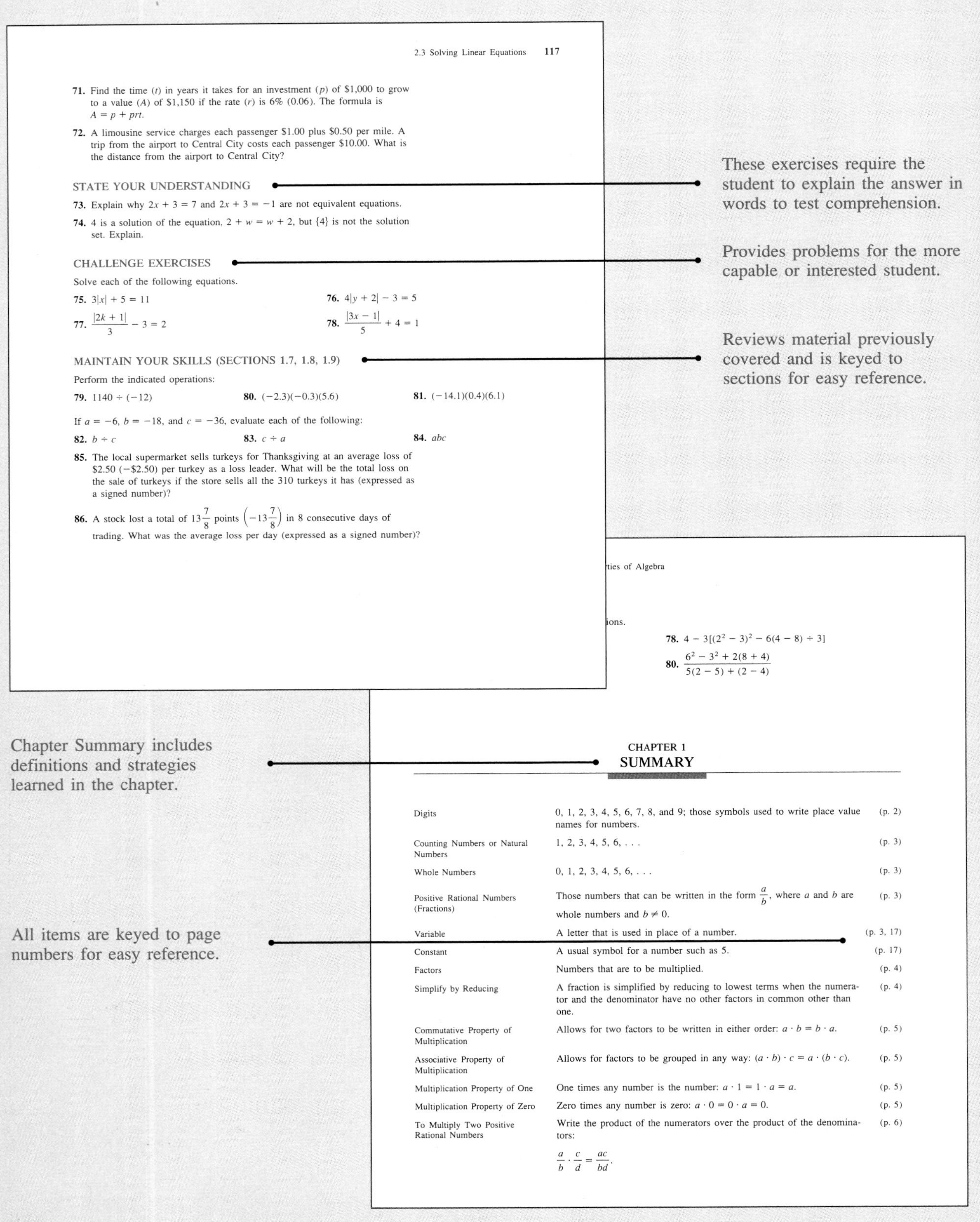

2.3 Solving Linear Equations 117

71. Find the time (t) in years it takes for an investment (p) of \$1,000 to grow to a value (A) of \$1,150 if the rate (r) is 6% (0.06). The formula is $A = p + prt$.

72. A limousine service charges each passenger \$1.00 plus \$0.50 per mile. A trip from the airport to Central City costs each passenger \$10.00. What is the distance from the airport to Central City?

STATE YOUR UNDERSTANDING

73. Explain why $2x + 3 = 7$ and $2x + 3 = -1$ are not equivalent equations.

74. 4 is a solution of the equation, $2 + w = w + 2$, but $\{4\}$ is not the solution set. Explain.

CHALLENGE EXERCISES

Solve each of the following equations.

75. $3|x| + 5 = 11$

76. $4|y + 2| - 3 = 5$

77. $\frac{|2k + 1|}{3} - 3 = 2$

78. $\frac{|3x - 1|}{5} + 4 = 1$

MAINTAIN YOUR SKILLS (SECTIONS 1.7, 1.8, 1.9)

Perform the indicated operations:

79. $1140 \div (-12)$

80. $(-2.3)(-0.3)(5.6)$

81. $(-14.1)(0.4)(6.1)$

If $a = -6$, $b = -18$, and $c = -36$, evaluate each of the following:

82. $b \div c$

83. $c \div a$

84. abc

85. The local supermarket sells turkeys for Thanksgiving at an average loss of \$2.50 (−\$2.50) per turkey as a loss leader. What will be the total loss on the sale of turkeys if the store sells all the 310 turkeys it has (expressed as a signed number)?

86. A stock lost a total of $13\frac{7}{8}$ points $\left(-13\frac{7}{8}\right)$ in 8 consecutive days of trading. What was the average loss per day (expressed as a signed number)?

ties of Algebra

ions.

78. $4 - 3[(2^2 - 3)^2 - 6(4 - 8) \div 3]$

80. $\frac{6^2 - 3^2 + 2(8 + 4)}{5(2 - 5) + (2 - 4)}$

CHAPTER 1
SUMMARY

Digits	0, 1, 2, 3, 4, 5, 6, 7, 8, and 9; those symbols used to write place value names for numbers.	(p. 2)
Counting Numbers or Natural Numbers	1, 2, 3, 4, 5, 6, . . .	(p. 3)
Whole Numbers	0, 1, 2, 3, 4, 5, 6, . . .	(p. 3)
Positive Rational Numbers (Fractions)	Those numbers that can be written in the form $\frac{a}{b}$, where a and b are whole numbers and $b \neq 0$.	(p. 3)
Variable	A letter that is used in place of a number.	(p. 3, 17)
Constant	A usual symbol for a number such as 5.	(p. 17)
Factors	Numbers that are to be multiplied.	(p. 4)
Simplify by Reducing	A fraction is simplified by reducing to lowest terms when the numerator and the denominator have no other factors in common other than one.	(p. 4)
Commutative Property of Multiplication	Allows for two factors to be written in either order: $a \cdot b = b \cdot a$.	(p. 5)
Associative Property of Multiplication	Allows for factors to be grouped in any way: $(a \cdot b) \cdot c = a \cdot (b \cdot c)$.	(p. 5)
Multiplication Property of One	One times any number is the number: $a \cdot 1 = 1 \cdot a = a$.	(p. 5)
Multiplication Property of Zero	Zero times any number is zero: $a \cdot 0 = 0 \cdot a = 0$.	(p. 5)
To Multiply Two Positive Rational Numbers	Write the product of the numerators over the product of the denominators: $\frac{a}{b} \cdot \frac{c}{d} = \frac{ac}{bd}$.	(p. 6)

Subtraction of Real Numbers	Add the opposite of the number to be subtracted: $a - b = a + (-b)$.	(p. 58)
Multiplication of Real Numbers	The product of two positive numbers is positive. The product of two negative numbers is positive. The product of a positive and a negative number in either order is negative.	(p. 64)
Multiplying by -1	The product of -1 and a number yields the number's opposite.	(p. 64)
Multiplication Property of 1	$1 \cdot a = a \cdot 1 = a$, for any real number a	(p. 64)
Multiplication Property of 0	$0 \cdot a = a \cdot 0 = 0$, for any real number a	(p. 64)
Division of Real Numbers	The quotient of two positive numbers is positive. The quotient of two negative numbers is positive. The quotient of a positive and a negative number in either order is negative.	(p. 70)

CHAPTER 1

REVIEW EXERCISES

Problems reviewing all objectives are included here to test student's level of proficiency. Exercises are keyed to section number and objective.

SECTION 1.1 Objective 1

Simplify the following by reducing to lowest terms.

1. $\frac{144}{240}$ **2.** $\frac{240}{256}$ **3.** $\frac{91}{208}$ **4.** $\frac{180}{260}$ **5.** $\frac{147}{189}$

SECTION 1.1 Objective 2

Multiply or divide as indicated.

6. $\frac{15}{27} \cdot \frac{18}{75}$ **7.** $0.4375 \div 0.25$ **8.** $19{,}872 \div 72$ **9.** $2\frac{3}{36} \cdot \frac{21}{45}$

10. $59 \div 236$

SECTION 1.1 Objective 3

Add or subtract as indicated.

11. $234 + 5789 + 12 + 102$ **12.** $6845.3 - 82.97$ **13.** $\frac{19}{34} - \frac{1}{2}$

14. $3\frac{3}{4} + 1\frac{5}{9}$ **15.** $46.4 + 2.1231 + 0.8769$

, $y = -4$, and $z = 3$.

$= -1$, $b = 1$, and $c = -1$.

96. Is -12 a solution of $8x - 4 = -100$?

97. Is -15 a solution of $12 - 5x = 87$?

98. Is -21 a solution of $8x - 6 = -162$?

99. Is 16 a solution of $25 - 3x = 63$?

100. Is -6 a solution of $6x + 12 = 9x + 30$?

This review checks the student's understanding of the concepts presented in the chapter.

CHAPTER 1

TRUE–FALSE CONCEPT REVIEW

Check your understanding of the language of algebra. Tell whether each of the following is true (always true) or false (not always true).

1. 3^4 is an example of a number written in exponential form.

2. In the expression 3^4, 3 is the value of the expression.

3. A variable is usually a letter used as a placeholder for a number.

4. The symbol π is a variable.

5. In an expression containing more than one operation, multiplication is always done before division.

6. To evaluate the expression $4x + 2$, one needs to know the replacement for x.

7. The solution set of an equation is the set of all replacements for the variable.

8. The opposite of a number is always negative.

9. The symbol "$-x$" should be read "negative x."

10. Rational numbers can be represented by a fraction.

11. Zero is an integer.

12. The absolute value of a number is always positive.

13. The sum of two integers is always larger than either of the integers being added.

14. To subtract two integers, add the opposite of the number being subtracted.

15. $-7 + 7 = 0$ is an example of the addition property of opposites.

16. The sign of the product of two signed numbers is determined by the sign of the number with the larger absolute value.

17. The sum of two negative numbers is positive.

18. To find the opposite of a number, multiply the number by -1.

19. The rules for division of signed numbers is the same as the rules for multiplication of signed numbers.

20. The order of operations for signed numbers is different from that for whole numbers.

CHAPTER 1 TEST

This test is used to evaluate student comprehension.

1. Add: 22.3 + 1.79 + 6.7
2. Find the solution for $A = \ell w$ when $\ell = 2.5$ and $w = 0.6$.
3. Is 8 a solution of the equation $8x - 4 = 6x + 12$?
4. Find the absolute value: $|3.4|$
5. Subtract: $\frac{5}{8} - \frac{1}{3}$
6. Add: $-8.56 + (-6.58) + 7$
7. Evaluate $2a - b + 15$ when $a = 27$ and $b = 14$.
8. Divide: $91 \div (-13)$
9. Multiply: (2.3)(17.7)
10. Find the opposite of 43.
11. Find the value of 7^4.
12. Subtract: $-75 - 22 - 15 - (-1)$
13. Add: $-21 + (-84) + 19$
14. Divide: $0.34 \div 0.4$
15. Multiply: $(-27)(-14)(3)$
16. Write $12 \cdot 12 \cdot 12 \cdot 12 \cdot 12 \cdot 12$ in exponential form.
17. Subtract: $187 - (-122)$
18. Evaluate $2a^2 - 5b^3 \div 12c$ when $a = 2$, $b = -3$, and $c = -5$.
19. Find the absolute value: $|-87|$
20. Is -2 a solution of the equation $4x + 17 = -9x - 9$?
21. Translate the following expression to algebraic form. Let x represent the number: the product of the square of a number and 5
22. Find the opposite of -13.5.
23. Find the value of $15 \div 5 + 12 \cdot 5^2 - 8 \cdot 4$.
24. Divide: $-144 \div (-48)$
25. Translate the following expression to algebraic form; let n represent the first number and p the second number: the first number divided by the product of 4 and the second number
26. Multiply: $(-5.6)(-2)(-4.4)$
27. A teacher uses positive and negative points to keep a record of students' progress. If Sara had the following record, what is her total: +87 (test), −40 (absent), +44 (quiz), and −26 (homework late)
28. What is the difference between a temperature of 60°F and a temperature of −15°F?
29. Use the formula $C = \frac{5}{9}(F - 32)$ to find the Celsius temperature that is equal to 14°F.

Students can work Cumulative Review Exercises after Chapters 4 and 8 to reinforce concepts and sharpen skills.

CHAPTERS 1–4 CUMULATIVE REVIEW

CHAPTER 1

1. Reduce to lowest terms: $\frac{108}{162}$

Perform the indicated operations:

2. 4.67×0.713
3. $0.7668 \div 0.27$
4. $8.72 + 0.7 + 0.52 + 0.537$
5. $27.2 - 19.517$
6. Write in exponential form: $2 \cdot 2 \cdot 2 \cdot 2 \cdot 7 \cdot 7 \cdot 9 \cdot 9 \cdot 9$
7. $2^3 - 3[10 - 2(5^2 - 3 \cdot 7)]$
8. $\frac{3 \cdot 14^2 + 4 \cdot 7 - 6^3}{4(7^2 - 3^3) \div (5^2 - 14)}$
9. Find the value of x if $x = 8^2 - 4 \cdot 7^2 + 5^3$

If $x = 4.75$, $y = 2.64$, and $z = 0.015$, evaluate:

10. $\frac{y}{z}$
11. $x^2 - y^3 + z$
12. Is 4 a solution of $5x - 12 = 2x + 1$?
13. Is 13 a solution of $3x + 4 = 4x - 9$?

Classify the following as either true or false:

14. $\sqrt{8}$ is an integer.
15. All whole numbers are rational numbers.
16. No real number is an irrational number.
17. $-12 < -5$

Determine the value of each of the following:

18. $(-6) + (+4) + (-3) + (+12)$
19. $-5.2 + 3.47$
20. $\frac{3}{10} + \left(-\frac{1}{2}\right)$
21. $-|4 + (-7)|$

Determine the value of each of the following:

22. $-4.5 - 2.7$
23. $-\frac{5}{8} - \frac{5}{6}$
24. $-21 - 47 - (-34)$
25. $-(-12) - (17) - (-41) - (30)$

Determine the value of each of the following:

26. $(-2)(5)(-7)$
27. $(0.6)(-0.3)(4)$
28. $(-3)(41)(0)(-17)$
29. $-4^2(-3)^2(-1)^3$

Determine the value of each of the following:

30. $(-4.41) \div (0.7)$
31. $\left(-\frac{5}{12}\right) \div \left(-\frac{10}{27}\right)$
32. $\frac{-5.4}{0.04}$
33. $\left(1\frac{3}{4}\right) \div \left(-\frac{3}{8}\right)$

Determine the value of each of the following:

34. $-20\left|\frac{12 - 3}{-4 - 6}\right| - 4(11 - 19)$
35. $5(3^2 - 4^3) - (-7)^2(-6)$

Evaluate if $a = -7.2$, $b = -8.4$, and $c = -4.3$

36. $2a^2 - 3bc$
37. $(a + 2)(b - c)$

CHAPTER

1

BASIC CONCEPTS AND PROPERTIES OF ALGEBRA

SECTIONS

The stock market is a good example of the application of signed numbers. Gains in stock prices can be expressed with positive numbers: if a stock price opens at $31\frac{1}{4}$ and closes at $31\frac{7}{8}$, there is a gain of $\frac{5}{8}$ $\left(+\frac{5}{8}\right)$. If the same stock had closed at $29\frac{1}{2}$ it would have recorded a loss of $\frac{3}{4}$ $\left(-\frac{3}{4}\right)$. In Exercise 73, Section 1.9, signed numbers are used to look at the net profit or loss incurred by Mr. Gentry when he sold two stocks, one at a loss and the other one for a profit. *(Michael Going/The Image Bank)*

PREVIEW

We review the operations of addition, subtraction, multiplication, and division of positive rational numbers. We also review the basic properties of operations that are used in both arithmetic and algebra—commutative properties, associative properties, distributive property, multiplication properties of one and zero, and the reciprocal property.

The study of algebra begins with the introduction of exponents and algebraic logic through order of operations. This leads to the uses of variables and expressions containing variables. This ''language of algebra'' is reinforced by translating from English to algebra and vice versa. We introduce elementary equations. Solutions of these equations and other equations play a major role throughout the study we present in this text. The study of numbers includes the set of real numbers. Absolute value leads to the rules for operations on signed numbers. We conclude by checking equations with positive and negative numbers and by reviewing the order of operations.

The material contained in this chapter is basic to the study of algebra and must be mastered before proceeding.

1.1

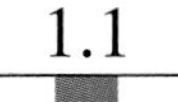

POSITIVE RATIONAL NUMBERS (FRACTIONS)

OBJECTIVES

1. Simplify positive rational numbers (fractions) by reducing.
2. Multiply and divide positive rational numbers.
3. Add and subtract positive rational numbers.

The study of mathematics includes the study of numbers and their properties. In this section, we review the properties and the operations on positive rational numbers, which are commonly called fractions.

There are ten symbols called digits that are used to write the place value names for numbers.

DEFINITION

Digits

The *digits* are the ten symbols

0, 1, 2, 3, 4, 5, 6, 7, 8, and 9.

Natural Numbers

DEFINITION

The *counting numbers* or *natural numbers* (those used when counting) are

1, 2, 3, 4, 5, 6, 7, 8, 9, 10, 11, 12, . . .

and continue without bound.

Whole Numbers

DEFINITION

The *whole numbers* are

0, 1, 2, 3, 4, 5, 6, 7, 8, 9, . . .

and continue without bound.

Positive Rational Numbers

DEFINITION

A *positive rational number* (fraction) is a number that can be written in the form

$$\frac{a}{b} \begin{array}{l} \longleftarrow \text{NUMERATOR} \\ \longleftarrow \text{DENOMINATOR} \end{array}$$

where a is a whole number and b is a natural number.

Letters are used to represent the numerator and the denominator of the fraction. This is very common in algebra.

Variable

DEFINITION

The letters used to represent numbers are called *variables*.

The letter x is frequently used as a variable; hence, when writing a multiplication problem, we do not use the $\times$ for a times sign. We will write:

$2 \cdot 5$	$2(5)$	$(2)5$	$(2)(5)$
ab	$a \cdot b$	$a(b)$	$(a)(b)$
$2a$	$2 \cdot a$	$2(a)$	$(2)(a)$

for multiplication. In each of these expressions, 2, 5, a, and b are called factors.

DEFINITION

Factors
Product

Numbers that are multiplied are called *factors,* and the answer is called a *product.*

Since each whole number can be written as a fraction whose denominator is 1, all whole numbers, except zero, are also positive rational numbers.

Decimals are other symbols used to write fractions; therefore, decimals are positive rational numbers as well.

DEFINITION

Reduced to Lowest Terms

A fraction (positive rational number) is *reduced to lowest terms* when the numerator and denominator have no factors in common other than 1.

EXAMPLE 1

Simplify the fraction $\frac{16}{36}$ by reducing to lowest terms.

$\frac{16}{36} = \frac{4 \cdot 4}{4 \cdot 9}$ — The numerator and the denominator have a common factor of 4.

$= \frac{\not{4} \cdot 4}{\not{4} \cdot 9}$ — Divide both the numerator and the denominator by 4.

$= \frac{4}{9}$ — Reduced form. ■

PROCEDURE

Reduce a Fraction

To simplify a fraction by reducing to lowest terms, divide the numerator and the denominator by any common factor until the only common factor remaining is 1.

Recall the names given to the parts of multiplication and division problems.

FACTOR		FACTOR		PRODUCT
9	$\cdot$	8	$=$	72

DIVIDEND		DIVISOR		QUOTIENT
72	÷	9	=	8

Two numbers can be multiplied in either order without changing the product. This fact is called the *commutative property of multiplication.*

Commutative Property of Multiplication

■ **PROPERTY**

$a \cdot b = b \cdot a$

As a result of the commutative property of multiplication, we know that $8 \cdot 9 = 9 \cdot 8$.

When three or more numbers are multiplied, the grouping or association can be changed. This fact is called the *associative property of multiplication.*

Associative Property of Multiplication

■ **PROPERTY**

$(a \cdot b) \cdot c = a \cdot (b \cdot c)$

As a result of the associative property of multiplication, we know that $(2 \cdot 6) \cdot 8 = 2 \cdot (6 \cdot 8)$.

A property involving multiplication by one is very useful in algebra. It is called the *multiplication property of one.*

Multiplication Property of One

■ **PROPERTY**

$a \cdot 1 = 1 \cdot a = a$

Thus, $1 \cdot 9 = 9 \cdot 1 = 9$.

The *multiplication property of zero* states that zero times any number is zero.

Multiplication Property of Zero

■ **PROPERTY**

$a \cdot 0 = 0 \cdot a = 0$

Thus, $0 \cdot 5 = 5 \cdot 0 = 0$.

To find the product of two positive rational numbers, write the product of their numerators over the product of their denominators.

RULE

If $\frac{a}{b}$ and $\frac{c}{d}$ are two positive rational numbers, then $\frac{a}{b} \cdot \frac{c}{d} = \frac{ac}{bd}$.

When multiplying rational numbers, be sure the final product is reduced to lowest terms.

EXAMPLE 2

Multiply: $\frac{9}{5} \cdot \frac{15}{36}$

$\frac{9}{5} \cdot \frac{15}{36} = \frac{9 \cdot 15}{5 \cdot 36}$ Multiply the numerators and the denominators.

$= \frac{3 \cdot \not{3} \cdot \not{3} \cdot \not{5}}{2 \cdot 2 \cdot \not{3} \cdot \not{3} \cdot \not{5}}$ Prime factor the numerator and the denominator.

$= \frac{3}{4}$ Divide both the numerator and the denominator by $3 \cdot 3 \cdot 5 = 45$. ■

To multiply mixed numbers, change the mixed numbers to improper fractions and multiply.

EXAMPLE 3

Multiply: $\left(2\frac{1}{2}\right)\left(3\frac{3}{4}\right)$

$\left(2\frac{1}{2}\right)\left(3\frac{3}{4}\right) = \frac{5}{2} \cdot \frac{15}{4}$ Change each mixed number to an improper fraction.

$= \frac{75}{8}$ Multiply the numerators and the denominators.

$= 9\frac{3}{8}$ Change the improper fraction to a mixed number. ■

To multiply two positive rational numbers written as decimals, we proceed as follows.

EXAMPLE 4

Multiply: 3.25(25.3)

```
  25.3
  3.25
------
 1 265
 5 06
75 9
------
82.225
```

Multiply, being sure to place the products in the proper columns. The product must have the same number of decimal places as the total number of places in both factors. ■

PROPERTY

Reciprocal Property

$$\frac{a}{b} \cdot \frac{b}{a} = 1$$

DEFINITION

Reciprocals

Two fractions are *reciprocals* of each other if their product is one.

$\frac{5}{8}$ and $\frac{8}{5}$ are reciprocals since $\frac{5}{8} \cdot \frac{8}{5} = 1$. Also $1\frac{3}{4}$ and $\frac{4}{7}$ are reciprocals since $1\frac{3}{4} = \frac{7}{4}$ and $\frac{7}{4} \cdot \frac{4}{7} = 1$.

To divide two fractions, multiply the dividend by the reciprocal of the divisor.

RULE

$$\frac{a}{b} \div \frac{c}{d} = \frac{a}{b} \cdot \frac{d}{c}$$

EXAMPLE 5

Divide: $\frac{15}{48} \div \frac{30}{24}$

$$\frac{15}{48} \div \frac{30}{24} = \frac{15}{48} \cdot \frac{24}{30}$$ Multiply by the reciprocal of $\frac{30}{24}$.

$$= \frac{15 \cdot 24}{48 \cdot 30}$$

$$= \frac{2 \cdot 2 \cdot 2 \cdot 3 \cdot 3 \cdot 5}{2 \cdot 2 \cdot 2 \cdot 2 \cdot 2 \cdot 3 \cdot 3 \cdot 5}$$

$$= \frac{1}{4}$$ Reduce. The factors of $2 \cdot 2 \cdot 2 \cdot 3 \cdot 3 \cdot 5$ are divided out of the numerator and denominator. ■

Since 0 has no reciprocal, we must be careful with division and zero.

■ PROPERTY

Zero Divided by Nonzero Number

$0 \div a = 0, a \neq 0$

Zero divided by any nonzero number is zero. So, $0 \div 1 = 0$, $0 \div 2 = 0$, $0 \div \frac{2}{5} = 0$, $0 \div 1.11 = 0$.

■ CAUTION

Division by Zero

Division by zero has no value.

The expressions $1 \div 0$, $2 \div 0$, $1.6 \div 0$, and $\frac{2}{3} \div 0$ have no value.

To divide mixed numbers, first change the mixed numbers to improper fractions.

EXAMPLE 6

Divide: $\left(2\frac{3}{4}\right) \div \left(1\frac{5}{8}\right)$

$$\left(2\frac{3}{4}\right) \div \left(1\frac{5}{8}\right) = \frac{11}{4} \div \frac{13}{8}$$ Change the mixed numbers to improper fractions.

$$= \frac{11}{4} \cdot \frac{8}{13}$$ Multiply by the reciprocal.

$= \dfrac{11 \cdot 4 \cdot 2}{4 \cdot 13}$ Multiply the numerators and denominators. Also $8 = 4 \cdot 2$.

$= \dfrac{22}{13}$ Reduce.

$= 1\dfrac{9}{13}$ Change to a mixed number. ■

Example 7 shows how to divide two positive rational numbers written as decimals.

EXAMPLE 7

Divide: $1.1772 \div 0.27$

```
        4.36
0.27)1.17 72
     1 08
        9 7
        8 1
        1 62
        1 62
```

To assure the proper placement of the decimal point, divide by a whole number. To do this, move the decimal point two places to the right (multiply by 100) in both the divisor and dividend. The decimal point in the quotient is directly above the new position in the dividend. ■

Recall the different names given to the parts of addition and subtraction problems.

ADDEND		ADDEND		SUM
5	+	3	=	8

MINUEND		SUBTRAHEND		DIFFERENCE
8	−	3	=	5

Two numbers can be added in either order. This is called the *commutative property of addition.*

■ PROPERTY

Commutative Property of Addition

$$a + b = b + a$$

Thus, $6 + 4 = 4 + 6$.

When three or more numbers are added, it is possible to group them in any way we wish. This is called the *associative property of addition.*

PROPERTY

Associative Property of Addition

$$(a + b) + c = a + (b + c)$$

Thus, $(12 + 8) + 23 = 12 + (8 + 23)$.

A property that involves both addition and multiplication is the *distributive property of multiplication over addition*. The common name for it is the *distributive property*.

PROPERTY

Distributive Property

$$a(b + c) = ab + ac$$

As a result of the distributive property, we know that $9(8 + 3) = 9 \cdot 8 + 9 \cdot 3$.

The sum of two fractions having a common denominator (the same denominators) is found by adding the numerators and putting that sum over the common denominator.

RULE

If $\frac{a}{b}$ and $\frac{c}{b}$ are two fractions then,

$$\frac{a}{b} + \frac{c}{b} = \frac{a + c}{b}.$$

EXAMPLE 8

Add: $\frac{6}{25} + \frac{9}{25}$

$$\frac{6}{25} + \frac{9}{25} = \frac{6 + 9}{25}$$ Add the numerators, and write the sum over the common denominator.

$$= \frac{15}{25}$$ Simplify.

$$= \frac{3}{5}$$ Reduce. ■

To add fractions that do not have common denominators, find the least common denominator and then build equivalent fractions by multiplying by forms of one.

EXAMPLE 9

Add: $\frac{5}{24} + \frac{1}{12} + \frac{2}{3}$

The least common denominator of 24, 12, and 3 is 24.

$$\frac{5}{24} + \frac{1}{12} + \frac{2}{3} = \frac{5}{24} + \frac{1}{12} \cdot \frac{2}{2} + \frac{2}{3} \cdot \frac{8}{8}$$

Multiply $\frac{1}{12}$ by $\frac{2}{2}$ and $\frac{2}{3}$ by $\frac{8}{8}$ to "build" the fractions to the common denominator.

$$= \frac{5}{24} + \frac{2}{24} + \frac{16}{24}$$

Multiply.

$$= \frac{23}{24}$$

Add. ■

To add mixed numbers, add the whole numbers and add the fractions.

EXAMPLE 10

Add: $2\frac{1}{2} + 3\frac{3}{4}$

$$2\frac{1}{2} = 2\frac{2}{4}$$
$$3\frac{3}{4} = 3\frac{3}{4}$$
$$5\frac{5}{4} = 5 + 1\frac{1}{4} = 6\frac{1}{4}$$

Write the mixed numbers in columns so that the whole number parts are lined up and the fraction parts are lined up. Rewrite $\frac{1}{2}$ with a denominator of 4 so that the fractions will have a common denominator.

Write $\frac{5}{4}$ as a mixed number, and then add. ■

To add two positive rational numbers written as decimals, insert zeros on the right where necessary so as to have the same number of decimal places. When writing in a column be sure to keep the decimal column straight and add as if they were whole numbers.

EXAMPLE 11

Add: $3.6 + 4.14 + 2.03$

3.6	3.60	Write the numbers in column form. Inserting a zero on the right in 3.6 helps keep the digits in the correct columns.
4.14	4.14	
2.03	2.03	
9.77	9.77	

■

To subtract two fractions with a common denominator, subtract the numerators, and write the difference over the common denominator.

■ RULE

If $\frac{a}{b}$ and $\frac{c}{b}$ are two fractions, then

$$\frac{a}{b} - \frac{c}{b} = \frac{a - c}{b}.$$

EXAMPLE 12

Subtract: $\frac{7}{12} - \frac{5}{12}$

$\frac{7}{12} - \frac{5}{12} = \frac{7 - 5}{12}$ Subtract the numerators, and write the difference over the common denominator.

$= \frac{2}{12}$ Subtract.

$= \frac{1}{6}$ Reduce. ■

EXAMPLE 13

Subtract: $\frac{23}{24} - \frac{1}{3}$

$\frac{23}{24} - \frac{1}{3} = \frac{23}{24} - \frac{8}{24}$ Write each fraction with a common denominator of 24. Multiply $\frac{1}{3}$ by $\frac{8}{8}$.

$= \frac{23 - 8}{24}$ Subtract the numerators, and write the difference over the common denominator.

$$= \frac{15}{24}$$

$$= \frac{5}{8} \qquad \text{Reduce.}$$

Example 14 reviews subtraction of mixed numbers.

EXAMPLE 14

Subtract: $3\frac{3}{7} - 1\frac{1}{5}$

$$\begin{array}{r} 3\frac{3}{7} = 3\frac{15}{35} \\ 1\frac{1}{5} = 1\frac{7}{35} \\ \hline 2\frac{8}{35} \end{array}$$

First, rewrite each of the fractions using the least common multiple of the denominators 7 and 5.

To subtract two positive numbers written as decimals, make sure the number of decimal places is the same, and subtract as though they were whole numbers.

EXAMPLE 15

Subtract: $0.675 - 0.49$

$$\begin{array}{r} 0.\overset{5}{\not{6}}\overset{17}{\not{7}}5 \\ 0.490 \\ \hline 0.185 \end{array}$$

Rewrite in column form, making sure to line up the decimal points. $(0.49 = 0.490)$

Exercise 1.1

A

Simplify each fraction by reducing to lowest terms:

1. $\frac{10}{25}$ **2.** $\frac{9}{12}$ **3.** $\frac{18}{36}$ **4.** $\frac{15}{35}$

5. $\frac{14}{28}$ **6.** $\frac{72}{96}$

Multiply:

7. $\frac{3}{4} \cdot \frac{8}{9}$

8. $\frac{9}{14} \cdot \frac{8}{27}$

9. $\frac{5}{14} \cdot \frac{7}{15}$

10. $\frac{20}{21} \cdot \frac{14}{25}$

Divide:

11. $\frac{5}{8} \div \frac{2}{3}$

12. $\frac{7}{12} \div \frac{5}{24}$

13. $\frac{7}{10} \div \frac{5}{14}$

14. $\frac{8}{15} \div \frac{16}{3}$

Add:

15. $\frac{3}{4} + \frac{2}{3} + \frac{1}{2}$

16. $\frac{5}{8} + \frac{2}{5} + \frac{1}{4}$

17. $$\begin{array}{r} 2\frac{1}{2} \\ 3\frac{3}{5} \\ 5\frac{5}{6} \\ \hline \end{array}$$

18. $$\begin{array}{r} 1\frac{3}{4} \\ 2\frac{1}{3} \\ 3\frac{3}{8} \\ \hline \end{array}$$

Subtract:

19. $\frac{3}{4} - \frac{1}{2}$

20. $\frac{7}{8} - \frac{2}{3}$

21. $$\begin{array}{r} 4\frac{3}{4} \\ 1\frac{4}{5} \\ \hline \end{array}$$

22. $$\begin{array}{r} 12\frac{1}{2} \\ 8\frac{5}{8} \\ \hline \end{array}$$

B

Simplify each fraction by reducing to lowest terms:

23. $\frac{48}{56}$

24. $\frac{64}{72}$

25. $\frac{75}{125}$

26. $\frac{36}{108}$

Multiply:

27. $3\frac{1}{2} \cdot 4\frac{5}{7}$

28. $6\frac{3}{4} \cdot 2\frac{2}{9}$

29. $5\frac{7}{12} \cdot 3\frac{3}{7}$

30. $10\frac{2}{3} \cdot 7\frac{1}{8}$

Divide:

31. $14\frac{14}{15} \div 5\frac{3}{5}$

32. $13\frac{21}{32} \div 2\frac{7}{8}$

33. $19\frac{1}{3} \div 4\frac{2}{15}$

34. $21\frac{3}{8} \div 5\frac{3}{4}$

Add:

35. $234 + 5246 + 56 + 340 + 1756$

36. $4500 + 123 + 349 + 34$

37. $\frac{11}{24} + \frac{5}{18} + \frac{5}{36}$

38. $14\frac{7}{20} + 5\frac{7}{45} + 22\frac{8}{15}$

Subtract:

39. $\begin{array}{r} 0.8 \\ \underline{0.27} \end{array}$

40. $\begin{array}{r} 0.9 \\ \underline{0.64} \end{array}$

41. 8.926 − 3.118

42. 52.09 − 17.83

C

Simplify each fraction by reducing to lowest terms:

43. $\frac{168}{224}$

44. $\frac{112}{168}$

45. $\frac{135}{216}$

46. $\frac{177}{413}$

Multiply:

47. $\begin{array}{r} 3.2 \\ \underline{2.5} \end{array}$

48. $\begin{array}{r} 8.1 \\ \underline{9.6} \end{array}$

49. $\begin{array}{r} 2.37 \\ \underline{0.842} \end{array}$

50. $\begin{array}{r} 8.34 \\ \underline{0.219} \end{array}$

Divide:

51. $0.7\overline{)1.82}$

52. $0.9\overline{)324}$

53. $0.18\overline{)0.603}$

54. $0.23\overline{)0.7475}$

Add:

55. 2.53 + 0.82 + 0.4 + 0.069

56. 0.036 + 0.5 + 1.27 + 0.31

57. 124.67 + 3.08 + 56.983 + 4.005 + 3

58. 135.67 + 45.3 + 17.983 + 100.01 + 57

Subtract:

59. 11.356 − 9.7

60. 9.63 − 0.413

61. 367.002 − 45.98

62. 2356.987 − 1321.678

63. Al worked $4\frac{3}{4}$ hours on Monday, $2\frac{2}{3}$ hours on Wednesday, and $8\frac{1}{2}$ hours on Friday. How many total hours did he work for the three days?

64. Maria paid $19.04 for 8.5 pounds of sausage. What was the cost per pound?

65. Renee normally earns $5.50 per hour. For overtime work, she is paid $1\frac{1}{2}$ times her normal hourly rate. What is the hourly rate when Renee works overtime?

66. An electric meter read 56,589 kilowatt hours (KWH) at the beginning of April and 57,474 KWH at the beginning of May. How much electricity was used during the month of April?

67. How many ounces of fruit juice are contained in 20 cases, if each case contains 24 12-ounce cans of juice?

68. Floors 'R' Us has 5 carpet remnants of lengths 2.7 m, 6.24 m, 1.08 m, 0.6 m, and 4.2 m. What is the total length of these pieces?

69. Cartons, each $15\frac{1}{2}$ inches tall, are to be stored in a closet 8 feet high. How many cartons can be stacked there?

70. Kahseem purchased 3 compact discs at a cost of $12.99 each. He gave the clerk $40. How much change did he receive?

STATE YOUR UNDERSTANDING

71. Two fractions are to be added, but they do not have common denominators. How do we "build" the fractions so that they do have common denominators?

72. The operations of addition and multiplication are both commutative. Explain why the operations of subtraction and division are not commutative. Use examples to support your answer.

73. The operations of addition and multiplication are both associative. Explain why the operations of subtraction and division are not associative. Use examples to support your answer.

CHALLENGE EXERCISES

74. Jose earns $7.20 per hour for the first 40 hours he works each week. He receives overtime pay of $1\frac{1}{2}$ times his normal rate for every hour worked over 40. Last week he worked $47\frac{3}{4}$ hours. What were his earnings for the week?

75. The Nut Store mixed 80 pounds of peanuts costing $1.99 per pound with 30 pounds of cashews costing $3.99 per pound and 40 pounds of almonds costing $2.79 per pound. The mixture was packaged in 2-pound bags that sold for $6.99 each. How much profit was earned on this transaction?

76. The fact that 5 10-pound weights is equivalent to 10 5-pound weights illustrates what fundamental property?

77. The cost of purchasing 6 cans of orange juice at $0.89/can and 5 cans of peas costing $0.89/can may be determined by simply multiplying $0.89 by 11. This is an application of which fundamental property?

1.2

EXPONENTS, ORDER OF OPERATIONS, AND EVALUATING EXPRESSIONS

OBJECTIVES

1. Find the value of expressions written in exponential form.
2. Write a product of like factors in exponential form.
3. Find the value of (evaluate) numerical expressions.

An example of *exponential form* of a number is 3^4. The "3" is called the *base,* and the "4" is called the *exponent.* Whole number exponents, greater than one, show how many times the base is used as a factor.

EXPONENT ↓

$$3^4 = \underbrace{3 \cdot 3 \cdot 3 \cdot 3}_{\text{FOUR FACTORS}} = 81$$

↓ BASE ↓ FOUR FACTORS ↓ POWER OR VALUE

The *power* or *value* of an exponential expression is the product.

Letters used to represent numbers are called variables. Letters such as x, y, and z are often used as variables. *Constants* are symbols that are assigned a fixed value. The symbol "4" is called a constant since it always names the number "four."

The following table shows examples of numbers written in exponential form, how to read them, and what they represent. The exponential form of writing a number is often used to write multiplications with repeated factors.

Exponential Form	Read	Factors	Value (Power)
3^4	3 to the fourth	$3 \cdot 3 \cdot 3 \cdot 3$	3^4 or 81
2^5	2 to the fifth	$2 \cdot 2 \cdot 2 \cdot 2 \cdot 2$	2^5 or 32
5^3	5 to the third or 5 cubed	$5 \cdot 5 \cdot 5$	5^3 or 125
$\left(\frac{1}{3}\right)^2$	$\frac{1}{3}$ squared	$\frac{1}{3} \cdot \frac{1}{3}$	$\left(\frac{1}{3}\right)^2$ or $\frac{1}{9}$

Any whole number can be used as an exponent. The numbers 0 and 1 are special exponents. The following definitions tell how to find the values in these cases.

RULE

Any number (except 0) with an exponent of 0 has a value of 1. That is,

$x^0 = 1, \quad x \neq 0.$

Thus,

$2^0 = 1, \quad 11^0 = 1, \quad 3^0 = 1, \quad (0.5)^0 = 1.$

The expression 0^0 has no meaning. This is similar to division by 0; $0 \div 0$ has no meaning.

RULE

Any number with an exponent of 1 has a value of the number itself. So,

$x^1 = x.$

Therefore,

$2^1 = 2, \quad 11^1 = 11, \quad 3^1 = 3, \quad (0.5)^1 = 0.5.$

Any number can be regarded as having an exponent of 1; for example, 125 can be written as 125^1.

PROCEDURE

To find the value of a number written in exponential form:

1. If the exponent is 0, the value is 1 (base not 0).
2. If the exponent is 1, the value is the same as the base.
3. If the exponent is a whole number larger than 1, then the exponent tells how many times the base is used as a factor, and the value is found by multiplying.

EXAMPLE 1

Exponential Form	Factors	Value (Power)
7^3	$7 \cdot 7 \cdot 7$	7^3 or 243
11^1	None (special case)	11^1 or 11
7^0	None (special case)	7^0 or 1
4^5	$4 \cdot 4 \cdot 4 \cdot 4 \cdot 4$	4^5 or 1024

■

EXAMPLE 2

Find the value of 25^3

See Appendix VI for calculator instructions.

ENTER	DISPLAY
[25]	25.
[×]	25.
[25]	25.
[×]	625.
[25]	25.
[=]	15625.

$25^3 = 15625$

If your calculator has a [y^x] key:

ENTER	DISPLAY
[25]	25.
[y^x]	25.
[3]	3.
[=]	15625.

Therefore, the value of 25^3 is 15625. ■

EXAMPLE 3

Find the value of x that will make the statement true: $x = 15^4$

$15^4 = 15 \cdot 15 \cdot 15 \cdot 15 = 50{,}625$ The exponent indicates four factors of 15.

Therefore, if x is replaced with 50,625, the statement will be true. ■

EXAMPLE 4

A congressional committee proposes to increase the national debt by 13×10^9 dollars. Write the proposed increase in numerical form.

$$13 \times 10^9 = 13 \cdot 10 \cdot 10 \cdot 10 \cdot 10 \cdot 10 \cdot 10 \cdot 10 \cdot 10 \cdot 10$$
$$= 13 \cdot 1{,}000{,}000{,}000$$
$$= 13{,}000{,}000{,}000 \text{ (thirteen billion)}$$

The exponent (9) indicates 9 factors of 10.

The proposed increase is \$13,000,000,000. ■

■ PROCEDURE

To write the indicated product of like factors in exponential form, count the number of factors to determine the exponent for the base.

EXAMPLE 5

Write the following product using an exponent: $6 \cdot 6 \cdot 6 \cdot 6 \cdot 6$

$6 \cdot 6 \cdot 6 \cdot 6 \cdot 6 = 6^5$

There are five factors of 6; therefore, the exponent is 5, and the base is 6. ■

■ DEFINITION

Numerical Expressions

Numerical expressions are written using the usual symbols for numbers together with *signs of operation.*

Some of the signs of operation that are used are addition, subtraction, multiplication, and division. For example, "$2 \cdot 3 + 4$" is a numerical expression.

■ RULE

Evaluate a Numerical Expression

To *find the value of (evaluate) a numerical expression,* perform the indicated operations.

When a numerical expression has two or more indicated operations, evaluating could lead to different values if we were able to perform the operations in any order. To avoid the possible confusion of different values, an agreement has been established that guarantees only one answer. So,

$16 - 8 + 3$

$8 + 3$ Subtract.

11	Add.
and	
$15 - 3 \cdot 4$	
$15 - 12$	Multiply.
3	Subtract.

The agreement follows:

■ PROCEDURE

In an expression with two or more operations, perform the operations in the following order:

1. GROUPING SYMBOLS—perform operations included by grouping symbols first (parentheses, fraction bar, brackets, and braces) according to steps 2, 3, and 4.
2. EXPONENTS—perform operations indicated by exponents.
3. MULTIPLY AND DIVIDE—perform multiplication and division as they appear from left to right.
4. ADD AND SUBTRACT—perform addition and subtraction as they appear from left to right.

For the purposes of step 1, grouping symbols are used:

Parentheses	**Brackets**	**Braces**	**Fraction Bar**
( )	[ ]	{ }	——

In the examples that follow, different combinations of operations are involved.

EXAMPLE 6

Evaluate: $24 - \frac{8}{2} + 4(5)$

$24 - \frac{8}{2} + 4(5) = 24 - 4 + 4(5)$	Divide. Multiplication and division are done from left to right as they appear.
$= 24 - 4 + 20$	Multiply.
$= 20 + 20$	Subtract. Addition and subtraction are done from left to right as they appear.
$= 40$	Add. ■

EXAMPLE 7

Evaluate: $8 \cdot 6 - 3^2 + \frac{51}{17}$

$8 \cdot 6 - 3^2 + \frac{51}{17} = 8 \cdot 6 - 9 + \frac{51}{17}$ Exponents are done first. The square of three (3^2) is 9.

$= 48 - 9 + 3$ Multiply and divide from left to right as they appear.

$= 39 + 3$ Subtract.

$= 42$ Add. ■

EXAMPLE 8

Evaluate: $16 \cdot 3 - \frac{48 - 6^2}{12 - 6}$

$16 \cdot 3 - \frac{48 - 6^2}{12 - 6} = 16 \cdot 3 - \frac{48 - 36}{12 - 6}$ Exponents are done first. The fraction bar is a grouping symbol, so subtract above and below the fraction bar before dividing.

$= 16 \cdot 3 - \frac{12}{6}$

$= 48 - 2$ Multiply and divide from left to right.

$= 46$ Subtract. ■

EXAMPLE 9

Find the value of x if $12^2 - 4 \cdot 8 + 8^3 = x$.

$12^2 - 4 \cdot 8 + 8^3 = 144 - 4 \cdot 8 + 512$ Exponents are done first. $12^2 = 144$ and $8^3 = 512$.

$= 144 - 32 + 512$ Multiply.

$= 624$ Subtract and add from left to right as they appear.

Therefore, x has the value 624. ■

EXAMPLE 10

Find the value of b if $(15 + 3)^3 + 4(12 - 7)^2 = b$.

$(15 + 3)^3 + 4(12 - 7)^2 = (18)^3 + 4(5)^2$ Perform operations inside the parentheses first.

$= 5832 + 4 \cdot 25$	Exponents are done next. $18^3 = 5832$ and $5^2 = 25$.
$= 5832 + 100$	Multiply.
$= 5932$	Add.

Therefore, the value of b is 5932. ■

EXAMPLE 11

Find the value of V if $V = \frac{4}{3}(3.14)(6.5)^3$ to the nearest tenth.

ENTER	DISPLAY	
4	4.	
÷	4.	
3	3.	
×	1.3333333	4 has been divided by 3.
3.14	3.14	
×	4.1866667	The quotient is multiplied by 3.14.
6.5	6.5	
×	27.213333	Multiplied by 6.5 once.
6.5	6.5	
×	176.88667	Multiplied by 6.5 again.
6.5	6.5	
=	1149.7633	Multiplied by 6.5 again.

Therefore, $V \approx 1149.8$ The symbol "$\approx$" means "is approximately equal to." ■

EXAMPLE 12

The price of a stereo component system is $698. The Merry Melody Music Store offers a deferred-payment plan of $50 down and 36 monthly payments of $22 each. The full price (P) when the payment plan is used is

$P = 50 + 36 \cdot 22$.

What is the deferred-payment (P) price? To find the deferred-payment price of the stereo, evaluate the numerical expression.

$50 + 36(22) = 50 + 792$	Multiply.
$= 842$	Add.

The deferred-payment price is $842. ■

Exercise 1.2

A

Fill in the missing parts:

	Exponential Form	Factors	Value (Power)
1.	4^2		
2.	5^3		
3.		$3 \cdot 3 \cdot 3 \cdot 3$	
4.		$(4)(4)(4)$	
5.	$\left(\frac{1}{4}\right)^3$		
6.	$\left(\frac{1}{8}\right)^2$		
7.	$\left(\frac{1}{2}\right)^5$		
8.		$7 \cdot 7 \cdot 7$	
9.			7^2 or 49
10.			5^2 or 25
11.		$1 \cdot 1 \cdot 1 \cdot 1 \cdot 1 \cdot 1$	
12.		$1 \cdot 1 \cdot 1$	

Write each of the following products using exponents:

13. $3 \cdot 3 \cdot 3 \cdot 3$ **14.** $5 \cdot 5 \cdot 5 \cdot 5 \cdot 5$ **15.** $8 \cdot 8 \cdot 8$ **16.** $9 \cdot 9 \cdot 9 \cdot 9 \cdot 9 \cdot 9$

Find the value:

17. $\frac{40}{8} + 2$ **18.** $\frac{40}{8 + 2}$ **19.** $\frac{36 - 8}{4 + 3}$ **20.** $\frac{60 - 22}{28 - 9}$

21. $30 - (16 - 1) \div 5$ **22.** $24 - (32 - 5) \div 9$

23. $4(8 - 3) \div 10$ **24.** $12(7 + 2) \div 12$

B

Fill in the missing parts:

	Exponential Form	Factors	Value (Power)
25.	$\left[\frac{2}{3}\right]^4$		

	Exponential Form	Factors	Value (Power)
26.	$\left[\frac{5}{6}\right]^3$		

In each of the following, find the value of the variable that will make each statement true:

27. $12^2 = x$ **28.** $14^3 = y$ **29.** $22^3 = z$ **30.** $15^4 = w$

31. $12^3 = x$ **32.** $16^4 = y$ **33.** $17^4 = z$ **34.** $19^3 = w$

35. $24 - 9 \cdot 2 + 6 \cdot 3$ **36.** $12 \cdot 3 - 5 \cdot 4 + 1$ **37.** $28 \div 4 + 3 - 1$

38. $44 \div 2 \cdot 11 + 11$ **39.** $(55 + 25) \div 8 - 3$ **40.** $(90 - 36) \div (9 - 3)$

41. $68 \div 4 + 5 \cdot 2$ **42.** $5 \cdot 15 \div 3 + 4 \cdot 11$ **43.** $12(4 + 2) - 24 \div 12$

44. $(5 \cdot 2 + 3)2 - 10$ **45.** $20 - [(28 - 11) - 2^3]$ **46.** $(8^2 - 56) \div (20 \div 5) + 5^2$

47. $21 + 18(15 - 4) - 3(14 - 9)$ **48.** $48 + 12(18 - 6) - 4(21 - 3)$

49. $(5^2 - 7) + 3 \cdot 8 \div 6 - 9$ **50.** $12(69 - 21) \div 24 + 14(2^3 + 3^2)$

51. $\dfrac{16(35 - 19)}{4^2}$ **52.** $\dfrac{18(48 - 3)}{9^2}$ **53.** $\dfrac{5(18 - 7) + 1}{2^3 - 1}$ **54.** $\dfrac{26(3^3 - 3) + 4 \cdot 6}{5^2 - 7}$

C

Fill in the missing parts:

	Exponential Form	Factors	Value (Power)
55.		$12 \cdot 12 \cdot 12 \cdot 22 \cdot 22 \cdot 22$	
56.		$7 \cdot 7 \cdot 7 \cdot 8 \cdot 8 \cdot 8 \cdot 9 \cdot 9$	
57.	$13^4 \cdot 5^3 \cdot 8^0$		
58.	$25^0 \cdot 8^3 \cdot 12^3$		

In each of the following, what number could replace the variable and make the statement true?

59. $9^3 \cdot 4^2 = x$ **60.** $5^3 \cdot 3^2 = y$

61. $8^3 \cdot 4^3 = z$ **62.** $3^4 \cdot 5^3 = w$

63. $2^3 \cdot 3^2 = v$ **64.** $9^3 \cdot 2^5 = a$

65. $135 + 15(21 - 3 \cdot 4) \div 45$ **66.** $184 - 21(36 - 5 \cdot 6) \div 18$

67. $\dfrac{12^2 + 8 \cdot 9 - 5^3}{4^4 - 3^3 - 6 \cdot 37}$ **68.** $\dfrac{15^2 - 10^2 + 7 \cdot 4}{9^2 - 8^2}$

69. $\dfrac{[32(20 - 12) - 6^2(30 - 5^2)]}{3(9^2 - 1) \div (8^2 - 4)}$ **70.** $\dfrac{5^2 \cdot 3 - 7 \cdot 2^2 + 4(3^2 - 5)}{8(2^3 + 1) - 5(4^2 - 3)}$

71. $16^3 - 12(8 + 7)^2 + (8 + 3)^4$

72. $25^2 + 15^2(8 - 3)^4 - 7(7 + 9)^3$

73. $4[15(12 + 5) - 8(14 + 3)] - 15$

74. $8[16(18 - 4) - 4(16 - 12)] - 14$

75. $9^2 + 18^2(15 + 2)^4 - 8^3(5 + 9)^4$

76. $25^3 + 50^2(12 + 6)^3 - 12^2(14 - 6)^5$

D

77. A fiscal subcommittee suggests cutting a state debt by 4×10^5 dollars. Write this value in numeral form.

78. A midwestern state had a tax surplus of 16×10^7 dollars for a given year. Write this value in numerical form.

79. The Earth is more than 7×10^8 years old. Write this number in numeral form.

80. A sample of bacteria contained 5×10^6 bacteria. Write this number in numeral form.

81. The distance from the Earth to the nearest star (Alpha Centauri) is approximately 255×10^{11} miles. Write this distance in numeral form.

82. A service station pumped 2.5×10^4 gallons of gas in a given month. Write this value in numeral form.

83. The ABC Publishers have sold 362×10^4 books in the year 1989. Write this number in numeral form.

84. During one week, the state lottery collected 52×10^5 dollars. Write this value in numeral form.

85. In a pyramid investment scheme, two investors put up \$2000 each. Each investor then finds two more investors, and these four new members each find two more, and so on. To complete the tenth level, 2^{10} investors are needed. Write this value in numeral form, and find the number of dollars invested at this level.

86. In a pyramid investment scheme, three investors put up \$1000 each. Each investor then finds three more investors, and these nine each find three more, and so on. To complete the sixth level, 3^6 investors are needed. Write this value in numeral form, and find the number of dollars invested at the sixth level.

87. Find the value of A if $A = 3.14(128.35)^2$.

88. Find the value of F if $F = \frac{9}{5} \cdot 30 + 32$.

89. Find the value of y if $y = (18.32)(27)$.

90. Find the value of x if $x = (16.14)(32) + 18.7$.

91. Jane buys a stereo for \$75 down and \$31 per month for 12 months. The total price (T) that she will pay is $T = 75 + 12(31)$. What is the total price of the stereo?

92. The E-Z Chair Company advertises recliners for \$10 down and \$15 per month for 24 months. What will be the total cost to the consumer under this plan? (*Hint:* The total cost will be $10 + 15 \cdot 24$.)

93. The local auto import dealer offers compact cars for \$210 down and \$138 per month for 48 months. What is the total cost to the consumer?

94. An auto dealer offers a midsize car for \$300 down and \$180 per month for 48 months. What is the total cost to the consumer?

95. Pete buys a car from his friend José. He agrees to pay José \$100 a month for 12 months and then increase his payment to \$125 per month for the final 24 months. How much does Pete pay José?

96. The court orders Jane to pay \$250 per month for child support for the next three years and then decreases the payment to \$175 per month for the following two years. How much did the court order Jane to pay for child support during the next five years?

STATE YOUR UNDERSTANDING

97. What is the power or value of an exponential expression?

98. Locate the error in the following example. Indicate why it is not correct. Determine the correct answer.

$$
\begin{aligned}
&2\{2^2 - 2[10^2 - 2(4^2 - 3^2)^2] + 5^2\} = \\
&2\{2^2 - 2[10^2 - 2(16 - 9)^2] + 5^2\} = \\
&2\{2^2 - 2[10^2 - 2(7)^2] + 5^2\} = \\
&2\{2^2 - 2[100 - 2(49)] + 5^2\} = \\
&2\{2^2 - 2[100 - 98] + 5^2\} = \\
&2\{2^2 - 2[2] + 5^2\} = \\
&2\{4 - 2[2] + 25\} = \\
&2\{2[2] + 25\} = \\
&2\{4 + 25\} = \\
&2\{29\} = \\
&58
\end{aligned}
$$

CHALLENGE EXERCISES

Determine the value:

99. $23 - \{5 - [16 - 4(4 - 1)1]\}$

100. $12 - 7(6 - 4)^2 \div 4 + 3$

101. $\{(250 \cdot 6) \div [(4 \cdot 16) \div (8 \cdot 2)]\} + 25 \cdot (8 \div 4)$

102. $2\{3^2 - 2[4^2 - 2(4^2 - 3^2)] + 5^2\} \div [4 - 3(2^2 - 3)^2]$

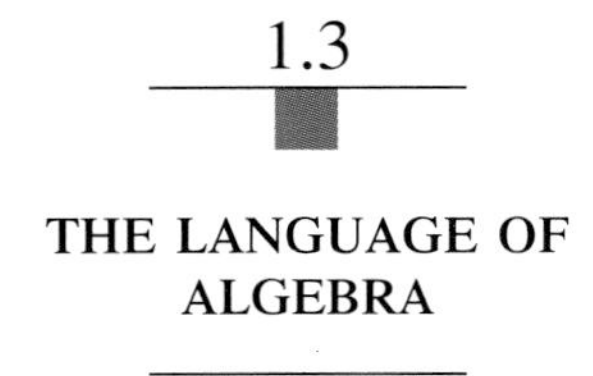

1.3 THE LANGUAGE OF ALGEBRA

OBJECTIVES

1. Evaluate algebraic expressions and formulas.
2. Translate an English phrase to algebraic form.
3. Determine whether a number is a solution of an equation.

In this section, terminology associated with algebra is introduced. Here are some of the words we will use.

A *set* is a collection of objects such as a chess set or a set of flatware or tableware. The objects contained in the set are called the *members* or *elements* of the set. In mathematics, a pair of braces, { }, is used to enclose the members of a set. If we write $A = \{1, 2, 3\}$, then A represents the set $\{1, 2, 3\}$. That is, A is the set whose members or elements are 1, 2, and 3. A set with no members or elements is called the *empty set* or *null set*. The symbol used to identify the empty or null set is "$\emptyset$."

Algebraic Expressions

DEFINITION

Algebraic expressions are formed by using constants and variables together with signs of operations.

Formula

DEFINITION

A *formula* is shorthand for stating a rule or a method of procedure.

A formula states that two algebraic expressions are equal. That is, they name the same number. To *evaluate* a formula or an expression is to find the value of that formula or expression given the replacement values for the variables or letters.

The following formulas are among those used most often in arithmetic and geometry:

$A = \ell w$ Area of a rectangle:

$P = 2\ell + 2w$ Perimeter of a rectangle:

$P = 4s$ Perimeter of a square:

w
ℓ
s
s s
s

$D = rt$	Distance (D) traveled at speed or rate (r) for a given amount of time (t).

These formulas illustrate some of the ideas of algebra. The variables A, ℓ, w, p, s, D, r, and t are used to represent numbers.

The variable D (for distance) does not represent the word *distance* but represents the number of miles (or number of feet or kilometers) traveled.

The formula $A = \ell w$ is read "A equals ℓ times w," which is shorthand for saying "The area is the product of the length (ℓ) and the width (w)."

The formula $V = \ell wh$ can be used to find the volume of a box. We can *evaluate* or solve for V when we know the value of ℓ, w, and h.

EXAMPLE 1

Find the value of V in $V = \ell wh$ if $\ell = 2$, $w = 3$, and $h = 5$.

$V = \ell wh$	Formula.
$V = 2 \cdot 3 \cdot 5$	Substitute.
$V = 30$	Multiply.

So V has a value of 30. ■

Solution of a Formula

In Example 1, the number 30 is called the *solution* for V. In any formula, we can find the solution (replacement) for a single variable when the other values are listed.

An algebraic expression can be evaluated when the replacement values for the variables involved are known.

EXAMPLE 2

Evaluate: $x + 14$ if $x = 25$

$x + 14 = 25 + 14$	Replace x with 25, and perform the indicated operations.
$= 39$	

■

■ PROCEDURE

To evaluate expressions and find solutions for formulas:

1. Replace the variables with the values given in the problem.
2. Perform the indicated operations.

EXAMPLE 3

Evaluate: $3xy + 2x$ if $x = 20$ and $y = 12$

$3xy + 2x = 3(20)(12) + 2(20)$ Replace x with 20 and y with 12.

$= 720 + 40$ Multiply first.

$= 760$ Add. ■

EXAMPLE 4

Find the solution for A in the formula $A = 0.5bh$ when $b = 30$ and $h = 15$.

$A = 0.5(30)(15)$ Substitute 30 for b and 15 for h.

ENTER	DISPLAY
.5	0.5
×	0.5
30	30.
×	15.
15	15.
=	225.

So $A = 225$. ■

Much of algebra is understanding what algebraic expressions mean and how to write them given a common language version of the expressions in a ''real-life'' situation. To translate an arithmetic or algebraic expression to an English phrase, consider the following table.

Arithmetic Form	Algebraic Form	English Phrase
	Addition	
$2 + 3$	$x + 3$	The sum of x and 3
	$2 + y$	y more than 2
	$x + y$	x increased by y
	Subtraction	
$10 - 3$	$x - 3$	3 less than x *or* x less 3
	$10 - y$	Subtract y from 10 *or* 10 minus y
	$x - y$	x diminished by y *or* the difference of x and y
	$82 - w$	The value after w is subtracted from 82 *or* the difference of 82 and w

Arithmetic Form	Algebraic Form	English Phrase
	Multiplication	
3×6	$(6)x$	6 times x
$3 \cdot 6$		
$3(6)$	$3(y)$, $3 \cdot y$, $3y$	y multiplied by 3 (*Note:* In $3y$ the parentheses or the $\cdot$ has been dropped. This is the usual way to write the product of a variable and a constant.)
$(3)(6)$	$x \cdot y$, xy, $(x)(y)$	The product of two numbers represented by x and y
$5 \cdot 5 \cdot 5$	x^3	x cubed *or* a number cubed *or* the cube of a number
	Division	
$3 \div 4$	$x \div 4$, $\frac{x}{4}$	x divided by 4 (*Note:* In algebra, this is most often written as $\frac{x}{4}$.)
$4\overline{)3}$		
$\frac{3}{4}$	$\frac{x}{y}$	x over y *or* x divided by y *or* y into x

EXAMPLE 5

Translate to English: $a - 24$

The English phrase could be any one of the following:
"A number minus 24"
"The difference of a number and 24"
"Subtract 24 from a number"

The minus sign means "less," "difference," or "subtract," and if a represents a number, we list three possibilities. ■

EXAMPLE 6

Translate to English: x^4

"A number raised to the fourth power."

The letter x represents a number with an exponent of 4. ■

The process of translating from an English phrase to an algebraic form (expression) is shown in the following table:

English Phrase	Variable	Operation	Algebraic Form
Three more than a given number	Let x represent the given number	"More than" indicates addition	$x + 3$
A number decreased by 5	Let y represent the number	"Decreased" indicates subtraction	$y - 5$
The product of 6 and a number	Let a represent the number	"Product" indicates multiplication	$6 \cdot a$ or $6a$
The square of a number	Let b represent the number	"Square" indicates that the base is used as a factor two times, or there is an exponent of 2	b^2
A number cubed	Let z represent the number	"Cube" indicates an exponent of 3	z^3

PROCEDURE

To translate from English phrases to algebraic form, represent each unknown number with a variable, and indicate the operation.

EXAMPLE 7

Translate to algebraic form: two less than a number

The algebraic form is $x - 2$. Let x represent the number; "less than" indicates subtraction. ■

EXAMPLE 8

Translate to algebraic form: five more than the product of two numbers

xy represents the product. Let x represent one of the numbers and y the other. "Product" indicates multiply.

The algebraic form is $xy + 5$. "More than" indicates addition. ■

DEFINITION

Equation

An *equation* is a statement about numbers that states that two expressions are equal.

The symbol "=" is used to show this relationship.

DEFINITION

Solution

The *solution set* of an equation is the set of all numbers that, when used to replace the variable, make the equation true. A *solution* of an equation is a number that is a member of the solution set.

To tell whether a given number is a solution of a given equation, replace the variable in the equation with the number. Perform the indicated operations. If the statement is true, the number is a solution. If the statement is false, the number is not a solution.

EXAMPLE 9

Is 2 a solution of the equation $2x + 4 = 8$?

$2x + 4 = 8$	Original equation.
$2(2) + 4 = 8$	Replace x with 2.
$4 + 4 = 8$	Multiply first.
$8 = 8$	The statement is true.

2 is a solution of the equation. ■

PROCEDURE

To determine whether a number is a solution of an equation:

1. Replace the variable with the given number.
2. Perform the indicated operations.
3. If the result is a true statement, the number is a solution. If the result is a false statement, the number is not a solution.

EXAMPLE 10

Is 5 a solution of the equation $8x - 12 = 25$?

$8x - 12 = 25$

$8(5) - 12 = 25$ Replace x with 5.

$40 - 12 = 25$ Multiply.

$28 = 25$ Subtract.

The statement is false, so 5 is not a solution. ■

EXERCISE 1.3

A

If $x = 4$, $y = 6$, and $z = 11$, evaluate:

1. $x + y$

2. $z - x$

3. $\frac{66}{y}$

4. xy

If $a = 7$, $b = 21$, and $c = 3$, evaluate:

5. $29 - b$

6. ac

7. $\frac{b}{a}$

8. a^4

9. b^2

Translate the following expressions to algebraic form; let x represent the given number:

10. 6 more than a given number

11. 15 less than a given number

12. the square of a number

Translate the following algebraic expressions to English:

13. $x - 20$

14. $a + 13$

15. $24y$

16. Is 7 a solution of $x + 4 = 3$?

17. Is 12 a solution of $x - 6 = 6$?

18. Is 8 a solution of $x + 13 = 22$?

B

If $x = 72$, $y = 18$, and $z = 6$, evaluate:

19. yz

20. $100 - x$

21. $\frac{x}{y}$

22. $\frac{z}{y}$

23. $x + y + z$

24. $x - y$

If $a = 50$, $b = 10$, and $c = 25$, evaluate:

25. $50 - a$ **26.** bc **27.** $\frac{a}{b^2}$ **28.** $\frac{b^2}{a}$

29. $c - b$ **30.** $a + b + c$ **31.** a^2 **32.** c^3

Translate the following expressions to algebraic form; let x represent the first number and y represent the second number:

33. the first number minus the second number

34. 12 less the second number

35. the product of the two numbers

36. the second number divided by the first number

37. the sum of the two numbers plus 18

38. twice the first number decreased by the second number

39. the square of the first number times the second number

40. the first number times the square of the second number

41. Is 8 a solution of $2x - 12 = 4$?

42. Is 9 a solution of $3x - 7 = 18$?

43. Is 14 a solution of $5x + 9 = 79$?

44. Is 16 a solution of $8x - 12 = 126$?

C

If $x = 6.15$, $y = 2.05$, and $z = 0.0123$, evaluate:

45. xy **46.** $x + y + z$ **47.** $\frac{x}{z}$ **48.** $\frac{x}{y}$

49. $10z$ **50.** $3.0495 - y$ **51.** x^3 **52.** y^4

53. z^2 **54.** z^4 **55.** $x^2 - y^3$

56. $x^2 - y - z$ **57.** $x^2 + 5y + z^3$ **58.** $x^2y^2z^5$

59. Is 13 a solution of $3x - 4 = 2x + 9$?

60. Is 12 a solution of $4x + 7 = 5x - 5$?

61. Is 7 a solution of $9x - 8 = 4x + 27$?

62. Is 9 a solution of $12x + 4 = 21x - 77$?

63. Is 7 a solution of $9x - 7 = 4x + 25$?

64. Is 11 a solution of $7x + 5 = 115 - 3x$?

D

65. Find D if $r = 50$ and $t = 4$:
$D = rt$ (Distance formula)

66. Find I if $p = 1000$, $r = 0.08$, and $t = 2$:
$I = prt$ (Simple interest formula)

67. Find C when $d = 8$; $\pi = 3.14$:
$C = \pi d$ (Circumference of a circle)

68. Find P when $s = 16$:
$P = 4s$ (Perimeter of a square)

69. Find the area of a rectangular plot of ground that has a length of 15 feet and a width of 12 feet.

70. If Reggie Pete earns $5.25 per hour, how much will he make in 35 hours?
[*Hint:* total earnings = (hourly wage)(hours worked), or $e = wh$]

71. How many eggs are there in 60 cartons, each of which contains 12 eggs?
[*Hint:* number of eggs = (number of cartons)(12 eggs per carton), or $E = C(12)$]

72. The number of cents in x quarters is given by the formula $c = 25x$. How many cents are in 19 quarters?

73. Portland and Seattle are 180 miles apart. How long will it take the Blazers to make the trip to Seattle if they average 45 miles per hour?
$\left(\textit{Hint:}\ \text{time} = \dfrac{\text{distance}}{\text{rate}},\ \text{or } t = \dfrac{D}{r}\right)$

74. If a can of Vinesweet peas costs 62¢, what will a case of 24 cans cost?
[*Hint:* cost = (price per can)(number of cans), or $c = pn$]

75. Ohm's law is $I = \dfrac{E}{R}$, where I is measured in amperes, E is the voltage, and R is the resistance in ohms. Find the amperage if the voltage is 115 and the resistance is 10 ohms.

76. $P = EI$ is a formula to determine the power (P) in watts given the volts (E) and the amperage (I). How much power is developed if $E = 120$ and $I = 25$?

77. The formula used to determine a person's IQ is $\text{IQ} = \dfrac{100(MA)}{CA}$, where MA is a person's mental age as measured by a test and CA is that person's chronological age. What is a person's IQ if her MA is 32 and her chronological age is 25?

78. The formula for finding the area of a circle is $A = \pi r^2$. What is the area of a circle with radius (r) of 2.5 cm if $\pi \approx 3.14$? (Recall that "$\approx$" means "approximately equal to.")

STATE YOUR UNDERSTANDING

79. What is an English translation of $2x - 3$?

80. Given the equation $5y + 13 = 43$. Describe how you would go about determining whether or not $y = 7$ is a solution of the equation.

CHALLENGE EXERCISES

81. Find the value of $\dfrac{ac + x - bd}{cy - ax + ab}$ when $a = 3$, $b = 5$, $c = 4$, $d = 1$, $x = 8$, and $y = 6$.

82. Find the value of $\dfrac{x^4 - 2x^2 + 1}{(x - 1)^2(x + 1)^2}$ when $x = 4$.

83. The electric current I, in amperes delivered by 4 batteries in series is given by $I = \dfrac{nE}{R + nr}$. Find I when $n = 4$, $E = 1.5$ volts, $R = 8$ ohms, and $r = 0.4$ ohms.

84. The electric current, I, in amperes through three resistances in parallel is given by the formula $I = E\left(\frac{1}{A} + \frac{1}{B} + \frac{1}{C}\right)$. Find I when $E = 220$ volts, $A = 0.20$ ohms, $B = 0.25$ ohms, and $C = 0.50$ ohms.

1.4

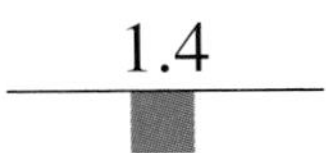

REAL NUMBERS AND THEIR PROPERTIES

OBJECTIVES

1. Identify natural numbers, whole numbers, integers, rational numbers, and irrational numbers.
2. Match real numbers with points on the number line.
3. Determine which of two real numbers is smaller.

Real Numbers

Part of the study of algebra includes the study of the set of real numbers. To illustrate the set of real numbers, we use a *number line*. It is called the real number line, and any number associated with a point on this line is called a *real number*. To construct a number line, draw a horizontal line, and select a point to associate with zero. Then select a convenient unit of measure, and mark off equal distances in either direction from the point labeled zero. Those points to the right of zero are labeled 1, 2, 3, . . . , and those to the left of zero are labeled −1, −2, −3, ... The three dots indicate that the number line continues without end in either direction. The symbol "−" is used for numbers to the left of zero. Such a number line is shown in Figure 1.1.

DEFINITION

Positive Numbers

Numbers to the right of zero are called *positive numbers* and can be written with a positive sign (+) in front.

Negative Numbers

Numbers to the left of zero are called *negative numbers* and are written with a negative sign (−) in front.

Positive and negative numbers are sometimes called *signed numbers.*

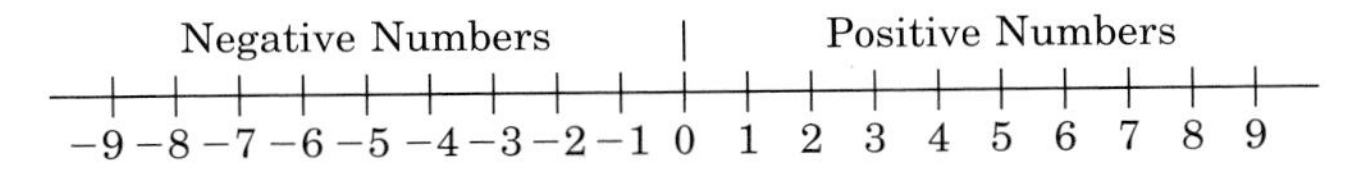

FIGURE 1.1

DEFINITION

Integers

The set

$\{\ldots, -5, -4, -3, -2, -1, 0, 1, 2, 3, 4, 5, \ldots\}$

make up the *integers.*

A few integers are pictured on the number line in Figure 1.1. Note that the whole numbers make up part of those numbers we call integers.

Zero is neither positive nor negative.

The set of *rational numbers* contains the positive rational numbers, zero, and negative rational numbers. Some rational numbers are shown on the number line in Figure 1.2.

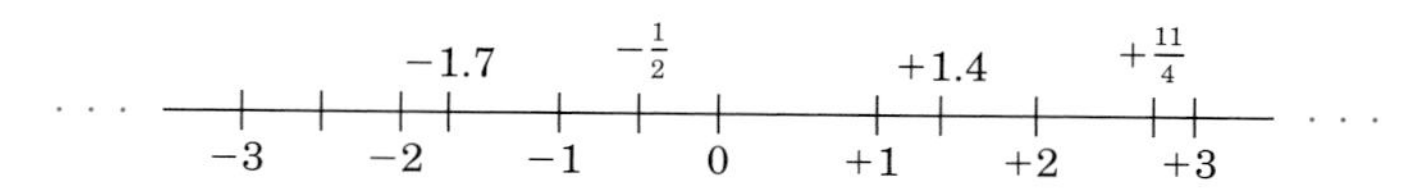

FIGURE 1.2

When a set contains more than a few numbers, we use *set builder notation.* An example of set builder notation follows.

If R is used to denote the set of real numbers, we have

Set Builder Notation

$R = \{x \mid x$ is a number associated with a point on the number line$\}$.

This is read "R is the set of all x such that x is a number associated with a point on the number line."

The following are some symbols that name real numbers and how to read them:

$+7$	Seven or positive seven
-0.5	Negative five tenths
-3	Negative 3
2	Two or positive two

The set of real numbers comprises more than one set of numbers. Some of the sets within the real numbers are:

Integers	$\{\ldots, -4, -3, -2, -1, 0, 1, 2, 3, 4, \ldots\}$
Rational numbers	$\{x \mid x$ is the quotient of two integers, with the divisor not zero$\}$

Some examples of rational numbers are:

$$\frac{3}{5}, -\frac{2}{3}, -8, 0, 5, \frac{15}{16}, 5\frac{1}{2}, -\frac{18}{5}, \quad \text{and} \quad -19.$$

Irrational numbers $\{x | x \text{ is not a rational number}\}$

Some examples of irrational numbers are:

$\sqrt{2}, -\sqrt{5}, \sqrt{10}, -\sqrt{15},$ and π.

The symbol π represents the ratio of the circumference of a circle to its radius.

The following chart shows the classifications of real numbers.

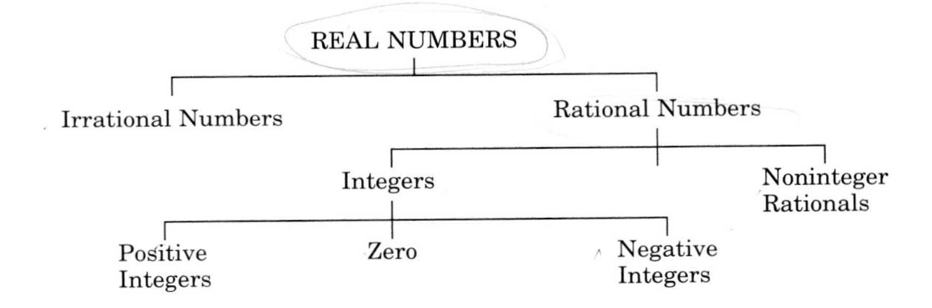

EXAMPLE 1

Given $\left\{-5, -\frac{15}{4}, -3, -\sqrt{5}, -2.5, -1, 0, 1, 3, \frac{21}{5}, \sqrt{29}, 34.7\right\}$, list

(1) real numbers,
(2) rational numbers,
(3) irrational numbers,
(4) positive numbers,
(5) negative numbers.

1. All are real numbers

2. $\left\{-5, -\frac{15}{4}, -3, -2.5, -1, 0, 1, 3, \frac{21}{5}, 34.7\right\}$

3. $\{-\sqrt{5}, \sqrt{29}\}$

4. $\left\{1, 3, \frac{21}{5}, \sqrt{29}, 34.7\right\}$

5. $\left\{-5, -\frac{15}{4}, -3, -\sqrt{5}, -2.5, -1\right\}$ ■

EXAMPLE 2

Match the points on the number line with the following numbers:
$\left\{-5, -\frac{15}{4}, -1.5, -\frac{1}{4}, 0, \frac{1}{3}, 3, \frac{21}{5}, 4.5\right\}$

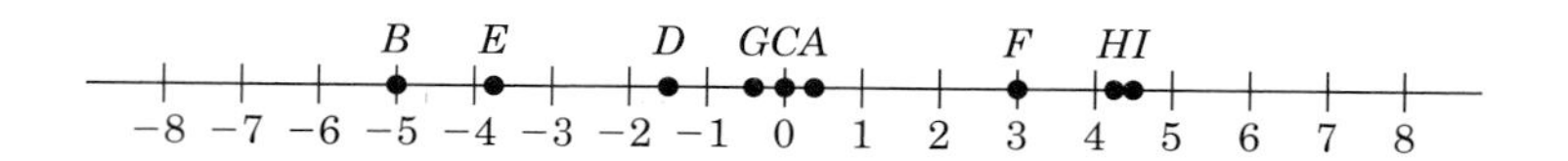

$A \leftrightarrow \frac{1}{3}$, $B \leftrightarrow -5$, $C \leftrightarrow 0$, $D \leftrightarrow -1.5$,

$E \leftrightarrow -\frac{15}{4}$, $F \leftrightarrow 3$, $G \leftrightarrow -\frac{1}{4}$, $H \leftrightarrow \frac{21}{5}$, $I \leftrightarrow 4.5$ ■

To write the relationship between two numbers when one is larger or smaller than the other, we use the following symbols:

Symbol	English Phrase
$<$	less than (points to the smaller number)
$>$	greater than (points to the smaller number)
$\leq$	less than or equal to
$\geq$	greater than or equal to
$\neq$	not equal to

When comparing whole numbers, you can determine by inspection which is the larger or smaller. When working with real numbers, we compare by graphing the numbers on the number line. The larger is on the right (see Figure 1.3).

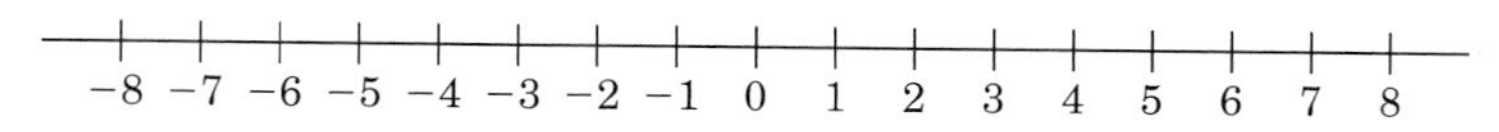

FIGURE 1.3

From the number line we see that:

2 is to the left of 5,	so $2 < 5$	or	$5 > 2$
0 is to the left of 3,	so $0 < 3$	or	$3 > 0$
1 is to the left of 8,	so $1 < 8$	or	$8 > 1$

and

-1 is to the left of 5,	so $-1 < 5$	or	$5 > -1$
-2 is to the left of 0,	so $-2 < 0$	or	$0 > -2$
-5 is to the left of -3,	so $-5 < -3$	or	$-3 > -5$

In general:

Less Than

Given any two real numbers x and y, if x is to the left of y on the number line, then $x < y$.

It also follows that if $x < y$, then $y > x$.

EXAMPLE 3

Are the following statements true or false?
(1) $-3 < 3$
(2) $0 > -1$
(3) $-5 > -2$
(4) Any negative number is less than 0

1. $-3 < 3$, True — -3 is to the left of 3 on the number line.

2. $0 > -1$, True — 0 is to the right of -1 on the number line.

3. $-5 > -2$, False — -5 is to the left of -2 on the number line.

4. Any negative number is less than zero, True — All negative numbers are to the left of zero. ■

EXERCISE 1.4

A

Given $A = \left\{-12, -9.8, -7, -\frac{15}{4}, -1, 0, +2, \pi, \frac{7}{2}, 9.3, +18\right\}$:

1. List the whole numbers.

2. List the integers.

3. List the positive numbers.

4. List the negative numbers.

5. List the rational numbers.

6. List the irrational numbers.

Match the points on the number line with the following numbers:

$$\left\{-8, -\frac{15}{2}, -5.8, -4, -1, 0.5, 1, \frac{7}{3}, 6, 7.5\right\}$$

A E J G D B I C F H

−8 −7 −6 −5 −4 −3 −2 −1 0 1 2 3 4 5 6 7 8

7. $A \leftrightarrow$

8. $B \leftrightarrow$

9. $C \leftrightarrow$

10. $D \leftrightarrow$

11. $E \leftrightarrow$

12. $F \leftrightarrow$

13. $G \leftrightarrow$

14. $H \leftrightarrow$

15. $I \leftrightarrow$

16. $J \leftrightarrow$

True or false:

17. $5 < 12$

18. $0 < 8$

19. $16 > 12$

20. $48 > 21$

B

True or false:

21. 2.5 is an integer.

22. −4 is a negative number.

23. −8 is a rational number.

24. 14 is a rational number.

25. −3.2 is a real number.

26. 7.6 is a real number.

27. 16 is an integer.

28. −17 is an integer.

29. 0 is a positive number.

30. 0 is a negative number.

31. $-5 < 12$

32. $-7 < 0$

33. $-15 < -4$

34. $-17 < -9$

35. $0 > 5$

36. $0 < -4$

37. $-8 < 3$

38. $9 > -2$

39. $-15 < -24$

40. $-21 > -14$

C

True or false:

41. $\sqrt{5}$ is a real number.

42. $\sqrt{5}$ is an irrational number.

43. $\sqrt{5}$ is a rational number.

44. $\sqrt{5}$ is an integer.

45. 3.9 is a real number.

46. 3.9 is an irrational number.

47. 3.9 is a rational number.

48. 3.9 is an integer.

49. −8.2 is an irrational number.

50. −8.2 is a real number.

51. −8.2 is an integer.

52. −8.2 is a rational number.

53. $-8 > -4$

54. $-12 > -7$

55. 0 is neither positive nor negative.

56. All integers are rational numbers.

57. All rational numbers are integers.

58. All irrational numbers are real numbers.

Translate the following expressions to algebra.

59. 32 is greater than 25.

60. 4 + 8 is less than 18.

61. 8 − 2 is less than 6 + 2.

62. 80 ÷ 4 is less than 7 · 6.

63. The product of 8 and 4 is greater than the sum of 9 and 6.

64. The product of 12 and 3 is less than the cube of 4.

65. 20 decreased by 2 times the sum of 3 and 1 is greater than 9.

66. The quotient of 40 and 2 decreased by the product of 3 and 5 is less than 7.

STATE YOUR UNDERSTANDING

67. What is the difference between a rational number and an irrational number?

68. Explain the difference between the following:
A. 4 less than 9 B. 4 is less than 9

CHALLENGE EXERCISES

Write an inequality for each of the following. Let x represent the unknown number. Find two numbers that will make each of the statements true.

69. Six times a number is less than 54.

70. A number increased by 8 is greater than 17.

71. A number decreased by 12 is less than 25.

72. Twice a number increased by 7 is less than 40.

OPPOSITES, ABSOLUTE VALUE, AND ADDING REAL NUMBERS

OBJECTIVES

1. Find the opposite (additive inverse) of a real number.
2. Find the absolute value of a real number.
3. Add real numbers.

Wherever quantities can be measured in opposite directions, positive and negative numbers can be used to show direction.

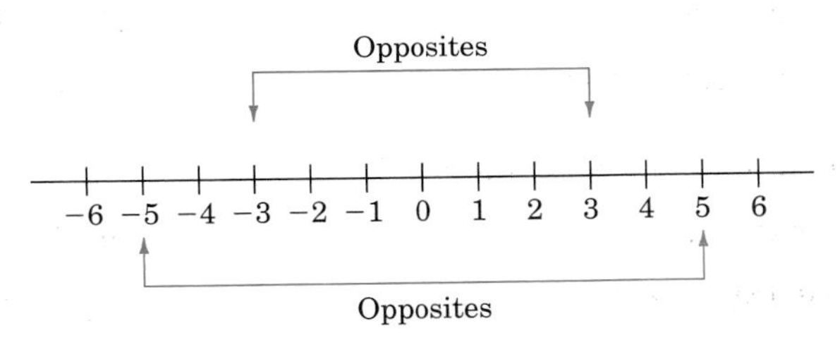

DEFINITION

Opposites

Opposites are pairs of numbers that are the same distance from zero on the number line but on opposite sides. Another name that is given to them is *additive inverse*.

The opposite of a number is indicated by writing a dash in front of the number.

$-(5) = -5$	The opposite of positive 5 is negative 5.
$-(-3) = 3$	The opposite of negative 3 is positive 3.
$-(0) = 0$	The opposite of 0 is 0.

The dash in front of a number can now be read in two different ways:

-19	The opposite of 19.
-19	Negative 19.

We avoid saying "minus 19" to prevent confusion with subtraction. The dash in front of a variable will be read only one way:

$-x$	The opposite of x.

It is sometimes read "the negative *of* x," but we avoid this to prevent confusion with negative numbers.

Note that the opposite of negative 3 is written $-(-3)$ and not as $--3$. This is done to avoid the possibility of writing the two "$-$" symbols so close together that they are thought to be one. Using the parentheses leaves no doubt about what is meant.

RULE

In general, if a represents a positive number:

$-(a) = -a$	The opposite of a positive number is a negative number.
$-(-a) = a$	The opposite of a negative number is a positive number.
$-(0) = 0$	The opposite of zero is zero.

EXAMPLE 1

Write the symbol for the opposite of 8.

The symbol is $-(8)$.

$-(8) = -8$

The opposite of 8 units to the right of zero.

The opposite of 8 is negative 8. ■

EXAMPLE 2

Write the symbol for the opposite of negative 9.

The symbol is $-(-9)$.

$-(-9) = 9$

The opposite of 9 units to the left of zero.

The opposite of -9 is 9. ■

EXAMPLE 3

Read the symbol -12.

The symbol -12 can be read "negative twelve" or "the opposite of twelve."

There are two ways of reading the symbol. ■

EXAMPLE 4

Write $-(-37)$ as a signed number.

$-(-37) = 37$

The symbol is read "the opposite of negative thirty seven." ■

EXAMPLE 5

For what replacement of x is $-x = 4$ true?

$-x = 4$

If the opposite of x is 4, then we want to know what number has an opposite of 4.

The replacement for x is -4 since $-(-4) = 4$.

The number must be negative since its opposite is positive. ■

EXAMPLE 6

Translate to algebraic form: two more than the opposite of a number. Let x represent the number.

$-x$

If x represents the number, $-x$ will represent the opposite of the number.

$-x + 2$

"Two more than the opposite of a number" means that 2 is added to the opposite of x. ■

When the units in a number are important but its location on the number line is not, we use what is called the *absolute value of the number*.

DEFINITION

Absolute Value

> The *absolute value* of a signed number is the distance between the number and zero on the number line, and this distance is always taken to be positive.

To indicate the absolute value of a number, we write it between two vertical bars. For instance, $|5|$ is read "the absolute value of 5."

RULE

> In general, if a is positive:
>
> $|a| = a$ The absolute value of a positive number is itself.
>
> $|-a| = a$ The absolute value of a negative number is its opposite.
>
> $|0| = 0$ The absolute value of zero is zero.

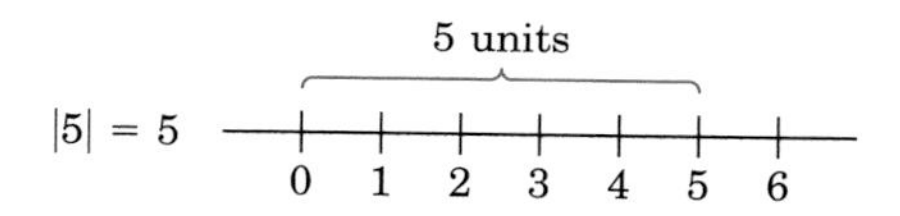

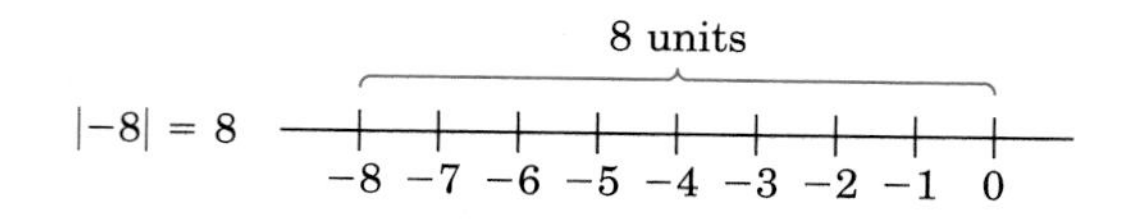

0 units

$|0| = 0$

−2 −1 0 1 2

EXAMPLE 7

Write the symbol for and find the value of the absolute value of six.

The symbol for the absolute value of six is $|6|$.

$|6| = 6$ The absolute value of a positive number is positive. ■

EXAMPLE 8

Write the symbol for and find the value of the absolute value of negative fourteen.

The symbol for the absolute value of negative fourteen is $|-14|$.

$|-14| = 14$ The absolute value of a negative number is positive. ■

EXAMPLE 9

Write $-|15|$ as a signed number.

$-|15| = -(15)$ The absolute value of a positive number is positive.

$= -15$ The opposite of a positive number is negative. ■

EXAMPLE 10

For what replacement of x is $|x| = 3$ true?

$|x| = 3$ There are two integers with an absolute value of 3.

If x is 3 or -3, the statement will be true since the absolute value of each is 3. ■

Recall that on the number line, the negative numbers are to the left of zero, and the positive numbers are to the right of zero. Thus, going to the left is the negative direction, and going to the right is the positive direction. We can represent positive and negative numbers on the number line using arrows to show the direction. The numbers -5 and 4 are represented on the number line below.

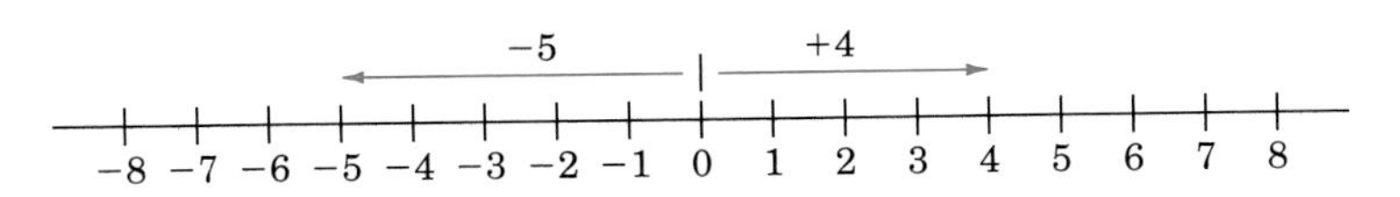

The arrows do not need to start at zero. As long as they contain the correct number of units and go in the correct direction, that is all that is necessary. For example,

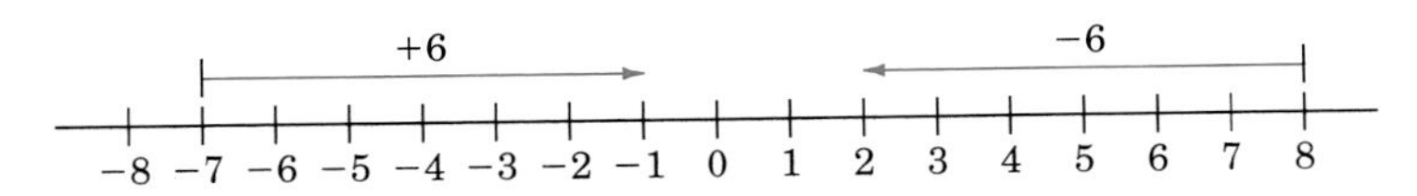

Addition of integers on the number line can be shown using the arrows. Starting at 0, draw an arrow (to the left or the right) that represents the first addend. The second arrow begins at the end of the first arrow and (either left or right) represents the second addend. Where the second arrow ends is the sum of the two signed numbers.

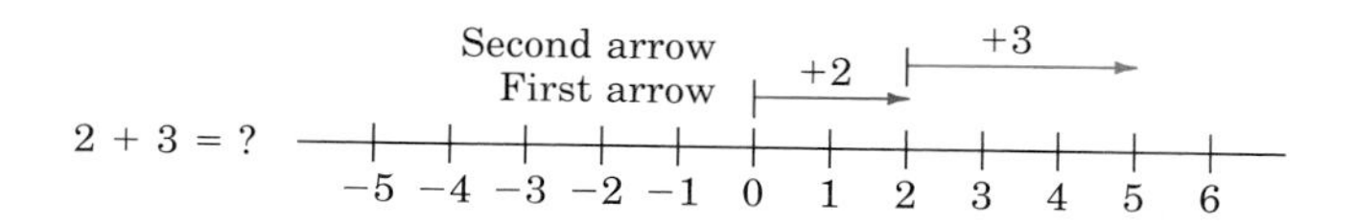

Since the second arrow ends at 5, we say that

$2 + 3 = 5.$

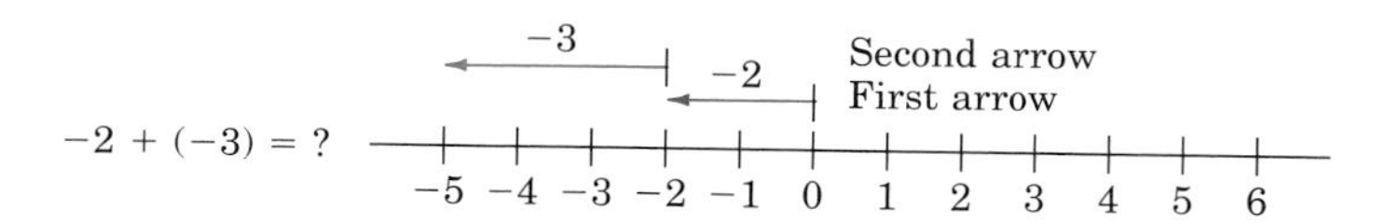

Since the second arrow ends at -5, we say that

$-2 + (-3) = -5.$

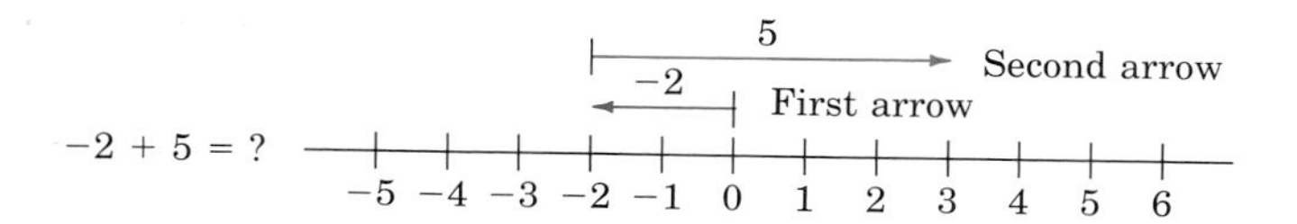

Since the second arrow ends at 3, we say that

$-2 + 5 = 3.$

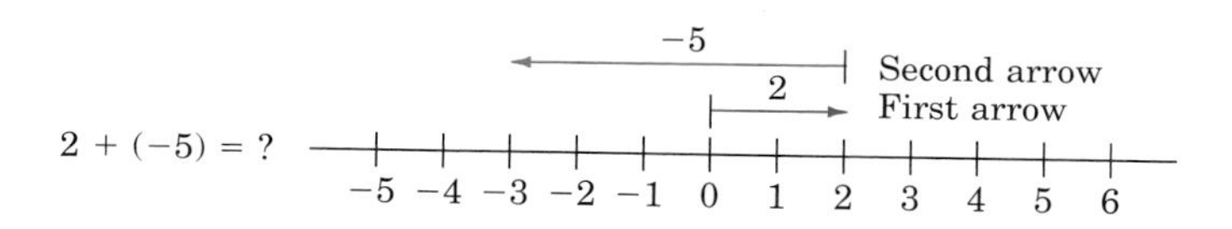

Since the second arrow ends at -3, we say that

$2 + (-5) = -3.$

We now state the rules for adding signed numbers so that the sums are the same as in the number line examples.

■ PROCEDURE

Addition of Real Numbers

1. To find the sum of two numbers of like sign, add their absolute values, and use the common sign.
2. To find the sum of a positive and a negative, find the difference of their absolute values, and use the sign of the number with the larger absolute value.

Using the rules for adding signed numbers, we see that:

$3 + 2 = 5$	The sum of two positive numbers is positive.
$-2 + (-1) = -3$	The sum of two negative numbers is the sum of their absolute values, and the sum is negative.
$3 + (-1) = 2$	Find the difference of their absolute values. Since 3 has the larger absolute value, the sum is positive.
$-10 + 13 = 3$	Find the difference of their absolute values. Since 13 has the larger absolute value, the sum is positive.
$1 + (-2) = -1$	Find the difference of their absolute values. Since -2 has the larger absolute value, the sum is negative.
$-5 + 3 = -2$	Find the difference of their absolute values. Since -5 has the larger absolute value, the sum is negative.

Here are some other observations about special sums of signed numbers.

PROPERTY

Addition Property of Zero

The sum of any number and zero is the number:

$a + 0 = 0 + a = a.$

This statement is called the Addition Property of Zero or the Additive Identity Property.

Thus, we have

$5 + 0 = 5$

$-6 + 0 = -6$

$0 + (-8) = -8$

$0 + 7 = 7.$

PROPERTY

Addition Property of Opposites

The sum of a number and its opposite is zero:

$a + (-a) = -a + a = 0.$

This is known as the Addition Property of Opposites or the Additive Inverse Property.

So,

$6 + (-6) = 0$

$-9 + 9 = 0.$

EXAMPLE 11

Add: $137 + (-115)$

$|137| - |-115| = 137 - 115$

$= 22$

To find the sum of a positive and a negative number, find the difference of their absolute values.

So, $137 + (-115) = 22.$

Since $|137|$ is larger than $|-115|$, the sum is positive. ■

EXAMPLE 12

Add: $329 + (-448)$

$|-448| - |329| = 448 - 329$

$= 119$

Subtract their absolute values.

So, $329 + (-448) = -119.$

Since $|-448|$ is larger than $|329|$, the sum is negative. ■

EXAMPLE 13

Add: $-13 + 72 + (-20) + (-23)$

$-13 + 72 + (-20) + (-23)$

Add from left to right.

$= 59 + (-20) + (-23)$

$= 39 + (-23)$

$= 16$

Alternative Method:

$-13 + 72 + (-20) + (-23)$

$= 72 + (-13) + (-20) + (-23)$

First add all the negative numbers, and then add that sum to the positive number.

$= 72 + (-33) + (-23)$

$= 72 + (-56)$

$= 16$

■

EXAMPLE 14

Add: $-\frac{3}{4} + \frac{7}{8} + \left(-\frac{1}{2}\right) + \frac{7}{8}$

$= \frac{1}{8} + \left(-\frac{1}{2}\right) + \frac{7}{8}$ Add from left to right.

$= -\frac{3}{8} + \frac{7}{8}$

$= \frac{4}{8}$

$= \frac{1}{2}$ ■

EXAMPLE 15

Add: $-47 + 38 + (-72) + 91$

ENTER	DISPLAY	
[47]	47.	
[+/−]	−47.	Pressing this key replaces the number with its opposite.
[+]	−47.	
[38]	38.	
[+]	−9.	
[72]	72.	
[+/−]	−72.	
[+]	−81.	
[91]	91.	
[=]	10.	

The sum is 10. ■

EXAMPLE 16

Is -27 a solution of $x + (-65) = -92$?

$x + (-65) = -92$ Replace x with -27, and perform the indicated operations. If the resulting statement is true, the answer is yes.

$-27 + (-65) = -92$

$-92 = -92$

The statement is true, so -27 is a solution of the equation. ■

EXAMPLE 17

If x represents a first number and z represents a second number, write an algebraic expression that represents the sum of the opposite of the first number and the opposite of the second number.

The opposite of the first number is $-x$, and the opposite of the second number is $-z$. — Write the opposite of each number.

$-x + (-z)$ — Sum indicates addition. ■

EXAMPLE 18

Add: $[-3 + 5 + (-2 + 7)] + (-10)$

$[-3 + 5 + (-2 + 7)] + (-10) = [-3 + 5 + 5] + (-10)$ — Add inside the innermost grouping symbol.

$= 7 + (-10)$ — Add inside the brackets.

$= -3$ ■

EXAMPLE 19

An airplane is being reloaded. If 577 pounds of baggage and mail are removed (-577 lb) and 482 pounds of baggage and mail are loaded ($+482$ lb), what net change in weight (w) should the cargo master report?

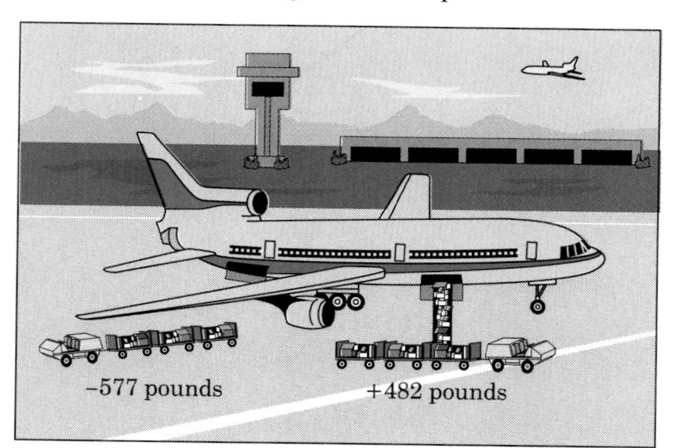

$w = (-577) + (+482)$

$= -95$

To find the net change, we add the numbers on the right side of the equation.

Therefore, the cargo master will report that the plane is 95 lb lighter. ■

EXERCISE 1.5

A

Find the opposite of each number:

1. -8 **2.** 12 **3.** 37

4. -28 **5.** 150 **6.** -340

Find the absolute value:

7. $|-2|$ **8.** $|7|$ **9.** $|-9|$ **10.** $|29|$

Find the value of each of the following:

11. $-|10|$ **12.** $-|-5|$ **13.** $-|-8|$ **14.** $-|12|$

What replacement(s) for the variable will make the following true?

15. $|y| = 5$ **16.** $|z| = 7$

Add:

17. $-16 + 12$ **18.** $-8 + 2$ **19.** $-6 + (-7)$ **20.** $-2 + (-11)$ **21.** $-7 + 7$

22. $14 + (-14)$ **23.** $-10 + (-3)$ **24.** $-15 + (-12)$ **25.** $7 + (-14)$ **26.** $-8 + 3$

What replacement for the variable will make the statement true?

27. $x = -5 + 2$ **28.** $y = -8 + 4$ **29.** $a = 12 + (-4)$

30. $B = 15 + (-16)$ **31.** $z = -8 + (-2) + (-3)$ **32.** $w = -2 + (-4) + 9$

B

Write as a signed number:

33. $-|25|$ **34.** $-|-36|$ **35.** $-|-47|$ **36.** $-|78|$

37. $|27 - 16|$ **38.** $-|82 - 37|$

What replacement(s) for the variable will make the following true?

39. $|x| = 2$ **40.** $|x| = 4$ **41.** $-|x| = 7$

42. $-x = -(-2)$ **43.** $-x = 2$ **44.** $-x = -2$

Add:

45. $6.3 + 3.7$ **46.** $-1.4 + 4.1$

47. $\frac{5}{12} + \left(-\frac{7}{12}\right)$ **48.** $-\frac{4}{15} + \frac{6}{15}$

What replacement for the variable will make the statement true?

49. $a = -25 + 32$ **50.** $b = -92 + (-84)$ **51.** $c = 76 + (-96)$

52. $d = -82 + 127$ **53.** $e = -21.1 + 16.8$ **54.** $f = -23.02 + (-16.7)$

55. $g = \left(-\frac{7}{8}\right) + \left(-\frac{3}{8}\right)$ **56.** $h = \left(-\frac{2}{3}\right) + \frac{3}{4}$

C

Write as a signed number:

57. $|-(-8.1)|$ **58.** $-\left|-\left(\frac{2}{3}\right)\right|$ **59.** $\left|-2\frac{1}{5}\right|$ **60.** $-|-(9.27)|$

61. $-(-|48|)$ **62.** $-(-|-98|)$ **63.** $-(-|-86|)$ **64.** $-(-|57|)$

What replacement(s) for the variable will make the following true?

65. $-|x| = -2$ **66.** $|x| = 5$

67. $|x| = -2$ **68.** $-|x| = -4$

Add:

69. $\frac{13}{9} + \left(-\frac{8}{3}\right) + \left(-\frac{7}{3}\right)$ **70.** $-\frac{11}{2} + \frac{19}{4} + \left(-\frac{13}{8}\right)$

What replacement for the variable will make the statement true?

71. $z = -47 + (-82) + 76$ **72.** $m = 48 + (-62) + (-81)$

73. $x = 0.5 + (-0.75) + (-0.125)$ **74.** $y = -2.15 + (-4.2) + (-0.723)$

75. $z = \left(\frac{2}{5}\right) + \left(-\frac{3}{10}\right) + \left(-\frac{1}{2}\right)$ **76.** $a = \left(-\frac{1}{4}\right) + \left(-\frac{2}{3} + \left(-\frac{1}{9}\right)\right)$

77. Is -72 a solution of $x + (-48) + 92 + (-85) + (-31) = -144$?

78. Is 79 a solution of $345 + (-217) + x + (-182) = 25$?

79. Is -38.25 a solution of $18.45 + y + (-65.34) + 7.9 + 0.75 = -76.89$?

80. Is 23.4 a solution of $-8.45 + (-5.82) + (-9.71) + y = -32.98$?

D

Write the following phrases with mathematical symbols. *Examples:* the opposite of positive 13: -13 or $-(13)$; the distance between -14 and 0: $|-14|$.

81. the opposite of seven **82.** the opposite of negative five

83. the opposite of positive eighteen

84. the opposite of negative forty-six

85. the distance between 8 and 0

86. the distance between -31 and 0

87. the opposite of the distance between 0 and -63

88. a temperature of 18° below zero

89. a loss of six and three-fourths points on the stock exchange

90. a gain of one-eighth of a point for a certain stock

Translate the following expressions to algebraic form. Let x represent the first number and y the second number.

91. the opposite of a first number multiplied times a second number

92. the opposite of the first number divided by the second number

93. the opposite of a first number less a second number

94. the opposite of a first number added to the opposite of a second number

95. A U-boat dives 1020 feet below the surface of the ocean and then rises 830 feet. What is the depth of the U-boat? Write as a signed number.

96. Stacey deposited $311.56 in her bank account. She wrote checks for $92.25 and $61.60. What is her balance?

97. At the beginning of the week, a certain stock sold for $42\frac{7}{8}$. The net changes for the first three days were $-\frac{1}{8}$, $+\frac{1}{2}$, $-\frac{1}{4}$. What was the price of the stock at the beginning of the fourth day?

98. Another stock priced at $112.50 a share fell $7.25, and then rose $2.75. What is the present value of the stock?

99. An airplane is being reloaded. If 2338 pounds of baggage and mail are removed (-2338 lb) and 6313 pounds of baggage and mail are loaded on ($+6313$ lb), what net change in weight should the cargo master report?

100. During the current fiscal year, Le Baroque Coffee House recorded the following quarterly earnings (positive numbers represent profit, negative numbers represent loss): $3456, −$507, −$498, $4007. What was the total profit (or loss) for the year?

101. The Pacific Northwest Book Depository handles most textbooks for local schools. On September 1, the inventory was 18,340 volumes. During the month, the company made the following transactions (positive numbers represent volumes received, negative numbers represent shipments): 1800, -356, -843, -500, 250, -650. What is the inventory at the end of the month?

102. The change in altitude of a plane in flight was measured every ten minutes. The figures between 3 P.M. and 4 P.M. were

3:00 P.M.	30,000 ft initially (+30,000)
3:10 P.M.	increase of 220 ft (+220)
3:20 P.M.	decrease of 200 ft (−200)
3:30 P.M.	increase of 55 ft (+55)
3:40 P.M.	decrease of 110 ft (−110)
3:50 P.M.	decrease of 25 ft (−25)
4:00 P.M.	increase of 40 ft (+40)

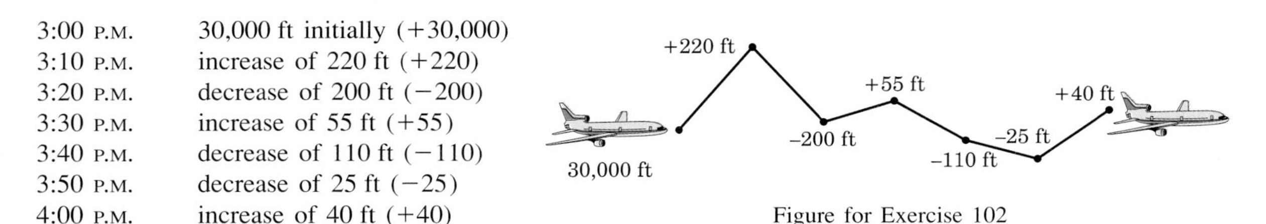

Figure for Exercise 102

What was the altitude of the plane? (*Hint:* Find the sum of the initial altitude and the six measured changes between 3 and 4 P.M.)

If w represents a first number and z represents a second number, write an algebraic expression that represents the following English phrases:

103. the sum of -8 and the opposite of the second number

104. the sum of the first number and the opposite of -9

105. the first number increased by the second number

106. the second number increased by the sum of the first number and 10

STATE YOUR UNDERSTANDING

107. How do we recognize opposites or additive inverses?

108. Explain how the number line can be used to add two numbers.

CHALLENGE EXERCISES

Translate the following into algebraic expressions.

109. Four more than the sum of -8 and x.

110. Twice the sum of a and b increased by 3 times the sum of $7a$ and $2b$.

111. Four times the sum of 2 and the cube of y, increased by 5 times the sum of $2y$ and negative seven.

112. The product of 8 and the square of k, increased by the sum of negative 3 and 4 times the square of k.

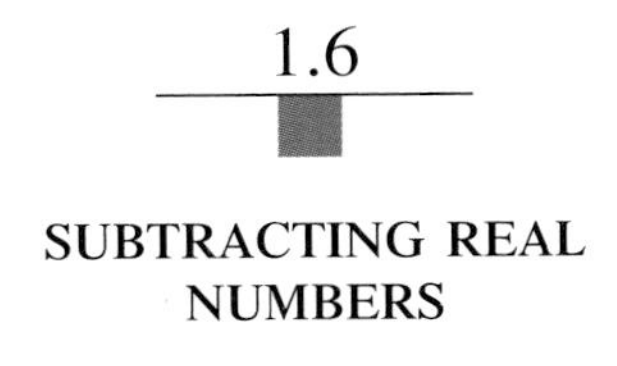

1.6 SUBTRACTING REAL NUMBERS

OBJECTIVE

1. Subtract real numbers.

Subtraction is the inverse (opposite) of addition: $10 - 4 = 6$ because $4 + 6 = 10$. This chart will help you understand the operation of subtraction.

$11 - 8$	What must be added to 8 to equal 11?	$11 - 8 = 3$ because $8 + 3 = 11$
$17 - 5$	What must be added to 5 to equal 17?	$17 - 5 = 12$ because $5 + 12 = 17$
$(-3) - (5)$	What must be added to 5 to equal -3?	$(-3) - (5) = -8$ because $5 + (-8) = -3$
$7 - (-2)$	What number must be added to -2 to equal 7?	$7 - (-2) = 9$ because $(-2) + 9 = 7$
$(-1) - (-13)$	What number must be added to -13 to equal -1?	$(-1) - (-13) = 12$ because $(-13) + 12 = -1$

A comparison of the two columns of the next chart leads us to a shortcut for subtraction.

Subtraction	Addition
$11 - 8 = 3$	$11 + (-8) = 3$
$17 - 5 = 12$	$17 + (-5) = 12$
$-3 - 5 = -8$	$-3 + (-5) = -8$
$20 - (-3) = 23$	$20 + 3 = 23$
$-15 - (-2) = -13$	$-15 + 2 = -13$

In each row of the chart, we see that the answer to the subtraction problem can be found by addition. The addition problem is formed by adding the opposite of the number to be subtracted. This is similar, in a way, to dividing fractions by changing the problem to the appropriate multiplication problem.

Subtraction can also be related to distance. If we want to know the distance on the number line between two numbers, it can be found by subtracting one number from another. For example, to find the distance from 8 to 11, we do the subtraction problem $11 - 8$.

RULE

Subtraction of Real Numbers

To subtract two real numbers, add the opposite of the number to be subtracted. In algebraic form:

$a - b = a + (-b)$.

EXAMPLE 1

Subtract: $145 - 133$

$$145 - 133 = 145 + (-133)$$
$$= 12$$

Add the opposite of 133, which is -133.

EXAMPLE 2

Subtract: $-0.21 - (-3.4)$

$$-0.21 - (-3.4) = -0.21 + 3.4$$
$$= 3.19$$

Add the opposite of -3.4, which is 3.4.

EXAMPLE 3

Subtract: $-891 - (-227)$

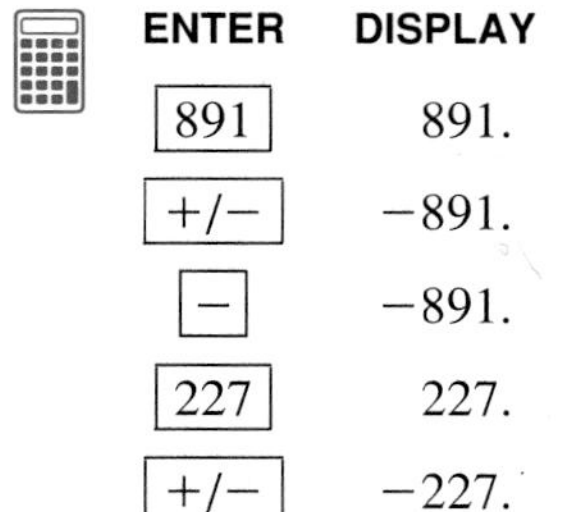

ENTER	DISPLAY
891	891.
+/−	−891.
−	−891.
227	227.
+/−	−227.
=	−664.

The difference is -664.

EXAMPLE 4

Subtract: $-18 - 23 - 14$

$$-18 - 23 - 14 = -18 + (-23) + (-14)$$
$$= -55$$

There are two subtractions; change each to add the opposite.

EXAMPLE 5

Subtract: $\frac{15}{4} - \left(-\frac{2}{3}\right)$

$\frac{15}{4} - \left(-\frac{2}{3}\right) = \frac{15}{4} + \frac{2}{3}$ — Add the opposite of $-\frac{2}{3}$.

$= \frac{45}{12} + \frac{8}{12}$ — Find the common denominator and add.

$= \frac{53}{12}$ ■

EXAMPLE 6

Subtract: $-5 - (8 - 5) - (-2 - 3)$

$= -5 - [8 + (-5)] - [-2 + (-3)]$ — Change subtract to add the opposite inside the parentheses.

$= -5 - 3 - (-5)$ — Add inside the parentheses.

$= -5 + (-3) + 5$ — Change subtract to add the opposite.

$= -3$ — Add. ■

EXAMPLE 7

Is -468 a solution of $x = -215 - 253$?

$x = -215 - 253$ — Substitute -468 for x, and see if the resulting statement is true.

$-468 = -215 - 253$

$-468 = -215 + (-253)$ — Add the opposite of 253.

$-468 = -468$ — Add.

The statement is true, so -468 is a solution. ■

EXAMPLE 8

If k represents a number, translate the following to an algebraic expression: -8 less than a number.

$k - (-8)$ — "Less than" indicates subtraction. ■

EXAMPLE 9

Viking II recorded a high temperature of $-22°C$ and a low temperature of $-107°C$ for one day at a point on the surface of Mars. What was the difference in temperature for the day?

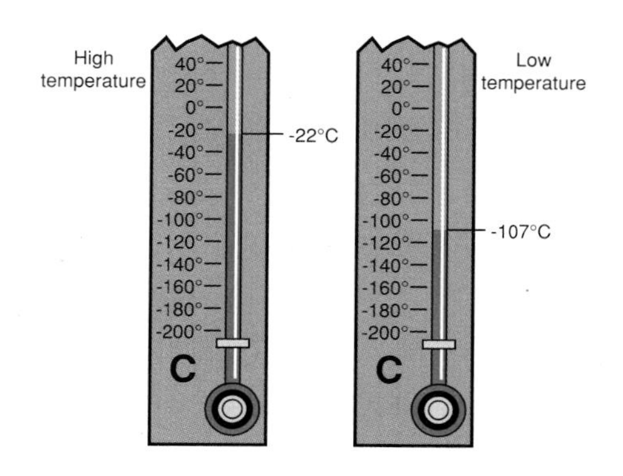

$C = -22 - (-107)$ To find the difference, subtract the lower temperature (smaller number) from the higher temperature (larger number).

$C = -22 + 107$ Add the opposite of -107.

$C = 85$

The change in temperature that day was 85°C. ■

EXERCISE 1.6

A

Subtract:

1. $8 - 4$ **2.** $9 - 3$ **3.** $-12 - 7$ **4.** $-15 - 9$

5. $-16 - (-4)$ **6.** $-15 - (-6)$ **7.** $11 - (-15)$ **8.** $16 - (-22)$

9. $-5 - (-9)$ **10.** $-4 - (-6)$ **11.** $-4 - 8 - 2$ **12.** $-7 - 4 - 5$

13. $-15 - (-4) - (-2)$ **14.** $-8 - (-3) - (-9)$

What replacement for the variable will make the statement true?

15. $y = 8 - (-4)$ **16.** $z = -8 - 9$

B

Subtract:

17. $-25 - 48$
18. $-37 - 52$
19. $-25 - (-48)$
20. $-37 - (-52)$
21. $25 - 48$
22. $37 - 52$
23. $25 - (-48)$
24. $37 - (-52)$
25. $-48 - (-25)$
26. $-52 - (-37)$
27. $214 - 385$
28. $416 - 846$
29. $214 - (385)$
30. $416 - (-846)$
31. $-214 - 385$
32. $-416 - 846$
33. $-214 - (-385)$
34. $-416 - (-846)$

What replacement for the variable will make the statement true?

35. $x = -97 - (-73)$
36. $y = 98 - (-98)$

C

Subtract:

37. $0.35 - 2.6$
38. $0.64 - 1.8$
39. $18.7 - (-3.8)$
40. $16.3 - (-1.7)$
41. $-\frac{3}{4} - \frac{3}{2}$
42. $-\frac{17}{4} - \frac{5}{8}$
43. $-\frac{9}{2} - \left(-\frac{5}{6}\right)$
44. $-\frac{17}{6} - \left(-\frac{13}{3}\right)$
45. $-\frac{7}{3} - \left(\frac{1}{4}\right)$
46. $-\frac{5}{3} - \frac{7}{4}$
47. $-0.0035 - (-0.821)$
48. $-0.015 - (-0.614)$
49. $-288 - 122 - (-621)$
50. $-622 - (-471) - 521$
51. $374 - 482 - 261 - 142$
52. $-(-325) - (192) - (-527) - (219)$
53. $-821 - 144 - 48 - (-831)$
54. $-724 - 482 - 921 - (-858)$

What replacement for the variable will make the statement true?

55. If $a = 0.76 - 5.8$, then $a = ?$
56. If $x = -12.9 - (-8.3)$, then $x = ?$
57. If $y = \frac{5}{6} - \left(-\frac{1}{4}\right)$, then $y = ?$
58. If $z = -\frac{5}{2} - \frac{9}{4}$, then $z = ?$

D

59. Mary had a balance of $298.75 in her checking account. What is the new balance after she writes a check for $78.47?
60. Referring to Exercise 59, what would Mary's new balance be if the check she wrote had been for $300?
61. Referring to Exercise 59, what would Mary's new balance be if she had written two checks, one for $92.11 and one for $17.26?

62. At 3 P.M. the temperature was 11°, and at 11 P.M. it was −12°. What was the change in temperature expressed as a signed number?

63. At sunrise the temperature in Coldtown was −9°F. At 2 P.M. the temperature was 15°F. What was the rise in temperature?

64. What is the difference in altitude between the highest and lowest points in the United States? The highest point, Mount McKinley, is 20,300 ft above sea level (20,300). The lowest point, Death Valley, is 282 ft below sea level (−282).

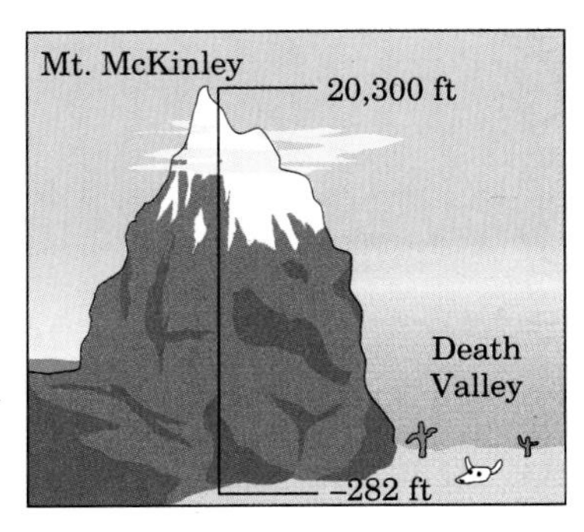

Figure for Exercise 64

65. At the beginning of the month, Norm's bank account had a balance of \$614.46, whereas at the end of the month, the account was overdrawn by \$4.86 (−\$4.86). If there were no deposits during the month, what was the total of the checks that Norm wrote for the month?

66. What is the difference in altitude of an airplane flying at 22,500 feet and a submarine submerged to 143 feet?

67. A truck travels due east 173 miles and then turns around and travels due west for 213 miles (−213). What is the distance the truck is from the starting point and in which direction?

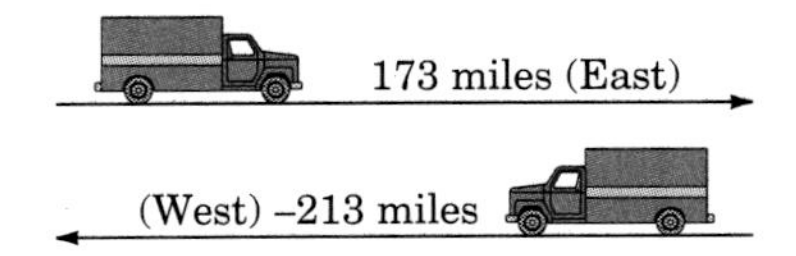

Figure for Exercise 67

If x represents a first number and y a second number, write an algebraic expression that represents the following English phrases:

68. the difference of the first number and the second number

69. 18 less than the first number

70. the opposite of the first number decreased by −5

71. −32 decreased by the opposite of the second number

STATE YOUR UNDERSTANDING

72. What is the difference between adding real numbers and subtracting real numbers?

73. Explain how the number line can be used to perform subtraction.

CHALLENGE EXERCISES

Translate the following into algebraic expressions.

74. Seven more than the difference of −5 and y.

75. Six times the difference of x and y, decreased by twice the difference of $4x$ and $5y$.

76. Three times the difference of a-squared and 4, subtracted from 5 times the product of negative 4 and the square of a.

77. The quotient of x and negative 2, decreased by the difference of $2x$ and x^2.

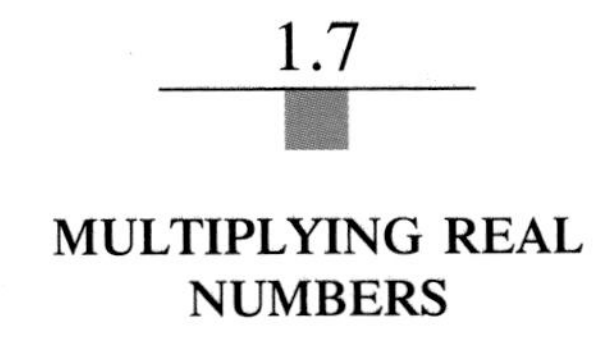

1.7 MULTIPLYING REAL NUMBERS

OBJECTIVE

1. Multiply real numbers.

We know that the product of two positive numbers is positive. Using that knowledge together with the properties of operations, we can develop rules for multiplying signed numbers.

$2 \cdot 0 = 0$	Multiplication property of zero.
$2[3 + (-3)] = 0$	Substitute $3 + (-3)$ for 0 since $0 = 3 + (-3)$.
$2 \cdot 3 + 2(-3) = 0$	Distributive property.
$6 + 2(-3) = 0$	Multiply the positive numbers.
$6 + (?) = 0$	

What number goes in the place of the question mark? We know that -6 must be added to 6 to get zero, so $2(-3)$ must be -6. A similar sequence of steps can be written for the product of any positive number times a negative number, so we conclude that the product of a positive number and a negative number is a negative number. Using the commutative property of multiplication, we further conclude that the product of a negative number and a positive number is also negative.

To find the product of two negative numbers, consider the next example.

$-2 \cdot 0 = 0$	Multiplication property of zero.
$-2[3 + (-3)] = 0$	Substitute $3 + (-3)$ for 0.
$-2 \cdot 3 + (-2)(-3) = 0$	Distributive property.
$-6 + (-2)(-3) = 0$	Multiply the negative and positive number.
$-6 + (?) = 0$	

What number goes in the place of the question mark? We know that 6 must be added to -6 to get zero, so $(-2)(-3)$ must be 6. Again a similar argument will hold for the product of any two negative numbers, so we conclude that the product of two negative numbers is positive.

We now state rules for multiplying signed numbers so that the products are the same as in the examples above.

$(+)(+) = +$
$(-)(-) = +$
$(+)(-) = -$
$(-)(+) = -$

RULE

1. The product of two positive numbers is positive.
2. The product of two negative numbers is positive.
3. The product of a positive number and a negative number in either order is negative.

$3 \cdot 4 = 12$	The product of two positive numbers is positive.
$(-3)(-4) = 12$	The product of two negative numbers is positive.
$-2(3) = -6$	The product of a negative number and a positive number is negative.
$5(-2) = -10$	The product of a positive number and a negative number is negative.
$-1(5) = -5$	Note that -1 times a positive number is its opposite.
$-1(-4) = 4$	Note that -1 times a negative number is its opposite.

PROPERTY

The product of -1 and a number yields the number's opposite.

$-1 \cdot a = -a$

The multiplication properties of one and zero given in Section 1.1 for the positive rational numbers can be extended to all real numbers.

PROPERTY

Multiplication Property of One (Multiplicative Identity Property)

$1 \cdot a = a \cdot 1 = a$

This property states that the product of any number and one is that number. So,

$$1(-4.6) = -4.6 \quad \text{and} \quad 1 \cdot \frac{13}{17} = \frac{13}{17}.$$

PROPERTY

Multiplication Property of Zero

$0 \cdot a = a \cdot 0 = 0$

This property states that the product of any number and zero is zero. So,

$$-3 \cdot 0 = 0 \quad \text{and} \quad 0 \cdot \frac{8}{11} = 0.$$

The following are examples of finding products of different combinations of positive and negative integers and rational numbers.

EXAMPLE 1

Multiply: $-6 \cdot 5$

$-6 \cdot 5 = -30$ The product of a negative number and a positive number is negative. ■

EXAMPLE 2

Multiply: $-\frac{1}{3} \cdot \frac{1}{5}$

$$-\frac{1}{3} \cdot \frac{1}{5} = -\frac{1}{15}$$ ■

EXAMPLE 3

Multiply: $-6(-7)$

$-6(-7) = 42$ ■

EXAMPLE 4

Multiply: $3(-2)(-5)$

$$\begin{aligned} 3(-2)(-5) &= -6(-5) \\ &= 30 \end{aligned}$$

Multiply from left to right. ■

EXAMPLE 5

Multiply: $-\frac{2}{7}\left(-\frac{3}{8}\right)$

$$-\frac{2}{7}\left(-\frac{3}{8}\right) = \frac{6}{56} = \frac{3}{28}$$ ■

EXAMPLE 6

Multiply: $0.3(-0.2)(-0.5)$

$0.3(-0.2)(-0.5) = -0.06(-0.5)$ Multiply from left to right.

$= 0.030$ The 0 to the right of the 3 is not needed.

$= 0.03$ ■

EXAMPLE 7

Multiply: $-6.27(38.4)$

ENTER	DISPLAY
6.27	6.27
+/−	−6.27
×	−6.27
38.4	38.4
=	−240.768

■

EXAMPLE 8

Is the following statement true or false: $-5^2 = (-5)^2$

$$-5^2 = -1 \cdot 5^2$$
$$= -1 \cdot 25$$
$$= -25$$

■ **CAUTION**

The expression -5^2 is read "the opposite of, 5 squared," *not* "the opposite of 5, squared" nor "negative 5, squared."

$$(-5)^2 = (-5)(-5)$$
$$= 25$$

The expression $(-5)^2$ is read "negative 5 squared."

Therefore, $-5^2 \neq (-5)^2$, and the statement is false. ■

EXAMPLE 9

What replacement for x will make the statement $x = -8(9)(-9)$ true?

$x = -8(9)(-9)$

$x = -72(-9)$

$x = 648$

To determine the replacement for x, perform the multiplication on the right side of the equation.

or

$x = -8(9)(-9)$

$x = -8(-81)$

$x = 648$

Since only multiplication is involved, we can use the associative property of multiplication to multiply in a different order.

EXAMPLE 10

If x represents a number, translate the following expression to an algebraic expression and simplify: the product of 12 and the opposite of a number

$12(-x) = 12(-1 \cdot x)$ — Product is the result of multiplication. If x is the number, then "$-x$" is the opposite of that number.

$= [12(-1)]x$ — Recall $-x = -1 \cdot x$. Use the associative property of multiplication and simplify.

$= -12x$

EXAMPLE 11

The formula for converting temperature measurement from Fahrenheit to Celsius is $C = \frac{5}{9}(F - 32)$. What Celsius measure is equal to 14°F?

$C = \frac{5}{9}(F - 32)$

$C = \frac{5}{9}(14 - 32)$ — Substitute 14 for F, and perform the indicated operations.

$C = \frac{5}{9}(-18)$

$C = -\frac{90}{9}$

$C = -10$

A temperature of 14°F is the same as −10°C.

EXERCISE 1.7

A

Multiply:

1. $(-3)(2)$
2. $(-4)(3)$
3. $(-5)(-6)$
4. $(-9)(-4)$
5. $(-8)6$
6. $7(-5)$
7. $(-12)(-3)$
8. $(-11)(-4)$
9. $-9(-6)$
10. $8(-4)$
11. -3^2
12. $(-3)^2$

What replacement for the variable will make the statement true?

13. $x = 9(-3)$

14. $x = -8(-2)$

15. $y = -7(-4)$

16. $y = -4 \cdot 6$

B

Multiply:

17. $(-11)(10)$

18. $(-13)(11)$

19. $(-12)(-12)$

20. $(-15)(-15)$

21. $(43)(-6)$

22. $(-21)(5)$

23. $(-14)(-15)$

24. $(-21)(-4)$

25. $(-4)(3)(-5)$

26. $6(0)(-8)$

27. $(-4)(-1)(-11)$

28. $(-7)(-2)(-2)$

29. $(-6)(-5)(8)$

30. $(-2)(-2)(-2)$

31. $(-1)(-3)(-4)(-2)$

32. $(-2)(-5)(-6)(-3)$

33. $(-5)(-6)(2)(4)$

34. $(-18)(-1)(2)(3)$

35. $(-5)^4$

36. 6^2

37. -5^4

38. $-(-7)^2$

What replacement for the variable will make the statement true?

39. $a = 16(-8)$

40. $a = -12(-18)$

41. $b = 14(-8)(-12)$

42. $b = 17(-9)(14)$

C

Multiply:

43. $\left(-\frac{2}{3}\right)\left(\frac{1}{2}\right)$

44. $\left(-\frac{4}{3}\right)\left(\frac{3}{5}\right)$

45. $\left(-\frac{1}{2}\right)\left(\frac{2}{3}\right)\left(\frac{6}{7}\right)$

46. $\left(-\frac{1}{3}\right)\left(\frac{3}{4}\right)\left(\frac{5}{6}\right)$

47. $(0.2)(-0.5)(6)$

48. $(-0.6)(0)(0.3)$

49. $(-0.19)(-0.27)$

50. $(-0.39)(-0.18)$

51. $\left(-\frac{4}{9}\right)\left(-\frac{3}{8}\right)\left(-\frac{4}{5}\right)$

52. $\left(-\frac{5}{7}\right)\left(-\frac{14}{15}\right)\left(-\frac{3}{8}\right)$

53. $(-0.8)(-0.1)(-0.1)(5)$

54. $(-1)(-1)(-1)(4.6)$

55. $(-1)^2(-2)^2(-3)^2 = ?$

56. $-(2)^2(-4)^2(-2)^2 = ?$

57. If $w = \left(-\frac{4}{5}\right) \cdot \left(-\frac{5}{8}\right) \cdot \left(-\frac{7}{9}\right)$, then $w = ?$

58. If $w = \frac{4}{9} \cdot \left(-\frac{9}{10}\right) \cdot \left(-\frac{15}{17}\right)$, then $w = ?$

59. If $z = (0.8)(0.7)(4)$, then $z = ?$

60. If $z = (2.3)(-0.2)(-5)(-1)$, then $z = ?$

61. If $x = (-12)^2(-5)$, then $x = ?$

62. If $x = -8(-15)^2$, then $x = ?$

D

63. For six consecutive weeks, Mr. Obese lost 3.5 lb every week. If each loss is represented by -3.5 lb, what was his change in weight for the six weeks?

64. If Mrs. Tomlison lost 2.1 lb every week for eight consecutive weeks, what is her total change in weight, expressed as a signed number?

65. A stock sustained a 2.83 decline (-2.83) for twelve consecutive days. What was the total change during the twelve-day period, expressed as a signed number?

66. Super Jock, the star fullback of a local high-school football team, lost 4 yards (-4 yd) in each of six carries. What was his net gain, expressed as a signed number?

67. The temperature at midnight was 81°, and it dropped 3° an hour for the next 5 hours. What was the temperature at 5:00 A.M.?

68. The value of a stock rose \$1.75 for three consecutive days and then it dropped \$1.50 for each of the next four days. What was the net change, expressed as a signed number?

Use the formula $C = \frac{5}{9}(F - 32)$ to convert the following Fahrenheit measures to Celsius measures:

69. -76°F **70.** -31°F **71.** 14°F **72.** 5°F

Translate the following expressions to algebra: Let x represent a first number and y a second number.

73. the product of a first number and -8

74. the product of the opposite of a first number and -9

75. the opposite of the product of -7 and the opposite of a second number

76. the opposite of the product of a first number and -12

77. the opposite of the product of a first number and a second number

78. the product of a first number and the opposite of a second number

STATE YOUR UNDERSTANDING

79. Given a number, what process must be done to find the opposite of that number?

80. Explain why $-a^2 \neq (-a)^2$.

CHALLENGE EXERCISES

Determine the value of the following expressions.

81. $13 - \{2 - [1 - 4(5 - 8)]\}$

82. $20 - \{5 - [3 - 5(6 - 2)]\}$

Determine the value of each of the following if $a = -5$ and $b = 4$.

83. $3(2a^2 - b)$

84. $(2a - b)(3b - 4a)$

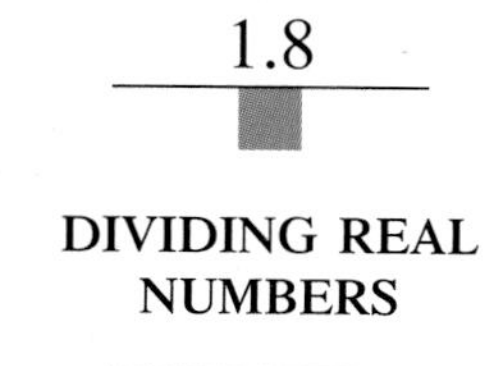

1.8 DIVIDING REAL NUMBERS

OBJECTIVE

1. Divide real numbers.

Division is the inverse (opposite) of multiplication: $12 \div 3 = 4$ because $3 \cdot 4 = 12$. This chart will help you understand the operation of division.

$72 \div 6$	What must 6 be multiplied by to equal 72?	$72 \div 6 = 12$ because $6 \cdot 12 = 72$
$8 \div 3$	What must 3 be multiplied by to equal 8?	$8 \div 3 = \frac{8}{3}$ because $3 \cdot \frac{8}{3} = 8$
$(-27) \div 3$	What must 3 be multiplied by to equal -27?	$(-27) \div 3 = -9$ because $3(-9) = -27$
$30 \div (-6)$	What must -6 be multiplied by to equal 30?	$30 \div (-6) = -5$ because $-6(-5) = 30$
$(-18) \div (-2)$	What must -2 be multiplied by to equal -18?	$(-18) \div (-2) = 9$ because $(-2)(9) = -18$

These examples lead to rules similar to those for multiplication.

RULE

1. The quotient of two positive numbers is positive.
2. The quotient of two negative numbers is positive.
3. The quotient of a positive number and a negative number in either order is negative.

$(+) \div (+) = +$
$(-) \div (-) = +$
$(+) \div (-) = -$
$(-) \div (+) = -$

CAUTION

Care must be taken if one of the numbers in a divison problem is zero.

Study the following:

$0 \div 7$	What must 7 be multiplied by to equal 0?	$0 \div 7 = 0$ because $7 \cdot 0 = 0$
$0 \div (-5)$	What must -5 be multiplied by to equal 0?	$0 \div (-5) = 0$ because $-5 \cdot 0 = 0$
$10 \div 0$	What must 0 be multiplied by to equal 10?	No such number exists because no number times 0 is 10. Therefore, we say division by zero is not defined.
$0 \div 0$	What must 0 be multiplied by to equal 0?	Many answers. Any number times 0 is 0. Therefore, we say it is not defined.

The following are examples of finding quotients of (dividing) different combinations of positive and negative integers and rational numbers.

EXAMPLE 1

Divide: $-8.6 \div 4.3$

$-8.6 \div 4.3 = -2$ The quotient of a negative number and a positive number is negative. ■

EXAMPLE 2

Divide: $-\frac{3}{4} \div \frac{1}{2}$

$$-\frac{3}{4} \div \frac{1}{2} = -\frac{3}{4} \cdot \frac{2}{1}$$

$$= -\frac{3}{2}$$ The quotient of a negative number and a positive number is negative. ■

Division can also be written as a fraction.

EXAMPLE 3

Divide: $\frac{-48}{-12}$

$\frac{-48}{-12} = 4$ The quotient of two negative numbers is positive. ■

EXAMPLE 4

Divide: $92.16 \div (-3.6)$

$92.16 \div (-3.6) = -25.6$ The quotient is negative. ■

EXAMPLE 5

Divide: $-8624 \div 32$

ENTER	DISPLAY
8624	8624.
+/−	−8624.
÷	−8624.
32	32.
=	−269.5

The quotient is -269.5. ■

EXAMPLE 6

If x represents a number, translate the following to an algebraic expression: the quotient of a number and -8

$\frac{x}{-8}$ Quotient means divide. ■

EXAMPLE 7

The coldest temperature in Eycee Northland on each of five days was $-15°$, $-6°$, $-4°$, $-2°$, and $-3°$. What was the average of the low temperatures during this five-day period?

$-15 + (-6) + (-4) + (-2) + (-3) = -30$

$-30 \div 5 = -6$

The average temperature in Eycee for the five days is the sum of the temperatures divided by five.

The average daily temperature was $-6°$, or $6°$ below zero. ■

EXERCISE 1.8

A

Divide:

1. $8 \div (-2)$
2. $9 \div (-3)$
3. $(-15) \div 3$
4. $(-24) \div 4$
5. $(-14) \div (-2)$
6. $(-21) \div (3)$
7. $(-33) \div 11$
8. $(-28) \div 14$
9. $16 \div (-4)$
10. $36 \div (-4)$
11. $-\dfrac{72}{6}$
12. $\dfrac{84}{-7}$

What replacement for the variable will make the statement true?

13. $x = 48 \div (-2)$
14. $x = -64 \div (-8)$
15. $y = -21 \div 3$
16. If $y = -36 \div 9$, then $y = ?$

B

Divide:

17. $125 \div (-25)$
18. $225 \div (-15)$
19. $(-260) \div 13$
20. $(-480) \div 24$
21. $(-300) \div (-12)$
22. $(-400) \div (-16)$
23. $252 \div (-14)$
24. $182 \div (-13)$
25. $(-276) \div (-23)$
26. $(-462) \div (-21)$
27. $(-240) \div 15$
28. $(-360) \div 18$
29. $368 \div 16$
30. $357 \div 17$
31. $-374 \div 22$
32. $312 \div -12$
33. $\dfrac{-465}{-15}$
34. $\dfrac{-594}{-18}$
35. $\dfrac{816}{-34}$
36. $\dfrac{968}{-22}$
37. $\dfrac{-2860}{-10}$
38. $\dfrac{-3150}{10}$
39. $\dfrac{-1050}{-25}$
40. $\dfrac{352}{-11}$

What replacement for the variable will make the statement true?

41. $a = -324 \div 9$
42. $b = (-624) \div (-16)$
43. $c = (-1197) \div (-21)$
44. $f = (480) \div (-15)$
45. $b = 253 \div (-11)$
46. $b = 504 \div (-21)$

C

Divide:

47. $4.95 \div (-0.9)$
48. $3.92 \div (-0.7)$
49. $(-15.5) \div (-0.05)$

50. $(-12.8) \div (-0.04)$

51. $\left(-\frac{3}{4}\right) \div \left(-\frac{3}{5}\right)$

52. $\left(-\frac{9}{10}\right) \div \left(\frac{9}{20}\right)$

53. $6.06 \div (-0.6)$

54. $(8.08) \div (-0.4)$

55. $\left(-\frac{14}{9}\right) \div \left(\frac{7}{8}\right)$

56. $\left(-\frac{15}{8}\right) \div \left(\frac{3}{4}\right)$

57. $\frac{0.875}{-0.007}$

58. $\frac{-15.9}{0.003}$

59. $\frac{-7.25}{-0.05}$

60. $\frac{52.4}{-0.004}$

What replacement for the variable will make the statement true?

61. $z = -48.4 \div 0.22$

62. $z = -44.4 \div 0.37$

63. $x = -576 \div (-4.8)$

64. $x = 487.5 \div (-3.9)$

65. $-117.3 \div (5.1)$

66. $278.4 \div (-8.7)$

D

67. Find the average of the following group of numbers:
$-12, \quad -16, \quad -24, \quad -36, \quad -42$

68. Find the average of the following group of numbers:
$-82, \quad -29, \quad 63, \quad -18, \quad 52, \quad 44$

69. A certain stock lost 26 points (-26) over a five-day period. What is the average change over this period? Express your answer as a signed number.

70. If a store lost a total of \$15,824 (−\$15,824) over a period of eight months, what was the average loss per month, expressed as a signed number?

71. The lowest temperature in Far North town on each of seven consecutive days was $-8°$, $-15°$, $-24°$, $-12°$, $-6°$, $-10°$, and $-9°$. What was the average of the low temperatures for this seven-day period?

72. On six consecutive plays, the local football team lost a total of 36 yards (-36). What was the average loss per play?

73. During the first week of July, a major midwest stock exchange registered the following changes: $11\frac{7}{8}$, $-\left(17\frac{1}{8}\right)$, $-\left(21\frac{3}{4}\right)$, $5\frac{1}{2}$, and $-\left(6\frac{7}{8}\right)$. What was the average change for the week?

74. The low temperatures for a certain week in Valdez, Alaska, were $-12°$, $-10°$, $5°$, $11°$, $7°$, $-9°$, $-13°$. What was the average low temperature for this seven-day period?

75. Over a five-day period, the AZ Bolt Company shipped out 20,175 bolts. What was the average number shipped out per day? We will assume shipping bolts out is a negative number.

76. A local supermarket listed the following losses from shoplifting for a six-month period: $-\$2175$, $-\$3110$, $-\$1925$, $-\$2250$, $-\$2570$, $-\$2820$. Find the average loss per month.

Translate the following expressions to algebra. Let x represent a first number and y a second number.

77. the quotient of the opposite of a first number and 18

78. the quotient of the opposite of -22 and a second number

79. the quotient of a first number and a second number

80. the quotient of a first number and the opposite of a second number

STATE YOUR UNDERSTANDING

81. When dividing two real numbers, care must be taken if one of the numbers is zero. Why?

82. Division by 0 is never permitted, but $4 \div 0$ is not the same as $0 \div 0$. Explain why $4 \div 0$ and $0 \div 0$ are not defined.

CHALLENGE EXERCISES

Translate the following expressions to algebra. Let x represent a first number and y represent a second number.

83. The difference of a first number decreased by 8 and the product of 8 and a second number.

84. The quotient of a first number less 4 and the sum of a second number and 12.

85. 12 times a first number, decreased by the difference of a second number and 8.

86. 35 less than a first number, decreased by the sum of a second number and 5.

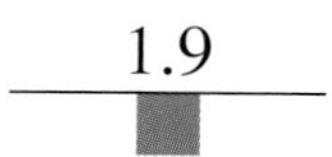

ORDER OF OPERATIONS WITH REAL NUMBERS

OBJECTIVES

1. Evaluate algebraic expressions with real numbers.
2. Determine whether a real number is a solution of an equation.

In Section 1.2, the agreement for the order of operations was given. At that time, only positive numbers, zero, and number expressions were used. We now take a second look at order of operations using signed numbers and algebraic expressions. The agreement for the order of operations is restated.

PROCEDURE

In an expression with two or more operations, perform the operations in the following order:

1. GROUPING SYMBOLS—perform operations included by grouping symbols first (parentheses, fraction bar, and so on) according to steps 2, 3, and 4.
2. EXPONENTS—perform operations indicated by exponents.
3. MULTIPLY AND DIVIDE—perform multiplication and division as they appear from left to right.
4. ADD AND SUBTRACT—perform addition and subtraction as they appear from left to right.

We consider some examples in which real numbers are used. Recall the definition of exponents.

EXAMPLE 1

Evaluate: $12 - 5(35 \div 7)^2$

$12 - 5(35 \div 7)^2 = 12 - 5(5)^2$	Perform the operation inside the parentheses.
$= 12 - 5 \cdot 25$	Find the value of 5^2.
$= 12 - 125$	Multiply.
$= -113$	Subtract. ■

EXAMPLE 2

Evaluate: $49 - 9[12(48 - 53) - 64] - 52$

$49 - 9[12(48 - 53) - 64] - 52$ $= 49 - 9[12(-5) - 64] - 52$	Perform the operation inside the parentheses.
$= 49 - 9[-60 - 64] - 52$	Multiply inside the brackets.
$= 49 - 9[-124] - 52$	Subtract inside the brackets.
$= 49 + 1116 - 52$	Multiply.
$= 1113$	Add. ■

Recall that fraction bars are also grouping symbols.

EXAMPLE 3

Evaluate: $\frac{2}{3} - \frac{\frac{5}{6}}{1 - \frac{4}{3}} \div \frac{6}{7}$

$\frac{2}{3} - \frac{\frac{5}{6}}{1 - \frac{4}{3}} \div \frac{6}{7} = \frac{2}{3} - \frac{\frac{5}{6}}{-\frac{1}{3}} \div \frac{6}{7}$	Subtract below the fraction bar. Recall that the fraction bar also means divide.
$= \frac{2}{3} - \left(-\frac{5}{2}\right) \div \frac{6}{7}$	Perform the division, starting on the left.
$= \frac{2}{3} - \left(-\frac{35}{12}\right)$	Perform the remaining division.
$= \frac{43}{12}$	Subtract. ■

Recall that to evaluate algebraic expressions, we replace the variables with their numeric values and then perform the indicated operations following the accepted order.

EXAMPLE 4

Evaluate: $5ab^2c + 24a^2bc^2$ when $a = -6$, $b = -7$, and $c = -4$

$5ab^2c + 24a^2bc^2$	
$= 5(-6)(-7)^2(-4) + 24(-6)^2(-7)(-4)^2$	Replace a with -6, b with -7, and c with -4.

$= 5(-6)(49)(-4) + 24(36)(-7)(16)$ Find values of the squares.

$= 5880 + (-96768)$ Multiply.

$= -90888$ Add. ■

EXAMPLE 5

If b represents a first number and c represents a second number, translate the following to an algebraic expression: the product of a first number and 3, and that product, decreased by the quotient of a second number and -10.

$b \cdot 3$ Product is the result of multiplication.

$\frac{c}{-10}$ Quotient is the result of division.

$b \cdot 3 - \frac{c}{-10} = 3b - \frac{c}{-10}$ "Decreased by" means subtract from. ■

EXAMPLE 6

To convert degrees Fahrenheit to degrees Celsius, we use the formula

$$C = \frac{5}{9}(F - 32).$$

What is the Celsius temperature that is equal to $-40°$F?

$$C = \frac{5}{9}(F - 32)$$

$$C = \frac{5}{9}(-40 - 32)$$

To find the Celsius temperature that is equal, substitute -40 for F, and perform the indicated operations.

$$C = \frac{5}{9}(-72)$$

$$C = -40$$

Therefore, $-40°\text{F} = -40°\text{C}$. ■

Recall that we check a solution to an equation by substituting a number for the variable. If the result is a true statement, the number is a solution of the equation. If the result is a false statement, the number is not a solution of the equation.

EXAMPLE 7

Is -16 a solution of $6x + 48 = -3x - 96$?

$$6x + 48 = -3x - 96$$

$$6(-16) + 48 = -3(-16) - 96 \qquad \text{Substitute } -16 \text{ for } x.$$

$$-96 + 48 = 48 - 96 \qquad \text{Multiply.}$$

$$-48 = -48 \qquad \text{Add.}$$

The statement is true; therefore, -16 is a solution. ■

EXERCISE 1.9

A

Evaluate:

1. $21 - (-8)2 - 5 \cdot 4$
2. $37 - 4(-2) - 8 \cdot 7$
3. $5[(12 - 32)5 - 18]$
4. $-8[8(15 + 16) - 32]$
5. $5^2 - (12 - 7)^2 \cdot 4$
6. $6^2 - (15 - 4)^2 \cdot 0$
7. $\frac{1}{2} - \left(\frac{2}{3} \div \frac{1}{2}\right) \cdot \frac{3}{4}$
8. $\frac{7}{8} - \left(\frac{5}{6} \div \frac{3}{4}\right) \cdot \frac{4}{5}$

Evaluate the following if $a = -7$, $b = -4$, and $c = 3$:

9. $15a - 3b - 5c$
10. $19b - 4c + 8a$
11. $a(3b - 4c) - 5a$
12. $c(4a + 7b) - 12c$
13. $5a^2 - 4b^2 - 3c^2$
14. $6c^2 - 5b^2 - 2a^2$
15. $-4ab - 3bc + 5abc$
16. $-8bc + 7a^2c^2 - 4ab$

Check the given solution in each equation:

17. $2x - 8 = 16$, 12
18. $5y + 4 = -32$, -7
19. $7b + 9 = 55$, 8
20. $8a - 9 = 55$, 8

B

Evaluate:

21. $-15 - \dfrac{8^2 - 4}{3^2 + 3}$
22. $-22 + \dfrac{9^2 - 6}{6^2 - 11}$
23. $\dfrac{12(8 - 24)}{5^2 - 3^2} \div (-12)$
24. $\dfrac{15(12 - 45)}{6^2 - 5^2} \div (-9)$
25. $\dfrac{1}{2} - \dfrac{\frac{4}{5}}{1 - \frac{1}{3}} \div \dfrac{2}{5}$
26. $\dfrac{7}{8} + \dfrac{6}{1 - \frac{5}{9}} \div \dfrac{1}{2}$
27. $(12.7 - 4.8)^2 + 8.16 \div 0.25$
28. $(14.3 - 7.9)^2 - 9.32 \div 0.4$
29. $-8|125 - 321| - 21^2 + 8(-7)$
30. $-9|482 - 632| - 17^2 + 9(-9)$

31. $-6(8^2 - 9^2)^2 - (-7)20$

32. $-5(6^2 - 7^2)^2 - (-8)19$

Evaluate when $a = -5$, $b = -3$, and $c = 6$:

33. $4a - 2b[(4a + 3b) - 3c] - 5a$

34. $3a - 2c[(3c - 2b) + 5a] - 4c$

35. $-2a^2b + 4ab^2 - 6c^2$

36. $3ab^2 - 4b^2c + 5a^2$

37. $2c(a - b)^3 \div c^2$

38. $-3a(b + c)^4 \div 5b^2$

Check the given solution in the equation:

39. $5x - 8 = 3x + 5, \dfrac{13}{2}$

40. $9z + 12 = 13z + 15, \dfrac{-3}{4}$

41. $8a - 22 = 12a + 17, \dfrac{-5}{4}$

42. $15b - 36 = 21b + 14, \dfrac{-20}{3}$

C

Evaluate:

43. $-30\left|\dfrac{15 - 4}{-8 - 7}\right| - 25(15 - 18)$

44. $-48\left|\dfrac{21 - 12}{-5 - 11}\right| - 30(21 - 13)$

45. $25.6 + \dfrac{18.5 - 20.25}{0.25} + (1.8)(32.7)$

46. $13.41 - \dfrac{27.5}{53.7 - 41.2} - 21.6 \div 1.8$

47. $(12.2)(4.5)^2 - 13.7 \div (2.5)^2$

48. $99.9 \div (0.3)^3 + (-4.2)^3 + 21.6$

49. $16.4(41.9 - 62.8)^2 + 21.6$

50. $(3.4)^2(-1.5)^3[-32.6 - 25.7] \div (-1.35)$

51. $-3.6[(5.5)^2 - (7.3)^2] - 2.3(-3.7)^2$

52. $-6.5[(7.8)^2 + (4.7)^2] - (-1.8)^2(8.6)^2$

53. $5\dfrac{2}{3} + \dfrac{4\frac{1}{2}}{2\frac{7}{8} - 3\frac{1}{6}} \div 3\dfrac{1}{3}$

54. $-\dfrac{53}{5} - \dfrac{6\frac{1}{3} - 3\frac{4}{5}}{8\frac{4}{15} - 7\frac{2}{3}} \div 3\dfrac{3}{4}$

Evaluate the following if $a = -31.2$, $b = -5.7$, and $c = 14.5$:

55. $3.75b - ac$

56. $ab - bc - ac$

57. $a^2 - b^2 - c^2$

58. $a^2 - b^2c - 25.7c$

59. $abc - 37.3c^2$

60. $(bc - 2a)^2$

Check the given solution in each equation:

61. $2.3x + 5.6 = 22.298, 7.26$

62. $13.3 - 2.5x = -9.95, 9.3$

63. $(5.3)^2 - 21.6x = 5.9x - 68.16, 3.5$

64. $(14.7 - 19.36)x = 5.96x - 228.33, -21.5$

D

65. What Celsius temperature is equivalent to 95.9°F? [*Hint:* Use the formula $C = \frac{5}{9}(F - 32)$.]

66. What Celsius temperature is equivalent to −4.9°F?

67. What Celsius temperature is equivalent to 131°F?

68. What Fahrenheit temperature is equivalent to 92.5°C? (*Hint:* Use the formula $F = \frac{9}{5}C + 32$.)

69. What Fahrenheit temperature is equivalent to −20.5°C?

70. What Fahrenheit temperature is equivalent to −55°C?

71. During a ''blue light'' special, B-Mart sold 112 pairs of socks at a loss of 36¢ a pair (−$0.36). The remainder of the day they sold 212 pairs at a profit of 22¢ a pair. Express the profit or loss on the sale of socks as a signed number.

72. Fly America sold 40 seats on flight 402 at a loss of $48 per seat (−$48). Fly America also sold 27 seats at a profit of $82 per seat. Express the profit or loss on the sale of the 67 seats on flight 402 as a signed number.

73. Mr. Gentry sold 200 shares of General Grocery stocks at a loss of $1.75 (−1.75) per share. He also sold 400 shares of Red Brick Company stocks at a gain of $0.80 per share (0.80). Did Mr. Gentry have a profit or a loss and how much?

74. The Thrift Mart had bread, milk, and potato chips on sale last week. They sold 170 loaves of bread at a loss of eight cents per loaf, 190 quarts of milk at a profit of seventeen cents a quart, and 82 bags of potato chips at a loss of eleven cents a bag. Determine the profit or loss on the sales of these three items.

STATE YOUR UNDERSTANDING

75. What is the order of operations for real numbers?

76. Locate the error in the following example. Indicate why it is not correct. Determine the correct answer.

$3 - [5 - 2(6 - 4^2)^3] =$
$3 - [5 - 2(6 - 16)^3] =$
$3 - [5 - 2(-10)^3] =$
$3 - [5 - (-20)^3] =$
$3 - [5 - (-8000)] =$
$3 - [5 + 8000] =$
$3 - 8005 =$
-8002

CHALLENGE EXERCISES

Determine the value of the following expressions.

77. $4 - 5[6(4 - 8) + 2 - 9]$

78. $4 - 3[(2^2 - 3)^2 - 6(4 - 8) \div 3]$

79. $\dfrac{8(-5) + 6(4)}{2(-5) - 6}$

80. $\dfrac{6^2 - 3^2 + 2(8 + 4)}{5(2 - 5) + (2 - 4)}$

CHAPTER 1
SUMMARY

Digits	0, 1, 2, 3, 4, 5, 6, 7, 8, and 9; those symbols used to write place value names for numbers.	(p. 2)
Counting Numbers or Natural Numbers	1, 2, 3, 4, 5, 6, . . .	(p. 3)
Whole Numbers	0, 1, 2, 3, 4, 5, 6, . . .	(p. 3)
Positive Rational Numbers (Fractions)	Those numbers that can be written in the form $\frac{a}{b}$, where a and b are whole numbers and $b \neq 0$.	(p. 3)
Variable	A letter that is used in place of a number.	(p. 3, 17)
Constant	A usual symbol for a number such as 5.	(p. 17)
Factors	Numbers that are to be multiplied.	(p. 4)
Simplify by Reducing	A fraction is simplified by reducing to lowest terms when the numerator and the denominator have no other factors in common other than one.	(p. 4)
Commutative Property of Multiplication	Allows for two factors to be written in either order: $a \cdot b = b \cdot a$.	(p. 5)
Associative Property of Multiplication	Allows for factors to be grouped in any way: $(a \cdot b) \cdot c = a \cdot (b \cdot c)$.	(p. 5)
Multiplication Property of One	One times any number is the number: $a \cdot 1 = 1 \cdot a = a$.	(p. 5)
Multiplication Property of Zero	Zero times any number is zero: $a \cdot 0 = 0 \cdot a = 0$.	(p. 5)
To Multiply Two Positive Rational Numbers	Write the product of the numerators over the product of the denominators: $\frac{a}{b} \cdot \frac{c}{d} = \frac{ac}{bd}$.	(p. 6)

Reciprocals	Two fractions are reciprocals of each other if their product is one. $\frac{a}{b} \cdot \frac{b}{a} = 1$, $\frac{a}{b}$ and $\frac{b}{a}$ are reciprocals.	(p. 7)
To Divide Two Positive Rational Numbers	Multiply the dividend by the reciprocal of the divisor: $\frac{a}{b} \div \frac{c}{d} = \frac{a}{b} \cdot \frac{d}{c}$.	(p. 7)
Properties of Zero Related to Division	**1.** Zero divided by a nonzero number is zero: $0 \div a = 0$ **2.** Division by zero has no value: $a \div 0$ is undefined	(p. 8)
Commutative Property of Addition	It is possible to add two numbers in either order: $a + b = b + a$.	(p. 9)
Associative Property of Addition	When three or more numbers are added, it is possible to group them in any way: $(a + b) + c = a + (b + c)$.	(p. 10)
Distributive Property	Involves both multiplication and addition: $a(b + c) = ab + ac$.	(p. 10)
To Add Two Fractions with a Common Denominator	Add the numerators, and put that sum over the common denominator: $\frac{a}{b} + \frac{c}{b} = \frac{a + c}{b}$	(p. 10)
To Build a Fraction	Multiply both numerator and denominator by the same factor: $\frac{a}{b} \cdot \frac{c}{c} = \frac{ac}{bc}$	(p. 11)
To Subtract Two Fractions with a Common Denominator	Subtract the numerators, and put the difference over the common denominator: $\frac{a}{b} - \frac{c}{b} = \frac{a - c}{b}$	(p. 12)
Exponential Form	3^4 is an example of exponential form. The base is 3, and the exponent is 4. The power or value is 81.	(p. 17)
Exponent of Zero	Any number (except 0) with an exponent of 0 has a value of 1: $a^0 = 1, a \neq 0$	(p. 18)
Exponent of One	Any number with an exponent of 1 has a value of the number itself: $a^1 = a$	(p. 18)
Numerical Expressions	Written using the usual symbols for numbers together with signs of operations.	(p. 20)
To Evaluate a Numerical Expression	Find the value of that expression.	(p. 20)
Order of Operations	**1.** Grouping symbols **2.** Exponents **3.** Multiply and divide from left to right **4.** Add and subtract from left to right	(p. 21)

Algebraic Expressions	Algebraic expressions are formed by using constants and variables and signs of operation.	(p. 28)
Formula	Shorthand for stating a rule or a method of procedure.	(p. 28)
Solution of a Formula or Evaluate a Formula	The value of a formula given the replacement values for the variables or letters.	(p. 29)
Equation	A statement about two numbers that states that the two expressions are equal.	(p. 33)
Solution of an Equation	A number that, when used to replace the variable, makes the equation true.	(p. 33)
Solution Set of an Equation	The set of all numbers that are solutions of the equation.	(p. 33)
Real Numbers	Those numbers that can be associated with a point on the real number line.	(p. 37)
Positive Numbers	Those numbers that are to the right of 0 on the number line.	(p. 37)
Negative Numbers	Those numbers that are to the left of 0 on the number line.	(p. 37)
Integers	Any number from the group described by $\ldots, -3, -2, -1, 0, 1, 2, 3, \ldots$	(p. 38)
Rational Numbers	The positive rational numbers together with those negative numbers that can be written as fractions.	(p. 3, 38)
Irrational Numbers	All the real numbers that are not rational.	(p. 39)
Symbols That Indicate a Relationship Between Two Numbers	$<$ less than $>$ greater than $\leq$ less than or equal to $\geq$ greater than or equal to $\neq$ not equal to	(p. 40)
Opposites (Additive Inverses)	Pairs of numbers that are the same distance from zero on the number line but on opposite sides.	(p. 44)
Absolute Value of a Number	The distance between the number and zero on the number line. The distance is always positive or zero.	(p. 46)
Addition of Real Numbers	**1.** To find the sum of two numbers of like sign, add their absolute values, and use the common sign. **2.** To find the sum of a positive and a negative, find the difference of their absolute values, and use the sign of the number with the larger absolute value.	(p. 48)
Addition Property of Zero (Additive Inverse Property)	The sum of any number and zero is the number: $a + 0 = 0 + a = a$.	(p. 49)
Addition Property of Opposites	The sum of a number and its opposite is zero: $a + (-a) = -a + a = 0$.	(p. 49)

Subtraction of Real Numbers	Add the opposite of the number to be subtracted: $a - b = a + (-b)$.	(p. 58)
Multiplication of Real Numbers	The product of two positive numbers is positive. The product of two negative numbers is positive. The product of a positive and a negative number in either order is negative.	(p. 64)
Multiplying by -1	The product of -1 and a number yields the number's opposite.	(p. 64)
Multiplication Property of 1	$1 \cdot a = a \cdot 1 = a$, for any real number a	(p. 64)
Multiplication Property of 0	$0 \cdot a = a \cdot 0 = 0$, for any real number a	(p. 64)
Division of Real Numbers	The quotient of two positive numbers is positive. The quotient of two negative numbers is positive. The quotient of a positive and a negative number in either order is negative.	(p. 70)

CHAPTER 1
REVIEW EXERCISES

SECTION 1.1 Objective 1

Simplify the following by reducing to lowest terms.

1. $\frac{144}{240}$ **2.** $\frac{240}{256}$ **3.** $\frac{91}{208}$ **4.** $\frac{180}{260}$ **5.** $\frac{147}{189}$

SECTION 1.1 Objective 2

Multiply or divide as indicated.

6. $\frac{15}{27} \cdot \frac{18}{75}$ **7.** $0.4375 \div 0.25$ **8.** $19{,}872 \div 72$ **9.** $2\frac{3}{36} \cdot \frac{21}{45}$

10. $59 \div 236$

SECTION 1.1 Objective 3

Add or subtract as indicated.

11. $234 + 5789 + 12 + 102$ **12.** $6845.3 - 82.97$ **13.** $\frac{19}{34} - \frac{1}{2}$

14. $3\frac{3}{4} + 1\frac{5}{9}$ **15.** $46.4 + 2.1231 + 0.8769$

SECTION 1.2 Objective 1

Find the value of each of the following.

16. $x = 12^4$

17. $x = 8^5$

18. $x = 4^6$

19. $x = 5^4$

20. $x = 4^8$

SECTION 1.2 Objective 2

Write each of the following in exponential form.

21. $7 \cdot 7 \cdot 7 \cdot 7 \cdot 7 \cdot 7$

22. $9 \cdot 9 \cdot 9 \cdot 9 \cdot 9 \cdot 9 \cdot 9$

23. $8 \cdot 8 \cdot 8 \cdot 8$

24. $1 \cdot 1 \cdot 1 \cdot 1 \cdot 1 \cdot 1 \cdot 1 \cdot 1$

25. $2 \cdot 2 \cdot 2 \cdot 2 \cdot 2 \cdot 2 \cdot 2 \cdot 2 \cdot 2 \cdot 2$

SECTION 1.2 Objective 3

Evaluate the following numerical expressions.

26. $48 - \dfrac{16}{4} + 8 \times 10$

27. $16 \cdot 18 - 9^2$

28. $32 \cdot 6 - \dfrac{96 - 8^2}{12 - 8}$

29. Find the value of x if $16^2 - 8 \cdot 9 + 64 = x$.

30. Find the value of y if $y = (20 + 4)^3 - 5(15 + 8)^2$.

SECTION 1.3 Objective 1

Evaluate the following algebraic expressions. In 31–33, $w = 8$, $x = 5$, $y = 2$, and $z = 4$.

31. $x + 21 + y$

32. $4wx + 3yz$

34. Find the solution for D in $D = rt$ if $r = 65$ and $t = 8$.

35. Find the solution for P in $P = 2\ell + 2w$ if $\ell = 16$ and $w = 8$.

SECTION 1.3 Objective 2

Translate the following English phrases to algebraic form.

36. 10 less than a number

37. 6 more than the product of a number and 6

38. 20 less than the quotient of a number and 12

Translate the following to English.

39. x^5

40. $4x - 5$

SECTION 1.3 Objective 3

Determine whether or not the given number is a solution to the given equation.

41. Is 12 a solution of $6x - 14 = 58$?

42. Is 9 a solution of $4x - 8 = 28$?

43. Is 28 a solution of $5y + 44 = 96$?

44. Is 16 a solution of $8a + 24 = 104$?

45. Is 9 a solution of $48 - 4x = 12$?

SECTION 1.4 Objective 1

Given $\{-15, -12.3, -9, -6.4, -0.1, 0, 1, 3.5, 12\}$.

46. List the integers.

47. List the rational numbers.

48. List the whole numbers.

49. List the real numbers.

50. List the irrational numbers.

SECTION 1.4 Objective 2

Match the real numbers with the points indicated on the number line:
$\{-3.5, -1, 1.3, 2.5, 4\}$.

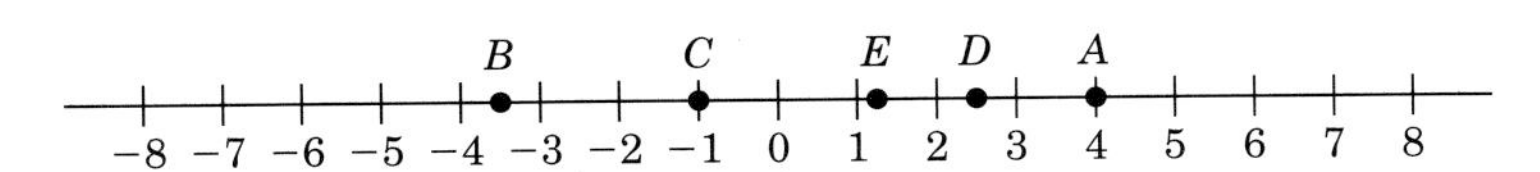

51. $A \leftrightarrow$

52. $B \leftrightarrow$

53. $C \leftrightarrow$

54. $D \leftrightarrow$

55. $E \leftrightarrow$

SECTION 1.4 Objective 3

Determine whether each of the following is true or false.

56. $-8 > -4$

57. $-12 < -18$

58. $-6 < 0$

59. $14 > -3$

60. $0 > -6$

SECTION 1.5 Objective 1

Find the opposite of each of the following.

61. 47

62. -38

63. $-(-21)$

64. For what replacement of x is $-x = -9$ true?

65. For what replacement of x is $-x = 21$ true?

SECTION 1.5 Objective 2

Find the value of each of the following.

66. $|21|$ **67.** $|-48|$ **68.** $-|41|$ **69.** $-|-72|$

70. For what replacement of x is $|x| = 5$ true?

SECTION 1.5 Objective 3

Add.

71. $-17 + 28 + (-18)$ **72.** $-68 + 19 + 29$ **73.** $-62 + (-49) + (-51)$

74. Is -84 a solution of $x + (-29) = -55$? **75.** Is -62 a solution of $-x + 71 = 133$?

SECTION 1.6 Objective 1

Subtract.

76. $-12 - (18 - 4)$ **77.** $-45 - (-16 - 12)$ **78.** $29 - 48 - 72$

79. Is -214 a solution of $x - 121 = -93$? **80.** Is -17 a solution of $48 - x = 31$?

SECTION 1.7 Objective 1

Multiply.

81. $-8(-9)(-10)$ **82.** $-12(7)(-4)$ **83.** $14(-8)(5)$

84. What replacement for x will make $x = (-14)(-17)$ true?

85. What replacement for y will make $y = (-2)(-5)(-7)(-8)$ true?

SECTION 1.8 Objective 1

Divide.

86. $-720 \div (-18)$ **87.** $-475 \div [-(-25)]$ **88.** $-(-364) \div [-(-28)]$

89. $-\frac{25}{42} \div \frac{125}{126}$ **90.** $-48.24 \div (-0.08)$

SECTION 1.9 Objective 1

Evaluate the following expressions.

91. $16 - 8(48 \div 8)^2$ **92.** $50 - 9[13(49 - 54) - 65] - 53$

93. $\frac{4}{5} - \frac{\frac{2}{3}}{2 - \frac{5}{3}} \div \frac{5}{6}$

94. Evaluate $4xy^2z - 3xyz$ when $x = 2$, $y = -4$, and $z = 3$.

95. Evaluate $3a^2bc^2 + 4ab^2c^2$ when $a = -1$, $b = 1$, and $c = -1$.

SECTION 1.9 **Objective 2**

Check the following solutions.

96. Is -12 a solution of $8x - 4 = -100$?

97. Is -15 a solution of $12 - 5x = 87$?

98. Is -21 a solution of $8x - 6 = -162$?

99. Is 16 a solution of $25 - 3x = 63$?

100. Is -6 a solution of $6x + 12 = 9x + 30$?

CHAPTER 1
TRUE–FALSE CONCEPT REVIEW

Check your understanding of the language of algebra. Tell whether each of the following is true (always true) or false (not always true).

1. 3^4 is an example of a number written in exponential form.

2. In the expression 3^4, 3 is the value of the expression.

3. A variable is usually a letter used as a placeholder for a number.

4. The symbol π is a variable.

5. In an expression containing more than one operation, multiplication is always done before division.

6. To evaluate the expression $4x + 2$, one needs to know the replacement for x.

7. The solution set of an equation is the set of all replacements for the variable.

8. The opposite of a number is always negative.

9. The symbol "$-x$" should be read "negative x."

10. Rational numbers can be represented by a fraction.

11. Zero is an integer.

12. The absolute value of a number is always positive.

13. The sum of two integers is always larger than either of the integers being added.

14. To subtract two integers, add the opposite of the number being subtracted.

15. $-7 + 7 = 0$ is an example of the addition property of opposites.

16. The sign of the product of two signed numbers is determined by the sign of the number with the larger absolute value.

17. The sum of two negative numbers is positive.

18. To find the opposite of a number, multiply the number by -1.

19. The rules for division of signed numbers is the same as the rules for multiplication of signed numbers.

20. The order of operations for signed numbers is different from that for whole numbers.

CHAPTER 1
TEST

1. Add: $22.3 + 1.79 + 6.7$
2. Find the solution for $A = \ell w$ when $\ell = 2.5$ and $w = 0.6$.
3. Is 8 a solution of the equation $8x - 4 = 6x + 12$?
4. Find the absolute value: $|3.4|$
5. Subtract: $\frac{5}{8} - \frac{1}{3}$
6. Add: $-8.56 + (-6.58) + 7$
7. Evaluate $2a - b + 15$ when $a = 27$ and $b = 14$.
8. Divide: $91 \div (-13)$
9. Multiply: $(2.3)(17.7)$
10. Find the opposite of 43.
11. Find the value of 7^4.
12. Subtract: $-75 - 22 - 15 - (-1)$
13. Add: $-21 + (-84) + 19$
14. Divide: $0.34 \div 0.4$
15. Multiply: $(-27)(-14)(3)$
16. Write $12 \cdot 12 \cdot 12 \cdot 12 \cdot 12 \cdot 12$ in exponential form.
17. Subtract: $187 - (-122)$
18. Evaluate $2a^2 - 5b^3 \div 12c$ when $a = 2$, $b = -3$, and $c = -5$.
19. Find the absolute value: $|-87|$
20. Is -2 a solution of the equation $4x + 17 = -9x - 9$?
21. Translate the following expression to algebraic form. Let x represent the number: the product of the square of a number and 5
22. Find the opposite of -13.5.
23. Find the value of $15 \div 5 + 12 \cdot 5^2 - 8 \cdot 4$.
24. Divide: $-144 \div (-48)$
25. Translate the following expression to algebraic form; let n represent the first number and p the second number: the first number divided by the product of 4 and the second number
26. Multiply: $(-5.6)(-2)(-4.4)$
27. A teacher uses positive and negative points to keep a record of students' progress. If Sara had the following record, what is her total: $+87$ (test), -40 (absent), $+44$ (quiz), and -26 (homework late)
28. What is the difference between a temperature of 60°F and a temperature of -15°F?
29. Use the formula $C = \frac{5}{9}(F - 32)$ to find the Celsius temperature that is equal to 14°F.

CHAPTER

2

LINEAR EQUATIONS AND INEQUALITIES

SECTIONS

The speed at which a free-falling object falls during its descent can be found using the formula, $s = v + gt$, where s is the speed, v the initial velocity, g the effect of gravity, and t the time the object has been falling. In Exercise 61, Section 2.3, you are asked to find the time it takes a skydiver to reach a falling speed of 172 feet per second. Substitute the given data in the formula and simplify. *(Heinz Fischer/The Image Bank)*

PREVIEW

A great many of the practical applications of algebra involve solving equations and inequalities. In Sections 2.1 and 2.2, we examine the properties of equality that are used to solve linear equations. Throughout the text, you will find a variety of equations and their applications. Section 2.3 exhibits equations that require more than one property to solve. Section 2.4 shows how to simplify expressions in order to be able to solve equations with more terms like those that appear in Section 2.5. Percents in Section 2.6 model one of the most common uses of equations in business. Finally, in Section 2.7, we look at inequalities and ways to show their solution, including graphing.

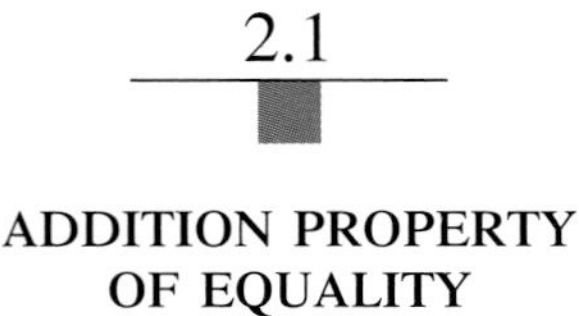

2.1 ADDITION PROPERTY OF EQUALITY

OBJECTIVE

1. Solve linear equations of the form $x + a = b$ or $x - a = b$, where a, b, and x are real numbers.

Recall from Section 1.3 that an equation is a statement about numbers that declares two expressions are equal.

The statements

$T = D + pm, \qquad 3x + 12 = 10, \qquad x - 75.6 = 2.1 \quad \text{and} \quad 4x = 8$

are equations. A formula is an equation that is used repeatedly and often contains variables which are the first letters in the words. For example, $V = \ell wh$ is the formula for the volume of a rectangular box and can be read "The volume, V, of a rectangular box is equal to its length, ℓ, times its width, w, times its height, h."

DEFINITION

Linear Equation

Equations of this form are *linear equations.*

Equations are classified in different ways. Here we are concerned with the classification based on whether the equation is true or false after the variables have been replaced by numbers.

DEFINITION

Conditional Equations

A *conditional equation* is true for some but not for all replacements of the variables. The numbers that make an equation true are called the *solutions* or *roots* of the equation.

$y - 2 = 7$ — The equation is true when y is replaced by 9 since $9 - 2 = 7$, but the equation is false when y is replaced by 21 since $21 - 2 \neq 7$.

The equation, $y - 2 = 7$, is a conditional equation.

DEFINITION

Identities

An *identity* is an equation that is true for all permitted variable replacements.

$3x = 3x$ — is true for all variable replacements and is an example of the reflexive property of equality discussed below

$6x + 3x = 9x$ — is true for all variable replacements and is an example of the use of the distributive property

DEFINITION

Contradictions

A *contradiction* is an equation that is never true for any variable replacements.

$17 - 3 = 3$ — is false

$0x = 12$ — is false when x is replaced by any number

The solution set of a contradiction is $\emptyset$.

Remember from Section 1.3 that the solution set of an equation is the set containing all solutions to the equation. And, to solve an equation means to find the solution set of the equation.

A large part of algebra is the study of methods for solving equations of different types. The equation

$3x = 12$

is easily solved by inspection since we know that $3(4) = 12$. The solution set is $\{4\}$.

Three properties of equality are fundamental to the study of equations.

PROPERTIES

Reflexive Property

$x = x$ (Any expression is equal to itself.) — Reflexive property of equality.

PROPERTIES

Symmetric Property

If $x = a$, then $a = x$. (The right and left sides of any equation may be interchanged.) — Symmetric property of equality.

Transitive Property

If $x = y$ and $y = x$, then $y = z$. (Two expressions that are equal to a third are equal to each other.) — Transitive property of equality.

Solutions of equations such as $x - 46 = 29$ can be found by trial and error, but a better approach is to find a systematic method to solve for x. Let us examine the result of adding a number to each side of an equation and subtracting a number from each side of an equation.

TABLE 2.1

Equation	Add or Subtract	Result
$12 = 12$	$12 + 3 = 12 + 3$	$15 = 15$
$12 = 12$	$12 - 7 = 12 - 7$	$5 = 5$
$21 = 21$	$21 + 6.5 = 21 + 6.5$	$27.5 = 27.5$
$21 = 21$	$21 - 5.6 = 21 - 5.6$	$15.4 = 15.4$
$6 = 4 + 2$	$6 + 33 = 4 + 2 + 33$	$39 = 39$

In each row of Table 2.1, we started with a true equation (identity) and ended with a different true equation. If

$$x - 46 = 29$$

is true for some replacement of x, then

$$x - 46 + 46 = 29 + 46$$

is also true for the same replacement of x, and furthermore by simplifying, $x + 0 = 75$ and $x = 75$ are true for the same replacement. The four equations:

$$x - 46 = 29, \quad x - 46 + 46 = 29 + 46, \quad x + 0 = 75, \quad \text{and} \quad x = 75$$

are called equivalent equations.

DEFINITION

Equivalent Equations

Equivalent equations are equations that have the same solution set.

PROCEDURE

The solution of a linear equation is found by writing a series of equivalent equations until the variable is isolated.

For the equation $x - 46 = 29$, we add 46 to each side because adding 46 is the inverse of (undoes) the indicated subtraction and leaves x isolated on the left side. (Similarly, subtracting 25 from each side of $x + 25 = 70$ will undo the indicated addition.)

EXAMPLE 1

Solve: $x + 25 = 70$

$x + 25 - 25 = 70 - 25$	Subtract 25 from each side.
$x = 70 - 25$	Simplify the left side: $x + 25 - 25 = x + 0 = x$
$x = 45$	Simplify the right side.

Check:

$45 + 25 = 70$ True	Substitute 45 in the original equation.

The solution set is $\{45\}$. ■

Table 2.1 and Example 1 illustrate two properties of algebra. We call these the addition and subtraction properties of equality.

PROPERTY

Addition and Subtraction Properties of Equality

If $x = y$ is true, then

$x + a = y + a$ is true	Addition property of equality.
and	
$x - a = y - a$ is true.	Subtraction property of equality.

Furthermore, the new equations, $x + a = y + a$ and $x - a = y - a$ are equivalent to the original.

EXAMPLE 2

Solve: $y - 1.8 = 11$

$y - 1.8 + 1.8 = 11 + 1.8$	Add 1.8 to each side to isolate y.
$y = 11 + 1.8$	Simplify the left side: $y - 1.8 + 1.8 = y + 0 = y$
$y = 12.8$	Simplify the right side.

Check:

$12.8 - 1.8 = 11$ True — Substitute in the original equation.

The solution set is $\{12.8\}$. ■

■ PROCEDURE

To solve an equation of the form $x + a = b$, add the opposite of a to each side or subtract a from each side to isolate x.

$$x + a - a = b - a$$
$$x = b - a$$

To solve an equation of the form $x - a = b$, add a to each side to isolate x.

$$x - a + a = b + a$$
$$x = b + a$$

All equations of the form $x + a = b$ can be solved by using the addition property of equality; that is, add the opposite of a to each side.

EXAMPLE 3

Solve: $y + \frac{3}{5} = \frac{7}{8}$

$y + \frac{3}{5} + \left(-\frac{3}{5}\right) = \frac{7}{8} + \left(-\frac{3}{5}\right)$ — Add $-\frac{3}{5}$ to each side.

Simplify both sides.

Left: $y + \frac{3}{5} + \left(-\frac{3}{5}\right) = y + 0 = y$.

$y = \frac{11}{40}$ — Right: $\frac{7}{8} + \left(-\frac{3}{5}\right) = \frac{35}{40} + \left(-\frac{24}{40}\right) = \frac{11}{40}$.

The solution set is $\left\{\frac{11}{40}\right\}$. The check is left for the student. ■

■ DEFINITION

Literal Equation

An equation that contains more than one variable is sometimes referred to as a *literal equation.*

The equation $y + a = k$ is a literal equation. Such equations can be solved using the same properties of equality.

EXAMPLE 4

Solve $y + a = k$ for y.

$y + a - a = k - a$	Subtract a from each side.
$y = k - a$	Simplify both sides.
The solution is $y = k - a$.	Set notation will not be used to describe the solution of a literal equation. ■

It is often desirable to alter a formula by solving for a different variable.

EXAMPLE 5

The selling price of a Walker stereo is found by adding the markup (the store's gross profit) to the cost (the price paid by the store). The formula is $C + M = S$. Solve for C.

$C + M - M = S - M$	Subtract M from each side.
$C = S - M$	Simplify. In words we can say, "cost equals selling price minus markup." ■

The same properties can be used to solve an equation that arises from a word problem.

EXAMPLE 6

Forty-four less than a number is seventy-eight. What is the number?

Simpler word form:

A number minus 44 equals 78.

Select a variable:

Let n represent the number.

Translate to algebra:

$n - 44 = 78$

Translate the sentence into symbols. "Less than" becomes subtraction, so "forty-four less than a number" becomes "$n - 44$." The word *is* is translated to "=" in the algebraic form.

Solve:

$$n - 44 + 44 = 78 + 44$$
$$n = 122$$

Check:

Is 44 less than 122 equal to 78?

$$\begin{array}{r} 122 \\ 44 \\ \hline 78 \end{array}$$

Subtract.

Yes.

The number is 122. ■

Numerous useful equations are used in business.

EXAMPLE 7

The selling price of a pair of shoes is \$48.95. If the markup on the shoes is \$14.82, what is their cost? Cost + markup = selling price. To find the cost of the pair of shoes, first translate the English sentence to an algebraic form.

Formula:

$C + M = S$

Since cost + markup = selling price, we use the formula $C + M = S$.

Substitute:

$C + 14.82 = 48.95$

Since $M = \$14.82$ and $S = \$48.95$, we can substitute for the variables M and S and solve.

Solve:

$$C + 14.82 - 14.82 = 48.95 - 14.82$$
$$C = 34.13$$

When stores use the word *cost* they mean the price paid by the store, not by the customer. The customer pays the selling price.

The shoes cost \$34.13. ■

Exercise 2.1

A

Solve:

1. $a - 3 = 29$
2. $b - 6 = 24$
3. $x + 2 = 65$
4. $y + 1 = 30$
5. $z - 3 = -16$
6. $w - 7 = -26$
7. $d + 6 = -41$
8. $f - 19 = 211$
9. $g - 7 = -47$
10. $h - 9 = -49$
11. $x + 22 = -48$
12. $y + 33 = 21$
13. $37 = w + 6$
14. $-12 = t + 15$
15. $-45 = s - 55$
16. $-77 = p - 62$
17. Solve for ℓ: $\ell + w = p$
18. Solve for a: $a - b = c$
19. Solve for x: $t = -s + x$
20. Solve for y: $t = r + y$

B

Solve:

21. $x - 72 = 30$
22. $x - 28 = 78$
23. $y + 139 = 95$
24. $y - 113 = 77$
25. $x + 12 = -23$
26. $x - 90 = 98$
27. $w + 72 = 0$
28. $z + 98 = 0$
29. $y + 323 = -338$
30. $y + 646 = -689$
31. $x - 341 = 263$
32. $x - 22 = 748$
33. $w + 477 = -662$
34. $w + 119 = -989$
35. $-221 = a - 477$
36. $-112 = b - 111$
37. $-28 = y - 42$
38. $-36 = z - 32$
39. Solve for a: $a - c = b$
40. Solve for y: $x + y = -z$

C

Solve:

41. $x - \frac{3}{4} = -\frac{3}{2}$
42. $y + \frac{5}{6} = -\frac{1}{3}$
43. $\frac{2}{5} = x - \frac{1}{4}$
44. $z + 99.9 = -43.1$
45. $-\frac{4}{5} = c - \frac{1}{2}$
46. $x - 6.02 = 82.44$
47. $x - 8.04 = 2.61$
48. $z + 99.9 = 43.1$
49. $z + 17.86 = 5.49$
50. $18.3 = t + 20$
51. $-22.7 = k + 30$
52. $t + 18.6 = -42.3$
53. $c - 42.8 = -50.3$
54. Solve for w: $w - p = r$
55. Solve for p: $p + r = w$
56. Solve for ℓ: $\ell + w = \frac{p}{2}$
57. Solve for t: $t - m = 5$

58. Solve for n: $17 = n - y$

59. Solve for p: $p - r = -s$

60. Solve for z: $z + r = -2$

D

61. Sixteen more than some number is 60. What is the number?

62. Forty-five more than a given number is 27. What is the number?

63. Thirty-three less than a number is 18. What is the number?

64. The difference between a number and eighty-three is 6. What is the number?

65. The difference between a number and 37 is -11.

66. A number decreased by 372.6 gives the result 421.2. What is the number?

67. The selling price of a stereo system is \$498.79. If the markup on the system is \$83.13, what is the cost (to the store) of the system? (cost + markup = selling price)

68. The selling price of a personal computer is \$995.95. If the cost is \$855.29, what is the markup?

69. The selling price of a compact disc player is \$449.98. If the cost is \$321.22, what is the markup?

70. After Rosa had gained $5\frac{1}{2}$ pounds, she weighed 110 pounds. How much did Rosa weigh before the weight gain?

71. The length of a rectangular garage is 1.8 meters more than its width. If the width is 7.3 meters, what is its length?

72. After the cost of an automobile was increased by \$879, it sold for \$14,524. What was the previous selling price?

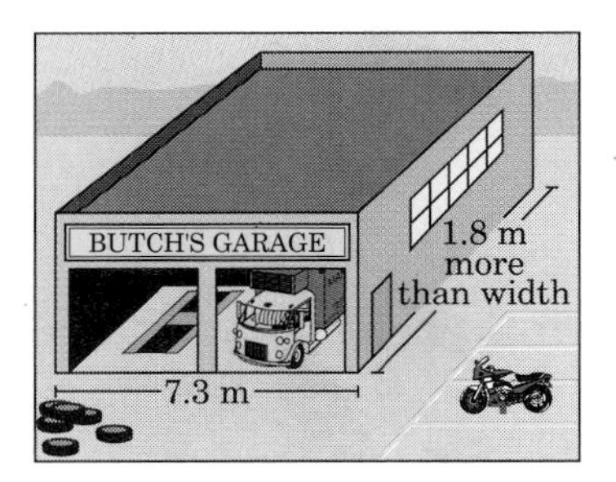

Figure for Exercise 71

STATE YOUR UNDERSTANDING

73. The equations $x + 3 = 7$ and $x + 3 = -1$ are not equivalent equations. Explain.

74. When solving the equation $x + 2.5 = 5.2$, you have the option of either subtracting 2.5 or adding -2.5. Why?

CHALLENGE EXERCISES

Solve the following equations.

75. $|x| = 4$

76. $|x| + 7 = 12$

77. $|y| - 5 = 13$

78. $|x| + 3 = 3$

MAINTAIN YOUR SKILLS (SECTIONS 1.5, 1.6, 1.9)

Perform the indicated operations:

79. $-33 - (-45)$

80. $-56 + (-33)$

81. $-3.45 - (-5.67)$

82. $-6.78 + 5.34$

What number can replace the variable to make the statement true?

83. $z = -65 - 45 + (-32)$

84. $x = -33 - 45 - (-23)$

85. $y = -(-45) + (-34) - (-54)$

86. $z = 34 - (-12) + (-73) - (-26)$

2.2

MULTIPLICATION PROPERTY OF EQUALITY

OBJECTIVE

1. Solve linear equations of the form $ax = b$ or $\frac{x}{a} = b$, where x, a, and b are real numbers.

Solutions of equations such as $3x = 39$ can be found by trial and error. Just look for a number that when multiplied by 3 has a product of 39, ($x = 13$). In this section, we advocate a systematic method that can be used for all such equations. Table 2.2 illustrates the effect of multiplying or dividing each side of an equation by the same nonzero number.

TABLE 2.2

Equation	Multiply or Divide	Result
$5 = 5$	Multiply by 7: $5 \cdot 7 = 5 \cdot 7$	$35 = 35$
$40 = 40$	Divide by 10: $\frac{40}{10} = \frac{40}{10}$	$4 = 4$
$\frac{5}{3} = \frac{5}{3}$	Multiply by 3: $3 \cdot \frac{5}{3} = 3 \cdot \frac{5}{3}$	$5 = 5$
$9 = 9$	Divide by 4: $\frac{9}{4} = \frac{9}{4}$	$2.25 = 2.25$

In each row of Table 2.2, we started with a true equation (identity) and ended with a different true equation. If

$$3x = 27.06$$

is true for some replacement of x then

$$\frac{3x}{3} = \frac{27.06}{3}$$

is also true for the same replacement of x, and furthermore by simplifying, $x = 9.02$ is also true. The three equations

$$3x = 27.06, \qquad \frac{3x}{3} = \frac{27.06}{3}, \quad \text{and} \quad x = 9.02$$

are equivalent equations.

So the solution set is $\{9.02\}$.

EXAMPLE 1

Solve: $5x = 13$

$$\frac{5x}{5} = \frac{13}{5}$$ Divide each side by 5 to undo the indicated multiplication and isolate the variable x.

$$x = \frac{13}{5}$$

Check:

$5 \cdot \frac{13}{5} = 13$ True Substitute $\frac{13}{5}$ in the original equation.

The solution set is $\left\{\frac{13}{5}\right\}$ or $\{2.6\}$. ■

■ PROCEDURE

To solve an equation of the form $ax = b$, divide each side by a:

$$\frac{ax}{a} = \frac{b}{a} \quad \text{so} \quad x = \frac{b}{a},\ a \neq 0.$$

Our next example involves a fraction. The fraction can be ''cleared'' by multiplying each side of the equation by the same expression so that the result is an equivalent equation with integer coefficients.

EXAMPLE 2

Solve: $\frac{x}{23} = 8$

$23 \cdot \frac{x}{23} = 23(8)$ Multiply each side by 23 to clear the fraction.

$x = 184$ Simplify each side.

Left side: $23 \cdot \frac{x}{23} = x$ since multiplying by 23 and dividing by 23 undo each other.

Check:

$\frac{184}{23} = 8$ True Substitute 184 in the original equation.

The solution set is $\{184\}$. ■

Table 2.2 and Examples 1 and 2 illustrate two properties of equality. We call these the multiplication and division properties of equality.

■ PROPERTY

Multiplication and Division Properties of Equality

If $x = y$ is true and $a \neq 0$, then

$ax = ay$ is true Multiplication property of equality.

and

$\frac{x}{a} = \frac{y}{a}$ is true. Division property of equality.

Furthermore, the new equations $ax = ay$ and $\frac{x}{a} = \frac{y}{a}$ are equivalent to the original.

Multiplication by zero is avoided because it does not lead to an equivalent equation. If each side of the equation $4x = 32$, which has solution set $\{8\}$, is multiplied by zero, we get $0 \cdot 4x = 0 \cdot 32$ or $0x = 0$, which has the solution set {all real numbers}. The equations $4x = 32$ and $0x = 0$ are not equivalent. Division by 0 is avoided since it is undefined.

EXAMPLE 3

Solve: $-8.3 = \frac{y}{7}$

$7(-8.3) = 7 \cdot \frac{y}{7}$ — Multiply each side by 7.

$-58.1 = y$ — Simplify both sides.
Left side: $7(-8.3) = -58.1$.
Right side: $7 \cdot \frac{y}{7} = y$.

$y = -58.1$ — Symmetric property of equality.

The solution set is $\{-58.1\}$. — The check is left for the student. ■

Division by a and multiplication by $\frac{1}{a}$ (the reciprocal of a) are equivalent operations: $b \div a = b \cdot \frac{1}{a} = \frac{b}{a}$, $a \neq 0$. Therefore, equations of the form $ax = b$ can also be solved by multiplying each side by the reciprocal of a.

For equations such as $\frac{5x}{6} = 12$, we multiply both sides by 6 (to clear the fraction) then isolate x by multiplication or division.

EXAMPLE 4

Solve: $\frac{5x}{6} = 12$

$\frac{5x}{6} \cdot 6 = 12 \cdot 6$ — Multiply each side by 6.

$5x = 72$ — Then divide each side by 5 $\left(\text{or multiply each side by } \frac{1}{5}\right)$.

$x = \frac{72}{5}$

The solution set is $\left\{\frac{72}{5}\right\}$ or $\{14.4\}$. — The check is left for the student. ■

■ PROCEDURE

To solve an equation of the form $\frac{x}{a} = b$, multiply each side by a:

$$a\left(\frac{x}{a}\right) = a(b) \quad \text{so} \quad x = ab.$$

■ CAUTION

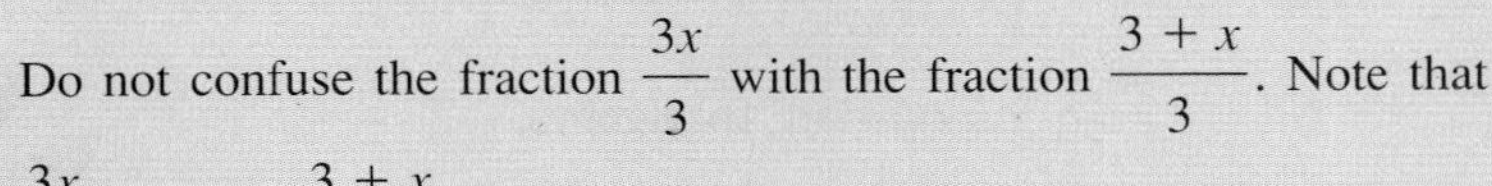

Do not confuse the fraction $\frac{3x}{3}$ with the fraction $\frac{3 + x}{3}$. Note that

$\frac{3x}{3} = x$, but $\frac{3 + x}{3} \neq x$.

Literal equations such as $RB = A$ can be solved using the same rules. We can find a formula for R.

EXAMPLE 5

Solve: $RB = A$ for R

$\frac{RB}{B} = \frac{A}{B}$	Divide each side by B.
$R = \frac{A}{B}$	Simplify.

The solution is $R = \frac{A}{B}$. ■

The formula for time (traveled) is $t = \frac{D}{r}$. Find the formula for distance.

EXAMPLE 6

Solve: $t = \frac{D}{r}$ for D

$r(t) = r \cdot \frac{D}{r}$	Multiply each side by r.
$rt = D$ or $D = rt$	The formula for distance is $D = rt$ (distance equals rate times time). ■

The same properties can be used to solve several equations that arise from number problems.

EXAMPLE 7

A certain number divided by 37 is 43. Find the number.

Simpler word form:

A number divided by 37 is 43.

Select a variable:

Let n represent the number.

Translate to algebra:

$\frac{n}{37} = 43$ — Translate the sentence into symbols.

Solve:

$37 \cdot \frac{n}{37} = 37(43)$ — Clear the fraction by multiplying each side by 37.

$n = 1591$

Check:

Is 1591 divided by 37 equal to 43?

$$\begin{array}{r} 43 \\ 37\overline{)1591} \\ \underline{148} \\ 111 \\ \underline{111} \end{array}$$

True

The number is 1591. ■

The formula for the area of a rectangle, $A = \ell w$, is often used for land measure.

EXAMPLE 8

What is the width (w) of a rectangular plot of ground on which a house is to be built if the length (ℓ) is 125 feet and the area (A) is 9375 square feet? Use the formula $A = \ell w$.

Formula:

$A = \ell w$

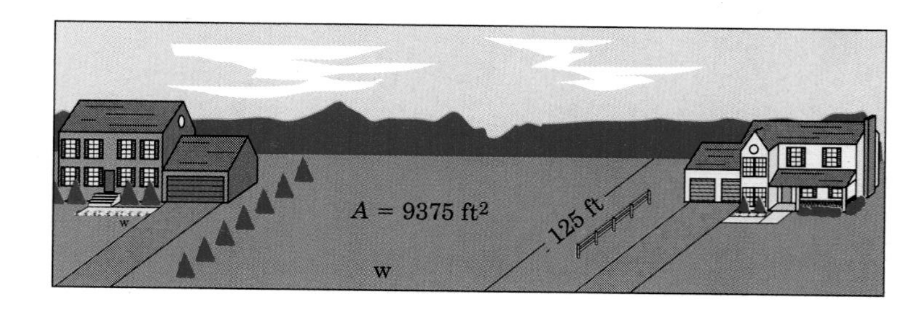

Substitute:

$9375 = 125\,w$ — To find the width of the building lot, substitute the area, $A = 9375$, and the length, $\ell = 125$, into the formula and solve.

Solve:

$$\frac{9375}{125} = \frac{125w}{125}$$

$$75 = w$$

Check:

If the width is 75 feet and the length is 125 feet, is the area 9375 square feet?

$A = (125 \text{ ft})(75 \text{ ft}) = 9375 \text{ sq ft.}$ True

The width of the lot is 75 feet. ■

EXERCISE 2.2

A

Solve:

1. $2x = 50$

2. $4x = 40$

3. $-19x = 38$

4. $-22y = -66$

5. $\frac{x}{2} = 50$

6. $\frac{x}{4} = 40$

7. $8y = -72$

8. $6y = -72$

9. $\frac{y}{6} = 11$

10. $\frac{y}{8} = -11$

11. $\frac{-12w}{8} = 72$

12. $\frac{z}{-6} = 13$

13. $\frac{-15w}{-8} = 30$

14. Solve for p: $pr = i$

15. Solve for i: $\frac{i}{r} = p$

16. Solve for r: $pr = i$

17. Solve for x: $\frac{x}{t} = 7$

18. Solve for y: $\frac{y}{12} = p$

19. Solve for a: $18 = \frac{a}{q}$

20. Solve for b: $bw = 13$

B

Solve:

21. $\frac{t}{22} = 88$

22. $\frac{t}{5} = 85$

23. $-27v = 405$

24. $28v = -728$

25. $\frac{w}{-6} = 30$ **26.** $\frac{w}{3} = -90$ **27.** $34z = 2380$ **28.** $-65b = 3250$

29. $\frac{y}{16} = 27$ **30.** $\frac{y}{25} = -13$ **31.** $\frac{y}{-8} = -14$ **32.** $\frac{b}{-21} = 13$

33. $\frac{z}{22} = 25$ **34.** $\frac{8w}{-3} = 15$ **35.** $\frac{-10p}{7} = -40$ **36.** $\frac{-12t}{-5} = 24$

37. Solve for x: $5x = m$

38. Solve for y: $10y = b$

39. Solve for x: $\frac{x}{8} = a$

40. Solve for w: $\frac{w}{-3} = c$

C

Solve:

41. $7w = -9.03$ **42.** $-4y = -14$ **43.** $\frac{z}{-1.1} = 3.5$ **44.** $\frac{z}{5.3} = -2.2$

45. $6x = -4$ **46.** $4y = 14$ **47.** $-16a = 18$ **48.** $\frac{s}{\frac{2}{3}} = \frac{7}{8}$

49. $\frac{t}{\frac{3}{4}} = -\frac{4}{5}$ **50.** $\frac{m}{-\frac{4}{3}} = -\frac{4}{5}$ **51.** $\frac{t}{\frac{4}{3}} = \frac{4}{5}$ **52.** $\frac{2x}{3} = \frac{5}{6}$

53. $\frac{3t}{5} = -\frac{7}{10}$

54. Solve for s: $4s = P$

55. Solve for n: $C = np$

56. Solve for C: $\frac{C}{D} = \pi$

57. Solve for A: $\frac{A}{w} = \ell$

58. Solve for t: $p = \frac{t}{2m}$

59. Solve for m: $a = \frac{m}{4b}$

60. Solve for m: $3mp = 4a$

D

61. A certain number divided by 37 is 53. What is the number?

62. A certain number divided by 53 is 73. What is the number?

63. Fifteen times a number is 180. What is the number?

64. A number multiplied by seventy-seven is -3234. What is the number?

65. A number divided by -13 is 49. What is the number?

66. The quotient of a number divided by 49 is 63. What is the number?

67. The product of a number and 1.05 is 0.063. What is the number?

68. A number times 63.5 is 1143. What is the number?

69. Find the length of a rectangle that has a width of 12 feet and an area of 252 square feet.

70. Find the width of a rectangular plot of ground that has an area of 6765 square meters and a length of 123 meters.

71. Find the principal (p) if the rate of interest (r) is 0.065, the interest (i) is \$78 and the time (t) is 1 year. Use the formula $i = prt$.

72. Find the rate of interest (r) if the interest (i) is \$195.36, the principal (p) is \$1,320 and the time (t) is 2 years. Use the formula $i = prt$.

73. Mark cut a board into 8 pieces of equal length. If each piece was $1\frac{1}{2}$ feet long, what was the length of the board?

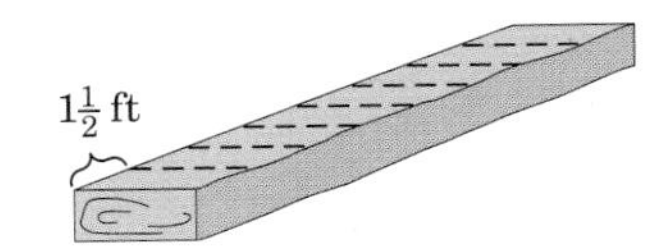

Figure for Exercise 73

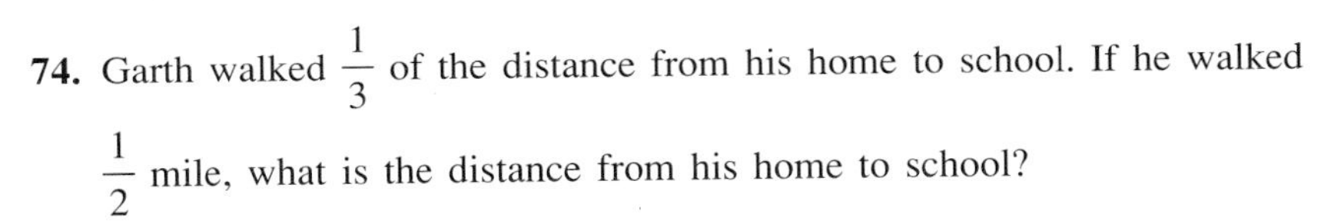

74. Garth walked $\frac{1}{3}$ of the distance from his home to school. If he walked $\frac{1}{2}$ mile, what is the distance from his home to school?

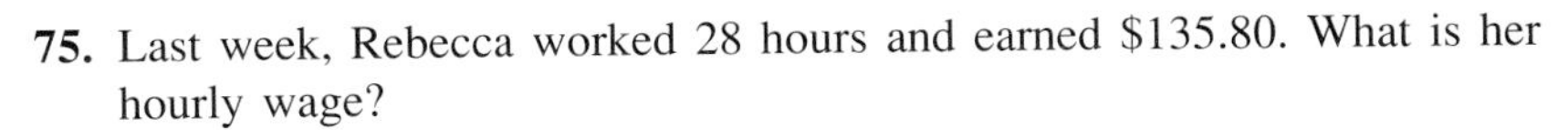

75. Last week, Rebecca worked 28 hours and earned \$135.80. What is her hourly wage?

STATE YOUR UNDERSTANDING

76. Explain why both sides of a conditional equation cannot be multiplied by 0.

77. Why is $\emptyset$ the solution set of $0 \cdot x = 12$?

CHALLENGE EXERCISES

Solve the following equations.

78. $6 \cdot |x| = 48$

79. $\dfrac{|y|}{3} = 1.5$

80. $2|x + 1| = 0$

81. $|y - 1| + 4 = 16$

MAINTAIN YOUR SKILLS (SECTIONS 1.6, 1.9)

Perform the indicated operations:

82. $-45 - (-75) + (-34) - (-55)$

83. $-|23| + (-19) - (-40) - 14$

84. $-(-67) - (-89) + 64 - (-43) - 62$

If $a = -34$, $b = 17$, and $c = -48$, evaluate each of the following:

85. $a - b$

86. $c - a$

87. $c - b$

88. The number of cases of soda that Hank's Supermarket had in stock on Friday was 135. If the number of cases sold is represented by negative numbers and the number of cases received is represented by positive numbers, how many cases does Hank's have in stock after the following transactions: 45, −67, 36, −78, 52, −45, 97, −102.

89. The Acme Corporation lost 1.3 million dollars (−1.3) during the first quarter of 1990. The company reported a net profit of 1.05 million dollars (+1.05) for the first two quarters of 1990. What was the profit (or loss) during the second quarter (expressed as a signed number)?

2.3

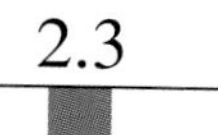

SOLVING LINEAR EQUATIONS

OBJECTIVE

1. Solve linear equations of the form $ax + b = c$, $ax - b = c$, $\frac{x}{a} + b = c$, and $\frac{x}{a} - b = c$, where a, x, b, and c are real numbers.

In the previous sections, we solved equations such as $x + 4 = 7$, $x - 8 = -11$, $-3x = 142$, and $\frac{x}{3} = 142$. Each of these can be solved in one step using one of the basic properties of equations. Other equations require the use of two or more of the basic properties to isolate the variable. Examples 1 and 2 are such equations.

EXAMPLE 1

Solve: $14y - 39 = 87$

$14y - 39 + 39 = 87 + 39$ The first step is to isolate $14y$ by adding 39 to each side.

$14y = 126$ Simplify.

$\frac{14y}{14} = \frac{126}{14}$ The second step is to isolate the variable by dividing each side by 14.

$y = 9$

Check:

$14(9) - 39 = 87$ Substitute 9 into the original equation.

$126 - 39 = 87$ True Multiply first, then subtract.

The solution set is $\{9\}$. ■

EXAMPLE 2

Solve: $-13 = -7 - 5x$

$7 - 13 = 7 - 7 - 5x$ Add 7 to each side to isolate $-5x$. If you prefer, you can rewrite the equation using the symmetric law of equality: $-7 - 5x = -13$.

$-6 = -5x$ Simplify.

$\dfrac{-6}{-5} = \dfrac{-5x}{-5}$ Divide each side by -5 to isolate x.

$\dfrac{6}{5} = x$ Simplify.

The solution set is $\left\{\dfrac{6}{5}\right\}$. The check is left for the student. ■

An equation that contains fractions is called a fractional equation. An equivalent equation that does *not* contain fractions can be derived from a fractional equation by using the multiplication property of equality. The equivalent equation is solved in the usual manner. Examples 3 and 4 are such equations.

EXAMPLE 3

Solve: $\dfrac{x}{7} + 32 = -6$

$7\left(\dfrac{x}{7} + 32\right) = 7(-6)$ The first step is to *clear the fraction* by multiplying each side by 7, which is the denominator of the fraction.

$7\left(\dfrac{x}{7}\right) + 7(32) = 7(-6)$ Simplify the left side by using the distributive property to write the multiplication as the sum of two products.

$x + 224 = -42$ Simplify.

$x = -266$ Subtract 224 from each side to isolate x.

The solution set is $\{-266\}$. The check is left for the student. ■

EXAMPLE 4

Solve: $-1 = \dfrac{x}{4} + 5$

$\dfrac{x}{4} + 5 = -1$	Symmetric property of equality. Interchange the left and right sides.
$4\left(\dfrac{x}{4} + 5\right) = 4(-1)$	Multiply each side by 4 to clear the fraction.
$4\left(\dfrac{x}{4}\right) + 4(5) = 4(-1)$	Distributive property.
$x + 20 = -4$	Simplify.
$x = -24$	Subtract 20 from each side.
The solution set is $\{-24\}$.	The check is left for the student. ■

If an equation contains more than a single fraction, all the fractions can be cleared by multiplying by the least common multiple of the denominators of the fractions.

EXAMPLE 5

Solve: $\dfrac{y}{3} + \dfrac{1}{4} = 2$

$12\left(\dfrac{y}{3} + \dfrac{1}{4}\right) = 12(2)$	Multiply each side by 12, the LCM of 3 and 4, to clear the fractions.
$12\left(\dfrac{y}{3}\right) + 12\left(\dfrac{1}{4}\right) = 12(2)$	Distributive property.
$4y + 3 = 24$	Simplify.
$4y = 21$	Subtract 3 from each side.
$y = \dfrac{21}{4}$	Divide each side by 4.
The solution set is $\left\{\dfrac{21}{4}\right\}$ or $\{5.25\}$.	The check is left for the student. ■

The same properties are used to solve literal equations.

EXAMPLE 6

Solve $ax - b = c$ for x.

$ax - b + b = c + b$	Add b to each side.

$$ax = c + b$$ Simplify.

$$\frac{ax}{a} = \frac{c + b}{a}$$ Divide each side by a.

$$x = \frac{c + b}{a}$$ Simplify. ■

EXAMPLE 7

Solve $RI + e = E$ for I.

$$RI + e - e = E - e$$ Subtract e from each side. When both uppercase and lowercase letters (E and e) are used in the same formula, they are different variables and represent different values.

$$RI = E - e$$ Simplify.

$$\frac{RI}{R} = \frac{E - e}{R}$$ Divide each side by R to isolate I.

$$I = \frac{E - e}{R}$$ Simplify the left side. Since I is isolated on the left, we have "solved for I." ■

Several algebraic formulas are used in the solution of business-related problems. The next example demonstrates this.

EXAMPLE 8

The formula for the balance of a loan (D) is $D = B - NP$, where P represents the monthly payment, N represents the number of payments, and B represents the amount of money borrowed. Find N when $D = \$375$, $B = \$630$, and $P = \$15$.

Formula:

$$D = B - NP$$

Substitute:

$$375 = 630 - N(15)$$ We substitute the given values into the formula: $D = 375$, $B = 630$, and $P = 15$.

Solve:

$$-255 = -15N$$ Subtract 630 from each side.

$$17 = N$$ Divide each side by -15.

Check:

If 17 payments have been made, is the balance \$375?

$$\begin{array}{l} \$630 \\ \underline{\$255} \\ \$375 \quad \text{True} \end{array}$$

Subtract 17(15) or 255 from the loan of \$630.

Answer:

Seventeen payments have been made. ■

Our last example concerning free-falling objects comes from physics.

EXAMPLE 9

The speed, s (in feet per second), of a free-falling skydiver is given by the formula

$$s = v + gt,$$

where v is the initial velocity (in feet per second), g is the force of gravity, and t is the time (in seconds). Find the time, t, it takes the skydiver to reach a speed of 125 feet per second if $v = -3$ feet per second and $g = 32$.

Formula:

$$s = v + gt$$

Substitute:

$$125 = -3 + 32t$$

We substitute the given values into the formula: $s = 125$, $v = -3$, and $g = 32$.

Solve:

$$128 = 32t$$

Add 3 to each side.

$$4 = t$$

Divide each side by 32.

It takes 4 seconds for the skydiver to reach a speed of 125 feet per second.

The check is left for the student. ■

EXERCISE 2.3

A

Solve:

1. $7x + 5 = 40$
2. $8y + 3 = 51$
3. $3w - 4 = -16$
4. $2x - 5 = -17$
5. $2y + 17 = -8$
6. $4x + 21 = -11$
7. $13 = 18w - 23$
8. $-14 = 37y - 88$
9. $0 = -3x + 15$
10. $19 = -8w + 3$
11. $\frac{x}{3} + 6 = 16$
12. $\frac{y}{5} - 4 = -16$
13. Solve $ax + b = c$ for x if $a = 3$, $b = 4$, and $c = 1$.
14. Solve $ax + b = c$ for x if $a = 6$, $b = 14$, and $c = 2$.
15. Solve for x: $tx + c = d$
16. Solve for y: $my + r = w$

Solve:

17. $22 = 4 + \frac{x}{7}$
18. $19 = 9 + \frac{y}{10}$
19. $15 = \frac{x}{6} - 5$
20. $3 = \frac{y}{11} - 8$

B

Solve:

21. $29y + 2 = 205$
22. $74w + 3 = 225$
23. $-2x + 5 = -17$
24. $9z - 15 = -87$
25. $67w + 34 = 369$
26. $50w - 16 = 384$
27. $-18y - 10 = -1$
28. $-12x - 10 = -2$
29. $4 = 22x - 150$
30. $2 = -3x - 61$
31. $\frac{w}{9} - 8 = -2$
32. $\frac{x}{7} + 6 = -5$
33. $\frac{p}{-2} - 14 = 6$
34. $\frac{t}{-3} + 23 = 17$
35. $\frac{x}{8} - 6 = \frac{1}{2}$
36. $\frac{y}{6} - 8 = \frac{5}{2}$
37. Solve $ax - b = c$ for x if $a = 12$, $b = 6$, and $c = -14$.
38. Solve $ax - b = c$ for x if $a = 3$, $b = 5$, and $c = -50$.
39. Solve for x: $gx - t = k$
40. Solve for y: $sy - r = pq$

C

Solve:

41. $2x - \frac{1}{3} = \frac{1}{2}$

42. $-6x + \frac{1}{5} = \frac{3}{10}$

43. $-2p + \frac{1}{6} = \frac{2}{3}$

44. $3y - \frac{1}{4} = \frac{3}{8}$

45. $\frac{x}{4} + \frac{1}{2} = \frac{1}{3}$

46. $\frac{y}{3} + \frac{1}{4} = \frac{1}{2}$

47. $22.8x + 56.6 = 170.6$

48. $44.5y + 9.4 = 320.9$

49. $-13y + 20 = 35.6$

50. $-5x + 17 = 30.5$

51. $-18x + 3.8 = 65$

52. $-14y + 3.2 = 90$

53. $2.3p - 19.3 = -83.7$

54. $9.45p + 55.5 = -246.9$

55. Solve $ax + b = c$ for x if $a = 7$, $b = 1.3$, and $c = -2.9$.

56. Solve $ax + b = c$ for x if $a = 3$, $b = \frac{4}{7}$, and $c = -\frac{3}{2}$.

57. Solve for w: $2\ell + 2w = P$

58. Solve for ℓ: $2\ell + 2w = P$

59. Solve for p: $c = 18p + 20$

60. Solve for t: $s = v + gt$

D

61. Find the time (t) it takes a free-falling skydiver to reach a falling speed of 173 feet per second if $v = -3$ and $g = 32$. The formula is $s = v + gt$.

62. Find the time (t) it takes a free-falling skydiver to reach a falling speed of 77 feet per second if $v = -3$ and $g = 32$. The formula is $s = v + gt$.

63. A formula for distance traveled is $2d = t^2a + 2tv$, where d represents distance, v represents initial velocity, t represents time, and a represents acceleration. Find a if $d = 240$, $v = 20$, and $t = 3$.

64. Use the formula in Exercise 63 to find a if $d = 240$, $v = 20$, and $t = 5$.

65. If 98 is added to six times some number, the sum is 266. What is the number?

66. If 73 is added to eleven times a number, the sum is 117. What is the number?

67. The difference of 15 times a number and 87 is -39. What is the number?

68. The difference of 24 times a number and 78 is 78. What is the number?

69. Using the formula $D = B - NP$ (see Example 8), find the number of payments made (N) if the balance (D) is \$205, the monthly payment (P) is \$35, and the amount borrowed (B) is \$450.

70. Use the formula in Exercise 69 to find the number of payments made if the balance is \$459, the monthly payment is \$57, and the amount borrowed is \$1200.

71. Find the time (t) in years it takes for an investment (p) of \$1,000 to grow to a value (A) of \$1,150 if the rate (r) is 6% (0.06). The formula is $A = p + prt$.

72. A limousine service charges each passenger \$1.00 plus \$0.50 per mile. A trip from the airport to Central City costs each passenger \$10.00. What is the distance from the airport to Central City?

STATE YOUR UNDERSTANDING

73. Explain why $2x + 3 = 7$ and $2x + 3 = -1$ are not equivalent equations.

74. 4 is a solution of the equation, $2 + w = w + 2$, but $\{4\}$ is not the solution set. Explain.

CHALLENGE EXERCISES

Solve each of the following equations.

75. $3|x| + 5 = 11$

76. $4|y + 2| - 3 = 5$

77. $\dfrac{|2k + 1|}{3} - 3 = 2$

78. $\dfrac{|3x - 1|}{5} + 4 = 1$

MAINTAIN YOUR SKILLS (SECTIONS 1.7, 1.8, 1.9)

Perform the indicated operations:

79. $1140 \div (-12)$

80. $(-2.3)(-0.3)(5.6)$

81. $(-14.1)(0.4)(6.1)$

If $a = -6$, $b = -18$, and $c = -36$, evaluate each of the following:

82. $b \div c$

83. $c \div a$

84. abc

85. The local supermarket sells turkeys for Thanksgiving at an average loss of \$2.50 (−\$2.50) per turkey as a loss leader. What will be the total loss on the sale of turkeys if the store sells all the 310 turkeys it has (expressed as a signed number)?

86. A stock lost a total of $13\frac{7}{8}$ points $\left(-13\frac{7}{8}\right)$ in 8 consecutive days of trading. What was the average loss per day (expressed as a signed number)?

2.4 SIMPLIFYING EXPRESSIONS

OBJECTIVES

1. Combine like terms.
2. Simplify algebraic expressions that contain like terms.

Term

DEFINITION

A *term* is a number, a variable, or the indicated product or quotient of numbers and variables.

Algebraic expressions such as $2a$, $5b^4$, $\frac{1}{x}$, $\sqrt{y}$, and $|2x|$ are terms. In this section, we limit our discussion of terms to those that are products of numbers and variables with whole number exponents. These include algebraic expressions such as $3x$, cz, $2m^2$, $\frac{2}{3}x$, and $2.5ab^2c^3$.

Coefficient

DEFINITION

In a term which is written as an indicated product of numbers and variables, the number is called the *coefficient* of the term.

The coefficient of $3x$ is 3, of cz is 1, of $2m^2$ is 2, of $\frac{2}{3}x$ is $\frac{2}{3}$, and of $2.5ab^2c^3$ is 2.5.

Like Terms

DEFINITION

Like terms are terms with common variables raised to the same power.

For example, $5xy^2$ and $-16xy^2$ are like terms, but $7x^2y^3$ and $7xy^4$ are not like terms.

Like Terms	Unlike Terms
$3xy$ and $-4xy$	$5x$ and $5w$
$-6x^2$ and $-15x^2$	$4x^2$ and $6x$
w and $5w$	$6ab$ and $6a^2b$

To combine like terms, we use the distributive property. When the terms have been combined, we say the expression has been simplified.

The following table contains some examples of combining terms.

Algebraic Expression	Distributive Property	Result
$16x + 22x$	$(16 + 22)x$	$38x$
$37t - 19t$	$(37 - 19)t$	$18t$
$54mn - 55mn$	$(54 - 55)mn$	$-1mn$ or $-mn$
$y + 48y$	$(1 + 48)y$	$49y$
$24b + 24w$	Unlike terms cannot be combined	
$16x^2 + 15x$	Unlike terms cannot be combined	
$14p - 7 + 25p - 9p$	$(14 + 25 - 9)p - 7$	$30p - 7$

The middle step in the table (distributive property) is often done mentally and not written out.

PROCEDURE

To combine like terms, add or subtract the coefficients, and write the common variable factors in an indicated product.

EXAMPLE 1

Simplify: $27ay - 32ay$

$27ay - 32ay = (27 - 32)ay$ Distributive property.

$= -5ay$ Subtract. ■

In example 1 the expression $-5ay$, which has but one term, is both simpler and equivalent to the original expression, $27ay - 32ay$.

EXAMPLE 2

Simplify: $3.4a^2b + 4.9a^2b - 1.7a^2b$

$3.4a^2b + 4.9a^2b - 1.7a^2b = (3.4 + 4.9 - 1.7)a^2b$ Distributive property.

$= 6.6a^2b$ Add and subtract. ■

It is frequently necessary to rearrange the terms of an expression so that the like terms are grouped. This is made possible by the commutative and associative properties of addition.

EXAMPLE 3

Simplify: $4x + 5y + 6x - 22y + x$

$= (4x + 6x + x) + (5y - 22y)$	Group the like terms using the commutative and associative properties.
$= (4 + 6 + 1)x + (5 - 22)y$	Distributive property.
$= 11x - 17y$	Add and subtract. ■

EXAMPLE 4

Simplify: $2x^2 + 17x - 3 + 9x^2 - 4x + 18$

$= (2x^2 + 9x^2) + (17x - 4x) + (-3 + 18)$	Group like terms. (This step could be done mentally.)
$= 11x^2 + 13x + 15$	Add and subtract. Here we have used the distributive property mentally. ■

EXERCISE 2.4

A

Simplify:

1. $13x + 9x$
2. $18y + 7y$
3. $21t - 8t$
4. $15m - 3m$
5. $42a - 48a$
6. $37b - 41b$
7. $13x^2 + 16x^2 + 5x^2$
8. $8y^3 + 12y^3 + 7y^3$
9. $45ab + 22ab - 31ab$
10. $54acx + 21acx - 31acx$
11. $8.9p + 11.2p - 6.6p$
12. $9.7r + 22.1r - 5.4r$
13. $78y + 34y + 6$
14. $22z + 19z + 12$
15. $8.2t^2 - 9.5t^2 - 3.2$
16. $7.3t^2 - 11.6t^2 - 2.8$
17. $\frac{4}{5}w + \frac{9}{2}w$
18. $\frac{2}{5}a^2m + \frac{1}{3}a^2m$
19. $\frac{7}{8}c^2d^2 + \frac{1}{2}c^2d^2 - \frac{3}{4}c^2d^2$
20. $\frac{5}{6}st^3 + \frac{1}{3}st^3 - \frac{1}{2}st^3$

B

Simplify:

21. $-15x - 14x + 5x - 4x$
22. $-22y + 14y - 65y + 8y$

23. $-32s - s - 11s - 8s$

24. $-43t - 5t - t - 12t$

25. $-74w^2 + 22w^2 - 15w^2 + 84w^2$

26. $-47mn + 55mn - 47mn + 46mn$

27. $67abd^2 - 15abd^2 - 23abd^2 - 36abd^2$

28. $81x^2yz - 44x^2yz - 51x^2yz - 18x^2yz$

29. $17x^2 - 19y + 14$

30. $-21t^2 + 4t - 13$

31. $12x + 7x + 5y - 3y$

32. $18a + 34a + 7b - 6b$

33. $12x + 7y + 5x - 3y$

34. $18a + 34b + 7a - 6b$

35. $14xy + 5xy - 17xy^2 - 8xy^2$

36. $23p^2q + 8p^2q - 9pq^2 - 2pq^2$

37. $14xy + 5xy^2 - 17xy - 8xy^2$

38. $23p^2q + 8pq^2 - 9p^2q - 2pq^2$

39. $25a - b + 16a - 5c + 26b - a + 2b + 8c + 17c$

40. $27x^2 - 19 + 28y^2 - 12x^2 + 23 - 9x^2 - 32y^2 + 8 + 5y^2$

C

Simplify:

41. $8x - a + 4x + 2$

42. $12a - x + 6a + 3$

43. $p - q + 7p - 9 + 2q$

44. $r - t + 10t + 5 - 4r$

45. $17a + 12b - c + 2c - 15b + 7a$

46. $18y + 2x - z - 7z + 9x - 19y$

47. $2.3x - 1.8p + 9.4 - 3.2x - 0.7 - 0.7p$

48. $3.8y - 1.3 + 4.2z - 1.9y - 3.3z - 3.3$

49. $-5.22m + 2.55n - 6.1 - 1.4n + 3.4m + 5.4$

50. $-8.77 + 7.88f - 8.78g - 2.3f + 2.9 + 9.2g$

51. $5a - 3b + 6c - 3b + 4a - 2a + 7c + 7b - c$

52. $8x - 4y + z - 8y + 6x - 10x + 11z + 2y - 3z$

53. $-\frac{1}{6}p + \frac{2}{3} + \frac{1}{3}p - q - \frac{5}{6} - \frac{1}{2}q$

54. $-\frac{1}{3}u + v + \frac{2}{3}w - w + \frac{1}{2}u - \frac{5}{8}v$

55. $\frac{1}{2}a - \frac{1}{3}b + c + a - \frac{1}{3}a + \frac{3}{4}b - \frac{3}{5}c$

56. $\frac{1}{3}x + 3y^2 - \frac{1}{5}xy - \frac{4}{5}y^2 + \frac{3}{4}x - \frac{6}{5}y^2 + \frac{1}{2}xy$

D

Translate each statement to an algebraic expression and simplify:

57. Eight added to twice a number is added to the number.

58. Nine added to six times a number is added to the number.

59. The sum of five times a number and eleven is added to the sum of seven times the number and one.

60. The sum of six times a number and five is added to the sum of twice the number and four.

61. The difference of a number and ten is added to the sum of the number and three.

62. The difference of a number and two is added to the sum of twice the number and nine.

63. Twelve is subtracted from the sum of five times a number and three.

64. Seven is subtracted from the difference of fifteen times a number and one.

STATE YOUR UNDERSTANDING

65. Explain how to identify like terms.

66. Explain how to combine like terms.

CHALLENGE EXERCISES

67. $3y + [6y + (y - 4)]$

68. $[4b - 4a - 5] + [(3a + b) + (2b - a)]$

69. $27x^2yz + 16xy^2z - 31x^2yz + 19xyz^2 - 27xy^2z$

70. $(3x^2 - 5x + 8) + (6x^2 + 7x - 4) + (x^2 + 2x + 10)$

MAINTAIN YOUR SKILLS (SECTIONS 1.2, 1.3)

Translate the following expressions to algebraic form. Let x represent the first number and y represent the second number.

71. Three more than the first number and the result multiplied by 5.

72. The product of the second number and 6 and that product increased by 7.

73. The product of the two numbers and that result decreased by the square of the second number.

74. The square of the first number decreased by three times the square of the second number.

Translate the following algebraic expressions to English.

75. $2x - 10$

76. $xy + 5$

Evaluate the following expressions if $x = 2.1$ and $y = 3.6$.

77. $2x - y$

78. $x^2 + 5y - 10$

2.5 MORE LINEAR EQUATIONS

OBJECTIVE

1. Solve linear equations that contain like terms.

When an equation contains like terms, we can use the addition and subtraction properties of equality to isolate the terms containing the variable on one side of the equation. Combine the like terms, and solve for the variable.

EXAMPLE 1

Solve: $8x - 12 = 11x + 12$

$8x - 11x - 12 = 11x - 11x + 12$	Subtract $11x$ from each side to isolate the variable on the left.
$-3x - 12 = 12$	Simplify each side by combining the like terms.
$-3x - 12 + 12 = 12 + 12$	Add 12 to each side.
$-3x = 24$	Simplify.
$\frac{-3x}{-3} = \frac{24}{-3}$	Divide each side by -3.
$x = -8$	Simplify.
The solution set is $\{-8\}$.	The check is left for the student. ■

It is sometimes possible to combine like terms on each side before isolating the variable.

EXAMPLE 2

Solve: $y + 13 + 2y - 15 = y - 14 + 36$

$(y + 2y) + (13 - 15) = y + (-14 + 36)$	Group the like terms using the commutative and associative properties. (Often done mentally.)
$3y - 2 = y + 22$	Combine the like terms on each side.
$3y - y - 2 = y - y + 22$	Subtract y from each side.
$2y - 2 = 22$	Combine like terms on each side.
$2y - 2 + 2 = 22 + 2$	Add 2 to each side.
$2y = 24$	Simplify.

$$\frac{2y}{2} = \frac{24}{2}$$ Divide each side by 2.

$$y = 12$$

The solution set is {12}. The check is left for the student. ■

■ PROCEDURE

To solve an equation containing like terms:

1. Simplify each side by combining like terms.
2. Use the addition and subtraction properties of equality to isolate the term containing the variable.
3. Use the multiplication and division properties of equality to solve the equation and check.

If some of the steps are done mentally, fewer steps need to be shown.

EXAMPLE 3

Solve: $3x - 15 = x - 2x - 14$

$3x - 15 = -x - 14$	Combine the like terms on the right.
$4x - 15 = -14$	Add x to each side.
$4x = 1$	Add 15 to each side.
$x = 0.25$	Divide each side by 4.

Check:

$3(0.25) - 15 = 0.25 - 2(0.25) - 14$	Substitute 0.25 for x in the original equation.
$0.75 - 15 = 0.25 - 0.5 - 14$	Multiply.
$-14.25 = -14.25$	True.

The solution set is $\{0.25\}$ or $\left\{\frac{1}{4}\right\}$. ■

Some word problems can be translated into equations that contain like terms.

EXAMPLE 4

When seven is added to twice a number, the result is three less than the number. What is the number?

Simpler word form:

Twice a number plus 7 equals the number minus 3.

Select a variable:

Let n represent the number.

Translate to algebra:

$2n + 7 = n - 3$

Translate the expression "seven is added to twice a number," then translate "three less than the number," and set them equal.

Solve:

$n + 7 = -3$ — Subtract n from each side.

$n = -10$ — Subtract 7 from each side.

Check:

Is twice -10 plus seven equal to -10 minus 3?

$-20 + 7 = -10 - 3$ True

Answer:

The number is -10. ■

EXERCISE 2.5

A

Solve:

1. $15x - 12x = -13 + 19$
2. $-5x + 3x = 11 + 15$
3. $12x - 19x = -1 + 50$
4. $14x - 16x = -8 + 24$
5. $7y - 4 = 4y + 20$
6. $8y - 13 = 3y + 12$
7. $6y - 5 = 2y - 5$
8. $3y - 14 = 2y - 14$
9. $21 - 2t = 8t - 19$
10. $14 - 6t = 9t - 16$
11. $9x - 26 = 5x + 22$
12. $15x - 23 = 11x + 21$
13. $5w + 5 = 13 - 2w + 6$
14. $8w + 6 = 16 - w + 17$
15. $4r - 5r - 15 = 2r$

16. $3r - 4r - 18 = 2r$

17. $2x - 17x - 15 = 0$

18. $6x - 16x - 10 = 0$

19. $19 - 2y - 5 = 5y + 7$

20. $9 - 5y - 5 = 3y - 12$

B

Solve:

21. $23x + 2 = 27x - 18$

22. $27x + 10 = 31x + 6$

23. $39x + 32 = 31x$

24. $39x - 32 = 55x$

25. $6t - 2 - 6t = 16 - t$

26. $13t - 6 - t = 15t + 15$

27. $5x + 7x = 4x$

28. $8x - 3x = 2x$

29. $2 - 3w = 1 - w$

30. $11 - 7w = 1 + 13w$

31. $20 - a - 3 = 7a - 23 - 2a$

32. $10 - a - 13 = 87a - 23 - 26a$

33. $2y - 17 - y + 15 = 0$

34. $7y - 34 - 2y + 44 = 0$

35. $43 - 2 + 7x = 42 + 10x$

36. $50 + x - 51 = 5x - 19$

37. $3 - v + 7 = -1 - 4 + 2v$

38. $30 - 26v + 18 = -10v - 32$

39. $5w - 4 = 5 + 4w$

40. $-2 + 8w = -2w + 8$

C

Solve:

41. $3x + 15 + 21 = 22 + 4x$

42. $2x + 15 + 15 = 23 - 5x$

43. $7 - 13y - 32 = 2y - 5y + 5$

44. $14 - 14y - 1 = 5y - 7y + 37$

45. $14 - 25z + 44 = 75z + 18$

46. $15 - 31z + 23 = 69z - 2$

47. $40p - 24 - 33p - 77 = 102$

48. $27p - 24 - 44p - 77 = -390$

49. $37q + 9 - 8q = 8q - 9 - 87$

50. $9q + 73 - q = 109 + q - 78$

51. $x^2 - 5x + 8 = x^2 + 5x - 22$

52. $x^2 - 8x + 12 = x^2 - 12x + 8$

53. $15 - y - y^2 = y - 3 - y^2$

54. $-6 + y - y^2 = 8 - y - y^2$

55. $13 - w^3 + 19 = 7w - w^3 + w$

56. $31 - w^3 - 57 = 12w - w^3 - 64w$

57. $940x - 112 = 922x + 212$

58. $843x - 221 = 850x + 122$

59. $p^2 + 6p + 9 - p^2 - 4p - 4 = 1$

60. $p^2 + 4p - 5 - p^2 - p + 6 = 13$

D

61. Three times a number is twenty-two more than twice the number. What is the number?

62. Seven times a number is sixteen more than three times the number. What is the number?

63. Twice a number is thirty less than four times the number. What is the number?

64. Five times a number is twenty-four less than three times the number? What is the number?

65. Dorina's mother is three times as old as Dorina. In five years, Dorina will be the same age her mother was fifteen years ago. How old is Dorina?

66. Chico's father is four times as old as Chico. In ten years, Chico will be the same age his father was twenty-three years ago. How old is Chico?

67. A house and the lot on which it was built cost $270,000. The cost of the house was five times the cost of the lot. What was the cost of the lot?

68. J.R. drove from Baltimore to Miami, a distance of 1,125 miles, in two days. The second day he traveled twice as far as he did the first day. How far did he drive each day?

69. A triangle has a perimeter of 75.6 centimeters. The longest side is twice the shortest side, and the remaining side is three-fourths of the longest side. Determine the length of each side.

STATE YOUR UNDERSTANDING

70. Write a process that may be used to solve equations containing like terms.

71. Why is it a good practice to combine like terms before using the properties of equality?

CHALLENGE EXERCISES

Solve the following equations.

72. $3(x - 4) + 6(2 - x) = 3(2 - x)$

73. $4(2y - 1) + 5(y + 2) = 10y + 3(y + 2)$

74. $\frac{1}{2}(4x - 3) - \frac{1}{12}(2x + 5) = \frac{3}{4}(2x + 5)$

75. Solve for b. $A = \frac{1}{2}h(a + b)$

MAINTAIN YOUR SKILLS (SECTION 1.4)

Given $C = \left\{-6, -\frac{5}{4}, -\sqrt{2}, -1, 0, 2, \frac{7}{3}, 8.61, 12\right\}$:

76. List the set of whole numbers.

77. List the set of negative integers.

78. List the set of rational numbers.

79. List the set of irrational numbers.

True or False:

80. -6.9 is a rational number.

81. $-3 < -18.6$

82. .3333 is an irrational number.

83. $0 > -14.6$

2.6

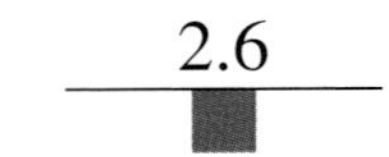

SOLVING PERCENT PROBLEMS

OBJECTIVE

1. Solve problems of the form $A = RB$ or $RB = A$, where R is a rate of percent.

Another way to name a fraction or decimal is to use percent (%). Percent means ''per hundred'' and thus 48% means

$$\frac{48}{100} \quad \text{or} \quad 48 \cdot \frac{1}{100} \quad \text{or} \quad 48(0.01).$$

If we need to change a percent to a decimal, we replace the percent symbol (%) by 0.01 and multiply. So $48\% = 48(0.01) = 0.48$.

In the formula for percent, $A = RB$, R is the *rate of percent* and can be identified by the ''%'' symbol. B is the *base unit* and often follows the word *of*. A is the *amount* that is compared to B and is sometimes called the *percentage*.

Our work with equations provides a method for solving percent problems. The word *of* in ''A is what percent of B?'' indicates multiplication. The word *is* describes the relationship ''is equal to.'' So we can write

A is what percent of B? $\qquad A = RB$

and

What percent of B is A? $\qquad RB = A$

When this formula is used, R *must be* expressed as a decimal or fraction.

EXAMPLE 1

12% of what number is 42?

$RB = A$	Formula.
$0.12B = 42$	Substitute $R = 12\% = 0.12$ and $A = 42$.
$B = 350$	Divide each side by 0.12.

So, 12% of 350 is 42. ■

The same formula is used to find A.

EXAMPLE 2

15% of 30 is what number?

$RB = A$	Formula.
$0.15(30) = A$	Substitute $R = 15\% = 0.15$ and $B = 30$.
$4.5 = A$	

So, 15% of 30 is 4.5. ■

And, again, the same formula is used to find R.

EXAMPLE 3

What percent of 50 is 12?

RB = A	Formula.
$R(50) = 12$	Substitute $B = 50$ and $A = 12$.
$R = \frac{12}{50} = 0.24 = 24\%$	Divide each side by 50.

So, 24% of 50 is 12. ■

Problems involving percent are very common in financial applications.

EXAMPLE 4

Stella paid cost plus 15% for a food processor. If the store charged her $129.95, what was the cost of the processor?

Simpler word form:

115% of what is 129.95?	Since Stella has to pay the total cost (100%) and 15% more, the rate of percent is 115%.

Formula:

$RB = A$	The basic percent formula.

Substitute:

$1.15B = 129.95$

Solve:

$B = 113$ — Divide each side by 1.15.

The cost of the food processor was \$113. — The check is left for the student. ■

Sports statisticians use percents to describe a variety of comparisons.

EXAMPLE 5

The Portland Mavericks baseball team won 60% of their games last year. How many games did they play if their win record was 66 games?

Simpler word form:

60% of what number is 66?

Select a variable:

Let B represent the number of games played. — The base is the number of games played.

Substitute:

$0.60B = 66$ — Substitute $R = 0.60$ and $A = 66$ in the formula $RB = A$.

Solve:

$$\frac{0.60B}{0.60} = \frac{66}{0.60}$$ Divide each side by 0.60.

$$B = 110$$ Simplify each side.

Check:

Is 60% of 110 equal to 66?

60% of $110 = 0.60(110) = 66.00$ — Yes.

Answer:

The Mavericks played 110 games. ■

EXERCISE 2.6

A

Solve:

1. What percent of 40 is 20?
2. What percent of 40 is 10?
3. 20% of 10 is what?
4. 80% of 10 is what?
5. 50% of what number is 13?
6. 25% of what number is 6?
7. 80% of 60 is what?
8. 30% of 60 is what?
9. 40% of what number is 48?
10. 70% of what number is 21?
11. What percent of 32 is 8?
12. What percent of 90 is 9?
13. 150% of 30 is what?
14. 200% of 35 is what?
15. What percent of 10 is 20?
16. What percent of 10 is 30?
17. What is 175% of 16?
18. What is 180% of 25?
19. 175% of what number is 49?
20. 180% of what number is 81?

B

Solve:

21. 39% of 60 is what number?
22. 110% of 80 is what number?
23. 30% of what number is 15?
24. 15% of what number is 30?
25. What percent of 48 is 16?
26. What percent of 16 is 48?
27. 2 is what percent of 12?
28. 12 is what percent of 2?
29. What percent of 200 is 18.6?
30. 6.2% of 1500 is what number?
31. 1 is what percent of 1000?
32. 1000 is what percent of 250?
33. What percent of 420 is 50.4?
34. 8.7% of what number is 104.4?
35. 12.3% of what number is 282.9?
36. What percent of 86.5 is 51.9?
37. $\frac{3}{4}$% of what number is 21? (*Hint:* $\frac{3}{4}\% = 0.75 \times 0.01 = 0.0075$)
38. $\frac{1}{2}$% of what number is 3?
39. $\frac{3}{5}$% of what number is 12.3?
40. $\frac{4}{5}$% of what number is 0.04?

C

Solve:

41. 43% of what number is 39? (to the nearest tenth)
42. 23.4% of what number is 75? (to the nearest tenth)
43. 53 is what percent of 42? (to the nearest tenth of one percent)
44. 6 is what percent of 18.8? (to the nearest tenth of one percent)
45. 9.3% of 940 is what number? (to the nearest tenth)

46. 92 is 61.1% of what number? (to the nearest tenth)

47. 81.7 is what percent of 28? (to the nearest whole-number percent)

48. 28 is what percent of 81.7? (to the nearest whole-number percent)

49. 62.3% of 18 is what number? (to the nearest tenth)

50. 3.45% of 760 is what number? (to the nearest tenth)

51. 72.6% of what number is 930? (to the nearest tenth)

52. What percent of 92.3 is 14.7? (to the nearest tenth of one percent)

53. 121.3% of 73 is what number? (to the nearest tenth)

54. 205.8% of 40 is what number? (to the nearest tenth)

55. What percent of 16 is 2.75? (to the nearest tenth of one percent)

56. What percent of 2.75 is 16? (to the nearest tenth of one percent)

57. 73.5 is 22% of what number? (to the nearest hundredth)

58. 22 is 73.5% of what number? (to the nearest hundredth)

59. 6525 is what percent of 8000? (to the nearest whole-number percent)

60. 8000 is what percent of 6525? (to the nearest whole-number percent)

D

61. Springdale College's basketball team finished the season with a record of 16 wins out of 24 games played. What percent of the games played were won?

62. Springdale's baseball team won 20 games out of 32 games for the season. What percent of the games played were won?

63. Verna bought a new car for $7449. She made a down payment of 16%. What was the down payment?

64. Basil bought a used car for $3225. He made a down payment of 22%. What was the down payment?

65. Felah pays $275 per month for her apartment. This is 24% of her monthly income. What is her monthly income?

66. Leo pays $355 per month for his apartment. This is 28% of his monthly income. What is his monthly income?

67. An appliance store has a sale on dryers. What is the sale price of a dryer that was priced at $424.95 and is marked at 16% off?

68. A furniture store has a sale on tables. What is the sale price of a table that was priced at $649.95 and is marked at 22% off?

69. John paid cost plus 15% for a pair of speakers. If the store charged him $362.25, what was the cost of the speakers?

70. Yen Thi paid cost plus 12% for a winter coat. If the store charged her $85.34, what was the cost of the coat?

71. The population of Ross County increased 8% since the last census. If its former population was 64,500, what is its present population?

72. The population of Emmitburg increased 14% since the last census. If its former population was 7300, what is its present population?

73. A discount market advertises television sets at cost plus 15%. What is the selling price on a set that cost the store $520?

74. The U-price Market advertises canned goods at cost plus 3%. What is the selling price on a case of canned goods that cost the store $6.48?

75. In preparing a mixture of concrete, Dolores uses 150 pounds of gravel, 50 pounds of cement, and 100 pounds of sand. What percent of the mixture is gravel?

76. In preparing a mixture of concrete, Manuel uses $2\frac{1}{4}$ yards of gravel, $\frac{2}{3}$ yard of cement, and $1\frac{1}{5}$ yards of sand. What percent of the mixture is gravel?

STATE YOUR UNDERSTANDING

77. Explain the meaning of each of the variables in the formula $A = RB$.

78. Describe how to recognize the amount, the base, and the rate in a statement of the form:

    X is Y percent of Z

CHALLENGE EXERCISES

79. Julie has worked 15 examples of which 9 are correct. By working all of the remaining examples correctly, she obtained a score of 85%. How many examples in all did she work?

80. A baseball team has won 20 out of 32 games played. By winning all of the remaining games it will have won 70% of the games played. How many games will be played?

81. The Stereo Shop purchased a CD player having a retail value of $295. If the store received one discount of 35% and then a second discount of 10%, how much did it pay for the CD player?

82. After receiving discounts of 25% and 15%, an appliance store paid $545 for a refrigerator. What was the original price of the refrigerator?

MAINTAIN YOUR SKILLS (SECTIONS 1.7, 1.8)

Perform the indicated operations:

83. $\left(-\frac{5}{12}\right)\left(-\frac{6}{7}\right)$

84. $\left(-\frac{7}{15}\right)\left(\frac{18}{21}\right)$

85. $(-4)(-64)$

86. $(-14)(-21)(18)$

87. $(-34.24) \div (-32)$

88. $\left(-\frac{12}{15}\right) \div \left(\frac{9}{20}\right)$

89. $(-900) \div (36)$

90. $(-0.3)(-12)(-3.5)$

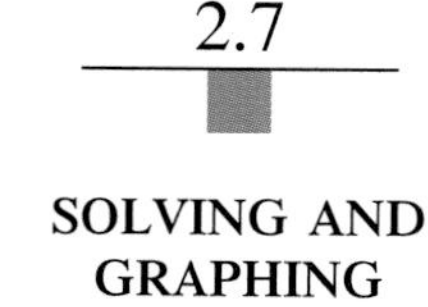

2.7 SOLVING AND GRAPHING INEQUALITIES

OBJECTIVE

1. Solve a linear inequality, and graph the solution set.

To write the relationship between two numbers that are not equal, we use the symbols for inequality that were defined in Section 1.4.

Symbol	Meaning
$<$	less than (points to the smaller number)
$>$	greater than (points to the smaller number)
$\leq$	less than or equal to
$\geq$	greater than or equal to
$\neq$	not equal to

These symbols are used to write mathematical sentences called *inequalities*. Recall that:

1. The numbers pictured on the number line are called *real numbers* and include whole numbers, integers, rational and irrational numbers.
2. On the number line, the smaller of two numbers appears to the left.

In fact, all the numbers to the left of a number, 5, for instance, are less than 5. To solve a linear inequality such as $ax + b < c$, we need properties similar to those for equations. The following table shows the result of adding a number to each side of an inequality.

Inequality	Add the Same Number	Result
$7 < 10$	$7 + 2 < 10 + 2$	$9 < 12$ True
$-8 \leq 5$	$-8 + 4 \leq 5 + 4$	$-4 \leq 9$ True
$12 > 6$	$12 + 3 > 6 + 3$	$15 > 9$ True

Inequalities are similar to equations in that if we add or subtract the same number from each side of the inequality, the result is an equivalent inequality.

PROPERTY

Addition Property of Inequality

If $x < y$ is true, then

$x + a < y + a$ is true. Addition Property of Inequality.

and

PROPERTY

Subtraction Property of Inequality

If $x < y$ is true, then

$x - a < y - a$ is true. Subtraction Property of Inequality.

Similar properties apply when $<$ is replaced by $>$ or $\leq$ or $\geq$.

EXAMPLE 1

Solve: $x + 4 < 6$

$x + 4 - 4 < 6 - 4$ Subtract 4 from each side.

$x < 2$ We can see that any number less than 2 is a solution of the original inequality. ■

When a solution set contains more than a few numbers, as in Example 1, there is more than one way to symbolize the solution. Here we shall use *set builder notation* to write the solution and the number line to illustrate the solution graphically.

Set Builder Notation

DEFINITION

Set builder notation is a compound expression used to represent a set that satisfies given conditions. The expression $S = \{x | x > 10\}$ is read:

$S =$	$\{x$	$\mid$	$x > 10\}.$
↑	↑	↑	↑
S IS THE SET	OF ALL x	SUCH THAT	x IS GREATER THAN 10

In this example, S is the set of all real numbers that are to the right of 10 on the number line.

The symbol " ϵ " is used to indicate that an element is a member of a set. The expression $x \epsilon A$ is read:

"x is an element of A," "x is a member of A," or "x is in A."

The set of real numbers greater than or equal to -6 is

$\{x | x \geq -6 \text{ and } x \text{ is a real number}\} = \{x | x \geq -6 \text{ and } x \epsilon R\}$.

However, when using the set of real numbers, the qualification, $x \epsilon R$ is understood and omitted so that $\{x | x \geq -6\}$ is understood to represent the same set.

EXAMPLE 2

Solve $x - 13 \geq -4$, and write the solution in set builder notation.

$x - 13 + 13 \geq -4 + 13$	Add 13 to each side.
$x \geq 9$	Simplify each side.

The solution set is $\{x | x \geq 9\}$. ■

To deduce two other properties of inequalities, study the following tables.

Inequality	Multiply Each Side	Result	
$2 < 8$	$3 \cdot 2 < 3 \cdot 8$	$6 < 24$	True
$-1 > -3$	$4 \cdot (-1) > 4 \cdot (-3)$	$-4 > -12$	True
$-5 < 7$	$-1 \cdot (-5) < -1 \cdot (7)$	$5 < -7$	False
$-2 > -4$	$-4 \cdot (-2) > -4 \cdot (-4)$	$8 > 16$	False

Inequality	Divide Each Side	Result	
$2 < 8$	$\frac{2}{2} < \frac{8}{2}$	$1 < 4$	True
$9 > -6$	$\frac{9}{3} > \frac{-6}{3}$	$3 > -2$	True
$2 < 8$	$\frac{2}{-2} < \frac{8}{-2}$	$-1 < -4$	False
$9 > -6$	$\frac{9}{-3} > \frac{-6}{-3}$	$-3 > 2$	False

Inequalities do not have the same properties for multiplication and division as do equations. Notice that when each side is multiplied or divided by the same positive number, the result is a true inequality. When each side is multiplied or divided by the same negative number, the result is a false inequality. However, if we reverse the inequalities in the latter cases, we get a true statement in each case. Therefore, we have the following multiplication and division properties.

PROPERTY

Multiplication Property of Inequality

If $x < y$ is true and $a > 0$ (a is positive), $ax < ay$ is true.

If $x < y$ is true and $a < 0$ (a is negative), $ax > ay$ is true.

Multiplication Property of Inequality.

Division Property of Inequality

If $x < y$ is true and $a > 0$ (a is positive), $\frac{x}{a} < \frac{y}{a}$ is true.

If $x < y$ is true and $a < 0$ (a is negative), $\frac{x}{a} > \frac{y}{a}$ is true.

Division Property of Inequality.

Similar properties apply if $<$ is replaced by $>$ or $\leq$ or $\geq$.

Stated in words, if the same positive number is multiplied times (divided into) both sides of an inequality, the result is an equivalent inequality. If the same negative number is multiplied times (divided into) both sides of an inequality, the result is an equivalent inequality if the inequality symbol is reversed.

EXAMPLE 3

Solve: $2x + 5 < 9$

$2x + 5 + (-5) < 9 + (-5)$	Add -5 to each side.
$2x < 4$	Simplify.
$\frac{2x}{2} < \frac{4}{2}$	Divide each side by 2.
$x < 2$	
The solution set is $\{x \mid x < 2\}$.	Remember that when the variable represents a real number $\{x \mid < 2\} = \{x \mid x < 2 \text{ and } x \in R\}$. ■

■ PROCEDURE

To solve an inequality of the form $x + a < b$, subtract a from both sides.

$$x + a - a < b - a$$
$$x < b - a$$

To solve an inequality of the form $x - a < b$, add a to both sides.

$$x - a + a < b + a$$
$$x < b + a$$

To solve an inequality of the form $ax < b$, divide both sides by a. If a is negative, reverse the order of the inequality.

$$\frac{ax}{a} < \frac{b}{a},\ a \text{ positive} \quad \text{or} \quad \frac{ax}{a} > \frac{b}{a},\ a \text{ negative}$$
$$x < \frac{b}{a} \qquad\qquad x > \frac{b}{a}$$

To solve an inequality of the form $\frac{x}{a} < b$, multiply both sides by a. If a is negative, reverse the order of the inequality.

$$a \cdot \frac{x}{a} < a(b),\ a \text{ positive} \quad \text{or} \quad a \cdot \frac{x}{a} > a(b),\ a \text{ negative}$$
$$x < ab \qquad\qquad x > ab$$

For inequalities with $>$, $\leq$, or $\geq$, use the appropriate properties.

To show the graph of the solution of an inequality, we use a closed or open dot for the points at the right or left end of the solution interval. The closed dot shows that the number on the end is included, and the open dot shows that the number on the end is not included. The remaining solutions are shown by a blue arrow to the right or left as appropriate.

CAUTION

When multiplying or dividing both sides of an inequality by a negative number, the order of the inequality must be reversed.

EXAMPLE 4

Solve $-2x + 6 \leq 9$, and graph the solution.

$-2x + 6 - 6 \leq 9 - 6$ — Subtract 6 from both sides or, equivalently, add -6 to each side.

$-2x \leq 3$ — Simplify both sides.

$-\frac{1}{2} \cdot -2x \geq -\frac{1}{2} \cdot 3$ — Multiply both sides by $-\frac{1}{2}$ or, equivalently, divide both sides by -2. This example illustrates the major difference between solving equations and solving inequalities. When you multiply or divide both sides by a negative number, *reverse the inequality.*

$x \geq -\frac{3}{2}$ — Simplify both sides.

The solution set is $\left\{x \mid x \geq -\frac{3}{2}\right\}$.

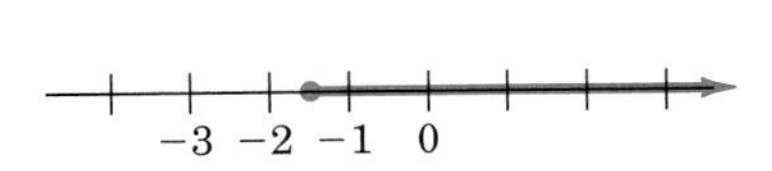

The closed dot at $-\frac{3}{2}$ shows that the number on the end is included. ■

EXAMPLE 5

Solve $\frac{3}{4}x + 2 \leq 7$, and graph the solution.

$4\left(\frac{3}{4}x + 2\right) \leq 4(7)$ — Multiply both sides by 4 to clear the fraction. Since 4 is positive, the inequality is not reversed.

$3x + 8 \leq 28$ — Simplify.

$$3x \le 20 \qquad \text{Subtract 8 from both sides to isolate } 3x.$$

$$x \le 6\frac{2}{3} \qquad \text{Divide both sides by 3.}$$

The solution set is $\left\{x \mid x \le 6\frac{2}{3}\right\}$.

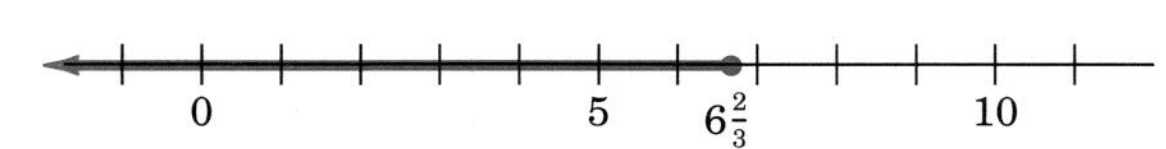

■

Since an elevator can hold a range of weights, we can describe this range using an inequality.

EXAMPLE 6

The maximum capacity of the elevator at a construction site is 2000 lb. If a sack of cement weighs 98 lb and the elevator operator weighs 236 lb, how many bags of cement can be lifted on the elevator at one time?

Simpler word form:

(weight of cement) + (weight of operator) is less than or equal to 2000 lb

Select a variable:

Let x represent the number of bags of cement (an integer).

Translate to algebra:

$98x + 236 \le 2000$

Solve:

$$98x + 236 - 236 \le 2000 - 236$$

$$98x \le 1764$$

$$\frac{98x}{98} \le \frac{1764}{98}$$

$$x \le 18$$

Check:

If 18 bags of cement are loaded on the elevator, will the total weight be less than or equal to 2000 lb?

$$\begin{array}{rl} 98(18) = 1764 & \text{(weight of cement)} \\ \underline{236} & \text{(weight of operator)} \\ 2000 & \end{array}$$

Eighteen or fewer bags of cement can be safely lifted on the elevator at one time. ■

EXERCISE 2.7

A

Solve:

1. $x - 7 > 13$
2. $x + 7 < -2$
3. $y + 2 > 1$
4. $y - 3 \geq -1$
5. $5x \leq -45$
6. $7x > -84$
7. $-4w > -84$
8. $13w \geq -65$
9. $-6x < 54$
10. $-11x > 132$
11. $-9x \leq -135$
12. $-9x \geq -108$

Solve and graph the solution:

13. $x + 8 \geq 16$
14. $x - 7 \geq 7$
15. $-4x > 8$
16. $-3x \leq -18$

B

Solve:

17. $4x + 1 \geq 5$
18. $2x - 5 \leq -3$
19. $6y - 9 \geq 9$
20. $8x + 2 \geq 26$
21. $9x + 6 > -12$
22. $7x - 9 < 12$
23. $5x + 12 < 32$
24. $6x + 15 \geq -21$
25. $-10x + 8 > 48$
26. $-11x - 7 < 15$
27. $-13x - 12 > 27$
28. $-12x + 16 \leq -20$

Solve and graph the solution:

29. $9x - 5 > -23$
30. $-8x + 21 \leq -3$
31. $3x + 7 < 13$
32. $4x - 5 > 11$
33. $-8x + 2 \geq 26$
34. $-9x + 6 \leq -12$
35. $4(x - 2) \leq 12$
36. $5x - 2 < 3(x + 4)$

C

Solve:

37. $\frac{1}{2}x + 4 < 5$

38. $\frac{1}{3}x - 2 > 3$

39. $\frac{1}{5}y - 9 \geq 1$

40. $\frac{1}{4}y + 5 \leq 2$

41. $-3w + 2 \geq 5$

42. $-6x - 5 \leq 13$

43. $\frac{x}{4} + 2 \leq -2$

44. $\frac{x}{5} - 6 \geq 4$

45. $\frac{x}{-2} + 3 < 4$

46. $\frac{x}{-3} + 4 > 1$

47. $-4w + 7 < 11$

48. $-7w - 12 > 2$

49. $5z - 8 \leq 10$

50. $-3t + 11 > 13$

51. $-7z + 3 \geq -11$

52. $4y - 3 \geq 3$

Solve and graph the solution:

53. $\frac{2}{3}y + 4 < 2$

54. $\frac{4}{5}y - 2 > 3$

55. $-\frac{5}{2}x + 3 \leq 5$

56. $-\frac{6}{5}x - 6 \geq 12$

D

57. Eight more than four times a number is less than or equal to twelve. What numbers satisfy this statement?

58. Six less than four times a number is greater than or equal to ten. What numbers satisfy this statement?

59. Five times a number less six is less than thirty-four. What numbers satisfy this statement?

60. Eight more than the quotient of a number and five is less than -2. What numbers satisfy this statement?

61. If $F = \frac{9}{5}C + 32$ is the formula to convert Celsius temperature to Fahrenheit temperature, what Celsius temperatures are less than or equal to 32°F? Graph the solution.

62. Using the formula in Exercise 61, find the Celsius temperatures that are greater than 86°F. Graph the solution.

63. The formula for the area of a rectangle is $A = \ell w$. What is the width, greater than zero, of a rectangle with a length of 18 inches if the area can be no more than 144 in^2? Graph the solution.

64. What is the length, greater than zero, of a rectangle with a width of 24 meters if the area must be greater than 792 m^2? Graph the solution.

65. The velocity, v, of a projectile in ft/sec is given by the formula $v = 72 - 32t$, where t represents the time in seconds. If $v > 0$, the object is rising. For what values of t does the projectile rise?

66. Sam received grades of 96, 82, 85, and 91 on his first four exams. What is the lowest grade he can receive on the fifth exam to have an average of at least 90?

STATE YOUR UNDERSTANDING

67. How do the properties for inequalities differ from the properties for equations? Explain in detail.

68. Explain what you would do to solve the following inequality. Describe everything you would do in detail.

$$2 - \frac{y}{3} \leq 5$$

CHALLENGE EXERCISES

69. Solve and graph.

$$3x^3 - 4x^2 - 2x - 5 - 4 - 3x + 4x^2 - 3x^3 \geq 2x + 5$$

70. The maximum capacity of the elevator at a construction site is 2,000 lb. If a sack of cement weighs 126 lb. and the operator weighs 212 lb, what is the maximum number of bags of cement that can be lifted on the elevator at one time?

Solution: $126x + 212 \leq 2000$
$x \leq 14.19$

71. A sponsor agreed to underwrite the cost of at least 350 tickets for a circus. If it agreed to purchase 100 less adult tickets than youth tickets, what was the least number of youth tickets it could buy?

72. Ms. Smith is 13 years younger than Mr. Moss, and Mr. Young is one-half as old as Mr. Moss. The sum of Ms. Smith's age and Mr. Moss' age is at least 50 years more than Mr. Young's age. Find Mr. Moss' minimum age.

MAINTAIN YOUR SKILLS (SECTIONS 1.2, 1.9)

Evaluate:

73. $-4|36 - 81| + (-12)(31) - (-10)^2$

74. $-3(5^2 - 7^2)^2 - (-8)(-9)$

Evaluate the following when $a = -15$, $b = -24$, and $c = -\frac{2}{3}$:

75. $ac - bc$

76. $c^2 - ac - bc$

77. $\frac{a}{b}(c)$

78. $\frac{a - b}{c}$

Is the equation true for the given solution?

79. $-3x - 6 = 5x - 17, \frac{11}{8}$

80. $45 - 18b = 2b - (-40), \frac{1}{4}$

CHAPTER 2
SUMMARY

Conditional Equations	True for some but not for all replacements of the variables.	(p. 92)
Identity	True for all replacements of the variables.	(p. 93)
Contradiction	False for all replacements of the variables.	(p. 93)
Linear Equations	Equations of the form $ax + b = c$.	(p. 92)
Equivalent Equations	Equations that have the same solution set.	(p. 94)
Literal Equation	An equation that contains more than one variable.	(p. 97)
Solving Equations	Use the properties listed below to write a series of equivalent equations until the variable is isolated on one side of the equation.	
Addition Property of Equality	If $x = y$, then $x + a = y + a$. Adding the same number to each side of an equation "undoes" an indicated subtraction.	(pp. 95, 96)
Subtraction Property of Equality	If $x = y$, then $x - a = y - a$. Subtracting the same number from each side of an equation can "undo" an indicated addition.	(pp. 95, 96)
Multiplication Property of Equality	If $x = y$, then $ax = ay$. Multiplying the same number times each side of an equation can "undo" an indicated division and can be used to "clear a fraction."	(pp. 103, 104)
Division Property of Equality	If $x = y$, then $\frac{x}{a} = \frac{y}{a}$. Dividing the same number into each side of an equation can "undo" an indicated multiplication.	(pp. 102, 103)
Like Terms	Terms with common variables raised to the same power.	(p. 118)

Combine Like Terms	To combine like terms, add or subtract the coefficients, and write the common variable factors in an indicated product. The distributive property is used to combine like terms and simplify expressions. Equations that contain like terms can be solved by simplifying and then using the appropriate properties of equality.	(pp. 119, 124)
Percent Formula	In the formula for percent, $A = RB$, R is the *rate of percent* and can be identified by the "%" symbol. B is the *base unit* and often follows the word *of*. A is the *amount* that is compared to B and is sometimes called the *percentage*.	(p. 128)
Inequalities	Mathematical sentences containing the symbols $<$, $>$, $\leq$, or $\geq$.	(p. 134)
Solving Linear Inequalities	Use the properties listed below to write a series of equivalent inequalities until the variable is isolated on one side of the inequality. The number line may be used to illustrate the solution.	
Addition Property of Inequality	If $x < y$, then $x + a < y + a$.	(p. 135)
Subtraction Property of Inequality	If $x < y$, then $x - a < y - a$.	(p. 135)
Multiplication Property of Inequality	If $x < y$ and a is positive, then $ax < ay$. If $x < y$ and a is negative, then $ax > ay$.	(p. 137)
Division Property of Inequality	If $x < y$ and a is positive, then $\frac{x}{a} < \frac{y}{a}$. If $x < y$ and a is negative, then $\frac{x}{a} > \frac{y}{a}$.	(p. 137)

CHAPTER 2
REVIEW EXERCISES

SECTION 2.1 Objective 1

Solve.

1. $a - 23 = 49$

2. $b - 16 = 44$

3. $-45 = w + 25$

4. $-76 = x + 67$

5. $y - \frac{3}{8} = \frac{1}{4}$

6. $x - \frac{7}{8} = \frac{1}{2}$

7. $4.28 + x = 2.33$

8. $-42.8 + t = -70.3$

9. Solve for r: $r - k = g$

10. Solve for s: $d + s = h$

SECTION 2.2 Objective 1

Solve.

11. $13z = 169$

12. $14c = 210$

13. $-23x = -207$

14. $-31y = 341$

15. $\frac{x}{-8} = 40$

16. $\frac{y}{-5} = -20$

17. $\frac{3w}{7} = 36$

18. $35 = \frac{-5p}{12}$

19. Solve for b: $v = ab$

20. Solve for c: $q = \frac{c}{p}$

SECTION 2.3 Objective 1

Solve.

21. $7x + 5 = 159$

22. $7x + 159 = 82$

23. $4 = -3x - 38$

24. $6 = \frac{t}{8} - 7$

25. $-15y + 37 = 70$

26. $\frac{w}{5} + \frac{1}{2} = \frac{13}{20}$

27. Solve for t: $4s + 5t = m$

28. Solve for g: $r = gt - a$

SECTION 2.4 Objective 1

Simplify.

29. $4x^2 + 12x^2 + x^2$

30. $-y - 8y + 24y$

31. $-8a - 9a - 3a - 8a$

32. $5y + 4y + 7y + 3y$

SECTION 2.4 Objective 2

Simplify.

33. $-8a - 9 - 3a - 8$

34. $x + y + x - y + 4x - y$

35. $5y^2 + 4y + 7y + 3$

36. $13a + 14b - c + 16a - 13b + 7c$

37. $11ab + 4a - 2b - ab - a - b$

38. $-47x^2 + 3x - 73 + 81x^2 - 9 - 33x$

39. $4xy - 3x + 9y - 5 - 3xy - 14x - 2y + 3 + 7xy - 9y$

40. Translate to an algebraic expression and simplify: Fourteen times a number minus fifteen is added to the same number.

6

$47x^2 + 3x - 73 + 81x^3 - 9 - 33x$

-47

SECTION 2.5 **Objective 1**

Solve.

41. $y - 14 + y = 4$

42. $x - 2x + 3 = 4 - 2x$

43. $6w - 21 - 3 = 2w - 4$

44. $5 - x + 3 = -4 - 9 + 2x$

45. $5y - 7 - 35y - 60 + 6 = -1$

46. $2x^2 - 13x - 7 - 2x^2 + 6 - 24 = 1$

SECTION 2.6 **Objective 1**

Solve.

47. What percent of 40 is 35?

48. 16 is 8% of what?

49. 14% of what is 19.18?

50. What percent of 160 is 83 to the nearest tenth of a percent?

51. 122 is 11% of what number (to the nearest tenth).

52. 34% of what is 25.2 (to the nearest tenth).

SECTION 2.7 **Objective 1**

Solve.

53. $x - 22 > -13$

54. $22x \leq -55$

55. $-12 + 3x < 9$

56. $12 - 3x < 9$

57. $\frac{1}{2}x - 16 \geq -4$

58. $16 - \frac{1}{2}x \leq -4$

59. $4x - 7 < 8x + 3$

60. $7 - 4x \leq 3 + 8x$

CHAPTER 2
TRUE–FALSE CONCEPT REVIEW

Check your understanding of the language of algebra. Tell whether each of the following statements is true (always true) or false (not always true).

1. The equation $3x - 3 = 3x + 4$ is an example of a contradiction.

2. The equations $x + 5 = 5$ and $\frac{x}{5} = 0$ are equivalent equations.

3. The equation $6x = 4x$ has no solution.

4. Every equation contains an equals (=) sign.

5. Every equation has one number in its solution set.

6. Both the addition property of equality and the subtraction property of equality can be used to solve the equation $x + 789 = 22$.

7. Literal equations are solved using the same properties of equality used for conditional equations.

8. The multiplication property of equality can be used to solve the equation $5x = 45$.

9. The solution set of the equation $\frac{x}{0} = 5$ is $\{0\}$.

10. Harley makes \$100 per week for a part-time job. He is given a 10% raise when business is good. Two months later, business drops off, so he is given a 10% cut. After the raise and the cut, he is back to making \$100 per week.

11. It is possible to decrease a county's population by 110%.

12. If $7\frac{2}{5}\%$ of 400 is A, then $A = 30$.

13. 700 is 7000% of 10.

14. 0.7% of 20 is 0.14.

15. Equations that contain rational numbers (fractions) can be solved by clearing the fractions using the multiplication property of equality and solving the resulting, equivalent, equation.

16. The largest integer in the solution set of $-2x + 5 \geq 13$ is -4.

17. The smallest integer in the solution set of $-2x + 5 \geq 13$ is -10.

18. Inequalities are solved using the same properties of equality that are used to solve equations.

19. The inequalities $6 - x < -6$ and $-6 > 6 - x$ are equivalent.

20. The inequalities $\frac{x}{-3} > -22$ and $x > 66$ are equivalent.

CHAPTER 2
TEST

1. 23% of 17 is what number?

2. Solve for s: $\frac{s}{9} = -\frac{7}{18}$

3. Solve for z: $13 - z = 28$

4. Simplify: $18x + 11y - w + 3w - 29y + 13x$

5. Solve for x: $b + ax = c$

6. Solve for t: $0.2t + 5.8 = 1.8$

7. Solve for a: $a + 8.5 = 2.3$

8. What percent of 96 is 120?

9. Solve for c: $c - \frac{3}{5} = \frac{1}{2}$

10. Simplify: $17x^2 - 9 + 18y^2 + 3xy - 8x^2 + 3y^2 - 14$

11. Solve for x: $\frac{x}{-11} = -11$

12. Solve for x: $\frac{x}{-11} > -11$

13. Graph the solution: $3x - 1 \leq 8$

14. 8.5% of what number is 51?

15. Solve for m: $6m + 4 < 21$

16. Solve for z: $-\frac{2}{3}z = -10$

17. Solve for x: $3x - 12 = 100 - 4x$

18. Solve for P: $AM + PM = R$

19. The selling price of a graphing calculator is \$59.95. The markup on the calculator is \$14.55. What is the cost (to the store) of the calculator?

20. A grocery store has put yellow arrow stickers on all products that are advertised at 12% off the marked price. What is the selling price of a box of Krunch Snacks that is marked \$1.25?

CHAPTER

3

ALGEBRA OF POLYNOMIALS AND RELATED EQUATIONS

SECTIONS

Distance, rate, and time problems lend themselves to solution by algebraic methods. Distance can be found by multiplying rate times time, $D = rt$. In Exercise 8, Section 3.6, two hikers are taking off from the same spot and traveling in the same direction. Since the number of days they hike is the same, we can represent the distance each travels in terms of their average speed and this common time. An equation can be written by observing that the difference between the distances of the two hikers is the total distance that they are apart. *(John Kelly/The Image Bank)*

PREVIEW

This chapter is devoted to the algebra of polynomials. The presentation covers operations of addition, subtraction, multiplication, and division and solving equations containing polynomials at each appropriate opportunity. We stress the FOIL method of multiplying binomials and identify special products. The associated patterns are used to ''shortcut'' the usual multiplication process.

Since polynomials play such a major role in the study of algebra, mastery of the skills presented in this chapter is necessary to be successful in the following chapters.

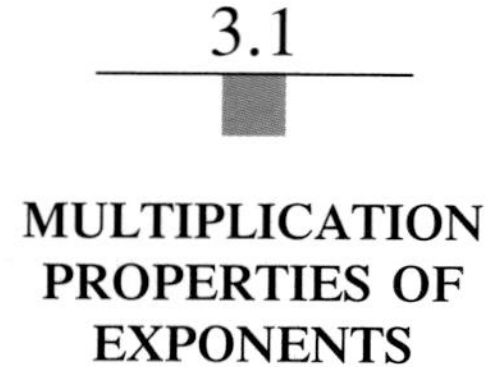

3.1 MULTIPLICATION PROPERTIES OF EXPONENTS

OBJECTIVES

1. Multiply powers with the same base.
2. Raise a power to a power.
3. Raise a product to a power.

To discover how to multiply powers with the same base, study the multiplications in the following table.

Problem	Factors	Regrouping	Result
$a^2 \cdot a^3$	$\underbrace{(a \cdot a)}_{2}\underbrace{(a \cdot a \cdot a)}_{3}$	$\underbrace{(a \cdot a \cdot a \cdot a \cdot a)}_{5}$	$a^{2+3} = a^5$
$x^4 \cdot x^2$	$\underbrace{(x \cdot x \cdot x \cdot x)}_{4}\underbrace{(x \cdot x)}_{2}$	$\underbrace{(x \cdot x \cdot x \cdot x \cdot x \cdot x)}_{6}$	$x^{4+2} = x^6$
$b^3 \cdot b^5$	$\underbrace{(b \cdot b \cdot b)}_{3}\underbrace{(b \cdot b \cdot b \cdot b \cdot b)}_{5}$	$\underbrace{(b \cdot b \cdot b \cdot b \cdot b \cdot b \cdot b \cdot b)}_{8}$	$b^{3+5} = b^8$

In each row of the table, we see that the number of factors is the sum of the exponents. From this, we conclude the following:

PROPERTY

First Property of Exponents

To multiply powers with the same base, add the exponents and keep the common base.

$a^m \cdot a^n = a^{m+n}$ Multiplying like bases.

EXAMPLE 1

Multiply: $y^2 \cdot y^3$

$y^2 \cdot y^3 = y^{2+3}$
$= y^5$

The bases are the same, so add the exponents. ■

The process is the same when multiplying more than two factors of the same base.

EXAMPLE 2

Multiply: $a^6 \cdot a^7 \cdot a^5 \cdot a$

$a^6 \cdot a^7 \cdot a^5 \cdot a = a^{6+7+5+1}$
$= a^{19}$

The bases are the same, so add the exponents. Recall that $a = a^1$. ■

What happens when the bases are unlike?

EXAMPLE 3

Multiply: $a^2 \cdot b^2$

a^2b^2

The bases are not the same. This expression is said to be in simplest form. ■

■ CAUTION

If the bases are unlike as in a^2b^4 the first property of exponents cannot be used.

EXAMPLE 4

Multiply: $2^2 \cdot 2^4 \cdot 2$

$2^2 \cdot 2^4 \cdot 2 = 2^{2+4+1}$
$= 2^7$

■ CAUTION

$2^2 \cdot 2^4 \cdot 2 \neq 8^7$. Do not multiply the bases.

■

EXAMPLE 5

Multiply: $a^0 \cdot a^3 \cdot a^7$

$$a^0 \cdot a^3 \cdot a^7 = 1 \cdot a^{3+7}$$
$$= a^{10}$$

■ **CAUTION**

The base cannot equal 0 when the exponent is zero. Here $a \neq 0$, therefore $a^0 = 1$.

■

The property also applies when the exponents are variables.

EXAMPLE 6

Multiply: $x^a \cdot x^2$

$x^a \cdot x^2 = x^{a+2}$ Add the exponents. ■

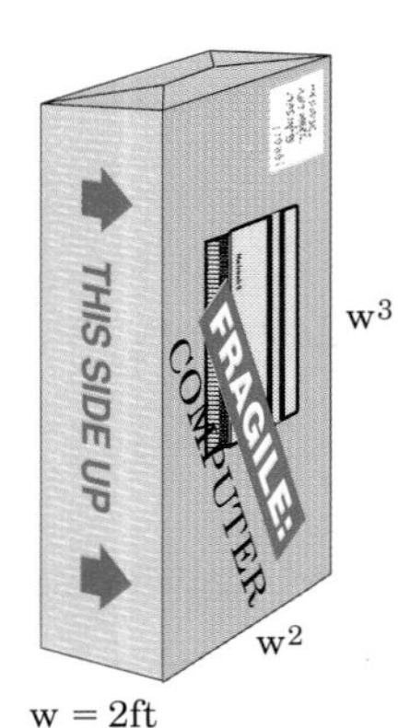

EXAMPLE 7

The volume of a crate containing a new computer is given by $V = \ell wh$ (ℓ is the length, w is the width, and h is the height). If the number of feet in the length is the square of the number of feet in the width ($\ell = w^2$) and the number of feet in the height is the cube of the number of feet in the width ($h = w^3$), express the volume in terms of w, using a single exponent, and then find the volume if $w = 2$ ft.

Formula:

$V = \ell wh$

Substitute:

$V = w^2 \cdot w \cdot w^3$

$V = w^6$

To express the volume of the crate in terms of the width (w), substitute w^2 for ℓ and w^3 for h in the formula, and multiply the factors by adding the exponents.

Solve:

$V = w^6$ New formula.

$V = 2^6$

$V = 64$

To find the volume of the crate when $w = 2$, substitute in the new formula and solve.

Check:

$V = \ell wh$ — Check by substituting 64 for V, 2^2 for ℓ, 2 for w, and 2^3 for h in the original formula.

$64 = (2^2)(2)(2^3)$

$64 = 64$

Answer:

The volume of the crate containing the new computer is w^6 in terms of the width and 64 cubic feet when the width is 2 feet. ■

The property of exponents for multiplying like bases leads us to a second property of exponents. Consider the multiplications in the following table.

Problem	Factors	Result
$(a^3)^2$	$a^3 \cdot a^3$	$a^{3+3} = a^6 = a^{3\cdot 2}$
$(a^5)^3$	$a^5 \cdot a^5 \cdot a^5$	$a^{5+5+5} = a^{15} = a^{5\cdot 3}$
$(a^2)^4$	$a^2 \cdot a^2 \cdot a^2 \cdot a^2$	$a^{2+2+2+2} = a^8 = a^{2\cdot 4}$

In each row of the table, we see that the total number of times the variable is used as a factor is the product of the exponents.

■ PROPERTY

Second Property of Exponents

To raise a power to a power, multiply the exponents and keep the common base.

$(a^m)^n = a^{mn}$ — Raise a power to a power.

EXAMPLE 8

Simplify: $(x^2)^5$

$(x^2)^5 = x^{2\cdot 5} = x^{10}$ — Raise a power to a power, multiply the exponents. ■

EXAMPLE 9

Simplify: $(x^0)^4$

$(x^0)^4 = x^{0\cdot 4} = x^0 = 1$ — Raise a power to a power, multiply the exponents. ■

The same property is used when the exponents are variables.

EXAMPLE 10

Simplify: $(y^a)^b$

$(y^a)^b = y^{a \cdot b} = y^{ab}$ Raise a power to a power, multiply the exponents. ■

A third property of exponents is developed in a similar manner.

Problem	Factors	Result
$(xy)^3$	$xy \cdot xy \cdot xy$	$x \cdot x \cdot x \cdot y \cdot y \cdot y \cdot = x^3y^3$
$(mn)^4$	$mn \cdot mn \cdot mn \cdot mn$	$m \cdot m \cdot m \cdot m \cdot n \cdot n \cdot n \cdot n = m^4n^4$
$(bc)^2$	$bc \cdot bc$	$b \cdot b \cdot c \cdot c = b^2c^2$

In each row of the table, in the result, we can see that the total number of times each variable is used as a factor is the number of times the product of the variables is used as a factor. From this, we conclude a third property of exponents.

■ PROPERTY

Third Property of Exponents

To raise a product to a power, raise each factor to the power.

$(ab)^m = a^mb^m$ Raise a product to a power.

EXAMPLE 11

Simplify: $(2z)^8$

$(2z)^8 = 2^8z^8$ Each factor is raised to the eighth power.

$= 256z^8$ ■

EXAMPLE 12

Simplify: $(a^2b^3)^4$

$(a^2b^3)^4 = (a^2)^4(b^3)^4$ Raise a product to a power.

$= a^8b^{12}$ Raise a power to a power. ■

In summary, the three properties of exponents discussed in this section are:

Property 1: To multiply like bases, add the exponents and keep the common base.

Property 2: To raise a power to a power, multiply the exponents and keep the common base.

Property 3: To raise a product to a power, raise each factor to that power.

EXERCISE 3.1

A

Multiply where possible:

1. $b^2 \cdot b^5$ **2.** $b^5 \cdot b^6$ **3.** $x^0 \cdot x^8$ **4.** $y^7 \cdot y^4$

5. $x^6 \cdot x^9$ **6.** $x^5 \cdot x^9$ **7.** $a^8 \cdot a^9$ **8.** $a^{12} \cdot a^8$

9. $x^2 \cdot y$ **10.** $x^{11} \cdot y^{12}$

Simplify:

11. $(x^2)^2$ **12.** $(y^3)^2$ **13.** $(a^1)^2$ **14.** $(b)^3$ **15.** $(y^4)^1$

16. $(a^5)^0$ **17.** $(3b)^2$ **18.** $(2b)^3$ **19.** $(mn)^4$ **20.** $(mn)^5$

B

Multiply where possible:

21. $x^4 \cdot x^5 \cdot x^2$ **22.** $a^7 \cdot a^5 \cdot a^3$ **23.** $y^2 \cdot y^4 \cdot y^6$

24. $z^3 \cdot z^2 \cdot z^5$ **25.** $b^{10} \cdot b^8 \cdot b^4$ **26.** $y^5 \cdot y^9 \cdot y^8$

27. $a^2 \cdot a^5 \cdot x^2 \cdot x^5$ **28.** $x^4 \cdot x^0 \cdot y^2$ **29.** $x^4 \cdot x^5 \cdot y^7 \cdot y$

30. $a^3 \cdot a \cdot b \cdot b^5$

Simplify:

31. $(x^5)^3$ **32.** $(y^3)^5$ **33.** $(a^{10})^3$ **34.** $(b^2)^{10}$

35. $(z^6)^4$ **36.** $(w^0)^5$ **37.** $(xy)^0$ **38.** $(xy)^{12}$

39. $(7a^2b)^2$ **40.** $(11a^3b^2)^2$ **41.** $(y^2z^2)^2$ **42.** $(y^3z^3)^2$

C

Multiply where possible:

43. $a \cdot a \cdot a^2 \cdot a^3$ **44.** $x^4 \cdot x \cdot x \cdot x^2 \cdot x^5$ **45.** $y^5 \cdot y^2 \cdot y^3 \cdot y^7$

46. $x^3 \cdot x^{10} \cdot x^6 \cdot x^{11}$ **47.** $a^0 \cdot a^4$ **48.** $x^0 \cdot x^0$

49. $x^a \cdot x^b$ **50.** $y^b \cdot y^3$ **51.** $a^3 \cdot a^m$

52. $b^2 \cdot b^n$

Simplify:

53. $(x^m)^2$ **54.** $(y^3)^n$ **55.** $(a^b)^c$ **56.** $(c^0)^e$

57. $(y^n)^n$ **58.** $(a^m)^m$ **59.** $(x^0y^5)^4$ **60.** $(a^5b^6)^3$

61. $(15x^2yz^4)^2$ **62.** $(3a^2bc^3)^5$

63. $(x^3y^2z^3)^2$ **64.** $(a^5b^4c^3)^3$

D

65. The volume of a cube is expressed by the formula $V = s^3$. Find the volume of a cube if $s = 5.1$ inches.

66. The area of a triangle is given by the formula $A = \frac{1}{2}bh$. Find the area of the triangle if the number of inches in the base is the square of the number of inches in the height ($b = h^2$) and $h = 3$.

67. Express the volume of a box in terms of the width (w) if the number of feet in the length is equal to the number of feet in the width ($\ell = w$) and the number of feet in the height is the square of the number of feet in the width ($h = w^2$). Find the volume if $w = 3$.

68. The area of a rectangle is given by the formula $A = \ell w$. Express the area in terms of the width (w) when the number of feet in the length (ℓ) is the cube of the number of feet in the width. Find the area when $w = 5$.

69. The volume of a right circular cylinder is given by the formula $V = \pi r^2 h$. Express the volume in terms of the radius (r) when the number of inches in the height is the cube of the number of inches in the radius. Find the volume when $r = 3$. Let $\pi \approx 3.14$.

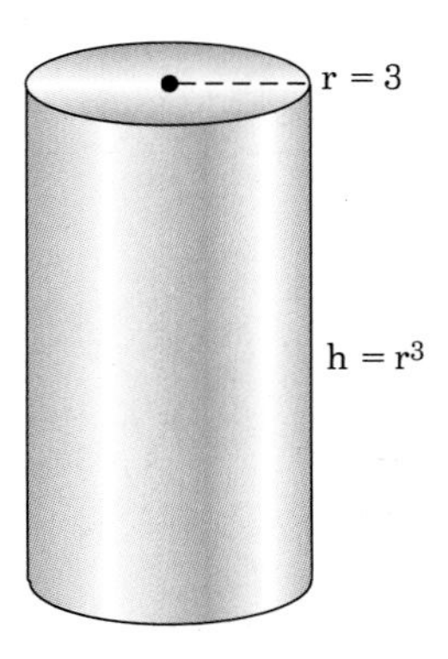

Figure for Exercise 69

70. Express the volume of a box in terms of the width if the number of feet in the length is the cube of the number of feet in the width and the number of feet in the height is the square of the number of feet in the width. Find the volume when $w = 1$.

71. The volume of a sphere is expressed by the formula $V = \frac{4}{3}\pi r^3$. Find the volume of a sphere if $r = 2$ cm and $\pi = 3.14$. (Round to the nearest tenth.)

72. Voltage is given by the formula $V = IR$. Find the voltage if the number of units in R is the cube of the number of units in I ($R = I^3$) and $I = 4$.

73. The area of a trapezoid is given by the formula $A = \frac{(B + b)}{2}h$. Find the area if the number of inches in the measure of B is the square of the number of inches in the measure of b ($B = b^2$) and the number of inches in the measure of h is the cube of the number of inches in the measure of b ($h = b^3$) and $b = 3$.

74. The formula for distance is $D = rt$. Find the distance, in miles, that an airplane flies if the number of units (mph) in the measure of r is the cube of the number of hours in the measure of t ($r = t^3$) and $t = 6$.

STATE YOUR UNDERSTANDING

75. Write the procedure for multiplying powers of the same base.

76. Write the procedure for raising a power to a power.

CHALLENGE EXERCISES

Multiply:

77. $3x^{2a} \cdot 2x^{3a}$

78. $2^5 \cdot 2^3$

79. $-3^2 \cdot 3^4$

Simplify:

80. $3x^0$

81. $x^m \cdot x$

82. -6^0

83. $(-x)^3 \cdot (-x)^2$

84. $(3^2 \cdot 2^3)^2$

85. $(2a^2bc^0)^3(-3a^3b^2c)^2$

MAINTAIN YOUR SKILLS (SECTIONS 1.9, 2.1)

Perform the indicated operations:

86. $-2[-14 - (5 - 13)] - 4(-15)$

87. $(-13)(-2)(-1) - (-5)(-4) + (-35)$

88. $\dfrac{(-3)(-12) - (-7)(9)}{21 - (-5)(-2)} - (-10) - (-5)$

89. $(-14)^2 - (12)^2 - (-14)(-12)$

Solve:

90. $y + 72 = 100$

91. $w + 24 = -432$

92. $-78 = -45 + x$

93. $\dfrac{5}{7} + x = \dfrac{5}{21}$

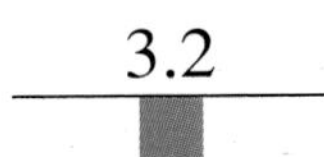

3.2 INTEGER EXPONENTS AND DIVISION PROPERTIES OF EXPONENTS

OBJECTIVES

1. Rewrite powers using positive exponents.
2. Multiply and divide powers with the same base.
3. Raise a quotient to a power.

To find a meaning for the expression 5^{-2}, study the following chart:

Expression	Reduce	Observation
$\frac{5^3}{5}$	$\frac{5 \cdot 5 \cdot 5}{5} = 5 \cdot 5 = 5^2$	$\frac{5^3}{5^1} = 5^{3-1} = 5^2$
$\frac{5^3}{5^2}$	$\frac{5 \cdot 5 \cdot 5}{5 \cdot 5} = 5 = 5^1$	$\frac{5^3}{5^2} = 5^{3-2} = 5^1$
$\frac{5^3}{5^3}$	$\frac{5 \cdot 5 \cdot 5}{5 \cdot 5 \cdot 5} = 1 = 5^0$	$\frac{5^3}{5^3} = 5^{3-3} = 5^0$
$\frac{5^3}{5^4}$	$\frac{5 \cdot 5 \cdot 5}{5 \cdot 5 \cdot 5 \cdot 5} = \frac{1}{5} = 5^?$	$\frac{5^3}{5^4} = 5^{3-4} = 5^{-1}$
$\frac{5^3}{5^5}$	$\frac{5 \cdot 5 \cdot 5}{5 \cdot 5 \cdot 5 \cdot 5 \cdot 5} = \frac{1}{5 \cdot 5} = 5^?$	$\frac{5^3}{5^5} = 5^{3-5} = 5^{-2}$

In the second column, we see that the exponent is decreasing by one from top to bottom. If this pattern is continued, we see that

$$5^{-1} = \frac{1}{5}$$

and

$$5^{-2} = \frac{1}{5^2}.$$

This pattern leads us to the following definition:

DEFINITION

Negative Exponent

$$x^{-n} = \frac{1}{x^n},\ x \neq 0, \quad n \text{ is a positive integer.}$$

Zero is excluded as the base since division by zero is not defined.

Algebraic expressions are not in simplest form if they include negative exponents.

EXAMPLE 1

Rewrite using positive exponents; simplify if possible: 10^{-2}

$10^{-2} = \frac{1}{10^2}$ Definition of a negative exponent.

$= \frac{1}{100}$ Simplify. ■

EXAMPLE 2

Rewrite using positive exponents; simplify if possible: a^{-5}

$a^{-5} = \frac{1}{a^5}$ Definition of a negative exponent. ■

The properties of Section 3.1 apply to negative exponents as well.

EXAMPLE 3

Simplify; write using positive exponents: $a^2 \cdot a^{-3} \cdot a^0$

$a^2 \cdot a^{-3} \cdot a^0 = a^{2+(-3)+0}$ Add the exponents.

$= a^{-1}$

$= \frac{1}{a}$ Definition of a negative exponent. ■

EXAMPLE 4

Simplify; write using positive exponents: $[x^{-5}]^{-2}$

$[x^{-5}]^{-2} = x^{(-5)(-2)}$ Raise a power to a power; multiply the exponents.

$= x^{10}$ ■

When a product is raised to a power, each factor is raised to that power.

EXAMPLE 5

Simplify; write using positive exponents: $(2y^3)^{-5}$

$(2y^3)^{-5} = 2^{-5}(y^3)^{-5}$ Raise a product to a power.

$= 2^{-5}y^{-15}$ Raise a power to a power.

$= \frac{1}{2^5}\frac{1}{y^{15}}$ Definition of negative exponent.

$= \frac{1}{32y^{15}}$ Simplify. ■

EXAMPLE 6

Simplify; write using positive exponents: $[x^2y^{-2}]^3$

$[x^2y^{-2}]^3 = [x^2]^3[y^{-2}]^3$ Raise a product to a power.

$= x^6y^{-6}$ Raise a power to a power.

$= x^6 \cdot \frac{1}{y^6}$ Definition of a negative exponent.

$= \frac{x^6}{y^6}$ ■

EXAMPLE 7

The radius of a red corpuscle is about 3.8×10^{-5} centimeters. Write this measurement in decimal form.

To find the radius of the red corpuscle in decimal form, rewrite 10^{-5} without exponents, and perform the indicated multiplication.

$3.8 \times 10^{-5} = 3.8 \times \frac{1}{10^5}$ Definition of a negative exponent.

$= 3.8 \times \frac{1}{100{,}000}$

$= 3.8 \times 0.00001$

$= 0.000038$

So, the radius is 0.000038 centimeters. ■

The third column of the table on page 158 leads us to a fourth property of exponents.

PROPERTY

Fourth Property of Exponents

To divide nonzero powers that have the same base, subtract the exponent in the denominator from the exponent in the numerator, and keep the common base.

$$\frac{a^m}{a^n} = a^{m-n},\ a \neq 0$$ Dividing like bases.

EXAMPLE 8

Divide and write the answer using positive exponents: $\frac{x^4}{x^7}$

$$\frac{x^4}{x^7} = x^{4-7}$$ Dividing like bases.

$$= x^{-3}$$

$$= \frac{1}{x^3}$$

EXAMPLE 9

Divide and write the answer using positive exponents: $\frac{a^{-6}}{a^{-4}}$

$$\frac{a^{-6}}{a^{-4}} = a^{-6-(-4)}$$

$$= a^{-2}$$

$$= \frac{1}{a^2}$$

More complex examples can include other properties of exponents.

EXAMPLE 10

Simplify; write using positive exponents: $\frac{(3x)^2}{3x^3}$

$$\frac{(3x)^2}{3x^3} = \frac{3^2x^2}{3x^3}$$ Raise a product to a power.

$$= \frac{3^2x^2}{3^1x^3}$$ $3 = 3^1$.

$= 3^{2-1}x^{2-3}$ Divide like bases.

$= 3^1x^{-1}$ Simplify.

$= 3 \cdot \frac{1}{x}$ Definition of a negative exponent.

$= \frac{3}{x}$ ■

The following table illustrates raising a quotient to a power.

Problem	Factors	Result
$\left(\frac{3}{5}\right)^2$	$\frac{3}{5} \cdot \frac{3}{5}$	$\frac{3 \cdot 3}{5 \cdot 5} = \frac{3^2}{5^2}$
$\left(\frac{x}{y}\right)^3$	$\frac{x}{y} \cdot \frac{x}{y} \cdot \frac{x}{y}$	$\frac{x \cdot x \cdot x}{y \cdot y \cdot y} = \frac{x^3}{y^3}$

When a quotient is raised to a power, the numerator is raised to the power, and the denominator is raised to the power. This leads to a fifth property of exponents.

■ **PROPERTY**

Fifth Property of Exponents

To raise a quotient to a power, raise the numerator and the denominator to that power.

$\left[\frac{a}{b}\right]^n = \frac{a^n}{b^n}$ Raise a quotient to a power.

EXAMPLE 11

Simplify; write using positive exponents: $\left(\frac{3x}{4y}\right)^3$

$\left(\frac{3x}{4y}\right)^3 = \frac{(3x)^3}{(4y)^3}$ Raise a quotient to a power.

$= \frac{3^3x^3}{4^3y^3}$ Raise a product to a power.

$= \frac{27x^3}{64y^3}$ Simplify. ■

The quotient property of exponents holds for negative exponents as well.

EXAMPLE 12

Simplify; write using positive exponents: $\left(\dfrac{5x^3y^2}{5^2xy^5}\right)^{-3}$

$$
\begin{aligned}
\left(\frac{5x^3y^2}{5^2xy^5}\right)^{-3} &= (5^{1-2}x^{3-1}y^{2-5})^{-3} && \text{Simplify inside the parentheses first.}\\
&= (5^{-1}x^2y^{-3})^{-3} \\
&= (5^{-1})^{-3}(x^2)^{-3}(y^{-3})^{-3} && \text{Raise a product to a power.}\\
&= 5^3x^{-6}y^9 && \text{Raise a power to a power by multiplying the exponents.}\\
&= 125\left(\frac{1}{x^6}\right)(y^9) \\
&= \frac{125y^9}{x^6} && \text{Simplify.}
\end{aligned}
$$

■

To summarize the properties for exponents when doing division, we have:

Property 4: To divide powers with like bases, subtract the exponent in the denominator from the exponent in the numerator.

Property 5: To raise a quotient to a power, raise the numerator and the denominator to that power.

EXERCISE 3.2

A

Simplify; write using positive exponents:

1. 6^{-1}
2. 11^{-2}
3. 6^{-3}
4. w^{-6}
5. z^{-4}
6. x^{-7}
7. $\dfrac{2^{-3}}{2^6}$
8. $3^2 \cdot 3^{-3}$
9. $\dfrac{x^7}{x^{10}}$
10. $y^5 \cdot y^0$
11. $\dfrac{y^{-3}}{y^{-5}}$
12. $x^8 \cdot x^{-3}$
13. $(x^{-2})^2$
14. $(a^{-3})^2$
15. $(w^{-3})^{-2}$
16. $\left(\dfrac{a^2}{c}\right)^5$
17. $\left(\dfrac{a^{-3}}{b}\right)^2$
18. $\left(\dfrac{a^2}{b^3}\right)^2$
19. $\left[\dfrac{a^{-2}}{b^{-2}}\right]^{-2}$
20. $\left[\dfrac{x^{-3}}{y^{-4}}\right]^{-1}$

B

Simplify; write using positive exponents:

21. $2^{-3} \cdot 2^4 \cdot 2^{-3}$

22. $3^0 \cdot 3^{-1} \cdot 3^{-1}$

23. $10^{-4} \cdot 10^2$

24. $\dfrac{8^{-5}}{8^3}$

25. $x^{-3} \cdot x^{-5} \cdot x^2$

26. $\dfrac{y^{11}}{y^{-15}}$

27. $y^{-5} \cdot y^{13} \cdot y^{-8}$

28. $\dfrac{x^0}{x^{-8}}$

29. $(x^{-2} \cdot x^0)^3$

30. $(a^3 \cdot a^{-1})^{-2}$

31. $(y^4 \cdot y^{-5})^{-1}$

32. $(c^4 \cdot c^{-8})$

33. $a^x \cdot a^y$

34. $\dfrac{a^m}{a^n}$

35. $\dfrac{x^a}{x^n}$

36. $x^a \cdot x^0$

Simplify; write using positive exponents:

37. $\left(\dfrac{a^3}{b^2}\right)^4$

38. $\left(\dfrac{a^{-5}}{b^4}\right)^3$

39. $\left(\dfrac{x^5y^2}{w^2z^3}\right)^3$

40. $\left(\dfrac{a^{-4}b^3}{c^3d^{-3}}\right)^4$

41. $\left(\dfrac{5a^3}{5^2a^2}\right)^{-2}$

42. $\left(\dfrac{6a^4}{a^3}\right)^{-3}$

43. $\left(\dfrac{4^3a^{-3}}{4a^{-4}}\right)^{-2}$

44. $\left(\dfrac{6b^{-7}}{6b^{-6}}\right)^{-2}$

C

Simplify; write using positive exponents:

45. $\dfrac{5^{-1} \cdot 5^3}{5^4}$

46. $\dfrac{4^3 \cdot 4^{-5}}{4^6}$

47. $\dfrac{3^0 \cdot 3^{-7}}{3^{-5}}$

48. $\dfrac{x^5 \cdot x^{-6}}{x^{-3}}$

49. $\dfrac{y^{-8} \cdot y^{-10}}{y^{-11}}$

50. $\dfrac{x^{10} \cdot x^{-8}}{x^3}$

51. $\dfrac{x^3 \cdot x^{-2}}{x^{-5} \cdot x^4}$

52. $\dfrac{y^{-6} \cdot y^3}{y^5 \cdot y^{-7}}$

Simplify; write using positive exponents:

53. $\left(\dfrac{y^{-3} \cdot y^{-2}}{y^2 \cdot y^4}\right)^{-1}$

54. $\left(\dfrac{y^3 \cdot y^4}{y^{-2} \cdot y^{-4}}\right)^{-1}$

55. $\left(\dfrac{a^7 \cdot a^{-3}}{a^{-6} \cdot a^4}\right)^{-2}$

56. $\left(\dfrac{b^{-9} \cdot b^5}{b^{-2} \cdot b^7}\right)^{-2}$

57. $\dfrac{a^m \cdot a^n}{a^n}$

58. $\dfrac{x^y \cdot x^z}{x^y}$

59. $\dfrac{x^2 \cdot x^{-3} \cdot x^5}{x^4 \cdot x^0}$

60. $\dfrac{x^{15}x^{-12}}{x^{-18}x^{20}}$

61. $\dfrac{x^{-18}x^{14}}{x^{17}x^{-13}}$

62. $\dfrac{x^{25}x^{-21}}{x^{-17}x^{25}}$

63. $\dfrac{x^{31}x^{-22}}{x^{-16}x^{24}}$

64. $\dfrac{x^{48}x^{-28}}{x^{-26}x^{19}}$

65. $\left(\dfrac{a^{-15}b^{-8}}{c^{-7}d^{-4}}\right)^{-2}$

66. $\left(\dfrac{x^{-12}y^{-7}}{w^{-9}z^{-12}}\right)^{-3}$

67. $\left(\dfrac{x^{7}y^{8}}{x^{4}y^{9}}\right)^{-3}$

68. $\left(\dfrac{a^{9}b^{4}}{a^{3}b^{4}}\right)^{-4}$

69. $\left(\dfrac{10x^{-3}y^{-4}}{10^{3}x^{-3}y^{-7}}\right)^{3}$

70. $\left(\dfrac{3^{4}a^{-8}b^{-5}}{3a^{6}b^{4}}\right)^{2}$

D

71. The diameter of an atom is approximately 5×10^{-9} inch. Write this distance in decimal form.

72. The length of a wave of yellow light is 2.28×10^{-5} inch. Write this length in decimal form.

73. The radius of a spherical raindrop is approximately 1.5×10^{-3} meter. Write this distance in decimal form.

74. Given the formula $H = qst$, find H if $t = 2$, $s = t^{-3}$, and $q = t^{8}$.

75. Voltage is given by the formula $V = IR$. Find the voltage if $I = 10^{-3}$ and $R = 10^{4}$.

76. The volume of a cylinder is given by the formula $V = \pi r^{2}h$. Find the volume if $r = 2$ and the numbers for h and r are related so that $h = r^{-3}$. Let $\pi \approx 3.14$.

77. In Exercise 76, find the volume if the numbers r and h are related so that $r = h^{2}$ and $h = 2$. Let $\pi \approx 3.14$.

78. The volume of a gas under a constant temperature is given by $V = kp^{-1}$, where k is the constant temperature, V is the volume, and p is the pressure. Find V when $k = 80$ and $p = 100$.

STATE YOUR UNDERSTANDING

79. What does a negative exponent indicate?

80. Explain the difference between $x^{3} \cdot y^{-2}$ and $x^{3} + y^{-2}$.

CHALLENGE EXERCISES

81. Use the Properties of Exponents to show that $\left(\dfrac{x}{y}\right)^{-a} = \left(\dfrac{y}{x}\right)^{a}$.

82. Simplify: $\left(\dfrac{x^{-3}}{4y}\right)^{-3}\left(\dfrac{3y^{4}}{x^{-2}}\right)^{-2}$

83. Simplify: $\dfrac{3^{4x} \cdot 9^{2x+1}}{27^{x-1}}$

84. Simplify: $\dfrac{4^{-2} \cdot 2^{3a+1} \cdot 8^{a}}{16^{2a}}$

MAINTAIN YOUR SKILLS (SECTIONS 1.5, 1.9, 2.2)

Perform the indicated operations:

85. $|-5 + 2| - 3 \cdot 5 + 2 \cdot 7$

86. $\left(-\frac{2}{3}\right) \div \left(-\frac{5}{6}\right) - \frac{2}{3} \cdot \frac{5}{6}$

87. $-5[3(-4) - (-4)(-9)] + (-5)(15)$

88. $[3(9 - 12)]^2 - [4(-3 + 7)]^2$

Solve:

89. $4x = -76$

90. $\frac{y}{-13} = -14$

91. $-82y = 3936$

92. $\frac{x}{-36} = -25$

3.3 SCIENTIFIC NOTATION

OBJECTIVES

1. Write a number in scientific notation.
2. Change scientific notation to place value notation.

Scientific applications of mathematics often have uses for very large and very small numbers. *Scientific notation* is the name given to the shorthand for displaying those numbers.

A number is written in scientific notation when it is expressed as the product of a number between 1 and 10 and a power of 10.

DEFINITION

Scientific Notation

A number written in scientific notation has the form:

$a \times 10^m$, $1 \le a < 10$ and m is an integer

The number 347 can be written in scientific notation in three steps:

Step 1: 3.47 is between 1 and 10 (the decimal has been moved two places to the left, 3.4.7.).

Step 2: 100 or 10^2 times 3.47 will restore the value of 347.

Step 3: $347 = 3.47 \times 10^2$

Since the decimal was moved two places to the left in Step 1, 3.47 is multiplied by 10^2.

The number 0.0756 can be written in scientific notation as follows:

Step 1: 7.56 is between 1 and 10 (the decimal has been moved two places to the right, 0.0 7 5 6).

Step 2: $\frac{1}{100} = \frac{1}{10^2}$ or 10^{-2} times 7.56 will restore the value to 0.0756 (two places to the left will restore the value 0 0 7.5 6).

Step 3: $0.0756 = 7.56 \times 10^{-2}$

Since the decimal was moved two places to the right in Step 1, 7.56 is multiplied by 10^{-2}.

It will help you remember the placement of the decimal and the correct power of 10 if you note that, in scientific notation:

A number whose value is greater than 1 has a positive or zero exponent of 10.

A number whose value is greater than 0 and less than 1 has a negative exponent of 10.

PROCEDURE

To write a number in scientific notation:

1. Move the decimal left or right so that the result is 1 or a number between 1 and 10.
2. Find a power of ten that will restore the original value after multiplying. (Note that the exponent of 10 is the number of positions and the direction the decimal point is moved in order to restore it to its original position; negative to the left, positive to the right.)
3. Write the original number as the product of the number in step 1 and the power of 10 in step 2.

Example 1 illustrates how to write a number larger than one in scientific notation.

EXAMPLE 1

Write in scientific notation: 38,900

3.89	Step 1: Write a number between 1 and 10. (Move decimal four places to the left.)
3.89 times 10000, or 10^4, will yield a product of 38,900.	Step 2: Find a power of 10 that will restore the original value. (Since the decimal was moved four places to the left, multiply by 10^4.)

$38{,}900 = 3.89 \times 10^4$ — Step 3: Write the original number as the product of the number in Step 1 and the power of 10 in Step 2. ■

Example 2 illustrates how to write a number smaller than one in scientific notation.

EXAMPLE 2

Write in scientific notation: 0.000207

2.07 — Step 1: (Move decimal four places to the right.)

2.07 times $\dfrac{1}{10{,}000}$ is 0.000207. — Step 2.

$\dfrac{1}{10{,}000} = 10^{-4}$ — Change the fraction to a power of 10. (Since the decimal was moved four places to the right, multiply by 10^{-4}.)

$0.000207 = 2.07 \times 10^{-4}$ — Step 3. ■

Example 3 is from astronomy.

EXAMPLE 3

The distance from the Earth to the sun is approximately 92,900,000 miles. Write this number in scientific notation.

To write the distance from the Earth to the sun in scientific notation, we follow the steps given in the rule.

92,900,000 — Original number.

9.29 — Step 1.

$9.29 \times 10{,}000{,}000$, or 10^7, is equal to 92,900,000 — Step 2.

$92{,}900{,}000 = 9.29 \times 10^7$ — Step 3.

So, the distance from the Earth to the sun in scientific notation is 9.29×10^7 miles. ■

To multiply a number by 10, we move the decimal point one place to the right. Therefore, to multiply a number by a positive power of 10, move the decimal point to the right the same number of places as the exponent of 10.

Multiplying a number by a negative power of 10 is equivalent to dividing the number by a positive power of 10. Therefore, to multiply a number by a negative power of 10, we move the decimal point to the left the same number of places as the negative exponent of 10.

To write a number in place value notation, the process is reversed.

■ PROCEDURE

> To change a number in scientific notation to place value notation, do the indicated multiplication.

EXAMPLE 4

Write in place value notation: 8.28×10^5

$8.28 \times 10^5 = 828{,}000$ — To multiply by 10^5, move the decimal point five places to the right. ■

Example 5 shows how to change from scientific notation to place value notation when the power of 10 is negative.

EXAMPLE 5

Write in place value notation: 8.28×10^{-6}

$8.28 \times 10^{-6} = 0.00000828$ — To multiply by 10^{-6}, move the decimal point six places to the left. ■

Numbers may be multiplied or divided when written in scientific notation.

■ PROCEDURE

> To perform multiplications or divisions with numbers written in scientific notation:
>
> 1. Multiply or divide the numbers that are between 1 and 10.
> 2. Multiply or divide the powers of 10.
> 3. Write the result in scientific notation or place value notation as indicated.

EXAMPLE 6

Simplify and write the result in place value notation: $\dfrac{3.6 \times 10^{-5}}{1.2 \times 10^{-4}}$

$$\frac{3.6 \times 10^{-5}}{1.2 \times 10^{-4}} = \frac{3.6}{1.2} \cdot \frac{10^{-5}}{10^{-4}}$$

$= 3 \cdot 10^{-5-(-4)}$ Subtract exponents.

$= 3 \cdot 10^{-1}$

$= 0.3$ ■

EXAMPLE 7

Simplify and write the result in scientific notation: $(3.1 \times 10^4)(4.2 \times 10^3)$

$(3.1 \times 10^4)(4.2 \times 10^3) = (3.1)(4.2)[(10^4)(10^3)]$ Associative and commutative properties.

$= (13.02)(10^7)$ Multiply.

$= (1.302 \times 10^1)(10^7)$ Write 13.02 in scientific notation.

$= 1.302 \times 10^8$ Multiply the powers of 10. Add the exponents. ■

EXERCISE 3.3

A

Write in scientific notation:

1. 50,000 **2.** 0.0005 **3.** 0.00004 **4.** 400,000

5. 9,000,000 **6.** 0.000009 **7.** 7000 **8.** 0.006

9. 210,000 **10.** 0.00023

Change to place value notation:

11. 9.3×10^2 **12.** 9.3×10^{-2} **13.** 8.91×10^4

14. 8.91×10^{-4} **15.** 5.4×10^7 **16.** 5.4×10^{-7}

17. 4.6×10^1 **18.** 4.6×10^{-1} **19.** 8.2×10^{-5}

20. 7.8×10^7

B

Write in scientific notation:

21. 377,000
22. 0.00122
23. 0.0000701
24. 80,600,000
25. 611,000,000
26. 0.000116
27. 0.000456
28. 291
29. 0.0000922
30. 8,230,000

Change to place value notation:

31. 3.21×10^{-1}
32. 1.23×10^{1}
33. 6.89×10^{4}
34. 9.68×10^{-2}
35. 7.43×10^{2}
36. 5.65×10^{-3}
37. 4.6×10^{5}
38. 3.33×10^{-3}
39. 8.62×10^{5}
40. 5.19×10^{-3}

C

Write in scientific notation:

41. 3784
42. 0.2894
43. 34,480
44. 0.07762
45. 0.000484
46. 2,876,000
47. 11,100,000
48. 0.000000000023
49. 296,200,000
50. 0.00000676

Change to place value notation:

51. 3.84×10^{7}
52. 4.38×10^{-6}
53. 2.36×10^{-9}
54. 6.35×10^{7}
55. 1.01×10^{-5}
56. 1.91×10^{9}
57. 9.11×10^{-8}
58. 4.44×10^{8}
59. 5.78×10^{7}
60. 7.07×10^{-8}

Simplify and write the result in scientific notation:

61. $\dfrac{4.8 \times 10^{-6}}{2.4 \times 10^{5}}$
62. $(9.2 \times 10^{5})(3.2 \times 10^{-6})$
63. $(1.39 \times 10^{7})(2.1 \times 10^{3})$
64. $\dfrac{9.1 \times 10^{5}}{1.3 \times 10^{9}}$

Simplify and write the result in place value notation:

65. $(2.7 \times 10)(4.1 \times 10)$
66. $\dfrac{3.6 \times 10^{-4}}{1.2 \times 10^{4}}$
67. $\dfrac{1.28 \times 10^{10}}{3.2 \times 10^{4}}$
68. $(6.8 \times 10^{3})(2.5 \times 10^{-5})$

D

Write in scientific notation:

69. In one year, light travels approximately 5,870,000,000,000 miles.

70. The king bolt bushings of an early-model automobile had a clearance of 0.0005 in.

71. A bat emits high-pitched sounds (too high for humans to hear) of approximately 50,000 cycles per second.

72. The age of a 22-year-old student is approximately 694,000,000 seconds.

73. A bearing, which is for the drive mechanism of an automobile, has a tolerance of 0.003 inch.

Write in place value notation:

74. The retirement fund of a certain company is $\$3.72 \times 10^6$.

75. The shortest wavelength of visible light is approximately 4×10^{-5} centimeters.

76. A certain x-ray has a length of 1.3×10^{-7} centimeters.

77. The distance from the Earth to the nearest star is approximately 2.55×10^{13} miles.

78. The average number of red blood cells in the human adult is approximately 2.0×10^{12}.

STATE YOUR UNDERSTANDING

79. Define scientific notation.

80. When is scientific notation useful?

CHALLENGE EXERCISES

Simplify the following.

81. $\dfrac{(3.25 \times 10^{-2})(2.4 \times 10^2)}{(4.8 \times 10^{-2})(2.5 \times 10^{-2})}$

82. $\dfrac{(4.5 \times 10^{-6})(5.6 \times 10)}{(6.3 \times 10^{-8})(5.0 \times 10^{-9})}$

83. The distance from Earth to the Sun is 9.3×10^7 miles, and the distance from Uranus to the Sun is 1.8×10^9 miles. Uranus is how many times further from the Sun than Earth? Give your answer to the nearest whole number?

84. The resistance of a wire is given by the formula $R = \dfrac{k}{d^2}$. Find R, if $k = 0.000210$ ohms $\cdot$ cm^2 and $d = 0.00840$ cm. Give the answer to the nearest hundredth. Use scientific notation to calculate your answer.

MAINTAIN YOUR SKILLS (SECTIONS 2.3, 2.6)

85. 225% of what number is 220.5?

86. What percent of 384 is 480?

87. 145% of 620 is what number?

88. At the local theater, there are a total of 690 seats. If there are 207 seats in the balcony, what percent of the seats are not in the balcony?

89. José's Furniture offers a chair at 40% off the list price. What is the selling price if the chair is listed at $385?

Solve:

90. $6x - 25 = 85$ **91.** $15 = 29 - 7x$ **92.** $\frac{x}{-3} - 11 = 14$

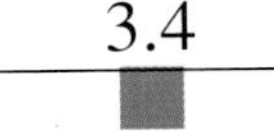

3.4 INTRODUCTION TO POLYNOMIALS

OBJECTIVES

1. Classify polynomials by degree.
2. Classify polynomials by the number of terms.
3. Write a polynomial in descending order of powers.

Recall that a term can be the product of numbers and variables. The numerical factor of the term is the coefficient.

The expression "$3 + 4x$" contains two terms: 3 and $4x$.

We have seen before that the use of variables (letters) to hold the place of numbers often leads to writing indicated operations. The expression

$3x - cz + 2m^2$

has six indicated operations and three terms. There are three multiplications, one addition, one subtraction, and one raising to a power. The terms are connected by the symbols + and −.

$3x$	$-$	cz	$+$	$2m^2$
FIRST TERM		SECOND TERM		THIRD TERM

DEFINITION

Monomial

A *monomial* is a term that contains a number or a number multiplied times one or more variables raised to whole number exponents.

The following terms are monomials:

$$4, \quad 9w, \quad 3x, \quad -\frac{3}{8}x^2y, \quad 4(xy)z^2, \quad -7w.$$

The numerical factors 4, 9, 3, $-\frac{3}{8}$, 4, and -7 are called coefficients. In the terms x, y^2, x^2y, a^3bc^4, and $rstv$, the numerical factor is understood to be one:

$x = 1(x)$, $y^2 = 1(y^2)$, $x^2y = 1(x^2y)$, $a^3bc^4 = 1(a^3bc^4)$, and $rstv = 1(rstv)$.

In the terms $-t$, $-abc$, $-c^3$, $-x^3y^5$, and $-z^7$, the numerical factor is understood to be negative one (-1):

$-t = -1(t)$, $-abc = -1(abc)$, $-c^3 = -1(c^3)$, $-x^3y^5 = -1(x^3y^5)$, and $-z^7 = -1(z^7)$.

Monomials are classified by degree.

DEFINITION

Degree of a Monomial

The degree of a monomial that contains only one variable is the exponent of the variable.

The degree of a monomial that contains more than one variable is the sum of the exponents of the variables.

If the monomial is a non-zero constant, the degree of the monomial is zero.

The monomial zero has no degree.

Monomial	Degree	
$3x$	1	$(3x = 3x^1)$
$4x^3$	3	
$3x^2y^2$	4	Add the exponents, $2 + 2 = 4$
5	0	$5 = 5(1) = 5x^0$

EXAMPLE 1

Find the coefficient and degree of each of the monomials:
$3x^2$, $-18x^5y$, rst, $-x^3$

$3x^2$

Coefficient is 3.	The coefficient is the numerical factor.
Degree is 2.	The degree is the exponent of the variable.

$-18x^5y$

Coefficient is -18.	The numerical factor is the coefficient.
Degree is 6.	When a term contains more than one variable, the degree is the sum of the exponents, $5 + 1 = 6$. (Recall that y has an exponent of 1, $y = y^1$.)

rst

Coefficient is 1.	$rst = 1(rst)$
Degree is 3.	$rst = r^1s^1t^1$ $(1 + 1 + 1 = 3)$

$-x^3$

Coefficient is -1.	$-x^3 = -1(x^3)$
Degree is 3.	

■

■ DEFINITION

Polynomial

A *polynomial* is an algebraic expression containing one or more monomials. If the polynomial has more than one term, the terms are separated by a plus or minus sign.

Polynomials

The following algebraic expressions are polynomials:

$$x + 2, \quad 3xy, \quad \frac{3}{4}y + 7, \quad a + 3b + c, \quad x^2 - 2x + 3.$$

The *degree of a polynomial* is the degree of the term (monomial) with the highest degree.

Degree of a Polynomial

Polynomial	Term of Polynomial	Degree of Term	Degree of Polynomial	Also Called
$\frac{1}{4}x + 6$	$\frac{1}{4}x$ 6	1 0	1	Linear
$x^2 + 3x + 4$	x^2 $3x$ 4	2 1 0	2	Quadratic

(continues on next page)

Polynomial	Term of Polynomial	Degree of Term	Degree of Polynomial	Also Called
$x^2y^3 + 3x^2y + 7$	x^2y^3 $3x^2y$ 7	5 3 0	5	No special name.

EXAMPLE 2

Classify the polynomials by degree: $-3x^2$, $2x^2 + 7x + 5$, $-x^2y^2 + 2$

$-3x^2$ The exponent of the variable is 2.

Degree is 2.

$2x^2 + 7x + 5$ The term with the highest degree, $2x^2$, has degree 2.

Degree is 2.

$-x^2y^2 + 2$ The term with the highest degree, $-x^2y^2$, has degree 4.

Degree is 4. ■

Polynomials are also classified by the number of terms, which is illustrated in the following table.

Monomial

Binomial

Trinomial

Name	Number of Terms	Examples
Monomial	One term	$3xy$ or $\frac{4}{3}y$
Binomial	Two terms	$x + 3$ or $x^2 - y^2$
Trinomial	Three terms	$a + b + c$ or $x^2 + 6x + 7$
	More than three terms	$x^3 + 2x^2 - x + 10$ or $a + b + c + d$

EXAMPLE 3

Classify the polynomials by the number of terms: $-3x^2$, $2x^2 + 7x + 5$, $-x^2y^2 + 2$

$-3x^2$ One term.
Monomial

$2x^2 + 7x + 5$ Three terms.
Trinomial

$-x^2y^2 + 2$ Two terms.
Binomial ■

It is customary to write a polynomial in one variable in descending powers of the variable.

Descending Order

To write a polynomial in *descending order* means to write the term with the largest exponent of the variable first, followed by the next largest, and so on.

Study the following table.

Polynomial	Arranged in Descending Order of Powers
$x^2 + x^3 + 2x + x^4 + 6$	$x^4 + x^3 + x^2 + 2x + 6$
$x^4 + 7 - 3x + 6x^2 - 2x^3$	$x^4 - 2x^3 + 6x^2 - 3x + 7$
$x^3 + 3x + 8$	$x^3 + 3x + 8$
$x^2 - 7 + 4x^4$	$4x^4 + x^2 - 7$
$x^5 - 1$	$x^5 - 1$

The commutative and associative properties of addition allow for the rearranging of terms so that any polynomial can be written in descending powers of a variable.

Writing a polynomial in descending powers is very useful in performing operations with polynomials.

■ DEFINITION

Standard Form of a Polynomial in x

A polynomial in x is an algebraic expression containing a finite number of terms of the form ax^n, where a is any real number and n is a whole number. If the terms are written in descending order, the polynomial in x is said to be in standard form.

EXAMPLE 4

Write the polynomial in descending powers: $3x^2 - 2x^7 + 4x^4 + 2$

$-2x^7 + 4x^4 + 3x^2 + 2$ In descending powers, the terms are written from the highest degree to the lowest degree from left to right. ■

EXERCISE 3.4

A

Give the coefficient and degree of each of the following monomials:

1. $5x^2$

2. $-3y$

3. 12

4. -35

5. $16a^4$

6. $-14b^9$

7. $-\frac{7}{9}x$

8. $\frac{23}{33}y$

9. $\frac{14}{3}x^2$

10. $\frac{3}{17}y^3$

Classify each of the following polynomials by the number of terms:

11. $3x + 3$

12. $x^2 - 5x$

13. $-15x^3y^4$

14. $-12xyz$

15. $a + b - 14$

16. $x^2 - y^2 + z^2$

B

Give the coefficient and degree of each of the following monomials:

17. $-14x^4y^6$

18. $45x^3y^7$

19. x^5yz

20. rs^3t^2

21. $-3a^3b^3c^3$

22. $-19x^4y^5z^6$

23. $-\frac{35}{23}xy^2$

24. $\frac{14}{15}a^2bc$

25. $-\frac{15}{2}$

26. $\frac{13}{3}abc$

Classify each of the following polynomials by the number of terms:

27. $2x^3 - 3x^2 - 4x$

28. $5c^7 - 3c^6 + 4c^5$

29. $-12x^8y^9z^6$

30. $34a^{11}b^8c^2$

31. $6xy - 4z$

32. $18x^2y^2 - xy$

C

Classify each of the following polynomials by degree:

33. $4x^3 - 2x^2 - 5x + 9$

34. $2c^3 - 3c^2 + 5c - 8$

35. $15x^6 - 2y^8 + 19$

36. $-45a^7 + 32b^9$

37. $8x^2y^3 - 14x^3y + 23xy^4$

38. $9a^4b^3 + 3a^2b^3 - 7ab^7$

39. $4x^7 - 3x^2 - 4x^9 + 3x^5 - 7x^6 + 12$

40. $13y^4 - 6y^8 + 2y^3 + 4y^5 - 3y - 23$

41. $9a^2b^3c^5 - 3a^3bc^7 + 12a^5b^3c^2 - 14a^7b^7$

42. $-13x^4y^5z^2 - 34x^4y^6z^2 + 16x^2y^6z^2 + 12xyz$

Write the following polynomials in descending powers: Rearrange

43. $8x^2 - 3x^5 + 2x - 4x^3 + 3 - x^4$

44. $9 - 5x^5 - 3x + 2x^2 - 7x^4 + 2x^3$

45. $7x - 14x^5 - 3x^2 - 6$

46. $32 + 7x^4 - 2x^5 + 2x$

47. $14x^9 - 3x^5 + 1$

48. $2x^7 - 3x + 7x^4$

STATE YOUR UNDERSTANDING

49. What is meant by the coefficient of a term?

50. What is meant by the degree of a term?

51. What is meant by the degree of a polynomial?

52. $4x^2 + \frac{5}{x} - 2$ contains three terms, but it is not a trinomial. Explain. $4x^2 + 5x^{-1} - 2$

No variables in denom. Is a binomial

CHALLENGE EXERCISES

Determine the number of terms in each of the following expressions. When the expression is a polynomial, classify it as a monomial, binomial, or trinomial.

53. $2x + \frac{3x + 1}{2}$

54. $(2a^2 + 3a - 7)$

55. $7x^2 - (2x + 3)(3x - 4)$

56. $\frac{3x - 5}{4x^2}$

57. $3\{4x + 2[3x + 7(x - 2)] + 1\}$

MAINTAIN YOUR SKILLS (SECTIONS 2.3, 2.6)

58. 13 is what percent of 134? (to the nearest tenth of a percent)

59. 114.6% of 34 is what number? (to the nearest hundredth)

60. 76.5% of what number is 52.4? (to the nearest tenth)

61. Jacob pays $345 a month for his apartment. If this represents 32% of his monthly income, what is his monthly income (to the nearest dollar)?

62. Solve $ax + b = c$ for x if $a = 24$, $b = 56$, and $c = -20$.

63. Solve $cx - dy = e$ for x.

Solve:

64. $4y - \frac{3}{7} = \frac{1}{4}$

65. $\frac{x}{5} + \frac{5}{8} = \frac{1}{3}$

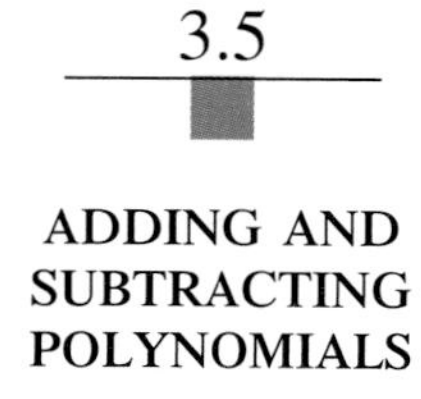

3.5 ADDING AND SUBTRACTING POLYNOMIALS

OBJECTIVES

1. Combine monomials (like terms).
2. Add polynomials.
3. Write the opposite of a polynomial.
4. Subtract polynomials.
5. Solve equations containing polynomials.

Like terms were combined in Chapter 2. The type of terms that were combined are what we now call monomials. As a result of that discussion, combining (adding and subtracting) monomials is a review of combining terms.

When combining terms, only like terms can be combined. This is also true for monomials:

PROCEDURE

To combine monomials, add or subtract the coefficients and write the common variable factors in an indicated product.

EXAMPLE 1

Simplify: $8.7p^2q + 4.5p^2q - 10p^2q$

$(8.7 + 4.5 - 10)p^2q$	Distributive property.
$3.2p^2q$	Simplify.

EXAMPLE 2

Simplify: $8ax + 3bx - 7cx$

$8ax + 3bx - 7cx$	There are no like monomials, so they cannot be combined.

To add polynomials, we will make use of the commutative and associative properties of addition discussed in Chapter 1. Their use is illustrated in the following.

$(5a + 3b) + (2a + 6b)$

$(5a + 2a) + (3b + 6b)$ Group like terms.

$7a + 9b$ Combine like terms.

If the polynomials contain a subtraction sign, rewrite it as addition before grouping like terms. For instance,

$(2x - 4y) + (5x + 2y)$

$[2x + (-4y)] + (5x + 2y)$ Write subtraction as addition.

$(2x + 5x) + (-4y + 2y)$ Group like terms.

$7x + (-2y)$ Combine like terms.

$7x - 2y$ Change from adding the opposite to subtraction.

PROCEDURE

To add polynomials:

1. Change subtractions to addition of opposites.
2. Group like terms.
3. Combine like terms.

EXAMPLE 3

Add: $(4a - 5b) + (a - 2b) + (3b - 3a)$

$[4a + (-5b)] + [a + (-2b)] + [3b + (-3a)]$ Change to addition by adding the opposite.

$[4a + a + (-3a)] + [-5b + (-2b) + 3b]$ Group like terms.

$2a + (-4b)$ Add.

$2a - 4b$ Change from adding the opposite to subtraction. ■

It is sometimes convenient to write the sum of polynomials in vertical form so that like terms will appear in the same column.

EXAMPLE 4

Add: $2a - 3x + 7y$, $4x - 2a + 8y$, and $3y - 5x + 3a$

$$\begin{array}{r} 2a - 3x + 7y \\ -2a + 4x + 8y \\ 3a - 5x + 3y \\ \hline 3a - 4x + 18y \end{array}$$

Use the commutative and associative properties of addition to change the order of the terms. Writing the polynomials in columns gives an easy way to show the grouping of like terms. ■

EXAMPLE 5

Add: $(3x + 10y - 3) + (4x - 7y) + (2y - 8)$

$(3x + 4x) + (10y - 7y + 2y) + (-3 - 8)$ — Group like terms, and mentally change subtraction to addition.

$7x + 5y - 11$ — Combine like terms. ■

Some formulas can be simplified by adding polynomials.

EXAMPLE 6

The city of Bonanza, Colorado, has a triangular park that it wishes to enclose with a chain-link fence. The lengths of the sides of the park are shown in the diagram. Express the length of the fence (F) needed in terms of x and simplify. Find the length of fence needed when x is 350 meters.

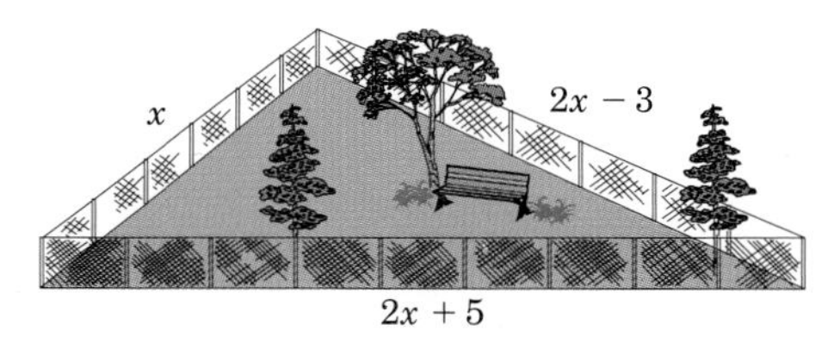

The perimeter of the park will be the length of the fence needed (F). We need the formula for the perimeter of a triangle. The length of one of the sides is represented by x.

$P = a + b + c$ — Formula for the perimeter of a triangle.

$F = x + (2x - 3) + (2x + 5)$ — Substitute into the formula.

$F = (x + 2x + 2x) + (-3 + 5)$ — Group like terms.

$F = 5x + 2$ — Add.

The simplified algebraic form for the fence length is $F = 5x + 2$. To find F if $x = 350$, substitute into this new formula.

$F = 5(350) + 2$ — Substitute 350 for x.

$F = 1750 + 2$

$F = 1752$

The city should order 1752 meters of fence. ■

Opposite of a Polynomial

The opposite of a polynomial is found by changing the sign of each term of the polynomial; that is, use the distributive property to multiply each term by -1.

$$\begin{aligned} -(a + b) &= -1(a + b) \\ &= (-1)a + (-1)b \\ &= -a + (-b) \\ &= -a - b \end{aligned}$$

So,

$$\begin{aligned} -(2x + 3y) &= -2x - 3y \\ -(x^2 - 3x + 9) &= -x^2 + 3x - 9 \end{aligned}$$

and

$$-(-2x - 3y - 7) = 2x + 3y + 7$$

EXAMPLE 7

Simplify $-(3x - 2y + 4)$

$-(3x - 2y + 4) = -3x + 2y - 4$ Multiply each term by -1, or change the sign of each term. ■

EXAMPLE 8

Simplify $-(-8a - 3b - 4c)$

$-(-8a - 3b - 4c) = 8a + 3b + 4c$ Multiply each term by -1, or change the sign of each term. ■

Subtraction of polynomials follows the same pattern as subtraction of signed numbers. That is, the subtraction is changed to addition by adding the opposite.

Subtract Polynomials

■ DEFINITION

$$a - b = a + (-b)$$

■ PROCEDURE

To subtract polynomials, add the opposite of the polynomial to be subtracted:

$$\begin{aligned} (a + b) - (d + e) &= (a + b) + [-(d + e)] \\ &= a + b + [(-d) + (-e)]. \end{aligned}$$

EXAMPLE 9

Subtract: $(2x - 5y) - (6x - 8y)$

$(2x - 5y) + [-(6x - 8y)]$	Add the opposite.
$(2x - 5y) + [-6x + 8y]$	
$(2x - 6x) + (-5y + 8y)$	Group like terms.
$-4x + 3y$	Combine like terms.

or

$(2x - 5y) - (6x - 8y)$	
$2x - 5y - 6x + 8y$	Remove the parentheses by adding the opposite, and combine like terms.
$-4x + 3y$	■

EXAMPLE 10

Subtract: $(x^2 - 2x + 6) - (3x^2 - 5x - 8)$

$x^2 - 2x + 6 - 3x^2 + 5x + 8$	Remove the parentheses by adding the opposite, and combine like terms.
$-2x^2 + 3x + 14$	■

Subtraction can also be written vertically.

EXAMPLE 11

Subtract:
$$\begin{array}{r} 0.18x + 0.05y + 9.1 \\ \underline{0.13x - 0.11y - 8.2} \end{array}$$

$$\begin{array}{r} 0.18x + 0.05y + 9.1 \\ \underline{-0.13x + 0.11y + 8.2} \\ 0.05x + 0.16y + 17.3 \end{array}$$

Subtract by adding the opposite. ■

Exercises with nested grouping symbols can be simplified by doing each operation working from inside out.

EXAMPLE 12

Combine: $4x - [3 + (x + 7)]$

$4x - [3 + (x + 7)] = 4x - [x + 10]$	Combine inside the brackets.
$= 4x - x - 10$	Remove the brackets by adding the opposite.
$= 3x - 10$	Combine like terms. ■

When an equation contains polynomials, use the addition and subtraction properties of equality to isolate the terms containing the variable on the same side of the equation. Combine like terms, and proceed as before.

EXAMPLE 13

Solve: $6x - 9 + 4x = -7 + 8x - 3$

$10x - 9 = 8x - 10$	Combine like terms on both sides.
$10x - 9 - 8x = 8x - 10 - 8x$	Subtract $8x$ from both sides.
$2x - 9 = -10$	
$2x - 9 + 9 = -10 + 9$	Add 9 to both sides.
$2x = -1$	
$x = -\dfrac{1}{2}$	

The solution set is $\left\{-\dfrac{1}{2}\right\}$. The check is left to the student. ■

EXAMPLE 14

Solve: $(3x - 5) - (4 - x) = 27 - 2x$

$3x - 5 - 4 + x = 27 - 2x$	Remove the parentheses by adding the opposite.
$4x - 9 = 27 - 2x$	Combine like terms.
$6x - 9 = 27$	Add $2x$ to each side.
$6x = 36$	Add 9 to each side.
$x = 6$	Divide.

The solution set is $\{6\}$. The check is left for the student. ■

EXAMPLE 15

The length of a rectangle is three less than twice the width (w).

$2w - 3$

w

Find the dimensions of the rectangle if the length is 10 feet longer than the width.

$(2w - 3) - (w) = 10$	The length $(2w - 3)$ minus the width (w) is 10 feet.
$w - 3 = 10$	Combine terms.
$w = 13$	Add 3 to both sides.
$2w - 3 = 2(13) - 3$	To find the length, substitute 13 for w in $2w - 3$.
$= 23$	

So, the dimensions of the rectangle are 13 ft wide and 23 ft long. ■

EXERCISE 3.5

A

Simplify:

1. $6a + 3a$
2. $3x + 5x$ 
3. $7a - 5a$
4. $22b - 3b$
5. $7y^2 + 11y^2 - 20y^2$
6. $4w^3 - 15w^3 + 9w^3$
7. $25xyz - 17xyz - 17xyz$
8. $33bc - 15bc - 19bc$
9. $3x^2 - 4y^2 + 6z^2$
10. $8a + 9b - 17c$
11. $-(2x - 3)$
12. $-(3x + 4)$
13. $-(-x + 5)$
14. $-(-2x - 6)$
15. $-(-3x - 7y - 12)$
16. $-(-7y + 9z - 17)$
17. $(7m - 8) - (3m - 9)$
18. $(5d + 2) - (3d + 5)$
19. $(19x - 5y) - (-21x + 8y)$
20. $(12a - 9b) - (16a + 3b)$

Solve:

21. $2x - 9x = 14$
22. $4x - 12x = 56$
23. $3a - 6a - 9a = -24$
24. $5x - (2x - 1) = 10$
25. $(4x + 7) - 5x = 4$
26. $(6x - 7) - 3x = 11$

B

Add:

27. $(x^2 - 8x + 3) + (4x^2 + 9x + 7) + (-8x^2 + 6x - 11)$
28. $(9a^2 - b^3 + c) + (3a^2 + 2b^3 - c) + (8b^2 + c)$
29. $(2a + 3b - c) + (a - b + c)$
30. $(4ab - 7c^2 - 12) + (3ab + 7c^2 - 8)$

31. $\begin{array}{r} 15y^2 - 8y + 19 \\ \underline{12y^2 - 11y - 31} \end{array}$

32. $\begin{array}{r} -13x^2 - 12x + 18 \\ \underline{-24x^2 + 30x + 12} \end{array}$

33. $\begin{array}{r} 10r + 9s - 8t \\ -r + 5s + 2t \\ \underline{-12r - 7s - 3t} \end{array}$

34. $\begin{array}{r} -5a^2 - 3b^2 + 2ac \\ -4a^2 + b^2 - 5ac \\ \underline{-a^2 + b^2 - 5ac} \end{array}$

35. $\begin{array}{r} a^2 - 7ab + 3b^2 \\ \underline{7a^2 + 3ab - 9b^2} \end{array}$

36. $\begin{array}{r} 34x^2 - 7xy + 16y^2 \\ \underline{30x^2 - 12xy + 17y^2} \end{array}$

Subtract:

37. $(2a^2 + 47a - 110) - (28a^2 - 100a + 79)$

38. $(17x^2 - 42x + 33) - (36x^2 + 25x + 71)$

39. $(21y^2 + 19y - 28) - (-8y^2 - 7y + 15)$

40. $(a^3 - 3ab^2 - a^2b) - (a^2b + b^3 - 3ab^2)$

41. $(5a^2 - 3ab - 4b^2) - (2a^2 + 2ab + b^2)$

42. $(3t^2 - 4t + 3) - (5t^2 - 2t - 2)$

43. $\begin{array}{r} 8x - 3y + 8 \\ \underline{7x + 6y + 3} \end{array}$

44. $\begin{array}{r} -11x^2y + 16xy^2 - 8x^2y^2 \\ \underline{-3x^2y - 8xy^2 - 4x^2y^2} \end{array}$

45. $\begin{array}{r} 212x^2 + 53x + 7 \\ \underline{200x^2 - 17x + 25} \end{array}$

46. $\begin{array}{r} 15y^2 - 17y + 8 \\ \underline{13y^2 - 16y + 27} \end{array}$

Solve:

47. $(8x + 3) + (-5x + 2) = -2x - 15$

48. $(-6x + 5) + (-x + 6) = 8 - 3x$

49. $(9x - 6) + (-5x - 4) = -6 + 2x$

50. $(3x - 5) - (x + 4) = x - 5$

51. $(-6x - 10) - (2x - 6) = 4 - 2x$

52. $3 + t - (t + 3) = 6 - t$

C

Simplify:

53. $(3x - 4) + (2x + 6) + (8x + 9) + (3x + 6)$

54. $(x^2 + y^2) + (3x^2 - 4y^2) + (7x^2 + 3y^2) + (x^2 - y^2)$

55. $\left(\frac{1}{4}x - 2y + \frac{3}{2}z\right) + \left(\frac{5}{4}x - \frac{3}{4}y + z\right) + \left(-\frac{1}{2}x + \frac{1}{2}y - \frac{1}{2}z\right)$

56. $\left(\frac{2}{3}mn - \frac{1}{2}n + 2\right) + \left(\frac{3}{4}mn + \frac{5}{6}n - 1\right) + \left(mn + \frac{1}{3}n + \frac{1}{2}\right)$

57. $(-8xy - 7yz + 5xz) + (6xz - 8xyz - 9yz) + (12yz - 8xz - 20xyz)$

58. $(4abc + 3ab - 5bc) + (6abc - 15bc + 4ab) + (7ab + 4bc - 8abc)$

59.
$$\begin{array}{r} 0.34r - 0.55s + 0.17t \\ 0.28r + 0.32s - 0.83t \\ -0.16r - 0.23s + 0.21t \\ \hline \end{array}$$

60.
$$\begin{array}{r} 0.08a - 0.15b - 0.21 \\ 0.61a - 0.31b + 0.28 \\ -0.21a + 0.50b - 0.07 \\ \hline \end{array}$$

61.
$$\begin{array}{r} 0.8bc - 3.5ad \\ -1.1bc - 0.8ad \\ 0.2bc + 2.3ad \\ \hline \end{array}$$

62.
$$\begin{array}{r} 6.22w - 8.91 \\ -9.13w + 5.83 \\ 0.07w + 1.35 \\ \hline \end{array}$$

63. $(6a^2 - 5b^2) - (2a^2 + 3ab - b^2)$

64. $(9x^2 + 7y^2) - (3x^2 - 3xy - y^2)$

65. $14x - [3x - (2x + 9)]$

66. $27x - [-2x - (-9x + 6)]$

67. $(5x - 2) - [-5x - (x - 12) - 4] + 6$

68. $(2x - 4) - [7x - (12 + x) + 7] - 8$

69. $[4 - (3x - 2)] - [-(4x - 6) - 9]$

70. $[12x - (8 - 6x)] - [-(5x + 2) - 12]$

Subtract:

71.
$$\begin{array}{r} 0.79a - 0.32b + 0.62 \\ -0.18a - 0.42b - 0.18 \\ \hline \end{array}$$

72.
$$\begin{array}{r} 0.10b + 1.36c + 2.01 \\ 1.16b - 2.03c + 1.89 \\ \hline \end{array}$$

Solve:

73. $(3x + 8) + (2x - 5) = (4x + 2) + (-3x + 5)$

74. $(7x - 8) + (4x + 7) = (5x + 5) + (x + 6)$

75. $(-8a - 12) + (7a - 9) + 12 = (-4a - 16) + (9a - 15) + 23$

76. $(-7c - 18) + (-9c - 19) + 12 = 22 + (-5c + 14) + (12c - 13)$

77. $(2x - 6) - (4x - 7) = (5x + 2) - (x + 1)$

78. $1 - (y - (6 - y)) = 2y - (3 - y)$

79. $-(6x - 2) - (3x - 7) = -(8x - 2) - (4x + 4)$

80. $(5 + 8a) - (2 - (3a - 5)) = 5 - a - (2a + 3)$

D

81. If four more than a number is added to seventeen less than the number, the result is 75. Find the number.

82. If eight less than a number is added to three more than twice the number, the result is 106. Find the number.

83. Seven less than twice a number is added to twelve times the number. If the result is 147, find the number.

84. If four less than three times a number is added to 85, the result is 48. Find the number.

85. If six more than four times a number is added to twelve less than eight times the number, the result is thirty-one less than twice the number. Find the number.

86. Nine more than three times a number added to three less than nine times the number gives a result of five less than seven times the number. Find the number.

87. If the sum of a certain number and six is increased by two more than the number, the result is eight less than the number. Find the number.

88. Barbara's house has an odd-shaped yard, as shown to the right. Find the perimeter of the yard in terms of x and simplify. Find the perimeter if x is 25 feet.

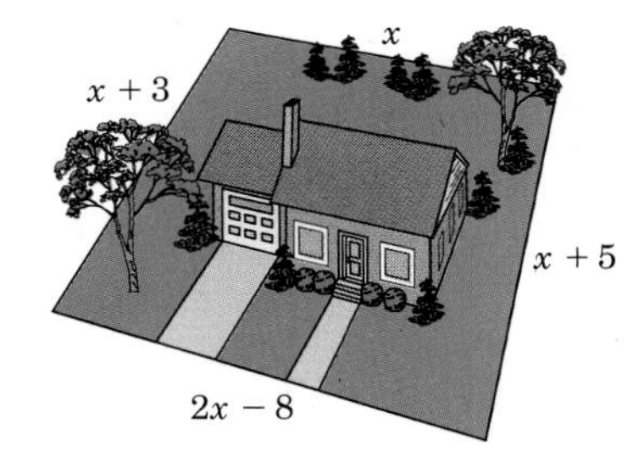

Figure for Exercise 88

89. A flower garden is in the shape of a triangle, as shown in the diagram. Find the length of each side if the total amount of fencing to enclose it is 41 feet.

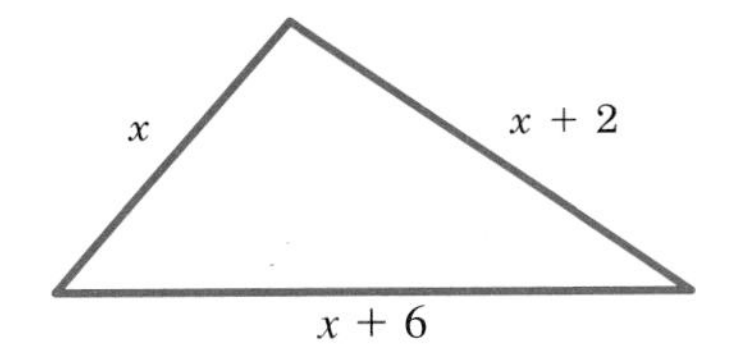

Figure for Exercise 89

90. A piece of wire is bent into the shape of a rectangle with the length seven more than twice the width. If the perimeter of the rectangle is 38 inches, find the length and width.

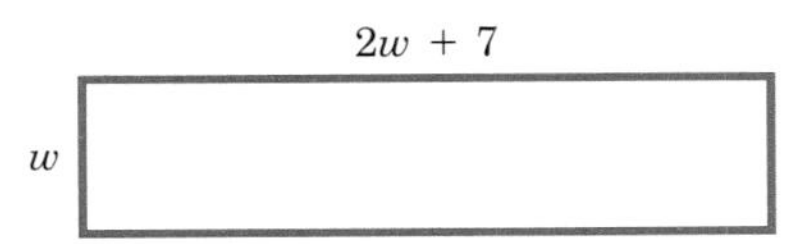

Figure for Exercise 90

91. Express the perimeter of the trapezoid shown in the diagram in terms of w and simplify. Find the length of the perimeter if w is 15 centimeters.

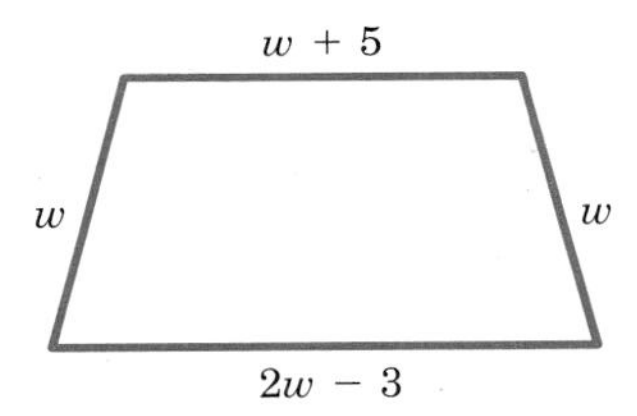

Figure for Exercise 91

92. One number is three less than six times another number. If the difference between the two numbers is 17, find the numbers.

93. The larger of two numbers is five more than twice the smaller. If the difference between the larger number and three times the smaller is -8, what are the two numbers? (Recall that we write the difference between two numbers, a and b, as $a - b$.)

94. The difference between one more than five times a number and three less than two times that same number is 7. Find the number.

95. Pete's score can be written as six more than eight times a number, whereas Amber's score can be written as eight less than nine times the same number. If the differences between Pete's score and Amber's score is 4, find their scores.

96. Mr. Braun had 18 fewer than three times the number of stamps that Mrs. Liber had. If the difference between the number of stamps Mr. Braun had and Mrs. Liber had is 22, find the number of stamps each one had.

97. In a Gobblebug game, Cindy's score was four more than five times Minh's score. If the difference in their scores was 4900, what did each score?

98. Cynthia and Max are comparing their present fortunes. Cynthia has \$3.82 less than twice as much as Max has. If the difference in the amount of money each has is \$2.33, how much does each have?

99. Two rockets are fired at the same time from ground level. The height (H in feet) of the first rocket at any time t (seconds) is $H_1 = 420t - 16t^2$. The height of the second and slower rocket is $H_2 = 212t - 16t^2$. In how many seconds will they be 1040 ft apart?

100. Mary earns 10 cents an hour more than one half of what Joan earns. If Joan earns \$2.95 an hour more than Mary, how much does each earn per hour?

101. The amperage in an electrical current at any time t is given by the equation $I_1 = 16 - 3t - t^2$. The amperage of a second system is given by $I_2 = 28 + 15t - t^2$. At what time is the difference $(I_2 - I_1)$ in amperage in the two systems equal to 60?

STATE YOUR UNDERSTANDING

102. What is the difference between adding and subtracting polynomials?

103. Explain why it is permissible to simplify $(3x - 4y) - (2x - 6y)$ as if it were $(3x - 4y) - 1(2x - 6y)$.

CHALLENGE EXERCISES

104. Simplify: $[(5x^2 - 3x - 2) - (x^2 + 2x - 5)] - [(7x - x^2) + (-4x^2 - 5x + 1)]$

105. What algebraic expression must be added to $3x^2 - 4x + 2$ to give $4x^2 + 5x - 3$ as the result?

106. What algebraic expression must be added to $-4y^2 + 3y - 6$ to have 0 as the sum? What is the relationship between the two expressions?

107. The perimeter of an isosceles triangle is $5x - 3y$ cm. The length of each of the equal legs is $2x + 5y$. What is the length of the base?

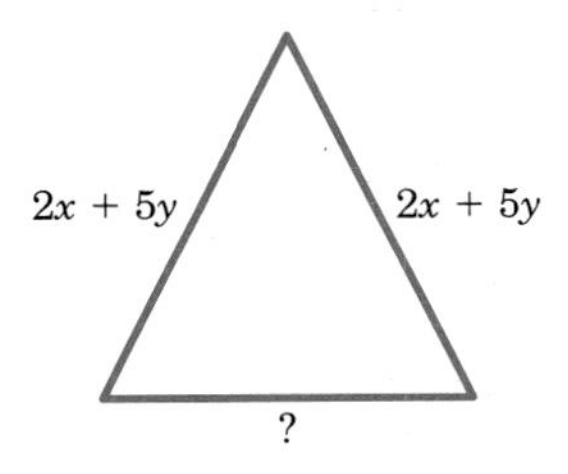

Figure for Exercise 107

MAINTAIN YOUR SKILLS (SECTIONS 2.3, 2.5, 2.7, 3.1)

Solve:

108. $-3.4x - 6.7 = 10.64$

109. $5.8x - 7.1 > -47.12$

110. $8.3 + 7.2x \leq 19.82$

111. $4.8x - 9.3 = -24.9$

112. $5xy - 13z = 15$, for y

113. $rst + ab = 16y$, for s

Multiply:

114. $x^8 \cdot x^9 \cdot x^3$

115. $y^{11} \cdot y^{13} \cdot y^6 \cdot y^2$

3.6 APPLICATION SOLUTIONS

OBJECTIVE

1. Solve applications often referred to as word problems, verbal problems, or story problems.

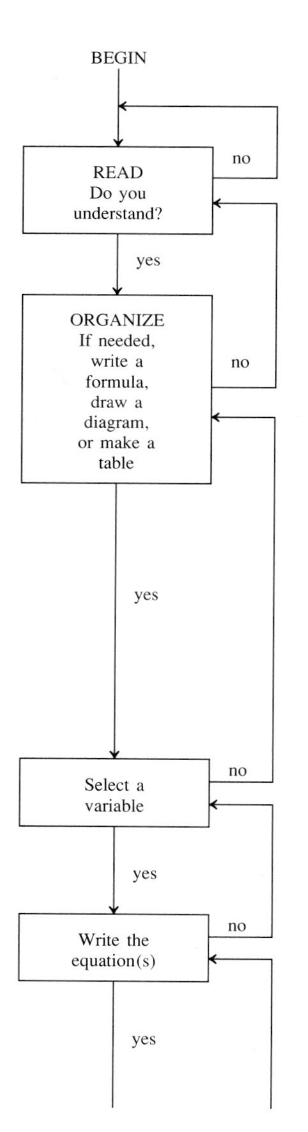

Translating words to algebra requires a lot of practice. The following is a list of suggested steps that can help you make the translation. On the left is a flow chart to follow. The path for the answer to each step is shown. On the right is a step-by-step description of the flow chart.

1. Read the problem. First read the problem through quickly once or twice to get the general idea. Then read the problem through slowly several times to sort out the information. Read especially to identify the question. (What problem are we asked to solve?)

2. Organize the information. The information in a problem can be arranged in different ways (from the original) to make it more useful. For instance,
   a. Make a drawing, and label it with the problem information. Look for relationships between values (for instance, sides of triangles, speeds of planes, and rates of work).
   b. Make a table, and label the columns and rows so that the information can be written in the spaces (see Example 2).
   c. Write a shorter, simpler word statement that expresses the relationships stated (often suggested) by the problem. This is done by comparing or relating the data given (see Example 2).
   d. Use a formula. Some formulas are contained within the problem. Other formulas are commonly found in geometry, business, science, social science, and other fields (see Example 1).

   Your organization may use any or all of these.

3. Select a variable (such as x or n) to represent one of the values that is asked for (unknown). Write an explicit definition of this basic variable. Use this variable to write algebraic expressions to represent other unknowns (if there are others). In some problems two variables might be used (see Chapter 6).

4. Translate the information into algebra.
   a. Use a drawing or table to write an equation. The left and right sides are two (different) expressions that represent the same value (see Examples 2 and 3).
   b. Write an equation from a simpler word statement (see Example 3).
   c. Substitute known values into an appropriate formula to get an equation (see Example 2).

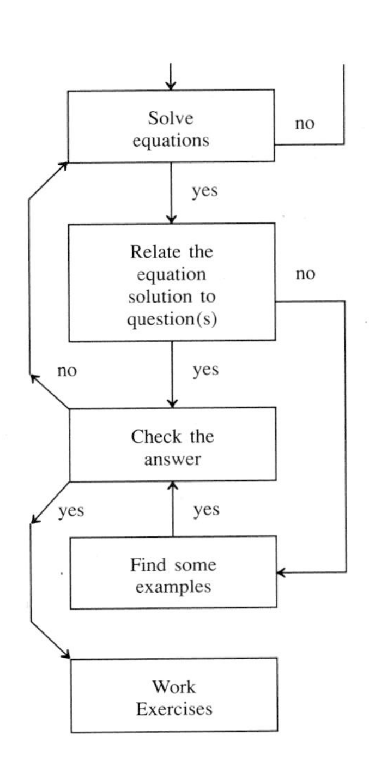

5. Use the laws of equations to solve the equation or system of equations.

6. Translate the solution of the equation(s) to the answer. Write in words the answer to the question(s) asked in the problem (see all the Examples).

7. Use the original problem (not the equation) to check. If the equation is incorrect, this kind of check will help you make this discovery (see Example 1).

8. Go to the set of Exercises and work similar problems. In this way, you will become better skilled at problem solving.

The order of steps 2, 3, and 4 may vary depending on the nature of the problem and your understanding of the concept involved in the solution.

Most of the applications that we have worked so far could be solved by using arithmetic or substituting into a formula. We are now ready to solve applications involving more complex situations that need careful translations to algebra.
Some examples of formulas used in applications follow:

speed (r) · time (t) = distance (d)

$$rt = d$$

rate (r) · principal (p) · time (t) = interest (i)

$$rpt = i$$

rate (percent) (R) · base (B) = amount (A)

$$RB = A$$

number of items (n) · cost per item (c) = total cost (t)

$$nc = t$$

value of a coin (v) · number of coins (n) = total value (V)

$$vn = V$$

$$\begin{pmatrix}\text{\% of a mixture } (R)\\ \text{that is sugar}\end{pmatrix} \cdot \begin{pmatrix}\text{total amount } (A)\\ \text{of mixture}\end{pmatrix} = \begin{pmatrix}\text{amount of } (S)\\ \text{sugar in mixture}\end{pmatrix}$$

$$RA = S$$

If only one relationship is involved in an application, a simple formula (equation) is used. It can be solved by substitution, as in Example 1.

EXAMPLE 1

If an alcohol and water solution is 35% alcohol, how much alcohol is in 5 gallons of the solution?

Simpler word form:

$$\underset{\left(\begin{array}{c}\text{\% of a solution}\\ \text{that is alcohol}\end{array}\right)}{R} \cdot \underset{\left(\begin{array}{c}\text{total amount}\\ \text{of solution}\end{array}\right)}{B} = \underset{\left(\begin{array}{c}\text{amount of alcohol in}\\ \text{the solution}\end{array}\right)}{A}$$

Select variable:

Let A represent the amount of alcohol in the solution.

Translate to algebra:

$(35\%)(5) = A$

Solve:

$(0.35)(5) = A$

$1.75 = A$

Check:

Is 1.75 gallons 35% of 5 gallons?

$\frac{1.75}{5} = 0.35 = 35\%$ Yes

Answer:

There are 1.75 gallons of alcohol in the solution. ■

If two or more like relationships are involved in the same problem, a chart is useful in organizing the data.

EXAMPLE 2

Car A and Car B start from the same place at the same time and drive in the same direction. If Car A travels at 40 mph and Car B travels at 52 mph, in how many hours will they be 60 miles apart?

Rate, time, and distance are involved ($rt = d$).

Simpler word form:

$$\begin{pmatrix}\text{distance traveled}\\ \text{by the faster car}\end{pmatrix} - \begin{pmatrix}\text{distance traveled}\\ \text{by the slower car}\end{pmatrix} = \begin{pmatrix}\text{distance}\\ \text{apart}\end{pmatrix}$$

The difference of the two distances is their distance apart.

Select variable:

Let t represent the time the cars are driven.

	Rate · Time = Distance		
Car A	40	t	$40t$
Car B	52	t	$52t$

Translate to algebra:

$52t - 40t = 60$

$52t$ (rate times time traveled) is the distance of Car B. $40t$ is the distance of Car A.

Solve:

$12t = 60$

$t = 5$

Check:

If the cars are driven for 5 hours, will they be 60 miles apart?

52 mph · 5 hours = 260 mi

40 mph · 5 hours = 200 mi

260 mi − 200 mi = 60 mi

Answer:

In 5 hours, the cars will be 60 miles apart. ■

These relationships and others like them will be used throughout the text. You may want to mark this section for easy reference.

EXAMPLE 3

Tang decided to raid his piggy bank to buy the latest hit CD by the Ape Brothers. He found that his bank contained an equal amount of nickels, dimes, and quarters. How many of each coin were in the bank if the total value was $16?

Simpler word form:

$$\begin{pmatrix}\text{value of}\\ \text{the nickels}\end{pmatrix} + \begin{pmatrix}\text{value of}\\ \text{the dimes}\end{pmatrix} + \begin{pmatrix}\text{value of}\\ \text{the quarters}\end{pmatrix} = \begin{pmatrix}\text{total value}\\ \text{of the coins}\end{pmatrix}$$

Select variable:

Since there are the same number of nickels, dimes, and quarters, let n represent the number of each coin.

We use a chart to organize the data.

	Value of Each Coin	· Number of Coins	= Value of Coins
Nickels	0.05	n	$0.05n$
Dimes	0.10	n	$0.10n$
Quarters	0.25	n	$0.25n$

Translate to algebra:

$0.05n + 0.10n + 0.25n = 16$

Solve:

$$0.40n = 16$$
$$n = 40$$

Check:

Is the value of 40 nickels, 40 dimes, and 40 quarters equal to $16?

$$\begin{aligned}(40)(\$0.05) &= \$\ \ 2.00\\ (40)(\$0.10) &= \$\ \ 4.00\\ (40)(\$0.25) &= \underline{\$10.00}\\ \text{total value} &= \$16.00\end{aligned}$$

Answer:

Tang found 40 nickels, 40 dimes, and 40 quarters in his bank. ■

EXAMPLE 4

Mr. Money Bags invested a like amount in each of two tax-free municipal bonds. One bond pays 9.36% simple interest, and the other pays 10.15% simple interest. If his income from the two bonds is \$4877.50, how much does he have invested in each bond?

Simpler word form:

$$\begin{pmatrix}\text{interest from}\\ \text{first bond}\end{pmatrix} + \begin{pmatrix}\text{interest from}\\ \text{second bond}\end{pmatrix} = \begin{pmatrix}\text{total}\\ \text{interest}\end{pmatrix}$$

Select variable:

Let m equal the amount of money invested in each bond.

We use a chart to organize the data.

	Rate	· Principal	= Interest
First bond	9.36%	m	$0.0936m$
Second bond	10.15%	m	$0.1015m$

Translate to algebra:

$$0.0936m + 0.1015m = 4877.50$$

Solve:

$$0.1951m = 4877.50$$
$$m = 25{,}000$$

Check:

If \$25,000 was invested at 9.36% interest and at 10.15% interest, is the total interest \$4877.50?

$$\begin{aligned}(\$25{,}000)(0.0936) &= \$2340.00\\ (\$25{,}000)(0.1015) &= \underline{\$2537.50}\\ \text{total interest} &= \$4877.50\end{aligned}$$

Answer:

Mr. Money Bags invested \$25,000 in each municipal bond. ■

EXERCISE 3.6

1. Two autos headed in the same direction start at the same time from the same place, one at 65 mph and the other at 58 mph. How long will it take them to be 126 miles apart?

2. Adam has the same number of pennies, nickels, and dimes in his piggy bank. The total value is $5.12. How many of each type of coin does he have?

3. A catering company has the same number of pounds of cashew nuts and peanuts. The cashew nuts cost $4.60 per pound, and the peanuts cost $2.15 per pound. If the total cost was $148.50, how many pounds of each type were there?

4. Car A starts from one town driving at 50 mph and Car B starts from another town driving at 55 mph. The two towns are 378 miles apart, and the cars are driving toward each other. How long will it take for them to meet?

5. Two cyclists start in opposite directions from the same place, one at 6 mph and the other at 9 mph. How long will it take for them to be 24 miles apart?

6. Two airplanes take off from Portland International Airport flying in opposite directions. The first plane flies at 340 mph and the second at 450 mph. How long will it take them to be 1185 miles apart?

7. Two cars leave from the same town at the same time traveling in the same direction. If the first car travels at a speed of 60 mph and the second car at 45 mph, how long will it take for the cars to be 80 miles apart?

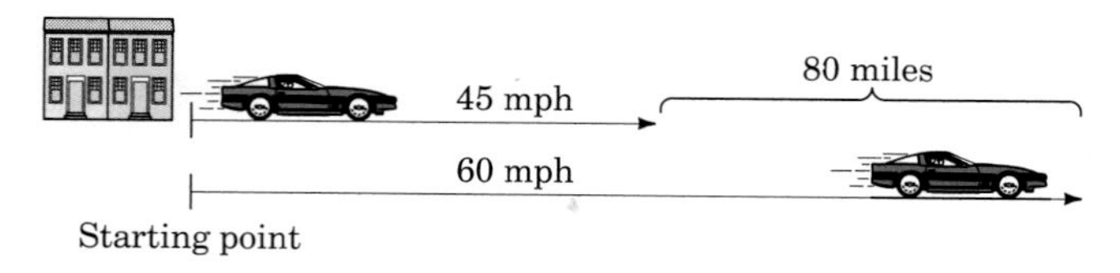

Figure for Exercise 7

8. Two hikers start out on the Pacific Crest Trail at the same location and walk in the same direction. If one hiker averages 13.4 miles per day and the other averages 9.6 miles per day, in how many days will they be 49.4 miles apart?

9. At the end of the Little League candy sale, it was noted that the receipts contained the same number of quarters and half dollars. If the value of these coins was $101.25, how many of each were there?

10. The cashier at the National Women's Bank found that a certain deposit contained the same number of pennies, nickels, and dimes. If the value of these coins was $12, how many of each were in the deposit?

11. The local Grandmother's Market ordered the same number of cases of green beans and pork 'n' beans. The green beans cost \$9.60 per case, and the pork 'n' beans cost \$15.12 per case. How many cases of each were ordered if the total cost to the store was \$543.84?

12. A shipment of rare crystal costing \$1410 was received by the Exquisite Gift Shoppe. The shipment contained equal numbers of water, wine, and sherbet goblets. If the water goblets cost \$21 each, the wine goblets cost \$19.50 each, and the sherbets cost \$18.25 each, how many of each were in the shipment?

13. The First National Bank offers 6.75% simple interest on its saving accounts. The Second National Bank offers 7.15% interest on its saving accounts. John Q. Public invests the same amount in each bank. If the difference between the interest earned at the two banks is \$10.24, how much does John Q. Public have invested in each bank?

14. In Average City, USA, it is found that 8.6% of the residents are black, 6.7% are Hispanic, 5.7% are Asian, and 0.15% are American Indian. If the total number of these minorities is 13,536, how many people live in Average City?

15. Mr. Bar Fly had two drinks after work. Each drink contained the same number of ounces of liquid. One was 30% alcohol and the other was 25% alcohol. How many ounces of liquid were in each drink if he consumed 4.4 ounces of alcohol?

16. Mr. Peanut packages two varieties of mixed nuts. Each package contains the same number of ounces of nuts. One package contains 1 ounce more cashews than the other. If the packages contain 20% and 12% cashews, how many ounces of nuts are in each package?

STATE YOUR UNDERSTANDING

17. What is the recommended first step in the procedure for solving word problems?

18. Why should the answer to a word problem be checked in the original problem instead of the equation.

CHALLENGE EXERCISES

19. Two planes take off from an airport at the same time and fly in opposite directions. One averages 60 mph faster than the other. At the end of $1\frac{1}{2}$ hours the planes are 810 miles apart. Find the average speed of each plane.

20. A boy left home at 7:30 AM and walked at 2.5 mph. At 10:15 AM his father left home in the family car to catch him and give him his lunch. What was the father's average speed if he overtook the boy at 10:30 AM?

21. Mike and Jill are partners. The agreement is that Mike's share of the profit is to be 25% less than Jill's portion. The profit for last year was \$56,000. How much did each partner receive?

22. The sum of three consecutive odd integers is 57. Find the numbers.

23. If three times the first of three consecutive even integers is subtracted from four times the third, the result is 42. Find the numbers.

MAINTAIN YOUR SKILLS (SECTIONS 1.4, 2.7)

True or false:

24. $-82 < 114$

25. $-82 < -114$

Solve:

26. $5x - 17 > 32$

27. $3x + 42 \leq 36$

28. $-7x - 12 > -33$

29. $-15x + 37 \geq 57$

30. The maximum weight that Joe's pickup truck can safely carry is 1500 pounds. If Joe weighs 210 pounds and wheat weighs 75 pounds per sack, how many sacks of wheat can Joe safely haul?

31. What is the integral length, greater than zero, of a rectangle with a width of 45 meters if the area must be greater than 2745 m^2?

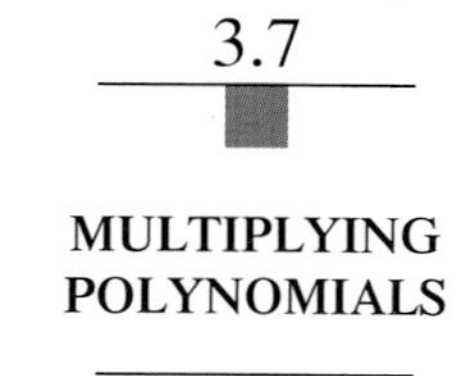

3.7 MULTIPLYING POLYNOMIALS

OBJECTIVES

1. Multiply polynomials.
2. Multiply binomials.
3. Solve equations that involve multiplying polynomials.

To multiply monomials, multiply the coefficients and the variables.

The *distributive property* is used to find the product of two polynomials. The following examples show the product of a monomial and a binomial and the product of a monomial and a trinomial.

$2(3x + 4) = 2(3x) + 2(4)$ Distributive property.

$= 2 \cdot 3 \cdot x + 8$

$= 6x + 8$ Simplify.

$$-3w(4w + 8y - 2) = -3w(4w) + (-3w)(8y) + (-3w)(-2) \quad \text{Distributive property.}$$
$$= -3 \cdot 4 \cdot w \cdot w + (-3 \cdot 8)w \cdot y + (-3)(-2)w$$
$$= -12w^2 + (-24wy) + 6w \quad \text{Simplify.}$$
$$= -12w^2 - 24wy + 6w$$
■

■ PROCEDURE

To multiply a monomial times a polynomial, use the distributive property to write the products of the monomial and each term of the polynomial. Then multiply the monomials and simplify.

EXAMPLE 1

Multiply: $5mn(3m - 9n)$

$$5mn(3m - 9n) = 5mn(3m) + 5mn(-9n) \quad \text{Distributive property.}$$
$$= 5 \cdot 3mn \cdot n + 5(-9)mn \cdot n$$
$$= 15m^2n - 45mn^2 \quad \text{Simplify.}$$
■

EXAMPLE 2

Multiply: $-0.5abc(0.8ab - 0.6bc - 0.2ac)$

$$= -0.5abc(0.8ab) + (-0.5abc)(-0.6bc) + (-0.5abc)(-0.2ac) \quad \text{Distributive property.}$$
$$= -0.4a^2b^2c + 0.3ab^2c^2 + 0.1a^2bc^2 \quad \text{Simplify.}$$
■

Products of other polynomials are found by repeated use of the distributive, commutative, and associative properties. For example, to multiply $(x + 3)$ and $(x^2 - 2x + 5)$, think of $(x + 3)$ as a monomial M, and

Think: $M(x^2 - 2x + 5) = M[x^2 + (-2x) + 5]$
$$= M(x^2) + M(-2x) + M(5)$$

Write: $(x + 3)(x^2 - 2x + 5) = (x + 3)[x^2 + (-2x) + 5]$
$$= (x + 3)x^2 + (x + 3)(-2x) + (x + 3)(5)$$

Now simplify:
$$= (x^3 + 3x^2) + [-2x^2 + (-6x)] + (5x + 15)$$
$$= x^3 + 3x^2 - 2x^2 - 6x + 5x + 15$$
$$= x^3 + x^2 - x + 15$$

PROCEDURE

To multiply two polynomials:

1. Use the distributive property to multiply each term of one polynomial times each term of the other.
2. Combine like terms.

EXAMPLE 3

Multiply: $(x - y)(x^2 + xy + y^2)$

$(x - y)(x^2 + xy + y^2) = (x - y)x^2 + (x - y)xy + (x - y)y^2$	Distributive property.
$= x^3 - x^2y + x^2y - xy^2 + xy^2 - y^3$	Distributive property.
$= x^3 - y^3$	Combine like terms. ■

The multiplication can also be written vertically. Be sure and multiply each term of the first polynomial by each term of the second polynomial. Line up like terms in the product for easy addition.

$$
\begin{array}{rrrrr}
 & & 2x^2 & - \ 3x & + \ 7 \\
 & & x^2 & + \ 5x & - \ 10 \\
\hline
 & & -20x^2 & + \ 30x & - \ 70 \\
 & 10x^3 & - \ 15x^2 & + \ 35x & \\
2x^4 & - \ 3x^3 & + \ 7x^2 & & \\
\hline
2x^4 & + \ 7x^3 & - \ 28x^2 & + \ 65x & - \ 70
\end{array}
$$

EXAMPLE 4

Multiply: $(0.3f^2 + 0.6f - 0.2)(0.4f - 0.25)$

$$
\begin{array}{rrrr}
 & 0.3f^2 & + \ 0.6f & - \ 0.2 \\
 & & 0.4f & - \ 0.25 \\
\hline
 & -0.075f^2 & - \ 0.15f & + \ 0.05 \\
0.12f^3 \ + & 0.24f^2 & - \ 0.08f & \\
\hline
0.12f^3 \ + & 0.165f^2 & - \ 0.23f & + \ 0.05
\end{array}
$$

Write the problem in vertical form.
Multiply by -0.25.
Multiply by $0.4f$.
Combine terms. ■

EXAMPLE 5

Multiply: $(a - b + 2c)(2a - 3b + c)$

$$
\begin{array}{rrrrrr}
 & & a & -\ b & +\ 2c & \\
 & & 2a & -\ 3b & +\ c & \\
\hline
 & & ac & -\ bc & +\ 2c^2 & \\
 -\ 3ab + 3b^2 & & & -\ 6bc & & \\
 2a^2 - 2ab & & +\ 4ac & & & \\
\hline
 2a^2 - 5ab + 3b^2 & & +\ 5ac & -\ 7bc & +\ 2c^2 &
\end{array}
$$

Write the problem in vertical form.
Multiply by c.
Multiply by $-3b$.
Multiply by $2a$. ■

EXAMPLE 6

Multiply: $(2x + 3)(3x - 2)(4x + 5)$

$$
\begin{aligned}
(2x + 3)(3x - 2)(4x + 5) &= (6x^2 + 5x - 6)(4x + 5) \\
&= 24x^3 + 50x^2 + x - 30
\end{aligned}
$$

First find the product of $(2x + 3)$ and $(3x - 2)$, then do the last multiplication. ■

Binomial multiplication can also be done by a shortcut method that is called **F O I L** multiplication. **F O I L** means the product of the First terms plus the product of the Outside terms plus the product of the Inside terms plus the product of the Last terms.

The binomial product is labeled as follows:

F L F L

$(2m + 3)(3m - 1)$

O I I O

to indicate that

F refers to the first term of each binomial.

O refers to the outside terms of the binomials.

I refers to the inside terms of the binomials.

L refers to the last term of each binomial.

To find the product of $(2m + 3)(3m - 1)$, we do the following:

O, F

$(2m + 3)(3m - 1) = (2m + 3)(3m + -1)$

I, L

$$
\begin{aligned}
&\quad \mathbf{F} \quad + \quad \mathbf{O} \quad + \quad \mathbf{I} \quad + \quad \mathbf{L} \\
&= (2m)(3m) + (2m)(-1) + (3)(3m) + (3)(-1) \\
&= 6m^2 + (-2m) + 9m + (-3) \\
&= 6m^2 + 7m - 3
\end{aligned}
$$

PROCEDURE

> The product of two binomials is the sum of the products of the first terms, outer terms, inner terms, and last terms. Combine and simplify where possible.

EXAMPLE 7

Multiply: $(x + 5)(x + 11)$

	F O I L	
$(x + 5)(x + 11)$	$= x \cdot x + x(11) + 5(x) + 5(11)$	Do this step mentally using FOIL.
	$= x^2 + 16x + 55$	Multiply and combine like terms. ■

EXAMPLE 8

Multiply: $(3y + 2)(3y - 7)$

	F O I L	
$(3y + 2)(3y - 7)$	$= 3y \cdot 3y + 3y(-7) + 2(3y) + 2(-7)$	FOIL, done mentally.
	$= 9y^2 - 15y - 14$	Multiply and combine terms. ■

EXAMPLE 9

Multiply: $(5x - 7y)(6x + 9y)$

$(5x - 7y)(6x + 9y) = 30x^2 + 3xy - 63y^2$ — Done mentally. Your goal should be to do as many as possible mentally. ■

To solve some equations, it is useful to remove parentheses first.

PROCEDURE

> To solve an equation that involves multiplication of polynomials:
>
> 1. Multiply the polynomials.
> 2. Combine like terms.
> 3. Solve as before.

EXAMPLE 10

Solve: $4(x - 3) + 7 = 2(5 - x)$

$4x - 12 + 7 = 10 - 2x$	Distributive property.
$4x - 5 = 10 - 2x$	Combine terms.
$6x - 5 = 10$	Add $2x$ to both sides.
$6x = 15$	Add 5 to both sides.
$x = \dfrac{5}{2}$ or $x = 2.5$	

The solution set is $\left\{\dfrac{5}{2}\right\}$. Check left for student. ■

EXAMPLE 11

Solve: $(x + 2)(x^2 + x + 1) = (x + 1)(x^2 - x + 2) + 3x^2 - 1$

$x^3 + 3x^2 + 3x + 2 = x^3 + x + 3x^2 + 1$	Multiply and combine terms.
$3x = x - 1$	Subtract x^3, $3x^2$, and 2 from each side.
$2x = -1$	Subtract x from each side.
$x = -\dfrac{1}{2}$	

The solution set is $\left\{-\dfrac{1}{2}\right\}$. Check is left for student. ■

EXAMPLE 12

Solve: $(y - 3)(y + 4) = (y + 6)(y - 8)$

$y^2 + y - 12 = y^2 - 2y - 48$	Multiply.
$y - 12 = -2y - 48$	Subtract y^2 from both sides.
$3y - 12 = -48$	Add $2y$ to both sides.
$3y = -36$	Add 12 to both sides.
$y = -12$	Divide both sides by 3.

The solution set is $\{-12\}$. Check is left for the student. ■

EXAMPLE 13

This year Mary is twice as old as her sister Jane. Five years from now, the product of their ages will be 85 more than the product of their ages now. How old is each today?

Simpler word form:

$$\begin{pmatrix}\text{Mary's}\\ \text{age in}\\ \text{5 years}\end{pmatrix}\cdot\begin{pmatrix}\text{Jane's}\\ \text{age in}\\ \text{5 years}\end{pmatrix}=\begin{pmatrix}\text{Mary's}\\ \text{age now}\end{pmatrix}\cdot\begin{pmatrix}\text{Jane's}\\ \text{age now}\end{pmatrix}+85$$

First, write the simpler word form.

Select a variable:

Let x represent Jane's age.

Second, if x represents Jane's age now, then $2x$ represents Mary's age now.

Third, make a chart.

	Age Now	Age in 5 Years
Jane	x	$x + 5$
Mary	$2x$	$2x + 5$

Translate to algebra:

$$(2x + 5)(x + 5) = (2x)(x) + 85$$

Fourth, translate the simpler word form to an equation.

Solve:

$$(2x + 5)(x + 5) = (2x)(x) + 85$$
$$2x^2 + 15x + 25 = 2x^2 + 85$$
$$15x + 25 = 85$$
$$15x = 60$$
$$x = 4$$

Fifth, solve the equation.

Answer:

Since x represents Jane's age today, she is now 4 years old. Since $2x$ represents Mary's age today, she is now 8 years old.

Sixth, answer the question. ■

EXERCISE 3.7

A

Multiply:

1. $5(2x - 3)$
2. $8(4k - 7)$
3. $2x(x - 3)$
4. $3a(4a - 5)$
5. $7c(4a - 3b + 4c)$
6. $9a(4a + 5b - 8c)$
7. $2x^2y(3x - y + 1)$
8. $-ab(3a^2b - 2ab^2 + 2b)$
9. $0.8xyz(-10x^2y^2 + 30x^2z^2 - 50y^2z^2)$
10. $0.25abc(20a^2bc - 40bc^3 + 80b^3c^3)$
11. $0.5x^2(-4x^3y^3 + 5x^2y^2 - 0.7xy)$
12. $20b^4(-0.2b^2c^2 + 0.9a^2b^3 - 0.6a^5c^2)$
13. $(x - 1)(x - 2)$
14. $(x - 3)(x - 2)$
15. $(z + 5)(z + 3)$
16. $(c - 4)(c + 1)$
17. $(z + 8)(z + 12)$
18. $(c - 9)(c + 7)$
19. $(a - 12)(a - 6)$
20. $(b + 9)(b + 12)$
21. $(y + 5)(y + 2)$
22. $(k - 8)(k - 5)$

Solve:

23. $2(x + 3) + 4(x - 5) = 16$
24. $4(5y - 9) - 2(6 - y) = 62$
25. $-5(2a - 4) - 3(4a + 4) = 24$
26. $-2(3x - 7) + 7(x + 4) = 32$

B

Multiply:

27. $(x + y)(x - y)(x + y)$
28. $(x - y)(x + y)(x - y)$
29. $(2x + 3)(x^2 - 5x - 6)$
30. $(3a + 2)(a^2 - 6a - 5)$
31. $(5z - 3)(z^2 - 4z + 7)$
32. $(3y - 5)(y^2 + 7y - 4)$
33. $\begin{array}{r} 4x^3 + 3xy + 2 \\ x - y \\ \hline \end{array}$
34. $\begin{array}{r} 3a^3 + 4ab - 1 \\ a - b \\ \hline \end{array}$
35. $(2x - 3)(3x - 4)(3x + 1)$
36. $(4x + 3)(2x - 5)(x + 1)$
37. $(a - 3b)(4a + b)(2a - b)$
38. $(x + 2y)(3x - y)(x + 2y)$
39. $(3y - 1)(y + 4)$
40. $(5c - 3)(c - 6)$
41. $(4a - 3)(a - 5)$
42. $(6y - 1)(y - 4)$
43. $(2b + 1)(3b + 2)$
44. $(4b + 3)(2b + 1)$

45. $(4x + 5)(2x - 3)$
46. $(5z + 2)(4z - 3)$
47. $(5 - 3s)(1 + 7s)$
48. $(6 - 9a)(4 + 3a)$
49. $(8 - 5c)(7 - 3c)$
50. $(12 + 4c)(6 + 8c)$

Solve:

51. $4x(x + 3) - 8 = 2x(2x - 9) - 6$
52. $2w(3w + 1) = w(w - 8) + 5w(w + 1) - 17$
53. $3x(4x - 5) - 12 = 6x(2x - 3)$
54. $6a(3a - 4) + 8 = 2a(9a - 6)$
55. $(x - 6)(x + 4) = x^2$
56. $(y + 3)(y - 5) = y^2 + 1$
57. $(x - 7)(x - 3) = x^2 - 1$
58. $(x - 5)(x - 4) = x^2 + 7$

C

Multiply:

59. $(x - 2)(x^2 + 2x + 4)$
60. $(x + 3)(x^2 - 3x + 9)$
61. $(2a - b)(4a^2 + 2ab + b^2)$
62. $(4x + y)(16x^2 - 4xy + y^2)$
63. $(x - y)(4x^3 + 3xy + 2)$
64. $(a - b)(3a^3 + 4ab - 1)$
65. $(0.6x^2 + 0.5x + 0.3)(0.2x - 0.3)$
66. $(0.8y^2 + 0.6y + 0.5)(0.3y - 0.1)$
67. $\left(\frac{1}{2}x - \frac{1}{4}y\right)\left(\frac{1}{3}x^2 + 4xy + \frac{1}{5}y^2\right)$
68. $\left(\frac{1}{5}a - \frac{1}{3}b\right)\left(\frac{1}{6}a^2 + \frac{1}{8}ab + \frac{1}{4}b^2\right)$
69. $(x + y + z)(x + y - z)$
70. $(a + b - 2c)(2a + 3b - c)$
71. $(5x - 9)(4x + 3)$
72. $(6x - 5)(8x + 7)$
73. $(7x - 4)(4x + 7)$
74. $(8a - 9)(9a - 8)$
75. $(5a - 3b)(4a + 2b)$
76. $(6y - 5z)(3y + 8z)$
77. $(0.5x + 0.2)(0.4x + 0.1)$
78. $(0.6m + 0.5)(0.3m + 0.2)$
79. $\left(\frac{1}{4}x - \frac{1}{3}\right)\left(\frac{1}{3}x - \frac{1}{4}\right)$
80. $\left(\frac{1}{5}a - \frac{1}{2}\right)\left(\frac{1}{2}a - \frac{1}{5}\right)$
81. $(0.8 - 0.7x)(0.1 + 0.3x)$
82. $(0.6 + 0.4x)(0.5 + 0.1x)$

Solve:

83. $(x - 2)(x^2 + 2x + 4) = x^3 + 2x + 8$
84. $(x + 2)(x^2 - 2x + 4) = x^3 + 4x + 6$
85. $(x + 4)(x^2 - 4x + 16) = x^3 - 5x + 9$
86. $(x - 5)(x^2 + 5x + 25) = x^3 + 6x - 1$
87. $(x - 7)(x + 5) = (x + 8)(x + 1)$
88. $(y - 6)(y + 4) = (y + 7)(y + 2)$
89. $(2a + 3)(a - 1) = 2a(a + 2) - 5$
90. $(3b + 4)(b - 2) = 3b(b + 2) - 5$
91. $(3x - 7)(2x + 3) = (6x + 1)(x - 5)$
92. $(2x + 5)(4x - 2) = (8x - 2)(x + 7)$
93. $(d - 1)(2d + 1) = 2d(d - 5) + 17$
94. $(3x - 5)(x + 1) = (5x + 1)(x - 6) - 2x^2$

D

95. Express the area of the following rectangle in terms of w and simplify.

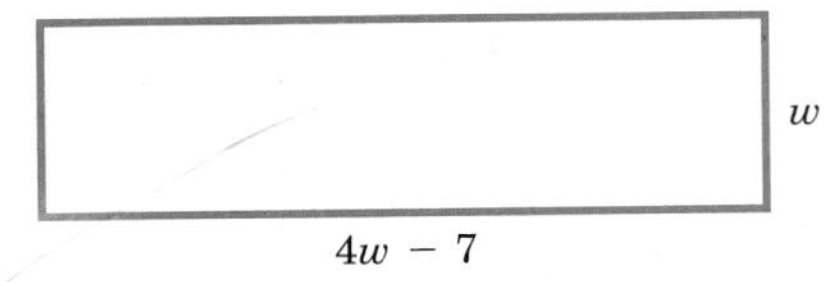

Figure for Exercise 95

Find the area when w is 10 meters.

96. Express the area of the following rectangle in terms of w and simplify.

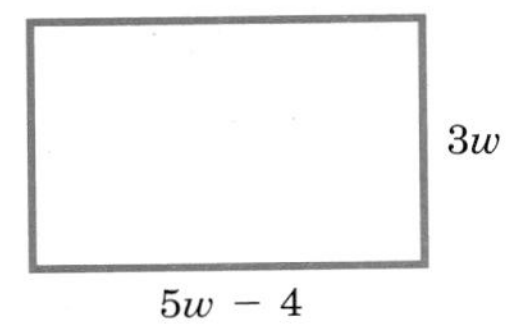

Figure for Exercise 96

Find the area when w is 4 feet.

97. Four less than a number is multiplied by three. This is subtracted from twice the number. If the difference is 7, find the number.

98. Mary earns one dollar an hour more than John earns per hour. If together they earn \$96.80 in an eight-hour day, how much does each earn per hour?

99. Two cars leave Miami along the same route, with Car A going 60 mph and Car B going 55 mph. If Car B leaves two hours before Car A, how long will it take Car A to be within 50 miles of Car B?

100. Spokane and Seattle are 512 miles apart. A car leaves Spokane and heads to Seattle, and at the same time a car leaves Seattle for Spokane. If one car travels 15 mph faster than the other one and they are 350 miles apart at the end of two hours, how fast is each car traveling?

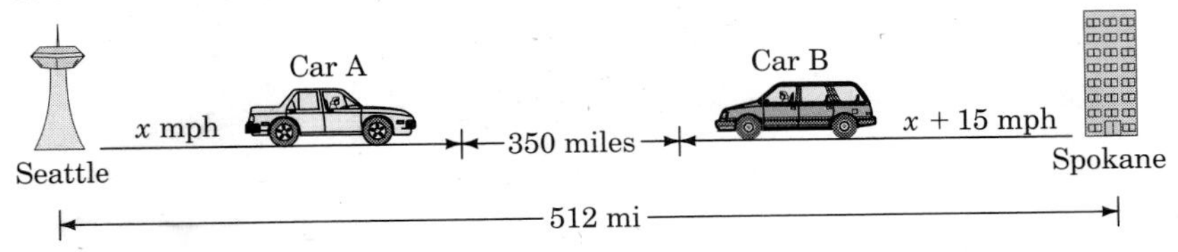

Figure for Exercise 100

101. If two more than a number is multiplied by five less than the same number, the result is four less than the square of the number. What is the number?

102. If six more than five times a number is multiplied by two less than the number, the result is five times the square of the number. Find the number.

103. If a number is multiplied by seven more than the number, the result is 21 less than the square of the number. Find the number.

104. If three less than a number is multiplied by two less than the same number, the product is ten less than the square of the number. What is the number?

105. The product of three times a number and one less than the number is equal to the product of eight less than the number and one more than three times the number. Find the number.

106. Mrs. DeLang is nine times as old as her daughter. In four years, she will be five times as old. Find their ages today.

107. Today Pete is 3 years younger than Jean. In 6 years, the product of their ages will be 138 more than the product is today. Find the ages of each today.

108. Annie is three times as old as Sally is today. Five years ago, the product of their ages was 275 less than the product today. Find the ages of each today.

109. Mr. Wisner invests $12,000. Part of the money is invested at 10% and part at 12%. The annual income is $1332. Find the amount invested at each rate.

110. Mr. Wisner's brother also invests $12,000. Part of the money is invested at 8% and part at 15%. His annual income is $1520. Find the amount invested at each rate.

STATE YOUR UNDERSTANDING

111. What is the FOIL procedure?

112. Whenever two polynomials are multiplied, the degree of the product is equal to the sum of the degrees of the two polynomials. Why is this true?

CHALLENGE EXERCISES

Multiply and simplify:

113. $(3a - 2b)(9a^2 + 6ab + 4b^2) - (2a + b)(4a^2 - 2ab + b^2)$

114. $5x(3x^n - 4x^{n-1} - 2x^{n-2})$

115. $(4x^n - 3)(2x^n + 1)$

116. $(x - 2)(3x^n - 4x^{n-1})$

117. $(x^{a+1} + 5)(x^{a-1} - 2)$

MAINTAIN YOUR SKILLS (SECTIONS 3.1, 3.2, 3.3)

Multiply or divide and write using positive exponents:

118. $\left(\frac{a^7}{a^2}\right)^5$ **119.** $\left(\frac{x^3 \cdot y^7}{x^{-4} \cdot y^6}\right)^{-2}$ **120.** $\left(\frac{3x^{-3}}{x^2y^4}\right)^{-2}$ **121.** $\left(\frac{4x^{-5}y^{-7}}{4^2x^3y^{-2}}\right)^{-5}$

Write in scientific notation:

122. 38.5 **123.** 123,000 **124.** 0.00000034

Change to place value notation:

125. 5.99×10^7

3.8

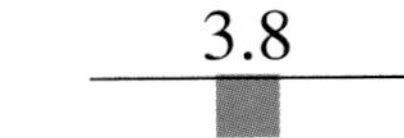

SPECIAL BINOMIAL PRODUCTS

OBJECTIVES

1. Find the product of conjugate binomials.
2. Find the square of a binomial.

Conjugate Binomials

Consider the binomials $x - 2$ and $x + 2$. They are the sum and difference of the same terms. They are called *conjugate binomials*. We can find their product using the **F O I L** method. The following chart, using the **F O I L** method, leads us to a further shortcut.

Conjugates	F	O	I	L	Products
$(x - 3)(x + 3)$	x^2	$+\ 3x$	$-\ 3x$	$-\ 9$	$x^2 - 9$
$(x + 5)(x - 5)$	x^2	$-\ 5x$	$+\ 5x$	$-\ 25$	$x^2 - 25$
$(b + 10)(b - 10)$	b^2	$-\ 10b$	$+\ 10b$	$-\ 100$	$b^2 - 100$
$(2x - 3)(2x + 3)$	$4x^2$	$+\ 6x$	$-\ 6x$	$-\ 9$	$4x^2 - 9$
$(4x + 7)(4x - 7)$	$16x^2$	$-\ 28x$	$+\ 28x$	$-\ 49$	$16x^2 - 49$

Difference of Two Squares

In each case, the sum of the **O** and **I** terms is zero. These terms may be omitted in the multiplication process. The product of the sum and difference of two terms is often called the *difference of two squares*. So,

$$(x + 2)(x - 2) = x^2 - 4$$

product of conjugates = difference of two squares.

PROCEDURE

To find the product of the sum and the difference of two terms, write the square of the first term minus the square of the second term.

RULE

$$(x - a)(x + a) = x^2 - a^2$$

EXAMPLE 1

Multiply: $(x - 7)(x + 7)$

$(x - 7)(x + 7) = x^2 - 49$ The product of conjugate pairs is the difference of two squares. ■

EXAMPLE 2

Multiply: $(3x + 7)(3x - 7)$

$(3x + 7)(3x - 7) = 9x^2 - 49$ ■

To find a shorter procedure for the square of a binomial, we first use the FOIL method.

$(x - 3)^2 = (x - 3)(x - 3)$

	F	O I	L
$=$	x^2	$-3x - 3x +$	9
		$2(-3x)$	
	SQUARE OF FIRST TERM	DOUBLE THE PRODUCT OF THE TERMS	SQUARE OF LAST TERM

$$(x - 3)^2 = x^2 - 6x + 9$$

Follow the pattern in the following chart:

	Square of First Term		Double the Product of the Terms		Square of Last Term		Product
$(x + 4)^2 =$	$(x)^2$	$+$	$2(4)(x)$	$+$	4^2	$=$	$x^2 + 8x + 16$
$(x - 6)^2 =$	$(x)^2$	$+$	$2(-6)(x)$	$+$	$(-6)^2$	$=$	$x^2 - 12x + 36$
$(3a - 1)^2 =$	$(3a)^2$	$+$	$2(-1)(3a)$	$+$	$(-1)^2$	$=$	$9a^2 - 6a + 1$
$(4y + 5)^2 =$	$(4y)^2$	$+$	$2(5)(4y)$	$+$	$(5)^2$	$=$	$16y^2 + 40y + 25$
$(b - 10)^2 =$	$(b)^2$	$+$	$2(-10)(b)$	$+$	$(-10)^2$	$=$	$b^2 - 20b + 100$

These steps can be done mentally, so:

$(x - 12)^2 = x^2 - 24x + 144.$

Perfect Square Trinomial

The square of a binomial is called a *perfect square trinomial.*

■ PROCEDURE

To square a binomial, write the sum of:

1. The square of the first term
2. Double the product of the first and last terms
3. The square of the last term

■ RULE

$(a + b)^2 = a^2 + 2ab + b^2$

$(a - b)^2 = a^2 - 2ab + b^2$

■ CAUTION

$(a + b)^2 \neq a^2 + b^2$

$(a - b)^2 \neq a^2 - b^2$

EXAMPLE 3

Multiply: $(x + 3)^2$

$$(x + 3)^2 = x^2 + 2(3 \cdot x) + 9$$
$$= x^2 + 6x + 9$$

Recall that $(x + 3)^2 = (x + 3)(x + 3)$. Write the sum of (1) the square of the first term, (2) double the product of the first and last terms, and (3) the square of the last term. ■

EXAMPLE 4

Multiply: $(3x - 7)^2$

$$(3x - 7)^2 = (3x)^2 + 2(3x)(-7) + (-7)^2$$
$$= 9x^2 - 42x + 49$$

To solve equations in which special products are involved, remove parentheses first by multiplying.

EXAMPLE 5

Solve for x: $(x - 4)(x + 4) = (x - 2)^2$

$(x - 4)(x + 4) = (x - 2)^2$

$x^2 - 16 = x^2 - 4x + 4$ Multiply.

$4x = 20$

$x = 5$

The solution set is $\{5\}$. The check is left for the student. ■

EXAMPLE 6

Hurtado owns a square corner lot in Frendtown. The city planning commission has sent him notice that he must give up a 15-foot strip on each of two adjacent sides of his lot to provide for new streets in his neighborhood. This will reduce the area of his lot by 3075 square feet. What was the original length of each side of his lot?

Simpler word form:

original area − new area = area lost

Select variable:

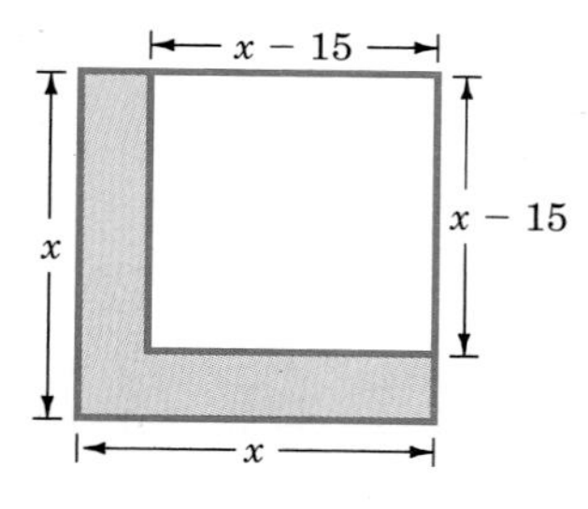

original area $= x^2$

new area $= (x - 15)^2$

Let x represent the original length of a side. Then $x - 15$ represents the length of a side of the new square.

Translate to algebra:

$x^2 - (x - 15)^2 = 3075$

Solve:

$$x^2 - (x - 15)^2 = 3075$$

$$x^2 - (x^2 - 30x + 225) = 3075$$

$$x^2 - x^2 + 30x - 225 = 3075$$
$$30x - 225 = 3075$$
$$30x = 3300$$
$$x = 110$$

Answer:

Hurtado's original lot was 110 ft on each side. ■

EXERCISE 3.8

A

Multiply:

1. $(x - 8)^2$
2. $(y - 4)^2$
3. $(a + 9)(a - 9)$
4. $(x + 5)(x - 5)$
5. $(b + 10)^2$
6. $(c + 12)^2$
7. $(d - 6)(d + 6)$
8. $(m - 3)(m + 3)$
9. $(z + 10)(z - 10)$
10. $(w + 11)(w - 11)$
11. $(x + y)^2$
12. $(w - z)^2$
13. $(x + y)(x - y)$
14. $(w + z)(w - z)$
15. $(a - 12)(a + 12)$
16. $(a + 12)^2$

B

Multiply:

17. $(2x + 3)^2$
18. $(3x + 2)^2$
19. $(2c - 5)(2c + 5)$
20. $(5b + 2)(5b - 2)$
21. $(3a - 4)^2$
22. $(4a + 3)^2$
23. $(6k + 7)(6k - 7)$
24. $(7m + 6)(7m - 6)$
25. $(5y - 9)(5y + 9)$
26. $(9y + 5)(9y - 5)$
27. $(8 - 3c)(8 + 3c)$
28. $(7 + 4a)(7 - 4a)$
29. $(12 - y)^2$
30. $(15 + z)^2$

Solve:

31. $(x - 2)^2 = x^2$
32. $(x + 2)^2 = x^2$
33. $(x + 3)^2 = x^2 - 3$
34. $(x - 5)(x + 3) = x^2 + 7$
35. $(x + 2)(x - 2) = x^2 + x$
36. $(a - 3)(a + 3) = a^2 + 2a$

C

Multiply:

37. $(x + y)^2$ **38.** $(m - n)^2$ **39.** $(b - c)(b + c)$

40. $(m + n)(m - n)$ **41.** $(3a + 2b)(3a - 2b)$ **42.** $(4x - 5y)(4x + 5y)$

43. $(3x - 2y)^2$ **44.** $(4c - 3d)^2$ **45.** $(4c + 3d)^2$

46. $(6k - 3b)^2$

Solve:

47. $(x + 2)(x - 2) = (x + 4)^2$ **48.** $(x - 5)(x + 5) = (x + 5)^2$

49. $(x - 8)^2 = (x - 4)(x + 4)$ **50.** $(x - 7)^2 = (x + 9)(x - 9)$

51. $(x - 2)^2 = (x + 2)^2$ **52.** $(x - 4)^2 = (x + 6)^2$

53. $(x + 4)^2 = (x + 3)^2 - 5$ **54.** $(x - 4)^2 = (x - 3)^2 + 5$

55. $(x - 5)^2 - (x - 4)^2 = 5$ **56.** $(x + 5)^2 - (x + 4)^2 = 5$

D

A method that is sometimes used to find certain products is to write the factors as conjugate pairs. In these cases, the answers can often be found mentally. For example,

$$\begin{aligned}(38)(42) &= (40 - 2)(40 + 2)\\ &= 40^2 - 2^2\\ &= 1600 - 4\\ &= 1596\end{aligned}$$

Use this method to find the following products:

57. (43)(37) **58.** (72)(68) **59.** (99)(101) **60.** (155)(165)

61. If the length of each of the sides of a square is increased by 2 ft, the area is increased by 44 sq ft. What is the length of each side of the original square?

62. If the length of each of the sides of a square is increased by 6 in., the area is increased by 108 sq in. What is the length of each side of the original square?

63. If one side of a square is increased by three centimeters and an adjacent side by one centimeter, the area of the new rectangle will be 27 square centimeters more than the area of the original square. Find the dimensions of the original square.

64. The height of a triangle and a rectangle are equal. The base of the triangle is twice its height, whereas the length of the rectangle is seven more than its height. The area of the rectangle is 14 square feet more than the area of the triangle. Find the length of the rectangle.

65. Abdul owns a square lot that is 656 sq ft too small to build a house on. If he purchased a strip 4 ft wide on two adjacent sides, he could then build his house. What is the present size of his lot?

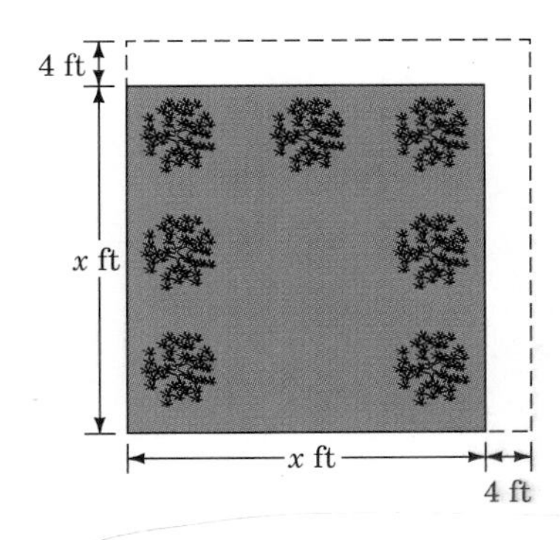

Figure for Exercise 65

66. The plans of a house indicate that the utility room is a square. If the lengths of two adjacent sides are increased by 2 feet, the area will be increased by 36 sq ft. What is the size of room that is indicated on the plan?

67. The difference of the squares of two consecutive odd integers is 32. Find the integers.

68. The product of the first two of three consecutive even integers is 76 less than the square of the third consecutive even integer. Find the integers.

69. A square pantry and a rectangular hallway occupy the same area. The length of the hallway is 7 feet greater than a side of the pantry, while the width of the hallway is $3\frac{1}{2}$ feet less than a side of the pantry. What are the dimensions of each?

STATE YOUR UNDERSTANDING

70. What is the pattern for squaring a binomial?

71. What is meant by conjugate binomials?

CHALLENGE EXERCISES

Multiply:

72. $(x + 5)^3$

73. $(a - 2)^4$

74. $(y^{2n} - y^n)^2$

75. $(x^n - 3)^3$

MAINTAIN YOUR SKILLS (SECTIONS 2.5, 3.5)

76. $6.5xyz + 3.4xy - 1.9xyz$

77. $9ab - 13ab - 24bc$

78. $\frac{7}{8}st - \frac{3}{4}st + \frac{5}{16}st$

79. $-\frac{5}{3}x^3 - \frac{8}{9}x^3 + \frac{5}{12}x^3$

80. $3.78mn - 7.89mn - 1.23mn + 5.67mn$

81. $5x^2y - 13x^2y + 17x^2y - 34xy$

Solve:

82. $15x - 9x + 13 = 49$

83. $21x - 54 = 13x + 19$

3.9

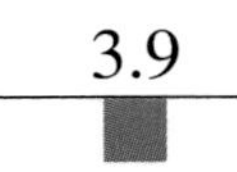

DIVIDING POLYNOMIALS

OBJECTIVES

1. Divide two monomials.
2. Divide a polynomial of more than one term by a monomial.
3. Divide polynomials.

To divide monomials, we need to recall (1) how to reduce fractions and (2) a property of exponents (dividing like bases). The division problem $8b^5 \div 2b^3$ can be performed in the following way:

$8b^5 \div 2b^3 = \dfrac{8b^5}{2b^3}$	Write as a fraction.
$= \dfrac{8}{2} \cdot \dfrac{b^5}{b^3}$	Consider the fraction formed by the coefficients and the fraction formed by the like bases.
$= 4 \cdot b^{5-3}$	Reduce the fraction formed by the coefficients and use the property of exponents for dividing like bases.
$= 4b^2$	Simplify.

EXAMPLE 1

Divide: $16x^5y^3 \div 8xy^2$

$= \dfrac{16x^5y^3}{8xy^2}$	Rewrite as a fraction.
$= \dfrac{16}{8} \cdot \dfrac{x^5}{x} \cdot \dfrac{y^3}{y^2}$	Rewrite as the product of fractions.
$= 2x^{5-1}y^{3-2}$	When dividing like bases, subtract the exponents.
$= 2x^4y$	

■

EXAMPLE 2

Divide: $(48x^{16}y^{10}) \div (6x^4y^9)$

$\dfrac{48x^{16}y^{10}}{6x^4y^9}$	Rewrite as a fraction.
$= \dfrac{48}{6} \cdot \dfrac{x^{16}}{x^4} \cdot \dfrac{y^{10}}{y^9}$	Rewrite as the product of fractions.

$$= 8x^{16-4}y^{10-9}$$

$$= 8x^{12}y$$

■

EXAMPLE 3

Divide: $\dfrac{35x^4yz^2}{15xy^3z^4}$

$$\frac{35x^4yz^2}{15xy^3z^4} = \frac{35}{15} \cdot \frac{x^4}{x} \cdot \frac{y}{y^3} \cdot \frac{z^2}{z^4}$$

Rewrite as a product of fractions.

$$= \frac{7}{3}x^{4-1}y^{1-3}z^{2-4}$$

Reduce the numerical fraction, and subtract exponents of like bases.

$$= \frac{7}{3}x^3y^{-2}z^{-2}$$

Simplify.

$$= \frac{7x^3}{3y^2z^2}$$

Definition of negative exponents. ■

To divide a polynomial of more than one term by a monomial we recall the rule for adding rational numbers (fractions).

$$\frac{a}{c} + \frac{b}{c} = \frac{a+b}{c}$$

Using the symmetric property of equality, we have:

$$\frac{a+b}{c} = \frac{a}{c} + \frac{b}{c}$$

This allows us to rewrite a division of a polynomial of more than one term by a monomial as the sum of two or more quotients.

EXAMPLE 4

Divide: $(15a^2b^3c - 25ab^2c^3 + 40abc) \div 5abc$

$$= \frac{15a^2b^3c - 25ab^2c^3 + 40abc}{5abc}$$

Write as a fraction.

$$= \frac{15a^2b^3c}{5abc} - \frac{25ab^2c^3}{5abc} + \frac{40abc}{5abc}$$

Divide each term by $5abc$.

$$= 3ab^2 - 5bc^2 + 8$$

Perform the three divisions. ■

Division of polynomials follows the same pattern as division of whole numbers.

Review the following division:

$$
\begin{array}{r l}
745 & \\
23\overline{)17135} & \\
\underline{16100} & 700 \times 23 \\
1035 & \\
\underline{920} & 40 \times 23 \\
115 & \\
\underline{115} & 5 \times 23 \\
0 &
\end{array}
$$

Now let's try it with polynomials.

EXAMPLE 5

Divide: $x + 3\overline{)x^2 + 7x + 12}$

$$
x + 3\overline{)x^2 + 7x + 12} \qquad x \cdot x = x^2
$$

Think: What times x will give a product of x^2?

$$
\begin{array}{r l}
x & \\
x + 3\overline{)x^2 + 7x + 12} & \\
\underline{x^2 + 3x \quad \downarrow} & x(x + 3) \\
4x + 12 & \\
 & 4 \cdot x = 4x
\end{array}
$$

Place the x over the x column and multiply x by $(x + 3)$, then subtract.

Think: What times x will give a product of $4x$?

$$
\begin{array}{r l}
x + 4 & \\
x + 3\overline{)x^2 + 7x + 12} & \\
\underline{x^2 + 3x } & x(x + 3) \\
4x + 12 & \\
\underline{4x + 12} & 4(x + 3) \\
0 &
\end{array}
$$

Place the 4 over the last column and multiply 4 by $(x + 3)$, then subtract.

Since there is no remainder, the quotient is $x + 4$. ■

EXAMPLE 6

Divide: $(12x^2 + 10x - 8) \div (3x + 4)$

$$
\begin{array}{r}
4x - 2 \\
3x + 4\overline{)12x^2 + 10x - 8} \\
\underline{12x^2 + 16x \quad \downarrow} \\
-6x - 8 \\
\underline{-6x - 8} \\
0
\end{array}
$$

Think: $\dfrac{12x^2}{3x} = 4x$

Multiply: $4x(3x + 4) = 12x^2 + 16x$.
Subtract.

Think: $\dfrac{-6x}{3x} = -2$

Multiply: $-2(3x + 4) = -6x - 8$
Subtract.

The quotient is $4x - 2$. ■

EXAMPLE 7

Divide: $(x^4 + 2x^3 + x^2 - 3x - 3) \div (x + 1)$

$$
\begin{array}{rlr}
 & \quad x^3 + x^2 \qquad\quad - 3 & \\
x + 1 & \overline{)x^4 + 2x^3 + x^2 - 3x - 3} & \\
 & \underline{x^4 + \ x^3} \quad \downarrow \quad \ \ | \quad\ \ | & x^3(x + 1) \\
 & \qquad x^3 + x^2 & \\
 & \qquad \underline{x^3 + x^2} \quad\ \downarrow \quad \downarrow & x^2(x + 1) \\
 & \qquad\qquad\quad - 3x - 3 & \\
 & \qquad\qquad\quad \underline{- 3x - 3} & -3(x + 1) \\
 & \qquad\qquad\qquad\qquad 0 &
\end{array}
$$

So the quotient is $x^3 + x^2 - 3$. ■

If the terms of the polynomials are not in descending order of powers or if some of the powers are missing, it is useful to rearrange them and insert zeros for the missing terms.

EXAMPLE 8

Divide: $(x^3 - y^3) \div (x - y)$

$$
\begin{array}{rl}
 & \quad x^2 \qquad + xy \ + y^2 \\
x - y & \overline{)x^3 + 0 \cdot x^2y + 0 \cdot xy^2 - y^3} \\
 & \underline{x^3 - \quad x^2y} \\
 & \qquad x^2y + 0 \cdot xy^2 \\
 & \qquad \underline{x^2y - \quad xy^2} \\
 & \qquad\qquad\quad xy^2 - y^3 \\
 & \qquad\qquad\quad \underline{xy^2 - y^3} \\
 & \qquad\qquad\qquad\qquad 0
\end{array}
$$

Since the terms containing x^2y and xy^2 are missing, we insert $0x^2y$ and $0xy^2$ for them. ■

If the division has a remainder, we will show it as: $\dfrac{\text{remainder}}{\text{divisor}}$. A remainder is identified when it has a degree that is smaller than the divisor.

EXAMPLE 9

Divide: $(x^2 + 7x - 8) \div (x + 3)$

$$
\begin{array}{rl}
 & \qquad x + \ 4 \\
x + 3 & \overline{)x^2 + 7x - \ 8} \\
 & \underline{x^2 + 3x} \\
 & \qquad 4x - \ 8 \\
 & \qquad \underline{4x + 12} \\
 & \qquad\quad -20
\end{array}
$$

So the quotient is $x + 4 + \dfrac{-20}{x + 3}$. ■

EXAMPLE 10

Divide: $(4x^4 + x^2 - x + 9) \div (2x - 3)$

$$
\begin{array}{r}
2x^3 + 3x^2 + 5x + 7 + \dfrac{30}{2x-3} \\
2x - 3\overline{)4x^4 + 0x^3 + x^2 - x + 9} \\
\underline{4x^4 - 6x^3} \qquad\qquad\qquad \\
6x^3 + x^2 \qquad\qquad \\
\underline{6x^3 - 9x^2} \qquad\qquad \\
10x^2 - x \qquad \\
\underline{10x^2 - 15x} \qquad \\
14x + 9 \\
\underline{14x - 21} \\
30
\end{array}
$$

Insert a placeholder for the term containing x^3.

There is a remainder of 30. ■

EXAMPLE 11

Divide: $(7x^2 + 9x + x^4 + 12 + 3x^3) \div (x^2 + 3)$

$$
\begin{array}{r}
x^2 + 3x + 4 \\
x^2 + 3\overline{)x^4 + 3x^3 + 7x^2 + 9x + 12} \\
\underline{x^4 \qquad + 3x^2} \qquad\qquad\quad \\
3x^3 + 4x^2 + 9x \qquad \\
\underline{3x^3 \qquad + 9x} \qquad \\
4x^2 \qquad + 12 \\
\underline{4x^2 \qquad + 12} \\
0
\end{array}
$$

Rewrite in descending order. Since the divisor $x^2 + 3$ has no term containing x, the product has no term containing x^3. Subtract 0 from $3x^3$. There is no remainder. ■

EXERCISE 3.9

A

Divide:

1. $\dfrac{24a^4}{4a^3}$
2. $\dfrac{-40b^7}{-8b^3}$
3. $\dfrac{42x^3y^5z^7}{-7x^2y^2z^2}$
4. $\dfrac{-56x^2y^7z^3}{4xy^5z^2}$
5. $(25y + 60y^2) \div 5y$
6. $(40a^3 - 88a^2) \div 4a^2$
7. $x - 1\overline{)8x - 8}$
8. $y + 2\overline{)6y + 12}$
9. $x + 1\overline{)x^2 + 4x + 5}$
10. $x + 2\overline{)x^2 + 5x + 6}$

11. $(a^2 - 2a - 3) \div (a + 1)$

12. $(b^2 - 3b + 4) \div (b + 1)$

13. $x - 5\overline{)x^2 - 7x + 10}$

14. $a + 3\overline{)a^2 + 10a + 21}$

15. $(x^2 + 8x + 15) \div (x + 5)$

16. $(x^2 - 8x + 12) \div (x - 6)$

B

Divide:

17. $\dfrac{-45a^3b^9c^5}{9a^2b^3c^4}$

18. $\dfrac{-150a^7b^2c^3}{-30a^5bc^2}$

19. $(40t^3 - 88t^2) \div (-4t^2)$

20. $(-84x^2 - 4x) \div (-4x)$

21. $(7x^2 + x - 8) \div (x - 1)$

22. $(16b^2 - 2b - 3) \div (2b - 1)$

23. $(8a^2 + 6ab - 9b^2) \div (4a - 3b)$

24. $(42x^2 + 23xy - 10y^2) \div (7x - 2y)$

25. $(x^2 + 8x - 10) \div (x - 5)$

26. $(x^2 - 9x + 5) \div (x - 6)$

27. $(2a^2 + a - 27) \div (a + 3)$

28. $(4x^2 + 12x + 7) \div (2x + 1)$

29. $(4b^2 - 4b - 5) \div (2b - 1)$

30. $(2y^2 - 3y - 15) \div (2y + 5)$

31. $(6x^2 + 3x - 1) \div (2x - 1)$

32. $(6y^2 + y - 7) \div (3y - 1)$

33. $(3x + 2x^2 - 50) \div (x - 5)$

34. $(5x - 4 + 3x^2) \div (x + 2)$

35. $(3a^2 - 1) \div (a + 2)$

36. $(5z^2 + 7) \div (z + 3)$

C

Divide:

37. $(48a^3b + 64ab^2 - 240a^2b) \div 16ab$

38. $(333w^2z^3 - 612w^2z + 99w^2) \div 9w^2$

39. $(-7x^2 + 2x^3 + 7x - 2) \div (x - 2)$

40. $(3x^3 - 25 + 16x^2) \div (x + 5)$

41. $(x^4 + x^2 + 3x + 1) \div (x + 1)$

42. $(2x^4 - 3x^3 - 5x^2 + 4x + 4) \div (x - 2)$

43. $(8x^3 + 32x + x^4 + 16 + 24x^2) \div (x + 2)$

44. $(81 + 54x^2 - 12x^3 + x^4 - 108x) \div (x - 3)$

45. $(x^3 - 1) \div (x - 1)$

46. $(x^4 - 1) \div (x - 1)$

47. $(x^3 - 1) \div (x + 1)$

48. $(x^5 - 1) \div (x + 1)$

49. $(a^5 - b^5) \div (a - b)$

50. $(a^7 + b^7) \div (a + b)$

51. $(x^3 + 6x^2 + 9x + 4) \div (x^2 + 2x + 1)$

52. $(x^3 - 7x^2 + 10x + 8) \div (x^2 - 3x - 2)$

53. $(x^4 - 10x^2 + 9) \div (x^2 + 4x + 3)$

54. $(x^4 - 9x^2 - 24x - 16) \div (x^2 + 3x + 4)$

55. $(8x^3 - 1) \div (2x - 1)$

56. $(27x^3 + 1) \div (3x + 1)$

D

57. A rectangle of width $y - 5$ inches has an area of $5y^2 - 23y - 10$ in^2. What is its length?

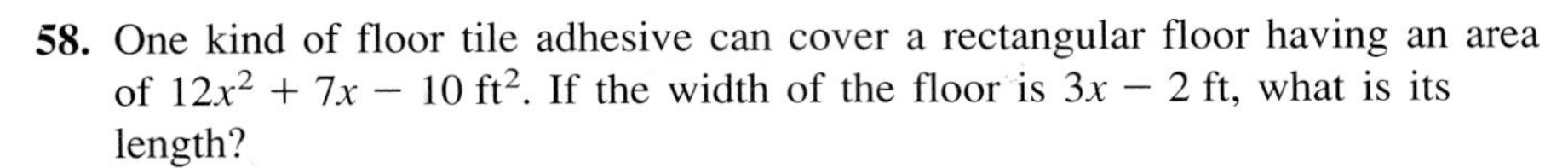

58. One kind of floor tile adhesive can cover a rectangular floor having an area of $12x^2 + 7x - 10$ ft^2. If the width of the floor is $3x - 2$ ft, what is its length?

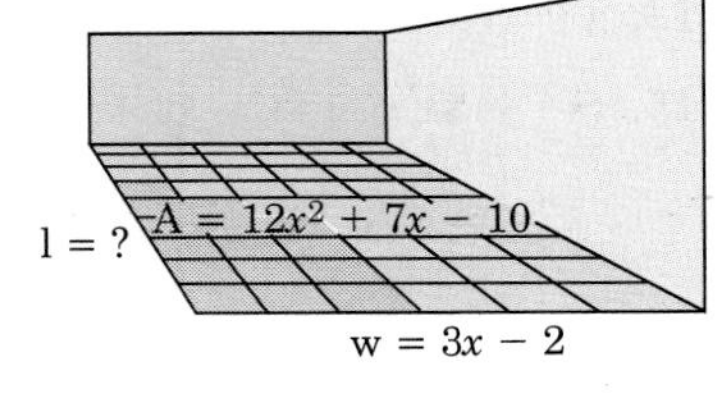

Figure for Exercise 58

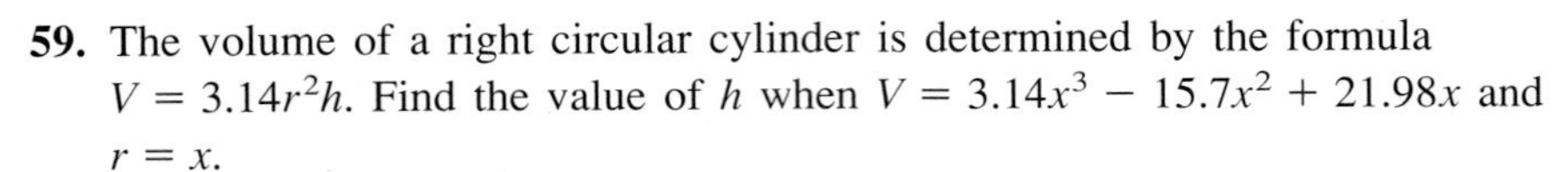

59. The volume of a right circular cylinder is determined by the formula $V = 3.14r^2h$. Find the value of h when $V = 3.14x^3 - 15.7x^2 + 21.98x$ and $r = x$.

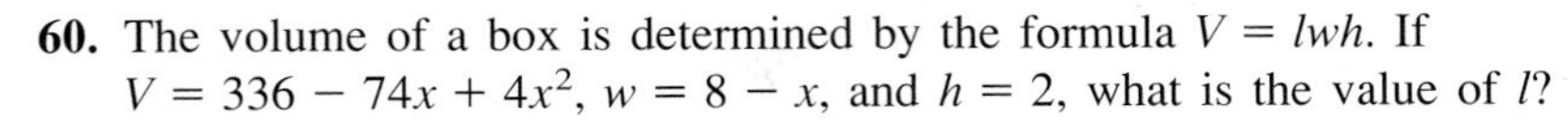

60. The volume of a box is determined by the formula $V = lwh$. If $V = 336 - 74x + 4x^2$, $w = 8 - x$, and $h = 2$, what is the value of l?

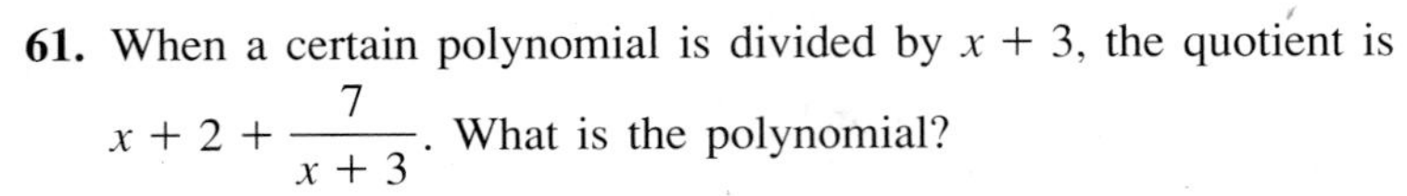

61. When a certain polynomial is divided by $x + 3$, the quotient is $x + 2 + \dfrac{7}{x + 3}$. What is the polynomial?

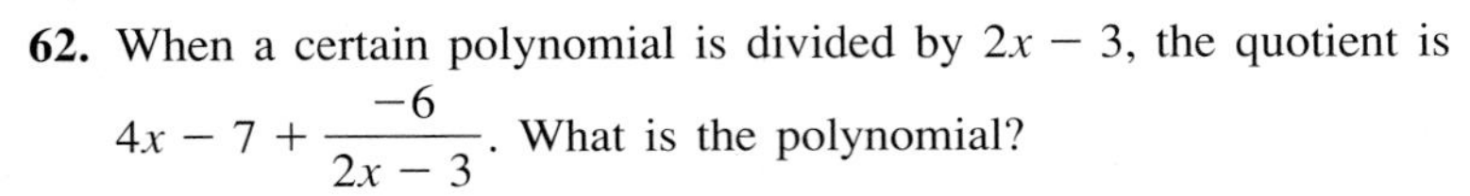

62. When a certain polynomial is divided by $2x - 3$, the quotient is $4x - 7 + \dfrac{-6}{2x - 3}$. What is the polynomial?

63. Find the number, n, so that when $6x^2 - x + n$ is divided by $2x - 5$, the remainder is 0.

64. Find the number, n, so that when $15x^2 - x + n$ is divided by $5x + 3$ the remainder is 8.

STATE YOUR UNDERSTANDING

65. How does one know when a division of polynomials problem has come to an end?

66. Why are placeholders inserted for missing terms when dividing by a polynomial?

CHALLENGE EXERCISES

Perform the indicated divisions. Simplify the answer completely.

67. $\dfrac{4(x - 3)^3 - 2(x - 3)^2 + (x - 3)}{(x - 3)}$

68. $(x^5 - 2x^3 - x^2 - 35x + 7) \div (x^2 - 7)$

69. $\left(2x^3 - 4x^2 + \dfrac{1}{2}x + \dfrac{1}{2}\right) \div (2x + 1)$

70. $(x^{2n} + 4x^n - 12) \div (x^n - 2)$

MAINTAIN YOUR SKILLS (SECTIONS 3.2, 3.5, 3.7)

Simplify:

71. $1.5x - 3.4x - 6.7x + 8x$

72. $16ab - 4.5ab + 12bc - 2.1bc$

73. $(-1.2xy)(-x^2y^3)(3xy^4)(-x^4y^2)$

74. $\dfrac{a^7b^{12}c^8}{a^6b^6c^5}$

Solve:

75. $6(-3x) - 7(2x) + 34 = 78$

76. $7(-3y) + 34 = 18 - 3(5y)$

77. A survey at the local theater found that 47% of the theatergoers were under 16 years old, 23% were 16 to 20 years old, and 5% were over 65. If the total number of these people was 600, how many people were in the survey?

78. At a performance of *The Mikado,* the same number of \$15 tickets and \$23 tickets were sold. If \$8550 was taken in from the sale of these tickets, how many of each were sold?

CHAPTER 3

SUMMARY

Multiplication Properties of Exponents	$a^m \cdot a^n = a^{m+n}$ Multiply like bases.	(p. 150)
	$(a^m)^n = a^{mn}$ Raise a power to a power.	(p. 153)
	$(ab)^m = a^mb^m$ Raise a product to a power.	(p. 154)
Negative Exponent	$x^{-n} = \dfrac{1}{x^n}$, n is a positive integer.	(p. 158)
Division Properties of Exponents	$\dfrac{a^m}{a^n} = a^{m-n}$ Divide like bases.	(p. 161)
	$\left[\dfrac{a}{b}\right]^n = \dfrac{a^n}{b^n}$ Raise a quotient to a power.	(p. 162)
Scientific Notation	$a \times 10^m$, $1 \le a < 10$ and m an integer.	(p. 166)
Monomial	A term that contains a number or a number multiplied times one or more variables.	(p. 173)
Degree of a Monomial	If a monomial contains only one variable, the degree of the monomial is the exponent of the variable. If the monomial contains more than one variable, the degree of the monomial is the sum of the exponents of the variables.	(p. 174)

Polynomial	An algebraic expression containing one or more monomials.	(pp. 175, 177)
Degree of a Polynomial	The degree of the term with the highest degree.	(pp. 175, 176)
Binomial	A polynomial that contains exactly two terms.	(p. 176)
Trinomial	A polynomial that contains exactly three terms.	(p. 176)
Combine Monomials (Combine Like Terms)	Add or subtract the numerical factors, and write the common variable factors in an indicated product.	(p. 180)
Add Polynomials	**1.** Rewrite subtractions as addition of opposites. **2.** Group like terms. **3.** Combine like terms.	(p. 181)
To Find the Opposite of a Polynomial	Change the sign of each term of the polynomial; that is, multiply each term by -1.	(p. 183)
Subtract Polynomials	Add the opposite of the polynomial being subtracted.	(p. 183)
Multiply a Monomial Times a Polynomial	Use the distributive property to write the products of the monomial and each term of the polynomial.	(p. 201)
Multiply Two Polynomials	**1.** Use the distributive property to multiply each term of one polynomial times each term of the other. **2.** Combine like terms.	(p. 202)
Multiply Two Binomials	Use the FOIL shortcut, that is, the products of the first terms, outer terms, inner terms, and last terms. Combine and simplify where possible.	(pp. 203, 204)
Multiply Conjugate Binomials	The product of the first terms minus the product of the last terms: $(a - b)(a + b) = a^2 - b^2$ The difference of two squares.	(pp. 211, 212)
Square of a Binomial	Square the first term, double the product of the first term and last term and the square of the last term: $(a + b)^2 = a^2 + 2ab + b^2$ $(a - b)^2 = a^2 - 2ab + b^2$	(pp. 212, 213)
Divide Monomials	Reduce the fraction formed by their coefficients, and use the exponents property for dividing like bases.	(p. 218)
Divide a Polynomial by a Monomial	Divide each term of the polynomial by the monomial.	(p. 219)
Write a Polynomial in One Variable in Descending Order	Write the term with the largest exponent of the variable first, followed by the next largest, and so on.	(pp. 177, 221)
Division of a Polynomial by Another Polynomial	Similar to long division with whole numbers.	(pp. 219, 220)
Termination of Division of Polynomials	The division of two polynomials is ended when all terms of the dividend have been used and the remainder is zero or when the degree of the divisor is smaller than the degree of the divisor.	(p. 221)

CHAPTER 3
REVIEW EXERCISES

SECTION 3.1 Objective 1

Multiply:

1. $a^5 \cdot a^2 \cdot a^3$
2. $x^2 \cdot x^0 \cdot x^5$
3. $x^a \cdot x^b$
4. $y^2 \cdot y^3 \cdot y^a$
5. $a \cdot a^x \cdot a^y$

SECTION 3.1 Objective 2

Simplify:

6. $(a^2)^4$
7. $(b^3)^5$
8. $(x^0)^5$
9. $(x^{10})^0$
10. $(x^a)^c$

SECTION 3.1 Objective 3

Simplify:

11. $(4a)^3$
12. $(5a^2)^3$
13. $(4x^2y^3)^3$
14. $(6ab^2c^3)^2$
15. $(2x^ay^b)^3$

SECTION 3.2 Objective 1

Rewrite using positive exponents:

16. a^{-4}
17. $(x^{-2}y^{-3})^2$
18. $(x^{-8})^{-3}$
19. $(4y^2)^{-2}$
20. $(2x^{-2}y^3)^{-3}$

SECTION 3.2 Objective 2

Multiply or divide as indicated:

21. $\dfrac{x^8}{x^5}$
22. $\dfrac{x^{-8}}{x^{-5}}$
23. $\dfrac{(4ab)^3}{4a^2b^2}$
24. $\dfrac{5x^3y^2z}{5^2x^5yz^3}$
25. $\dfrac{3^{-2}x^{-2}y^2}{3x^2y^2}$

SECTION 3.2 Objective 3

Simplify:

26. $\left[\frac{3}{4}\right]^3$

27. $\left[\frac{4x^2y^4}{16xy^2}\right]^{-3}$

28. $\left[\frac{15x^2y^3}{25xy}\right]^2$

29. $\left[\frac{4^{-1}x}{3^{-1}y^{-4}}\right]^3$

30. $\left[\frac{2^{-1}a^2}{3^{-1}b^2}\right]^{-2}$

SECTION 3.3 Objective 1

Write the following in scientific notation:

31. 150,000,000

32. 0.00000024

33. 0.0000035

34. 315

35. 1.3

SECTION 3.3 Objective 2

Write the following in place value notation:

36. 8.2×10^5

37. 6.1×10^{-8}

38. 5.4×10^{-3}

39. 7.1×10^4

40. 1.6×10^9

SECTION 3.4 Objective 1

Classify the following polynomials by degree:

41. $4x^5 + 5x^4 - 3x + 2$

42. $3x^4 + 4x^2 + 3$

43. 5

44. $6a$

45. $4x^2 + 3xy + 5y^2$

SECTION 3.4 Objective 2

Classify the following polynomials by the number of terms:

46. $5x$

47. $6x + 2$

48. $3x^2 + 8x - 1$

49. $8x^3 + 4x^2 + 5x$

50. $6x^3 + 4x^2 + 7x - 1$

SECTION 3.4 Objective 3

Write the following polynomials in descending order:

51. $2 - x$

52. $4x + x^2 - 3$

53. $x^3 + x^2 + 1 + x$

54. $8x^5 + 2x^3 - 3x^9 + 4x^2$

55. $-3x^5 + 4x^3 - 6x^2 + 5x - 4$

SECTION 3.5 Objective 1

Add:

56. $8xy + 7xy - 9xy$

57. $48x^3 - 32x^3 + 47x^3$

58. $2.5x^2y^3 - 3x^2y^3 + 0.4x^2y^3$

59. $15x^2y - 13xy^2$

60. $21abc + 36abc - 21ab$

SECTION 3.5 Objective 2

Add:

61. $(12a^3 + 4a^2 - 3a + 1) + (3a^3 - 2a^2 - 2a - 3)$

62. $(25x^2 - 3x + 1) + (4x - 3)$

63. $(2a^2 - 3a + 4) + (4a^2 - a - 1)$

64.
$$\begin{array}{r} 12x^2 + 5x - 9 \\ \underline{15x^2 - 3x + 2} \end{array}$$

65.
$$\begin{array}{r} 13a^2bc^2 - 7ab^2c + 7abc \\ \underline{-5a^2bc^2 + 9ab^2c - 11abc} \end{array}$$

SECTION 3.5 Objective 3

Find the opposite of each of the following:

66. $4x - 3y$

67. $x^2 - 7x - 5$

68. $-ab - b - c$

69. $-(x^2 + 3x + 2)$

70. 0

SECTION 3.5 Objective 4

Subtract:

71. $(4a - 3b) - (3a + 2b)$

72. $(6x^2 - 3x - 4) - (2x^2 + 4x - 4)$

73. $(x^2 - 3 + 4x) - (5 + x^2 + 4x)$

74.
$$\begin{array}{r} 3.5x^2 - 2.8x - 1.3 \\ \underline{0.5x^2 + 3.2x - 1.3} \end{array}$$

75.
$$\begin{array}{r} 12x^2 + 6x - 9 \\ \underline{3x^2 - 6x - 9} \end{array}$$

SECTION 3.5 Objective 5

Solve each of the following equations:

76. $(6x - 2) - (3x + 4) = 15$

77. $(3a - 12) - (4a - 9) = 3a + 5$

78. $4x - 7 - 3x + 5 = 2x + 3$

79. $5y - 12 - 3y - 8 = 4y + 16$

80. $(4x^2 - 3x + 12) - (2x^2 - 7x + 4) = 2x^2 + 6x + 10$

SECTION 3.6 Objective 1

Solve the following applications:

81. A collection of nickels and dimes contains as many dimes as nickels. If their total value is \$3.75, how many coins of each kind are in the collection?

82. An alcohol and water solution is 45% alcohol. How much water is in 8 gallons of the solution?

83. Car A and Car B start from the same place at the same time traveling in opposite directions. Car A travels at 38 mph, and Car B travels at 42 mph. In how many hours will they be 48 miles apart?

84. Ms. Smith has some money invested, part at 5% and part at 6%. The amount invested at 6% is \$2000 more than the amount invested at 5%. If the total interest from both investments for one year is \$560, how much is invested at each rate?

85. There are the same number of nickels, dimes, and quarters that make up a total of \$12. How many coins of each kind does it take to make up this amount of money?

SECTION 3.7 Objective 1

Multiply:

86. $(x + 2)(x^2 - 3x - 4)$

87. $(3x + 1)(9x^2 - 3x + 1)$

88. $(2x - 3)(4x^2 + 6x + 9)$

89. $(3x + y)(2x^2 + 3xy - y^2)$

90. $(4a - 2b)(3a^2 + 2ab + b^2)$

SECTION 3.7 Objective 2

Multiply:

91. $(3x + 4)(x - 5)$

92. $(4x + 3)(6x + 1)$

93. $(4a + 2b)(3a + b)$

94. $(5a - 1)(3a + 2)$

95. $(5x + 1)(x - 5)$

SECTION 3.7 Objective 3

Solve:

96. $5(x - 7) + 2 = 4(3 - x)$

97. $(x + 2)(x^2 + 3x - 5) = (x + 1)(x^2 + 4x - 2)$

98. $(y + 4)(y + 5) = (y + 4)(y + 1)$

99. $(x + 2)(x^2 - 2x + 1) = x^3 - 7$

100. $(2x - 5)(3x + 2) = (6x + 1)(x - 1)$

SECTION 3.8 Objective 1

Multiply:

101. $(3x + 2)(3x - 2)$

102. $(4a - 3b)(4a + 3b)$

103. $(5x - 2y)(5x + 2y)$

104. $(6ab + 5cd)(6ab - 5cd)$

105. $(25 - 3a)(25 + 3a)$

SECTION 3.8 Objective 2

Multiply:

106. $(3x + 4)^2$

107. $(5a - 2)^2$

108. $(6a - 5)^2$

109. $(4a + 3b)^2$

110. $(7x - 6y)^2$

SECTION 3.9 Objective 1

Divide:

111. $\dfrac{12x^5y^2}{4xy^2}$

112. $\dfrac{136x^3y^4z}{34x^5yz^2}$

113. $\dfrac{152a^4b^2c^3}{36ab^4c}$

114. $\dfrac{272xy^2z}{34x^2yz^2}$

115. $\dfrac{228a^4bc}{19a^4b^3}$

SECTION 3.9 Objective 2

Divide:

116. $\dfrac{48a^3b^2c^3 - 36ab^2c}{4abc}$

117. $\dfrac{75x^2y - 48xy^2}{15xy}$

118. $\dfrac{25a^2b^3c^2 - 40a^2b^2c^2 + 50abc}{5abc}$

119. $\dfrac{45x^2y^2z^3 + 36xy^2z^4 - 48xyz}{12x^2yz^2}$

120. $\dfrac{110a^3b^2c^2 + 88a^2b^3c^3 - 165a^2bc}{22a^3b^3c^3}$

SECTION 3.9 Objective 3

Divide:

121. $(8x^2 - 8x - 6) \div (2x - 3)$

122. $(15x^2 + 17x + 4) \div (5x + 4)$

123. $2x + 1\overline{)4x^3 + 18x^2 + 10x + 3}$

124. $3x - 5\overline{)6x^4 - 10x^3 + 15x^2 - 28x + 5}$

125. $2x + 3\overline{)6x^3 + 25x^2 - 36}$

CHAPTER 3
TRUE–FALSE CONCEPT REVIEW

Check your understanding of the language of algebra. Tell whether or not each of the following is true (always true) or false (not always true).

1. All multiplication problems involving exponents can be done by adding the exponents.
2. To raise a power to a power, we can always multiply the exponents.
3. All the laws of exponents can be applied to expressions even if they involve a (nonzero) term with an exponent of zero.
4. The expression $(-3)^{-2}$ represents a negative rational number.
5. In scientific notation, the number 5 is written 5×10^0.
6. In scientific notation, the number 623.5 is written 0.6235×10^{-3}.
7. Every algebraic expression is a polynomial.
8. The polynomial $x^{23} - 2x^{14} - 45x^5 + 3x - 102$ is written in descending powers.
9. A polynomial contains two or more terms.
10. A binomial is a polynomial of degree 2.
11. A trinomial is a polynomial that contains three terms.
12. The degree of a polynomial is the same as the degree of the term with the highest degree.
13. The coefficient of $-13x^3y^2$ is -13.
14. Like terms are terms containing the same variables.
15. The sum of $14x$ and $31x$ is $45x^2$.
16. The distributive property is used to add like terms.
17. The opposite of a polynomial can be found by multiplying the polynomial by negative 1.
18. To group the like terms in the expression $(3a - 4b + 1) + (2a - 5b - 10)$, the associative and commutative properties of addition are used repeatedly.
19. When multiplying two polynomials, one polynomial is multiplied by each term of the other.
20. The FOIL system is a shortcut for multiplying two polynomials.
21. The product of $(3x + 2)(2x - 3)$ is $6x^2 - 6$.
22. Two binomials that are the sum and difference of terms containing the same variables are conjugate binomials.
23. The polynomials $x^2 - y^2$ and $x^2 + y^2$ are conjugate binomials.
24. The square of a binomial is always a trinomial.
25. The quotient of $x^4 - y^4$ and $x - y$ is $x^3 - y^3$.
26. The product of a binomial and a trinomial is always a five-term polynomial.

CHAPTER 3
TEST

1. Simplify: $-5x + 3x + 11x$
2. Write $m^{-6}n^3$ with positive exponents.
3. Multiply: $(2x + 9)(2x - 9)$
4. Divide: $(x^2 - 15x + 56) \div (x - 8)$
5. Write 3.6×10^{-8} in place value notation.
6. Write $\frac{2}{x^{-5}}$ with positive exponents.
7. Add: $(12a - 3b + 4) + (8a - 9b - 12)$
8. Write 8,920,000 in scientific notation.
9. Multiply: $a^3 \cdot a^2 \cdot a$
10. Classify the following polynomial by its number of terms: $4x^3 - 3x^2y^2 + 16xy^2$
11. Solve: $(3m - 11) - (5m + 6) = 20$
12. Simplify: $(x^3y^2)^3$
13. Simplify: $\frac{a^{-3}a^6}{a^{-5}}$
14. Multiply: $(2x - 7)(x + 6)$
15. Write the following polynomial in descending powers: $4x - 8x^9 - 3x^2 + 6 + 5x^5 - 7x^3$
16. Solve: $6x - 5 = 8x - 7$
17. Multiply: $w^9 \cdot w^5 \cdot w^7$
18. Multiply: $-4xyz(-12x - 7y + 2z)$
19. Subtract: $\begin{array}{r} 4x - 5y + 8 \\ \underline{6x + 2y - 7} \end{array}$
20. Add: $(2x - 3y) + (4x + 6y) + (2y - 5x)$
21. Multiply: $(m - 10)^2$
22. Multiply: $(2a - 3b)(4a^2 + 3ab - 2b^2)$
23. Solve: $(x - 3)(x + 4) = x^2 + 4$
24. Multiply: $(y - 5)(y^2 + 3y + 2)$
25. Solve: $12x + 3 - 8x = -9 + 7x + 6$
26. Solve: $12x(x - 5) = (3x - 20)(4x + 3)$
27. Multiply: $(4x - 9)(9x + 4)$
28. Multiply: $4a(a^2 - 3a - 2)$
29. Subtract: $(14x^2 - 5x) - (6x - 5)$
30. Simplify: $\left(\frac{x^5}{y^4}\right)^4$
31. Solve: $(x - 3)(x + 3) = (x - 2)(x - 5) + 2$
32. Divide: $(x^3 - 3x + 6) \div (x + 1)$
33. Six less than five times a number is subtracted from three times the number. If the result is 12, find the number.
34. Two space shuttles, *Traveler* and *Voyager*, leave the same center at the same time, heading in the same direction. If *Traveler* averages 7000 mph and *Voyager* averages 8600 mph, how long will it take until they are 5600 miles apart?

CHAPTER

4

FACTORING POLYNOMIALS AND RELATED EQUATIONS

SECTIONS

The formula for the area of a rectangle is $A = lw$. If we know the sum of the length and the width, the area can be written as a quadratic equation. From the perimeter formula, $P = 2w + 2l$, we can see that half the perimeter is the sum of the length and the width. In Exercise 69, Section 4.7, the length of fence to enclose a rectangular pen for turkeys is given. Using the formulas above, write the area in terms of the length or width and then solve for the dimensions. *(Denny Tillman/The Image Bank)*

PREVIEW

Chapter 4 leads off with the zero-product property to allow the solution of equations solved by factoring. A thorough discussion of factoring follows. At the completion of the chapter, you will be able to factor a trinomial with integer coefficients or know that it is prime. Also included are special polynomials: differences of squares, perfect square trinomial, and the sum and differences of two cubes. The final section mixes up all the types of factoring to provide practice in selecting the correct method.

Factoring is used in Chapter 5 to simplify rational expressions and in Chapter 8 to solve other quadratic equations.

4.1

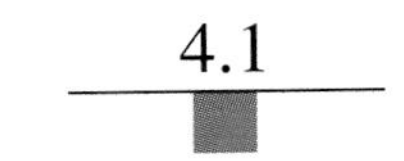

SOLVING EQUATIONS OF THE FORM $A \cdot B = 0$

OBJECTIVE

1. Solve a quadratic equation.

In Chapter 2, we saw how equations are classified by whether they are always true (identity), sometimes true (conditional), or never true (contradiction). Equations are also classified by degree. The degree of an equation in one variable is the largest exponent of any term in the equation.

Linear Equation (Degree One)

DEFINITION

A *linear equation* is an equation of degree one that can be written in the form $ax + b = c$, $a \neq 0$.

Quadratic Equation (Degree Two)

DEFINITION

A *quadratic equation* is an equation of degree two that can be written in the standard form $ax^2 + bx + c = 0$, $a \neq 0$.

Equations with degree greater than two are the subject of advanced algebra courses.

For equations in one variable, the degree of the equation also indicates the number of solutions of the equation. As we have seen, a linear equation (degree one) has one solution. A quadratic equation (degree two) has two solutions.

To solve a quadratic equation that is factored, we use the zero-product property.

PROPERTY

Zero-Product Property

> If two expressions have a product of zero, then at least one of the expressions is equal to zero. If $A \cdot B = 0$, then $A = 0$ or $B = 0$.

For example,

If $aw = 0$, then $a = 0$ or $w = 0$.
If $(x - 3)(y - 2) = 0$, then $x - 3 = 0$ or $y - 2 = 0$.
If $4s(s + 9) = 0$, then $4s = 0$ or $s + 9 = 0$.
If $(r + 3)(r - 5) = 0$, then $r + 3 = 0$ or $r - 5 = 0$.

If any factor contains a variable, then the equation has a solution when that factor is equal to zero. If there are two unlike factors with variables, there are two solutions.

EXAMPLE 1

Solve: $(2x - 1)(x + 6) = 0$

$2x - 1 = 0$ or $x + 6 = 0$ — Since the equation is in the form $A \cdot B = 0$, set each factor equal to zero using the zero-product property and solve.

$2x = 1$ or $x = -6$

$x = \frac{1}{2}$

Check: — Check in the original equation.

$\left[2\left(\frac{1}{2}\right) - 1\right]\left[\frac{1}{2} + 6\right] = 0$

$(1 - 1)\left(\frac{13}{2}\right) = 0$ — True since $1 - 1 = 0$.

$[2(-6) - 1][-6 + 6] = 0$ — True since $-6 + 6 = 0$.

The solution set is $\left\{-6, \frac{1}{2}\right\}$. ■

EXAMPLE 2

Solve: $(3x - 5)(2x - 1) = 0$

$3x - 5 = 0$ or $2x - 1 = 0$ — Zero-product property.

$$3x = 5 \quad \text{or} \quad 2x = 1$$

$$x = \frac{5}{3} \quad \text{or} \quad x = \frac{1}{2}$$

The solution set is $\left\{\frac{1}{2}, \frac{5}{3}\right\}$. The check is left for the student. ■

If the factors are identical, there is a single solution. Mathematicians call this single solution a *double root*. A double root is sometimes referred to as a *root of multiplicity two*.

EXAMPLE 3

Solve: $(m + 7)(m + 7) = 0$

$m + 7 = 0$ or $m + 7 = 0$ Zero-product property.
$m = -7$ or $m = -7$ A double root.

The solution set is $\{-7\}$. ■

Quadratic equations can be used to express some physical situations. For example, under certain conditions, the height of an object that is projected upward (a ball, a rocket, a bullet, etc.) can be found by the formula

$h = 16t(9 - t)$,

where h represents the height in feet and t represents the time in seconds.

EXAMPLE 4

How long will it take a ball, thrown from ground level, to return to the ground? The height of the ball at any time, t, is $h = 16t(9 - t)$.

Formula:

$h = 16t(9 - t)$

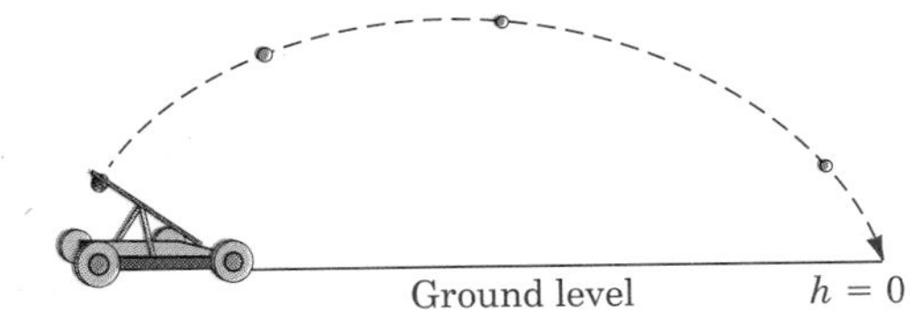

Substitute:

$0 = 16t(9 - t)$

At ground level, the height of the ball is 0 feet ($h = 0$).

Solve:

$16t = 0$ or $9 - t = 0$ Zero-product property.
$t = 0$ $t = 9$

The ball is on the ground initially, 0 seconds, and returns to the ground in 9 seconds. ■

EXERCISE 4.1

A

Solve:

1. $(x - 7)(x - 1) = 0$
2. $(x - 2)(x - 8) = 0$
3. $(y + 3)(y - 3) = 0$
4. $(y - 8)(y + 8) = 0$
5. $(w + 3)(w - 6) = 0$
6. $(w - 9)(w + 2) = 0$
7. $(t - 14)(t - 11) = 0$
8. $(t - 15)(t - 20) = 0$
9. $(r + 12)(r + 16) = 0$
10. $(r + 17)(r + 18) = 0$
11. $(x + 17)(x - 27) = 0$
12. $(x - 23)(x + 14) = 0$
13. $(y + 24)(y - 24) = 0$
14. $(y + 31)(y - 31) = 0$
15. $z(z - 21) = 0$
16. $z(z + 8) = 0$
17. $t(t + 33) = 0$
18. $t(t - 71) = 0$
19. $(x - 25)(x + 17) = 0$
20. $(x + 9)(x - 37) = 0$

B

Solve:

21. $(x - 6.2)(x - 3.4) = 0$
22. $(x - 2.6)(x - 8.5) = 0$
23. $(y + 3.4)(y - 0.6) = 0$
24. $(y + 2.2)(y - 9.1) = 0$
25. $(w + 3.3)(w - 3.3) = 0$
26. $(w - 14.7)(w + 14.7) = 0$
27. $t(t + 14.2) = 0$
28. $t(t - 21.8) = 0$
29. $\left(v + \frac{1}{3}\right)\left(v - \frac{2}{5}\right) = 0$
30. $\left(v - \frac{6}{7}\right)\left(v + \frac{7}{6}\right) = 0$
31. $\left(w - \frac{1}{2}\right)\left(w + \frac{2}{3}\right) = 0$
32. $\left(z + \frac{6}{7}\right)\left(2z - \frac{1}{3}\right) = 0$
33. $4s(s - 25) = 0$
34. $17s(s + 12) = 0$
35. $-2.7x(x - 2.7) = 0$
36. $-3.8x(x + 3.8) = 0$
37. $(30 - y)(40 - y) = 0$
38. $(27 - y)(33 + y) = 0$
39. $(18 - z)(z + 11) = 0$
40. $(2z + 3)(15 - z) = 0$

C

Solve:

41. $(2s - 5)(s - 6) = 0$
42. $(x + 8)(3x - 4) = 0$

43. $(2y - 9)(2y + 7) = 0$

44. $(3y - 1)(3y + 2) = 0$

45. $(5w - 11)(5w - 11) = 0$

46. $(4w + 3)(4w + 3) = 0$

47. $(8t + 7)(6t + 15) = 0$

48. $(8t - 6)(6t - 3) = 0$

49. $(8t + 6)(6t + 3) = 0$

50. $(5t + 1)(2t - 4) = 0$

51. $(9r + 6)(3r + 13) = 0$

52. $(4r - 33)(5r + 33) = 0$

53. $(7r + 91)(2r - 5) = 0$

54. $(3r - 105)(8r - 20) = 0$

55. $(6s - 46)(7s - 46) = 0$

56. $(10x - 7)(10x - 9) = 0$

57. $(10x - 1)(10x + 2) = 0$

58. $(12x + 5)(12x - 7) = 0$

59. $(100y - 663)(100y + 229) = 0$

60. $(100y + 511)(100y - 7) = 0$

D

61. Under certain conditions, the height of a falling object is given by the formula $16t(5 - t) = h$. At what time is the object on (or at) the ground ($h = 0$)? (See Example 4.)

62. Under certain conditions, the height of a falling object is given by the formula $16t(9 - 2t) = h$. At what time is the object on (or at) the ground?

63. Under certain conditions, the height of a falling object is given by the formula $16t(6 - 3t) = h$. At what time is the object on (or at) the ground?

64. One number is eight less than a second number, and their product is zero. What are the two numbers? (There are two sets of answers.) [*Hint:* Let x represent the first number. Let $x - 8$ represent the second number. The product of the numbers is 0, so $x(x - 8) = 0$.]

65. One number is 13 more than a second number. If their product is 0, what are the two numbers? (See hint in Exercise 64.)

66. One number is seven more than a second number, and a third number is five less than the second number. The product of all three numbers is zero. Find the three numbers. (There are three sets of answers.)

67. The formula for the height at any point of the arch of a bridge is $h = \frac{1}{100}d(140 - d)$, where h is the height in feet and d is the distance from one end of the arch in feet as shown in diagram (a).

How far is one end of the arch from the other in diagram (b)?

68. If the formula for the height, in feet, of the arch of a tunnel is $h = \frac{1}{121}d(161 - 2d)$, how far is one end of the arch from the other?

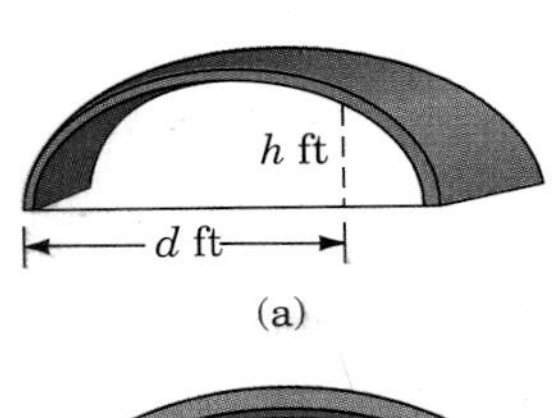

(a)

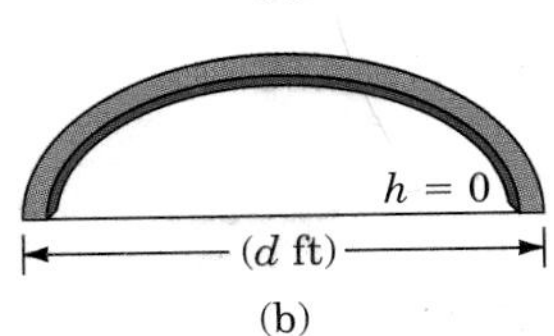

(b)

Figure for Exercise 67

STATE YOUR UNDERSTANDING

69. Explain the Zero-product Property.

70. Explain the difference between a linear (first degree) equation and a quadratic equation.

CHALLENGE EXERCISES

Solve the following equations.

71. $(2x - 5)(3x + 7)(5x + 8) = 0$

72. $x\left(\frac{1}{4} - \frac{2}{3}x\right)\left(\frac{5}{6} + \frac{7}{8}x\right) = 0$

73. $2\left(\frac{1}{x} + 2\right)\left(\frac{3}{x} - 5\right)\left(\frac{x}{4} + \frac{1}{2}\right) = 0$

74. $2\left(4 - \frac{1}{x}\right)\left(\frac{3}{5} + \frac{2}{x}\right)(4x - 1) = 0$

MAINTAIN YOUR SKILLS (SECTIONS 2.3, 2.7)

Solve:

75. $23 - 8x - 24x = -15x + 34$

76. $1.6x - 45 - 3.5x = 60 - 5.4x$

77. $5.5y - 7.2y - 78 = 8.8y + 583.5$

78. $-9a - 13a - 6a + 20a = 15a - 345$

79. $3x - 12 \geq -4x + 2$

80. $12x + 18 < 6x - 12$

81. $20x - 28 \leq 30x + 21$

82. $6x + 3 - 5x > 5x + 6 - 7x$

4.2 MONOMIAL FACTORS

OBJECTIVES

1. Factor a polynomial by writing it as the product of a monomial and a polynomial.
2. Solve related equations.

In Section 4.1, we solved quadratic equations where the left member was written as an indicated product. Recall that a product is the result of multiplying two (or more) factors.

$$\underbrace{y(y - 7)}_{\text{FACTORED FORM}} = \underbrace{y^2 - 7y}_{\text{PRODUCT}}$$

In the equation $x^2 - 4x = 0$, if we write $x^2 - 4x$ in factored form, we can solve the equation. The process of rewriting a product as two or more factors is called "factoring." Factoring is the inverse of multiplying so, in a sense, factoring is related to division.

DEFINITION

Factor

To *factor* a polynomial is to write it as an indicated product. The indicated product is called a "factored form" of the polynomial.

To learn how to factor a polynomial as a product of a monomial times another polynomial let's recall how we multiplied a monomial times a binomial. This is shown in the following table.

Factored Form	Use the Distributive Property	Product
$2(x + y)$	$2 \cdot x + 2 \cdot y$	$2x + 2y$
$a(b - c)$	$a \cdot b - a \cdot c$	$ab - ac$
$5(x + 2)$	$5 \cdot x + 5 \cdot 2$	$5x + 10$
$b(b - c)$	$b \cdot b - b \cdot c$	$b^2 - bc$
$x(x + y - z)$	$x \cdot x + x \cdot y - x \cdot z$	$x^2 + xy - xz$

In the table we might have included $2x + 2y = 4\left(\frac{1}{2}x + \frac{1}{2}y\right)$, but instead we shall require the factors to have integer coefficients.

DEFINITION

Polynomial Over the Integers

A *polynomial over the integers* is a polynomial that has only integers for coefficients.

When each term of a polynomial has a common factor, the greatest common factor (GCF) can be "factored out." The GCF of two or more monomials is itself a monomial.

DEFINITION

GCF of a Polynomial

The *GCF of a polynomial* is the product of:

1. The greatest integer factor, or its opposite, that is common to the coefficients, and
2. The variable factor(s) with the largest exponent(s) that is (are) common to each monomial.

To factor a common monomial factor from a polynomial, reverse the multiplication process.

Polynomial	Common Factor Located	Factored Form
$2x + 2y$	$2 \cdot x + 2 \cdot y$	$2(x + y)$
$ab - ac$	$a \cdot b - a \cdot c$	$a(b - c)$
$5x + 10$	$5 \cdot x + 5 \cdot 2$	$5(x + 2)$
$b^2 - bc$	$b \cdot b - b \cdot c$	$b(b - c)$
$x^2 + xy - xz$	$x \cdot x + x \cdot y - x \cdot z$	$x(x + y - z)$
$x^2 + x$	$x \cdot x + x \cdot 1$	$x(x + 1)$

In each case, the polynomial is rewritten so that each term is the product of the GCF and another monomial. The common factor is then factored out by using the distributive property to write the factored form.

EXAMPLE 1

Factor: $xy + xz$

$xy + xz = x(y + z)$

The only common factor (other than 1) is x, which is factored out. Check by multiplying. ■

EXAMPLE 2

Factor: $3a^3b + 6ab^3$

$3a^3b + 6ab^3 = 3ab \cdot a^2 + 3ab \cdot 2b^2$

The two terms of this polynomial have a common integral factor of 3 and common variable factors of a and b. The GCF is $3ab$.

$= 3ab(a^2 + 2b^2)$

Factor using the distributive property. ■

■ PROCEDURE

To factor a polynomial as the product of a monomial and another polynomial:

1. Find the GCF of all the terms.
2. Rewrite the polynomial so that each term is the product of the GCF and another monomial.

> 3. Factor out the GCF using the distributive property to write the factored form.

If the GCF of a polynomial is 1 and the polynomial has only factors of 1 and itself, it is called a prime polynomial.

DEFINITION

Prime Polynomial

> A *prime polynomial* is a polynomial that has exactly two factors that are polynomials over the integers, 1 and the polynomial itself.

EXAMPLE 3

Factor: $3x + 4y + 5z$

$3x + 4y + 5z$ is prime.	The only common factor is 1. ■

EXAMPLE 4

Factor: $-4r^2st - 20rs^2t - 36rst^2$

$= 4rst(-r) + 4rst(-5s) + 4rst(-9t)$	The greatest integer factor is 4 (or -4), and the common variable factors are r, s, and t. So, the GCF is $4rst$ (or $-4rst$).
$= 4rst(-r - 5s - 9t)$	Factored form using $4rst$ as the GFC.
or	
$= -4rst(r + 5s + 9t)$	Another factored form using $-4rst$ as the GCF. Either form is acceptable. ■

Examples 5 and 6 illustrate the use of factoring to solve an equation of the form $ax^2 + bx = 0$.

EXAMPLE 5

Solve: $x^2 - 8x = 0$

$x(x - 8) = 0$	Factor the left side.
$x = 0$ or $x - 8 = 0$	Zero-product property.
$x = 0$ or $x = 8$	The check, using the original equation, is left for the student.

The solution set is $\{0, 8\}$. ■

EXAMPLE 6

Solve: $15y^2 = 5y$

$15y^2 - 5y = 0$	Write the equation in standard form by subtracting $5y$ from each side.
$5y(3y - 1) = 0$	Factor the left side.
$5y = 0$ or $3y - 1 = 0$	Zero-product property.
$y = 0$ $\quad$ $3y = 1$	Solve.
$y = \frac{1}{3}$	The check is left for the student.

The solution set is $\left\{0, \frac{1}{3}\right\}$. ■

Now let's use this technique to solve a word problem.

EXAMPLE 7

Sam is seven years older than his sister Kate. How old is each when the product of their ages is twice the square of Kate's age?

Simpler word form:

Sam's age times Kate's age is twice the square of Kate's age.

Select variable:

Let x represent Kate's age now, then Sam's age is $x + 7$.	Sam is seven years older than Kate.

Translate to algebra:

$(x + 7)x = 2x^2$

Solve:

$x^2 + 7x = 2x^2$	
$-x^2 + 7x = 0$	
$x^2 - 7x = 0$	Multiply both sides by -1.
$x(x - 7) = 0$	Factor.
$x = 0$ or $x - 7 = 0$	Zero-product property.
$x = 0$ or $x = 7$	

Check:

Kate's age: $x = 7$
Sam's age: $x + 7 = 14$

Disregard $x = 0$ since Kate cannot be zero years old.

Product: $7(14) = 98$
Twice the square of Kate's age:
$2(7)^2 = 98$

Answer:

Kate is 7 years old, and Sam is 14 years old. ■

Exercise 4.2

A

Factor:

1. $7x + 7w$
2. $19a - 19b$
3. $ax + ay$
4. $bt - bq$
5. $6x + 12$
6. $8x + 18$
7. $9x - 15y$
8. $12b - 18$
9. $8ab - 14b$
10. $12xy + 16x$
11. $6bc + 12b$
12. $8ab - 16b$
13. $12xy - 36y$
14. $20ab + 12b$
15. $18st + 6t$
16. $39mn - 13n$

Solve:

17. $x^2 - 7x = 0$
18. $x^2 + 4x = 0$
19. $a^2 - 5a = 0$
20. $a^2 + 14a = 0$

B

Factor:

21. $15xyz + 7wxz$
22. $20abc + 9bcd$
23. $6a + 5b$
24. $13a + 14b$
25. $12bc - 16bd$
26. $16ab - 24ac$
27. $4ab + 8bc - 6cd$
28. $8xy + 12xz - 16yz$
29. $10x^2y + 25x^2z$
30. $18a^3b^2c - 3a^3b^2d$

Solve:

31. $3x^2 - 3x = 0$
32. $6x^2 - 6x = 0$
33. $5y^2 - 25y = 0$

34. $11y^2 - 66y = 0$

35. $-2w^2 + 8w = 0$

36. $5w^2 + 15w = 0$

37. $-7t^2 + 21t = 0$

38. $5t^2 + 25t = 0$

39. $-8a^2 + 24a = 0$

40. $15b^2 - 45b = 0$

C

Factor:

41. $x^2y^2 - x^3y^3 + x^4y^4$

42. $b^2c^3 - b^4c^4 + b^4c^5$

43. $\pi r^2h + \pi R^2h$

44. $4\pi r^3 + 2\pi R^3$

45. $16x^2y^2z^2 + 24x^2yz^2 + 32x^2yz$

46. $24a^3b^3c^3 + 18a^2bc^2 + 30a^2b^2c$

47. $3m^2n^2 - 6mn - 12m^2n - 15mn^2$

48. $4x^2y^2 + 8xy - 12x^2y - 16y^2$

49. $8r^2s^3t^5 - 48rs^2t^3 - 20r^2s^2t^2$

50. $6a^7b^2c^3 + 16a^4b^2c^7 - 18a^2b^2c^5$

Solve:

51. $4x^2 - 3x = 0$

52. $6x^2 - 5x = 0$

53. $9y^2 - 3y = 0$

54. $12z - 3z^2 = 0$

55. $9w^2 = 21w$

56. $4w^2 = 48w$

57. $15y^2 = -40y$

58. $6y^2 = -18y$

59. $t^2 = -17t$

60. $t^2 = 19t$

D

61. For a given square there are numerically as many square inches in the area as there are inches in the perimeter. What is the length of the side of the square?

62. If one side of a square is lengthened three inches to form a rectangle, then the area of the new rectangle is twice the area of the original square. Find the length of the side of the square.

63. Thelma is twice as old as her brother Pete. How old is each if the product of their ages is equal to six times Pete's age?

64. Jerry is three times as old as his niece Wendy. How old is each if the product of their ages is equal to twenty-four times Wendy's age?

65. A rectangle is twice as long as it is wide. If there are numerically as many square feet in the area as there are feet in the perimeter, what are the dimensions of the rectangle?

66. A rectangle is three times as long as it is wide. If there are numerically as many square yards in the area as there are yards in the perimeter, what are the dimensions of the rectangle?

67. If there are numerically as many square meters in the area of a circle as there are meters in the circumference, what is the radius of the circle?

68. If there are numerically twice as many square centimeters in the area of a circle as there are centimeters in the circumference, what is the radius of the circle?

69. Given the equation $s = 36t - 4t^2$, where s is the height of a falling object and t is the time that the object falls, how long does it take the object to hit the ground ($s = 0$)?

70. Given the equation $s = 64t - 16t^2$, where s is the height of a falling object and t is the time that the object falls, how long does it take the object to hit the ground ($s = 0$)?

STATE YOUR UNDERSTANDING

71. Describe how to determine the greatest common factor (GCF).

72. Once the greatest common factor (GCF) has been identified, the other factor can be determined by dividing by the GCF. Why is this the case?

CHALLENGE EXERCISES

Factor the following expressions.

73. $4x^2(x - 2y) + 8x(x - 2y) - 14(x - 2y)$

74. $(x - y)(x + 4y) + 2x(x - y)$

75. $(2a + 3b)(3a - b) + (2a + 3b)(2a + 7b)$

76. $2(4a - 5b)^3 - (4a - 5b)^2$

Solve the following equations.

77. $2x(x + 7) - 3(x + 7) = 0$

78. $3a(4a - 3) + 5(4a - 3) = 0$

MAINTAIN YOUR SKILLS (SECTIONS 2.5, 2.7, 3.5, 3.6)

Solve:

79. $3(-4a) - 5(-2a) - (3)(-6) = 5(2a) - 14$

80. $2.34x - 3(3.12x) + 4(-5.31) = -1(1.12x)$

81. Solve for x if $a = 5.3$, $b = -2.1$, and $c = 3.8$: $ax + bx + cx - ab = bc$

82. $2(-3x) - 2(-3) > (-2)(-5x) - 17$

83. $3(-7x) - 2(-6x) + 14 \leq 2x - 5$

84. $1.57 - 2.1(-3.4x) + 8 \geq 16 - 5.2(1.1x)$

85. A horse rider and a jogger leave the same time going in opposite directions. If the horse rider averages 20 mph and the jogger averages 12 mph, how long will it take for them to be 8 miles apart?

86. George mixed equal amounts of two windshield cleaners containing 25% cleanser and 33% cleanser to use in his car. If the mixture contains 116 ounces of cleanser, how many ounces of each mixture did he use?

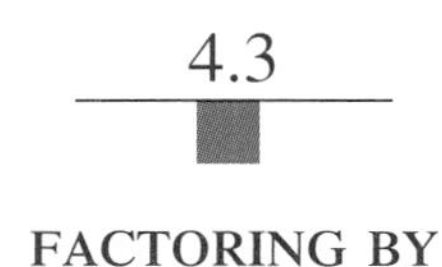

4.3 FACTORING BY GROUPING

OBJECTIVES

1. Factor a polynomial by grouping terms.
2. Solve related equations.

To see how polynomials can be factored by grouping terms, we review multiplication of polynomials using the distributive property.

$$(a + b)(x + y) = (a + b)x + (a + b)y$$
$$= ax + bx + ay + by$$

If we reverse this multiplication process, we can factor polynomials of four or more terms. Study the following table.

Polynomial	Group Terms with Common Factor	Common Factors Factored Out	Common Polynomial Factored Out
$ax + bx + ay + by$	$(ax + bx) + (ay + by)$	$x(a + b) + y(a + b)$	$(a + b)(x + y)$
$bz - dz + by - dy$	$(bz - dz) + (by - dy)$	$z(b - d) + y(b - d)$	$(b - d)(z + y)$
$x^2 + xy - zx - zy$	$(x^2 + xy) + (-zx - zy)$	$x(x + y) - z(x + y)$	$(x - z)(x + y)$
$8ax + 8bx - ay - by$	$(8ax + 8bx) + (-ay - by)$	$8x(a + b) - y(a + b)$	$(8x - y)(a + b)$
$6x^2 + 4x + 15x + 10$	$(6x^2 + 4x) + (15x + 10)$	$2x(3x + 2) + 5(3x + 2)$	$(2x + 5)(3x + 2)$
$2x^2 + 3x + 3xy + 2y$	$(2x^2 + 3x) + (3xy + 2y)$	$x(2x + 3) + y(3x + 2)$	There is no common factor, so the polynomial is prime

EXAMPLE 1

Factor: $2x + ax + 2y + ay$

$2x + ax + 2y + ay = (2x + ax) + (2y + ay)$	Group the first two terms and the last two terms.
$= x(2 + a) + y(2 + a)$	Factor x from the first two terms and y from the last two terms.
$= (2 + a)(x + y)$	Factor the common polynomial $(2 + a)$ from the expression by using the distributive property.

Check:

$(2 + a)(x + y) = 2x + ax + 2y + ay$	Multiply using FOIL. ■

■ PROCEDURE

Factor by Grouping

To factor a polynomial by grouping:

1. Group terms that have a common monomial factor.
2. Factor out the common monomial factor from each group.
3. Factor out the common polynomial factor (if one exists).

When factoring by grouping, exercise caution when the terms involve subtraction.

■ CAUTION

$$x^3 + 3x^2 - ax - 3a \neq (x^3 + 3x^2) - (ax - 3a)$$

EXAMPLE 2

Factor: $x^3 + 3x^2 - ax - 3a$

$x^3 + 3x^2 - ax - 3a = (x^3 + 3x^2) + (-ax - 3a)$	Group the terms.
$= x^2(x + 3) - a(x + 3)$	Factor x^2 from the first two terms and $-a$ from the last two terms.
$= (x^2 - a)(x + 3)$	Factor the common polynomial $(x + 3)$ from the expression. ■

EXAMPLE 3

Factor: $8ab - 3 - 2a + 12b$

$8ab - 3 - 2a + 12b = 8ab - 2a + 12b - 3$

The first two terms have no common factor (except 1). Use the commutative and associative properties to rearrange the terms so that the first two terms *will* have a common factor. This can be done in different ways.

$= 2a(4b - 1) + 3(4b - 1)$

$= (2a + 3)(4b - 1)$

Factor $2a$ out of the first two terms and 3 out of the last two.

or

$8ab - 3 - 2a + 12b = (8ab + 12b) + (-2a - 3)$

A different arrangement of terms.

$= 4b(2a + 3) + (-1)(2a + 3)$

$= [4a + (-1)][2a + 3]$

$= (4a - 1)(2a + 3)$

The different arrangement of terms produces the same set of factors. ■

To find all common factors, you need to recognize factors that are opposites. The factors $(a - b)$ and $(b - a)$ are not the same factor.

$a - b \neq b - a$

However, either can be used as a common factor by writing it as a product of -1 and the other one.

$a - b = -1(b - a)$ or $b - a = -1(a - b)$

EXAMPLE 4

Factor: $ax - bx + by - ay$

$ax - bx + by - ay = x(a - b) + y(b - a)$

$a - b$ and $b - a$ are opposites. By writing $b - a$ as $(a - b)(-1)$, we have a common factor.

$= x(a - b) + (-1)y(a - b)$

$= x(a - b) + (-y)(a - b)$ $\quad (-1)y = -y$

$= (a - b)(x - y)$ ■

Sometimes the grouping of terms results in common monomial factors in a group but not a common polynomial factor. This may indicate that the polynomial is prime.

EXAMPLE 5

Factor: $x^3 + 2x^2 + 4x - 8$

$x^3 + 2x^2 + 4x - 8 = (x^3 + 2x^2) + (4x - 8)$ Group the terms.

$= x^2(x + 2) + 4(x - 2)$ The factors $(x + 2)$ and $(x - 2)$ are not the same. This polynomial is prime. ■

Example 6 illustrates the use of factoring by grouping to solve an equation.

EXAMPLE 6

Solve: $6x^2 - 14x - 15x + 35 = 0$

$2x(3x - 7) + (-5)(3x - 7) = 0$ The common factor -5 is used for the group $(-15x + 35)$ instead of 5 because we want the polynomial factors $(3x - 7)$ to be the same.

$$2x(3x - 7) - 5(3x - 7) = 0$$
$$(2x - 5)(3x - 7) = 0$$

$2x - 5 = 0$ or $3x - 7 = 0$

$x = \frac{5}{2}$ or $x = \frac{7}{3}$

The solution set is $\left\{\frac{7}{3}, \frac{5}{2}\right\}$. Check left for the student. ■

EXERCISE 4.3

A

Factor by grouping:

1. $cx + cy + 3x + 3y$

2. $4a + 4b + ta + tb$

3. $x^2 - 10x + xy - 10y$

4. $d^2 - 8d + bd - 8b$

5. $x^2 + x + xw + w$
6. $m^2 + m + cm + c$
7. $y^2 - 3y + yd - 3d$
8. $3r^2 + 6r + rs + 2s$
9. $z^2 + 7z + 4z + 28$
10. $t^2 - 9t + 3t - 27$
11. $3x^2 - 12x + xy - 4y$
12. $5x^2 + 15x + 2x + 6$

Solve:

13. $x^2 + 2x + 5x + 10 = 0$
14. $x^2 + 7x + 2x + 14 = 0$
15. $w^2 - 13w + 2w - 26 = 0$
16. $w^2 - 2w + 3w - 6 = 0$
17. $y^2 + 7y + y + 7 = 0$
18. $y^2 + 11y + y + 11 = 0$
19. $m^2 + 3m + 8m + 24 = 0$
20. $w^2 + 2w + 12w + 24 = 0$

B

Factor by grouping:

21. $x^5 + x^3 + x^2 + 1$
22. $x^5 + 2x^3 + 4x^2 + 8$
23. $2x^2 - 4x + xy - 2y$
24. $7y^2 - 14y + by - 2b$
25. $ax + a + b + bx$
26. $xy + xz + wz + wy$
27. $4ab - 2c - 8a + bc$
28. $12xy - z - 4x + 3yz$
29. $24a^2b - 8a + 15abc - 5c$
30. $4ab + 2ac + 12bd^2 + 6cd^2$

Solve:

31. $3x^2 + 9x + 2x + 6 = 0$
32. $5x^2 + 10x + 3x + 6 = 0$
33. $3y^2 - 24y + 2y - 16 = 0$
34. $2y^2 - 14y + 3y - 21 = 0$
35. $2w^2 + w - 12w - 6 = 0$
36. $3w^2 + w - 12w - 4 = 0$
37. $6t^2 - 12t - t + 2 = 0$
38. $t^2 - 9t - 3t + 27 = 0$
39. $p^2 + 8p - 2p - 16 = 0$
40. $5x^2 - 15x - 3x + 9 = 0$

C

Factor by grouping:

41. $9x + 6 - 6ax - 4a$
42. $20a + 12 - 25ax - 15x$
43. $8abx + 15cy + 6cx + 20aby$
44. $9cxy + 8dz + 12cz + 6dxy$
45. $12a^2b^2c^2 + 4bc + 9a^2bc + 3$
46. $4x^2y^2z + 6xy + 10xyz + 15$
47. $10a^2b^2c + 15a^3b - 6b^2cd - 9abd$
48. $9r^3s^5 + 18r^4s^4 + 5st^2w^3 + 10rt^2w^3$

49. $a^3 + 2a^2 + ab^2 + 2b^2$

50. $ab^2 + b^2c + ad + cd$

Solve:

51. $8x^2 + 6x + 12x + 9 = 0$

52. $12x^2 + 8x + 15x + 10 = 0$

53. $6y^2 + 9y + 8y + 12 = 0$

54. $12w^2 + 15w + 16w + 20 = 0$

55. $6p^2 + 8p - 3p - 4 = 0$

56. $10x^2 - 4x - 25x + 10 = 0$

57. $15t^2 - 20t + 21t - 28 = 0$

58. $9t^2 + 33t - 15t - 55 = 0$

59. $6w^2 + 12w - 10w - 20 = 0$

60. $6w^2 + 21w - 12w - 42 = 0$

STATE YOUR UNDERSTANDING

61. How many different ways can a "four-term" polynomial be grouped? Justify your answer.

62. Explain how you decide which terms to group when factoring by grouping.

CHALLENGE EXERCISES

Factor the following polynomials.

63. $8a - 6bx + 4ax - 12b$

64. $12x^3 - 3xy - 24x^2y + 6y^2$

65. $8acd + 4bcd - 16ace - 8bce$

66. $5a^3x + 5a^3y + 10a^2b^3x + 10a^2b^3y$

67. $42x^3y^2 - 14x^2y^2z + 28x^3yz - 21x^2y^3$

MAINTAIN YOUR SKILLS (SECTION 3.5)

Combine:

68. $(13a + 5b - 6c) + (6a - 7b - 12c)$

69. $(21x - 13y + 6) + (-5x - 2y + 17)$

70. $(3y - 4z) + (4x - 7y) + (-7x - 5z)$

71. $(3x - 5) - (2x + 6) - (12 - 7x)$

Solve:

72. $(5x - 3) + (6 - 4x) = 3x + 16$

73. $(7y - 15) - (5y + 13) = 5y + 5$

74. $(3x - 7) + (2x - 9) - (6x + 11) = 49$

75. $(15a - 16) - (21 - 5a) = (12a + 34) + (3a - 1)$

4.4

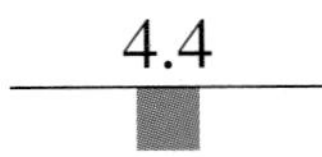

FACTORING TRINOMIALS OF THE FORM $x^2 + bx + c$

OBJECTIVES

1. Factor a trinomial of the form $x^2 + bx + c$.
2. Solve related equations.

DEFINITION

Leading Coefficient

The *leading coefficient* of a polynomial is the coefficient of the term of highest degree.

In the polynomial $4x^2 + 7x - 12$, the leading coefficient is 4, since $4x^2$ is the term of highest degree.

In this section, the leading coefficient of each trinomial is one (1).

Factoring quadratic trinomials with a leading coefficient of one can often be done by observation or by trial and error. The factors, if any, are two binomials.

Recall that the product of two binomials is often a trinomial. For example,

$(x + 3)(x + 9) = x^2 + 12x + 27$ and

$(x + 6)(x + 6) = x^2 + 12x + 36.$

In both examples, the middle term is $12x$. In one case, the sum is $3x + 9x$, and in the other case, the sum is $6x + 6x$. So the coefficient of the middle term is the sum of the factors of the third term.

To factor a trinomial such as $x^2 + 12x + 35$, we observe that the third term, 35, can be obtained only from the products $1 \cdot 35$ and $5 \cdot 7$ if we restrict ourselves to positive integers. The sum of the factors must be the coefficient of the middle term, $12x$. So the factors are:

$x^2 + 12x + 35 = (x + 5)(x + 7).$

EXAMPLE 1

Factor: $x^2 + 18x + 72$

$1 \cdot 72,\ 2 \cdot 36,\ 3 \cdot 24,\ 4 \cdot 18,\ 6 \cdot 12,\ 8 \cdot 9$	The factors of 72 using positive integer factors.
$6 + 12 = 18$	The factors 6 and 12 have the sum 18, the coefficient of the middle term, $18x$.
$x^2 + 18x + 72 = (x + 6)(x + 12)$	Factored form. ■

We now provide some clues to help arrive at the correct factors more quickly. We have observed that the sum of the factors of the third term (also called the constant term) is the coefficient of the middle term. That is, for

$x^2 + bx + c$

to be factored, the factors of c must have the sum b. The following chart shows how to choose positive or negative factors of c.

Sign of Middle Term (b)	Sign of Third Term (c)	Signs of the Two Factors of c
+	+	Both positive
−	+	Both negative
+	−	One positive One negative
−	−	One positive One negative

EXAMPLE 2

Factor: $x^2 - 7x - 30$

$(-1)(30)$, $(1)(-30)$, $(-2)(15)$
$(2)(-15)$, $(-3)(10)$, $(3)(-10)$
$(-5)(6)$, $(5)(-6)$

We use one positive factor and one negative factor since the constant term, -30, is negative.

$3 + (-10) = -7$ The correct factors of -30.

$x^2 - 7x - 30 = (x + 3)(x - 10)$ ■

The list of possible factors can also be trimmed when the pair are opposite in sign by noting that the factor with largest absolute value must have the same sign as the middle term.

EXAMPLE 3

Factor: $x^2 - 3x - 10$

$(1)(-10)$, $(2)(-5)$

These are the only factors of -10 we need to consider. The factors, $(-1)(10)$ and $(-2)(5)$, are disregarded since their sum is not negative.

$2 + (-5) = -3$ The correct factors of -10.

$x^2 - 3x - 10 = (x + 2)(x - 5)$ ■

PROCEDURE

To factor a trinomial of the form $x^2 + bx + c$:

1. Find the integers m and n such that $mn = c$ and $m + n = b$.
2. Write the factors:

   $x^2 + bx + c = (x + m)(x + n)$.

EXAMPLE 4

Factor: $x^2 - 7x - 20$

$(1)(-20)$, $(2)(-10)$, $(4)(-5)$ — Possible factors where the negative factor has the larger absolute value.

$1 + (-20) = -19$

$2 + (-10) = -8$ — None of the pairs of factors have the correct sum.

$4 + (-5) = -1$

$x^2 - 7x - 20$ is prime.

When c has a large number of possible factors, it might be easier to systematically list the products and their corresponding sums.

EXAMPLE 5

Factor: $x^2 - 5x - 84$

$mn = -84$	$m + n = -5$	
$(1)(-84)$	$1 + (-84) = -83$	List the factors whose product is -84, and find the sum. Here the negative factor must have the larger absolute value.
$(2)(-42)$	$2 + (-42) = -40$	
$(3)(-28)$	$3 + (-28) = -25$	
$(4)(-21)$	$4 + (-21) = -17$	
$(6)(-14)$	$6 + (-14) = -8$	
$(7)(-12)$	$7 + (-12) = -5$	The correct sum.

$x^2 - 5x - 84 = (x + 7)(x - 12)$

EXAMPLE 6

Factor: $z^2 + 5z - 104$

$mn = -104$	$m + n = 5$	List the factors whose product is -104, and find the sum.
$(-1)(104)$	$-1 + 104 = 103$	
$(-2)(52)$	$-2 + 52 = 50$	
$(-4)(26)$	$-4 + 26 = 22$	
$(-8)(13)$	$-8 + 13 = 5$	The correct sum.

$z^2 + 5z - 104 = (z - 8)(x + 13)$ ■

Equations of the form $x^2 + bx + c = 0$ can be solved by factoring the left side.

EXAMPLE 7

Solve: $x^2 + 10x + 21 = 0$

$(x + 7)(x + 3) = 0$ Factor the left side.

$x + 7 = 0$ or $x + 3 = 0$ Zero-product property.

$x = -7$ or $x = -3$

The solution set is $\{-7, -3\}$. Check left for student. ■

Some number problems lead to quadratic equations.

EXAMPLE 8

The product of four more than a positive number and twelve more than the same number is 345. What is the number?

Simpler word form:

(number plus 4) times (number plus 12) equals 345

Select a variable:

Let n represent the number.

Translate to algebra:

$(n + 4)(n + 12) = 345$

■ **CAUTION**

Since the right side is *not* equal to zero, we must first isolate all terms on the left side.

Solve:

$n^2 + 16n + 48 = 345$ — Multiply on left.

$n^2 + 16n - 297 = 0$ — Write the equation in standard form by subtracting 345 from each side.

$(n + 27)(n - 11) = 0$ — Factor the left side.

$n + 27 = 0$ or $n - 11 = 0$

$n = -27$ or $n = 11$ — The negative root is rejected since a positive number was specified in the original problem.

Answer:

The number is 11. — The check is left for the student. ■

Exercise 4.4

A

Factor:

1. $x^2 + 6x + 8$
2. $x^2 + 9x + 8$
3. $y^2 - 15y + 14$
4. $y^2 - 9y + 14$
5. $y^2 + 6y - 27$
6. $y^2 - 6y - 27$
7. $w^2 + 4w - 21$
8. $t^2 + 9t + 18$
9. $z^2 - 11z + 30$
10. $t^2 - 9t + 18$
11. $x^2 - 5x - 66$
12. $y^2 - 3y - 28$

Solve:

13. $x^2 + 16x + 15 = 0$
14. $x^2 + 8x + 15 = 0$
15. $y^2 - 13y + 22 = 0$
16. $y^2 - 23y + 22 = 0$
17. $w^2 - 11w - 26 = 0$
18. $w^2 + 11w - 26 = 0$
19. $x^2 - 10x + 24 = 0$
20. $r^2 - 16r + 55 = 0$

B

Factor:

21. $x^2 + 22x + 21$
22. $x^2 + 10x + 21$
23. $w^2 - 16w + 39$
24. $w^2 - 40w + 39$
25. $w^2 - 8w - 33$
26. $w^2 + 8w - 33$
27. $t^2 - 2t - 35$
28. $t^2 + 2t - 35$
29. $w^2 - 11w + 28$
30. $p^2 - 12p + 27$

Solve:

31. $x^2 + 10x + 24 = 0$
32. $a^2 - 8a + 15 = 0$
33. $y^2 + 26y - 27 = 0$
34. $y^2 - 26y - 27 = 0$
35. $w^2 - 15w + 26 = 0$
36. $w^2 + 22w + 40 = 0$
37. $r^2 - 6r - 55 = 0$
38. $w^2 - 4w - 45 = 0$
39. $x^2 - x - 90 = 0$
40. $y^2 + 12y - 45 = 0$

C

Factor:

41. $x^2 - 6x - 40$
42. $x^2 - 3x - 40$
43. $y^2 + 3y - 40$
44. $y^2 + 6y - 40$
45. $w^2 - 39w - 40$
46. $w^2 - 41w + 40$
47. $t^2 + 21t + 38$
48. $t^2 + 37t - 38$
49. $b^2 + 19b + 34$
50. $k^2 + 2k - 63$

Solve:

51. $x^2 + 23x + 42 = 0$
52. $x^2 - x - 42 = 0$
53. $y^2 + 30y + 56 = 0$
54. $y^2 + 15y + 56 = 0$
55. $w^2 - w - 56 = 0$
56. $w^2 - 55w - 56 = 0$
57. $r^2 - 2r - 63 = 0$
58. $r^2 + 2r - 63 = 0$
59. $x^2 - 2x - 120 = 0$
60. $y^2 - 17y - 84 = 0$

D

61. The product of two positive consecutive integers is 272. Find the integers. (*Hint:* Let x and $x + 1$ represent the consecutive integers.)

62. Find two positive consecutive even integers whose product is 624.

63. The length of a rectangular flower bed is 4 feet greater than its width. If the area of the flower bed is 21 ft^2, determine its length and its width.

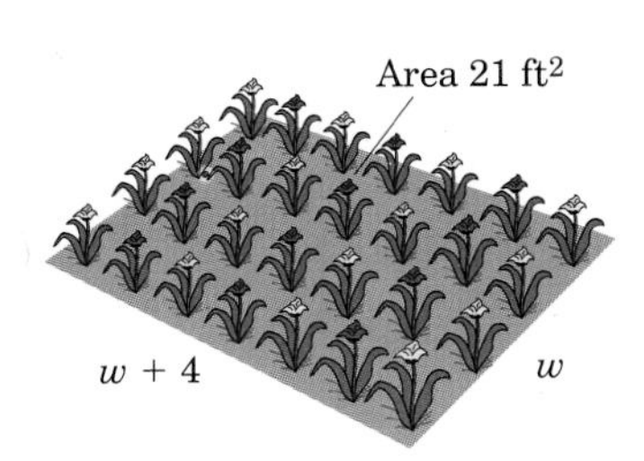

Figure for Exercise 63

64. The weight, w, in megagrams of the fuel in a booster rocket is given by $w = 144 - 7t - t^2$, where t is the time in seconds. How long after launch does the rocket run out of fuel? (*Hint:* What is the weight of the fuel when it is all gone?)

65. In exercise 64, how long after launch is there 100 megagrams of fuel remaining in the rocket?

66. A rectangular deck has dimensions of 4 m by 6 m. The owner wishes to double the area of the deck by increasing each side by the same amount. How much should each side be increased?

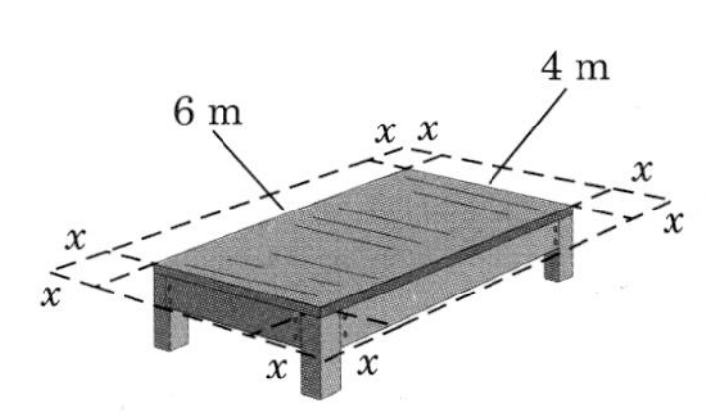

Figure for Exercise 66

67. If three times the smaller of two consecutive odd integers is subtracted from the square of the larger, the difference is 10. Determine the consecutive odd integers.

68. When the sides of a $12'$ by $15'$ solar panel were increased by the same amount, the area was increased by 90 ft^2. How much was added to each side?

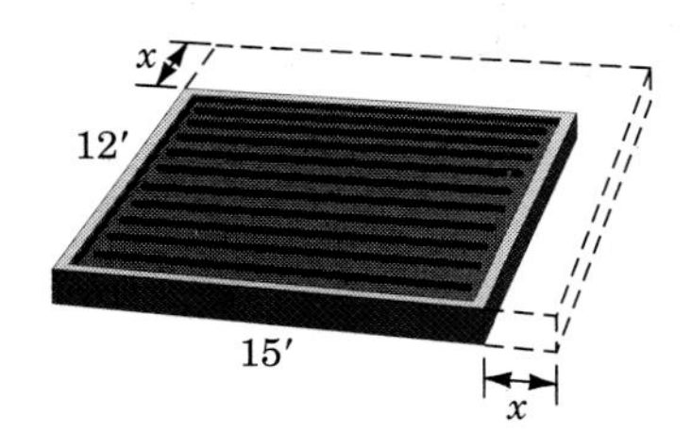

Figure for Exercise 68

STATE YOUR UNDERSTANDING

69. Explain how to determine the signs of the last terms of the factors of a trinomial.

70. How are the last terms of the factors of $x^2 + bx + c$ related to b and c?

CHALLENGE EXERCISES

Solve the following equations.

71. $5x^2 + 7x - 30 = 4x^2 - 5x - 2$

72. $(x - 9)(x - 4) = 6$

Express the area of the shaded region in factored form.

73.

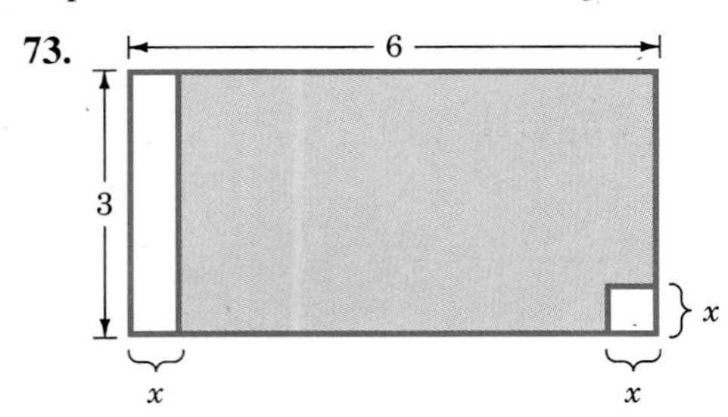

74.

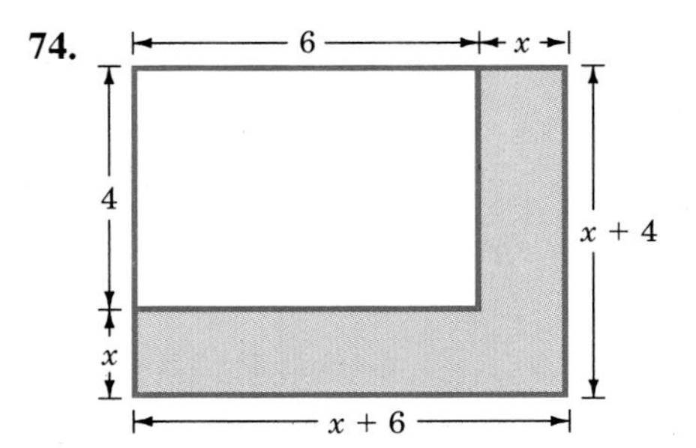

MAINTAIN YOUR SKILLS (SECTION 3.5)

Combine:

75. $(-3xy - 5xz + 7yz) + (-7xy + 3xz - 2yz) - (6xy + 2xz - 8yz)$

76. $(22x^2 - 3x + 17) - (12x^2 + 5x - 32) - (x^2 + 25x - 8)$

77. $(1.34y^2 - 4.6y + 2.3) + (3.6y^2 - 5.6y - 1.23) - (2.3y^2 - 5.67)$

78. $(2.5a^2 - 1.15b^2) - (1.3a^2 - 3.65ab + 2.7b^2) + (-3.45ab + 2.1b^2)$

Solve:

79. $(4x - 2) - (3x + 5) + (2x - 24) = (4x + 12) - 23$

80. $(-3x + 36) - (7x + 23) - (-4x + 15) = (2x + 9) - (12x - 17)$

81. $(3x^2 - 5x + 19) - (x^2 - 7x - 10) = 2x^2 + 4x - 33$

82. $(7x^2 + 2x - 17) + (3x^2 - 14x + 55) = 14 + (10x^2 - 3x + 15)$

4.5

FACTORING TRINOMIALS OF THE FORM $ax^2 + bx + c$

OBJECTIVES

1. Factor a trinomial of the form $ax^2 + bx + c$.
2. Solve related equations.

With a change or two, the procedure for factoring trinomials with leading coefficient of 1 can be extended to factor trinomials of the form $ax^2 + bx + c$. Recall the FOIL shortcut for multiplying binomials.

$$\begin{array}{l} \text{O I I O} \\ (2x + 3)(3x + 2) = 2x \cdot 3x + 2x \cdot 2 + 3 \cdot 3x + 3 \cdot 2 \\ \text{F L F L} \qquad\qquad\quad \text{F} \qquad\quad \text{O} \qquad\quad \text{I} \qquad\quad \text{L} \\ \qquad\qquad\qquad\qquad = 6x^2 + 13x + 6 \end{array}$$

We can reverse this procedure to factor a trinomial of form $ax^2 + bx + c$. For instance, to factor $12x^2 + 19x + 5$:

1. We know that the product of the two first terms is $12x^2$, so the two terms are

$(12x)(x)$, $(2x)(6x)$, or $(3x)(4x)$.

2. The product of the last terms is 5. The possible pairs of factors of 5 are

$(1)(5)$ — We use positive factors since the middle term is positive.

3. The sum of the inner and outer products (the middle term of the trinomial) is $19x$. We try the following combinations of factors.

Possible Factors	Middle Term	
$(12x + 1)(x + 5)$	$61x$	List all of the possible pairs of factors then check for the correct middle term.
$(12x + 5)(x + 1)$	$17x$	
$(6x + 1)(2x + 5)$	$32x$	
$(6x + 5)(2x + 1)$	$16x$	
$(3x + 1)(4x + 5)$	$19x$	This is the required middle term.
$(3x + 5)(4x + 1)$		

So, $12x^2 + 19x + 5 = (3x + 1)(4x + 5)$.

Use the chart in Section 4.4 to determine whether positive or negative factors of c are needed.

EXAMPLE 1

Factor: $10x^2 + 11x + 3$

Possible factors of $10x^2$:

$(10x)(x)$ $(2x)(5x)$

The product of the first terms is $10x^2$.

Possible factors of 3:

$(1)(3)$

Since the middle term is positive and the last term is positive, we use $(1)(3)$.

Possible Factors	Middle Term	
$(10x + 1)(x + 3)$	$31x$	
$(10x + 3)(x + 1)$	$13x$	
$(5x + 3)(2x + 1)$	$11x$	These are the correct factors.
$(5x + 1)(2x + 3)$		

$10x^2 + 11x + 3 = (5x + 3)(2x + 1)$ ■

EXAMPLE 2

Factor: $8x^2 + 14x - 15$

Possible factors of $8x^2$:

$(x)(8x)$ $(2x)(4x)$

The product of the first terms is $8x^2$.

Possible factors of -15:

$(-1)(15)$ $(1)(-15)$ $(3)(-5)$ $(-3)(5)$

The product of the last terms is -15; therefore, one of the last terms is negative.

Possible Factors	Middle Term	
$(x - 5)(8x + 3)$	$-37x$	No. If the signs are switched, the middle term becomes $37x$, so we do not try $(x + 5)(8x - 3)$.
$(x + 3)(8x - 5)$	$19x$	No. Again no need to switch signs.
$(x - 15)(8x + 1)$	$-119x$	No.
$(x + 1)(8x - 15)$	$-7x$	No.
$(2x + 3)(4x - 5)$	$2x$	No.
$(2x - 5)(4x + 3)$	$-14x$	No, notice that this is the opposite of what is needed. If we switch the signs of the last terms, the sum will be $14x$.
$(2x + 5)(4x - 3)$	$14x$	

$8x^2 + 14x - 15 = (2x + 5)(4x - 3)$ ■

EXAMPLE 3

Factor: $6x^2 + 4x + 5$

Possible factors of $6x^2$:

$(x)(6x) \quad (2x)(3x)$

The product of the first terms is $6x^2$.

Possible factors of 5:

$(1)(5)$

The product of the last terms is 5. Since the middle term is positive, the factors are positive.

Possible Factors	Middle Term
$(x + 5)(6x + 1)$	$31x$
$(x + 1)(6x + 5)$	$11x$
$(3x + 1)(2x + 5)$	$17x$
$(3x + 5)(2x + 1)$	$13x$

The list of possible factors of the trinomial has been exhausted.

So, $6x^2 + 4x + 5$ is a prime polynomial. ■

It is possible for the terms of the trinomial to have a GCF other than 1. When this happens, the GCF should be factored out first.

EXAMPLE 4

Factor: $42x^2 - 7x - 84$

$42x^2 - 7x - 84 = 7(6x^2 - x - 12)$

The trinomial has a common factor of 7.

■ **CAUTION**

> If 7 is not factored out, it will be a factor of one of the binomial factors.

$42x^2 - 7x - 84 = 7(2x - 3)(3x + 4)$ $\quad$ $6x^2 - x - 12 = (2x - 3)(3x + 4)$ ■

Trinomials of the form $ax^2 + bx + c$ can also be factored by grouping by replacing the middle term (bx) with two terms whose sum or difference is bx. Let's call these new terms mx and nx. To find these terms, we look for two numbers whose product is ac and whose sum is b. This method is effective when a and c have several factors.

EXAMPLE 5

Factor: $6x^2 + x - 15$

$mn = ac = 6(-15)$

mn is the product of the first and last terms. $m + n$ is the coefficient of the middle term. We choose the positive factor to be the one with larger absolute value since the middle term is positive.

$mn = -90$	$m + n = 1$
$-1(90)$	$-1 + 90 = 89$
$-2(45)$	$-2 + 45 = 43$
$-3(30)$	$-3 + 30 = 27$
$-5(18)$	$-5 + 18 = 13$
$-6(15)$	$-6 + 15 = 9$
$-9(10)$	$-9 + 10 = 1$

The correct numbers.

$6x^2 + x - 15 = 6x^2 - 9x + 10x - 15$ Replace x with $-9x + 10x$.

$= 3x(2x - 3) + 5(2x - 3)$ Factor by grouping.

$= (3x + 5)(2x - 3)$ ■

EXAMPLE 6

Factor: $6x^2 + 29x + 30$

$mn = ac = 6(30)$

Since 6 and 30 have several factors, we factor by grouping.

$mn = 180$	$m + n = 29$
1(180)	181
2(90)	92
3(60)	63
4(45)	49
5(36)	41
6(30)	36
9(20)	29

These are the correct numbers.

$6x^2 + 29x + 30 = 6x^2 + 9x + 20x + 30$ Replace $29x$ by $9x + 20x$.

$= 3x(2x + 3) + 10(2x + 3)$

$= (3x + 10)(2x + 3)$ ■

EXAMPLE 7

Factor: $40x^2 - 33x - 18$

$mn = ac = (40)(-18)$

mn is the product of the first and last terms. $m + n$ is the coefficient of the middle term. We choose the negative factor to be the one with larger absolute value since the middle term is negative.

$mn = -720$	$m + n = -33$	
$1(-720)$	$1 + (-720) = -719$	
$2(-360)$	$2 + (-360) = -358$	
$3(-240)$	$3 + (-240) = -237$	
$4(-180)$	$4 + (-180) = -176$	
$5(-144)$	$5 + (-144) = -139$	
$6(-120)$	$6 + (-120) = -114$	
$8(-90)$	$8 + (-90) = -82$	
$9(-80)$	$9 + (-80) = -71$	
$10(-72)$	$10 + (-72) = -62$	
$12(-60)$	$12 + (-60) = -48$	
$15(-48)$	$15 + (-48) = -33$	The correct numbers.

$40x^2 - 33x - 18 = 40x^2 + 15x - 48x - 18$ Replace $-33x$ with $15x - 48x$.

$= 5x(8x + 3) - 6(8x + 3)$ Factor by grouping.

$= (5x - 6)(8x + 3)$ ■

Equations of the form $ax^2 + bx + c = 0$ can be solved by factoring the left side and using the zero-product property.

EXAMPLE 8

Solve: $-3x^2 - 4x - 1 = 0$

$3x^2 + 4x + 1 = 0$ Multiply both sides by -1 so that the leading coefficient is positive.

$(3x + 1)(x + 1) = 0$ Factor.

$3x + 1 = 0$ or $x + 1 = 0$

$x = -\frac{1}{3}$ or $x = -1$

The solution set is $\left\{-1, -\frac{1}{3}\right\}$. Check left for student. ■

A variable electrical current can be expressed by the formula $i = at^2 + bt + c$, where i is in amperes; t is in seconds; and a, b, and c are constants.

EXAMPLE 9

A variable electrical current is given by $i = 2t^2 - 5t + 16$. If t is in seconds, in how many seconds is the current (i) equal to 79 amperes?

Formula:

$i = 2t^2 - 5t + 16$

Substitute:

$79 = 2t^2 - 5t + 16$ $\qquad i = 79$

Solve:

$0 = 2t^2 - 5t - 63$ $\qquad$ Write the equation in standard form by subtracting 79 from each side.

Factor the right side. The factors of $2t^2$ are $(2t)(t)$. The factors of -63 are $(-1)(63)$, $(1)(-63)$, $(-3)(21)$, $(3)(-21)$, $(-7)(9)$, and $(7)(-9)$. Since the difference of 9 and 7 is 2 (which is close to the coefficient of the middle term), we try these first.

Possible Factors	Middle Term	
$(2t - 7)(t + 9)$	$11t$	
$(2t - 9)(t + 7)$	$5t$	No, but if the signs are switched, it will be correct.
$(2t + 9)(t - 7)$	$-5t$	

$0 = (2t + 9)(t - 7)$

$2t + 9 = 0 \quad \text{or} \quad t - 7 = 0$

$t = -\frac{9}{2} \quad \text{or} \quad t = 7$ $\qquad$ Since we assume time to be positive, we reject the negative root.

Check:

$79 = 2(7)^2 - 5(7) + 16$ $\qquad$ Check $t = 7$ in the original formula.

$79 = 98 - 35 + 16$

Answer:

The current will be 79 amperes in 7 seconds. ■

EXERCISE 4.5

A

Factor:

1. $2x^2 + 17x + 26$ **2.** $2x^2 + 53x + 26$ **3.** $2x^2 + 19x + 35$

4. $2x^2 + 17x + 35$
5. $3x^2 - 19x + 28$
6. $3x^2 - 25x + 28$
7. $3x^2 - 20x + 28$
8. $3x^2 - 85x + 28$
9. $2x^2 + 71x + 35$
10. $2x^2 + 37x + 35$
11. $3x^2 - 8x - 28$
12. $3x^2 - 25x - 28$
13. $11x^2 + 80x + 21$
14. $11x^2 + 40x + 21$
15. $7x^2 + 57x - 64$
16. $7x^2 - 40x - 63$

Solve:

17. $2x^2 + 15x + 7 = 0$
18. $3x^2 - 4x + 1 = 0$
19. $2x^2 + 9x - 5 = 0$
20. $3x^2 - 11x - 4 = 0$

B

Factor:

21. $20x^2 + 23x - 7$
22. $20x^2 + 68x - 7$
23. $20x^2 + 4x - 7$
24. $20x^2 - 13x - 7$
25. $16x^2 + 10x + 1$
26. $16x^2 + 8x + 1$
27. $4x^2 - x - 14$
28. $4x^2 - 7x - 15$
29. $4x^2 - 59x - 15$
30. $4x^2 + 28x - 15$
31. $15x^2 - 23x - 22$
32. $15x^2 + 49x - 22$
33. $15x^2 + 7x - 22$
34. $15x^2 - 61x - 22$
35. $6x^2 - 43x + 55$
36. $6x^2 + 41x + 55$

Solve:

37. $10x^2 - 13x - 3 = 0$
38. $14x^2 + 19x - 3 = 0$
39. $12x^2 + 5x - 2 = 0$
40. $20x^2 - x - 12 = 0$

C

Factor:

41. $8x^2 - 34x + 21$
42. $8x^2 + 17x - 21$
43. $8x^2 - 59x + 21$
44. $8x^2 - 38x - 21$
45. $30x^2 + 425x + 70$
46. $36x^2 - 102x - 84$
47. $12x^2 + 62x + 28$
48. $24x^2 + 100x + 56$
49. $35x^2 - 39x + 10$
50. $35x^2 + 43x - 10$
51. $33x^2 - 34x - 35$
52. $33x^2 - 158x - 35$
53. $18x^2 - 37xy + 15y^2$
54. $18x^2 - 59xy + 15y^2$
55. $20x^2 - 19xy - 28y^2$
56. $18x^2 + 27xy - 35y^2$

Solve:

57. $2x^2 - 33x - 35 = 0$

58. $11x^2 - 10x - 21 = 0$

59. $18x^2 - 33x + 15 = 0$

60. $33x^2 - 382x - 35 = 0$

D

61. A variable electrical current is given by $i = 2t^2 - 5t + 25$. If t is in seconds, in how many seconds will the current (i) be equal to 100 amperes?

62. How long will it take the electrical current in Exercise 61 to reach 298 amperes?

63. The height of a given triangle is one less than three times the base. The area is 22 square feet. Find the length of the base.

64. The height of a given triangle is three more than four times the base. The area of the triangle is 81 square inches. Find the base and the height.

65. The area of a rectangle is 350 ft^2. If the length of the rectangle is 5 ft longer than three times the width, what are the dimensions of the rectangle?

66. The width of a rectangle is 10 less than twice its length. If the area of the rectangle is 72 square meters, what are its dimensions?

67. Eight times the square of Jon's age now less twenty-five times his age two years ago is 188. Find his age now.

68. The smaller of two positive numbers is five less than the larger number. The product of the two numbers is 66. Find the numbers.

69. The sum of the first n positive integers is given by $S = \frac{1}{2}n(n + 1)$. If S is 78, find the value for n.

70. The height of a body thrown upward with velocity of 48 feet per second has the formula $H = 48t - \frac{gt^2}{2}$, where t is the time and $g = 32$ feet/sec/sec. For what values of t will the object be at a height of 32 feet?

STATE YOUR UNDERSTANDING

71. In factoring trinomials, when is it helpful to rewrite $ax^2 + bx + c$ as $ax^2 + mx + nx + c$ and then factor by grouping?

72. When factoring $ax^2 + bx + c$ by rewriting it as $ax^2 + mx + nx + c$, how are the values for m and n determined?

CHALLENGE EXERCISES

Factor the following expressions.

73. $-6x^2 + 5x + 6$

74. $15x^4 - 11x^2 - 14$

75. $-90x^6 - 57x^4 + 12x^2$

76. $12x^2 - 16x + 27$

MAINTAIN YOUR SKILLS (SECTIONS 3.7, 3.8)

Multiply:

77. $(4y - 17)(3y + 5)$

78. $(6x + 9y)(4x - 3y)$

79. $(5x - 14)(5x + 14)$

80. $(6a - 5)^2$

Solve:

81. $(x + 4)(x - 3) - (x + 5)(x - 7) = 17$

82. $23 - (3x + 2)(x - 5) + (x + 12)(x - 8) = (12 - 2x)(3 + x)$

83. $(x - 5)(x + 14) - (x - 12)(x + 24) = 16$

84. A rectangular garden is three feet longer than it is wide. If the length and the width of the garden are both increased by six feet, the area is increased by 150 square feet. Find the original dimensions of the garden.

4.6

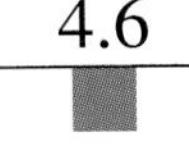

FACTORING: SPECIAL CASES

OBJECTIVES

1. Factor the difference of two squares.
2. Factor the sum or difference of two cubes.
3. Factor a perfect square trinomial.
4. Solve related equations.

Look again at the product of conjugate binomials.

$(3x - 5)(3x + 5) = 9x^2 - 25$

The product is the square of $3x$ minus the square of 5. To reverse the process (factor), we need to recognize the difference of two squares. Recall that a binomial is called the difference of two squares if the two terms are squares separated by a minus sign. Study the following table.

Polynomial	$(a)^2 - (b)^2$	$(a + b)(a - b)$
$x^2 - 16$	$(x)^2 - (4)^2$	$(x + 4)(x - 4)$
$4a^2 - 9$	$(2a)^2 - (3)^2$	$(2a + 3)(2a - 3)$
$25y^2 - 1$	$(5y)^2 - (1)^2$	$(5y + 1)(5y - 1)$
$16x^2 - 7$	7 is not the square of an integer	Prime
$64y^2 - 49$	$(8y)^2 - (7)^2$	$(8y + 7)(8y - 7)$
$36a^2 - 25b^2$	$(6a)^2 - (5b)^2$	$(6a + 5b)(6a - 5b)$

PROCEDURE

$a^2 - b^2 = (a - b)(a + b)$

To factor a binomial that is the difference of two squares, write the product of the sum and difference of the terms that are squared (conjugate binomials).

$a^2 - b^2 = (a + b)(a - b)$

EXAMPLE 1

Factor: $x^2 - 169$

$x^2 - 169 = (x)^2 - (13)^2$

The first term is the square of x, and the second term is the square of 13.

$= (x - 13)(x + 13)$

The factors are the sum and difference of the squared terms. Check by multiplying. ■

Even powers of a variable are perfect squares. Since $x^a \cdot x^a = x^{2a}$, $x^8 = (x^4)^2$ and $y^{10} = (y^5)^2$.

EXAMPLE 2

Factor: $a^2b^{16} - 196$

$a^2b^{16} - 196 = (ab^8)^2 - (14)^2$

$= (ab^8 - 14)(ab^8 + 14)$

CAUTION

b^{16} is $b^8 \cdot b^8$ and *not* $b^4 \cdot b^4$

■

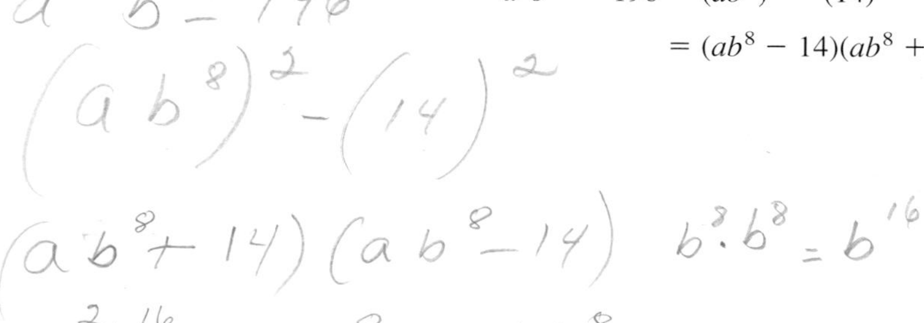

CAUTION

The sum of two squares, $a^2 + b^2$, cannot be factored by these methods and is a prime polynomial.

Two other special cases we consider are the sum and difference of two cubes.

DEFINITION

A binomial is called the sum or difference of two cubes if:

1. The two terms are cubes of other terms.
2. The two terms are separated by a plus or minus sign.

Consider the following products:

$$\begin{aligned}(a + b)(a^2 - ab + b^2) &= (a + b)a^2 + (a + b)(-ab) + (a + b)b^2 \\ &= a^3 + a^2b - a^2b - ab^2 + ab^2 + b^3 \\ &= a^3 + b^3.\end{aligned}$$

Similarly:

$$(a - b)(a^2 + ab + b^2) = a^3 - b^3.$$

We reverse these products to factor the sum or difference of two cubes. Study the following table.

Sum of Two Cubes

Polynomial	$(a)^3 + (b)^3$	$(a + b)(a^2 - ab + b^2)$
$x^3 + 1$	$(x)^3 + (1)^3$	$(x + 1)(x^2 - x + 1)$
$125y^3 + 27$	$(5y)^3 + (3)^3$	$(5y + 3)(25y^2 - 15y + 9)$
$x^{12} + 216y^6$	$(x^4)^3 + (6y^2)^3$	$(x^4 + 6y^2)(x^8 - 6x^4y^2 + 36y^4)$

and

Difference of Two Cubes

Polynomial	$(a)^3 - (b)^3$	$(a - b)(a^2 + ab + b^2)$
$y^3 - 8$	$(y)^3 - (2)^3$	$(y - 2)(y^2 + 2y + 4)$
$64x^3 - 729$	$(4x)^3 - (9)^3$	$(4x - 9)(16x^2 + 36x + 81)$
$a^{15} - 8b^9$	$(a^5)^3 - (2b^3)^3$	$(a^5 - 2b^3)(a^{10} + 2a^5b^3 + 4b^6)$

PROCEDURE

To factor the sum (difference) of two cubes:

1. The first factor is a binomial. This binomial is the sum (difference) of the terms that are cubed.
2. The second factor is a trinomial:
   i. first term: the square of the first term of the binomial in step 1.
   ii. second term: minus (plus) the product of the two terms in step 1.
   iii. third term: plus the square of the second term of the binomial in step 1.
3. Write the product of the two expressions found in steps 1 and 2.

$$a^3 + b^3 = (a + b)(a^2 - ab + b^2)$$

$$a^3 - b^3 = (a - b)(a^2 + ab + b^2)$$

EXAMPLE 3

Factor: $s^3 - 1000$

$s^3 - 1000 = (s)^3 - (10)^3$ — The first term is the cube of s. The second term is the cube of 10. The binomial is the difference of cubes.

$s - 10$ — The first factor is the binomial $(s - 10)$, the difference of the terms that are cubed.

Now write the trinomial factor.

First term: square of the first term of the binomial, s.

$s^2 + 10s + 100$ — Second term: plus the product of the terms of the binomial, s and 10.

Third term: plus the square of the second term of the binomial, 10.

$s^3 - 1000 = (s - 10)(s^2 + 10s + 100)$ — Write the product of the two factors. ■

The last special case we will consider is the perfect square trinomial. Recall that it is the result of squaring a binomial.

$$(x + 3)^2 = x^2 + 2(3x) + 9$$
$$= x^2 + 6x + 9$$

SQUARE OF FIRST TERM (x^2) — TWICE THE PRODUCT OF THE TERMS ($6x$) — SQUARE OF SECOND TERM (9)

When we recognize this pattern, the perfect square trinomial can be factored by writing the square of the sum of the two terms. Study the following table.

Trinomial	$(a)^2$	$2(a)(b)$	$(b)^2$	$(a + b)^2$
$x^2 + 10x + 25$	$(x)^2$	$2(x)(5)$	$(5)^2$	$(x + 5)^2$
$x^2 - 14x + 49$	$(x)^2$	$2(x)(-7)$	$(-7)^2$	$(x - 7)^2$
$x^2 - 11x + 16$	$(x)^2$	$-11x$ is not $2(x)(4)$	$(-4)^2$	Not a perfect square
$4a^2 + 12a + 9$	$(2a)^2$	$2(2a)(3)$	$(3)^2$	$(2a + 3)^2$
$16y^2 - 40y + 25$	$(4y)^2$	$2(4y)(-5)$	$(-5)^2$	$(4y - 5)^2$
$x^2 - 20x + 90$	$(x)^2$		90 is not the square of an integer	Not a perfect square

Notice that when the middle term of the trinomial is negative, the factor b is chosen to be negative.

PROCEDURE

To factor a perfect square trinomial:

1. Write the sum of
   i. the positive number whose square is a^2, and
   ii. the positive or negative number whose square is b^2; the sign chosen is the same sign as $2ab$.
2. Write the square of the sum:

   $a^2 + 2ab + b^2 = (a + b)^2$

   $a^2 - 2ab + b^2 = (a - b)^2$

EXAMPLE 4

Factor: $c^2 - 22c + 121$

$c^2 - 22c + 121 = (c)^2 + 2(-11)(c) + (-11)^2$ The first term is the square of c. The third term is the square of either 11 or -11. Since the middle term is negative, we choose -11.

$= [c + (-11)]^2 = (c - 11)^2$ Check by multiplication. ■

EXAMPLE 5

Factor: $16a^2 + 56ab + 49b^2$

$$16a^2 + 56ab + 49b^2 = (4a)^2 + 2(4a)(7b) + (7b)^2$$
$$= (4a + 7b)^2$$

Check to see that twice the product of $4a$ and $7b$ is $56ab$. ■

EXAMPLE 6

Factor: $t^2 + 14t + 196$

$$t^2 + 14t + 196 = (t)^2 + 2(7)(t) + (14)^2$$

The trinomial does *not* fit the pattern for a perfect square trinomial. The middle term is not $2(14)(x)$.

So $t^2 + 14t + 196$ is not a perfect square trinomial and cannot be factored by this method. In fact, it is a prime polynomial. ■

When factoring any polynomial, first check for a GCF other than 1.

EXAMPLE 7

Factor: $4x^2 + 24x + 36$

$$4x^2 + 24x + 36 = (2x)^2 + 2(6)(2x) + (6)^2$$
$$= (2x + 6)^2$$

A perfect square trinomial. $(2x + 6)$ is *not* a prime polynomial.

■ CAUTION

> The common factor should always be factored out first.

$$4x^2 + 24x + 36 = 4(x^2 + 6x + 9)$$

A common factor of 4.

$$= 4(x + 3)^2$$

$x^2 + 6x + 9$ is a perfect square trinomial. ■

Whenever one side of an equation is 0 and the other side is factored, we can use the zero-product property to solve.

EXAMPLE 8

Solve: $4x^2 - 28x + 49 = 0$

$(2x - 7)^2 = 0$ — Factor the left side, a perfect square trinomial.

$2x - 7 = 0$ or $2x - 7 = 0$ — Zero-product property.

$x = \dfrac{7}{2}$ or $x = \dfrac{7}{2}$ — A double root.

The solution set is $\left\{\dfrac{7}{2}\right\}$. ■

EXAMPLE 9

The square of Chin's age two years from now minus four times his age five years from now is zero. How old is Chin today?

Simpler word form:

$$\begin{pmatrix}\text{the square of}\\ \text{Chin's age two}\\ \text{years from now}\end{pmatrix} - \begin{pmatrix}\text{four times Chin's}\\ \text{age five years}\\ \text{from now}\end{pmatrix} \text{ is equal to 0.}$$

Select variable:

Let x represent Chin's age today; then $x + 2$ represents his age in two years and $x + 5$ represents his age five years from now.

Translate to algebra:

$(x + 2)^2 - 4(x + 5) = 0$

Solve:

$$x^2 + 4x + 4 - 4x - 20 = 0$$

$$x^2 - 16 = 0$$

$$(x - 4)(x + 4) = 0$$

$x = 4$ or $x = -4$ — The solution $x = -4$ is rejected because it does not measure age.

Answer:

Chin is 4 years old today. ■

EXERCISE 4.6

A

Factor:

1. $x^2 - 9$ **2.** $x^2 - 36$ **3.** $y^2 - 49$

4. $y^2 - 100$ **5.** $y^3 - 125$ **6.** $w^3 - 8$

7. $s^3 + 27$ **8.** $t^3 + 1000$ **9.** $w^2 + 4w + 4$

10. $w^2 + 12w + 36$ **11.** $t^2 - 16t + 64$ **12.** $t^2 + 20t + 100$

Solve:

13. $x^2 - 64 = 0$ **14.** $x^2 - 121 = 0$ **15.** $y^2 - 196 = 0$

16. $y^2 - 169 = 0$ **17.** $w^2 + 16w + 64 = 0$ **18.** $w^2 + 22w + 121 = 0$

19. $r^2 - 28r + 196 = 0$ **20.** $r^2 + 26r + 169 = 0$

B

Factor:

21. $9x^2 - 4$ **22.** $4x^2 - 9$ **23.** $16y^2 - 25$

24. $25y^2 - 81$ **25.** $8x^3 - 1$ **26.** $27y^3 - 8$

27. $w^6 + 1$ **28.** $r^9 + 8$ **29.** $4w^2 + 28w + 49$

30. $4w^2 + 12w + 9$ **31.** $9t^2 - 30t + 25$ **32.** $25t^2 - 30t + 9$

Solve:

33. $25x^2 - 36 = 0$ **34.** $36x^2 - 49 = 0$ **35.** $9y^2 - 196 = 0$

36. $25y^2 - 144 = 0$ **37.** $4w^2 + 20w + 25 = 0$ **38.** $4w^2 + 44w + 121 = 0$

39. $25r^2 - 80r + 64 = 0$ **40.** $64r^2 - 48r + 9 = 0$

C

Factor:

41. $4x^2 + 49$ **42.** $1 + 81r^2$ **43.** $49a^2b^2 - 4c^2$

44. $x^2 - 25w^2y^6$ **45.** $64x^3 - 27$ **46.** $27y^3 - 125$

47. $y^6 + 216$ **48.** $t^{21} + 1$ **49.** $169x^2 - 286x + 121$

50. $256a^2 - 288ab + 81b^2$ **51.** $81y^2 + 126yz + 49z^2$ **52.** $121a^2 + 220a + 100$

Solve:

53. $11x^2 + 3 = 2x^2 + 19$

54. $29x^2 + 11 = 4x^2 + 20$

55. $3x^2 + 2 = 2x^2 + 27$

56. $15x^2 - 15 = 6x^2 - 11$

57. $y^2 - 5y + 10 = 26 - 5y$

58. $5y^2 + 2y = 2y + 1 + 4y^2$

59. $3w^2 + 2w + 5 = 2w^2 - 6w - 11$

60. $9w^2 + 2w + 20 = 50w - 44$

61. $z^2 - z - 1 = 77z - 8z^2 - 170$

62. $4z^2 + 220 = 60z - 5$

D

63. The square of Ada's age six years from now minus twelve times her age fifteen years from now is zero. How old is Ada today?

64. The square of Manuel's age five years from now minus ten times his age nine years from now is 419. How old is Manuel today?

65. Twice the square of Krista's age today less seven times her age three years ago is 21. How old is Krista now?

66. A variable electrical current is given by $i = t^2 - 18t + 100$. If t is in seconds, in how many seconds will the current (i) equal 19 amperes?

67. A variable electrical current is given by $i = t^2 - 10t + 37$. If t is in seconds, in how many seconds will the current (i) equal 12 amperes?

68. Find two numbers that have a product of 456 by factoring $625 - 169$.

69. Find two numbers that have a product of 91 by factoring $100 - 9$.

70. The height above the ground of an object that is thrown directly upward from a height of 7 feet at any time (t) is given by $s = 7 + 56t - 16t^2$, where s is the distance in feet. In how many seconds will the object reach a height above the ground of 56 feet?

71. The height above the ground of an object that is thrown directly upward from a height of 8 feet at any time (t) is given by $s = 8 + 72t - 16t^2$, where s is the distance in feet. In how many seconds will the object reach a height above the ground of 89 feet?

72. The square of twice a number added to fifteen is 79. What is the number?

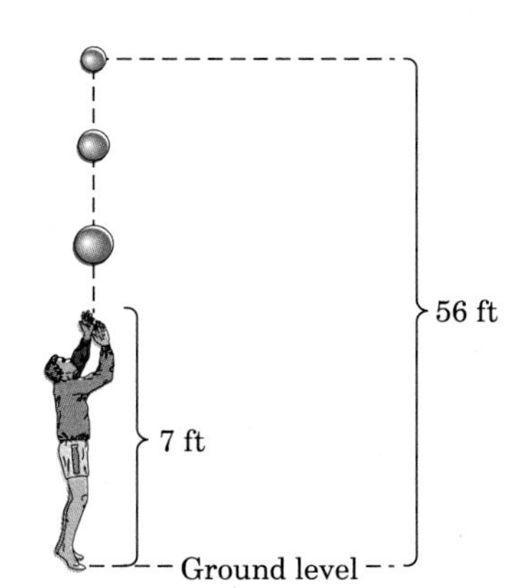

Figure for Exercise 70

STATE YOUR UNDERSTANDING

73. Explain how to recognize a perfect square trinomial.

74. Explain how to factor the sum or difference of two cubes.

CHALLENGE EXERCISES

Factor the following expressions

75. $x^{2n} - 225$

76. $4x^2 - 72x + 324$

77. $(x - y)^2 - 49$

78. $36 - (x - y)^2$

79. $(3x - 2y)^2 - (x - y)^2$

MAINTAIN YOUR SKILLS (SECTIONS 3.7, 3.8)

Multiply:

80. $(x + 7)(2x - 1)$

81. $(3x + 4)(x - 11)$

82. $(x - 2y)(x + 2y)$

83. $(3a - 7b)(3a - 7b)$

84. $(x - 2y)^2$

85. $(5y + 7)^2$

86. $(2y - 5)(7y + 8)$

87. $(9a + 7)(5a - 8)$

4.7

FACTORING POLYNOMIALS: A REVIEW

OBJECTIVES

1. Factor polynomials in general.
2. Solve related equations.

In the last five sections, you have learned to factor different types of polynomials. In this section, you need to recognize which type of polynomial is to be factored. Some problems will involve more than one type of polynomial.

EXAMPLE 1

Factor: $ax^2 - a$

$ax^2 - a = a(x^2 - 1)$ Common monomial factor.

$= a(x - 1)(x + 1)$ Difference of two squares. ■

A polynomial will be considered completely factored when each factor, other than monomial factors, is a prime polynomial. In Example 1, $a(x^2 - 1)$ is factored but not completely factored. However, $ax^2 - a = a(x - 1)(x + 1)$ is completely factored.

The following procedure will help when you are completely factoring a polynomial.

PROCEDURE

1. If there is a common factor, use the distributive property to factor out the greatest common monomial factor (GCF).
2. If the polynomial or any factor is a binomial, see if it can be factored as
   i. the difference of two squares.
   ii. the sum or difference of two cubes.
3. If the polynomial or any factor is a trinomial, try to factor it by the appropriate method.
4. If the polynomial contains four or more terms, try to factor it by grouping.
5. Check that all factors are prime, that is, that none of the remaining factors can be factored further by steps 1 through 4.

If the polynomial cannot be factored by any of these methods, for the purposes of this text we will say it cannot be factored. Other methods of factoring are studied in more advanced mathematics courses.

EXAMPLE 2

Factor: $7x^2 - 21x - 378$

$7x^2 - 21x - 378 = 7(x^2 - 3x - 54)$ — The GCF of the trinomial is 7.

Try to factor the trinomial in parentheses.

$mn = -54$	$m + n = -3$
$1(-54)$	-53
$2(-27)$	-25
$3(-18)$	-15
$6(-9)$	-3

So, $x^2 - 3x - 54 = (x - 9)(x + 6)$.

$7x^2 - 21x - 378 = 7(x - 9)(x + 6)$ ■

EXAMPLE 3

Factor: $5y^2 - 1125$

$5y^2 - 1125 = 5(y^2 - 225)$ — The GCF of the binomial is 5.

$5y^2 - 1125 = 5(y + 15)(y - 15)$ — The binomial in parentheses is the difference of two squares. ■

EXAMPLE 4

Factor: $6x^2 + 27x - 105$

$6x^2 + 27x - 105 = 3(2x^2 + 9x - 35)$

The GCF of the trinomial is 3.

$mn = 2(-35) = -70$	$m + n = 9$
$-1(70)$	69
$-2(35)$	33
$-5(14)$	9

Try to factor the trinomial in parentheses.

$$6x^2 + 27x - 105 = 3[(2x^2 - 5x + 14x - 35)]$$
$$= 3[x(2x - 5) + 7(2x - 5)]$$
$$= 3(x + 7)(2x - 5)$$

■

EXAMPLE 5

Factor: $ab^2 - 16a + b^2c - 16c$

$ab^2 - 16a + b^2c - 16c = (ab^2 - 16a) + (b^2c - 16c)$

Try factoring by grouping.

$= a(b^2 - 16) + c(b^2 - 16)$

$= (a + c)(b^2 - 16)$

The binomial factor on the right can be further factored as the difference of two squares.

$= (a + c)(b + 4)(b - 4)$

■

$16a^2+56ab+49b^2$

$(4a+7b)(4a+7b)$

$28ab$
$28ab$
$56ab$

EXAMPLE 6

Factor: $48x^4y^2 - 8x^3y^3 - 120x^2y^4$

$48x^4y^2 - 8x^3y^3 - 120x^2y^4 = 8x^2y^2(6x^2 - xy - 15y^2)$

The GCF is $8x^2y^2$.

$mn = -90$	$m + n = -1$
$1(-90)$	-89
$2(-45)$	-43
$3(-30)$	-27
$5(-18)$	-13
$6(-15)$	-9
$9(-10)$	-1

Try to factor the trinomial in parentheses.

$$48x^4y^2 - 8x^3y^3 - 120x^2y^4 = 8x^2y^2(6x^2 + 9xy - 10xy - 15y^2)$$
$$= 8x^2y^2[3x(2x + 3y) - 5y(2x + 3y)]$$
$$= 8x^2y^2(3x - 5y)(2x + 3y)$$

■

EXAMPLE 7

Factor: $432x^3y + 2y$

$432x^3y + 2y = 2y(216x^3 + 1)$ — The GCF is $2y$. The binomial is the sum of two cubes.

$= 2y(6x + 1)(36x^2 - 6x + 1)$ ■

When solving equations using factoring and the zero-product property, we must first write an equivalent equation with one side equal to zero.

EXAMPLE 8

Solve: $5x^2 + 6x - 10 = 3x^2 + 13x + 5$

$2x^2 - 7x - 15 = 0$ — Write the equation in standard form by adding $-3x^2 - 13x - 5$ to both sides.

$(2x + 3)(x - 5) = 0$ — Factor.

$2x + 3 = 0 \quad \text{or} \quad x - 5 = 0$ — Zero-product property.

$x = -\frac{3}{2} \quad \text{or} \quad x = 5$

The solution set is $\left\{-\frac{3}{2}, 5\right\}$. ■

EXAMPLE 9

Wendel frames an 8-by-10-in. photo with a frame of uniform width at the U-Frame-It Shop. If the area of the photo and frame is 120 square inches, what is the width of the frame?

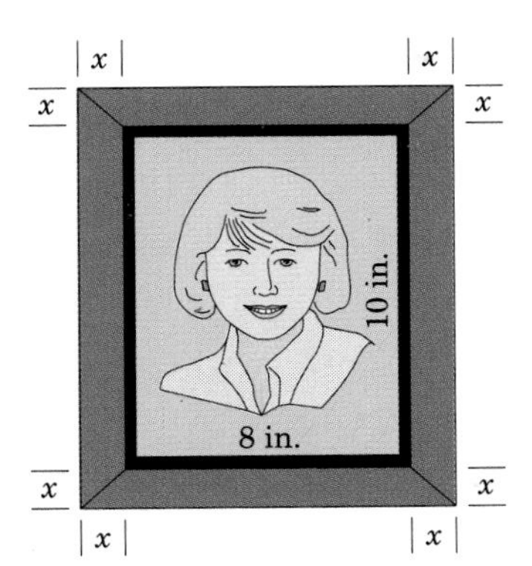

Formula:

$A = \ell w$

Select variable:

Let x represent the width of the frame. From the diagram, we see that the dimensions of the framed photo are represented by $(2x + 8)$ and $(2x + 10)$.

Substitute:

$120 = (2x + 10)(2x + 8)$

Solve:

$120 = 4x^2 + 36x + 80$

$0 = 4x^2 + 36x - 40$

$0 = 4(x^2 + 9x - 10)$

$0 = 4(x + 10)(x - 1)$

$x = -10$ or $x = 1$ — The solution $x = -10$ is rejected since the width of the frame cannot be negative.

Check:

$2x + 8 = 2(1) + 8 = 10$ — Total width.

$2x + 10 = 2(1) + 10 = 12$ — Total length.

$(10)(12) = 120$

Answer:

The width of the frame that Wendel used was 1 inch. ■

EXERCISE 4.7

A

Factor:

1. $6a^2 + 8ab$
2. $21c^2 - 35cd$
3. $x^2 - x - 56$
4. $x^2 - 10x - 56$
5. $16a^2b^4 - 25$
6. $16x^2y^2 - 9$
7. $a^2 - 46a + 529$
8. $t^2 + 52t + 676$
9. $5c^2 - 5cd + 3cd - 3d^2$
10. $2mn - 2mp - 3n^2 + 3np$
11. $10x^2 - 25x + 10$
12. $18x^2 - 60x + 18$
13. $7x^3 - 7$
14. $10y^3 + 10$

Solve:

15. $3y^2 - 507 = 0$
16. $7y^2 - 1372 = 0$
17. $2r^2 - 28r + 98 = 0$
18. $3r^2 + 48r + 192 = 0$

B

Factor:

19. $x^2 + 15x - 34$
20. $c^2 - 9c - 70$

21. $8x^4y - 16x^2y^3 - 4x^2y$

22. $7a^4b^4 + 7a^3b^4 - 42a^2b^3$

23. $6s^2 - st - 35t^2$

24. $5c^2 + 19cd - 4d^2$

25. $24m^2 + 8mt - 9mt - 3t^2$

26. $14s^2 + 7st - 10st - 5t^2$

27. $200k^2 - 2$

28. $3k^2 - 1200$

29. $2aw^2 + 44aw + 242a$

30. $3xw^2 - 72xw + 432x$

31. $16y^3w + 54wz^3$

32. $125a^4b^3c - 64ac^4$

Solve:

33. $r^2 + 9 = 6r$

34. $3t^2 + 19t = 14$

35. $6w^2 = -10w$

36. $140w^2 = 120w$

37. $3t^2 = -12 - 20t$

38. $64b^2 - 49 = 0$

39. $y^2 - 7 = (5 + y)(5 - y)$

40. $2y^2 + 13 = (16 + y)(16 - y)$

C

Factor:

41. $5x^3 - 5xy^2 + 2x^2y - 2y^3$

42. $3c^3 - 12cd^2 - c^2d + 4d^3$

43. $9a^3 - 9a$

44. $63c^3s - 28cs^3$

45. $8x^3 + 72x^2 + 112x$

46. $3a^4 - 18a^3 - 120a^2$

47. $42t^2 + 133t - 49$

48. $-75a^2 + 65ab + 30b^2$

49. $72a^3b^3 - 66a^2b^2c - 30abc^2$

50. $60c^4d - 28c^3d^2 - 8c^2d^3$

51. $2x^3 + 3x^2 - 50x - 75$

52. $12x^3 + 16x^2 - 3x - 4$

53. $4a^3b^2c - 32b^2c^4$

54. $135x^4y^3z^2 - 5xw^3z^2$

Solve:

55. $18p^2 = 98$

56. $50p^2 = 32$

57. $(x - 4)^2 = 25$

58. $16x^2 + 12x - 3 = -2x + 12$

Solve for x:

59. $2x^2 - 5ax - 3a^2 = 0$

60. $4x^2 - 9b^2 = 0$

Solve for y:

61. $(y - a)(y - a) + (y + a)(y + a) = 10a^2$

62. $(y + 10m)(y - 10m) = -2(y + 2m)(y - 2m)$

D

63. Twice the square of a number decreased by seven times the number is 15. Find the positive value for the number.

64. The product of six less than five times a number and three more than twice the number is -11. Find both values for the number.

65. The U-Frame-It Shop frames a 5-by-7-in. photo with a frame of uniform width. If the area of the photo and the frame is 99 square inches, what is the width of the frame?

66. The U-Frame-It Shop frames a 7-by-9-in. photo with a frame of uniform width. If the area of the photo and the frame is 99 square inches, what is the width of the frame?

67. An oil painting at the Sea-n-Breeze Gallery is rectangular and has an area of 143 square inches. If the length is 2 inches longer than the width, what are the dimensions of the oil painting?

68. A water color at the McClain Gallery is rectangular and has an area of 1080 square inches. If the length is 6 inches longer than the width, what are the dimensions of the water color?

69. A turkey farmer buys a roll of fence that measures 400 ft in length. Find the dimensions of the rectangular area the fence can enclose if it contains 9900 square feet.

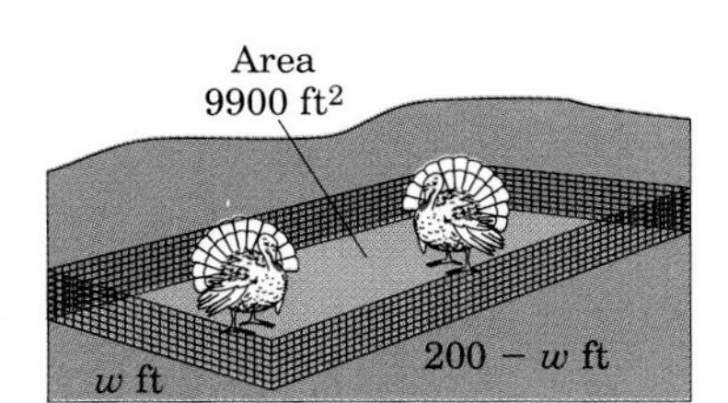

Figure for Exercise 69

70. A chicken farmer buys a roll of fence that measures 600 ft in length. Find the dimensions of the rectangular area the fence can enclose if it contains 19,475 square feet.

71. The length of a rectangle is three inches more than the width. If both the width and length are increased by two inches, the area of the new rectangle is 54 square inches. Find the dimensions of the original rectangle.

72. A rectangle has a length of five centimeters less than twice the width. If both the length and width are decreased by two centimeters, the area of the new rectangle is 20 square centimeters. Find the length of the original rectangle.

STATE YOUR UNDERSTANDING

73. Describe the procedure used to factor completely.

74. Describe the procedure used to solve a quadratic equation.

CHALLENGE EXERCISES

75. $x^6 - y^6$

76. $x^4 - 13x^2 + 36$

77. $x^4 - 8x^2 + 16$

78. $x^4 - x^2 - 12$

79. $4(x^2 - 25)^2 - 12(x^2 - 25) + 9$

MAINTAIN YOUR SKILLS (SECTIONS 3.7, 3.8, 3.9)

Multiply:

80. $(a - b + 3)(2a + 3b - 5)$

81. $(2m - 3n + 7)(4m - 2n - 2)$

82. $(2x - 3y)(2x + 3y)(3x + 4y)$

Solve:

83. $(3x - 1)(x^2 - 5x - 8) = 3(x^3 - 2) - 16x^2$

84. $(5y - 2)(2y^2 + 3y - 6) = -10y(4 - y^2) - (3 - 11y^2)$

Divide:

85. $(2x^2 - 7x - 15) \div (x - 5)$

86. $(14x^2 + 43x - 21) \div (2x + 7)$

87. $(x^3 + 3x^2 - 22x + 42) \div (x + 7)$

CHAPTER 4
SUMMARY

Linear Equation	A linear equation is an equation that can be written in the form $ax + b = c$, $a \neq 0$.	(p. 235)
Quadratic Equation	A quadratic equation is an equation that can be written in the form $ax^2 + bx + c = 0$, $a \neq 0$.	(p. 235)
Zero-Product Property	If $A \cdot B = 0$, then $A = 0$ or $B = 0$.	(p. 236)
Polynomial Over the Integers	A polynomial over the integers is a polynomial that has only integers for coefficients of its terms. $3x^2 - 7x + 17$ is a polynomial over the integers.	(p. 241)
Prime Polynomial	A prime polynomial is a polynomial that has exactly two factors that are polynomials over the integers, 1 and the polynomial itself. $x^2 + 1$ is a prime polynomial.	(p. 243)
Common Monomial Factor	A monomial that is a factor of each term of the polynomial. The polynomial can be factored using the common monomial factor as a factor.	(p. 242)
Factor by Grouping	Terms of a polynomial are grouped so that they have common monomial factors in each group. $ax + ay + bx + by = a(x + y) + b(x + y) = (a + b)(x + y)$	(p. 249)

Leading Coefficient	The leading coefficient of a polynomial is the coefficient of the term of highest degree. The leading coefficient of $3x^2 + 4x - 10$ is 3.	(p. 254)
Factoring $x^2 + bx + c$	Trinomials of the form $x^2 + bx + c$ can be factored by writing all the possible pairs of factors (mn) of c. The factors are $(x + m)(x + n)$, where $m + n = b$.	(p. 256)
Factoring $ax^2 + bx + c$	Trinomials of the form $ax^2 + bx + c$, $a > 1$, can be factored by writing all the possible pairs of factors $(ex + f)(gx + h)$, where $eg = a$ and $fh = c$. Multiply these pairs to find the correct pair of factors. Or replace the term bx with the terms $mx + nx$, where $mn = ac$ and $m + n = b$, and factor by grouping.	(p. 261)
Factoring the Difference of Two Squares	$a^2 - b^2 = (a - b)(a + b)$	(p. 270)
Factoring the Sum of Difference of Two Cubes	$a^3 + b^3 = (a + b)(a^2 - ab + b^2)$ $a^3 - b^3 = (a - b)(a^2 + ab + b^2)$	(p. 272)
Factoring a Perfect Square Trinomial	$x^2 \pm 2bx + b^2 = (x \pm b)^2$ The sign chosen is the sign of the middle term. $a^2x^2 \pm 2abx + b^2 = (ax \pm bx)^2$ The sign chosen is the sign of the middle term.	(p. 273)

CHAPTER 4
REVIEW EXERCISES

SECTION 4.1 Objective 1

Solve:

1. $(x - 12)(x + 3) = 0$

2. $(x + 6)(x - 1) = 0$

3. $(y - 21)(y - 2) = 0$

4. $(x + 14)(x - 1) = 0$

5. $(x - 3.5)(x + 2.1) = 0$

6. $\left(y + \frac{3}{5}\right)\left(y - \frac{2}{17}\right) = 0$

7. $6z(4z - 9) = 0$

8. $-14t(12t - 5) = 0$

9. $(3x - 7)(5x + 8) = 0$

10. $(10y - 3)(9y + 8) = 0$

SECTION 4.2 Objective 1

Factor:

11. $4y - 8z$

12. $2abc - 16abd$

13. $21ab + 7ad$

14. $34xy + 17xz$

15. $12x^2 - 36x$

16. $24x^2y - 32xy^2$

17. $4x^2 - 8x^3 + 6x^2y^2$

18. $45a^3bc^2 - 30a^2bc^2 + 60a^2b^2c^3$

19. $5a^2b^2 - 10ab - 15a^2b - 20ab^2$

20. $7m^2n^2 + 14mn - 21m^2n - 28n^2$

SECTION 4.2 Objective 2

Solve:

21. $x^2 + 9x = 0$

22. $y^2 - 13y = 0$

23. $9x^2 - 27x = 0$

24. $-13x^2 - 39x = 0$

25. $32x = -8x^2$

26. If one side of a square is lengthened 5 inches to form a rectangle, the area of the new rectangle is twice the area of the original square. Find the length of the side of the square.

SECTION 4.3 Objective 1

Factor by grouping:

27. $ct - cs + dt - ds$

28. $5a - 10b + ax - 2bx$

29. $w^6 + w^5 + 4w + 4$

30. $w^6 - 5w^4 + 3w^2 - 15$

31. $32x^2y - 16x + 10xyz - 5z$

32. $7ab + 14ac + 9bd^2 + 18cd^2$

33. $30abx + 35cx + 7cy + 6aby$

34. $4x^3y^5 + 8x^4y^4 + 7ys^2t^3 + 14xs^2t^3$

35. $s^3 + 5s^2 + s^2t + 5st + st^2 + 5t^2$

36. $7xy^2 + x^2y + 7xyz + x^2z + 7yz + xz$

SECTION 4.3 Objective 2

Solve:

37. $x^2 + 3x + 2x + 6 = 0$

38. $m^2 - 7m + 14m - 98 = 0$

39. $4x^2 + 12x + 5x + 15 = 0$

40. $6x^2 - x + 12x - 2 = 0$

41. $88t^2 - 55t + 24t - 15 = 0$

42. $63r^2 - 28r + 45r - 20 = 0$

SECTION 4.4 Objective 1

Factor:

43. $x^2 - 12x + 35$

44. $x^2 + 12x + 27$

45. $y^2 - 18y + 32$

46. $y^2 - 18y + 65$

47. $a^2 + a - 156$

48. $a^2 - 18a - 19$

49. $x^2 - 7x - 98$

50. $x^2 - 5x - 104$

51. $x^2 + x - 110$

52. $x^2 + 24x + 128$

SECTION 4.4 Objective 2

Solve:

53. $x^2 - 6x - 7 = 0$

54. $x^2 + 17x + 60 = 0$

55. $y^2 - 19y + 78 = 0$

56. $y^2 - 4y - 96 = 0$

57. $x^2 + 33x + 242 = 0$

58. The product of a number plus five and the same number less four is 36. Find the number.

SECTION 4.5 Objective 1

Factor:

59. $3x^2 - 16x - 12$

60. $4x^2 - x - 5$

61. $5x^2 + 29x - 6$

62. $6x^2 - 31x + 5$

63. $10x^2 + 17x - 63$

64. $12x^2 - 13x - 35$

65. $2x^2 + 21x + 54$

66. $3x^2 - 14x + 16$

67. $24x^2 - 22x - 35$

68. $54x^2 - 33x - 35$

SECTION 4.5 Objective 2

Solve:

69. $3x^2 + 13x - 30 = 0$

70. $2x^2 + 21x - 98 = 0$

71. $6x^2 - 7x - 5 = 0$

72. $12x^2 - 7x - 12 = 0$

73. $168x^2 + 38x - 45 = 0$

74. The height of a given triangle is five less than four times the base. The area is 108 square feet. Find the length of the base.

SECTION 4.6 Objective 1

Factor:

75. $w^2 - 400$

76. $t^2 - 225$

77. $16a^2 - 25b^2$

78. $81b^2c^2 - 16d^2$

79. $100y^2 - 121z^2$

80. $36x^2y^2 + 81z^2$

81. $49a^2b^2c^2 - 9d^2$

82. $x^2y^2 - 196w^2z^2$

83. $x^4 - 9$

84. $x^6 - 4$

SECTION 4.6 Objective 2

Factor:

85. $s^3 - 216$

86. $t^3 - 343$

87. $x^3 + 1000$

88. $w^3 + 64$

89. $8d^3 - 125$

90. $27b^3 - 1$

91. $64a^3b^3c^3 + 125d^3e^3$

92. $1000s^3t^3 + 27v^3$

93. $t^6 + 125$

94. $y^{15} - 8z^{15}$

SECTION 4.6 Objective 3

Factor:

95. $x^2 - 8x + 16$

96. $x^2 + 14x + 49$

97. $x^2 - 18x + 81$

98. $x^2 - 22x + 121$

99. $9x^2 + 42x + 49$

100. $16x^2 - 24x + 9$

101. $25x^2 + 80xy + 64y^2$

102. $25x^2 - 20xy + 4y^2$

103. $81a^2b^2 - 18abc + c^2$

104. $9m^2n^2 - 48mn + 64$

SECTION 4.6 Objective 4

Solve:

105. $49x^2 - 81 = 0$

106. $64y^2 - 9 = 0$

107. $36x^2 - 25 = 0$

108. $16x^2 = 121$

109. $x^2 - 50x + 625 = 0$

110. $9x^2 - 84x + 196 = 0$

111. $25y^2 + 144 = 120y$

112. The square of three times a number added to 30 is 471. What is the number?

SECTION 4.7 Objective 1

Factor:

113. $3x^2 - 12x - 135$

114. $10x^2 - 45x - 280$

115. $12x^3 - 147x$

116. $80x^2y - 45y$

117. $42a^2b^3 + 28ab^3 - 63a^2b^2 - 42ab^2$

118. $10x^4y - 25x^3y + 10x^2y$

119. $56x^3 + 7$

120. $216x^4 - 64x$

121. $14x^5 - 1134x$

122. $8a^7b - 8ab$

SECTION 4.7 Objective 2

Solve:

123. $4x^2 + 16x - 84 = 0$

124. $18x^2 - 2 = 0$

125. $12x^2 - 14x - 110 = 0$

126. $12a^2 - 44a = 0$

127. $24x^3 + 66x^2 - 63x = 0$

128. A cattle rancher buys four rolls of fence that each measures 500 ft in length. Find the dimensions of the rectangular area the fence can enclose if it contains 171,600 square feet.

CHAPTER 4
TRUE–FALSE CONCEPT REVIEW

Check your understanding of the language of algebra. Tell whether or not each of the following is true (always true) or false (not always true).

1. A quadratic equation is an equation that contains four terms.
2. The zero-product law states that if $A \cdot B = 0$, then $B = 0$.
3. Every quadratic equation has two distinct solutions.
4. If $(x - 5)(x + 7) = 0$, then $x = 5$ or $x = -7$.
5. $x(x - 3) + 5$ is a factored polynomial.
6. A common monomial factor is a factor of each term of a polynomial.
7. Factoring a polynomial is the inverse of multiplying two polynomials.
8. $3(x + y) - y(z + 5)$ is the factored form of $3x + 3y - 2z - 10$.
9. The leading coefficient of the trinomial, $ax^2 + bx + c$, is the coefficient of x^2.
10. $(x + a)(x + b) = x^2 + cx + d$ if $ab = c$ and $a + b = d$.
11. A perfect square trinomial is the square of a binomial.
12. The factors of the difference of two squares are the sum and difference of the terms that are squared.

13. The sum of two cubes cannot be factored using the methods of this chapter.

14. $16a^2 + 20x + 25$ is a perfect square trinomial.

15. $x^9 - 1 = (x^3 - 1)(x^3 + 1)$

16. If $(ax + b)(cx + d) = ex^2 + fx + g$, then $b + d = f$.

17. A polynomial is said to be completely factored when all its factors are prime polynomials.

18. $x^3 - y^3 = (x + y)(x - y)(x + y)$

19. $x^2 + 81y^2$ is a prime polynomial.

20. A polynomial that is the sum of two squares is prime.

CHAPTER 4
TEST

1. Factor: $10a^2bc^3 - 12ab^2c^2 + 16a^3bc^4$
2. Factor: $75a^3 - 270a^2b + 243ab^2$
3. Factor: $ax - bx + 2a - 2b$
4. Factor: $2a^2 - 3ab + 2ac - 3bc$
5. Solve: $x^2 + 20x + 36 = 0$
6. Factor: $x^2 - 11x + 18$
7. Solve: $x(x - 2) = 0$
8. Factor: $a^2 - a - 72$
9. Solve: $x^2 - 24x + 144 = 0$
10. Solve: $y^2 - 3y - 28 = 0$
11. Factor: $22ab - 44abc$
12. Factor: $64x^2y^2 - 49a^2$
13. Solve: $x^2 + 5x = 0$
14. Factor: $r^2 - 19r + 18$
15. Factor: $6x^2 - 5x - 6$
16. Factor: $14x^2y + 28xy - 77wy$
17. Factor: $16x^2 + 24xy + 9y^2$
18. Factor: $y^2 + 5y - 66$
19. Factor: $8x^2 - 4x + 2xy - y$
20. Factor: $12x^2 + 7x - 12$
21. Factor: $3x^3 - 375y^3$
22. Factor: $12x^2 - 25x + 12$
23. Factor: $36b^2c^2 - 1$
24. Factor: $6t^2 + 7t - 20$
25. Factor: $128a^2x - 160abx + 50b^2x$
26. Solve: $9y^2 - 100 = 0$
27. Solve: $6x^2 - 37x + 6 = 0$
28. Factor: $36x^2 + 84xy + 49y^2$
29. Find two numbers that have a sum of 9 and a product of -90.
30. A strip of metal 8 inches wide is to be bent into a trough of rectangular shape (open top). The area of the rectangular cross section is to be 8 square inches. Find the depth and width of the trough.

CHAPTERS 1–4

Cumulative Review

CHAPTER 1

1. Reduce to lowest terms: $\frac{108}{162}$

Perform the indicated operations:

2. 4.67×0.713

3. $0.7668 \div 0.27$

4. $8.72 + 0.7 + 0.52 + 0.537$

5. $27.2 - 19.517$

6. Write in exponential form: $2 \cdot 2 \cdot 2 \cdot 2 \cdot 7 \cdot 7 \cdot 9 \cdot 9 \cdot 9$

7. $2^3 - 3[10 - 2(5^2 - 3 \cdot 7)]$

8. $\frac{3 \cdot 14^2 + 4 \cdot 7 - 6^3}{4(7^2 - 3^3) \div (5^2 - 14)}$

9. Find the value of x if $x = 8^2 - 4 \cdot 7^2 + 5^3$

If $x = 4.75$, $y = 2.64$, and $z = 0.015$, evaluate:

10. $\frac{y}{z}$

11. $x^2 - y^3 + z$

12. Is 4 a solution of $5x - 12 = 2x + 1$?

13. Is 13 a solution of $3x + 4 = 4x - 9$?

Classify the following as either true or false:

14. $\sqrt{8}$ is an integer.

15. All whole numbers are rational numbers.

16. No real number is an irrational number.

17. $-12 < -5$

Determine the value of each of the following:

18. $(-6) + (+4) + (-3) + (+12)$

19. $-5.2 + 3.47$

20. $\frac{3}{10} + \left(-\frac{1}{2}\right)$

21. $-|4 + (-7)|$

Determine the value of each of the following:

22. $-4.5 - 2.7$

23. $-\frac{5}{8} - \frac{5}{6}$

24. $-21 - 47 - (-34)$

25. $-(-12) - (17) - (-41) - (30)$

Determine the value of each of the following:

26. $(-2)(5)(-7)$

27. $(0.6)(-0.3)(4)$

28. $(-3)(41)(0)(-17)$

29. $-4^2(-3)^2(-1)^3$

Determine the value of each of the following:

30. $(-4.41) \div (0.7)$

31. $\left(-\frac{5}{12}\right) \div \left(-\frac{10}{27}\right)$

32. $\frac{-5.4}{0.04}$

33. $\left(1\frac{3}{4}\right) \div \left(-\frac{3}{8}\right)$

Determine the value of each of the following:

34. $-20\left|\frac{12 - 3}{-4 - 6}\right| - 4(11 - 19)$

35. $5(3^2 - 4^3) - (-7)^2(-6)$

Evaluate if $a = -7.2$, $b = -8.4$, and $c = -4.3$

36. $2a^2 - 3bc$

37. $(a + 2)(b - c)$

CHAPTER 2

Solve:

38. $y + \frac{1}{3} = \frac{2}{5}$

39. $x - 17.9 = -41.7$

40. Solve for c: $c + m = s$

41. Solve for x: $y = x - 7$

Solve:

42. $-2.7x = 1.35$

43. $\frac{5}{6}x = -\frac{5}{9}$

44. Solve for ℓ: $A = \ell w$

45. Solve for I: $P = \frac{I}{rt}$

Solve:

46. $5x + \frac{2}{3} = \frac{5}{8}$

47. $44.5x - 40.9 = 83.7$

48. Solve for C: $F = \frac{9}{5}C + 32$

49. Solve for g: $s = v_0 t - \frac{1}{2}gt^2$

Simplify:

50. $12x^2 - 4x - x^2 + 7x$

51. $9x^3y - 2x^3y^2 + 3x^3y^2 - 4x^2y^3 - 5x^3y^2 - x^2y^3$

52. $-\frac{1}{2}a + \frac{2}{3}b - \frac{1}{4}c + \frac{1}{3}a - \frac{1}{6}b + \frac{2}{3}c$

53. $7.63 - 8.54u - 3.6v - 5.9 + 3.8u$

Solve:

54. $4x - 5 + 11 = 30 - 2x$

55. $y^2 - 8y + 5 = y^2 + 8y - 5$

56. $8x + 11 - 3x = 4x - 12 + x$

57. $308x - 195 = 315x + 148$

58. 87% of what number is 61? (to the nearest tenth)

59. What number is 42.7% of 125,000? (to the nearest hundred)

60. 35 is what percent of 48? (to the nearest tenth of one percent)

61. The employment at a machine shop is down 5%. If the previous figures showed 140 employees, how many people are currently employed there?

Solve and graph:

62. $2x - 3 > 11$

63. $\frac{2}{3}x + 16 \leq 12$

64. $\frac{7}{2} - \frac{2}{3}x \geq 4$

65. $2x - 1 < 5x + 6 - 3x$

CHAPTER 3

Simplify:

66. $x \cdot x^2 \cdot x^5 \cdot x^3$

67. $a^m \cdot a^4$

68. $(-3a^2b)^2$

69. $(4x^2y^0z^3)^3$

Simplify and write answers using positive exponents:

70. $\frac{4^5 \cdot 4^{-2}}{4^8 \cdot 4^{-4}}$

71. $\left(\frac{x^3 \cdot x^{-2}}{x^{-2} \cdot x^5}\right)^{-2}$

72. $\left(\dfrac{x^2y^{-3}}{x^{-5}y^{-4}}\right)^{-3}$

73. $\left(\dfrac{5^3a^2b^0c^{-4}}{5a^2b^{-3}c^2}\right)^2$

Write in scientific notation:

74. 4,715,000,000

75. 0.000693

Write in place value notation:

76. 8.74×10^{-6}

77. 6.03×10^{23}

Simplify and write the answer in scientific notation:

78. $(3.2 \times 10^5)(4.65 \times 10^{-8})$

79. $\dfrac{6.5 \times 10^4}{2.6 \times 10^{-4}}$

Determine the degree:

80. $8x^3 - 7x^4 + 5x^2 - 6x^5 - 12x + 11$

81. $14x^2y^3z^3 + 3x^3y^4z^2 - 6x^4y^5z$

82. Write Exercise 80 in descending powers:

Simplify:

83. $(8x^3 - 7x - 3) + (2x^2 + 4x + 5) - (-5x^3 + 4x^2 + x - 4)$

84. $[7x - (3x - 8)] - [-(4x - 3) - 6]$

Solve:

85. $(4x - 5) + (2x + 11) = (3x - 4) - (5x - 2)$

86. $(3x - 4) - (8x - 3) = 5 - 2x - (4x - 11)$

87. One driver traveling at 45 miles per hour leaves Boise for Reno at the same time another driver traveling at 55 mph leaves Reno for Boise. The cities are 425 miles apart. How long will it take for the drivers to meet each other?

88. Equal weights of cheddar cheese, selling at \$2.79 per pound, and mozarella cheese, selling at \$1.69 per pound, were shipped to Just A Cheese Store. The total value of the shipment was \$215.04. How many pounds of each cheese were shipped?

89. Equal volumes of 25% sulfuric acid and 55% sulfuric acid are mixed to help generate hydrogen in a chemistry laboratory. If a total of 32 ml of pure sulfuric acid are needed, how much of each solution is required?

90. A bank contains equal numbers of nickels, dimes, and quarters. The value of the coins is \$35.60. How many of each coin are in the bank?

Multiply:

91. $(7y - 2)(3y + 5)$

92. $(2x - 5)(x^2 - 4x + 7)$

Solve:

93. $(x - 7)(x + 4) = (5 - x)(6 - x)$

94. $(x + 2)(x^2 - 2x + 4) = x^3 - 4x - 16$

Multiply:

95. $(2x + 7)^2$

96. $(6x - 5)(6x + 5)$

Solve:

97. $(3x - 5)^2 = (3x + 5)^2$

98. $(x - 8)^2 = (x - 12)(x + 12)$

Divide:

99. $(18x^4y^4 + 24x^3y^3 - 6x^2y^2) \div (-6x^2y^2)$

100. $(6x^2 + x - 35) \div (2x + 5)$

101. $(10x^2 - 7y^2 - 38xy) \div (x - 4y)$

102. $(x^4 - 6x^2 + 8) \div (x - 3)$

CHAPTER 4

Solve:

103. $(4x - 3)(x + 2) = 0$

104. $x(3x + 2) = 0$

105. $(5x + 2)^2 = 0$

106. $(4x - 11)(4x + 11) = 0$

Factor:

107. $2x^3 - 8x^2 + 10x$

108. $32x^2y^3z^4 - 16x^3y^2z^3 + 24x^4y^2z$

Solve:

109. $30x^2 - 15x = 0$

110. $y^2 = -12y$

Factor by grouping:

111. $8ab + 4b - 6a - 3$

112. $xy - 2xz - y + 2z$

Solve:

113. $x^2 + 6x - 7x - 42 = 0$

114. $6x^2 - 3x + 8x - 4 = 0$

Factor:

115. $x^2 - 12x + 32$

116. $y^2 + 7y - 44$

Solve:

117. $x^2 - 10x - 75 = 0$

118. $x^2 + 21x - 72 = 0$

Factor:

119. $18x^2 - 3x - 10$

120. $35x^2 + 29x + 6$

Solve:

121. $8x^2 + 13x - 6 = 0$

122. $20x^2 - 9x - 20 = 0$

Factor:

123. $81x^2 - 49$

124. $36x^2 - 60x + 25$

125. $8x^3 - 125$

Solve:

126. $8x^2 + 5x + 30 = 4x^2 - 7x + 21$

127. $3x^2 - 6 = 2x^2 + 75$

Factor:

128. $ax^2 - 9a - 2bx^2 + 18b$

129. $64y^4 - 4$

130. $3x^2 - 48x + 192$

131. $256y^5 + 108y^2$

132. $120x^2 - 230x + 75$

Solve:

133. $20x^2 = 605$

134. $(x - 6)(x + 2) = 20$

135. Solve for x: $3x^2 + 10ax - 8a^2 = 0$

CHAPTER 5

RATIONAL EXPRESSIONS AND RELATED EQUATIONS

Architects and engineers must be aware of the fact that building materials contract and expand with changes in temperature. When planning a new facility, they allow for these changes in the design. In Exercise 77, Section 5.6, the amount of contraction of a 20-foot beam for each temperature decrease of 5 degrees is given. Using a proportion, find the amount of contraction that will occur in a 13-foot beam with a decrease of 7 degrees. *(© Eric Kamp/Phototake NYC)*

PREVIEW

Algebraic expressions that contain fractions are governed, for the most part, using properties parallel to those for rational numbers. In Sections 5.1 and 5.2, we examine the properties that are used to reduce and multiply rational expressions. Section 5.3 shows how to find the least common multiple of two or more rational expressions. A comparison shows that the method used here is a generalization of the method used for finding common multiples of sets of whole numbers. Section 5.4 establishes that rational expressions are combined in a manner akin to that for rational numbers. Ratios and proportions in Section 5.6 model one of the most common uses of equations that contain rational expressions. Lastly, in Section 5.7, we investigate variation as an important application of equations incorporating products and quotients.

5.1

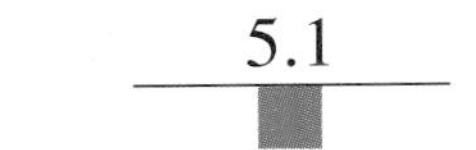

MULTIPLYING AND REDUCING RATIONAL EXPRESSIONS

OBJECTIVES

1. Reduce rational expressions.
2. Multiply rational expressions.
3. Convert measurements to equivalent measurements.

The procedures for reducing and multiplying rational expressions are very similar to those for fractions.

Rational Expressions

DEFINITION

A *rational expression* is the indicated quotient of two polynomials. $\frac{P}{Q}$ is a rational expression, where P and Q are polynomials and $Q \neq 0$.

As with fractions of arithmetic, the denominator of a rational expression cannot be zero. To guarantee this we restrict the domain of the variable(s) in the denominator.

Fraction	Replacements Not Allowed (Restrictions)
$-\frac{3}{4y}$	y cannot be replaced by zero, $y \neq 0$

Fraction	Replacements Not Allowed (Restrictions)
$\frac{x+7}{x-8}$	x cannot be replaced by eight, $x \neq 8$
$\frac{m^2-8m}{m^2+3}$	No restriction since $m^2 + 3$ is never equal to zero when m represents a real number

In arithmetic, you learned to reduce and multiply fractions (positive rational numbers). Rational expressions (ratios of polynomials) are treated similarly. For example, just as

$$\frac{24}{30} = \frac{4 \cdot 6}{5 \cdot 6} = \frac{4}{5} \cdot 1 = \frac{4}{5},$$

so,

$$\frac{24xy}{30x} = \frac{4y \cdot 6x}{5 \cdot 6x} = \frac{4y}{5} \cdot 1 = \frac{4y}{5}.$$

The rule for reducing is based on the rule for multiplying the fractions of algebra. In this form, the rule is called the basic principle of fractions.

RULE

Basic Principle of Fractions

$$\frac{AC}{BC} = \frac{A}{B} \cdot \frac{C}{C} = \frac{A}{B} \cdot 1 = \frac{A}{B}, \quad BC \neq 0$$

EXAMPLE 1

Reduce: $\frac{7a}{21b}$

$\frac{7a}{21b} = \frac{7 \cdot a}{7 \cdot 3b}$ Write the numerator and denominator in factored form.

$= \frac{a}{3b}$ Eliminate the common factor of 7, that is, reduce or "cancel." ■

PROCEDURE

To reduce a rational expression:

1. Write the numerator and denominator in factored form.

2. Eliminate all common factors (except 1) by dividing or "canceling."

$$\frac{AC}{BC} = \frac{A}{B}, \quad BC \neq 0$$

EXAMPLE 2

Reduce: $\dfrac{2x + 4}{x^2 + 5x + 6}$

$$\frac{2x + 4}{x^2 + 5x + 6} = \frac{2\overset{1}{\cancel{(x + 2)}}}{(x + 3)\underset{1}{\cancel{(x + 2)}}}$$

Factor both the numerator and denominator.

$$= \frac{2}{x + 3}$$

Eliminate the common factor $(x + 2)$ by dividing out or canceling. ■

EXAMPLE 3

Reduce: $\dfrac{x - y}{y - x}$

$$\frac{x - y}{y - x} = \frac{-1\overset{1}{\cancel{(y - x)}}}{1\underset{1}{\cancel{(y - x)}}}$$

The expression $x - y$ is the opposite of $y - x$. So $x - y = -(y - x) = -1(y - x)$.

$$= \frac{-1}{1}$$

Eliminate the common factor $(y - x)$.

$$= -1$$

or

$$\frac{x - y}{y - x} = \frac{1\overset{1}{\cancel{(x - y)}}}{-1\underset{1}{\cancel{(x - y)}}}$$

The expression $y - x$ is the opposite of $x - y$. So $y - x = -(x - y) = -1(x - y)$.

$$= \frac{1}{-1}$$

Eliminate the common factor $(x - y)$.

$$= -1$$

■

The rule for finding the product of rational expressions is similar to that for fractions of arithmetic.

As

$$\frac{3}{5} \cdot \frac{4}{7} = \frac{3 \cdot 4}{5 \cdot 7} = \frac{12}{35}$$

so,

$$\frac{3x}{5y} \cdot \frac{4x}{7} = \frac{(3x)(4x)}{(5y)(7)} = \frac{12x^2}{35y}.$$

PROCEDURE

To multiply two rational expressions, multiply the numerators and multiply the denominators.

$$\frac{A}{B} \cdot \frac{C}{D} = \frac{AC}{BD}, \quad BD \neq 0$$

Unless exercise directions ask for restrictions, we will assume that no variable will be replaced by a number that makes any denominator zero.

EXAMPLE 4

Multiply: $\dfrac{3a^2}{y} \cdot \dfrac{2ab}{5y}$

$$\frac{3a^2}{y} \cdot \frac{2ab}{5y} = \frac{3a^2 \cdot 2ab}{y \cdot 5y}$$ Multiply numerators and denominators.

$$= \frac{6a^3b}{5y^2}$$ Final product. ■

EXAMPLE 5

Multiply: $\dfrac{y+2}{y-9} \cdot \dfrac{y-11}{2y+3}$

$$\frac{y+2}{y-9} \cdot \frac{y-11}{2y+3} = \frac{(y+2)(y-11)}{(y-9)(2y+3)}$$ Multiply numerators and denominators.

$$= \frac{y^2 - 9y - 22}{2y^2 - 15y - 27}$$ Final product. ■

Multiplication of fractions often yields a product that can be reduced. The basic principle of fractions allows us to save time by first reducing and then multiplying, as we did with fractions of arithmetic.

EXAMPLE 6

Multiply: $\dfrac{6s+24}{3s-12}\cdot\dfrac{4s}{8s}$

$$\frac{6s+24}{3s-12}\cdot\frac{4s}{8s}=\frac{6(s+4)}{3(s-4)}\cdot\frac{4s}{8s}=\frac{2\cdot 3(s+4)}{3(s-4)}\cdot\frac{2\cdot 2\cdot s}{2\cdot 2\cdot 2\cdot s}\qquad\text{Factor.}$$

$$=\frac{\overset{1}{\cancel{2}}\cdot\overset{1}{\cancel{3}}(s+4)}{\underset{1}{\cancel{3}}(s-4)}\cdot\frac{\overset{1}{\cancel{2}}\cdot\overset{1}{\cancel{2}}\cdot\overset{1}{\cancel{s}}}{\underset{1}{\cancel{2}}\cdot\underset{1}{\cancel{2}}\cdot\underset{1}{\cancel{2}}\cdot\underset{1}{\cancel{s}}}\qquad\text{Reduce.}$$

$$=\frac{s+4}{s-4}\qquad\text{Multiply.}$$

Reducing first, as in Example 6, is less difficult because if we multiply first, we often get a polynomial having many terms that is difficult to factor.

PROCEDURE

To multiply two rational expressions:

1. Factor numerators and denominators.
2. Reduce if possible.
3. Multiply.

EXAMPLE 7

Multiply: $\dfrac{am}{m+n}\cdot\dfrac{m^2-n^2}{bm}$

$$\frac{am}{m+n}\cdot\frac{m^2-n^2}{bm}=\frac{a\overset{1}{\cancel{m}}}{\underset{1}{\cancel{(m+n)}}}\cdot\frac{\overset{1}{\cancel{(m+n)}}(m-n)}{b\underset{1}{\cancel{m}}}\qquad\text{Factor the numerator.}$$

$$=\frac{a(m-n)}{b}\qquad\text{Reduce.}$$

$$=\frac{am-an}{b}\qquad\text{Multiply.}$$

CAUTION

Reducing before performing the operation works only for multiplication. It does not work for addition, subtraction, or division. The reason is that the basic principle of fractions involves AC and BC (A times C and B times C).

EXAMPLE 8

Multiply: $\dfrac{y^2 + y - 2}{y^2 - 2y - 8} \cdot \dfrac{y^2 - 16}{y^2 + 2y - 3}$

$$= \frac{\overset{1}{\cancel{(y+2)}}\overset{1}{\cancel{(y-1)}}}{\underset{1}{\cancel{(y-4)}}\underset{1}{\cancel{(y+2)}}} \cdot \frac{(y+4)\overset{1}{\cancel{(y-4)}}}{(y+3)\underset{1}{\cancel{(y-1)}}} \qquad \text{Factor and reduce.}$$

$$= \frac{y+4}{y+3} \qquad \text{Multiply.}$$

■

Conversions of units of measure can be done using the procedure for multiplying rational expressions.

Equivalent Measures

Since 12 inches = 1 foot, these are two *equivalent measures*. If we consider the indicated division $\dfrac{12 \text{ inches}}{1 \text{ foot}}$ and think of these as numbers, we ask ‘‘how many units of 1 foot will it take to make 12 inches?’’ Since they measure the same length, the answer is 1.

$$\frac{12 \text{ inches}}{1 \text{ foot}} = 1 \quad \text{and} \quad \frac{1 \text{ foot}}{12 \text{ inches}} = 1$$

A fraction showing the indicated division of any two equivalent measures can always be thought of as a name for 1.

$$\frac{1 \text{ hour}}{60 \text{ minutes}} = 1 \qquad \frac{7 \text{ days}}{1 \text{ week}} = 1 \qquad \frac{2 \text{ pints}}{1 \text{ quart}} = 1$$

This idea is used along with the multiplication property of 1 to build a given measure to its equivalent measure in different units. For example, convert 48 inches to feet:

$$48 \text{ inches} = (48 \text{ inches})(1)$$

$$= 48\ \cancel{\text{inches}} \cdot \frac{1 \text{ foot}}{12\ \cancel{\text{inches}}}$$

$$= \frac{48}{12} \text{ feet}$$

$$= 4 \text{ feet.}$$

In some cases, it may be necessary to multiply by several names for one. For instance, convert 7200 seconds to hours:

$$7200 \text{ seconds} = \frac{7200 \cancel{\text{seconds}}}{1} \cdot \frac{1 \cancel{\text{minute}}}{60 \cancel{\text{seconds}}} \cdot \frac{1 \text{ hour}}{60 \cancel{\text{minutes}}}$$

$$= \frac{7200}{60(60)} \text{ hours}$$

$$= 2 \text{ hours.}$$

EXAMPLE 9

Convert 3 ft^2 to in^2. [*Hint:* $\text{ft}^2 = (\text{foot})(\text{foot})$.]

$$3 \text{ ft}^2 = \frac{3 \ (\text{foot})(\text{foot})}{1}$$

$$= \frac{3 \ \cancel{(\text{foot})}\cancel{(\text{foot})}}{1} \cdot \frac{12 \text{ inches}}{1 \ \cancel{\text{foot}}} \cdot \frac{12 \text{ inches}}{1 \ \cancel{\text{foot}}}$$

There are 12 inches in one foot. Multiply two times to eliminate both foot measures.

$$= \frac{3 \cdot 12 \cdot 12 \ (\text{inches})(\text{inches})}{1}$$

Multiply.

$$= 432 \text{ in}^2$$

■

EXAMPLE 10

Convert 60 miles per hour to feet per second.

$$60 \text{ mph} = \frac{60 \text{ miles}}{1 \text{ hour}}$$

$$= \frac{60 \ \cancel{\text{miles}}}{1 \ \cancel{\text{hour}}} \cdot \frac{1 \ \cancel{\text{hour}}}{60 \ \cancel{\text{minutes}}} \cdot \frac{1 \ \cancel{\text{minute}}}{60 \text{ seconds}} \cdot \frac{5280 \text{ feet}}{1 \ \cancel{\text{mile}}}$$

Change hours to minutes and minutes to seconds. Change miles to feet.

$$= \frac{60 \cdot 5280 \text{ feet}}{60 \cdot 60 \text{ seconds}}$$

Eliminate common measure factors.

$$= \frac{88 \text{ feet}}{1 \text{ second}}$$

Multiply.

$$= 88 \text{ feet per second}$$

■

EXAMPLE 11

The French Concorde can fly at a speed of 3234 feet per second (fps) (about three times the speed of sound). Convert this speed to miles per hour.

$$3234 \text{ fps} = \frac{3234 \cancel{\text{feet}}}{1 \cancel{\text{second}}} \cdot \frac{60 \cancel{\text{seconds}}}{1 \cancel{\text{minute}}} \cdot \frac{60 \cancel{\text{minutes}}}{1 \text{ hour}} \cdot \frac{1 \text{ mile}}{5280 \cancel{\text{feet}}}$$

Change seconds to minutes and minutes to hours. Change feet to miles.

$$= \frac{3234 \cdot 60 \cdot 60 \text{ miles}}{5280(1 \text{ hour})}$$

Eliminate common measure factors.

$$= \frac{2205 \text{ miles}}{1 \text{ hour}}$$

Multiply.

$= 2205$ miles per hour ■

EXERCISE 5.1

A

Reduce:

1. $\dfrac{xy}{y^2}$ **2.** $\dfrac{m^2n}{mp}$ **3.** $\dfrac{-15r^2s^4}{-3rs^2}$

4. $\dfrac{-27y^3z^5}{-9y^2z}$ **5.** $\dfrac{45a^5b^6c^7}{81a^3b^2c^5}$ **6.** $\dfrac{-2a^2}{abc}$

Multiply and reduce:

7. $\dfrac{a}{7} \cdot \dfrac{b}{8}$ **8.** $\dfrac{x}{5} \cdot \dfrac{x}{2y}$ **9.** $\dfrac{4m}{r} \cdot \dfrac{5m}{7}$

10. $\dfrac{4}{t} \cdot \dfrac{-v}{11t}$ **11.** $\dfrac{5}{-8w} \cdot \dfrac{5z}{6w}$ **12.** $\dfrac{w}{w-7} \cdot \dfrac{w-4}{w}$

13. $\dfrac{x+3}{x} \cdot \dfrac{x}{x+2}$ **14.** $\dfrac{20}{3y-9} \cdot \dfrac{7y-21}{5}$ **15.** $\dfrac{2x+5}{12} \cdot \dfrac{4}{4x+10}$

16. $\dfrac{3z-2}{8} \cdot \dfrac{6}{9x-6}$ **17.** $1 \text{ ft}^2 = 1(\text{ft})(\text{ft}) = ? \text{ in}^2$

18. $1 \text{ ft}^3 = 1(\text{ft})(\text{ft})(\text{ft}) = ? \text{ in}^3$

19. $1 \text{ yd}^3 = 1(\text{yd})(\text{yd})(\text{yd}) = ? \text{ ft}^3$

20. $1 \text{ yd}^2 = 1(\text{yd})(\text{yd}) = ? \text{ ft}^2$

B

Reduce:

21. $\dfrac{18a^2b}{24a^2b^2}$

22. $\dfrac{18x^3y^2}{-42xy^3}$

23. $\dfrac{bx + by}{bz}$

24. $\dfrac{cs - ct}{6c}$

25. $\dfrac{2cm + 2m}{3c + 3}$

26. $\dfrac{4x^2 - 4x}{ax - a}$

Multiply and reduce:

27. $\dfrac{x}{3} \cdot \dfrac{3x - 2}{5}$

28. $\dfrac{8ab}{3cd} \cdot \dfrac{2a + 3b}{5c - 4d}$

29. $-7 \cdot \dfrac{8m + 4}{2m - 1}$

30. $4a \cdot \dfrac{2b + 3c}{2b - 7}$

31. $\dfrac{19x}{13y} \cdot \dfrac{2x - 3}{4y - 5}$

32. $\dfrac{6x^2}{7ab} \cdot \dfrac{21a^2}{12x}$

33. $\dfrac{3s^3t^2}{8st} \cdot \dfrac{4s^2t}{6st^2}$

34. $\dfrac{m^2 - n^2}{m + n} \cdot \dfrac{m}{m^2 - mn}$

35. $\dfrac{2a^2 - 3a - 2}{a^2 - 2a - 15} \cdot \dfrac{3a^2 - 20a - 7}{3a^2 - 5a - 2}$

36. $\dfrac{2b^2 + 7b - 15}{b^2 + 2b - 15} \cdot \dfrac{5b^2 - 14b - 3}{2b^2 - 13b + 15}$

37. $\dfrac{4 \text{ ft}}{1 \text{ lb}} = \dfrac{? \text{ in.}}{1 \text{ lb}}$

38. $\dfrac{5 \text{ m}}{1 \text{ kg}} = \dfrac{? \text{ cm}}{1 \text{ kg}}$ (See Appendix IV)

39. $\dfrac{2 \text{ mi}}{1 \text{ hr}} = \dfrac{? \text{ ft}}{1 \text{ hr}}$

40. $\dfrac{5 \text{ km}}{1 \text{ min}} = \dfrac{? \text{ cm}}{1 \text{ min}}$

C

Reduce:

41. $\dfrac{6a^2 - 12ab}{6a^2 + 12ab}$

42. $\dfrac{5xy + 5xz}{10xy + 10x^2}$

43. $\dfrac{p^2 - q^2}{3q - 3p}$

44. $\dfrac{9a^2 - 1}{2 - 6a}$

45. $\dfrac{6x^2 + 5x - 4}{2x^2 + 11x - 6}$

46. $\dfrac{15a^2 + a - 6}{9a^2 - 4}$

Multiply and reduce:

47. $\frac{x^2 + 2x - 63}{3x^2 - 22x + 7} \cdot \frac{3x^2 - 7x + 2}{x^2 + 7x - 18}$

48. $\frac{m^2 - n^2}{m^2n - mn^2} \cdot \frac{m^2n^2}{(m + n)^2}$

49. $\frac{x^2 + 5x - 14}{x^2 - x - 2} \cdot \frac{3x^2 + x - 2}{2x^2 + 13x - 7}$

50. $\frac{x^2 - 3x - 4}{x^2 - 5x + 4} \cdot \frac{4x^2 + 4x - 3}{2x^2 + 5x + 3}$

51. $\frac{2a^3 + 3a^2 - 5a}{9a^2 - 8a - 1} \cdot \frac{27a^2 + 21a + 2}{2a^3 - a^2 - 15a}$

52. $\frac{6b^2 + 11bc - 2c^2}{6b^2c + 5bc^2 - c^3} \cdot \frac{b^2c - c^3}{2b^2 + 7bc + 6c^2}$

53. $\frac{ab - 3b^2}{2a^2 - 5ab} \cdot \frac{a^2 + 2ab + b^2}{a^2b - 2ab^2 - 3b^3} \cdot \frac{2a^2 - 5ab}{a^2 - b^2}$

54. $\frac{2x^2 - 7x - 4}{9x^2 + 12x - 5} \cdot \frac{15x^2 + x - 2}{2x^2 + 3x + 1} \cdot \frac{3x^2 + 8x + 5}{5x^2 - 18x - 8}$

55. $\frac{x^2 - y^2}{3 - x} \cdot \frac{x^2 + 3xy}{4y - 4x} \cdot \frac{xy - 3y}{x^2 + 4xy + 3y^2}$

56. $\frac{2a + b}{b^2} \cdot \frac{3a^2 - 3ab}{ab + 2b^2} \cdot \frac{5ab - 10b^2}{a^2 - 3ab + 2b^2}$

57. $\frac{10 \text{ mi}}{1 \text{ hr}} = \frac{? \text{ ft}}{1 \text{ sec}}$

58. $\frac{100 \text{ lb}}{1 \text{ ft}^2} = \frac{? \text{ oz}}{1 \text{ in}^2}$

59. $\frac{10 \text{ g}}{1 \ell} = \frac{? \text{ cg}}{1 \text{ c}\ell}$

60. $\frac{100 \text{ rpm}}{1 \text{ sec}} = \frac{? \text{ rpm}}{1 \text{ hr}}$

D

61. The probability that a certain repeated event out of t outcomes will take place six times in succession is given by

$$P = \frac{1}{t} \cdot \frac{1}{t} \cdot \frac{1}{t} \cdot \frac{1}{t} \cdot \frac{1}{t} \cdot \frac{1}{t}.$$

Simplify the formula. Find the probability if $t = 6$ (the probability that if you toss a die six times the same number will come up each time).

62. The probability that a certain repeated event out of t outcomes will happen twice in succession is

$$P = \frac{1}{t} \cdot \frac{1}{t}.$$

Simplify the formula. Find the probability if $t = 38$ (the probability that 16 will be landed on twice in a row on a roulette wheel).

63. In physics, it is important to know how far a metal bar will stretch if hung from a stationary platform with a weight attached to it. The formula is $d = \frac{P\ell}{AE}$, where P = load in pounds, A = cross-sectional area in square inches, ℓ = length of the bar in inches, E = modulus of elasticity in pounds per square inches, and d = change of length in inches. Find d if P = 60,000 lb, A = 1 sq in., ℓ = 10 in., E = 5,000,000 lb/sq in.

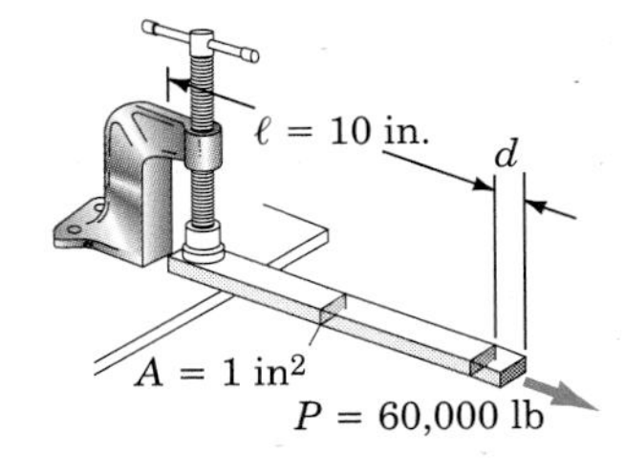

Figure for Exercise 63

64. Using the formula from Exercise 63, find d if P = 48,000 lb, A = 2 sq in., ℓ = 40 in., E = 6,000,000 lb/sq in.

65. A Vermont farmer collected 112 pints of maple syrup. How many gallons did he collect?

66. Joshua went on a fast to draw attention to the plight of some political prisoners. He fasted for 10,080 minutes. How many days did he fast?

67. Water weighs 64 pounds per ft^3. How many ounces does one in^3 weigh (to the nearest tenth of an ounce). [*Hint*: ft^3 = (foot)(foot)(foot).]

68. Ron Guidry of the New York Yankees throws the baseball at an average speed of 94 miles per hour. How many feet per second is this (to the nearest tenth)?

69. Renee read in the owner's manual of her new car that she could expect fuel economy to be 11.9 kilometers per liter. What is this in miles per gallon, to the nearest whole number? Use 1 mi = 1.6 km and 1 L = 1.06 qt.

STATE YOUR UNDERSTANDING

70. The polynomial in the denominator of a rational expression cannot equal zero. Why is this the case?

71. Explain how to determine the restrictions on a rational expression.

72. Explain the error in the following. What is the correct answer?

$$\frac{y^2 + 2y}{y^2 + 3y} = \frac{y^2 + 2y}{y^2 + 3y}$$

$$= \frac{1 + 2}{1 + 3}$$

$$= \frac{3}{4}$$

CHALLENGE EXERCISES

73. $\dfrac{8y^3 - 1}{7 - y} \cdot \dfrac{y^2 - 49}{4y^2 - 1}$

74. $\dfrac{x^2 + 7x + 12}{27x^3 + 1} \cdot \dfrac{4x^2 + 8x - 21}{x^2 + 8x + 16} \cdot \dfrac{9x^2 - 3x + 1}{6x^2 + x - 15}$

75. $\dfrac{2x + 4y - ax - 2ay}{x^2 - 4y^2} \cdot \dfrac{a^2 + 4a + 4}{a^2 - 4}$

76. $\dfrac{x^4 - 16}{x^3 - 3x^2 + 4x - 12} \cdot \dfrac{10x}{32 - 8x^2}$

MAINTAIN YOUR SKILLS (SECTIONS 3.5, 3.7, 4.1, 4.2)

Perform the indicated operations:

77. $(3x - 4y + 6) - (2x + 8y + 9)$

78. $(2y - 8) - (3y + 9) + (2y + 7)$

79. $(4xy^2)(6x^2y^2)(-3x^2y^4)$

80. $3xy^2(2x^2 - 5xy - 3y^2)$

Factor:

81. $8x^3 - 12x^2 + 10x$

82. $24a^4b^3c^3 - 36a^3b^4c^3 + 60a^3b^3c^4$

Solve:

83. $7x^2 - 42x = 0$

84. $54x = 18x^2$

5.2 DIVIDING RATIONAL EXPRESSIONS

OBJECTIVE

1. Divide rational expressions.

To divide fractions of arithmetic, we multiply by the reciprocal of the divisor. That is, invert the divisor and multiply.

$$\frac{2}{3} \div \frac{5}{4} = \frac{2}{3} \cdot \frac{4}{5} = \frac{8}{15}$$

Reciprocals

$\dfrac{c}{x}$ **and** $\dfrac{x}{c}$

Any number times its reciprocal has a product of 1. So,

Fraction	Reciprocal	Fraction	Reciprocal
$\frac{5}{4}$	$\frac{4}{5}$	7	$\frac{1}{7}$
$\frac{2}{x}$	$\frac{x}{2}$	$\frac{1}{x}$	x

Therefore,

$$\frac{3}{x} \div \frac{y}{2} = \frac{3}{x} \cdot \frac{2}{y} = \frac{6}{xy}.$$

PROCEDURE

To divide two rational expressions, multiply by the reciprocal of the divisor.

$$\frac{A}{B} \div \frac{C}{D} = \frac{A}{B} \cdot \frac{D}{C} = \frac{AD}{BC}, \quad BCD \neq 0$$

So,

$$\frac{x-3}{x^2+2} \div \frac{5}{x-6} = \frac{x-3}{x^2+2} \cdot \frac{x-6}{5} = \frac{x^2-9x+18}{5x^2+10}.$$

EXAMPLE 1

Divide: $\dfrac{2x+y}{4a} \div \dfrac{x}{b}$

$$\frac{2x+y}{4a} \div \frac{x}{b} = \frac{2x+y}{4a} \cdot \frac{b}{x}$$ Multiply by the reciprocal of the divisor.

$$= \frac{(2x+y)(b)}{4a \cdot x}$$ Multiply numerators and denominators.

$$= \frac{2bx+by}{4ax}$$ This product is the quotient of the original problem. ■

EXAMPLE 2

Divide: $\dfrac{x^2-2}{2x+1} \div \dfrac{2x+1}{x^2+7}$

$$= \frac{x^2-2}{2x+1} \cdot \frac{x^2+7}{2x+1}$$ Multiply by the reciprocal of the divisor.

$$= \frac{x^4+5x^2-14}{4x^2+4x+1}$$ This is the quotient. ■

After inverting the divisor, it is sometimes possible to reduce before multiplying.

EXAMPLE 3

Divide: $\dfrac{w^2 - 9}{6} \div \dfrac{w^2 - 3w}{2w + 2}$

$= \dfrac{w^2 - 9}{6} \cdot \dfrac{2w + 2}{w^2 - 3w}$ Invert the divisor.

$= \dfrac{(w + 3)\overset{1}{\cancel{(w - 3)}}}{\underset{1}{\cancel{2}} \cdot 3} \cdot \dfrac{\overset{1}{\cancel{2}}(w + 1)}{w\underset{1}{\cancel{(w - 3)}}}$ Factor and reduce.

$= \dfrac{(w + 3)(w + 1)}{3 \cdot w}$

$= \dfrac{w^2 + 4w + 3}{3w}$ Multiply. ■

EXAMPLE 4

Divide: $\dfrac{x^2 + 5x - 6}{x - 4} \div (1 - x)$

$= \dfrac{x^2 + 5x - 6}{x - 4} \cdot \dfrac{1}{1 - x}$ Invert the divisor.

$= \dfrac{(x + 6)(x - 1)}{x - 4} \cdot \dfrac{1}{1 - x}$

$= \dfrac{(x + 6)(-1)\overset{1}{\cancel{(1 - x)}}}{x - 4} \cdot \dfrac{1}{\underset{1}{\cancel{1 - x}}}$ Factor and reduce. The expression $x - 1$ is the opposite of $1 - x$. So, $x - 1 = -1(1 - x)$.

$= \dfrac{(-1)(x + 6)}{x - 4}$

$= \dfrac{-x - 6}{x - 4}$

Alternate form

$\dfrac{-x - 6}{x - 4} = \dfrac{-x - 6}{x - 4} \cdot \dfrac{(-1)}{(-1)}$ Multiply both the numerator and denominator by -1.

$= \dfrac{x + 6}{4 - x}$ Multiply. ■

EXERCISE 5.2

A

Divide and reduce:

1. $\frac{3c}{y} \div \frac{4x}{d}$
2. $3w \div \frac{5x}{7}$
3. $\frac{8y}{9} \div 3z$
4. $\frac{3ab}{5} \div \frac{-2c}{a}$
5. $\frac{4yz}{-3} \div \frac{-w}{8y}$
6. $\frac{15a}{-4b} \div \frac{7b}{-3a}$
7. $\frac{-4a}{3b} \div \frac{5b^2}{7a}$
8. $\frac{7z^3}{-8y} \div \frac{5y^4}{9z^2}$
9. $\frac{c}{d^2} \div \frac{c^2}{f}$
10. $\frac{y}{3} \div \frac{w}{z^2}$
11. $\frac{p^5q^7}{x^4y^6} \div \frac{p^3q^3}{x^2y^5}$
12. $\frac{a^3b^2c^5}{x^5y^2} \div \frac{a^2bc^4}{x^4y^4}$
13. $\frac{m+n}{5} \div \frac{m+n}{7}$
14. $\frac{3}{6p+q} \div \frac{13}{6p+q}$
15. $\frac{5x-10}{12} \div \frac{4x-8}{8}$
16. $\frac{x^2-x}{15} \div \frac{x-1}{5}$
17. $\left(\frac{2x+1}{4}\right) \div \left(\frac{2x+1}{3}\right)$
18. $\left(\frac{6}{2y+2}\right) \div \left(\frac{3}{y+1}\right)$
19. $\left(\frac{2a^2+4a}{18}\right) \div \left(\frac{3a^2+6a}{30}\right)$
20. $\left(\frac{4}{3x+6y}\right) \div \left(\frac{6}{9x+6y}\right)$

B

Divide and reduce:

21. $\frac{5a+3b}{x} \div \frac{3a+c}{9ab}$
22. $\frac{4x-2y}{3a-b} \div \frac{7}{4a}$
23. $\frac{2x+8}{5} \div \frac{3}{4x-7}$
24. $\frac{a}{2b+3c} \div \frac{b}{5a-c}$
25. $\frac{2ab}{3x+4y} \div \frac{4}{a-b}$
26. $\frac{4xy}{2a+3} \div \frac{5}{x+y}$
27. $\frac{12x^3y}{8xy^7} \div \frac{7x^5y^7}{6x^2y}$
28. $\frac{21w^2z^3}{98pq^2} \div \frac{9w^3z^3}{14p^2q^2}$
29. $\frac{p^2-2p}{15} \div \frac{p-2}{5}$
30. $\frac{y^2+2y+1}{6y^2} \div \frac{y+1}{9y^3}$
31. $\frac{10a^2-5a}{11a^3+22a^2} \div \frac{2a-1}{a^2+2a}$
32. $\frac{6pt^2-18pt}{10t^2+5t} \div \frac{p^2t-3p^2}{2t^3+t^2}$

33. $\dfrac{x^2 + 4x - 21}{x^2 + 8x + 7} \div \dfrac{x^2 - 2x - 3}{x^2 - 6x - 7}$

34. $\dfrac{y^2 + 7y + 10}{y^2 + 6y + 5} \div \dfrac{y^2 + 6y + 8}{y^2 - 7y - 8}$

35. $\dfrac{4x^2 - 4}{x^2 - 3x + 2} \div \dfrac{8x^2 + 32x + 24}{x^2 + 3x - 4}$

36. $\dfrac{3x^2 - 12}{x^2 + 7x + 10} \div \dfrac{6x^2 + 30x - 84}{x^2 + 11x + 30}$

C

Divide and reduce:

37. $\dfrac{am}{2a + 4m} \div \dfrac{3}{-7m}$

38. $\dfrac{4b - c}{-bc} \div \dfrac{8c}{-3}$

39. $\dfrac{4x - 9}{3x + 2} \div \dfrac{5 - 4x}{2x + 3}$

40. $\dfrac{-(7 - 8a)}{3a + 2} \div \dfrac{7a + 9}{2a + 3}$

41. $\dfrac{6a - 7y}{9z + 7} \div \dfrac{4z - 8}{2a + 5y}$

42. $\dfrac{9m - n}{8a + 5b} \div \dfrac{3a - b}{m + 9n}$

43. $\dfrac{3z + 7}{8a - 5} \div \dfrac{2z - 1}{2a + 3}$

44. $\dfrac{a - 5}{b + 2} \div \dfrac{2z - 1}{6a + 1}$

45. $\dfrac{5a^2 + 34a - 7}{2a^2 + 3a - 2} \div \dfrac{a^3 + 5a^2 - 14a}{a^3 - 4a}$

46. $\dfrac{r^2 - 9r + 18}{r^2 - 9} \div \dfrac{2r^3 - 11r^2 - 6r}{2r^2 + r}$

47. $\dfrac{2x^2 - 13x + 15}{6x^2 + 5x - 21} \div \dfrac{x^2 - 6x + 5}{3x^2 + 4x - 7}$

48. $\dfrac{12x^2 - 4x - 5}{10x^2 - 7x - 6} \div \dfrac{18x^2 - 3x - 10}{10x^2 + 13x - 30}$

49. $\dfrac{3x^2 - 22x - 45}{2x^2 - 9x - 56} \div \dfrac{x^2 - 14x + 49}{2x^2 - 7x - 49}$

50. $\dfrac{64x^2 - 25}{16x^2 + 18x + 5} \div \dfrac{64x^2 - 80x + 25}{16x^2 - 2x - 5}$

51. $\dfrac{3a + a^2}{12 + 3a} \div \dfrac{9a - a^3}{12 - a - a^2}$

52. $\dfrac{b^2 + 16b + 64}{2b^2 - 128} \div \dfrac{3b^2 + 30b + 48}{b^2 - 6b - 16}$

STATE YOUR UNDERSTANDING

53. Explain the procedure for dividing rational expressions.

54. In addition to the restrictions that $B \neq 0$ and $D \neq 0$, the division problem $\dfrac{A}{B} \div \dfrac{C}{D}$ has the additional restriction $C \neq 0$. Why is this the case?

CHALLENGE EXERCISES

55. $\dfrac{x^2 - 10x + 25}{10x^2 + 13x - 3} \cdot \dfrac{6 + x - 2x^2}{5x^2 - 24x - 5} \div \dfrac{x^2 - 7x + 10}{1 - 25x^2}$

56. $\dfrac{x^4 - y^4}{36a^2 - 16b^2} \div \dfrac{x^4 + 2x^2y^2 + y^4}{6ax - 4bx + 15ay - 10by} \cdot \dfrac{10b + 15a}{2x^2 + 3xy - 5y^2}$

57. $\dfrac{x^6 - y^6}{12x^2 - 28xy + 15y^2} \div \dfrac{9x^2 - 9xy + 9y^2}{12x - 18y} \div \dfrac{3x^3 - 3y^3}{25y - 30x}$

58. $\dfrac{12x^2 - 22x + 8}{8x^3 - 27} \div \left(\dfrac{4x^2 - 8x + 3}{x^2 - 4} \div \dfrac{4x^2 - 12x + 9}{8 - 2x - 3x^2}\right)$

MAINTAIN YOUR SKILLS (SECTIONS 2.3, 4.2, 4.3)

59. Solve $p = 2r - s^2$ for r if $p = 9$ and $s = 2$.

60. Solve $p = 2r - s^2$ for r.

61. Solve $i = prt$ for r if $p = 15{,}000$, $t = 2$, and $i = 2400$.

62. Solve $i = prt$ for t if $p = 10{,}000$, $r = 0.12$, and $i = 3600$.

Factor:

63. $9t^2x + 36t^2y - 18t^2z + 72t^2$

64. $54m^2n^2 - 45m^2n + 63mn^2 - 90mn$

65. $7x^2 - 7x + 25x - 25$

66. $4a^2b^2 - 6ab - 10abc + 15c$

5.3

LEAST COMMON MULTIPLE

OBJECTIVES

1. Find the least common multiple of two or more polynomials.
2. Solve equations containing rational expressions.

In the arithmetic of fractions, you learned that the least common multiple of two or more whole numbers is the smallest positive number that is a multiple of each of the numbers. It is also the smallest number other than zero that is divisible by each of the given numbers. We find the LCM of two or more whole numbers by first finding the prime factors of the numbers. The LCM is built using the prime factors.

The following table shows the procedure.

Find the LCM of	Prime Factors	LCM
4	$2 \cdot 2 = 2^2$	
8	$2 \cdot 2 \cdot 2 = 2^3$	$2^3 \cdot 3 = 24$
12	$2 \cdot 2 \cdot 3 = 2^2 \cdot 3$	
15	$3 \cdot 5 = 3 \cdot 5$	
50	$2 \cdot 5 \cdot 5 = 2 \cdot 5^2$	$2 \cdot 3 \cdot 5^2 = 150$
75	$3 \cdot 5 \cdot 5 = 3 \cdot 5^2$	

The LCM of two or more whole numbers is the product of the highest power of each different factor. The LCM of two or more polynomials is similar.

DEFINITION

LCM of Polynomials

The *least common multiple* of two or more polynomials is the polynomial with the least number of factors that is divisible by each of the given polynomials.

Find the LCM of	Factored	LCM
x^2	$x \cdot x$ or x^2	$x^3 \cdot y^3 = x^3y^3$
y^2	$y \cdot y$ or y^2	
x^3y^3	$x \cdot x \cdot x \cdot y \cdot y \cdot y$ or $x^3 \cdot y^3$	The different factors are x and y. The highest power of each is 3.
x	x	$x(x + y) = x^2 + xy$
$x + y$	$x + y$	The only factors are x and $x + y$. The highest power of each is 1.
$x^2 + xy$	$x(x + y)$	

Find the LCM of	Factored	LCM
$x^2 - y^2$ $x - y$ $x + y$	$(x - y)(x + y)$ $x - y$ $x + y$	$(x + y)(x - y) = x^2 - y^2$ The different factors are $(x - y)$ and $(x + y)$. The highest power of each is 1.
$x^2 + 2xy + y^2$ $x^2 + xy$ $x^2 - y^2$	$(x + y)^2$ $x\,(x + y)$ $(x - y)\,(x + y)$	$x(x - y)(x + y)^2$ The different factors are x, $x - y$, and $x + y$. The highest power of x and $x - y$ is 1; the highest power of $x + y$ is 2.
$x^2 + 5x + 6$ $x^2 + 6x + 9$ $x^2 + 4x + 4$	$(x + 2)(x + 3)$ $(x + 3)^2$ $(x + 2)^2$	$(x + 3)^2(x + 2)^2$ The different factors are $(x + 3)$ and $(x + 2)$, and 2 is the highest power of each.

PROCEDURE

To find the LCM of two or more polynomials:

1. Factor each polynomial completely.
2. Write the product of the highest power of each different factor.

EXAMPLE 1

Find the LCM of $4x^2$, $3y^2$, and $8xy$.

$4x^2 = 2^2 \cdot x^2$ — Factor each polynomial completely.

$3y^2 = 3 \cdot y^2$

$8xy = 2^3 \cdot x \cdot y$

$\text{LCM} = 2^3 \cdot 3 \cdot x^2 \cdot y^2$ — Write the product of the highest power of each different factor.

$= 24x^2y^2$ ■

EXAMPLE 2

Find the LCM of $a^2 - 1$, $a^2 + a - 2$, and $a^2 + 3a + 2$.

$$a^2 - 1 = (a - 1)(a + 1)$$
$$a^2 + a - 2 = (a + 2)(a - 1)$$
$$a^2 + 3a + 2 = (a + 2)(a + 1)$$

Factor each polynomial.

$$\text{LCM} = (a + 1)(a - 1)(a + 2)$$

Write the product of the highest power of each different factor and leave in factored form. ■

EXAMPLE 3

Find the LCM of $x^2 + 2xy + y^2$ and $x^2 - y^2$.

$$x^2 + 2xy + y^2 = (x + y)^2$$
$$x^2 - y^2 = (x - y)(x + y)$$
$$\text{LCM} = (x + y)^2(x - y)$$

The highest power of the factor $(x + y)$ is 2. ■

The LCM can be used to solve equations containing fractions. The LCM of two or more denominators is called the Least Common Denominator (LCD). Consider the following equation:

$$\frac{1}{x} + \frac{1}{x + 1} = \frac{5}{6}.$$

The LCD of *all* the denominators is $6x(x + 1)$. Using the multiplication property of equality, we have

$$6x(x + 1)\left(\frac{1}{x} + \frac{1}{x + 1}\right) = 6x(x + 1)\left(\frac{5}{6}\right)$$
$$6x(x + 1)\left(\frac{1}{x}\right) + 6x(x + 1)\left(\frac{1}{x + 1}\right) = 6x(x + 1)\left(\frac{5}{6}\right)$$
$$6(x + 1) + 6x \cdot 1 = x(x + 1)(5) \qquad \text{Reduce and multiply.}$$
$$6x + 6 + 6x = 5x^2 + 5x$$
$$12x + 6 = 5x^2 + 5x$$
$$0 = 5x^2 - 7x - 6$$
$$0 = (5x + 3)(x - 2)$$

$$5x + 3 = 0 \quad \text{or} \quad x - 2 = 0$$
$$5x = -3 \qquad\qquad x = 2$$
$$x = -\frac{3}{5}$$

The equation has two solutions. The solution set is $\left\{-\frac{3}{5}, 2\right\}$. ■

■ PROCEDURE

To solve an equation in which rational expressions are involved:

1. Find the LCD of all the denominators.
2. Multiply both sides of the equation by the LCM.
3. Solve the resulting equation as before.
4. Check.

EXAMPLE 4

Solve: $\frac{1}{3} - \frac{1}{x} = \frac{1}{2} + \frac{1}{4x}$

$$12x\left(\frac{1}{3} - \frac{1}{x}\right) = 12x\left(\frac{1}{2} + \frac{1}{4x}\right)$$

The LCD is $12x$. We multiply both sides of the equation by $12x$. This will eliminate the fractions.

$$12x \cdot \frac{1}{3} - 12x \cdot \frac{1}{x} = 12x \cdot \frac{1}{2} + 12x \cdot \frac{1}{4x}$$

Note $x \neq 0$.

$$4x - 12 = 6x + 3$$

The fractions have been eliminated.

$$-2x = 15$$
$$x = -\frac{15}{2}$$

Since no denominator is zero when $x = -\frac{15}{2}$, this is the solution if no errors occurred. The check is done in the next example.

The solution set is $\left\{-\frac{15}{2}\right\}$. ■

A calculator may be used to check equations. Since not all rational numbers can be written as exact decimals, the calculator values are often approximations.

EXAMPLE 5

Check Example 4.

$$x = -\frac{15}{2} = -7.5$$

Left side: $\frac{1}{3} - \frac{1}{x}$

[3] [1/x] [−] [7.5] [+/−] [1/x] [=] with the result 0.4666666

Right side: $\frac{1}{2} + \frac{1}{4x}$

[2] [1/x] [+] [(] [4] [×] [7.5] [+/−] [)] [1/x] [=] with the result 0.4666666

Since the left and right sides have the same (approximate) values, the equation checks. ■

EXAMPLE 6

Solve: $\frac{1}{x} - \frac{1}{x+3} = \frac{1}{18}$

$$18x(x+3)\left(\frac{1}{x} - \frac{1}{x+3}\right) = 18x(x+3)\left(\frac{1}{18}\right)$$

$$18x(x+3) \cdot \frac{1}{x} - 18x(x+3) \cdot \frac{1}{x+3} = 18x(x+3) \cdot \frac{1}{18}$$

The LCD is $18x(x + 3)$. We multiply both sides of the equation by $18x(x + 3)$. This will eliminate the fractions.

Note $x \neq -3, 0$.

$$18(x+3) - 18x = x(x+3)$$

$$18x + 54 - 18x = x^2 + 3x$$

The fractions are eliminated.

$$0 = x^2 + 3x - 54$$

The result is a quadratic equation.

$$0 = (x+9)(x-6)$$

Factor and use the zero-product property.

$x + 9 = 0 \quad \text{or} \quad x - 6 = 0$

$x = -9 \quad \text{or} \quad x = 6$

The check is left for the student.

The solution set is $\{-9, 6\}$. ■

In the next two examples, we see that some applications lead to equations that contain rational expressions.

EXAMPLE 7

Mike and Michelle can deliver papers on a route in the same amount of time. On Sunday they share the route and finish the delivery in two hours. How long does it take each to complete the route alone?

Simpler word form:

$$\begin{pmatrix}\text{Fraction of}\\ \text{route Mike}\\ \text{can complete}\\ \text{in one hour}\end{pmatrix} + \begin{pmatrix}\text{Fraction of}\\ \text{route Michelle}\\ \text{can complete}\\ \text{in one hour}\end{pmatrix} = \begin{pmatrix}\text{Fraction of}\\ \text{route they}\\ \text{can do together}\\ \text{in one hour}\end{pmatrix}$$

Select variable:

Let t represent the number of hours it takes for either to complete the route alone.

	Number of Hours to Do the Job Alone	Fractional Part of Job Done in 1 Hour
Mike	t	$\frac{1}{t}$
Michelle	t	$\frac{1}{t}$
Working Together	2	$\frac{1}{2}$

Make a table.

Since they each can deliver the route in the same amount of time, t represents the number of hours each takes to do the job.

Translate to algebra:

$$\frac{1}{t} + \frac{1}{t} = \frac{1}{2}$$

Solve:

$$2t\left(\frac{1}{t}+\frac{1}{t}\right)=2t\left(\frac{1}{2}\right)$$

$$2+2=t$$

$$4=t$$

Check:

If Mike can do the whole route in 4 hours, then he can do $\frac{1}{4}$ in one hour or $\frac{1}{2}$ in two hours. The same goes for Michelle. So, the whole job will be done in two hours.

Answer:

Mike or Michelle can complete the paper route in four hours, working alone. ■

EXAMPLE 8

Mr. Black travels the 600 miles from Pittsburgh to Chicago at a certain average speed. On the return trip, he increases his average speed by 15 mph. If the return trip takes two hours less time, what is his average speed from Pittsburgh to Chicago?

To find Mr. Black's average speed from Pittsburgh to Chicago, we write:

Simpler word form:

$$\begin{pmatrix}\text{Time from}\\ \text{Pittsburgh}\\ \text{to Chicago}\end{pmatrix}-\begin{pmatrix}\text{Time from}\\ \text{Chicago to}\\ \text{Pittsburgh}\end{pmatrix}=2\text{ hours}$$

Select variable:

Let v represent the average speed from Pittsburgh to Chicago.

	Rate	×	**Time** $\left(T=\frac{D}{R}\right)$	=	**Distance (Given)**
Pittsburgh to Chicago	v		$\frac{600}{v}$		600
Chicago to Pittsburgh	$v+15$		$\frac{600}{v+15}$		600

Translate to algebra:

$$\frac{600}{v} - \frac{600}{v + 15} = 2$$

Solve:

$$v(v + 15)\left(\frac{600}{v} - \frac{600}{v + 15}\right) = v(v + 15) \cdot 2$$

$$600v + 9000 - 600v = 2v^2 + 30v$$

$$0 = 2v^2 + 30v - 9000$$

$$0 = v^2 + 15v - 4500$$

$$0 = (v + 75)(v - 60)$$

$$v = -75 \quad \text{or} \quad v = 60$$

Answer:

The negative solution is rejected because speed is not measured with negative numbers. So the average speed from Pittsburgh to Chicago is 60 mph. ■

EXERCISE 5.3

A

Find the LCM of the following (leave in factored form):

1. $3a$, $5a$ **2.** $2z$, $7z$ **3.** $12x$, $2x$

4. $15a^2$, $3a$ **5.** $6x$, $4x^2$, 8 **6.** $12x$, 6, $3x^2$

7. $8(a + b)$, $6(a + b)$ **8.** $6(x - y)$, $4(x - y)$ **9.** $3ab$, $4a^2$, $6b^2$

10. $5xy$, $10x^2$, $20y^2$

Solve:

11. $\frac{1}{y} - \frac{1}{3y} = 2$ **12.** $\frac{2}{m} - \frac{1}{2m} = 1$ **13.** $\frac{1}{a} + \frac{1}{4a} = 1$ **14.** $\frac{1}{x} + \frac{1}{3x} = 2$

15. $\frac{1}{y} - \frac{1}{3y} = \frac{1}{3}$ **16.** $\frac{1}{z} + \frac{1}{4z} = \frac{1}{4}$ **17.** $\frac{1}{x} - \frac{1}{2} = \frac{5}{6}$ **18.** $\frac{1}{x} - \frac{3}{4} = \frac{3}{8}$

19. $\frac{1}{x - 3} = \frac{4}{5}$ **20.** $\frac{1}{x - 5} = \frac{4}{3}$

B

Find the LCM of the following (leave in factored form):

21. $2ab,\ 4a^2,\ 8b^2$

22. $6xy,\ 8x^2,\ 12b^2$

23. $3x + 3y,\ 2x + 2y$

24. $4a - 4b,\ 3a - 3b$

25. $x - y,\ x^2 - y^2$

26. $a - b,\ a^2 - b^2$

27. $4a^2 - 9,\ 14a + 21$

28. $9x^2 - 1,\ 6x - 2$

29. $2b^2 - 3b,\ 10b^2 - 15b$

30. $x^2 + 2x,\ 6x + 12$

Solve:

31. $\frac{1}{8a} - \frac{1}{6a} = \frac{1}{4}$

32. $\frac{2}{x} - \frac{1}{2x} = 6$

33. $\frac{1}{x-2} + \frac{3}{x-2} = \frac{2}{3}$

34. $\frac{2}{b-4} + \frac{6}{b-4} = \frac{3}{5}$

35. $\frac{1}{a+5} - \frac{2}{a+5} = \frac{1}{2}$

36. $\frac{2}{x-1} + \frac{3}{x-1} = \frac{2}{3}$

37. $\frac{-6}{b-4} + \frac{2}{b-4} = \frac{4}{3}$

38. $\frac{-12}{y+4} - \frac{8}{y+4} = \frac{9}{5}$

39. $\frac{6}{y-3} + \frac{4}{3-y} = \frac{2}{3}$

40. $\frac{-2}{3-x} + \frac{8}{x-3} = \frac{1}{2}$

C

Find the LCM of the following (leave in factored form):

41. $b + c,\ b - c,\ b^2 - c^2$

42. $m^2 + 3m + 2,\ m + 2,\ m + 1$

43. $y^2 - 9,\ y^2 - 6y + 9$

44. $a^2 + 10a + 25,\ a^2 - 25$

45. $a^2 - b^2,\ 3a - 3b,\ a^2 - 2ab + b^2$

46. $x^2 - y^2,\ (x - y)^2,\ 3x + 3y$

47. $x^2 + 5x + 6,\ x^2 + 4x + 4$

48. $b^2 + 6b + 8,\ b^2 + 8b + 16$

Solve:

49. $\frac{4}{a-1} + \frac{2}{1-a} = \frac{3}{5}$

50. $\frac{3}{b-2} + \frac{4}{2-b} = \frac{2}{3}$

51. $\frac{2}{a+3} = \frac{5}{a-4}$

52. $\frac{2}{x-5} = \frac{5}{x-2}$

53. $\frac{10}{x+4} - \frac{3}{x-2} = 0$

54. $\frac{3}{x-1} + \frac{1}{x-3} = 0$

55. $\frac{y}{y-1} = \frac{4}{5}$

56. $\frac{3}{x+1} - \frac{2}{3} = \frac{5}{x+1}$

57. $\frac{2}{x+1} + \frac{1}{x+2} = \frac{3}{x+2}$

58. $\frac{6x-1}{x^2-4} = \frac{2}{x+2} + \frac{4}{x-2}$

59. $\frac{x-7}{2x-6} + \frac{1}{3} = \frac{x-6}{x-3}$

60. $\frac{1}{x} + \frac{1}{x+2} = \frac{3}{x+2}$

D

61. The sum of the reciprocals of two consecutive integers is $\frac{11}{30}$. What are the integers?

62. The sum of the reciprocals of two consecutive integers is $\frac{3}{2}$. What are the integers?

63. The numerator of a fraction is five less than the denominator. If $\frac{2}{3}$ is added to the fraction, the sum is $\frac{5}{6}$. What is the original fraction?

64. The denominator of a fraction is two more than the numerator. The fraction increased by $\frac{1}{2}$ is $\frac{11}{10}$. What is the original fraction?

65. Chico could drive 594 miles in 11 hours time if he drove twice as fast as his usual average speed. What is Chico's usual average speed?

66. Tanya could bicycle 28 miles in 7 hours less time if she cycled twice as fast as usual. What is Tanya's usual speed?

67. The express bus averages 10 mph faster than the regular bus on the 200-mile trip from Kansas City to St. Louis. If the express bus takes one less hour to complete the trip, what is the average speed of each bus?

68. A pedestrian walks 12 miles at a certain average speed. On the return trip, she increases her average speed 1 mph and saves one hour. What is her speed in each direction?

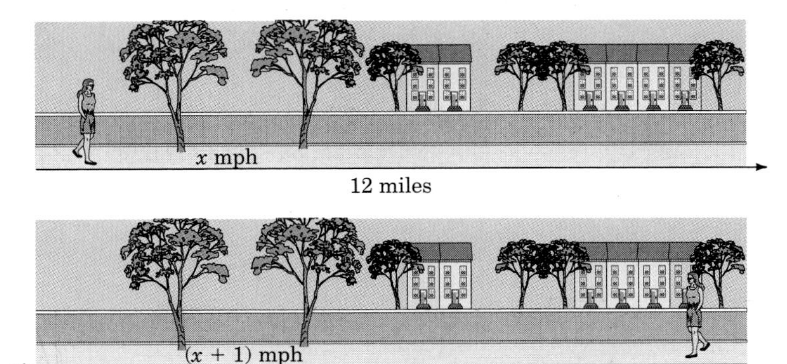

Figure for Exercise 68

69. Pedro can install a car stereo twice as fast as Luis. If it takes them 6 hours to install the stereo together, how long would it take Pedro or Luis to do it alone? (*Hint:* If a job can be done in t hours, then $\frac{1}{t}$ of it can be done in one hour.)

70. Jane can paint a wall in 15 minutes less time than Trong. If they work together, they can paint the wall in 10 minutes. How long will it take each to paint the wall alone?

STATE YOUR UNDERSTANDING

71. Describe the procedure for determining the least common multiple of polynomials.

72. Identify the error in the following problem. What is the correct answer? Determine the LCM of $x^2 - 6x + 8$, $x^2 + x - 6$, and $x^2 - 8x + 16$.

$x^2 - 6x + 8 = (x - 2)(x - 4)$

$x^2 + x - 6 = (x - 2)(x + 3)$

$x^2 - 8x + 16 = (x - 4)(x - 4)$

The LCM is $(x - 2)(x + 3)(x - 4)$.

CHALLENGE EXERCISES

Determine the least common multiple.

73. $3x^3 - 3y^3$; $6x^2 - 12xy + 6y^2$: $4x^2 - 4y^2$

74. $2a^2 - ab - 6b^2$: $x^2 + 3xy - 4y^2$: $2ax - 3by - 2ay + 3bx$

75. Perform the indicated operations.

$$\frac{4x + 12}{6x^2} \cdot \frac{9x}{12x + 36} - \frac{14x^2}{2x^2 + 10x} \div \frac{21x^2}{x + 5}$$

76. Solve for a. $\dfrac{1}{c} - \dfrac{1}{ab} = \dfrac{1}{ac}$

MAINTAIN YOUR SKILLS (SECTIONS 2.5, 3.5, 4.4)

Simplify:

77. $(4x - 8) - (3x - 12) + (6x + 9)$

78. $(14a - 3b + 2c) - (3a - 4b) + (-2b - 3c)$

79. $(2x^2 - 3x + 7y) - (-3x - 7y + 4x^2) + (4x - 7y^2)$

80. Solve $2s = an + n\ell$ for a.

Factor:

81. $x^2 - x - 56$

82. $x^2 + 17x + 60$

83. $x^2 - 17x + 66$

84. $x^2 + 10x - 56$

5.4 ADDING AND SUBTRACTING RATIONAL EXPRESSIONS

OBJECTIVES

1. Add and subtract rational expressions with like denominators.
2. Add and subtract rational expressions with opposite denominators.
3. Add and subtract rational expressions with unlike denominators.
4. Solve related equations.

In arithmetic, the sum or difference of two fractions with a common denominator is the sum or difference of their numerators over the common denominator.

$$\frac{2}{5} + \frac{1}{5} = \frac{2 + 1}{5} = \frac{3}{5}$$

$$\frac{5}{7} - \frac{3}{7} = \frac{5 - 3}{7} = \frac{2}{7}$$

In the same way, the sum or difference of two or more rational expressions can be found.

Indicated Sum or Difference	Combine the Numerators
$\frac{1}{x} + \frac{1}{x} = \frac{1 + 1}{x}$	$\frac{2}{x}$
$\frac{2}{y} - \frac{1}{y} = \frac{2 - 1}{y}$	$\frac{1}{y}$
$\frac{4}{x - 2} + \frac{3}{x - 2} = \frac{4 + 3}{x - 2}$	$\frac{7}{x - 2}$
$\frac{a}{b + 3} + \frac{c}{b + 3} = \frac{a + c}{b + 3}$	$\frac{a + c}{b + 3}$

PROCEDURE

To add or subtract two rational expressions that have a common denominator:

1. Find the sum or difference of the numerators.
2. Write the sum or difference over the common denominator.
3. Reduce if possible.

EXAMPLE 1

Add; reduce if possible: $\frac{1}{5x} + \frac{4}{5x}$

$\frac{1}{5x} + \frac{4}{5x} = \frac{1 + 4}{5x}$ Find the sum of the numerators.

$= \frac{5}{5x}$ Write the sum over the common denominator.

$= \frac{1}{x}$ Reduce.

EXAMPLE 2

Subtract; reduce if possible: $\frac{6}{x - 3} - \frac{2}{x - 3}$

$\frac{6}{x - 3} - \frac{2}{x - 3} = \frac{6 - 2}{x - 3}$ Find the difference of the numerators.

$= \frac{4}{x - 3}$ Write the difference over the common denominator.

In the special case of rational expressions with opposite denominators, one of the denominators can be multiplied by negative one (-1). The product of an expression and -1 is the opposite of the expression. For instance,

$-1(a) = -a$ and $-1(a + b) = -a - b$.

If a denominator is multiplied by -1, the numerator must also be multiplied by -1 to keep the fractions equivalent. The following table illustrates some equivalent fractions.

Fraction	Fraction Times (1)	Let $\frac{-1}{-1} = 1$	Result
$\frac{5}{6}$	$\frac{5}{6} \cdot 1$	$\frac{5}{6} \cdot \frac{(-1)}{(-1)}$	$\frac{-5}{-6}$
$\frac{a}{b}$	$\frac{a}{b} \cdot 1$	$\frac{a}{b} \cdot \frac{(-1)}{(-1)}$	$\frac{-a}{-b}$
$\frac{5}{-6}$	$\frac{5}{-6} \cdot 1$	$\frac{5}{-6} \cdot \frac{(-1)}{(-1)}$	$\frac{-5}{6}$

Fraction	Fraction Times (1)	Let $\frac{-1}{-1} = 1$	Result
$\frac{-a}{b}$	$\frac{-a}{b} \cdot 1$	$\frac{-a}{b} \cdot \frac{(-1)}{(-1)}$	$\frac{a}{-b}$
$\frac{1}{x-2}$	$\frac{1}{x-2} \cdot 1$	$\frac{1}{x-2} \cdot \frac{(-1)}{(-1)}$	$\frac{-1}{2-x}$
$\frac{3}{4-a}$	$\frac{3}{4-a} \cdot 1$	$\frac{3}{4-a} \cdot \frac{(-1)}{(-1)}$	$\frac{-3}{a-4}$

The table shows that if we take the opposite of both the numerator and denominator of a fraction of algebra, the value of the fraction is not changed.

In general,

PROPERTY

$$\frac{a}{b} = \frac{-a}{-b} \quad \text{and} \quad \frac{-a}{b} = \frac{a}{-b}.$$

The forms for the opposite of a fraction are closely related to these formulas:

$$\frac{-a}{b} = \frac{(-1)a}{b} = \frac{(-1)}{1} \cdot \frac{a}{b} = (-1)\frac{a}{b} = -\frac{a}{b}$$

In general,

PROPERTY

$$-\frac{a}{b} = \frac{-a}{b} = \frac{a}{-b}$$

We can use these properties of rational expressions when adding and subtracting fractions of algebra that have opposite denominators.

EXAMPLE 3

Add; reduce if possible: $\frac{6}{x} + \frac{8}{-x}$

$$\frac{6}{x} + \frac{8}{-x} = \frac{6}{x} + \frac{-8}{x}$$

Write the opposite of the numerator and the denominator of the second fraction.

$$= \frac{6 + (-8)}{x}$$

Find the sum of the numerators.

$$= -\frac{2}{x}$$

■

EXAMPLE 4

Subtract; reduce if possible: $\dfrac{2}{c-4} - \dfrac{5}{4-c}$

$$\frac{2}{c-4} - \frac{5}{4-c} = \frac{2}{c-4} - \frac{-5}{c-4}$$

Write the opposite of the numerator and the denominator of the second fraction.

$$= \frac{2-(-5)}{c-4}$$

$$= \frac{7}{c-4}$$

■

When the denominators of two rational expressions contain opposite factors, we can use the properties once again.

EXAMPLE 5

Add; reduce if possible: $\dfrac{x}{(x-2)(x+2)} + \dfrac{2}{(2+x)(2-x)}$

$$= \frac{x}{(x-2)(x+2)} + \frac{-2}{-(2+x)(2-x)}$$

$$= \frac{x}{(x-2)(x+2)} + \frac{-2}{(x-2)(x+2)}$$

Since $-(2+x)(2-x) = (x-2)(x+2)$

$$= \frac{x-2}{(x-2)(x+2)}$$

$$= \frac{1}{x+2}$$

Reduce.

■

By using both the procedures for finding the LCD and the procedures for adding or subtracting fractions, we can add and subtract fractions with unlike denominators. The following table illustrates the technique for simple denominators.

Problem	LCD of Denominators	Build Fractions	New Problem	Answer
$\frac{1}{4}+\frac{1}{3}$	12	$\frac{1}{4}\cdot\frac{3}{3}+\frac{1}{3}\cdot\frac{4}{4}$	$\frac{3}{12}+\frac{4}{12}$	$\frac{7}{12}$
$\frac{3}{2x}+\frac{5}{x^2}$	$2x^2$	$\frac{3}{2x}\cdot\frac{x}{x}+\frac{5}{x^2}\cdot\frac{2}{2}$	$\frac{3x}{2x^2}+\frac{10}{2x^2}$	$\frac{3x+10}{2x^2}$
$\frac{3}{a}-\frac{4}{b}$	ab	$\frac{3}{a}\cdot\frac{b}{b}-\frac{4}{b}\cdot\frac{a}{a}$	$\frac{3b}{ab}-\frac{4a}{ab}$	$\frac{3b-4a}{ab}$

■ PROCEDURE

To add or subtract two rational expressions that do not have a common denominator:

1. Find the LCD.
2. Rename (build) each fraction of algebra with the LCD as the denominator.
3. Add or subtract.
4. Reduce if possible.

EXAMPLE 6

Find the sum: $\frac{x}{a}+\frac{y}{a+b}$

$$\frac{x}{a}+\frac{y}{a+b}=\frac{x}{a}\cdot\frac{a+b}{a+b}+\frac{y}{a+b}\cdot\frac{a}{a}$$

The LCD is $a(a+b)$. Build each fraction to have $a(a+b)$ as the denominator.

$$=\frac{ax+bx}{a(a+b)}+\frac{ay}{a(a+b)}$$

$$=\frac{ax+bx+ay}{a(a+b)}$$

Add. ■

To combine more than two rational expressions, we build all of them so that they have a common denominator.

EXAMPLE 7

Perform the indicated operations: $\dfrac{7}{3x} - \dfrac{5}{x} + \dfrac{3}{2x}$

$$\frac{7}{3x} - \frac{5}{x} + \frac{3}{2x} = \frac{7}{3x} \cdot \frac{2}{2} - \frac{5}{x} \cdot \frac{6}{6} + \frac{3}{2x} \cdot \frac{3}{3}$$

The LCD is $6x$. Build each fraction to have $6x$ as the denominator.

$$= \frac{14}{6x} - \frac{30}{6x} + \frac{9}{6x}$$

$$= \frac{-7}{6x} = -\frac{7}{6x}$$

■

EXAMPLE 8

Subtract: $\dfrac{5x}{x+y} - \dfrac{4y}{x-y}$

$$\frac{5x}{x+y} - \frac{4y}{x-y} = \frac{5x}{x+y} \cdot \frac{x-y}{x-y} - \frac{4y}{x-y} \cdot \frac{x+y}{x+y}$$

The LCD is $(x+y)(x-y)$. Build each fraction.

$$= \frac{5x(x-y)}{(x+y)(x-y)} - \frac{4y(x+y)}{(x+y)(x-y)}$$

$$= \frac{(5x^2 - 5xy) - (4xy + 4y^2)}{(x+y)(x-y)}$$

$$= \frac{5x^2 - 5xy - 4xy - 4y^2}{(x+y)(x-y)}$$

$$= \frac{5x^2 - 9xy - 4y^2}{(x+y)(x-y)}$$

The result cannot be reduced. ■

■ CAUTION

It is a common error to try to multiply the fractions in an addition or subtraction problem to clear the denominators. This procedure (multiplying by the LCD) is valid *only* in an equation using the multiplication property of equality.

The next two examples illustrate the difference in the procedures for combining fractions and for solving equations involving fractions.

Combine

$$\frac{1}{x-2} + \frac{1}{5} - \frac{3}{2x-4} = \frac{1}{x-2} + \frac{1}{5} - \frac{3}{2(x-2)}$$

$$= \frac{10}{10(x-2)} + \frac{2(x-2)}{10(x-2)} - \frac{15}{10(x-2)}$$

$$= \frac{10 + 2(x-2) - 15}{10(x-2)}$$

$$= \frac{10 + 2x - 4 - 15}{10(x-2)}$$

$$= \frac{2x-9}{10(x-2)}$$

Solve

$$\frac{1}{x-2} + \frac{1}{5} - \frac{3}{2x-4} = 0$$

The LCM of the denominators is $10(x-2)$.

$$10(x-2)\left(\frac{1}{x-2} + \frac{1}{5} - \frac{3}{2(x-2)}\right) = 10(x-2) \cdot 0$$

$$10 + 2(x-2) - 15 = 0$$

$$10 + 2x - 4 - 15 = 0$$

$$2x - 9 = 0$$

$$2x = 9$$

$$x = \frac{9}{2}$$

The solution set is $\left\{\frac{9}{2}\right\}$.

To find the LCD, remember to factor the denominators.

EXAMPLE 9

Subtract: $\dfrac{2a}{a^2 + ab} - \dfrac{3}{a+b}$

$$\frac{2a}{a^2 + ab} - \frac{3}{a+b} = \frac{2a}{a(a+b)} - \frac{3}{a+b}$$

Factor the denominators.

$$= \frac{2a}{a(a+b)} - \frac{3}{a+b} \cdot \frac{a}{a}$$ The LCD is $a(a+b)$. Build the fractions.

$$= \frac{2a}{a(a+b)} - \frac{3a}{a(a+b)}$$

$$= \frac{-a}{a(a+b)}$$

$$= -\frac{1}{a+b}$$ Reduce. ■

The use of the LCD to clear an equation of fractions does not always yield an equivalent equation. For this reason, it is necessary to either:

1. Check the solutions obtained using the original equation, or

2. List the values of the variable that could cause division by zero in any fraction in the original equation. These values must be rejected since they are not solutions.

EXAMPLE 10

Solve: $\frac{3}{x-1} - \frac{x}{x-1} = \frac{2x}{x-1}$

$$(x-1)\left(\frac{3}{x-1}\right) - (x-1)\left(\frac{x}{x-1}\right) = (x-1)\left(\frac{2x}{x-1}\right)$$ Multiply each side by the LCD, $x - 1$, to clear the fractions.

$$3 - x = 2x$$ Simplify.

$$3 = 3x$$

$$1 = x$$ The solution set seems to be $\{1\}$, but replacing x by 1 in the original equation gives us undefined expressions.

Check:

$\frac{3}{0} = \frac{1}{0} + \frac{2}{0}$ is meaningless. Because $x = 1$ results in division by zero we conclude that there is no solution.

The solution set is $\emptyset$. ■

EXAMPLE 11

Solve: $\frac{4}{x} - \frac{1}{3} = \frac{5}{x + 14}$ — Multiply each side by the LCD $3x(x + 14)$.

$$(3x)(x + 14) \cdot \frac{4}{x} - (3x)(x + 14) \cdot \frac{1}{3} = (3x)(x + 14) \cdot \frac{5}{x + 14}$$

$3(x + 14)4 - (x)(x + 14) = (3x)5$ — Simplify.

$12x + 168 - x^2 - 14x = 15x$ — Remove parentheses.

$0 = x^2 + 17x - 168$ — Collect all terms on the right and factor.

$0 = (x + 24)(x - 7)$

$x + 24 = 0$ or $x - 7 = 0$ — Zero-product property.

$x = -24$ or $x = 7$ — The check is left for the student.

The solution set is $\{-24, 7\}$. ■

EXERCISE 5.4

A

Add or subtract; reduce if possible:

1. $\frac{6}{3 + x} - \frac{4}{3 + x}$
2. $\frac{5}{a - 7} + \frac{6}{a - 7}$
3. $\frac{4m}{2a - 3b} - \frac{5m}{2a - 3b}$
4. $\frac{5y}{5c + 3b} - \frac{4y}{5c + 3b}$
5. $\frac{x}{y - 4} + \frac{z}{4 - y}$
6. $\frac{3d}{2z - 5} - \frac{7d}{5 - 2z}$

Add or subtract; reduce if possible:

7. $\frac{m}{2y} - \frac{3m}{8y}$
8. $\frac{b}{3c} - \frac{5b}{9c}$
9. $\frac{2}{3a} + \frac{5}{6a}$
10. $\frac{7}{4b} + \frac{3}{20b}$
11. $\frac{3}{p^2} + \frac{4}{3p}$
12. $\frac{7}{6c} + \frac{5}{c^3}$
13. $\frac{a}{b} - \frac{b}{a}$
14. $\frac{x}{2} - \frac{2}{x}$
15. $\frac{4}{x} + \frac{7}{y}$
16. $\frac{3}{a} - \frac{5}{b}$

Solve:

17. $\frac{x}{2} = \frac{5}{2}$
18. $\frac{x}{5} = \frac{-7}{5}$
19. $\frac{2x}{3} = \frac{5}{3}$
20. $\frac{4x}{7} = \frac{-8}{7}$

B

Add or subtract; reduce if possible:

21. $\dfrac{2a}{4x+y} - \dfrac{3b}{4x+y}$

22. $\dfrac{5c}{2d-e} + \dfrac{4b}{2d-e}$

23. $\dfrac{10x}{5x-3y} - \dfrac{-6}{3y-5x}$

24. $\dfrac{12f}{2c-3d} - \dfrac{6}{3d-2c}$

25. $\dfrac{x}{x-y} - \dfrac{y}{x-y}$

26. $\dfrac{w}{w-z} + \dfrac{z}{z-w}$

27. $\dfrac{6}{5x^2y} + \dfrac{1}{10xy}$

28. $\dfrac{6}{5xy^2} - \dfrac{3}{2x^2y}$

29. $\dfrac{5}{3a^2b} + \dfrac{1}{4ab^2}$

30. $\dfrac{6}{5xy^2} + \dfrac{3}{2x^2y}$

31. $\dfrac{4}{a+b} - \dfrac{7}{a-b}$

32. $\dfrac{9}{x+4} - \dfrac{6}{x-4}$

33. $\dfrac{-18}{b} + \dfrac{2}{a} - \dfrac{1}{c}$

34. $\dfrac{14}{3m} + \dfrac{5}{6n} - \dfrac{1}{p}$

35. $\dfrac{8}{5a} - \dfrac{3}{2a^2} - \dfrac{3}{10}$

36. $\dfrac{3}{2b^2} + \dfrac{5}{3b} + \dfrac{5}{6b}$

Solve:

37. $\dfrac{x-3}{x+5} = \dfrac{2}{3}$

38. $\dfrac{2x+1}{x+7} = \dfrac{3}{4}$

39. $\dfrac{x+4}{x+6} = \dfrac{-2}{x+6}$

40. $\dfrac{x+3}{x+7} = \dfrac{-4}{x+7}$

C

Add or subtract; reduce if possible:

41. $\dfrac{2}{x+5} - \dfrac{7}{x-1}$

42. $\dfrac{5}{x-3} - \dfrac{6}{x+1}$

43. $\dfrac{3x}{(x+7)(x-3)} - \dfrac{9}{(x-3)(x+7)}$

44. $\dfrac{5a}{(a-5)(a-2)} - \dfrac{25}{(a-2)(a-5)}$

45. $\dfrac{5}{(x+7)(x-3)} + \dfrac{2}{(x+7)(3-x)}$

46. $\dfrac{6}{(a+4)(4-a)} + \dfrac{7}{(a+4)(a-4)}$

47. $\dfrac{1}{x} + x - 3$

48. $\dfrac{2}{b} + b + 4$

49. $\dfrac{1}{x-y} - \dfrac{x}{x^2-xy} + \dfrac{x^2}{x^3-x^2y}$

50. $\dfrac{a^2}{ab^2+b^3} + \dfrac{2a}{ab+b^2} + \dfrac{1}{a+b}$

51. $\dfrac{3}{(x+4)(x-5)} - \dfrac{4}{(x+4)(5-x)}$

52. $\dfrac{x+6}{(x-7)(x-3)} + \dfrac{x-8}{(x-7)(3-x)}$

53. $\dfrac{x}{x+y} + \dfrac{y}{x-y} + \dfrac{4}{x^2-y^2}$

54. $\dfrac{a}{a+b} - \dfrac{b}{a-b} - \dfrac{b}{a^2-b^2}$

55. $\dfrac{x}{x+y} + \dfrac{y}{x-y} + \dfrac{2}{x}$

56. $\dfrac{a}{a-b} - \dfrac{b}{a+b} - \dfrac{4}{a}$

Solve:

57. $\dfrac{2x-5}{5x} - \dfrac{x-2}{-5x} = 2$

58. $\dfrac{3a-7}{-4a} + \dfrac{a+5}{4a} = 4$

59. $\dfrac{x}{x-2} = \dfrac{-2}{2-x}$

60. $\dfrac{-x}{x-3} = \dfrac{3}{3-x}$

D

61. The formula to find the total resistance (R) of three resistors wired in parallel is

$$\frac{1}{R} = \frac{1}{r_1} + \frac{1}{r_2} + \frac{1}{r_3}.$$

Simplify the right side of the formula if all three resistors have the same resistance.

62. Using the formula in Exercise 61, simplify the right side of the formula if $r_1 = r_2 = r_3 = 3x$.

63. Express the right side of the formula as a single fraction if $a = b = c = 2x$.

$$\frac{1}{t} = \frac{1}{a} + \frac{1}{b} + \frac{1}{c}.$$

64. Using the formula in Exercise 63, express the right side of the formula as a single fraction if $a = b = c = 6x$.

65. The area of a trapezoid can be expressed as $A = \dfrac{ah}{2} + \dfrac{bh}{2}$. Solve for h.

66. Using the formula in Exercise 65, find h if $A = 9$ sq in., $a = 7$ in., and $b = 11$ in.

67. In optics, the focal length is given by $\dfrac{1}{f} = \dfrac{1}{a} + \dfrac{1}{b}$. Solve for f.

68. Using the formula in Exercise 67, find f if $a = 40$ mm and $b = 60$ mm.

69. If three times the reciprocal of twice a number is subtracted from five times the reciprocal of three times the same number, the result is $\dfrac{1}{30}$. Find the number.

70. If the reciprocal of seven times a number is added to six times the reciprocal of five times the same number, the result is $\frac{47}{105}$. Find the number.

STATE YOUR UNDERSTANDING

71. Why must caution be exercised when an equation is cleared of fractions?

72. Identify the error in the following problem. What is the correct answer?

Add: $\frac{2}{x^2 - 5x + 6} + \frac{3}{x^2 - x - 6} = \frac{2}{(x - 2)(x - 3)} + \frac{3}{(x + 2)(x - 3)}$

The LCM of the denominators is $(x + 2)(x - 2)(x - 3)$

$= (x + 2)(x - 2)(x - 3) \cdot \left[\frac{2}{(x - 2)(x - 3)}\right] + (x + 2)(x - 2)(x - 3) \cdot \left[\frac{3}{(x + 2)(x - 3)}\right]$

$= 2(x + 2) + 3(x - 2)$

$= 2x + 4 + 3x - 6$

$= 5x - 2$

CHALLENGE EXERCISES

73. Solve for x. $\frac{1}{a} - \frac{1}{x} = \frac{1}{b} + \frac{1}{x}$

74. The denominator of a fraction is three times the numerator. If eight is added to the numerator and six is subtracted from the denominator, the resulting fraction is 8/9. Find the original fraction.

75. One automobile averages six miles per hour more than another. It travels 350 miles in the same time the other travels 308 miles. Determine their rates.

76. A boat can travel 20 mph in still water. On the Monongahela River it can travel 60 miles downstream in the same time that it takes to go 40 miles upstream. What is the rate of the river?

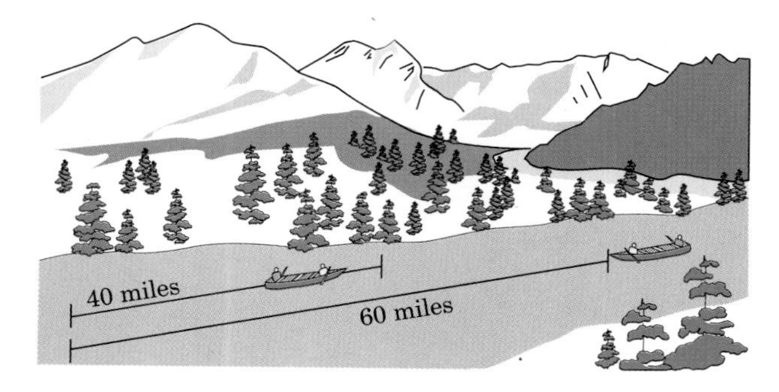

Figure for Exercise 76

MAINTAIN YOUR SKILLS (SECTIONS 2.5, 3.5, 4.3, 4.6)

Combine:

77. $2x - (3 - 4x) + (5 - 6x)$

78. $3(y + 7) - 6(y - 9) + 2y$

Factor:

79. $36x^2 - 108x + 81$

80. $25x^2 + 10x + 1$

Solve:

81. $10x - 12 - (-5x + 6) = 14 - (5x - 8)$

82. $-12x(x - 2) + 16 = 4x(4 - 3x) + 7x - 4$

83. $8x^2 - 20x + 6x - 15 = 0$

84. $15x^2 + 40x - 21x - 56 = 0$

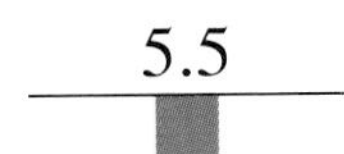

5.5 COMPLEX FRACTIONS

OBJECTIVE

1. Simplify complex fractions.

DEFINITION

Complex Fraction

A *complex fraction* is a fraction that contains one or more fractions in its numerator or its denominator.

To simplify a complex fraction means to find an equivalent expression that is a single fraction. This can be achieved by treating the complex fraction as a division exercise.

$$\frac{x}{y} = x \div y.$$

Using this idea, the complex fraction $\dfrac{\frac{1}{x}}{\frac{1}{y}}$ may be simplified.

$$\frac{\frac{1}{x}}{\frac{1}{y}} = \frac{1}{x} \div \frac{1}{y} = \frac{1}{x} \cdot \frac{y}{1} = \frac{y}{x}$$

All complex fractions can be simplified in this manner.

EXAMPLE 1

Simplify: $\dfrac{\dfrac{a}{b}}{\dfrac{3a}{2b}}$

$$\frac{\frac{a}{b}}{\frac{3a}{2b}} = \frac{a}{b} \div \frac{3a}{2b}$$ Divide the numerator by the denominator.

$$= \frac{a}{b} \cdot \frac{2b}{3a}$$ Invert the divisor.

$$= \frac{2}{3}$$

■

EXAMPLE 2

Simplify: $\dfrac{1 + \dfrac{3}{x}}{1 - \dfrac{4}{y}}$

$$\frac{1 + \frac{3}{x}}{1 - \frac{4}{y}} = \left(1 + \frac{3}{x}\right) \div \left(1 - \frac{4}{y}\right)$$

$$= \left(\frac{x + 3}{x}\right) \div \left(\frac{y - 4}{y}\right)$$ Combine terms in each set of parentheses.

$$= \left(\frac{x + 3}{x}\right)\left(\frac{y}{y - 4}\right)$$

$$= \frac{xy + 3y}{xy - 4x}$$

■ **CAUTION**

The terms in parentheses *must* be combined *before* inverting and multiplying.

■

An alternative method of simplifying complex fractions is based on the multiplication property of one.

$1 \cdot a = a$

EXAMPLE 3

Simplify: $\dfrac{\dfrac{1}{x} + 2}{\dfrac{1}{x} + 3}$

$$\frac{\dfrac{1}{x} + 2}{\dfrac{1}{x} + 3} = \frac{x}{x} \cdot \frac{\dfrac{1}{x} + 2}{\dfrac{1}{x} + 3}$$

Multiply by one (1) in the form $\dfrac{x}{x}$. This form is chosen because x is the LCD of the denominators of the fractions in the numerator and denominator of the complex fraction.

$$= \frac{x\left(\dfrac{1}{x} + 2\right)}{x\left(\dfrac{1}{x} + 3\right)}$$

$$= \frac{1 + 2x}{1 + 3x}$$

PROCEDURE

To simplify a complex fraction either:

1. Divide the numerator of the complex fraction by the denominator, or
2. Multiply the numerator and the denominator of the complex fraction by the LCD of the fractions within the complex fraction.

Example 4 shows both methods used on the same complex fraction.

EXAMPLE 4

Simplify: $\dfrac{\dfrac{3}{x} + 4}{2 - \dfrac{5}{x}}$

Treat as a Division Problem	Multiply the Numerator and Denominator by the LCD
$\dfrac{\dfrac{3}{x}+4}{2-\dfrac{5}{x}}=\left(\dfrac{3}{x}+4\right)\div\left(2-\dfrac{5}{x}\right)$	$\dfrac{\dfrac{3}{x}+4}{2-\dfrac{5}{x}}=\left(\dfrac{\dfrac{3}{x}+4}{2-\dfrac{5}{x}}\right)\left(\dfrac{x}{x}\right)$
$=\left(\dfrac{3+4x}{x}\right)\div\left(\dfrac{2x-5}{x}\right)$	$=\dfrac{\left(\dfrac{3}{x}+4\right)x}{\left(2-\dfrac{5}{x}\right)x}$
$=\left(\dfrac{3+4x}{x}\right)\left(\dfrac{x}{2x-5}\right)$	$=\dfrac{3+4x}{2x-5}$
$=\dfrac{3+4x}{2x-5}$	

■

EXAMPLE 5

Simplify: $\dfrac{4+\dfrac{1}{2x}}{3-\dfrac{1}{3x}}$

$$\frac{4+\dfrac{1}{2x}}{3-\dfrac{1}{3x}}=\frac{6x\left(4+\dfrac{1}{2x}\right)}{6x\left(3-\dfrac{1}{3x}\right)}$$

Multiply both the numerator and denominator by the LCD ($6x$).

$$=\frac{24x+3}{18x-2}$$

■

EXAMPLE 6

Simplify: $\dfrac{\dfrac{1}{a}-\dfrac{2}{b}}{\dfrac{3}{a}-\dfrac{5}{b}}$

$$\frac{\dfrac{1}{a}-\dfrac{2}{b}}{\dfrac{3}{a}-\dfrac{5}{b}}=\frac{ab\left(\dfrac{1}{a}-\dfrac{2}{b}\right)}{ab\left(\dfrac{3}{a}-\dfrac{5}{b}\right)}$$

Multiply both the numerator and denominator by the LCD (ab).

$$=\frac{b-2a}{3b-5a}$$

■

If we simplify Example 6 using division, we must perform the subtractions first.

EXAMPLE 7

Simplify by division: $\dfrac{\dfrac{1}{a} - \dfrac{2}{b}}{\dfrac{3}{a} - \dfrac{5}{b}}$

$$\frac{\dfrac{1}{a} - \dfrac{2}{b}}{\dfrac{3}{a} - \dfrac{5}{b}} = \left(\frac{1}{a} - \frac{2}{b}\right) \div \left(\frac{3}{a} - \frac{5}{b}\right)$$

Divide the numerator by the denominator.

$$= \frac{b - 2a}{ab} \div \frac{3b - 5a}{ab}$$

Combine terms inside the parentheses.

$$= \frac{b - 2a}{ab} \cdot \frac{ab}{3b - 5a}$$

Invert the divisor.

$$= \frac{b - 2a}{3b - 5a}$$

■

EXAMPLE 8

Greg drives from Memphis to Atlanta at an average speed of 40 mph. On the return trip, he averages 60 mph. What is his average speed for the entire trip?

Simpler word form:

$$\text{Average speed} = \frac{\text{Total distance}}{\text{Total time}}$$

The average speed (rate) for the trip, driving both ways, is the total distance divided by the total time: $r = \dfrac{D}{t}$.

Select variable(s):

Let D represent the distance from Memphis to Atlanta.
Let r represent the average speed.

	Rate (Given)	**Time** $\left(t = \frac{D}{r}\right)$	**Distance**
Memphis to Atlanta	40	$\frac{D}{40}$	D
Atlanta to Memphis	60	$\frac{D}{60}$	D
Total for round trip	r	$\frac{D}{40} + \frac{D}{60}$	$2D$

Make a table.

The columns for the rate and the distance are filled in first.

Translate to algebra:

$$r = \frac{2D}{\frac{D}{40} + \frac{D}{60}}$$

Now, to find the average speed (rate), write the fraction for the total distance, ($2D$), over the total time $\left(\frac{D}{40} + \frac{D}{60}\right)$.

Solve:

$$r = \frac{120(2D)}{120\left(\frac{D}{40} + \frac{D}{60}\right)}$$

$$r = \frac{240D}{3D + 2D}$$

$$r = \frac{240D}{5D}$$

$$r = 48$$

Answer:

The average speed was 48 mph. (Note that Greg spent *more time* driving at 40 mph.) ■

EXERCISE 5.5

A

Simplify:

1. $\dfrac{\frac{2}{a}}{\frac{5}{b}}$

2. $\dfrac{\frac{3}{c}}{\frac{3}{d}}$

3. $\dfrac{\frac{5}{a}}{\frac{-4}{b}}$

4. $\dfrac{\frac{7}{x}}{\frac{-2}{y}}$

5. $\dfrac{\frac{1}{4}}{\frac{x}{y}}$

6. $\dfrac{\frac{3}{5}}{\frac{a}{b}}$

7. $\dfrac{\frac{2}{y+1}}{\frac{6}{y+1}}$

8. $\dfrac{\frac{7}{a+3}}{\frac{10}{a+3}}$

9. $\dfrac{\frac{x^2}{y^2}}{\frac{x}{y}}$

10. $\dfrac{\frac{x}{y^2}}{\frac{x^2}{y}}$

11. $\dfrac{\frac{4x^3}{5a}}{\frac{3x}{4a^2}}$

12. $\dfrac{\frac{10a^3b}{3xy}}{\frac{5ab^3}{2xy^2}}$

13. $\dfrac{\frac{5x^2y}{4ab}}{\frac{15xy}{8a^2b^2}}$

14. $\dfrac{\frac{8a^4b^2}{5wz^2}}{\frac{12a^2b}{25w^2z^3}}$

15. $\dfrac{\frac{-33a^2b^4}{28r^2s^6}}{\frac{22ab^3}{21r^3s^3}}$

16. $\dfrac{\frac{48xyz}{76bcd}}{\frac{36x^2y^2}{19abc}}$

B

Simplify:

17. $\dfrac{\frac{x+1}{x-1}}{\frac{x+1}{x}}$

18. $\dfrac{\frac{a+3}{a+2}}{\frac{a+3}{6}}$

19. $\dfrac{\frac{p+2}{p-1}}{\frac{p+2}{p-3}}$

20. $\dfrac{\frac{2x-3}{x+5}}{\frac{x-7}{x+5}}$

21. $\dfrac{\frac{x^2-1}{x}}{\frac{x+1}{x}}$

22. $\dfrac{\frac{b-1}{b}}{\frac{b^2-1}{4}}$

23. $\dfrac{\frac{x-5}{x+3}}{\frac{2}{x^2+5x+6}}$

24. $\dfrac{\frac{5}{y^2-3y-28}}{\frac{7}{y-7}}$

25. $\dfrac{2+\frac{3}{a}}{3-\frac{1}{a}}$

26. $\dfrac{4 - \dfrac{6}{x}}{2 + \dfrac{5}{x}}$

27. $\dfrac{1 + \dfrac{1}{x - 1}}{1 - \dfrac{1}{x - 1}}$

28. $\dfrac{1 - \dfrac{1}{a - 3}}{1 + \dfrac{1}{a - 3}}$

29. $\dfrac{2 + \dfrac{3}{2x + 1}}{3 - \dfrac{1}{2x + 1}}$

30. $\dfrac{5 - \dfrac{1}{2b - 3}}{7 - \dfrac{1}{2b - 3}}$

31. $\dfrac{\dfrac{2}{5x - 1} + 3}{\dfrac{6}{5x - 1} + 2}$

32. $\dfrac{\dfrac{-5}{2a + 3} + 4}{\dfrac{-2}{2a + 3} + 5}$

33. $\dfrac{\dfrac{x}{5x + 3} - 5}{\dfrac{-x}{5x + 3} + 7}$

34. $\dfrac{\dfrac{-a}{3a - 5} - 6}{\dfrac{a}{3a - 5} - 4}$

35. $\dfrac{\dfrac{2a + 3}{a} - \dfrac{4}{a}}{\dfrac{3a - 5}{a} + \dfrac{6}{a}}$

36. $\dfrac{\dfrac{3z - 1}{z} + \dfrac{5}{z}}{\dfrac{2z + 7}{z} - \dfrac{3}{z}}$

C

Simplify:

37. $\dfrac{\dfrac{x}{y} - \dfrac{y}{x}}{\dfrac{x}{y} + \dfrac{y}{x}}$

38. $\dfrac{\dfrac{5}{a} - \dfrac{2}{b}}{\dfrac{4}{a} - \dfrac{3}{b}}$

39. $\dfrac{\dfrac{3}{c} - \dfrac{d}{4}}{\dfrac{d}{5} + \dfrac{3}{c}}$

40. $\dfrac{\dfrac{b}{3} + \dfrac{3}{a}}{\dfrac{5}{a} + \dfrac{b}{5}}$

41. $\dfrac{\dfrac{b}{c} - \dfrac{c}{b}}{\dfrac{b}{c} + \dfrac{c}{b}}$

42. $\dfrac{\dfrac{x}{y} - \dfrac{y}{x}}{\dfrac{y}{x} - \dfrac{x}{y}}$

43. $\dfrac{\dfrac{1}{3x} - \dfrac{1}{2y}}{\dfrac{4}{x} + \dfrac{3}{4y}}$

44. $\dfrac{\dfrac{2}{5a} + \dfrac{1}{3b}}{\dfrac{2}{3a} + \dfrac{5}{2b}}$

45. $\dfrac{\dfrac{1}{x - 1} + \dfrac{1}{x + 1}}{\dfrac{1}{x - 1} - \dfrac{1}{x + 1}}$

46. $\dfrac{\dfrac{1}{a + 3} - \dfrac{1}{a - 3}}{\dfrac{1}{a + 3} + \dfrac{1}{a - 3}}$

47. $\dfrac{x + 2 + \dfrac{1}{x + 2}}{x - 2 + \dfrac{1}{x - 2}}$

48. $\dfrac{2y - 1 + \dfrac{1}{y - 1}}{3y + 2 + \dfrac{1}{y - 1}}$

49. $\dfrac{\dfrac{4xy^2 - 8xy}{5a - 10a^2}}{\dfrac{5xy - 10x}{3a - 6a^2}}$

50. $\dfrac{\dfrac{7ab^2 - 5a^2b}{5a^2b - 7ab^2}}{\dfrac{2cd^2 - 3c^2d}{3c^2d - 2cd^2}}$

51. $\dfrac{\dfrac{x^2 - x - 6}{2x^2 - 3x - 2}}{\dfrac{x^2 + 5x + 6}{2x^2 + 3x + 1}}$

52. $\dfrac{\dfrac{4x^2 - 9}{3x^2 + x - 4}}{\dfrac{2x^2 - 5x + 3}{6x^2 + 17x + 12}}$

D

53. Larry averaged 60 mph when he drove from Salem to Eugene. On the return trip, traffic was slowed because of a rock concert at Albany, so he averaged only 30 mph. What was his average speed for the round trip?

54. An Olympic swimmer swam a certain distance at an average speed of 50 meters per minute. A prep swimmer swam the same distance at 35 meters per minute. What was the average rate of the two swimmers? (Round to the nearest tenth.)

55. The formula for current produced by cells in parallel is $I = \dfrac{E}{1 + \dfrac{R}{n}}$.

Simplify the right side of the formula.

56. Using the formula for Exercise 55, determine I to the nearest hundredth of an amp if $E = 1.5$ volts, $R = 0.1$ ohms, and $n = 4$.

57. Two objects in space with different masses and different velocities sometimes combine. The formula for finding their common velocity after combining is given by

$$v' = \dfrac{\dfrac{m_1}{v_2} + \dfrac{m_2}{v_1}}{\dfrac{m_1 + m_2}{v_1 v_2}}$$

Express v' as a simple fraction.

58. Using the formula from Exercise 57, find v' if $m_1 = 2000$, $m_2 = 3000$, $v_1 = 1800$ mph, and $v_2 = 1600$ mph.

59. Using the formula from Exercise 57, find v' if $m_1 = 1500$, $m_2 = 4000$, $v_1 = 2200$ mph, and $v_2 = 1800$ mph.

60. The average speed for a round trip can be determined by the formula $\bar{v} = \dfrac{2}{\dfrac{1}{v_1} + \dfrac{1}{v_2}}$, where $\bar{v}$ is the average speed for the roundtrip, v_1 is the speed in one direction and v_2 is the speed in the other direction. Express $\bar{v}$ as a simple fraction. Use this formula to verify your answers for Exercises 53 and 54. (This is called the harmonic mean.)

STATE YOUR UNDERSTANDING

61. Explain the basic differences between the two procedures for simplifying complex fractions.

62. Explain the error in the following problem. What is the correct answer?

$$\text{Simplify: } \frac{\dfrac{3}{x} + 2}{\dfrac{5}{x-1}} = \frac{x \cdot \left(\dfrac{3}{x} + 2\right)}{(x-1) \cdot \dfrac{5}{x-1}}$$

$$= \frac{3 + 2x}{5}$$

CHALLENGE EXERCISES

Simplify the following:

63. $\dfrac{\dfrac{1}{x} - 1}{\dfrac{1}{x^2} - 1}$

64. $\dfrac{\dfrac{2}{x} - \dfrac{4}{x^2} - \dfrac{16}{x^3}}{2 - \dfrac{9}{x} + \dfrac{4}{x^2}}$

65. $\dfrac{3 - \dfrac{4}{x+1}}{\dfrac{4}{x^2 - 1} - \dfrac{3}{x - 1}}$

66. $\dfrac{3}{\dfrac{3}{4 - \dfrac{1}{x}} - 2}$

MAINTAIN YOUR SKILLS (SECTIONS 3.7, 4.5)

Multiply:

67. $(3x - 7y)(2x + 5y)$

68. $(2x - 11)(x + 4)$

69. $(7ab + 3)(3ab - 4)$

70. $(9y - 1)(2y - 5)$

Factor:

71. $2x^2 + 7x - 15$

72. $2x^2 + x - 15$

73. $8x^2 + 12x - 20$

74. $10x^2 - 53x - 11$

5.6 RATIO AND PROPORTION

OBJECTIVE

1. Find the missing number that will make a given proportion true.

Two important ideas in mathematics are ratio and proportion. A ratio is the comparison of two numbers or quantities by division. Here are some examples of ratios.

$$\frac{2}{3}, \frac{4}{9}, \frac{x}{y}, \frac{208 \text{ miles}}{8 \text{ gallons}}, \frac{5 \text{ inches}}{2 \text{ feet}}, \frac{10 \text{ rpm}}{30 \text{ rpm}}$$

DEFINITION

Ratio

A *ratio* is a fraction that expresses the comparison of two numbers or measurements.

If the dimensions in a ratio are not the same, the ratio is sometimes called a "rate." The fraction $\frac{60 \text{ miles}}{1 \text{ hour}}$ is a rate of speed. If the dimensions in a ratio are the same, they may be dropped ("canceled"). So,

$$\frac{10 \text{ rpm}}{30 \text{ rpm}} = \frac{10}{30} = \frac{1}{3}.$$

When two ratios are written with an equal sign between them, the resulting equation is called a *proportion*. Here are two examples of proportions:

$$\frac{5 \text{ inches}}{2 \text{ feet}} = \frac{20 \text{ inches}}{8 \text{ feet}}, \qquad \frac{4}{9} = \frac{2}{3}.$$

For a proportion to be true, the ratios must represent the same value. To test whether or not the proportion

$$\frac{a}{b} = \frac{c}{d}$$

is true, multiply both sides by the product of the denominators, bd.

$$bd \cdot \frac{a}{b} = bd \cdot \frac{c}{d}$$

$$ad = bc$$

If $ad = bc$ is true, the proportion is true.

PROPERTY

> If $ad = bc$, then $\frac{a}{b} = \frac{c}{d}$, and if $\frac{a}{b} = \frac{c}{d}$, then $ad = bc$.

The proportion $\frac{14}{8} = \frac{35}{20}$ is true because

$$20 \cdot 14 = 8 \cdot 35$$

$$280 = 280.$$

The proportion $\frac{4}{9} = \frac{2}{3}$ is not true because

$$4 \cdot 3 \neq 2 \cdot 9$$

$$12 \neq 18.$$

When a proportion contains a variable, the missing number can be found by forming the equation $ad = bc$ and then solving by using the laws of equations.

$$\frac{3}{x} = \frac{4}{16}$$

$$3 \cdot 16 = 4 \cdot x$$

$$48 = 4x$$

$$12 = x$$

Proportions can also be solved by multiplying each side by the LCD. The preceding method is a shortcut called ‘‘cross multiplication’’.

PROCEDURE

> To solve the proportion (find the missing number) $\frac{a}{b} = \frac{c}{d}$:
>
> 1. Cross multiply to form the equation $ad = bc$.
> 2. Solve.

EXAMPLE 1

Solve the following proportion: $\frac{4}{9} = \frac{8}{x}$

$\frac{4}{9} = \frac{8}{x}$ Cross multiply; that is, form the equation $ad = bc$.

$4x = 72$

$x = 18$ ■

Proportions are solved the same way even if they contain decimals or fractions.

EXAMPLE 2

Solve the following proportion: $\frac{0.6}{x} = \frac{1.2}{0.84}$

$\frac{0.6}{x} = \frac{1.2}{0.84}$ Cross multiply; that is, form the equation $ad = bc$.

$0.504 = 1.2x$

$0.42 = x$ ■

EXAMPLE 3

Solve the following proportion: $\frac{\frac{3}{4}}{1\frac{2}{3}} = \frac{\frac{1}{2}}{x}$

$\frac{\frac{3}{4}}{1\frac{2}{3}} = \frac{\frac{1}{2}}{x}$

$\frac{3}{4}x = \frac{5}{6}$

$x = 1\frac{1}{9}$ ■

Many situations occur in which two quantities are compared by ratios. Assuming that the comparison of the quantities is constant, the given comparison can be used to discover the missing part of a second one.

EXAMPLE 4

If 2 pounds of seed cost \$0.36, what will 12 pounds cost?

	Pounds of Seed	Cost in Dollars
Case 1	2	0.36
Case 2	12	y

$$\frac{2 \text{ lb seed}}{12 \text{ lb seed}} = \frac{0.36}{y}$$

$$\frac{2}{12} = \frac{0.36}{y}$$

$$2y = 4.32$$

$$y = 2.16$$

Form two ratios of like quantities. These ratios are equal whenever the quantities change at the same rate.

So, 12 pounds of seed will cost \$2.16. ■

The ratio of the distance on a map to the corresponding distance on the earth is a very useful application of proportions.

EXAMPLE 5

A map of the western United States is scaled so that $\frac{3}{4}$ inch represents 100 miles. How many miles is it between San Diego and Seattle if the distance on the map is 9 inches?

Select variable:

Let D represent the distance between San Diego and Seattle.

Distance on Map	Distance Represented
$\frac{3}{4}$ in.	100 miles
9 in.	D miles

Make a table to organize the information given in the problem.

Translate to algebra:

$$\frac{\frac{3}{4}}{9} = \frac{100}{D}$$

Solve:

$$\frac{3}{4}D = 900$$

$$D = 1200$$

Answer:

The distance between San Diego and Seattle is 1200 miles. ■

EXERCISE 5.6

A

Solve:

1. $\frac{6}{8} = \frac{12}{x}$
2. $\frac{x}{42} = \frac{5}{7}$
3. $\frac{5}{2} = \frac{w}{8}$
4. $\frac{15}{4} = \frac{30}{x}$
5. $\frac{3}{y} = \frac{9}{12}$
6. $\frac{6}{11} = \frac{a}{22}$
7. $\frac{14}{9} = \frac{28}{x}$
8. $\frac{11}{5} = \frac{y}{30}$
9. $\frac{20}{25} = \frac{x}{5}$
10. $\frac{12}{x} = \frac{24}{18}$
11. $\frac{5}{y} = \frac{15}{27}$
12. $\frac{21}{x} = \frac{63}{12}$
13. $\frac{8}{15} = \frac{16}{y}$
14. $\frac{36}{c} = \frac{12}{25}$
15. $\frac{15}{48} = \frac{x}{16}$
16. $\frac{x}{16} = \frac{25}{80}$
17. $\frac{x}{4.8} = \frac{15}{1.6}$
18. $\frac{y}{8} = \frac{25}{1.6}$
19. $\frac{p}{0.25} = \frac{0.2}{0.05}$
20. $\frac{1.1}{0.05} = \frac{q}{3}$

B

Solve:

21. $\frac{5}{2} = \frac{w}{9}$
22. $\frac{5}{8} = \frac{7}{w}$
23. $\frac{x}{40} = \frac{\frac{3}{4}}{5}$
24. $\frac{\frac{1}{2}}{y} = \frac{\frac{2}{3}}{10}$

25. $\frac{8}{9} = \frac{\frac{1}{3}}{y}$

26. $\frac{4}{5} = \frac{x}{\frac{1}{2}}$

27. $\frac{5}{7} = \frac{\frac{2}{7}}{x}$

28. $\frac{x}{\frac{3}{5}} = \frac{25}{6}$

29. $\frac{2\frac{1}{2}}{3\frac{1}{3}} = \frac{4\frac{1}{4}}{x}$

30. $\frac{2\frac{1}{2}}{x} = \frac{3\frac{1}{5}}{4}$

31. $\frac{4}{\frac{2}{3}} = \frac{x}{1}$

32. $\frac{\frac{4}{5}}{5} = \frac{4}{x}$

33. $\frac{\frac{2}{3}}{x} = \frac{8}{12}$

34. $\frac{\frac{1}{4}}{k} = \frac{15}{60}$

35. $\frac{\frac{3}{8}}{\frac{6}{5}} = \frac{x}{16}$

36. $\frac{\frac{2}{15}}{\frac{1}{2}} = \frac{12}{x}$

37. $\frac{2\frac{1}{2}}{2} = \frac{4\frac{1}{4}}{x}$

38. $\frac{2\frac{1}{2}}{x} = \frac{3\frac{1}{3}}{6}$

39. $\frac{\frac{2}{3}}{\frac{2}{7}} = \frac{p}{\frac{3}{5}}$

40. $\frac{\frac{10}{7}}{\frac{8}{3}} = \frac{\frac{9}{7}}{t}$

C

Solve:

41. $\frac{6.5}{26} = \frac{y}{0.04}$

42. $\frac{0.014}{x} = \frac{7}{50}$

43. $\frac{8.4}{12} = \frac{14}{w}$

44. $\frac{0.55}{w} = \frac{11}{10}$

45. $\frac{0.05}{0.9} = \frac{y}{4.5}$

46. $\frac{1.2}{2.7} = \frac{3.4}{w}$

47. $\frac{0.055}{1} = \frac{R}{100}$

48. $\frac{0.112}{x} = \frac{7}{100}$

49. $\frac{5.6}{3.2} = \frac{28}{x}$

50. $\frac{9.5}{7.2} = \frac{x}{36}$

51. $\frac{12}{25} = \frac{2x}{14}$

52. $\frac{3a}{16} = \frac{18}{25}$

53. $\frac{27}{2x} = \frac{1.5}{2.3}$

54. $\frac{98}{5x} = \frac{1.4}{3.3}$

55. $\frac{125}{375} = \frac{x}{5}$

56. $\frac{155}{124} = \frac{y}{7}$

57. $\frac{235}{376} = \frac{c}{12}$

58. $\frac{483}{805} = \frac{18}{b}$

59. $\frac{225}{210} = \frac{b}{14}$

60. $\frac{168}{c} = \frac{2415}{667}$

D

61. A picture that measures 8 cm wide and 10 cm high is to be enlarged so that the height will be 25 cm. What will be the width of the enlargement?

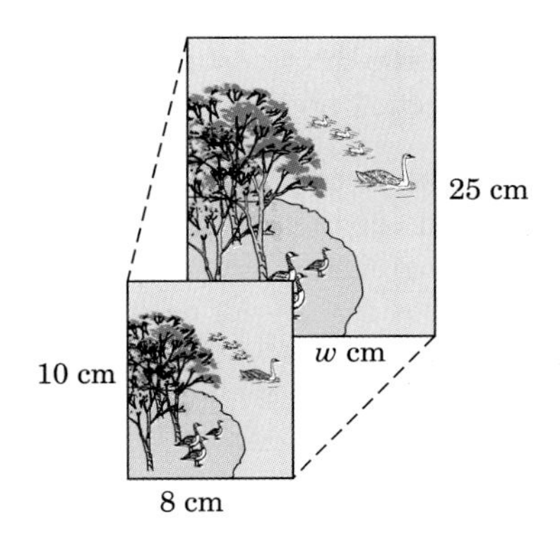

Figure for Exercise 61

	Width	Height
Case 1	8 cm	10 cm
Case 2	w	25 cm

62. Merle is knitting an afghan. The knitting gauge is 4 rows to 1 inch. How many rows must she knit to complete $19\frac{1}{2}$ inches of the afghan?

	Rows	Inches
Case 1	4	1
Case 2	r	$19\frac{1}{2}$

63. At a certain time of day, a building 300 m tall casts a shadow of 225 m, whereas a second building casts a shadow of 150 m. How tall is the second building?

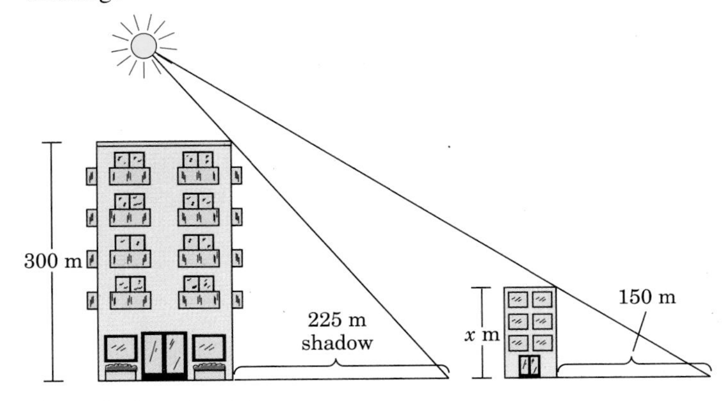

Figure for Exercise 63

64. If 100 lb of pesticide will cover 50 acres of potatoes, how much pesticide is needed for 135 acres of potatoes?

65. The Reedville Elementary School expects an enrollment of 1365 students. The district assigns teachers at the rate of 3 teachers for every 65 students. The district now employs 56 teachers. How many additional teachers does the district need to hire?

66. A homeowner in Texas pays $360 in property taxes on his home, which is assessed at $80,000. His neighbor's house is assessed at $145,000. How much property tax will his neighbor pay?

67. One can of frozen lemonade concentrate mixed with three and one-half cans of water makes one liter of lemonade. At the same rate, how many cans of water are needed to make five liters of lemonade?

68. A carpet store advertises 15 square yards of carpet installed for $275. At this price, what will 33 square yards of carpet cost?

69. The counter on a VCR registers 750 after the recorder has been running for 25 minutes. What will the counter register after 35 minutes?

70. During the first 560 miles of their vacation trip, the Scabery's auto used 17.5 gallons of gasoline. At this rate, how many gallons of gasoline to the nearest tenth of a gallon will be needed to finish the remaining 410 miles of the trip?

71. If it takes three men 18 hours to roof a house, how many of these jobs could they do in 72 hours?

72. In the first 15 games of a 24-game schedule, Maria's basketball team scored a total of 1150 points. At this rate, how many points can they expect to score in the remaining games?

73. A 16-ounce can of tomato sauce costs $0.53 and a 29-ounce can costs $0.82. Is the price per ounce the same in both cases? If not, then to the nearest cent, what would the price of the 29-ounce can have to be to equalize their prices?

74. A map of the world is scaled so that $\frac{5}{8}$ inch represents 100 miles. How many miles is it between Moscow and Paris if the distance on the map is $7\frac{1}{2}$ inches?

75. A doctor requires that nurse Ida give 10 milligrams of a certian drug to a patient. The drug is in a solution that contains 32 milligrams in one cubic centimeter (cc). How many cc's should Ida use for the injection?

76. If Marvin receives $550 for $\frac{3}{4}$ of a ton of raspberries, how much will he receive for $2\frac{1}{4}$ tons?

77. A 20-foot beam of structural steel contracts 0.0053 inch for each drop of 5 degrees in temperature. To the nearest ten-thousandth of an inch, how much will a 13-foot beam of structural steel contract for a drop of 7 degrees in temperature?

78. Betty prepares a mixture of nuts that has hazelnuts and walnuts in a ratio of 4 to 3. How many pounds of each will she need to make 84 pounds of the mixture?

79. The estate of the late Mrs. Hazel Geoffry is to be divided among her three nieces in the ratio of 6 to 4 to 3. How much of the estate, which is valued at $101,725, will each niece receive?

STATE YOUR UNDERSTANDING

80. Explain the difference between a ratio and a proportion.

81. If $\frac{a}{b} = \frac{c}{d}$, is it correct to say $\frac{b}{a} = \frac{d}{c}$? Why?

CHALLENGE EXERCISES

82. If $6x - 6y = 2x - 4y$, what is the ratio of x to y?

83. If $6ax + 2y = 4ay + 3x$, what is the ratio of x to y?

Solve:

84. $\frac{2x - 3}{x} = \frac{2}{x - 1}$

85. $\frac{3x + 5}{2x - 1} = \frac{x + 3}{x - 2}$

MAINTAIN YOUR SKILLS (SECTION 4.7)

Factor:

86. $16x^3y^2 - 10x^2y^2 + 3xy^2$

87. $x^2 - 7x - 144$

88. $50a^3 - 648a$

89. $25x^2 - 36y^2$

90. $12x^2 + 11x + 2$

91. $512 - x^3$

Solve:

92. $x^2 + 13x + 36 = 0$

93. $3x^2 - 4x = 15$

5.7 VARIATION

OBJECTIVES

1. Solve problems involving direct variation.
2. Solve problems involving inverse variation.

In our physical world, most things are in a state of change or variation. Measurements of changes in temperature, rainfall, light, heat, and fuel supplies are recorded. Changes in height, weight, and blood pressure can be important to a person's health. Many relationships between such measurements can be expressed in formulas. Here we are concerned about only two types of variation: *direct* and *inverse*.

Direct Variation

DEFINITION

y varies directly as x when the quotient of y and x is a constant.

$\frac{y}{x} = k$

$y = kx$

FORMULA

Direct variation:

$$\frac{y}{x} = k,\ k \neq 0 \quad \text{or} \quad y = kx,\ k \neq 0$$

Several traditional formulas are examples of direct variation.

Distance (D) traveled at a constant speed (r) varies directly as time (t).

$$D = rt \quad \text{or} \quad \frac{D}{t} = r$$

The weight (w) of a gold ingot varies directly as the volume (v).

$$w = kv \quad \text{or} \quad \frac{w}{v} = k$$

EXAMPLE 1

Write an equation expressing the following relationship: Weight varies directly as volume. Use k as the constant of variation.

$\frac{w}{v} = k$ In direct variation, the quotient is constant. Let w represent weight and v represent the volume. ■

EXAMPLE 2

The weight (w) of a piece of aluminum varies directly as the volume (v). If a piece of aluminum containing $1\frac{1}{2}$ cubic feet weighs 254 pounds, find the weight of a piece containing 5 cubic feet.

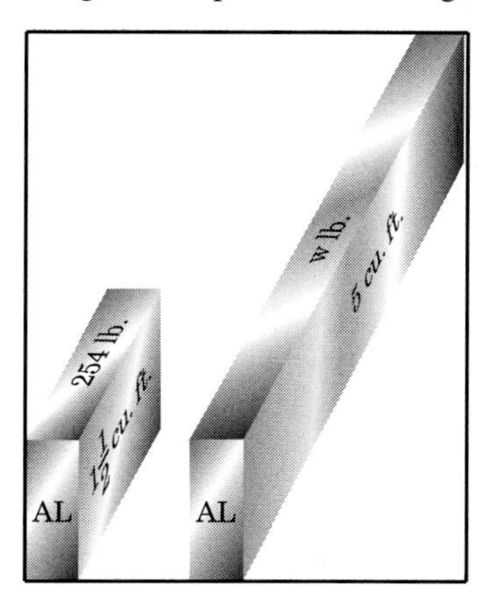

Write an equation expressing the direct variation relationship using k as the constant of variation. The weight varies directly as volume.

$$\frac{w}{v} = k$$

Formula for direct variation.

$$\frac{254}{1\frac{1}{2}} = k$$

Find the constant of variation by substituting the known values for w and v.

$$k = \frac{508}{3}$$

Solve the equation.

So,

$$\frac{w}{v} = \frac{508}{3}$$

Replace k with its calculated value.

$$\frac{w}{5} = \frac{508}{3}$$

To find the weight of a piece that contains 5 cubic feet, substitute 5 for v in the formula.

$$3w = 2540$$

Solve.

$$w = \frac{2540}{3}$$

$$w = 846\frac{2}{3}$$

So, the weight of a piece containing 5 cubic feet is $846\frac{2}{3}$ pounds. ■

Inverse Variation

DEFINITION

y varies inversely as x when the product of y and x is a constant.

$xy = k$

FORMULA

Inverse variation:

$xy = k, k \neq 0$ or $y = \frac{k}{x}, k \neq 0$

Several examples of inverse variation can be found in the physical sciences.

Pressure (P) varies inversely as volume (V) when the temperature (T) is constant.

$PV = T$

Rate (r) varies inversely as time (t) when the distance (D) is constant.

$rt = D$

When two pulleys are connected by a belt, the number of revolutions per minute (rpm) each turns varies inversely as their respective diameters.

Smaller Pulley		Larger Pulley
$dr = k$	and	$DR = k$

where d and D represent the diameters and r and R represent the revolutions per minute.

If the smaller pulley has a diameter of 10 inches and revolves 80 times per minute, how many rpm will the larger one make if it has a diameter of 16 inches?

First find k using the information about the smaller pulley. Then use this value of k to find the rpm of the larger pulley.

Smaller Pulley

$$dr = k$$
$$(10)(80) = k$$
$$800 = k$$

We now substitute 800 for k in the formula for the larger pulley.

Larger Pulley

$DR = k$
$DR = 800$
$16R = 800$
$R = 50$

So, the larger pulley will make 50 revolutions per minute.

EXAMPLE 3

Write an equation expressing the following relationship: If the distance is constant, speed varies inversely as time. Let k be the constant of variation.

$rt = k$ In inverse variation, the product is constant. Let r represent the rate of speed and t represent time. ■

EXAMPLE 4

The base (b) of a rectangle with constant area k varies inversely with its height. One rectangle has a base of 10 and a height of 6. Find the height of another rectangle with the same area that has a base of 12.

$bh = k$ Inverse variation, therefore, the product is a constant.

$10 \cdot 6 = k$

$k = 60$ Substitute.

So,

$bh = 60$ Substitute the calculated value for k.

$12h = 60$

$h = 5$ Solve.

Therefore, the height of the second rectangle is 5. ■

DEFINITION

Joint Variation

Three quantities *vary jointly* when one of the variables varies directly as the product of the other two.

FORMULA

$w = kxy$

Joint variation:

$$\frac{w}{xy} = k,\ k \neq 0 \quad \text{or} \quad w = kxy,\ k \neq 0$$

EXAMPLE 5

The interest (i) paid on borrowed money varies directly as the principal (P) and the time (t). If \$80 of interest is earned on a loan of \$600 in two years, how much interest is earned on \$1000 borrowed for three years?

$$\frac{i}{Pt} = k$$

The interest i varies directly as both P and t. This is called "joint variation." In this case, we say i varies jointly as P and t.

$$\frac{80}{(600)(2)} = k$$

Substitute to find the value of k.

$$k = \frac{1}{15}$$

$$\frac{i}{(1000)(3)} = \frac{1}{15}$$

Substitute the value of k and the second set of known values into the formula.

$$i = 200$$

The interest earned is \$200. ■

EXERCISE 5.7

A

Write an equation expressing the variation relationship given in each of the following. Use k as the constant of variation:

1. If time is held constant, the distance (d) varies inversely as rate (r).
2. The circumference (C) varies directly as the diameter (d).
3. The perimeter (P) of a square varies directly as the length of its side (s).
4. If the area of a rectangle is held constant, the length (ℓ) varies inversely as its width (w).
5. The volume (V) of a cone varies jointly as the area of its base (B) and its height (h).
6. The volume (V) of a cylinder varies jointly as the square of its radius (r) and its height (h).
7. The volume (V) of a gas varies directly as the temperature (T).
8. The volume (V) of a gas varies inversely as the pressure (P).
9. The current (I) passing through a wire varies directly as the voltage (E).

10. The current (I) passing through a wire varies inversely as the resistance (R).

11. The amount of illumination (I) from a light source varies inversely as the square of the distance (D).

12. The volume (V) of a sphere varies directly with the cube of the radius (r).

B

Find the constant of variation for each of the stated conditions:

13. y varies directly as x, and $y = 12$ when $x = 8$.

14. y varies directly as x, and $y = 16$ when $x = 12$.

15. y varies inversely as x, and $y = 18$ when $x = 4$.

16. y varies inversely as x, and $y = 22$ when $x = 3$.

17. y varies directly as the square of x, and $y = 14$ when $x = 7$.

18. y varies directly as the square of x, and $y = 50$ when $x = 5$.

19. y varies inversely as the square of x, and $y = 3$ when $x = 4$.

20. y varies inversely as the square of x, and $y = 6$ when $x = \frac{1}{2}$.

21. y varies inversely as the cube of x, and $y = 6$ when $x = 2$.

22. y varies directly as the cube of x, and $y = 36$ when $x = 2$.

23. y varies inversely as the cube of x, and $y = 5$ when $x = 3$.

24. y varies directly as the cube of x, and $y = 8$ when $x = 5$.

25. y varies jointly as x and z, and $y = 12$ when $x = 4$ and $z = 12$.

26. y varies jointly as x and z, and $y = 18$ when $x = 16$ and $z = 4$.

27. y varies jointly as x and the square of z, and $y = 25$ when $x = 3$ and $z = 5$.

28. y varies jointly as x and the square of z, and $y = 40$ when $x = 18$ and $z = 8$.

C

Solve by first obtaining the constant of variation:

29. If a varies directly as b and $a = 14$ when $b = 6$, find a when $b = 16$.

30. If a varies directly as m and $a = 16$ when $m = 12$, find a when $m = 23$.

31. If y varies inversely as x and $y = 21$ when $x = 16$, find y when $x = 24$.

32. If y varies inversely as x and $y = 36$ when $x = 12$, find y when $x = 6$.

33. If m varies directly as the square of n and $m = 30$ when $n = 5$, find m when $n = 16$.

34. If m varies directly as the square of n and $m = 72$ when $n = 4$, find m when $n = 2$.

35. If m varies directly as the square of n and $m = 11$ when $n = 3$, find m when $n = 6$.

36. If z varies directly as the cube of w and $z = 42$ when $w = 5$, find z when $w = 2$.

37. If z varies directly as the cube of w and $z = 81$ when $w = 3$, find z when $w = 2$.

38. If z varies inversely as the cube of w and $z = 21$ when $w = 2$, find z when $w = 4$.

39. If y varies jointly as x and z, and $y = 22$ when $x = 4$ and $z = 2$, find y when $x = 8$ and $z = 1$.

40. If y varies jointly as x and z, and $y = 31$ when $x = 2$ and $z = 10$, find y when $x = 16$ and $z = 2$.

41. If y varies jointly as x and the square of z, and $y = 36$ when $x = 12$ and $z = 3$, find y when $x = 8$ and $z = 6$.

42. If y varies jointly as z and the square of x, and $y = 144$ when $x = 6$ and $z = 12$, find y when $x = 4$ and $z = 9$.

D

43. The weight of a metal ingot varies directly as the volume v. If an ingot containing $1\frac{2}{3}$ cubic feet weighs 350 pounds, what will an ingot of $2\frac{1}{3}$ cubic feet weigh?

44. The time it takes to make a certain trip varies inversely as the speed. If it takes five hours at 50 mph, how long will it take at 60 mph?

45. A car salesman's salary varies directly as his total sales. If he receives $372 for sales of $3100, how much will he receive for sales of $4500?

46. The amount of quarterly income a person receives varies directly as the amount of money invested. If Dan earns $40 on a $2000 investment, how much would he earn on a $4300 investment?

47. The number of amperes varies directly as the number of watts. For a reading of 50 watts, the number of amperes is $\frac{5}{11}$. What are the amperes when there are 75 watts?

48. The weight of wire varies directly as its length. If 1000 ft of wire weighs 45 lb, what will one mile of wire weigh?

49. The length of a rectangle with a constant area varies inversely as the width. If one rectangle has a length of 12′ and a width of 8′, what will be the length of a rectangle with a width of 6′?

50. The force needed to raise an object with a crowbar varies inversely as the length of the crowbar. If it takes 40 lb of force to lift a certain object with a 2-ft-long crowbar, what force will be necessary if you use a 3-ft crowbar?

? lb
2 ft
40 lb
3 ft

Figure for Exercise 50

51. As a rule of thumb, realtors suggest that the price you can afford to pay for a house varies directly as your annual salary. If a person earning $18,500 can purchase a $46,250 home, what price home can a person earning $24,000 annually afford?

52. Assuming that each person works at the same rate, the time it takes to complete a job varies inversely as the number of people assigned to it. If it takes 5 people 12 hours to do a job, how long will it take 3 people?

STATE YOUR UNDERSTANDING

53. If a varies directly with b, what happens to a when b is doubled? is tripled? is multipled by 10? is divided by 4?

54. If a varies inversely with b, what happens to a when b is doubled? is tripled? is multipled by 10? is divided by 4?

CHALLENGE EXERCISES

55. The distance a freely falling body has traveled varies directly with the square of the time. If a body has fallen 64 ft in 4 sec, how far will it fall in 6 sec?

56. The Kinetic Energy of a moving automobile varies directly with the square of its speed. How is the Kinetic Energy affected when the speed is doubled?

57. Newton's Universal Law of Gravitation states that the gravitational force, F, between two objects varies inversely with the square of the distance, r, between them. How will tripling the distance between two objects affect the gravitational attraction of the two objects?

58. According to Kepler's Third Law, the square of the time it takes a planet to orbit the sun varies directly with the cube of its average distance from the sun. If Uranus is approximately nineteen times as far from the sun as Earth, how long, to the nearest year, does it take Uranus to complete one orbit of the sun? (Use your calculator.)

MAINTAIN YOUR SKILLS (SECTIONS 4.7, 5.1)

Factor completely:

59. $25x^2 - 100$

60. $36x^2 + 12x + 1$

61. $18a^2 - 66a - 24$

62. The length of a rectangular flower bed is three more than twice the width. The area of the bed is 152 sq ft. What are the dimensions of the flower bed?

Multiply:

63. $\dfrac{-3a^4}{2c} \cdot \dfrac{5b^2}{4c^2}$

64. $\dfrac{3s + 2t}{5s - t} \cdot \dfrac{3s - 2t}{3s + 5t}$

Divide:

65. $\dfrac{35m^2}{6n} \div \dfrac{8n^3}{5m}$

66. $\dfrac{4c - 5d}{c + 2d} \div \dfrac{2c - d}{4c}$

CHAPTER 5
SUMMARY

Rational Expression	The indicated quotient (fraction) of two polynomials.	(p. 297)
Basic Principle of Fractions	$\dfrac{AC}{BC} = \dfrac{A}{C}$, $BC \neq 0$	(p. 298)
Reducing Rational Expressions	Write both the numerator and denomintor in factored form, and eliminate all common factors using the basic principle of fractions.	(pp. 298, 299)
Multiplying Rational Expressions	Multiply the numerators and multiply the denominators. The basic principle of fractions allows reducing before multiplication.	(pp. 300, 301)
Dividing Rational Expressions	Multiply the dividend by the reciprocal of the divisior.	(p. 309)
Least Common Multiple (LCM)	The LCM is used to:	(p. 314)
	1. Clear the fractions from an equation containing rational expressions, and	(p. 317)
	2. Rewrite rational expressions with a common denominator so that they may be added and subtracted.	
Adding and Subtracting Rational Expressions	Find the LCM of the denominators, rewrite each expression with the LCM as the denominators, then add or subtract the numerators.	(p. 329)
Complex Fraction	A fraction or rational expression that contains one or more fractions in its numerator or denominator or both.	(p. 337)

Simplifying Complex Fractions	Either: **1.** Divide the numerator by the denominator after performing the operations in both, or **2.** Multiply the numerator and denominator by the LCM of the denominators of the fractions within the complex fraction.	(p. 339)
Ratios	Ratios are the comparison, by division, of two numbers or measurements. Ratios can be manipulated as other rational expressions.	(p. 347)
Proportions	A proportion is an equation with a single ratio on each side.	(p. 347)
Solving Proportions	When three parts of a proportion are given, the fourth part can be found by "cross multiplication," that is, if $\frac{a}{b} = \frac{c}{d}$, then $ad = bc$.	(p. 348)
Direct Variation	Two variables vary directly if their quotient is constant. If $\frac{y}{x} = k$, then y varies directly as x.	(p. 356)
Inverse Variation	Two variables vary inversely if their product is constant. If $xy = k$, then y varies inversely as x.	(p. 358)
Joint Variation	Three variables vary jointly if one of them varies directly as the product of the other two. If $\frac{w}{xy} = k$, then w varies jointly as x and y.	(p. 359)

CHAPTER 5

REVIEW EXERCISES

SECTION 5.1 Objective 1

Reduce; you need not state variable restrictions:

1. $\dfrac{3a - 2}{9a^2 - 4}$

2. $\dfrac{4x + 3}{16x^2 - 9}$

3. $\dfrac{4x + 12}{4x^2 + 28x + 48}$

4. $\dfrac{12x + 12}{6x^2 + 12x + 6}$

5. $\dfrac{6x^2 + 5x - 6}{8x^2 + 14x + 3}$

SECTION 5.1 Objective 2

Multiply:

6. $\dfrac{4a + 8}{a - 2} \cdot \dfrac{a^2 - 4a + 4}{8a + 16}$

7. $\dfrac{x - 4}{x + 6} \cdot \dfrac{x^2 - 36}{3x - 12}$

8. $\dfrac{x^2 + 7x + 12}{x^2 - 9} \cdot \dfrac{2x - 6}{x^2 + 8x + 16}$

9. $\dfrac{x^3 - 4x^2}{x^4 - x^3 - 12x^2} \cdot \dfrac{x^2 + 5x + 6}{x^2 + 3x + 2}$

10. $x(x^2 - 9)\left[\dfrac{x - 2}{x^2 + 3x}\right]$

SECTION 5.1 Objective 3

11. Convert 45 miles per hour to feet per second.

12. Convert 90 miles per hour to feet per second.

13. Convert 220 feet per second to miles per hour.

14. Convert 264 feet per second to miles per hour.

15. A wheel makes 500 revolutions per second. How many revolutions will it make in one hour?

SECTION 5.2 Objective 1

Divide:

16. $\dfrac{5a^2 - 6a + 1}{a^2 - 1} \div \dfrac{16a^2 - 9}{4a^2 + 7a + 3}$

17. $\dfrac{3x^2 - 3}{6x^2 + 18x + 12} \div \dfrac{x^2 - 3x + 2}{2x^2 - 8}$

18. $\dfrac{2x^2 + 17x + 21}{x^2 + 2x - 35} \div \dfrac{2x^2 - 7x - 15}{x^2 - 25}$

19. $\dfrac{18a^3 + 21a^2 - 60a}{21a^2 - 25a - 4} \div \dfrac{16a^3 + 28a^2 - 30a}{28a^2 - 17a - 3}$

20. $\dfrac{56y^3 + 54y^2 - 20y}{8y^2 - 2y - 15} \div \dfrac{63y^3 + 129y^2 - 42y}{6y^2 + 5y - 21}$

SECTION 5.3 Objective 1

Find the LCM of the following (leave in factored form):

21. $x^2 + 5x + 6,\ x^2 - 9$

22. $2x + 10,\ x^2 - 25$

23. $x^2 + 5x + 4,\ x^2 + 4x + 3$

24. $2x - 6,\ x^2 - 9,\ x^2 - 6x + 9$

25. $x^2 + 2x,\ x^2 - 4,\ x^2 - 4x + 4$

SECTION 5.3 Objective 2

Solve:

26. $\dfrac{x}{x + 1} + \dfrac{14}{5x + 5} = -\dfrac{4}{5}$

27. $\dfrac{x}{x - 2} - \dfrac{9}{2x - 4} = -\dfrac{3}{2}$

28. $\dfrac{8}{x^2 - 4} - \dfrac{1}{x - 2} = -\dfrac{3}{x + 2}$

29. $\dfrac{a}{a - 4} + \dfrac{4}{4 - a} = \dfrac{a}{2}$

30. $\dfrac{3}{x^2 + 5x + 6} + \dfrac{2x}{x + 2} = \dfrac{x}{x + 3}$

SECTION 5.4 Objective 1, 2

Add or subtract as indicated:

31. $\dfrac{5x}{5x + 4} + \dfrac{4}{5x + 4}$

32. $\dfrac{6x}{6x - 5} - \dfrac{5}{6x - 5}$

33. $\dfrac{a^2}{a + 4} - \dfrac{8a + 16}{-a - 4}$

34. $\dfrac{y^2}{y - 4} + \dfrac{8y - 16}{4 - y}$

35. $\dfrac{a^2}{a^2 + a - 6} + \dfrac{5a}{a^2 + a - 6} + \dfrac{6}{a^2 + a - 6}$

SECTION 5.4 Objective 3

Add or subtract as indicated:

36. $\dfrac{4}{x} + \dfrac{x + 1}{x + 2}$

37. $\dfrac{x}{x - 4} - \dfrac{5}{x}$

38. $\dfrac{4}{x^2 - 25} - \dfrac{3}{x^2 - 6x + 5}$

39. $\dfrac{4x}{x^2 + 8x + 15} - \dfrac{1}{x + 5}$

40. $\dfrac{x + 1}{x^2 + 2x + 1} + \dfrac{x + 2}{x^2 - 1} - \dfrac{3}{x + 1}$

SECTION 5.4 Objective 4

Solve:

41. $\dfrac{5}{x} + \dfrac{3}{2} = \dfrac{1}{x}$

42. $1 + \dfrac{8}{x} = \dfrac{20}{x^2}$

43. $\dfrac{5}{x - 2} - \dfrac{8}{x + 3} = \dfrac{19}{x^2 + x - 6}$

44. $\dfrac{8}{x} - \dfrac{3}{x^2 - 4} = \dfrac{3x + 2}{x(x - 2)}$

45. $\dfrac{2}{x + 5} + \dfrac{x}{x - 5} = \dfrac{7x - 1}{x^2 - 25}$

SECTION 5.5 Objective 1

Simplify:

46. $\dfrac{\dfrac{x+3}{x^2-9}}{\dfrac{5}{x-3}}$

47. $\dfrac{\dfrac{x+5}{x^2+7x+12}}{\dfrac{5}{x+4}}$

48. $\dfrac{1-\dfrac{1}{x}}{1-\dfrac{1}{x^2}}$

49. $\dfrac{1-\dfrac{1}{x}-\dfrac{6}{x^2}}{1-\dfrac{4}{x^2}}$

50. $\dfrac{1+\dfrac{1}{1+x}}{1+\dfrac{1}{1+\dfrac{1}{1+x}}}$

SECTION 5.6 Objective 1

Solve:

51. $\dfrac{3.6}{4.2}=\dfrac{x}{1.4}$

52. $\dfrac{0.25}{40}=\dfrac{30}{x}$

53. $\dfrac{49}{5x}=\dfrac{0.7}{130}$

54. $\dfrac{5}{1.5}=\dfrac{28}{x}$

55. $\dfrac{3\frac{1}{2}}{5}=\dfrac{6\frac{3}{4}}{x}$

56. An airplane travels 1625 miles in 10 hours. At the same rate, how far will it fly in 16 hours?

57. If a wheel makes 1100 revolutions in 2 minutes, how many revolutions will it make in 4.5 minutes?

SECTION 5.7 Objective 1

58. If y varies directly as x, and y is 12 when x is 21, what is the value of y when x is 28?

59. If y varies directly as x, and y is 4.2 when x is 0.3, what is the value of x when y is 14?

60. If a varies directly as the square of b, and a is 8 when b is 9, what is the value of a when b is 12?

61. If m varies directly as the square of n, and m is 8 when n is 8, what is the value of n when m is 18?

62. The cost of potatoes varies directly with the weight of the potatoes. If 10 pounds of potatoes cost \$1.68, what is the cost of 25 pounds?

SECTION 5.7 Objective 2

63. If y varies inversely as x, and y is 16 when x is 12, what is the value of y when x is 8?

64. If a varies inversely as b, and a is 48 when b is 18, what is the value of b when a is 135?

65. If x varies inversely as the square of y, and x is 16 when y is 12, what is the value of x when y is 24?

66. If y varies inversely as the square of x, and y is 15 when x is 9, what is the value of x when y is 5.4?

67. The intensity (I) of light varies inversely as the square of the distance (d) from the source. If at a distance of 10 feet the intensity is 300 candlepower, what is the intensity at 20 feet?

CHAPTER 5
TRUE–FALSE CONCEPT REVIEW

Check your understanding of the language of algebra. Tell whether each of the following statements is true (always true) or false (not always true).

1. Fractions of algebra are also called rational expressions.

2. The fraction $\dfrac{x}{x-3}$ is not a real number if x is replaced by zero.

3. The procedures for adding and multiplying rational expressions are the same for adding and multiplying fractions in arithmetic.

4. It is usually easier to multiply rational expressions if you factor and reduce before multiplying.

5. It is usually easier to divide two rational expressions by dividing their numerators and dividing their denominators.

6. By reducing and multiplying, we can say that $\dfrac{4}{x} \cdot \dfrac{x}{x-2} = \dfrac{4}{x-2}$ is always true.

7. When a fraction has been reduced to lowest terms, its value is less than it was originally.

8. Fractions can be reduced before multiplying without changing the product.

9. The sum of two fractions can be found by adding their numerators and their denominators and then reducing.

10. One of the applications of multiplying fractions is to change measures to equivalent measures.

11. One use of the basic principle of fractions is to reduce fractions.

12. To add or subtract two fractions, first multiply both fractions by the LCM of their denominators.

13. The only reason to check equations that contain fractions is to make sure that there are no errors.

14. It is possible to make up an equation that has no solution.

15. The LCM of two or more polynomials is used only to find the common denominator for fractions so that they can be added or subtracted.

16. The LCM of two or more polynomials consists of the highest power of each factor that occurs in any of the polynomials.
17. A complex fraction can be simplified by treating it as a division problem.
18. A complex fraction can be simplified by multiplying the numerator and denominator by the LCM of the denominators of the fractions within the complex fraction.
19. When a difference of any pair of numbers is equal, the numbers have the same ratio.
20. Two ratios with different values can make up a proportion.
21. We can use cross multiplication to test whether a proportion is true or false.
22. We can use cross multiplication to help solve a proportion.
23. If two quantities vary directly, as one increases, the other also increases in the same ratio.
24. If two quantities vary inversely, as one increases, the other also increases in the same ratio.
25. It is possible for a single quantity to vary directly as two other (different) quantities.

CHAPTER 5 TEST

1. Find the missing numerator: $\frac{14}{8x} = \frac{?}{32x^2y}$
2. Subtract: $\frac{-5w}{w+10} - \frac{2w}{w+10}$
3. Solve: $\frac{15}{16} = \frac{y}{8}$
4. Reduce: $\frac{18(a^2-49)}{3a-21}$
5. Multiply: $\frac{w}{7} \cdot \frac{4}{y} \cdot \frac{z}{5}$
6. Divide: $\frac{7}{y} \div \frac{w}{6}$
7. Add: $\frac{7r}{r-5} + \frac{2r}{5-r}$
8. Multiply; reduce if possible: $\frac{33t^2}{18pq} \cdot \frac{45q}{22t}$
9. Find the LCM: $w - 5$, $w^2 - 25$, and $w^2 - 10w + 25$
10. Add; reduce if possible: $\frac{1}{2y} + \frac{1}{3y} + \frac{1}{y}$
11. Convert 90 miles per hour to an equivalent measure in feet per second.
12. Solve: $\frac{4}{y-3} - \frac{3}{3-y} = 1$
13. Add and subtract; reduce if possible: $\frac{x+1}{2x-2} + \frac{1}{x-1} - \frac{x}{2x+2}$
14. Multiply; reduce if possible: $\frac{x^2-9}{x^2-1} \cdot \frac{x^2+2x+1}{x^2-2x-3}$
15. Divide; reduce if possible: $\frac{x^2y^2-xy}{4x-4y} \div \frac{3xy-3}{8x-8y}$
16. Simplify: $\frac{x - \frac{9}{x}}{\frac{3}{x} + 1}$

17. Solve: $\dfrac{y}{y+3} = \dfrac{y+4}{y+9}$

18. Subtract; reduce if possible:

$$\frac{3x}{x^2 + x - 2} - \frac{2}{x + 2}$$

19. A map of Arabia is scaled so that $\dfrac{5}{6}$ inch represents 100 miles. How many miles is it between Mecca and Riyadh if the distance of the map is 4 inches?

20. The area of surface that can be painted varies directly as the number of quarts of paint available. If one quart of paint will cover 210 square feet, how many quarts are needed to cover 1155 square feet?

21. The frequency (f) of air vibration in the pipes of a pipe organ is inversely proportional to the length (ℓ) of the pipe. If the frequency of air vibration in a 12-foot pipe is 42 vibrations per second, what is the frequency in an 18-foot pipe?

CHAPTER

6

LINEAR EQUATIONS, SYSTEMS, AND GRAPHS

SECTIONS

Systems of linear equations can be used to solve many types of mixture problems. In Exercise 65, Section 6.7, the Grow It Then Mow It seed company is marketing a combination of rye grass and blue grass seed. The key points to remember in writing the system of equations is that the weight of the mixture must equal the original weight of the rye grass seed plus the original weight of the blue grass seed. We assume no weight loss during the mixing process. The income generated by the mixture is the same as that generated if the seed were sold separately. *(Peter Hendrie/The Image Bank)*

PREVIEW

In this chapter, we introduce and solve equations in two variables with solutions written as ordered pairs. We also introduce the rectangular coordinate system so that these equations can be graphed. Properties of slope and intercept are discussed. The slope and the y-intercept are used to draw graphs of linear equations. Linear inequalities follow naturally and utilize the concept of half-planes. We then solve systems of linear equations by graphing, substitution, and linear combinations. We place less emphasis on solutions by graphing since nonintegral solutions can only be estimated by this method.

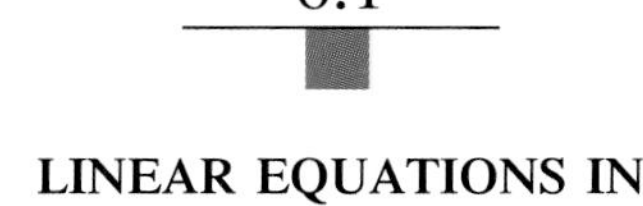

6.1 LINEAR EQUATIONS IN TWO VARIABLES

OBJECTIVE

1. Solve an equation in two variables given the value of one of the variables.

A problem involving two unknowns can often be solved using two variables. The use of two variables creates a need for two replacements to determine a solution. An equation that contains two variables has many solutions.

Equation	Solutions (Values of x and y)	Check
$x + 2y = 16$	$x = 8$ and $y = 4$	$8 + (2)(4) = 16$
$x + 2y = 16$	$x = 6$ and $y = 5$	$6 + (2)(5) = 16$
$x + 2y = 16$	$x = 20$ and $y = -2$	$20 + 2(-2) = 16$
$3x - 4y = 12$	$x = 2$ and $y = -\frac{3}{2}$	$3(2) - 4\left(-\frac{3}{2}\right) = 12$
$3x - 4y = 12$	$x = 0$ and $y = -3$	$3(0) - 4(-3) = 12$
$3x - 4y = 12$	$x = 8$ and $y = 3$	$3(8) - 4(3) = 12$

In the preceding table, three solutions (pairs of numbers) are given for each equation. In fact, a solution can be found by replacing either x or y with any number and solving for the value of the second variable.

EXAMPLE 1

Find the solution of $3x - 4y = 12$ when $x = 6$.

$3x - 4y = 12$

$3(6) - 4y = 12$ Substitute $x = 6$ and solve.

$$18 - 4y = 12$$
$$-4y = -6$$
$$y = \frac{3}{2}$$

So $x = 6$ and $y = \frac{3}{2}$ is a solution. ■

A solution can be found whether x or y is given.

EXAMPLE 2

Find the solution of $3x - 4y = 12$ when $y = 6$.

$3x - 4(6) = 12$ Substitute $y = 6$ and solve.

$$3x - 24 = 12$$
$$3x = 36$$
$$x = 12$$

So $x = 12$ and $y = 6$ is a solution. ■

The solutions to equations with two variables are written as ordered pairs.

■ DEFINITION

Ordered Pair (x, y)

A solution of an equation in two variables, x and y, is written as the *ordered pair* (x, y). The pair is called "ordered" since the value of x is always written first.

When writing several solutions, it is easier to use a table to list the ordered pairs.

Equation	Solutions in a Chart: x	Solutions in a Chart: y	Ordered Pairs
$3x - 4y = 12$			
	2	$-\frac{3}{2}$	$\left(2, -\frac{3}{2}\right)$
	0	-3	$(0, -3)$
	8	3	$(8, 3)$
	6	$\frac{3}{2}$	$\left(6, \frac{3}{2}\right)$
	12	6	$(12, 6)$

Equation	Solutions in a Chart		Ordered Pairs
$x + 2y = 16$	x	y	
	8	4	(8, 4)
	6	5	(6, 5)
	20	−2	(20, −2)
	−6	11	(−6, 11)

PROCEDURE

To solve an equation in two variables, given the value of one of them:

1. Substitute the known value of one of the variables in the equation.
2. Solve the resulting equation.
3. Write the solution as an ordered pair.

EXAMPLE 3

Given the equation $y = 3x - 4$, find the solutions when $x = 1$, $x = -2$, and $x = 0$.

$x = 1$	$x = -2$	$x = 0$
$y = 3x - 4$	$y = 3x - 4$	$y = 3x - 4$
$y = 3(1) - 4$	$y = 3(-2) - 4$	$y = 3(0) - 4$
$y = 3 - 4$	$y = -6 - 4$	$y = 0 - 4$
$y = -1$	$y = -10$	$y = -4$

There is a value of y for each value of x. To find the three values of y, substitute each value of x into the equation.

The solutions are $(1, -1)$, $(-2, -10)$, and $(0, -4)$. ■

EXAMPLE 4

Given the equation $2x + 3y = 6$, find the solutions when $y = 0$, $y = 2$, and $y = -2$. Write the solutions in a table.

$y = 0$	$y = 2$	$y = -2$
$2x + 3y = 6$	$2x + 3y = 6$	$2x + 3y = 6$
$2x + 3(0) = 6$	$2x + 3(2) = 6$	$2x + 3(-2) = 6$
$2x + 0 = 6$	$2x + 6 = 6$	$2x - 6 = 6$
$2x = 6$	$2x = 0$	$2x = 12$
$x = 3$	$x = 0$	$x = 6$

There is a value of x for each value of y. To find the three values of x, substitute each value of y into the equation.

The solutions are

x	y
3	0
0	2
6	−2

■

Problems with two unknowns can be solved using equations with two variables.

EXAMPLE 5

An apparel department is having a sale on jeans and tops. The jeans will sell for \$15 a pair and the tops for \$12 each. One sale came to \$66. If the customer bought three tops, how many pairs of jeans did she buy?

Simpler word form:

Cost of jeans + Cost of tops = 66

Select variables:

	Cost of Each	Number Sold	Total Cost of Each
Jeans	15	J	$15J$
Tops	12	T	$12T$

We organize the data in a table. Let J represent the number of jeans bought and T represent the number of tops bought.

Translate to algebra:

$$15J + 12T = 66$$

Substitute:

$$15J + 12(3) = 66$$

$T = 3$ since the customer bought three tops.

Solve:

$$15J + 36 = 66$$

$$15J = 30$$

$$J = 2$$

Answer:

The customer bought two pairs of jeans. ■

EXAMPLE 6

Georgia enters the post office with \$9.72 to spend on 30¢ and 18¢ stamps. If she needs twenty-four 30¢ stamps, how many 18¢ stamps can she buy?

Simpler word form:

Value of 30¢ stamps + Value of 18¢ stamps = Total value

Select variables:

Let t represent the number of 30¢ stamps; then, $0.30t$ represents the *value* of the 30¢ stamps. Let f represent the number of 18¢ stamps; then, $0.18f$ represents the *value* of the 18¢ stamps.

We use two variables to represent the number of stamps:
t = number of 30¢ stamps
f = number of 18¢ stamps

Translate to algebra:

$0.30t + 0.18f = \$9.72$

Substitute:

$0.30(24) + 0.18f = 9.72$

Georgia needs twenty-four 30¢ stamps, so $t = 24$.

Solve:

$$7.20 + 0.18f = 9.72$$
$$0.18f = 2.52$$
$$f = 14$$

Answer:

Georgia can buy fourteen 18¢ stamps. ■

EXERCISE 6.1

A

Given the equation $x + y = 18$, solve for y given the following values for x:

1. $x = 6$ **2.** $x = -3$ **3.** $x = 13$

4. $x = 18$ **5.** $x = -7$ **6.** $x = -5$

Given the equation $x + y = 15$, solve for x given the following values for y:

7. $y = 6$ **8.** $y = 3$ **9.** $y = -5$

10. $y = -17$ **11.** $y = 15$ **12.** $y = -20$

Given the equation $2x - y = 6$, solve for y given the following values for x:

13. $x = 0$ **14.** $x = -6$ **15.** $x = 22$

16. $x = 34$ **17.** $x = -28$ **18.** $x = -16$

Given the equation $2x - y = 6$, solve for x given the following values for y:

19. $y = 0$ **20.** $y = -6$ **21.** $y = -5$

22. $y = 34$ **23.** $y = -28$ **24.** $y = -16$

B

Given the equation $x + 2y = 4$, solve for y given the following value of x:

25. $x = \frac{1}{2}$ **26.** $x = -\frac{3}{4}$ **27.** $x = -\frac{1}{2}$

28. $x = \frac{7}{3}$ **29.** $x = -\frac{3}{5}$ **30.** $x = \frac{2}{7}$

Given the equation $x - 3y = 6$, solve for x given the following value of y:

31. $y = \frac{1}{6}$ **32.** $y = -\frac{5}{6}$ **33.** $y = -\frac{1}{9}$

34. $y = -\frac{5}{12}$ **35.** $y = \frac{7}{9}$ **36.** $y = -\frac{3}{4}$

Given the equation $2x + y = 8$, solve for y given the following value of x:

37. $x = -1.5$ **38.** $x = -6.4$ **39.** $x = -11.6$

40. $x = 3.75$ **41.** $x = 2.25$ **42.** $x = -3.33$

Given the equation $3x - 2y - 15 = 0$, solve for y given the following values for x:

43. $x = 0$ **44.** $x = 5$ **45.** $x = -10$

46. $x = -5$ **47.** $x = 1$ **48.** $x = 2$

C

Given the equation $3x - 2y + 15 = 0$, solve for x given the following values for y:

49. $y = 0$ **50.** $y = 3$ **51.** $y = -3$

52. $y = -6$

53. $y = 6$

54. $y = -9$

Given the equation $2x - 3y + 6 = 0$, solve for y given the following value of x:

55. $x = -1.2$

56. $x = -2.7$

57. $x = \frac{2}{3}$

58. $x = -\frac{5}{3}$

59. $x = -\frac{7}{2}$

60. $x = 2.76$

Given the equation $4x - 3y + 10 = 0$, solve for x given the following value of y:

61. $y = -6.5$

62. $y = 11.5$

63. $y = -\frac{3}{2}$

64. $y = \frac{7}{3}$

65. $y = -\frac{2}{5}$

66. $y = -\frac{5}{12}$

Given the equation $7x + 5y = 35$, solve for y given the following value of x:

67. $x = -8$

68. $x = 12$

69. $x = -3.5$

70. $x = 5.5$

71. $x = -\frac{5}{8}$

72. $x = \frac{5}{3}$

D

73. A first number is two less than twice a second ($x = 2y - 2$). If the first number is 47, what is the second?

74. A first number is 12 more than six times a second ($x = 6y + 12$). If the first number is 48, what is the second?

75. A certain number is eight more than three times a second number. If the second number is 44, what is the first number?

76. A certain number is three less than eight times a second number. If the second number is 21, what is the first number?

77. A load of cartons, some of which weighed 4 lb each and some 5 lb each, were to be loaded on a truck. The total weight of the boxes was 100 lb. If there were 15 of the 4-lb boxes, how many 5-lb boxes were in the load?

78. In Exercise 77, if there are 16 of the 5-lb boxes, how many 4-lb boxes are in the load?

79. Munch cereal provides 20 g of carbohydrate per unit, and Good AM cereal provides 30 g of carbohydrate. Gaston wants 180 g of carbohydrates (from cereal) per day. If he has already eaten two units of Good AM, how many units of Munch should he eat?

80. The perimeter of a rectangle is 60 ft ($2\ell + 2w = 60$). If the length is 24 ft, how wide is it?

81. The area of a trapezoid is 66 sq in. The formula for the area of a trapezoid is $A = \frac{1}{2}(b_1 + b_2) \cdot h$. If the lengths of the two bases are 10 in. and 12 in., find the height.

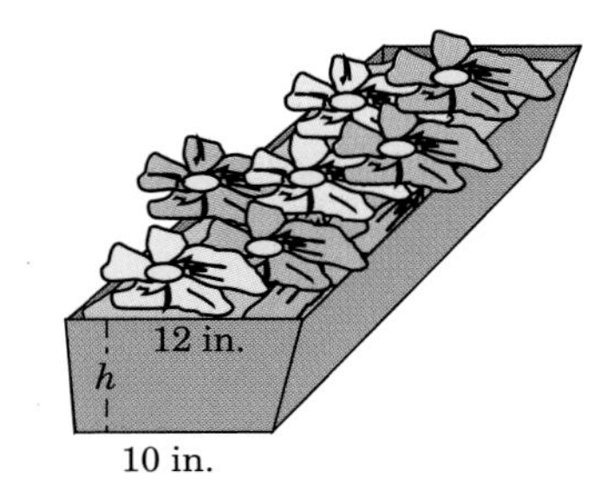

Figure for Exercise 81

82. The perimeter of an isosceles triangle is given by the formula $P = 2a + b$. Find a if $P = 38$ cm and $b = 10$ cm.

STATE YOUR UNDERSTANDING

83. Find the error in the following problem. What is the correct answer?
Find the solution for $5x - 2y = 10$ when $x = 4$.

$$5(4) - 2y = 10$$
$$20 - 2y = 10$$
$$-2y = -10$$
$$y = 5$$

The solution is (5, 4).

84. Explain why an equation in two variables has an infinite number of solutions.

CHALLENGE EXERCISES

For the equation $3y + 2 = 0$, find:

85. y when $x = 4$.

86. x when $y = -2/3$.

For the equation $4x - 5 = 0$, find:

87. x when $y = -3$.

88. y when $x = 1.25$.

MAINTAIN YOUR SKILLS (SECTIONS 5.4, 5.5)

Perform the indicated operations:

89. $\frac{3}{a} + \frac{5}{a}$

90. $\frac{1}{2x} + \frac{1}{5x} - \frac{1}{4}$

91. $\frac{1}{x^2 - 4} + \frac{3}{x - 2} - \frac{1}{x + 2}$

92. $\frac{x + 2}{x - 3} - \frac{x + 6}{x + 5} + \frac{1}{x^2 + 2x - 15}$

93. $\frac{x + 1}{x^2 + 2x - 3} + \frac{x + 4}{x + 3} - \frac{5}{x - 1}$

Simplify:

94. $\dfrac{\frac{a - 5}{b}}{\frac{a + 3}{b}}$

95. $\dfrac{\frac{a - b}{a + b}}{\frac{a^2 - b^2}{b + a}}$

96. $\dfrac{\frac{1}{2} - \frac{1}{b}}{\frac{1}{3} - \frac{1}{b}}$

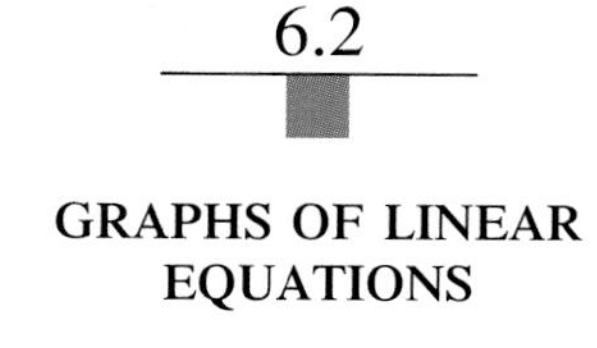

6.2 GRAPHS OF LINEAR EQUATIONS

OBJECTIVES

1. Graph a point given its coordinates.
2. Find the coordinates of a point given its graph.
3. Draw the graph of a linear equation.

To draw the graph of the solution of an equation in one variable, we used a number line. To draw the graph of a solution of an equation in two variables, we use two number lines placed at right angles, called a rectangular coordinate system.

DEFINITION

Rectangular Coordinate System

A *rectangular coordinate system* is formed by placing two number lines at right angles.

The horizontal line is called the x-axis.

The vertical line is called the y-axis.

The intersection of the axes is called the origin.

A quadrant is one of the four regions formed by the intersection of the axes.

A rectangular coordinate system with its parts labeled is shown in Figure 6.1.

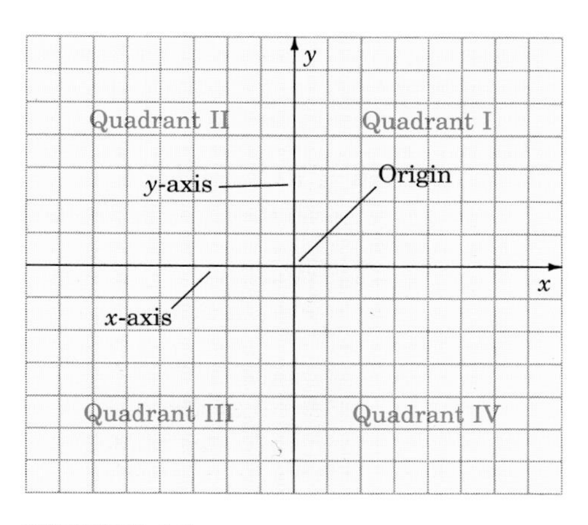

FIGURE 6.1

The x-axis (horizontal line) is the number line we worked with before. The y-axis (vertical line) is a number line with the positive numbers above the x-axis and the negative numbers below the x-axis. The ordered pairs associated with a given point show the distances from the axes. This is shown in Figure 6.2.

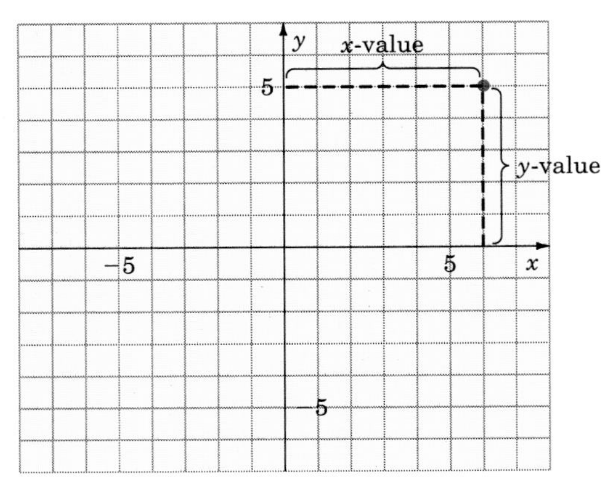

FIGURE 6.2

EXAMPLE 1
Plot the graph of the ordered pair (6, 3).

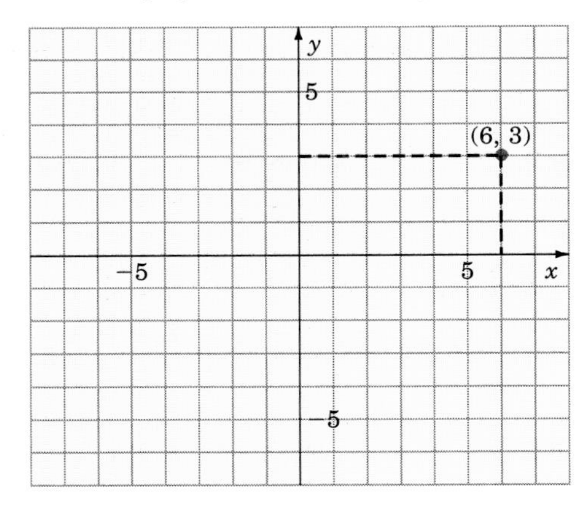

Count six units to the right of the origin on the x-axis and three units up. The point is at the corner of a rectangle and opposite the origin. (This is the reason it is called the rectangular coordinate system.)

PROCEDURE

To plot a point, given its coordinates:

1. Count (from the origin) right or left (positive or negative) the number of x-units.
2. From this point on the x-axis, count up or down (positive or negative) the number of y-units.
3. Label this point with its x- and y-values.

EXAMPLE 2

Plot the following points: $A(3, -2)$, $B(-5, 3)$, $C(-4, -1)$, $D(1.5, 5)$, $E(3, -4.5)$, and $F(0, 0)$.

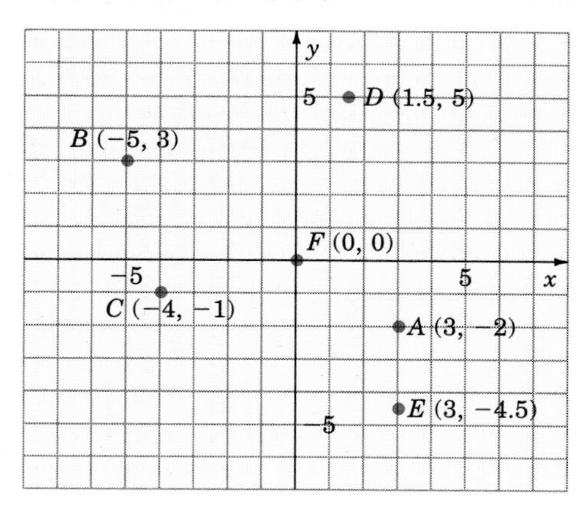

The points are shown as dots on the graph at the left.

To find the coordinates of a point when its graph is given, find the distance the point is from the y-axis (x-value) and from the x-axis (y-value).

EXAMPLE 3

Write the coordinates of the points shown on the rectangular coordinate system.

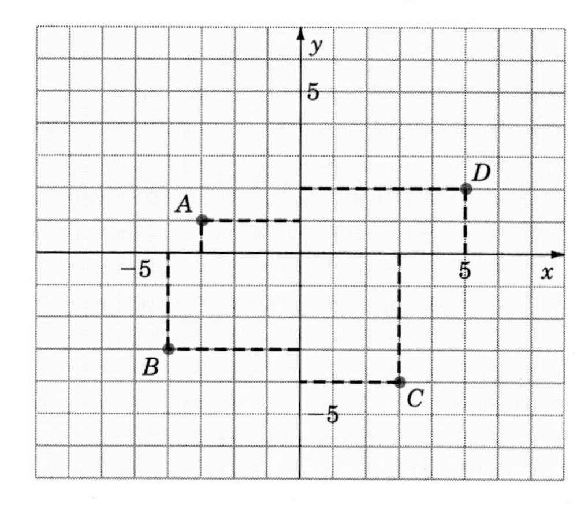

A: $(-3, 1)$ — The point A is three units to the left and one unit above the origin.

B: $(-4, -3)$ — The point B is four units to the left and three units below the origin.

C: $(3, -4)$ — Three units to the right and four units below the origin.

D: $(5, 2)$ — Five units to the right and two units above the origin.

PROCEDURE

To identify the coordinates of a point given its graph:

1. Find the distance of the point from the y-axis. This is the x-value. The x-value is positive if the point is to the right of the y-axis. The x-value is negative if the point is to the left of the y-axis.
2. Find the distance of the point from the x-axis. This is the y-value. The y-value is positive if the point is above the x-axis. The y-value is negative if the point is below the x-axis.
3. Write the ordered pair (x, y).

There are several different names used for the coordinates of a point. You should become familiar with all of them.

DEFINITION

x-Value
y-Value

The *x-value* (or x-coordinate) is called the abscissa.

The *y-value* (or y-coordinate) is called the ordinate.

Names used for ordered pairs are:

(x-value, y-value)

(abscissa, ordinate)

(x-coordinate, y-coordinate)

(independent variable, dependent variable)

(distance from y-axis, distance from x-axis)

A rectangular coordinate system can be used to draw the graph of a linear equation.

DEFINITION

Linear Equation

A *linear equation* is an equation that can be written in the form:

$ax + by = c$ (a and b not both zero)

A linear equation is a first-degree equation. The equation is called linear because the graph is a straight line.

To graph an equation, we find several ordered pairs that are solutions, plot the corresponding points on the coordinate system, and then draw a straight line through the points.

EXAMPLE 4

Sketch the graph of $2x - y = 8$.

First, solve the equation for y. Find several pairs by assigning values to x.

$y = 2x - 8$

$y = 2(-3) - 8 = -14$

$y = 2(-1) - 8 = -10$

$y = 2(0) - 8 = -8$

$y = 2(2) - 8 = -4$

$y = 2(4) - 8 = 0$

x	y
-3	-14
-1	-10
0	-8
2	-4
4	0

Plot the corresponding points on a coordinate system, and draw a straight line.

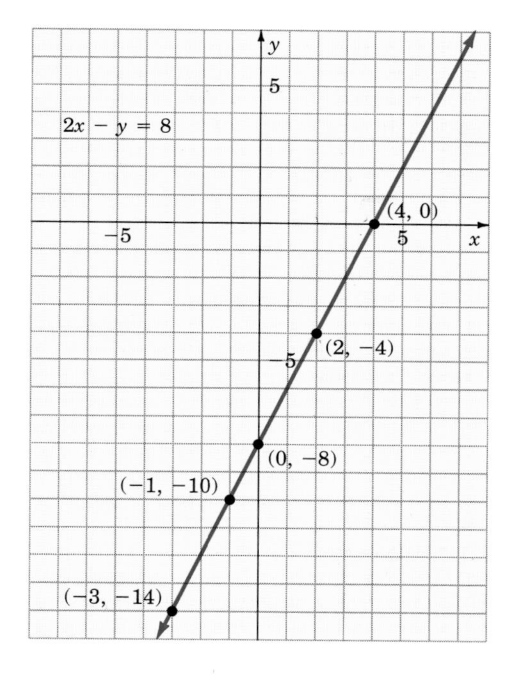

■

The significance of the graph of an equation is that every ordered pair of numbers that makes the equation true represents a point that lies on the line, and every point on the line has coordinates that make the equation true.

Since two points determine a straight line, you need to find only two solutions to draw the graph; however, using three or more solutions acts as a check and helps avoid errors.

■ CAUTION

Always use three or more points to draw the graph of a linear equation to avoid errors.

PROCEDURE

To draw a graph of a line given its equation:

1. Determine three or more points by assigning values to x (or y), and solve for y (or x) (make a table of values).
2. Plot the points on the rectangular coordinate system.
3. Draw the line that passes through all the plotted points.

Two points that are easy to find are those where the x and y coordinates are zero.

EXAMPLE 5

Draw the graph of $x - 5y = 5$.

x	y
0	-1
5	0
-5	-2

Find two points by letting $x = 0$ and $y = 0$. A third point is found by letting $x = -5$.

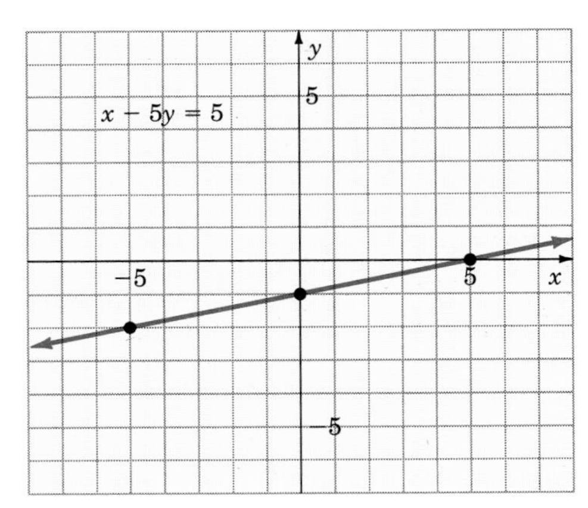

Draw the graph.

■

EXAMPLE 6

A company can depreciate the value of a small electronic control unit for tax purposes. One procedure is to use "straight-line depreciation." The procedure can be written in equation form

$y = -20x + 250,$

where y represents the value of the unit in any given year, and x represents the number of years the unit is in operation.

One of these units costs \$250 and has a scrap value of \$50 at the end of 10 years. Graph the equation, and list the coordinates that show the value of the unit each year.

Make a table of values.

x	y
0	250
4	170
8	90

Plot the points with these coordinates, and draw the line joining these points.

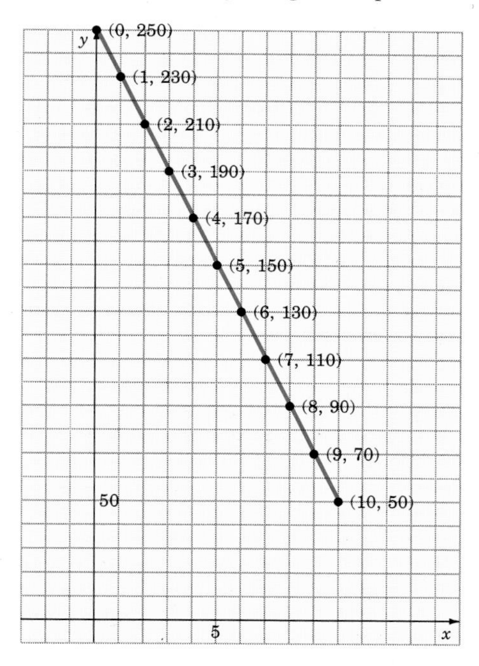

From the graph, we will be able to read the value of the electronic unit at any future time up to 10 years by looking at the x- and y-coordinates.

$$y = -20(0) + 250 = 250$$

$$y = -20(4) + 250 = 170$$

$$y = -20(8) + 250 = 90$$

To make it possible to draw the graph on one page, we use the following scale:

On the x-axis,

one unit = 1 year

On the y-axis,

one unit = 10 dollars

The graph stops when $x = 10$ because the unit is scrapped after 10 years. ■

Exercise 6.2

A

Construct a rectangular coordinate system, and plot the following points:

1. (1, 1)

2. (3, 4)

3. (−3, −2)

4. (−1, −4)

5. (−6, 2)

Name each of the following points using ordered pairs:

POINT		POINT	
6.	*A*	**7.**	*E*
8.	*G*	**9.**	*H*
10.	*D*	**11.**	*B*
12.	*F*	**13.**	*C*
14.	*K*	**15.**	*J*

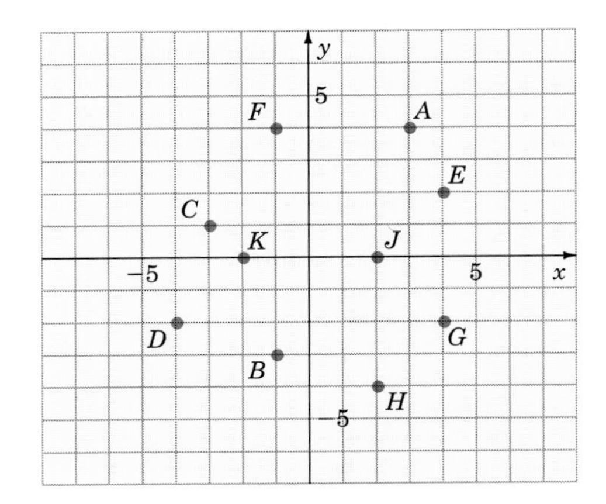

Construct a rectangular coordinate system, and plot the following points:

16. (4, 6)
17. (3, −2)
18. (−5, 1)
19. (−3, −4)
20. (−4, −1)

Name each of the following points using ordered pairs:

POINT		POINT	
21.	*E*	**22.**	*D*
23.	*A*	**24.**	*H*
25.	*J*	**26.**	*B*
27.	*G*	**28.**	*K*
29.	*C*	**30.**	*F*

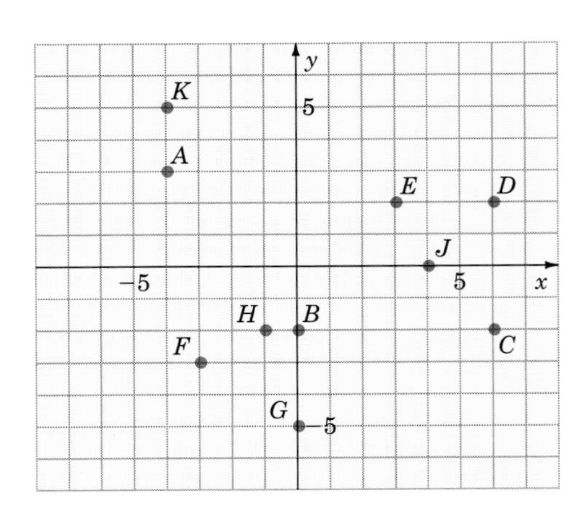

B

Construct a rectangular coordinate system, and plot the following points:

31. (1.5, 2)
32. (2.5, 1)

33. $\left(4\frac{1}{2}, -1\right)$

34. $\left(3\frac{1}{2}, -3\right)$

35. $(-0.5, -3)$

Name each of the following points using ordered pairs:

	POINT		POINT
36.	H	**37.**	A
38.	J	**39.**	K
40.	F	**41.**	C
42.	E	**43.**	D
44.	G	**45.**	B

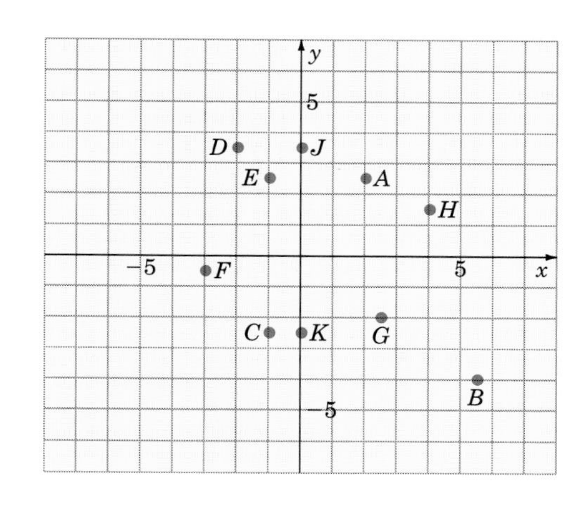

Complete the table of values, and graph each equation:

46. $y = \frac{1}{2}x$

x	y
-6	
0	
6	

47. $y = -\frac{1}{2}x$

x	y
-4	
0	
4	

48. $y = 3x$

x	y
-2	
0	
2	

49. $y = -3x$

x	y
-1	
0	
1	

50. $x - y = -4$

x	y
0	
	0
2	

51. $x - y = 5$

x	y
0	
	0
3	

52. $x + y = -2$

x	y
0	
	0
1	

53. $x + y = 4$

x	y
0	
	0
	−1

54. $x - y = 4$

x	y
0	
	0
2	

55. $x - y = -2$

x	y
0	
	0
1	

56. $x + 0y = -4$

x	y
−4	
	0
	−3

57. $0x + y = -2$

x	y
6	
	−2
−2	

58. $x - y = -3$

x	y
0	
	0
1	

59. $x + 2y = 6$

x	y
0	
	0
2	

60. $2x + y = 2$

x	y
	0
0	
-1	

C

Graph:

61. $-2x + y = 4$

62. $2x + 3y = 6$

63. $3x + 4y = 12$

64. $5x - 3y = 15$

65. $4x - 5y = 20$

66. $6x + 5y = 30$

67. $3x + 5y = -15$

68. $2x - 3y = 6$

69. $3x - 5y = -15$

70. $4x - 3y = 12$

71. $x - 4y = 8$

72. $-x + 2y = 6$

73. $5x + 2y = 10$

74. $5x - 7y = 10$

75. $4x - 5y = 18$

76. $6x + 5y = 15$

77. $x - 7y = -14$

78. $3x - 4y = 12$

79. $5x - 4y = 20$

80. $2x - y = 5$

81. $3x - 2y = 6$

82. $0.3x - 0.2y = 1.2$

83. $0.4x + 0.3y = 1.2$

84. $0.2x + 0.5y = 1$

85. $0.1x + 0.2y = 1$

86. $\frac{x}{4} + \frac{y}{7} = 1$

87. $\frac{x}{3} - \frac{y}{5} = 2$

88. $\frac{x}{3} - \frac{y}{2} = 2$

89. $\frac{x}{5} + \frac{y}{2} = 1$

90. $\frac{x}{4} - \frac{y}{3} = -2$

D

91. The sum of two numbers is 14. This can be written in equation form as $x + y = 14$ if x and y represent the two numbers. Draw the graph of this equation. (The graph will illustrate all such numbers.)

$y = mx + b$

92. The difference of two numbers is -4. Draw a graph whose coordinates are all such pairs of numbers.

93. The difference between three times a number and twice another number is twelve. Draw a graph whose coordinates are all such pairs of numbers.

94. The difference between four times a number and twice another number is ten. Draw a graph whose coordinates are all such pairs of numbers.

95. A business purchases an automobile for \$8000 and will use straight-line depreciation over a period of 5 years to a value of \$500. This can be described as $y = -1500x + 8000$, where y is the value of the automobile, and x is the number of years from 0 to 5. Graph this equation.

96. A building purchased for \$45,000 is depreciated over a 12-year period to \$1800. The formula for this is $y = 45{,}000 - 3600x$, where y is the value of the building, and x is the number of years from 0 to 12. Graph this equation. (*Hint:* Let one unit on the y-axis represent \$7000.)

97. If \$1,000 is deposited in an account earning simple interest at 6% annually, the amount, A, in the account after t years is given by the formula $A = 1000 + 60t$. Graph this equation for the first ten years.

98. A service station makes a profit of 20 cents on each gallon of regular gasoline and 25 cents on each gallon of premium gasoline. If x gallons of regular and y gallons of premium are sold the profit is \$500. This is represented by the equation $0.20x + 0.25y = 500$. Graph this equation.

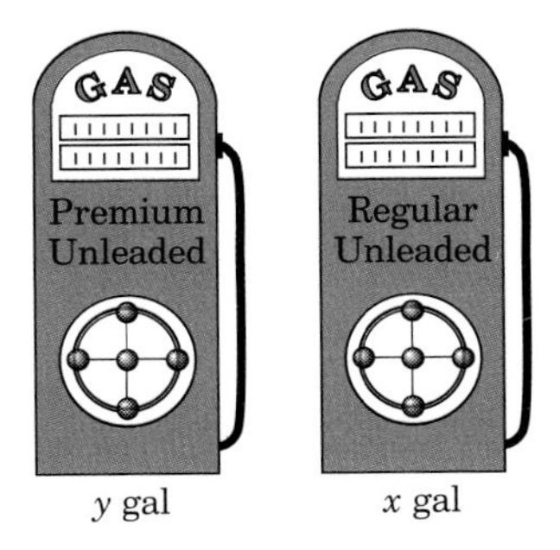

Figure for Exercise 98

STATE YOUR UNDERSTANDING

99. Describe the procedure used to determine the coordinates of a point given its graph.

100. What is another name for a first degree equation in two variables, and why is it given this name?

CHALLENGE EXERCISES

Graph the following linear equations.

101. $x = 4$

102. $y = -3$

103. $2y - 12 = 0$

104. $5x + 10 = 0$

MAINTAIN YOUR SKILLS (SECTION 5.1)

Reduce:

105. $\dfrac{45a^3b^4}{30a^4b^3}$

106. $\dfrac{3x^2 - 27x}{5xy - 45y}$

107. $\dfrac{3x^2 - 7x + 2}{5x^2 - 3x - 14}$

108. $\dfrac{6x^2 - 11x - 35}{3x^2 - 7x - 20}$

Find the missing numerator:

109. $\dfrac{5}{4y} = \dfrac{?}{60x^2y^2}$

110. $\dfrac{3x - 1}{2x + 5} = \dfrac{?}{2x^2 + x - 10}$

111. $\dfrac{24 \text{ mi}}{1 \text{ hr}} = \dfrac{? \text{ ft}}{1 \text{ sec}}$

112. $\dfrac{3a}{a - 3b} = \dfrac{?}{a^2 - 9b^2}$

6.3

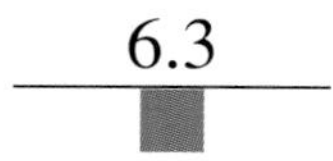

PROPERTIES OF STRAIGHT LINES

OBJECTIVES

1. Find the x-intercept and y-intercept given the equation of a line.
2. Find the slope of a line given two points on the line.
3. Find the slope of a line given its equation.
4. Draw the graph of the equation of a line using the slope and the y-intercept.

Straight lines, which are not parallel to one of the axes, intersect (cross) both the axes.

DEFINITION

x-intercept $(a, 0)$
y-intercept $(0, b)$

The *x-intercept* and *y-intercept* are the points where a line crosses (intersects) the x- and y-axes.

In Figure 6.3, the x-intercept is at $(a, 0)$, and the y-intercept is at $(0, b)$.

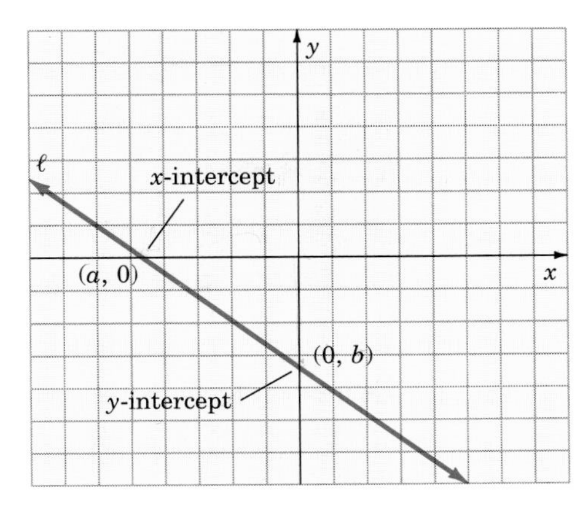

FIGURE 6.3

Since every point on the x-axis has a y-value of zero, the x-intercept can be found by substituting zero for y in the equation. Similarly, to find the y-intercept, substitute zero for x, and solve for the y-value.

EXAMPLE 1

Find the x- and y-intercepts of the graph of $5x + 7y = 35$.

x-intercept

$5x + 7(0) = 35$ — Substitute 0 for y in the equation, and solve for x.

$5x = 35$

$x = 7$

The x-intercept is $(7, 0)$.

y-intercept

$5(0) + 7y = 35$ — Substitute 0 for x in the equation, and solve for y.

$7y = 35$

$y = 5$

The y-intercept is $(0, 5)$ ■

■ PROCEDURE

To find the y-intercept:

1. Substitute 0 for x in the equation.
2. Solve for y.
3. Write the ordered pair $(0, y)$.

To find the x-intercept:

1. Substitute 0 for y in the equation.
2. Solve for x.
3. Write the ordered pair $(x, 0)$.

When we move from one point on a line to a second point, as shown in Figure 6.4, there is a vertical change and a horizontal change. We call the vertical change the "rise" and the horizontal change the "run."

Both rise and run can be positive or negative:

Positive rise	up
Negative rise	down
Positive run	right
Negative run	left

The word *slope* is used to describe these changes.

DEFINITION

Slope

$m = \dfrac{\text{rise}}{\text{run}}$

The *slope* of a line (m) is the ratio of the change in the vertical (rise) to the change in the horizontal (run) between any two points on a line.

The slope of the line in Figure 6.4 can be written in fraction form:

$$\text{slope} = m = \frac{\text{rise}}{\text{run}} = \frac{\text{change in } y}{\text{change in } x}, \quad \text{provided the change in } x \text{ is not zero.}$$

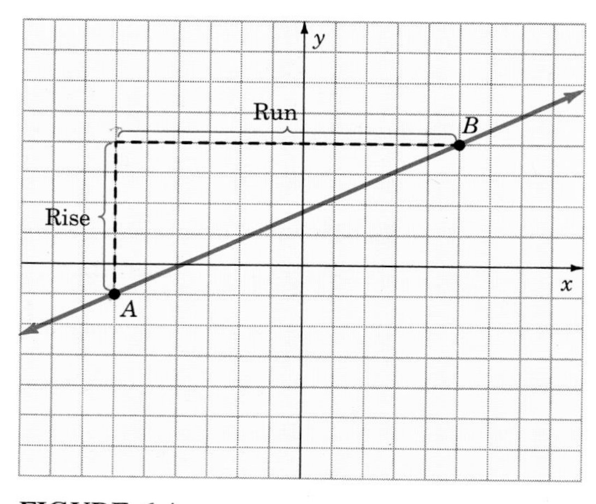

FIGURE 6.4

The slope of a line describes the slant (pitch, grade) of a line and is constant regardless of the points chosen.

To compute the slope, we find the rise (change in y) and the run (change in x) between the two points. We then divide the change in y by the change in x.

EXAMPLE 2

Find the slope of the line through the points A and B in the graph.

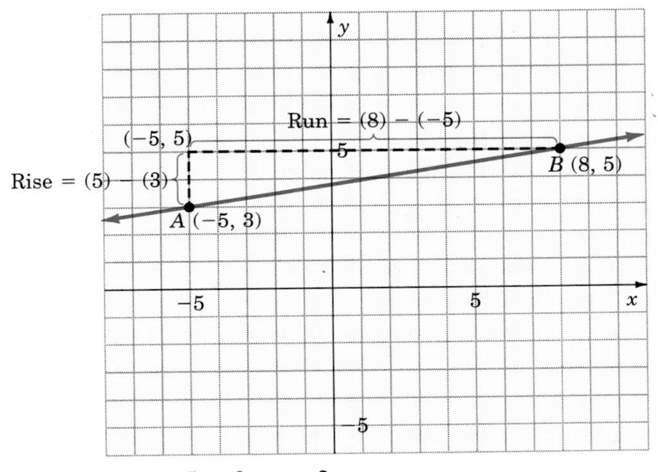

Going from A to B, the rise, from $y = 3$ to $y = 5$, is $5 - 3 = 2$.

The run, from $x = -5$ to $x = 8$, is $8 - (-5) = 13$.

$$\text{slope} = m = \frac{5-3}{8-(-5)} = \frac{2}{13}$$

Calculate the slope by dividing the rise by the run. ■

If the slope in Example 2 is calculated from B to A, the "rise" is actually a fall of 2, or -2, and the run is to the left, or -13. The slope is

$$m = \frac{3-5}{-5-(8)} = \frac{-2}{-13} = \frac{2}{13} \text{ as before.}$$

■ PROCEDURE

In general, given two points, $A(x_1, y_1)$ and $B(x_2, y_2)$, the slope of the line is

$$\text{slope} = m = \frac{y_2 - y_1}{x_2 - x_1}, \text{ where } x_2 - x_1 \neq 0.*$$

Figure 6.5 illustrates the formula for finding the slope.

*We can label *any* point either (x_1, y_1) or (x_2, y_2). The slope could just as easily be calculated as $m = \dfrac{y_1 - y_2}{x_1 - x_2}$.

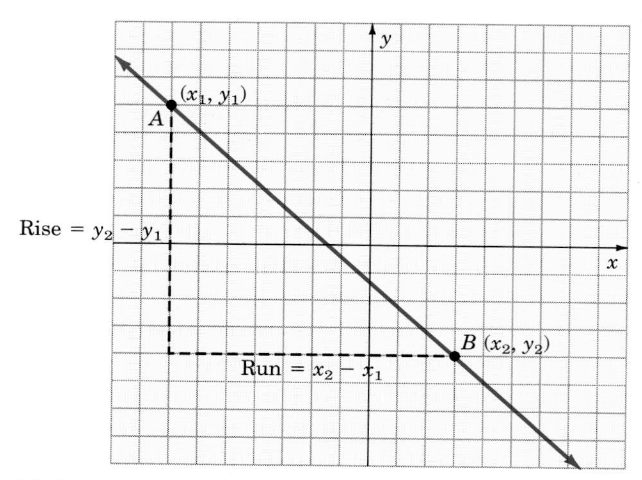

FIGURE 6.5

EXAMPLE 3

Find the slope of a line that contains the points (6, −8) and (4, 12).

Let $(x_1, y_1) = (6, -8)$ and $(x_2, y_2) = (4, 12)$.

$$m = \frac{y_2 - y_1}{x_2 - x_1} = \frac{12 - (-8)}{4 - 6}$$ Substitute in the formula and simplify.

$$= \frac{20}{-2} = -10$$

The slope is −10. ■

The slope of a line is negative if the line slants down as it goes to the right (Figure 6.6) and is positive if the line slants up as it goes to the right (Figure 6.7).

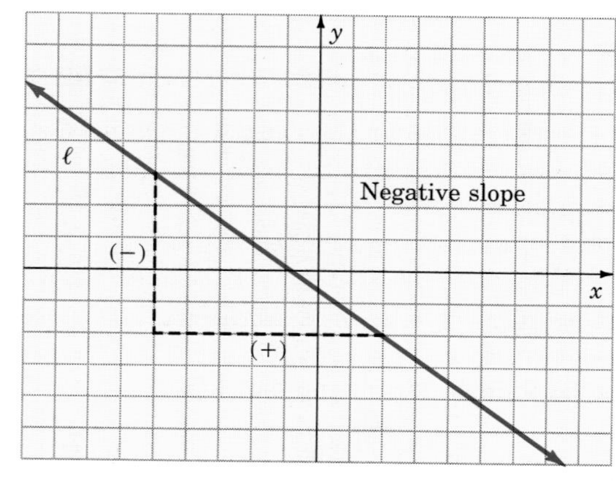

FIGURE 6.6

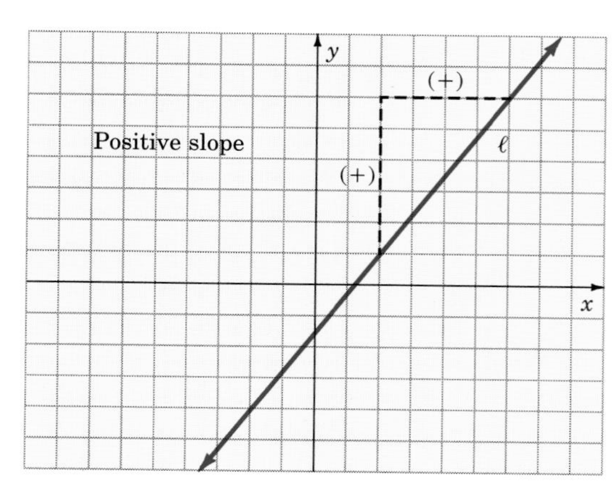

FIGURE 6.7

The slope of a line is 0 when the line is horizontal (parallel to the x-axis) because the rise is 0 (Figure 6.8). For example, consider the equation $y = 3$. This equation is equivalent to $y = 0x + 3$. For any value of x, y is always equal to 3. In general, the equation of a horizontal line is $y = k$, where k is a constant. The lines $y = 1$, $y = -2$, and $y = -4$ also have slope of 0.

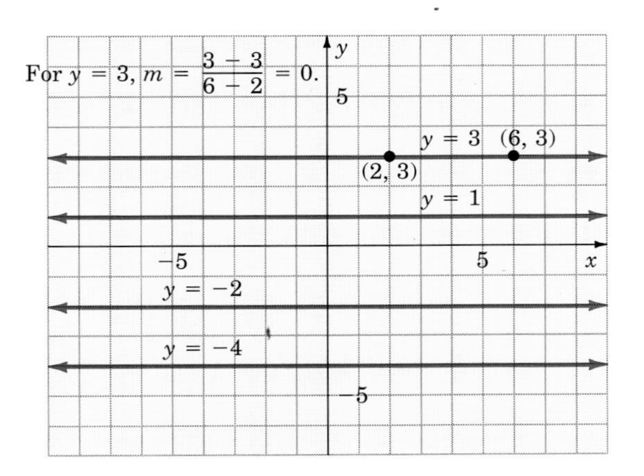

FIGURE 6.8

When a line is vertical (parallel to the y-axis), the line has no slope because the run is 0 (Figure 6.9). For example, the graph of the equation $x = -5$ has any ordered pair $(-5, y)$ as a solution. In general, the equation of a vertical line is $x = k$, where k is a constant.

$$\text{For } x = -5,\ m = \frac{6 - 2}{(-5) - (-5)} = \frac{4}{0}, \text{ which is not defined.}$$

The lines $x = -2$, $x = 3$, and $x = 7$ also have no slope.

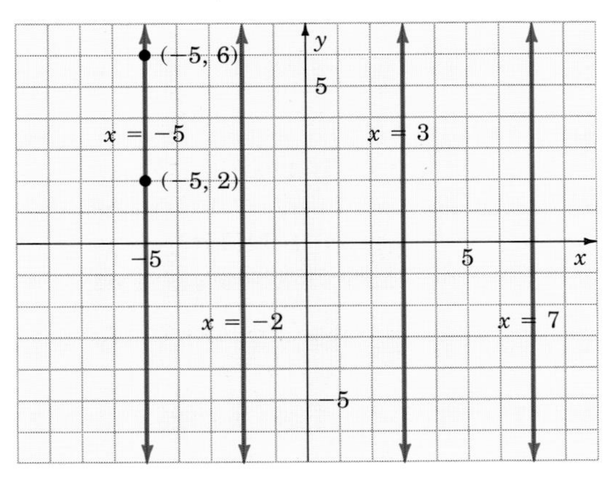

FIGURE 6.9

The following table shows a comparison of the slope of a line and the equation of the line when the equation has been solved for y.

Equation	Two Points on Line	$m = \dfrac{y_2 - y_1}{x_2 - x_1}$	y-Intercept	Equation Solved for y
$3x + 4y = 12$	$(0, 3), (4, 0)$	$\dfrac{0 - 3}{4 - 0} = -\dfrac{3}{4}$	$(0, 3)$	$y = -\dfrac{3}{4}x + 3$
$2x - 5y = -1$	$(2, 1), (-3, -1)$	$\dfrac{-1 - 1}{-3 - 2} = \dfrac{2}{5}$	$\left(0, \dfrac{1}{5}\right)$	$y = \dfrac{2}{5}x + \dfrac{1}{5}$
$5x - 8y = 10$	$(2, 0), (10, 5)$	$\dfrac{5 - 0}{10 - 2} = \dfrac{5}{8}$	$\left(0, -\dfrac{5}{4}\right)$	$y = \dfrac{5}{8}x - \dfrac{5}{4}$
$\dfrac{1}{2}x + \dfrac{1}{3}y = 1$	$(6, -6), (-2, 6)$	$\dfrac{6 - (-6)}{-2 - 6} = -\dfrac{3}{2}$	$(0, 3)$	$y = -\dfrac{3}{2}x + 3$

We make two observations from the table when the equation is solved for y:

1. The slope of the line (m) is identical to the coefficient of x.

2. The ordinate of the y-intercept is the constant term.

EXAMPLE 4

Find the slope and the y-intercept of the line whose equation is $7x - 5y = 10$.

$-5y = -7x + 10$ Solve the equation for y.

$$y = \frac{7}{5}x - 2$$

$m = \dfrac{7}{5}$ The slope is the coefficient of x.

y-intercept at $(0, -2)$ The ordinate of the y-intercept is the constant. ■

■ DEFINITION

Slope-Intercept Form

The *slope-intercept form* of a linear equation is the equation found by solving for y.

$$y = mx + b$$

PROCEDURE

To find the slope of a line and the y-intercept from the equation:

1. Write the equation in the form $y = mx + b$. (Solve for y in terms of x.
2. The coefficient of x is m, the slope.
3. The y-intercept is $(0, b)$.

EXAMPLE 5

Find the slope and the y-intercept of the line whose equation is $4x + 6y = 7$.

$6y = -4x + 7$ — Write the equation in the slope-intercept form.

$y = -\frac{2}{3}x + \frac{7}{6}$ — $y = mx + b$

$m = -\frac{2}{3}$, y-intercept at $\left(0, \frac{7}{6}\right)$ ■

We can use the slope-intercept form of the equation to draw the graph of the line. Starting with the y-intercept, we can find as many additional points as we want using the slope.

EXAMPLE 6

Draw the graph of $y = -\frac{2}{3}x - 2$.

$m = -\frac{2}{3}$, y-intercept at $(0, -2)$

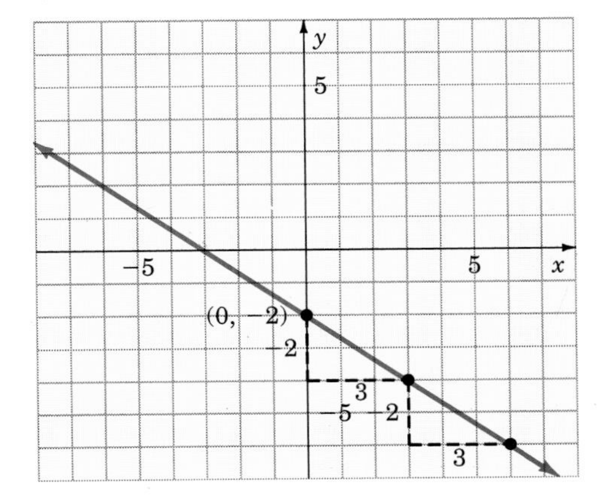

First, plot the y-intercept.

Now, using the slope, we can find another point. For every two units of fall (negative rise), there is a corresponding three units of run. The point we find is $(3, -4)$. We repeat the process to find the point $(6, -6)$.

Draw the line through the points. ■

PROCEDURE

To draw the graph of a line using the slope and the y-intercept:

1. Write the equation in the form $y = mx + b$.
2. Plot the y-intercept $(0, b)$.
3. Using the slope m, starting at the y-intercept, locate two or more additional points.
4. Draw the line through the points.

EXAMPLE 7

Using the slope and y-intercept, graph the line whose equation is $4x + 3y = 9$.

$3y = -4x + 9$

$y = -\dfrac{4}{3}x + 3$

First, write the equation in slope-intercept form.

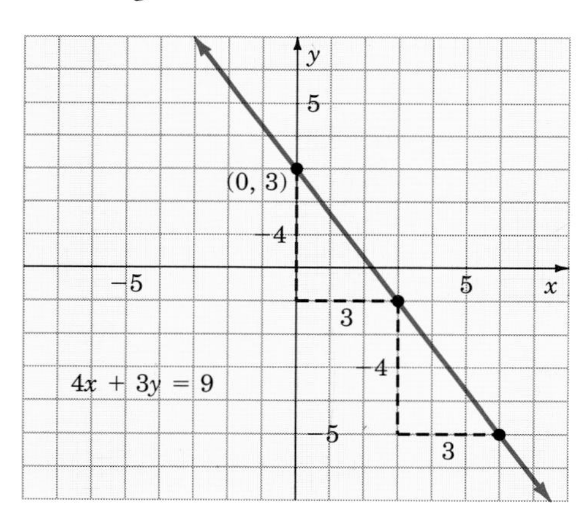

The slope is $-\dfrac{4}{3}$.

The y-intercept is $(0, 3)$.

Now plot the point $(0, 3)$. From there count down 4 units and right 3 units. This gives the point $(3, -1)$. Repeating the process, we get the point $(6, -5)$.

Draw the line through the points. ■

The familiar formula for distance, $d = rt$, is an example of a linear equation when either r or t is a constant.

EXAMPLE 8

A truck driver averages 80 km/hr. The formula for distance driven is

$d = 80t,$

where d represents the distance (in kilometers), and t represents the time (in hours). Draw a graph showing the relationship between distance and time.

$d = 80t$

$d = 80t + 0$

First, find the slope and y-intercept by writing the equation in slope-intercept form.

The slope is $m = 80$.
The y-intercept is $(0, 0)$.

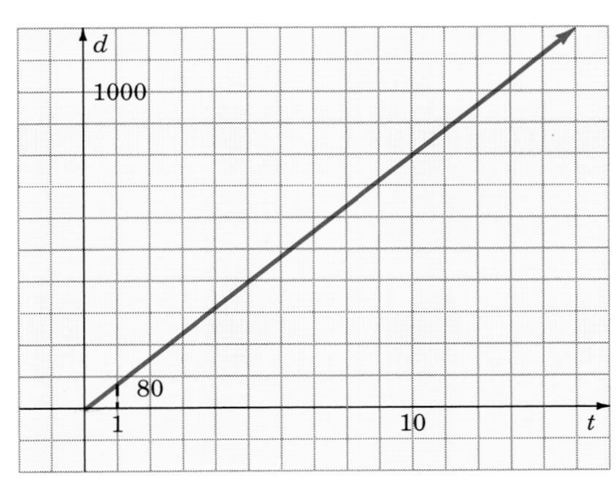

To draw the graph, let each unit on the t-axis represent 1 hour and each unit on the d-axis represent 100 kilometers.

Since the slope is $80 = \frac{80}{1}$, we know that for every 1 unit of run, we have 80 units of rise. We start at (0, 0) and count 80 units up and 1 unit right to find a second point and draw the graph.

CAUTION

The graph does *not* extend downward to the left since negative numbers do not apply here. Furthermore, a hasty glance at the graph may give the impression that the line has slope $m = 1$. A more careful look reveals that the t-axis and d-axis do not have the same scale and that the slope is actually $m = 80$.

EXERCISE 6.3

A

Find the x-intercept and the y-intercept of the graph of each of the following equations:

1. $3x + y = 9$

2. $7x - y = -14$

3. $x - 6y = -12$

4. $10x - 9y = 90$ **5.** $4x + 7y = 28$ **6.** $8x - 3y = 24$

Find the slope of the line containing the given pair of points:

7. $(2, 6), (3, 5)$ **8.** $(4, 7), (8, 10)$ **9.** $(-11, 3), (7, 0)$

10. $(0, 7), (4, -6)$ **11.** $(1, 2), (7, 8)$ **12.** $(11, 6), (8, 2)$

Find the slope of the line and the y-intercepts of the graph of each of the following equations:

13. $y = -\frac{1}{2}x + 6$ **14.** $y = \frac{3}{8}x + \frac{5}{8}$

15. $5y = 3x - 15$ **16.** $2y = 6x - 3$

Draw the graph of the line given its slope and y-intercept:

17. $m = \frac{2}{3}, (0, 5)$ **18.** $m = \frac{1}{2}, (0, 4)$ **19.** $m = -\frac{3}{2}, (0, -1)$

20. $m = -\frac{3}{2}, (0, 2)$ **21.** $m = 3, (0, -5)$ **22.** $m = 2, (0, 1)$

B

Find the x-intercept and the y-intercept of the graph of each of the following equations:

23. $2x + 5y = 11$ **24.** $3x + 7y = 18$ **25.** $6x - 7y = 12$

26. $4x - 5y = 17$

27. $10x - 7y = 20$

28. $7x + 20y = 10$

Find the slope of the line containing the given pair of points:

29. $(6, -5), (-3, -2)$

30. $(-3, 1), (6, -5)$

31. $(-2, 5), (5, 1)$

32. $(-3, -1), (7, 2)$

33. $(-13, -3), (-1, 4)$

34. $(-15, -15), (3, -12)$

Find the slope of the line and the y-intercept of the graph of each of the following equations:

35. $3x + 7y = 14$

36. $6x + 5y = 20$

37. $-2x - 5y = 8$

38. $-4x + 7y = 16$

39. $2y = 6$

40. $3x + 9 = 0$

Draw the graph of the line given its slope and y-intercept.

41. $m = -\frac{7}{4}, (0, 6)$

42. $m = \frac{1}{3}, (0, -3)$

43. $m = -3, (0, 4)$

44. $m = \frac{4}{3}, (0, 3)$

45. $m = -\frac{1}{3}, (0, -2)$

46. $m = 1, (0, -6)$

C

Find the x-intercept and the y-intercept of the graph of each of the following equations:

47. $\frac{1}{3}x + \frac{1}{2}y = -6$

48. $\frac{1}{2}x + \frac{1}{5}y = -3$

49. $\frac{1}{7}x + \frac{1}{3}y = \frac{1}{2}$

50. $\frac{3}{4}x - \frac{2}{3}y = 12$

51. $0.5x - 2.4y = 1$

52. $0.4x + 2y = 1.5$

Find the slope of the line containing the given pair of points:

53. $(5.6, -3.2), (4.1, 3.7)$

54. $(4.2, 2), (2.5, -3.1)$

55. $(0.36, 0.2), (0.4, 0.2)$

56. $(11.5, -3.4), (11.5, -2.5)$

57. $\left(\frac{2}{3}, -\frac{1}{2}\right), \left(\frac{1}{6}, \frac{3}{4}\right)$

58. $\left(\frac{1}{5}, -\frac{3}{2}\right), \left(-\frac{1}{2}, -\frac{2}{3}\right)$

Graph using the slope and the y-intercept.

59. $x - 4y = -8$

60. $6x + 5y = 15$

61. $2x + y = 4$

62. $5x - y = 2$

63. $3x + 5y = -3$

64. $6x - y = -2$

65. $x + \frac{2}{3}y = -1$

66. $\frac{1}{3}x + y = 2$

67. $\frac{1}{4}x + y = -2$

68. $x - \frac{3}{4}y = 3$

69. $-x + \frac{1}{2}y = 1$

70. $\frac{1}{2}x + \frac{2}{3}y = 3$

D

71. A part of the road from Reno to Lake Tahoe, Nevada, has a 6% grade. The equation $100y = 6x$ describes the grade or slope. What is the slope of the road?

72. A fire road in the Mt. Shasta area has a 7.5% grade. The equation $100y = 7.5x$ describes the grade or slope. What is the slope of the road?

73. The equation that describes the line of ascent of a jetliner is $1000y = 425x$. Find the rate of climb or slope.

74. The equation that describes the line of ascent of a corporate jet from Ajax Aircraft is $500y = 250x$. Find the rate of climb or slope.

75. A home in Aspen, Colorado, has a steep roof. For every four feet of run, it has three feet of rise. What is the pitch or slope of the roof?

76. A home in Hibbing, Minnesota, has a steep roof. For every two and one-half feet of run it has one and one-half feet of rise. What is the pitch or slope of the roof?

Figure for Exercise 76

77. The U-Drive-It auto leasing firm uses the following formula to compute the cost (c) of driving an auto x miles:

$$c = 0.17x + 10.$$

What is the cost per mile (slope of line)? What is the cost if the car is not driven during a day (y-value of the y-intercept)?

78. The U-Rent auto leasing firm uses the following formula to compute the cost (c) of driving an auto x miles:

$$c = 0.31x + 4.$$

What is the cost per mile (slope of line)? What is the cost if the car is not driven during a day (y-value of the y-intercept)?

79. A tourist averages 50 mph as he drives from coast to coast. The formula for distance (d) in terms of time (t) is $d = 50t$. Draw the graph.

80. The relationship between simple interest (I) at 10% for one year and principal (P) is given by $I = 0.10P$. Draw the graph.

81. The cost of setting up a production line for transistor radios is \$10,000. The additional cost to manufacture a radio is \$5. The cost (y) of producing x radios is given by the equation

$$y = 10{,}000 + 5x.$$

Graph the equation for 0 to 800 radios. (*Hint:* Use 1 unit equal to \$4000 on the y-axis and 1 unit equal to 100 radios on the x-axis.)

82. The cost of setting up a production line for a small cassette player is \$12,000. The additional cost to manufacture each cassette is \$15. The cost (y) of producing x cassettes is given by the equation

$y = 12{,}000 + 15x.$

Graph the equation for 0 to 800 cassettes. (*Hint:* Use 1 unit equal to \$5000 on the y-axis and 1 unit equal to 100 cassettes on the x-axis.)

STATE YOUR UNDERSTANDING

83. What do you know about the slope of this line? Explain.

84. Describe the slope-intercept form of the equation of a line.

CHALLENGE EXERCISES

Determine the slope of the line containing the given pair of points:

85. $(-7, -2)$ and $(3, -2)$

86. $(-4, -5)$ and $(-4, 7)$

Find the slope of the line and the y-intercept of the graph of each of the following equations:

87. $y - 3 = 0$

88. $x + 2 = 0$

MAINTAIN YOUR SKILLS (SECTIONS 4.4, 4.5, 5.1)

Multiply and reduce:

89. $\dfrac{5x^2}{6y^2} \cdot \dfrac{8y}{15x^3}$

90. $\dfrac{x^2 - 1}{16a^3} \cdot \dfrac{12a}{x + 1}$

91. $\dfrac{x^2 + 2x + 1}{x^2 + 3x + 2} \cdot \dfrac{x^2 + 4x + 4}{x^2 - x - 2}$

92. $\dfrac{15a^2}{a^3 - 9a} \cdot \dfrac{a^2 - 7a + 12}{3a^2 - 12a}$

Solve:

93. $5x^2 + 32x - 21 = 0$

94. $x^2 + 4x - 32 = 0$

95. The product of a number and 12 more than three times the number is 63. What is the number?

96. Today is Mary's birthday. She is now three years older than the square of her age five years ago. How old is Mary today?

6.4

GRAPHS OF LINEAR INEQUALITIES

OBJECTIVE

1. Draw the graph of a linear inequality.

■ **DEFINITION**

Linear Inequality

A *linear inequality* is an expression that can be written in one of the following forms:

$ax + by < c, \quad ax + by > c$

$ax + by \leq c, \quad ax + by \geq c$

where a and b are not both zero.

The graph of a linear inequality consists of an entire region rather than a straight line. To describe the graph of a linear inequality, we use the phrases *half-plane* and *boundary line*.

■ **DEFINITION**

Half-Plane
Boundary Line

A *half-plane* is formed on either side of a straight line when it is graphed on a rectangular coordinate system.

A *boundary line* is the graph of the straight line that divides the plane into two half-planes.

The graph of an inequality is one of the two half-planes. In some cases, the boundary line is also part of the graph. See Example 2.

EXAMPLE 1

Draw the graph of $x + y < 1$.

$x + y = 1$

$y = -x + 1$

First, draw the graph of the boundary line. The equation is found by replacing $<$ by $=$.

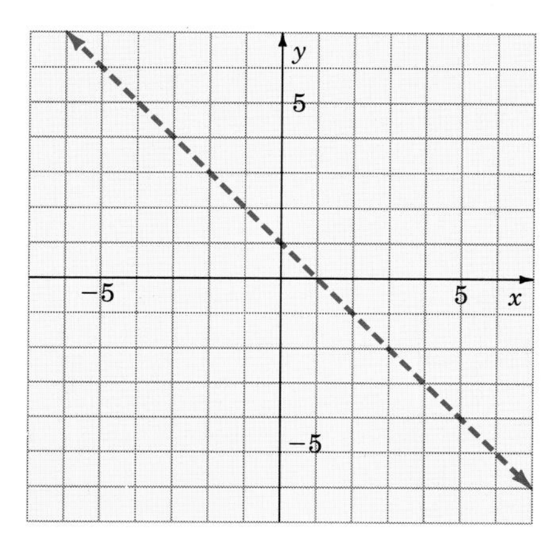

We use a broken line since the boundary line is not part of the graph.

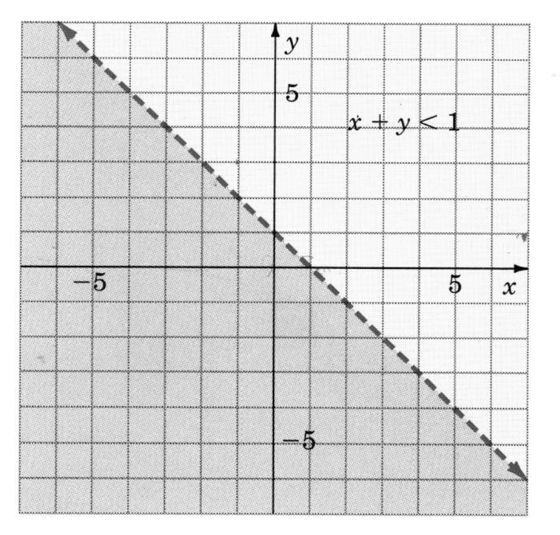

To find which half-plane is the graph, we test a point from either. The easiest point to test is the origin, (0, 0).

$x + y < 1$

$0 + 0 < 1$

$0 < 1$ True

So the graph is the half-plane below the line. It is shaded blue to show which points are solutions to the inequality.

■

Some things to remember about inequalities:

1. If a point in a half-plane is a solution of the inequality, all points in the half-plane are solutions of the inequality.
2. If a point in a half-plane is not a solution of the inequality, no points in the half-plane are solutions of the inequality.
3. The boundary line dividing the half-planes is not included in the solution if the symbols $<$ and $>$ are used. To show this, the line is drawn broken.
4. The boundary line is a solution of the inequality if the symbols $\leq$ and $\geq$ are used. To show this, the line is drawn solid.

■ PROCEDURE

To draw the graph of an inequality:

1. Draw the graph of the boundary line. Draw the line broken for $<$ and $>$ and solid for $\leq$ and $\geq$.

> 2. Test a point in one of the half-planes. If the inequality is true, shade that half-plane; if false, shade the other half-plane.

EXAMPLE 2

Graph the inequality $x - y \leq 1$.

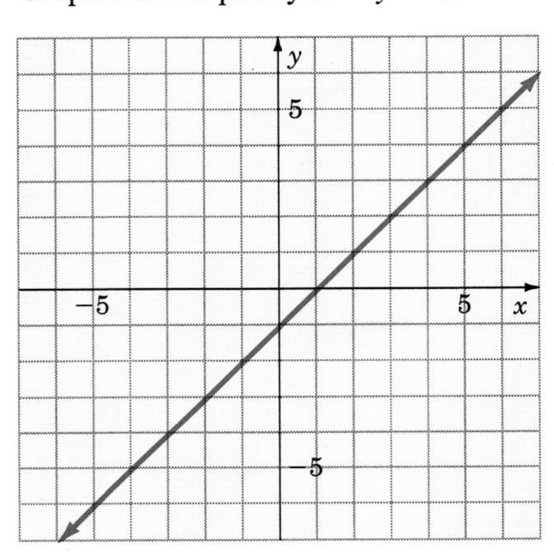

First, graph the boundary line $x - y = 1$. The graph is a solid line since the equality symbol is included, $\leq$.

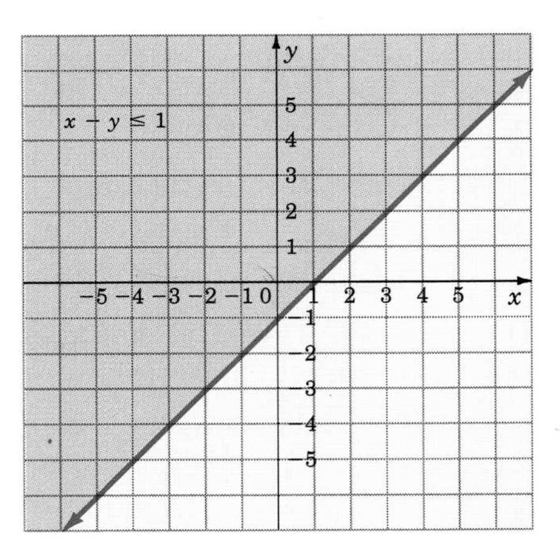

Next, check a point in one of the half-planes. The easiest point to check is the origin, (0, 0).

$x - y \leq 1$

$0 - 0 \leq 1$

$0 \leq 1$ True

The graph is the set of all points above the line and is shaded blue. ■

If we graph two inequalities on the same coordinate system, we can show the intersection of the solution sets by double shading the region that is contained in both solution sets.

EXAMPLE 3

Graph $x - 2y < -1$ and $x + 2y > 3$. Use double shading to indicate the region that is the intersection of both graphs.

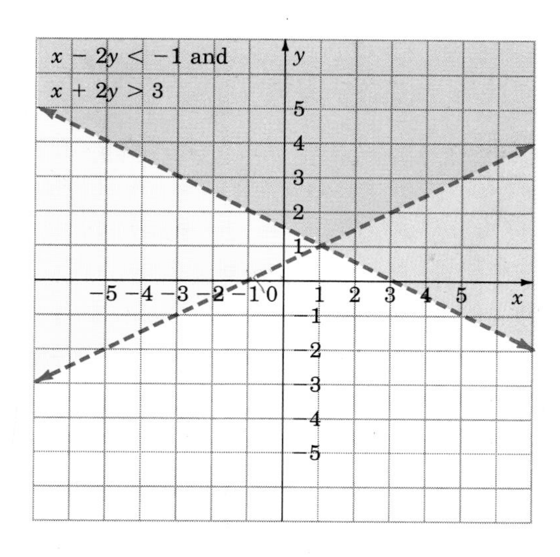

First, graph $x - 2y < -1$ using (0, 0) as a test point.

$x - 2y < -1$

$0 - 0 < -1$

$0 < -1$ False

Shade the region above the line yellow.

Second, graph $x + 2y > 3$ using (0, 0) as a test point.

$x + 2y > 3$

$0 + 0 > 3$

$0 > 3$ False

Shade the region above the line blue.

The common area is double shaded (shaded green). ■

EXERCISE 6.4

A

Graph each inequality:

1. $x - y < -4$

2. $x + y < 4$

3. $x + 3y > 2$

4. $x + 3y > -4$

5. $3x - y \leq 0$

6. $3x + y \leq 0$

7. $x + 2y \geq 7$

8. $x - 2y \geq -4$

9. $2x - y > 5$

10. $-3x + 2y \leq -1$

11. $4x + 3y \leq 12$

12. $x + 2y \geq -2$

B

Graph each inequality:

13. $x + y < 3$

14. $x + y > 5$

15. $4x - y < 4$

16. $4x - y < -2$

17. $3x - y \leq 5$

18. $2x + y \geq 4$

19. $x + 2y > 2$

20. $x - 3y < -3$

21. $x - 2y \leq -1$

22. $5x - y > 3$

23. $3x - 4y \geq -2$

24. $x + 3y < 5$

25. $5x + 2y < 2$

26. $4x - y \leq -2$

C

Graph both of these inequalities on the same graph. Shade the region that is part of both graphs:

27. $x + y < -2$ and $x - y > 3$

28. $x + y < 4$ and $x - y > -2$

29. $x - y > 3$ and $x + y < -1$

30. $x - 2y \leq 3$ and $2x + y > 1$

31. $x + 3y < 2$ and $2x - y > 2$

32. $2x - y > -2$ and $x + y > 5$

33. $5x - y > 0$ and $5x + y < 0$

34. $3x + y < 0$ and $3x - y > 0$

35. $2x + 3y \leq 6$ and $3x + 2y \geq 6$

36. $5x - 2y \geq 10$ and $5x + 2y \leq 10$

37. $3x - 4y > 8$ and $2x + 3y > -4$

38. $2x - 3y < 10$ and $3x + 4y < 8$

39. $3x - 2y < 6$ and $2x + 5y \geq 6$

40. $2x - 3y \leq -2$ and $4x + y > -2$

D

Write a linear inequality and graph it.

41. The ordinate of a point is less than its abscissa.

42. The ordinate of a point is greater than three times its abscissa.

43. The ordinate of a point is less than or equal to four less than its abscissa.

44. The abscissa of a point is greater than or equal to seven more than twice its ordinate.

45. The ordinate of a point is less than one-half of its abscissa.

46. The sum of the coordinates is greater than or equal to -7.

47. The sum of twice the abscissa and three times the ordinate is less than or equal to 6.

48. The difference of one-fourth the ordinate and one-half the abscissa is greater than zero.

STATE YOUR UNDERSTANDING

49. Discuss when the boundary line of an inequality is included and when it is not included in the graph of an equality.

50. If two inequalities are graphed on the same coordinate system, when is the point of intersection included in the graph? When is it not included?

CHALLENGE EXERCISES

Graph both inequalities on the same graph. Use double shading to indicate the region that is part of both graphs:

51. $2x + y > 2$ and $2x + y < 6$

52. $x + 2 \geq 0$ and $y - 5 < 0$

53. $3x - y \leq 6$ and $y > 3x$

54. $4x - 3y \geq 12$ and $y \geq \frac{4}{3}x$

MAINTAIN YOUR SKILLS (SECTION 3.3)

Combine:

55. $-3.45a + 7.65a - 8.3a - 5.2a + 9.21a$

56. $-0.16ab + 1.05b - 3.01b + 1.17ab + 2.14b$

57. $4[2(3a + 4) - 5(3 - 5a) + 16] - 2(-4a + 7)$

58. $(3x^2 - 4x + 16) - (5x^2 + 8x - 1) - (-3x^2 + 4x - 17)$

59. $-4a(b - 3c + 3) + b(2a - 3c + 7) - 5c(-a - 4b + 12)$

60. $3\{2 - 4[3(2 - b) - 5(3b + 7) + 8] - 3b\}$

61. $(x - 3y)^2 - 4(x + 2y)(3x - 5y) - 2(x^2 - 6xy - 3y^2)$

62. $34 - 5\{2 + 3[-2(a + b) + (2a - b) + 7] - 3(a - 4b)\}$

6.5

SYSTEMS OF EQUATIONS: SOLVING BY GRAPHING

OBJECTIVE

1. Solve a system of linear equations in two variables by graphing.

A pair of linear equations such as

$$\begin{cases} x + y = 4 \\ x + 2y = 6 \end{cases}$$

is called a system of linear equations.

DEFINITION

System of Equations

A *system of equations* is a set of two or more equations each of which contains more than one variable.

A system of two linear equations has three possible outcomes. The following words are used to describe these outcomes.

DEFINITION

Dependent and Consistent System

A *dependent and consistent system* is a system where graphs of the equations are the same line. The solution set of the system is the same as the solution set of either equation.

EXAMPLE 1

Solve the system: $\begin{cases} 3x - y = 2 & \text{equation of } \ell_1 \\ 6x - 2y = 4 & \text{equation of } \ell_2 \end{cases}$

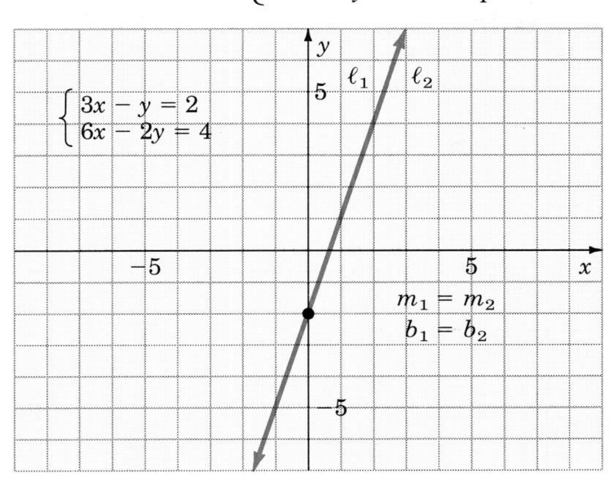

Draw the graph of each equation on the same set of axes.

$y = 3x - 2 \qquad \ell_1$
$y = 3x - 2 \qquad \ell_2$

Note that the slopes are the same and the y-intercepts are the same.

$m_1 = m_2$ and $b_1 = b_2$

The solution set is $\{(x, y) \mid 3x - y = 2\}$. ■

The dependent and consistent system can be recognized without graphing by writing the equations in the slope-intercept form. If the slopes and the intercepts are the same, the graphs are the same line. The solution set can be written using either one of the equations.

■ DEFINITION

Independent and Inconsistent System

An *independent and inconsistent system* is a system where the graphs of the equations are parallel lines. The solution set is $\emptyset$.

EXAMPLE 2

Solve the system: $\begin{cases} 3x - 7y = 9 & \text{equation of } \ell_1 \\ 3x - 7y = 2 & \text{equation of } \ell_2 \end{cases}$

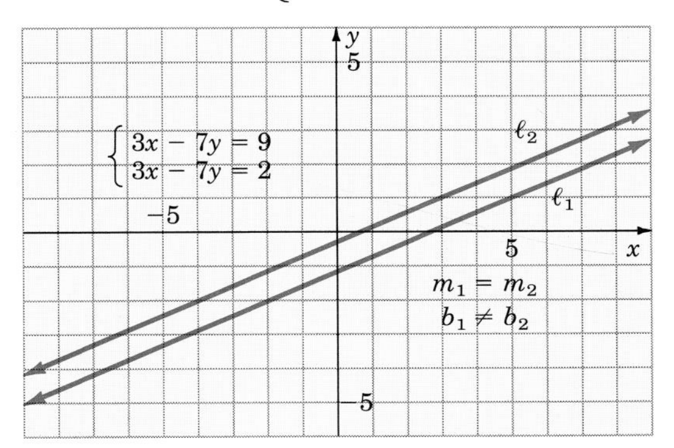

Rewrite each equation in slope-intercept form:

$y = \frac{3}{7}x - \frac{9}{7} \qquad \ell_1$

$y = \frac{3}{7}x - \frac{2}{7} \qquad \ell_2$

$m_1 = m_2$ and $b_1 \neq b_2$

The lines are parallel.

The solution set is $\emptyset$. ■

DEFINITION

Independent and Consistent System

An *independent and consistent system* is a system where the graphs of the equations intersect in exactly one point. The solution set contains the ordered pair of the point of intersection.

EXAMPLE 3

Solve the system: $\begin{cases} 3x - 2y = 6 & \text{equation of } \ell_1 \\ 3x + 2y = 6 & \text{equation of } \ell_2 \end{cases}$

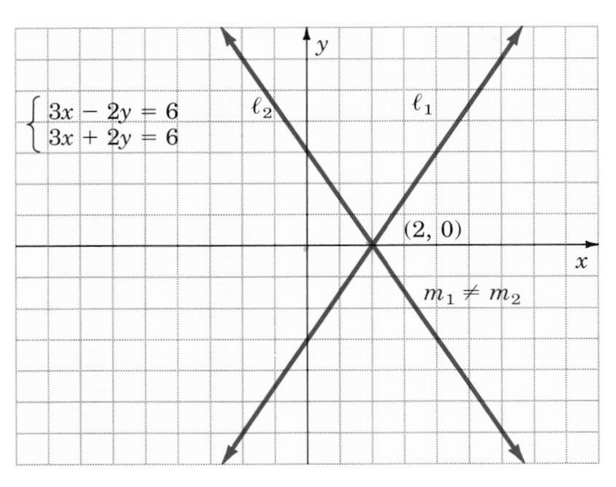

Write each equation in slope-intercept form.

$y = \frac{3}{2}x - 3 \quad \ell_1$

$y = -\frac{3}{2}x + 3 \quad \ell_2$

$m_1 \neq m_2$, so the lines intersect in exactly one point.

The solution set is $\{(2, 0)\}$. ■

The following table summarizes the types of systems and their solution sets.

Type of System	Slopes of Lines	*y*-Intercepts	Common Solutions (Points)
Dependent and consistent	Same	Same	Unlimited
Independent and inconsistent	Same	Different	No solution
Independent and consistent	Different	Same or different	One solution

PROCEDURE

To solve a system of equations in two variables by graphing:

1. Write each equation in the form $y = mx + b$ to determine the type of system.
2. If the system is independent and consistent, draw the graph of the system.
3. Write the solution set.
   a. Unlimited: use set builder notation and one of the equations (dependent and consistent system)
   b. $\emptyset$ (no solution, independent and inconsistent)
   c. The solution set containing the ordered pair (independent and consistent)

EXAMPLE 4

Solve the system: $\begin{cases} 2x + y = 8 & (\ell_1) \\ x - 2y = 9 & (\ell_2) \end{cases}$

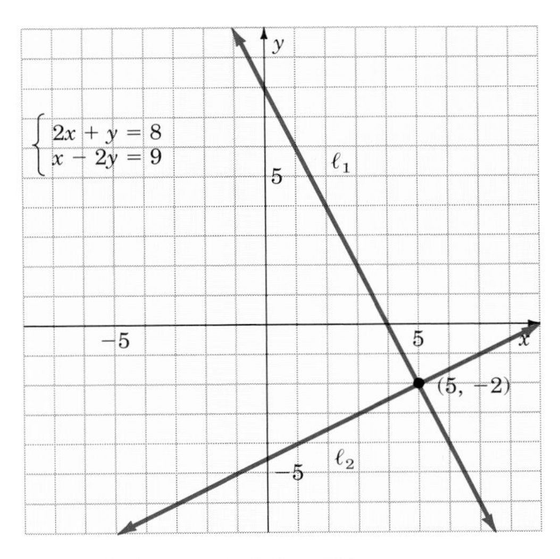

Write each equation in the form $y = mx + b$.

$y = -2x + 8$

$y = \frac{1}{2}x - \frac{9}{2}$

The slopes are different, so the system is independent and consistent. There is a single ordered pair for a solution.

Graph the equations to find the solution.

The solution set is $\{(5, -2)\}$. ■

Where the coordinates of the points of intersection are fractions, their values may not be easily identified. More precise methods for finding these values are given in the next two sections.

Example 5 is an application of linear systems from industry.

EXAMPLE 5

An electronics firm has two plants, each of which manufactures both color and black-and-white TV sets. In one day, the Deerberg plant can turn out 10 color sets and 50 black-and-white sets. The Elkland plant can produce 50 color sets and 10 black-and-white sets in a day. To fill an order for 100 color sets and 260 black-and-white sets, how many days at each plant are needed?

	Deerberg	Elkland	Order
Color sets	10	50	100
B & W sets	50	10	260

Use a table to help organize the material.

Simpler word form:

$$\begin{pmatrix}\text{Number of color}\\ \text{sets manufactured}\\ \text{at Deerberg}\end{pmatrix} + \begin{pmatrix}\text{Number of color}\\ \text{sets manufactured}\\ \text{at Elkland}\end{pmatrix} = \begin{pmatrix}\text{Total}\\ \text{color}\\ \text{sets}\end{pmatrix}$$

$$\begin{pmatrix}\text{Number of B \& W}\\ \text{sets manufactured}\\ \text{at Deerberg}\end{pmatrix} + \begin{pmatrix}\text{Number of B \& W}\\ \text{sets manufactured}\\ \text{at Elkland}\end{pmatrix} = \begin{pmatrix}\text{Total}\\ \text{B \& W}\\ \text{sets}\end{pmatrix}$$

Select variables:

If we let x represent the number of days the Deerberg plant operates and y represent the number of days the Elkland plant operates, we can add to the table.

	Deerberg (1 day)	Elkland (1 day)	Deerberg (x days)	Elkland (y days)	Order
Color sets	10	50	$10x$	$50y$	100
B & W sets	50	10	$50x$	$10y$	260

Translate to algebra:

With the information in the preceding table, we can write the two equations:

$$\begin{cases} 10x + 50y = 100 \\ 50x + 10y = 260 \end{cases}$$

Solve:

$$\begin{cases} x + 5y = 10 & (\ell_1) \\ 5x + y = 26 & (\ell_2) \end{cases}$$

Divide both sides of each equation by 10.

Sketch the graph of each equation.

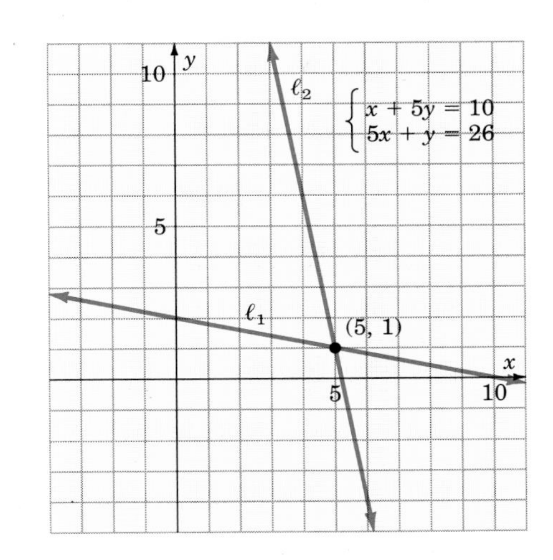

Answer:

The two lines intersect at the point (5, 1), which is the common solution. Therefore, if the Deerberg plant is run 5 days and the Elkland plant 1 day, the order can be filled. ■

EXERCISE 6.5

A

Solve by graphing:

1. $\begin{cases} x + y = 5 \\ x - y = 1 \end{cases}$

2. $\begin{cases} x + y = 12 \\ x - y = 2 \end{cases}$

3. $\begin{cases} x + y = 3 \\ x - y = 7 \end{cases}$

4. $\begin{cases} x + y = -3 \\ x - y = 7 \end{cases}$

5. $\begin{cases} x + y = -2 \\ x - y = -6 \end{cases}$

6. $\begin{cases} x + y = -5 \\ x - y = -1 \end{cases}$

7. $\begin{cases} x + y = 7 \\ x - y = -5 \end{cases}$

8. $\begin{cases} x + y = 4 \\ x - y = 6 \end{cases}$

9. $\begin{cases} x + y = 5 \\ -x - y = -6 \end{cases}$

10. $\begin{cases} x - y = 7 \\ y - x = 3 \end{cases}$

11. $\begin{cases} x - y = 2 \\ 3y - 3x = -6 \end{cases}$

12. $\begin{cases} x + y = 5 \\ 2x + 2y = 12 \end{cases}$

13. $\begin{cases} x + y = 8 \\ 2x + y = 12 \end{cases}$

14. $\begin{cases} x + y = 3 \\ x - 2y = 12 \end{cases}$

15. $\begin{cases} 7x - y = 7 \\ x - y = 1 \end{cases}$

16. $\begin{cases} 2x + y = 8 \\ 2x - y = -4 \end{cases}$

B

Solve by graphing:

17. $\begin{cases} y = x + 11 \\ y = -x + 3 \end{cases}$

18. $\begin{cases} y = x - 7 \\ y = x + 3 \end{cases}$

19. $\begin{cases} y = x + 5 \\ y = -x + 3 \end{cases}$

20. $\begin{cases} x = 2y + 5 \\ x = y + 1 \end{cases}$

21. $\begin{cases} x = -y + 3 \\ x = 3y - 1 \end{cases}$

22. $\begin{cases} y = -3x + 2 \\ y = 5x - 6 \end{cases}$

23. $\begin{cases} 2x + 4y = 2 \\ 3x + 6y = 3 \end{cases}$

24. $\begin{cases} 4x - 4y = 8 \\ y - x = -2 \end{cases}$

25. $\begin{cases} 2x + 3y = -1 \\ 3x + 2y = 1 \end{cases}$

26. $\begin{cases} 2x - 3y = -11 \\ x + 3y = 8 \end{cases}$

27. $\begin{cases} x + 3y = 15 \\ 2x + y = 10 \end{cases}$

28. $\begin{cases} 3y - 5x = 23 \\ 4x + y = 2 \end{cases}$

29. $\begin{cases} 2x - y = 3 \\ y - 2x = 4 \end{cases}$

30. $\begin{cases} 6x + 3y = -3 \\ 4x + 2y = 2 \end{cases}$

31. $\begin{cases} 4x + 3y = 11 \\ 2x - y = 3 \end{cases}$

32. $\begin{cases} 5x + 4y = -3 \\ 3x - y = 5 \end{cases}$

33. $\begin{cases} 2x - 3y = 6 \\ 6y - 4x = 12 \end{cases}$

34. $\begin{cases} 5x + 3y = 2 \\ x - 2y = 3 \end{cases}$

35. $\begin{cases} -2x + y = -8 \\ 2x + 3y = 0 \end{cases}$

36. $\begin{cases} 2x + 4y = 10 \\ 4x - 5y = 20 \end{cases}$

37. $\begin{cases} 2x + 4y = -2 \\ x - 2y = -1 \end{cases}$

38. $\begin{cases} x - 3y = 9 \\ 5x + 3y = 27 \end{cases}$

C

Solve by graphing; estimate the solution to the system to the nearest half-unit:

39. $\begin{cases} 2x + 4y = 9 \\ x + y = 1 \end{cases}$

40. $\begin{cases} x - 3y = 15 \\ 5x + 3y = 6 \end{cases}$

41. $\begin{cases} x + y = 0 \\ 2x + 4y = 1 \end{cases}$

42. $\begin{cases} 3x - 4y = 10 \\ 2x - y = 11 \end{cases}$

43. $\begin{cases} x + y = 3 \\ 2x - y = 5 \end{cases}$

44. $\begin{cases} 2x + 5y = 2 \\ 4x - 5y = 13 \end{cases}$

D

45. The sum of two numbers is 12. The difference of the same numbers is 2. Write a system of equations for the two numbers, and solve by graphing.

46. The difference of two numbers is seven. The sum of twice the first and three times the second is nine. Write a system of equations for the two numbers, and solve by graphing.

47. Find the number of days for each of the plants in Example 5 to fill an order for 300 color sets and 300 black-and-white sets.

48. Two cross-country runners leave the same starting point one hour apart. The first to leave runs at an average speed of 5 mph and the second at an average speed of 7 mph. How far will they have run when the second runner overtakes the first? The table summarizes what we know.

Figure for Exercise 48

	Rate	Time	Distance
First	5	1 hour longer	Same distance for both runners
Second	7		

If we let t represent the time for the *second* runner and d represent the distance of *each* runner, we can fill in the table.

	Rate	Time	Distance
First	5	$t + 1$	d
Second	7	t	d

Now, since $d = rt$, we have the system:

$$\begin{cases} d = 5(t + 1) \\ d = 7t \end{cases}$$

Construct a rectangular coordinate system, and find the solution by graphing.

49. If three adults and four children go to the museum, the total cost is \$27. When one adult accompanies three children the cost is \$14. Write a system of equations and determine the cost for one adult and the cost for one child by graphing.

50. In a shipment of 16 cartons of health aid products, some cartons weighed 6 lb and some weighed 8 lb. The weight of the shipment was 110 lb. Write a system of equations and use a graphical solution to determine the number of cartons of each weight in the shipment.

51. Twenty-three people are enrolled in an algebra class. There are three more women than men in the class. Write a system of equations and determine the number of men and the number of women enrolled in the class by graphing.

52. A computer requires 10.5 seconds to do two series of calculations. The second series requires one-half the time as the first. Write a system of equations and determine the time required for each series by graphing.

STATE YOUR UNDERSTANDING

53. Identify the three different types of linear equations, and describe each system.

54. State at least one disadvantage of solving a system of equations by graphical methods.

CHALLENGE EXERCISES

Solve by graphing:

55. $\begin{cases} x = -2 \\ y = 3 \\ 2x + 3y = 5 \end{cases}$

56. $\begin{cases} 3x - 4y = 4 \\ x + 2y = 8 \\ y = \frac{5}{2}x - 8 \end{cases}$

57. $\begin{cases} y = 2x - 1 \\ 3x + 2y = 6 \\ y = -1 \end{cases}$

58. $\begin{cases} 4x - y = 3 \\ x + y = 2 \\ 2y = 8x - 6 \end{cases}$

MAINTAIN YOUR SKILLS (SECTIONS 5.1, 5.2, 5.4)

Perform the indicated operations; reduce if possible:

59. $\dfrac{36x^4y^2}{25xy^3} \div \dfrac{9xy}{10x^3y^2}$

60. $\dfrac{a^2 + 5a + 6}{a^2 - 2a - 35} \cdot \dfrac{a^2 - 3a - 28}{a^2 + 7a + 12}$

61. $\dfrac{x}{x + 1} + \dfrac{2}{x + 2}$

62. $\dfrac{a - b}{a + b} - \dfrac{a + b}{a - b}$

63. $\dfrac{10}{t - 3} - \dfrac{15}{3 - t}$

64. $\dfrac{3x}{x - 3} + \dfrac{x + 3}{2x}$

65. $\dfrac{2m}{m - 1} - \dfrac{m + 4}{m - 3}$

66. $\dfrac{2}{x - 1} - \dfrac{1}{x + 1} + \dfrac{3}{x^2 - 1}$

6.6 SYSTEMS OF EQUATIONS: SOLVING BY SUBSTITUTION

OBJECTIVE

1. Solve a system of linear equations in two variables by substitution.

The second method for solving a system of equations is substitution. This method is practical when one of the variables has a coefficient of one. We solve the equation for that variable, and then substitute the value in the other equation.

EXAMPLE 1

Solve the system: $\begin{cases} 3x - y = -8 & (1) \\ x + 2y = 9 & (2) \end{cases}$

$3x - y = -8 \quad (1)$ Solve equation (1) for y.

$-y = -3x - 8$

$y = 3x + 8$

$x + 2y = 9 \quad (2)$ Next, substitute $(3x + 8)$ for y in equation (2), and solve for x.

$x + 2(3x + 8) = 9$

$x + 6x + 16 = 9$

$7x = -7$

$x = -1$

$3x - y = -8$ Now solve for y by replacing x by -1 in equation (1).

$3(-1) - y = -8$

$-y = -5$

$y = 5$

Check: Check the values of x and y in equation (2). We need not check in both equations since we used equation (1) to solve for y.

$x + 2y = 9$

$-1 + 2(5) = 9$ True.

The solution set is $\{(-1, 5)\}$.

PROCEDURE

To solve a system of linear equations in two variables by substitution:

Step 1. Solve either equation for x (or y).

Step 2. Eliminate x (or y) by substituting this expression for x (or y) in the other equation.

Step 3. Solve.

Step 4. Use this value of y (or x) to find the other value.

Step 5. Check the value of x and y in the equation not used in Step 4.

Step 6. Write the solution set $\{(x, y)\}$.

If both x and y are eliminated in Step 3, the new equation is either true (identity) or false (contradiction), which tells us the following about the system:

New Equation	Type of System	Example	Solution Set
Contradiction	Independent and inconsistent	$\begin{cases} 5x - y = 8 \\ 15x - 3y = 19 \end{cases}$	$\emptyset$
Identity	Dependent and consistent	$\begin{cases} 2x - y = 7 \\ 4x - 2y = 14 \end{cases}$	Either $\{(x, y) \mid 2x - y = 7\}$ or $\{(x, y) \mid 4x - 2y = 14\}$

EXAMPLE 2

Solve the system: $\begin{cases} x - 2y = 3 & (1) \\ 2x - 4y = 7 & (2) \end{cases}$

Step 1

$x - 2y = 3$ Solve equation (1) for x.

$x = 2y + 3$

Step 2

$2x - 4y = 7$ Substitute $(2y + 3)$ for x in equation (2).

$2(2y + 3) - 4y = 7$

Step 3

$4y + 6 - 4y = 7$ $6 = 7$	Solve for y. This is a contradiction. When a contradiction is reached, we know that there is no solution. Steps 4 and 5 are unnecessary.

The solution set is $\emptyset$. ■

EXAMPLE 3

Solve the system: $\begin{cases} x + y = 3 & (1) \\ 2x - y = 5 & (2) \end{cases}$

Step 1

$x + y = 3$ — Solve equation (1) for y.

$y = -x + 3$

Step 2

$2x - y = 5$ (2) — Replace y with $(-x + 3)$ in equation (2).

$2x - (-x + 3) = 5$

Step 3

$2x + x - 3 = 5$ — Solve for x.

$3x = 8$

$x = \dfrac{8}{3}$

Step 4

$x + y = 3$ (1) — Replace x in equation (1) with $\dfrac{8}{3}$, and solve for y.

$\dfrac{8}{3} + y = 3$

$y = 3 - \dfrac{8}{3}$

$y = \dfrac{1}{3}$

Step 5

$2x - y = 5$ (2) — Check $x = \dfrac{8}{3}$ and $y = \dfrac{1}{3}$ in equation (2).

$2\left(\dfrac{8}{3}\right) - \dfrac{1}{3} = 5$

$\dfrac{16}{3} - \dfrac{1}{3} = 5$ — True.

Step 6

The solution set is $\left\{\left(\frac{8}{3}, \frac{1}{3}\right)\right\}$.

■

If the system of equations has coefficients that are fractions or decimals, it is often easier to first write an equivalent system. The equivalent system is formed by multiplying each equation by a constant.

EXAMPLE 4

Solve the system: $\begin{cases} 0.2x - 0.1y = 0.3 & (1) \\ 2.5x - 0.3y = 4.7 & (2) \end{cases}$

$0.2x - 0.1y = 0.3 \quad (1)$

$10(0.2x - 0.1y) = 10(0.3)$

$2x - y = 3 \quad (1A)$

To eliminate the decimals, we multiply each of the equations by 10.

$2.5x - 0.3y = 4.7 \quad (2)$

$10(2.5x - 0.3y) = 10(4.7)$

$25x - 3y = 47 \quad (2A)$

$\begin{cases} 2x - y = 3 & (1A) \\ 25x - 3y = 47 & (2A) \end{cases}$

We now have a new, but equivalent, system to solve.

Step 1

$2x - y = 3 \quad (1A)$

$y = 2x - 3$

Solve equation (1A) for y.

Step 2

■

$25x - 3y = 47 \quad (2A)$

$25x - 3(2x - 3) = 47$

Substitute $(2x - 3)$ for y in equation (2A), and solve for x.

Step 3

$25x - 6x + 9 = 47$

$19x = 38$

$x = 2$

Step 4

$0.2x - 0.1y = 0.3 \quad (1)$

$0.2(2) - 0.1y = 0.3$

$0.4 - 0.1y = 0.3$

$-0.1y = -0.1$

$y = 1$

We now go back to the original system, replace x with 2 in equation (1), and solve for y.

Step 5

$$2.5x - 0.3y = 4.7 \quad (2)$$

Check $x = 2$ and $y = 1$ in equation (2).

$$2.5(2) - 0.3(1) = 4.7$$

$$5 - 0.3 = 4.7$$

True.

Step 6

The solution set is $\{(2, 1)\}$. ■

Example 5 shows the application of a system of equations to investment earnings.

EXAMPLE 5

Mary invested \$20,000, part at 8% interest and the remainder at 9%. If the total return on the investments is \$1690, how much is invested at each rate?

Simpler word form:

Total amount of money = \$20,000

$$\begin{pmatrix}\text{Amount of}\\ \text{interest}\\ \text{at } 8\%\end{pmatrix} + \begin{pmatrix}\text{Amount of}\\ \text{interest}\\ \text{at } 9\%\end{pmatrix} = \begin{pmatrix}\text{Total}\\ \text{amount of}\\ \text{return}\end{pmatrix}$$

Assign variables:

Let x be the amount invested at 8% and y the amount invested at 9%.

Translate to algebra:

$$\begin{cases} x + y = 20{,}000 & (1)\\ 0.08x + 0.09y = 1690 & (2)\end{cases}$$

Solve:

Step 1

$$y = 20{,}000 - x$$

Solve equation (1) for y.

Step 2

$$8x + 9y = 169{,}000 \quad (2A)$$

Multiply equation (2) by 100.

$$8x + 9(20{,}000 - x) = 169{,}000$$

Substitute $20{,}000 - x$ for y in equation (2A).

Step 3

$$8x + 180{,}000 - 9x = 169{,}000$$
$$-x = -11{,}000$$
$$x = 11{,}000$$

Solve for x.

Step 4

$$11{,}000 + y = 20{,}000$$
$$y = 9000$$

Substitute $x = 11{,}000$ in equation (1).

Step 5

Check:

$$0.08(11{,}000) + 0.09(9000) = 1690$$
$$880 + 810 = 1690$$

Substitute $x = 11{,}000$ and $y = 9000$ in equation (2).

Step 6

Answer:

The solution to the system is (11,000, 9000), which means that \$11,000 is invested at 8% and \$9000 is invested at 9%. ■

EXAMPLE 6

A landscaper wants 600 pounds of grass seed that is 45% bluegrass. He has two mixtures available. One contains 40% bluegrass and the other 70% bluegrass. How many pounds of each must he use?

	First Mix	Second Mix	Desired Mix
% of bluegrass (in decimal form)	0.40	0.70	0.45
Number of pounds	?	?	600

Use a chart to help organize the material.

Simpler word form:

Total number of pounds = 600

$$\begin{pmatrix}\text{amount of}\\ \text{bluegrass in}\\ \text{first mix}\end{pmatrix} + \begin{pmatrix}\text{amount of}\\ \text{bluegrass in}\\ \text{second mix}\end{pmatrix} = \begin{pmatrix}\text{total}\\ \text{amount of}\\ \text{bluegrass}\end{pmatrix}$$

Assign variables:

Let x be the number of pounds of the first mix and y be the number of pounds of the second mix.

Fill in the missing parts using the variables.

	First Mix	Second Mix	Desired Mix
% of bluegrass (in decimal form)	0.40	0.70	0.45
Number of pounds	x	y	600
Number of pounds of bluegrass seed	$0.40x$	$0.70y$	$0.45(600) = 270$

Translate to algebra:

$$\begin{cases} x + y = 600 & (1) \\ 0.40x + 0.70y = 270 & (2) \end{cases}$$

Solve:

Step 1

$y = 600 - x \quad (1)$

Solve equation (1) for y.

Step 2

$4x + 7y = 2700 \quad (2A)$

Multiply equation (2) by 10.

$4x + 7(600 - x) = 2700$

Substitute $600 - x$ for y in equation (2A).

Step 3

$$4x + 4200 - 7x = 2700$$
$$-3x + 4200 = 2700$$
$$-3x = -1500$$
$$x = 500$$

Step 4

$$500 + y = 600$$
$$y = 100$$

Substitute $x = 500$ in equation (1).

Step 5

Check:

$$0.40x + 0.70y = 270 \quad (2)$$
$$0.40(500) + 0.70(100) = 270$$
$$200 + 70 = 270$$

Step 6

Answer:

The solution to the system is (500, 100), which means that by using 500 pounds of 40% mix and 100 pounds of 70% mix, the landscaper can get 600 pounds of 45% bluegrass. ■

EXERCISE 6.6

A

Solve by substitution:

1. $\begin{cases} y = x - 5 \\ x + y = 3 \end{cases}$

2. $\begin{cases} y = x + 6 \\ x + y = 10 \end{cases}$

3. $\begin{cases} x = y - 1 \\ x + y = 3 \end{cases}$

4. $\begin{cases} y = x - 7 \\ y + x = 3 \end{cases}$

5. $\begin{cases} x = 11 - y \\ 2x - 3y = 12 \end{cases}$

6. $\begin{cases} y = -x - 3 \\ 7x + 2y = 4 \end{cases}$

7. $\begin{cases} x = y + 1 \\ 3x + 4y = 17 \end{cases}$

8. $\begin{cases} y = -x - 1 \\ 5x - 2y = -19 \end{cases}$

9. $\begin{cases} x + y = -8 \\ -3x + 2y = 9 \end{cases}$

10. $\begin{cases} 3x + 4y = -29 \\ x - y = 2 \end{cases}$

11. $\begin{cases} x + y = 3 \\ 2x - 3y = 1 \end{cases}$

12. $\begin{cases} x + y = -9 \\ 2y = 3x + 2 \end{cases}$

13. $\begin{cases} x - y = 12 \\ 5y = 4x - 12 \end{cases}$

14. $\begin{cases} x - y = 9 \\ 4x = -3y + 22 \end{cases}$

15. $\begin{cases} x = y + 2 \\ 4x - 2 = 3y \end{cases}$

16. $\begin{cases} y = x - 7 \\ -3y = 6 - 6x \end{cases}$

17. $\begin{cases} x + y = -4 \\ 2x - 5y = -29 \end{cases}$

18. $\begin{cases} x - y = 2 \\ 3x - 5y = -6 \end{cases}$

19. $\begin{cases} x + y = 9 \\ 8x - y = -18 \end{cases}$

20. $\begin{cases} x - y = 11 \\ x - 6y = -9 \end{cases}$

B

Solve by substitution:

21. $\begin{cases} x + 3y = 5 \\ 4x + 5y = 13 \end{cases}$

22. $\begin{cases} x + 2y = 7 \\ 3x - 2y = -11 \end{cases}$

23. $\begin{cases} x + 8y = 6 \\ -2x - 4y = 3 \end{cases}$

24. $\begin{cases} x + 5y = 14 \\ 2x - y = -5 \end{cases}$

25. $\begin{cases} 2a + 3b = 1 \\ a - 4b = 6 \end{cases}$

26. $\begin{cases} 2p - q = -15 \\ p + 2q = 5 \end{cases}$

27. $\begin{cases} 4x + y = 12 \\ 2y = 24 - 8x \end{cases}$

28. $\begin{cases} 5x = y + 8 \\ 10x - 2y = 16 \end{cases}$

29. $\begin{cases} 6x - 9y = 11 \\ -2x + 3y = -5 \end{cases}$

30. $\begin{cases} 2x - y = 7 \\ 2y - 4x = 7 \end{cases}$

31. $\begin{cases} \dfrac{x}{2} + \dfrac{y}{8} = \dfrac{1}{4} \\ 5x + 3y = 20 \end{cases}$

32. $\begin{cases} x + 5y = 14 \\ \dfrac{x}{5} - \dfrac{y}{10} = -\dfrac{1}{2} \end{cases}$

33. $\begin{cases} 2x + 3y = 23 \\ 2y - x = -1 \end{cases}$

34. $\begin{cases} 5x - 2y = 10 \\ 2x - y = 3 \end{cases}$

35. $\begin{cases} 2x - y = -7 \\ 3x + y = -3 \end{cases}$

36. $\begin{cases} 2x + 8y = -6 \\ -x + 3y = 10 \end{cases}$

37. $\begin{cases} y = \dfrac{2}{3}x - 4 \\ 5y - x = 1 \end{cases}$

38. $\begin{cases} x = \dfrac{2}{5}y - 6 \\ 2y + 7x = 6 \end{cases}$

39. $\begin{cases} y = \dfrac{1}{2}x + 4 \\ 2x + 5y = 11 \end{cases}$

40. $\begin{cases} x = \dfrac{2}{3}y - 8 \\ 8x - y = 1 \end{cases}$

C

Solve by substitution:

41. $\begin{cases} x + 2y = \dfrac{3}{4} \\ 2x + y = -\dfrac{3}{8} \end{cases}$

42. $\begin{cases} x - 2y = \dfrac{3}{2} \\ 2x + y = \dfrac{9}{8} \end{cases}$

43. $\begin{cases} 6x - y = 7 \\ 2x + 3y = 5 \end{cases}$

44. $\begin{cases} 2x + 3y = -4 \\ 3x + y = 6 \end{cases}$

45. $\begin{cases} 4x - 8y = 12 \\ 3x - 9 = 6y \end{cases}$

46. $\begin{cases} y = 2x + 5 \\ 6x - 3y = 1 \end{cases}$

47. $\begin{cases} 2x + y = 3 \\ 4x + 3y = 10 \end{cases}$

48. $\begin{cases} x = 4y + 1 \\ 3x - 8y = 8 \end{cases}$

49. $\begin{cases} \dfrac{x}{4} + \dfrac{y}{2} = \dfrac{7}{8} \\ 2x = 10 - 6y \end{cases}$

50. $\begin{cases} \dfrac{x}{4} + \dfrac{y}{3} = -\dfrac{7}{24} \\ 3y = 18 + 39x \end{cases}$

51. $\begin{cases} 2x + 3y = 3 \\ 6x - 3y = 1 \end{cases}$

52. $\begin{cases} 2x + 2y = -12 \\ 5y = 3x - 6 \end{cases}$

53. $\begin{cases} \dfrac{3}{4}x + 2y = 4 \\ 2x + 7y = 9 \end{cases}$

54. $\begin{cases} \dfrac{2}{3}y = x + 10 \\ 2y + 3x = 6 \end{cases}$

55. $\begin{cases} 0.1x + 0.1y = 0.5 \\ 3x - 2y = 5 \end{cases}$

56. $\begin{cases} 0.3x + 0.1y = 0.3 \\ 2x - 3y = 10 \end{cases}$

57. $\begin{cases} 0.3x - 0.1y = 0.9 \\ 0.5x - 2y = 7 \end{cases}$

58. $\begin{cases} \dfrac{2}{3}x - y = -6 \\ 2x - y = 14 \end{cases}$

59. $\begin{cases} 0.03x - 0.01y = -0.03 \\ 5x + 3y = -19 \end{cases}$

60. $\begin{cases} 0.01x + 0.05y = -0.1 \\ 5x - y = -24 \end{cases}$

D

61. If the landscaper in Example 6 needs 600 pounds of grass seed that is 55% bluegrass for a different soil, how many pounds of each of his mixtures should be used?

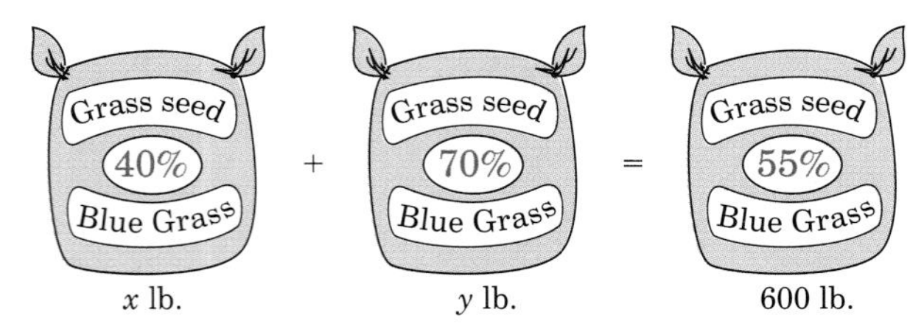

Figure for Exercise 61

62. Ellie needs some change for her weekend garage sale. She has enough nickels and wants three times as many quarters as dimes. If she takes $20.40 to the bank, how many quarters and how many dimes can she get?

63. The Little Theater group put on a play last week. There were twenty more adults attending than children. The adult tickets were $8 each and the child tickets $3 each. If they collected $1260, how many adult tickets were purchased?

64. Krista Wiess invested $15,000, part at 9% and part at 12%. If the interest for the year on the two investments was $1680, how much was invested at each rate?

65. Joe Cool invested $18,000, part at 12% and part at 15% interest. If the annual return on each investment was the same, find the amount invested at each rate.

66. Melba Sharp and Jose Remus both work for the Sweeter-Than-Sweet Ice Cream Company. Melba earns $8 per hour and Jose earns $6.50 per hour.

The owner paid them a total of \$388 for a combined total of 50 hours worked. How much did each earn?

67. The cost of driving Ms. Chinn's Cadillac is 32¢ per mile. The cost of driving her Datsun pickup is 22¢ per mile. During the month of December, the cars were driven a total of 660 miles at a cost of \$186.20. How many miles was each car driven during December?

68. For the grand opening of the new Opera House, 2600 tickets were sold. Some of the tickets sold for \$25 each, and the rest sold for \$15 each. How many tickets of each price were sold if the sales totaled \$46,000?

69. Jon bought 50 stamps. Some were 30¢ stamps, and some were 25¢ stamps. They cost him \$13.80. How many of each type of stamp did Jon buy?

x stamps

y stamps

Figure for Exercise 69

70. A candy store made up a mixture of almonds and cashew nuts. There was a total of 20 pounds. The almonds were \$3.00 a pound, and the cashews were \$4.50 a pound. The total value of the mixture was \$72. How many pounds of each type of nut was in the mixture?

71. Cindy Sloe borrowed money at 18% and 20%. If the annual interest payment is \$396, and she borrowed a total of \$2100, how much did she borrow at each rate?

72. Alpine Community College pays its administrators an average of \$2100 per month and its faculty an average of \$1950 per month. If the monthly payroll for the 120 employees in the two categories is \$237,000, how many administrators and how many faculty members does Alpine employ?

STATE YOUR UNDERSTANDING

73. Summarize the strategy for solving a system of equations by substitution.

74. Explain the error in the following example. Determine the correct solution.
Solve by substitution:

$$3x + y = 4$$
$$2x + 4y = 6$$

Solution:

$y = 4 - 3x$ (Solving the first equation for y.)
$3x + (4 - 3x) = 4$ (Substituting)
$3x + 4 - 3x = 4$ (Solving)
$4 = 4$ (Identity)

The system is dependent and consistent.

$$\{(x, y) \mid 3x + y = 4\}$$

CHALLENGE EXERCISES

Solve by substitution:

75. $\begin{cases} 2x + 3y = 5 \\ 6x - 5y = 1 \end{cases}$

76. $\begin{cases} 3x + 5y = 12 \\ 4x - 2y = 7 \end{cases}$

77. Determine the values of a and b so that the graph of $ax + by = 11$ will contain the points (4, 1) and (10, −3). Write the equation.

78. Determine the values of a and c so that the graph of $ax - 3y = c$ will contain the points (−1, −1) and (2, 4). Write the equation.

MAINTAIN YOUR SKILLS (SECTION 6.3)

Find the x-intercept and the y-intercept for the graph of each of the following:

79. $2y - 7x = 14$

80. $9x - 4y = 36$

81. $3x - 5y = 7$

82. $4y - 11x = 33$

Find the slope of the line containing the following points:

83. (3, 7), (2, 2)

84. (−2, −1), (−5, −2)

85. (3, 0), (1, −6)

86. (8, −7), (−1, 5)

6.7

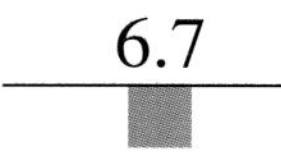

SYSTEMS OF EQUATIONS: SOLVING BY LINEAR COMBINATIONS

OBJECTIVE

1. Solve a system of linear equations in two variables by linear combinations.

If we are given a system of equations in which none of the coefficients of x or y is 1, we recommend solving the system by linear combinations. To solve a system by linear combinations, we first write an equivalent system in which the coefficients of one of the variables are opposites. Then add the equations to eliminate the variable.

PROPERTY

Linear Combination

If the system $\begin{cases} A_1x + B_1y = C_1 \\ A_2x + B_2y = C_2 \end{cases}$ has a single solution (a, b), then the linear combination $(A_1 + A_2)x + (B_1 + B_2)y = C_1 + C_2$ also has (a, b) for a solution.

Now let's use linear combinations to solve a system.

EXAMPLE 1

Solve the system: $\begin{cases} x - 2y = -3 & (1) \\ 3x + 2y = 7 & (2) \end{cases}$

$$\begin{array}{r} x - 2y = -3 \\ 3x + 2y = 7 \\ \hline 4x \quad = 4 \\ x = 1 \end{array}$$

Since the coefficients of y are opposites, adding the equations will eliminate y.

$1 - 2y = -3$

Substitute 1 for x in equation (1) to find y.

$-2y = -4$

$y = 2$

Check:

$3(1) + 2(2) = 7$

$3 + 4 = 7$

Check in equation (2), the equation not used to find y.

The solution set is $\{(1, 2)\}$. ■

EXAMPLE 2

Solve the system: $\begin{cases} 2x + y = 4 & (1) \\ x - 3y = 9 & (2) \end{cases}$

$3(2x + y) = 3(4)$

$6x + 3y = 12 \quad (3)$

Write an equivalent system where the coefficients of y are opposites by multiplying equation (1) by 3.

$\begin{cases} 6x + 3y = 12 & (3) \\ x - 3y = 9 & (2) \end{cases}$

New system.

$$\begin{array}{r} 6x + 3y = 12 \\ x - 3y = 9 \\ \hline 7x \quad = 21 \end{array}$$

Add equations (2) and (3).

$x = 3$

Solve for x.

$2(3) + y = 4$

Substitute 3 for x in equation (1) to find y.

$6 + y = 4$

$y = -2$

Check:

$3 - 3(-2) = 9$

$3 + 6 = 9$

Use equation (2) to check.

The solution set is $\{(3, -2)\}$. ■

PROCEDURE

To solve a system of equations in two variables by addition when each equation is written in the form

$Ax + By = C$:

Step 1. Multiply one or both of the equations by factors so that the coefficients of y (or x) are opposites.

Step 2. Add the two equations so that y (or x) is eliminated.

Step 3. Solve.

Step 4. Find the value of the other variable by substitution in one of the original equations.

Step 5. Check the answer in the original equation not used in Step 4.

Step 6. Write the solution set $\{(x, y)\}$.

If both x and y are eliminated in Step 2 and the new equation is either true (identity) or false (contradiction), it tells us the following about the system:

Result of Adding the Equations	Type of System	Example	Solution Set
Contradiction	Independent and inconsistent	$\begin{cases} 5x - 3y = 8 \\ 15x - 9y = 19 \end{cases}$	$\emptyset$
Identity	Dependent and consistent	$\begin{cases} 2x - 3y = 7 \\ 4x - 6y = 14 \end{cases}$	Either $\{(x, y) \mid 2x - 3y = 7$ or $\{(x, y) \mid 4x - 6y = 14)$

EXAMPLE 3

Solve the system: $\begin{cases} 2x + 3y = 5 & (1) \\ 3x + 2y = 5 & (2) \end{cases}$

Step 1

$$3(2x + 3y) = 3(5)$$
$$-2(3x + 2y) = -2(5)$$

Write an equivalent system where the coefficients of x are opposites.
Multiply equation (1) by 3 and equation (2) by -2.

$$\begin{cases} 6x + 9y = 15 & (3) \\ -6x - 4y = -10 & (4) \end{cases}$$ Equivalent system.

Step 2

$$\begin{array}{r} 6x + 9y = 15 \\ -6x - 4y = -10 \\ \hline 5y = 5 \end{array}$$

Add equations (3) and (4) to eliminate the variable x, a linear combination of equations (3) and (4).

Step 3

$$y = 1$$ Solve for y.

Step 4

$$\begin{aligned} 2x + 3y &= 5 \\ 2x + 3(1) &= 5 \\ 2x + 3 &= 5 \\ 2x &= 2 \\ x &= 1 \end{aligned}$$

Substitute 1 for y in equation (1), and solve for x. We use one of the original equations to reveal any errors that might have been introduced while writing the equivalent system.

Step 5

$$3x + 2y = 5$$ Check in equation (2).

$$3(1) + 2(1) = 5$$

$$3 + 2 = 5$$ True.

Step 6

The solution set is $\{(1, 1)\}$. ■

EXAMPLE 4

Solve the system: $\begin{cases} 9x - 4y = 36 & (1) \\ 18x - 8y = 11 & (2) \end{cases}$

Step 1

$$-2(9x - 4y) = -2(36)$$ Multiply equation (1) by -2.

$$\begin{cases} -18x + 8y = -72 & (3) \\ 18x - 8y = 11 & (2) \end{cases}$$ Equivalent system.

Step 2

$$\begin{array}{r} -18x + 8y = -72 \\ 18x - 8y = 11 \\ \hline 0 = -61 \end{array}$$

Add equations (2) and (3).

The result is a contradiction, so there is no solution.

Step 6

The solution set is $\emptyset$. ■

It is possible to find the value of each variable using linear combinations. This is useful when the solution involves fractions.

EXAMPLE 5

Solve the system: $\begin{cases} 3x + 2y = -8 & (1) \\ 2x - 3y = -6 & (2) \end{cases}$

Step 1

$3(3x + 2y) = 3(-8)$

$2(2x - 3y) = 2(-6)$

Multiply equation (1) by 3 and equation (2) by 2 so that the coefficients of y are opposite.

$\begin{cases} 9x + 6y = -24 & (3) \\ 4x - 6y = -12 & (4) \end{cases}$

Equivalent system.

Step 2

$$\begin{array}{r} 9x + 6y = -24 \\ \underline{4x - 6y = -12} \\ 13x \qquad = -36 \end{array}$$

Add the equations.

Step 3

$13x = -36$

$x = -\dfrac{36}{13}$

Solve for x.

We now solve for y by repeating Steps 1, 2, and 3.

Step 1 (repeated)

$-2(3x + 2y) = -2(-8)$

$3(2x - 3y) = 3(-6)$

Multiply equation (1) by -2 and equation (2) by 3 so that the coefficients of x are opposites.

$\begin{cases} -6x - 4y = 16 & (5) \\ 6x - 9y = -18 & (6) \end{cases}$

Equivalent system.

Step 2

$$\begin{array}{r} -6x - 4y = 16 \\ \underline{6x - 9y = -18} \\ -13y = -2 \end{array}$$

Add equations (5) and (6).

Step 3

$y = \dfrac{2}{13}$

Solve for y. Now go back to Step 5.

Step 5 Check $x = -\frac{36}{13}$ and $y = \frac{2}{13}$ in equations (1) and (2) since we did not use Step 4.

$$3x + 2y = -8 \quad (1)$$

$$3\left(-\frac{36}{13}\right) + 2\left(\frac{2}{13}\right) = -8$$

$$-\frac{108}{13} + \frac{4}{13} = -8$$

$$-\frac{104}{13} = -8 \qquad \text{True.}$$

$$2x - 3y = -6 \quad (2)$$

$$2\left(-\frac{36}{13}\right) - 3\left(\frac{2}{13}\right) = -6$$

$$-\frac{72}{13} - \frac{6}{13} = -6 \qquad \text{True.}$$

Step 6

The solution set is $\left\{\left(-\frac{36}{13}, \frac{2}{13}\right)\right\}$. ■

If the equations in a system are not written in standard form, $Ax + By = C$, rewrite them before continuing.

EXAMPLE 6

Solve the system: $\begin{cases} 2x = 5y - 1 & (1) \\ 2y = 8 - 3x & (2) \end{cases}$

$$\begin{cases} 2x - 5y = -1 & (1A) \\ 3x + 2y = 8 & (2A) \end{cases}$$

Write each equation in the form $Ax + By = C$.

Step 1

$$\begin{cases} 4x - 10y = -2 & (3) \\ 15x + 10y = 40 & (4) \end{cases}$$

Multiply equation (1*A*) by 2 and equation (2*A*) by 5. The multiplication is done mentally.

Step 2

$$19x = 38$$

Add the equations.

Step 3

$$x = 2$$

Solve for x.

Step 4

$2x = 5y - 1 \quad (1)$ Solve for y using equation (1).

$2(2) = 5y - 1$

$5 = 5y$

$1 = y$

Step 5

$2y = 8 - 3x \quad (2)$ Check using equation (2).

$2(1) = 8 - 3(2)$

$2 = 8 - 6$ True.

Step 6

The solution set is $\{(2, 1)\}$. ■

The application in Example 5, courtesy of Exxon Co., shows how systems of equations can be used to solve mixture problems.

EXAMPLE 7

A tank contains a salt solution that is 10% salt by weight. A second tank contains a solution that is 50% salt by weight. How many gallons of each solution should be mixed to make 100 gallons of a 20% salt solution?

Simpler word form:

(Number of gallons of 10% solution) + (Number of gallons of 50% solution) = 100 gallons

(Number of gallons of salt in 10% solution) + (Number of gallons of salt in 50% solution) = (Number of gallons of salt in final 20% solution)

Select variables:

If we let a represent the number of gallons from the first tank (10% salt solution) and b represent the number of gallons from the second tank (50% salt solution), we can organize the information in a table:

	Number of Gallons from a	Number of Gallons from b	Total
Solution	a	b	100
Salt	$0.10a$	$0.50b$	$0.20(100) = 20$

Translate to algebra:

The system of equations is

$$\begin{cases} a + \quad b = 100 & (1) \\ 0.10a + 0.50b = 20 & (2) \end{cases}$$

The total number of gallons of solution is 100, so $a + b = 100$. The total amount of salt is 20.

Solve:

Step 1

$$-10(0.10x + 0.50b) = -10(20)$$

Multiply equation (2) by -10. Equivalent system.

$$\begin{cases} a + b = 100 & (1) \\ -a - 5b = -200 & (3) \end{cases}$$

Step 2

$$-4b = -100$$

Add equations (1) and (3).

Step 3

$$b = 25$$

Solve for b.

Step 4

$$a + b = 100$$
$$a + 25 = 100$$
$$a = 75$$

Substitute 25 for b in equation (1), and solve for a.

Step 5

$$0.10a + 0.50b = 20$$

Check in equation (2).

$$0.10(75) + 0.50(25) = 20$$
$$7.5 + 12.5 = 20$$

True.

Step 6

The solution set for the system is $\{(75, 25)\}$.

Answer:

To make the mixture, 75 gallons are needed from the first tank and 25 gallons from the second tank.

■

EXERCISE 6.7

A

Solve by linear combinations:

1. $\begin{cases} x + y = 4 \\ x - y = 2 \end{cases}$

2. $\begin{cases} x + y = -1 \\ x - y = 3 \end{cases}$

3. $\begin{cases} 3x - 2y = 10 \\ 3x + 2y = 14 \end{cases}$

4. $\begin{cases} 2x + 3y = 6 \\ 2x - 3y = 6 \end{cases}$

5. $\begin{cases} 2x - y = 3 \\ -2x + y = 2 \end{cases}$

6. $\begin{cases} x + 5y = 4 \\ -x - 5y = -4 \end{cases}$

7. $\begin{cases} x + y = 12 \\ x - y = 2 \end{cases}$

8. $\begin{cases} x + 2y = 5 \\ 2x - y = 0 \end{cases}$

9. $\begin{cases} x + 3y = 10 \\ x - y = -2 \end{cases}$

10. $\begin{cases} 3x + y = -8 \\ x + y = -4 \end{cases}$

11. $\begin{cases} x + 2y = 15 \\ x - y = 0 \end{cases}$

12. $\begin{cases} x + 3y = 5 \\ x + y = 1 \end{cases}$

13. $\begin{cases} 2x + y = 5 \\ 4x + y = 6 \end{cases}$

14. $\begin{cases} x + 4y = 2 \\ x - 8y = -7 \end{cases}$

15. $\begin{cases} x + y = 35 \\ x - 5y = 17 \end{cases}$

16. $\begin{cases} 5x + 2y = -2 \\ x - y = 8 \end{cases}$

17. $\begin{cases} 2x + 5y = 13 \\ 3x + y = 0 \end{cases}$

18. $\begin{cases} 4x - y = 19 \\ 2x + 3y = -1 \end{cases}$

19. $\begin{cases} 5x + 2y = -7 \\ x + 3y = 9 \end{cases}$

20. $\begin{cases} 2x - 7y = 2 \\ 3x + y = -20 \end{cases}$

B

Solve by linear combinations:

21. $\begin{cases} 3x - 2y = -11 \\ x + 2y = -1 \end{cases}$

22. $\begin{cases} 2x + 3y = 2 \\ 3x - 2y = 16 \end{cases}$

23. $\begin{cases} 5x - 2y - 7 = 0 \\ -3x + 4y + 7 = 0 \end{cases}$

24. $\begin{cases} 6x + 2y + 3 = 0 \\ -2x - y - 2 = 0 \end{cases}$

25. $\begin{cases} x + 5y + 1 = 0 \\ 2x + 7y - 1 = 0 \end{cases}$

26. $\begin{cases} 2x + y - 1 = 0 \\ 6x + 5y - 13 = 0 \end{cases}$

27. $\begin{cases} \frac{3}{4}x + y - 6 = 0 \\ 2x + y - 11 = 0 \end{cases}$

28. $\begin{cases} 3x + y = 7 \\ x - \frac{5}{2}y = -\frac{1}{2} \end{cases}$

29. $\begin{cases} 2x + 4y = 5 \\ 3x + 6y = 6 \end{cases}$

30. $\begin{cases} 4x - 3y = -4 \\ 10x + 9y = 1 \end{cases}$

31. $\begin{cases} 5x - 3y = 5 \\ 15x - 15y = 19 \end{cases}$

32. $\begin{cases} 4x - 2y = -1 \\ 24x - 40y = 15 \end{cases}$

33. $\begin{cases} 2x - 2y = 3 \\ 15x - 4y = 0 \end{cases}$

34. $\begin{cases} 8x - 3y = -5 \\ 40x + 16y = 3 \end{cases}$

35. $\begin{cases} 4x - 3y = -20 \\ 5x - 7y = -64 \end{cases}$

36. $\begin{cases} 7x - 5y = 19 \\ 3x - 2y = 6 \end{cases}$

37. $\begin{cases} 4x + 9y = 57 \\ 9x - 4y = -90 \end{cases}$

38. $\begin{cases} 5x - 7y = 18 \\ -8x + 5y = 27 \end{cases}$

39. $\begin{cases} 5y = 4x + 3 \\ 9x - 8y = 16 \end{cases}$

40. $\begin{cases} 14x - 5y = 2 \\ 9x - 8y = 97 \end{cases}$

C

Solve by linear combinations:

41. $\begin{cases} 3x = 5y + 10 \\ 7x - 3y - 14 = 0 \end{cases}$

42. $\begin{cases} 7x - 9 = 13y \\ 3y + 7 = 9x \end{cases}$

43. $\begin{cases} x = -y + 5 \\ 3x = 2y + 5 \end{cases}$

44. $\begin{cases} 4x - y = 12 \\ 2y = 24 - 8x \end{cases}$

45. $\begin{cases} 2x + y + 2 = 0 \\ 5x = y + 23 \end{cases}$

46. $\begin{cases} x - 3y - 7 = 0 \\ -5x + 2y - 4 = 0 \end{cases}$

47. $\begin{cases} x = 3 + 4y \\ 8y = 6x - 15 \end{cases}$

48. $\begin{cases} 6x + 3y - 9 = 0 \\ 6y + 12x = 18 \end{cases}$

49. $\begin{cases} \frac{1}{2}x + \frac{2}{3}y = \frac{7}{3} \\ 3x - 2y = -16 \end{cases}$

50. $\begin{cases} \frac{3}{4}x - \frac{3}{5}y = \frac{3}{20} \\ \frac{1}{2}x + \frac{2}{3}y = \frac{7}{6} \end{cases}$

51. $\begin{cases} 12x + 35y = 57 \\ 16x + 27y = 17 \end{cases}$

52. $\begin{cases} 7x + 9y = 1 \\ 15x + 17y = -7 \end{cases}$

53. $\begin{cases} 9x + 16y = -65 \\ 12x + 11y = -4 \end{cases}$

54. $\begin{cases} 15x - 14y = 32 \\ 25x + 35y = 170 \end{cases}$

55. $\begin{cases} 8x - 15y = 33 \\ 28x + 25y = 193 \end{cases}$

56. $\begin{cases} 9x - 28y = -11 \\ 5x - 12y = 1 \end{cases}$

57. $\begin{cases} 5x + 7y = 1 \\ 7x + 5y = -1 \end{cases}$

58. $\begin{cases} 21x - 24y = -2 \\ 12x - 9y = 2 \end{cases}$

59. $\begin{cases} 14x + 27y = 26 \\ 21x + 10y = 85 \end{cases}$

60. $\begin{cases} -18x - 25y = 89 \\ 24x + 7y = 13 \end{cases}$

D

61. How many ounces each of 85% pure gold and 70% pure gold are needed to make up 15 ounces of 75% pure gold?

62. Greg has 80 coins, all nickels and dimes. The value of the coins is \$6.75. Write a system of equations, and find how many nickels and how many dimes he has.

63. At a recent play, the Round Table Players sold 850 tickets. They sold only regular seats and boxseats. The regular seats cost \$12 each, and the boxseats cost \$18 each. If the total receipts were \$11,550, how many boxseats were sold?

64. The Buy Me I'm Beautiful furniture company produces chairs at a cost of \$110 each and sofas at a cost of \$322 each. If the cost of production of these two items last month was \$12,450 and 65 items were produced, how many sofas and how many chairs were made?

65. The Grow It Then Mow It seed company sells rye grass seed for \$4.80 per pound and bluegrass seed for \$5.40 per pound. They decide to market 500 pounds of a mixture of the two seeds. If the price of the mixture is \$5.25 per pound, how many pounds of each seed did they use?

66. Mr. Tall Texan owns two oil wells. One well pumps oil that sells for \$36 a barrel, and the other pumps oil that sells for \$42 a barrel. If Mr. Texan pumped 3600 barrels of oil last year for a sale of \$136,200, how many barrels were pumped from each well?

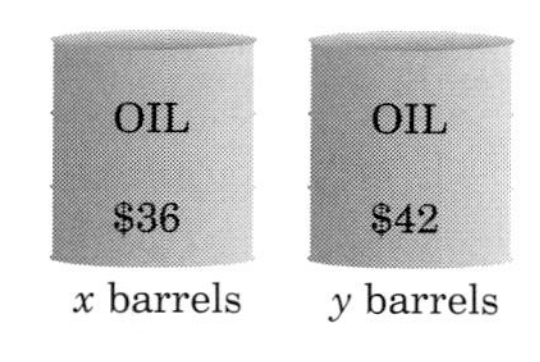

Figure for Exercise 66

67. Marlene Swift sold a record 210 subscriptions during the magazine sale at P.S. 76. The subscriptions were for either \$27 per year or \$48 for two years. If her sales amounted to \$6930, how many one-year and how many two-year subscriptions did Marlene sell?

68. Janet Hardsell makes a 3.5% commission for all sales she makes while in the store and a 6.6% commission for all sales she makes in the field. Last month her sales totaled \$45,500. How much did she sell in the store and in the field if her total commission was \$2600?

69. A cereal company intends to add enough dried fruit to each box of cereal so that each box will contain 21 grams of protein and 338 grams of carbohydrates. The approximate food values are as follows:

	Protein	Carbohydrates
Dried fruit	2%	75%
Cereal	10%	85%

How much cereal and how much dried fruit should each box contain?

70. Ida's mother is on a diabetic diet. The doctor's prescription for her diet includes 99 g of carbohydrate, 91 g of protein, and 30 g of fat. Her breakfast and lunch for Saturday have already provided 66 g of carbohydrate, 61 g of protein, and 18 g of fat. The bread, vegetables, and dessert at dinner will provide 33 g of carbohydrate and 9 g of protein. The approximate food values of liver and bacon are (in grams per 100 grams of weight):

	Carbohydrate	Protein	Fat
Liver	0	25	8
Bacon	0	20	40

How many grams of bacon and how many grams of liver should Ida fix for her mother's dinner to balance her diet for the day? (Round all weights to the nearest gram.)

STATE YOUR UNDERSTANDING

71. Summarize the strategy for solving a system of equations by linear combinations.

72. When is it practical to solve by substitution? When is it better to use linear combinations?

CHALLENGE EXERCISES

Solve by linear combinations:

73. $\begin{cases} \dfrac{1}{a} + \dfrac{1}{b} = 4 \\ \dfrac{1}{a} - \dfrac{1}{b} = -6 \end{cases}$ Hint: Let $x = \dfrac{1}{a}$ and $y = \dfrac{1}{b}$

Rewrite the system as

$$\begin{cases} x + y = 4 \\ x - y = -6 \end{cases}$$

74. $\begin{cases} \dfrac{3}{a} + \dfrac{4}{b} = 6 \\ \dfrac{9}{a} + \dfrac{8}{b} = 11 \end{cases}$

75. Determine the values of a and b so that the graph of $ax + by = 1$ will contain the points $(2, 3)$ and $(-3, -5)$. Write the equation.

76. Determine the values of b and c so that the graph of $\frac{1}{2}x + by = c$ will contain the points (4, 9) and (6, 6). Write the equation.

MAINTAIN YOUR SKILLS (SECTIONS 2.5, 3.8, 4.6, 6.3, 6.4)

Solve:

77. $12x - 3 = 27$

78. $12x^2 = 108$

79. $3[2 - 3(x + 5)] - 18 = 0$

80. $4x^2 - 20x + 25 = 0$

81. $(x - 3)(x + 2) - (x + 5)(x - 1) = 4$

82. Graph the line with slope $\frac{4}{5}$ and y-intercept $(0, -3)$.

83. Graph the line with slope $-\frac{3}{2}$ and y-intercept $\left(0, \frac{3}{2}\right)$.

84. Sketch the graph of the inequality $5x - 3y \geq 15$.

CHAPTER 6
SUMMARY

Ordered Pair	A solution to an equation in two variables, x and y, is written as the ordered pair (x, y).	(p. 374)
Solve an Equation in Two Variables	To solve an equation in two variables given the value of one of the variables, substitute the known value in the equation, and solve for the other.	(p. 375)
Rectangular Coordinate System	A rectangular coordinate system is formed by placing two number lines at right angles. The intersection is called the origin.	(p. 381)
Plotting a Point	To plot a point, count from the origin right or left (positive or negative) the number of x-units then up or down (positive or negative) the number of y-units.	(p. 382)
Writing the Coordinates of a Point.	Write the number of units the point is from the y-axis (positive, right; negative, left) and the number of units the point is from the x-axis (positive, up; negative, down).	(pp. 383, 384)
Names for an Ordered Pair	(x, y) (x-value, y-value) (abscissa, ordinate) (x-coordinate, y-coordinate) (independent variable, dependent variable) (distance from y-axis, distance from x-axis)	(p. 384)
Graph of a Straight Line	To draw the graph of a straight line, given its equation, find three or more solutions to the equation, plot the solutions on a coordinate system, and connect the points with a straight line.	(p. 386)
x-intercept $(a, 0)$	The point where the line crosses the x-axis. Find the intercept by setting $y = 0$.	(pp. 396, 398)
y-intercept $(0, b)$	The point where the line crosses the y-axis. Find the intercept by setting $x = 0$.	(pp. 396, 397)
Slope of a Line	The slope of a line (m) is the rate of change in the vertical (rise) to the change in the horizontal (run) between any two points on a line.	(p. 398)
Formula for Slope	$\text{slope} = m = \frac{\text{rise}}{\text{run}} = \frac{\text{change in } y}{\text{change in } x}$, providing the change in x is not zero $m = \frac{y_2 - y_1}{x_2 - x_1}$, given the points $A\ (x_1, y_1)$ and $B\ (x_2, y_2)$ and $x_2 - x_1 \neq 0$	(p. 399)
Slope of a Horizontal Line	The slope of a horizontal line is zero.	(p. 401)
Slope of a Vertical Line	A vertical line has no slope.	(p. 401)

Slope-Intercept Form	$y = mx + b$ is the slope-intercept form of a linear equation, m is the slope, and $(0, b)$ is the y-intercept.	(p. 402)
Linear Inequality	$ax + by < c, \quad ax + by > c$ $ax + by \leq c, \quad ax + by \geq c$	(p. 412)
Half-Plane	A half-plane is formed on either side of a straight line when it is graphed on a rectangular coordinate system.	(p. 412)
Boundary Line	A line that divides a coordinate system into two half-planes.	(p. 412)
Graph of an Inequality	The graph of an inequality of the form $ax + by < c$ or $ax + by > c$ is a half-plane. To find the graph of the inequality, test a point in either half-plane. If the inequality is true for the point, all points in that half-plane form the graph; if the inequality is false, all points in the other half-plane form the graph. The graph of an inequality of the form $ax + by \leq c$ or $ax + by \geq c$ is a half-plane and the boundary line.	(pp. 413, 414)
System of Linear Equations	$\begin{cases} A_1x + B_1y = C_1 \\ A_2x + B_2y = C_2 \end{cases}$	(pp. 422, 423)
Dependent System	The graphs of the two equations are the same line. The solution set is $\{(x, y) \mid A_1x + B_1y = C_1\}$ or $\{(x, y) \mid A_2x + B_2y = C_2\}$.	(pp. 422, 423)
Independent System	The graphs of the two equations are not the same line. The solution set is $\emptyset$ or a single ordered pair, $\{(a, b)\}$.	(pp. 423, 424)
Consistent System	The two equations have at least one point in common. The solution set is a single ordered pair, $\{(a, b)\}$, or all points on the line, $\{(x, y) \mid A_1x + B_1y = C_1\}$.	(pp. 422, 424)
Inconsistent System	The two lines have no points in common (the graphs are parallel lines). The solution set is $\emptyset$.	(pp. 423, 424)
Solve a System by Substitution	To solve a system of linear equations by substitution: **1.** Solve either equation for x (or y). **2.** Substitute the expression for x (or y) in the other equation. **3.** Solve. **4.** Use this value of y (or x) to find the other value.	(p. 436)
Solve a System by Linear Combinations	To solve a system of linear equations by linear combinations: **1.** Multiply one or both of the equations by factors so that the coefficients of y (or x) are opposites. **2.** Add the two equations so that y (or x) is eliminated. **3.** Solve. **4.** Use this value of y (or x) to find the other value.	(p. 448)

CHAPTER 6
REVIEW EXERCISES

SECTION 6.1 Objective 1

Given the equation $3x - 5y = 15$, solve for y given the following values of x:

1. $x = 0$ **2.** $x = 5$ **3.** $x = 2$

4. $x = -6$ **5.** $x = -3.5$

Given the equation $6x + 2y = 16$, solve for x given the following values of y:

6. $y = 0$ **7.** $y = -4$ **8.** $y = 5$

9. $y = -\frac{4}{3}$ **10.** $y = -\frac{5}{2}$

SECTION 6.2 Objective 1

Construct a rectangular coordinate system, and plot the following points:

11. $(3, 5)$ **12.** $(-3, 2)$ **13.** $(5, -2)$ **14.** $(-6, -5)$ **15.** $(6, -5)$

Construct a rectangular coordinate system, and plot the following points:

16. $(-1.5, -2)$ **17.** $(-2, 2.5)$ **18.** $(-3, 3.5)$ **19.** $(0, -0.5)$ **20.** $(4.5, 6.5)$

SECTION 6.2 Objective 2

Identify the coordinates of the points on the following graph:

21. A **22.** B

23. C **24.** D

25. E **26.** F

27. G **28.** H

29. I **30.** J

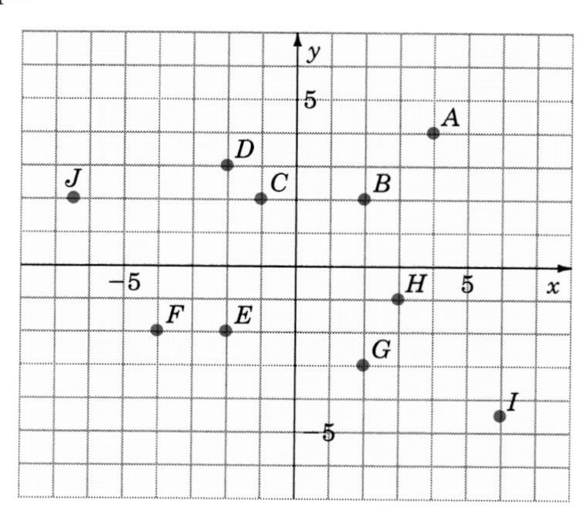

SECTION 6.2 Objective 3

Graph:

31. $y = -\dfrac{6}{5}x$

32. $y = \dfrac{8}{3}x$

33. $x + y = 8$

34. $x - y = 8$

35. $x - 6y = 6$

36. $x + 6y = 6$

37. $2x - 5y = 10$

38. $2x + 5y = 10$

39. $6x - 5y = 18$

40. $5x + 3y = 15$

SECTION 6.3 Objective 1

Find the x-intercept and the y-intercept of the graph of each of the following equations:

41. $x + y = 10$

42. $x - 3y = 9$

43. $5x + 6y = 15$

44. $3x - 4y = 20$

45. $4x + 7y = -28$

46. $2x - 9y = 18$

47. $\frac{1}{5}x - \frac{1}{9}y = -1$

48. $\frac{1}{6}x + \frac{1}{4}y = -\frac{1}{3}$

49. $0.2x + 1.2y = 2.4$

50. $0.7x - 1.4y = 2.8$

SECTION 6.3 Objective 2

Find the slope of the line containing the given pair of points:

51. $(-2, 3), (5, 4)$

52. $(0, -3), (5, -1)$

53. $(0, -1), (6, -5)$

54. $(-3, -5), (-1, -1)$

55. $(-2, 7), (5, 10)$

56. $(-6, 8), (3, -2)$

57. $\left(\frac{2}{3}, \frac{1}{2}\right), \left(\frac{1}{3}, -\frac{3}{2}\right)$

58. $\left(\frac{2}{3}, -\frac{3}{4}\right), \left(1, -\frac{1}{4}\right)$

59. $(2.5, -1.6), (3, -2.5)$

60. $(-1.5, -3), (-1.5, 0.3)$

SECTION 6.3 Objective 3

Find the slope of the graph of each of the following equations:

61. $2x - 3y = 7$

62. $3x + 4y = -9$

63. $2x - y = 8$

64. $5x + y = 8$

65. $5x + 3y = 10$

66. $-2x + 6y = 7$

67. $\frac{1}{3}x - \frac{1}{5}y = -1$

68. $\frac{2}{5}x - \frac{1}{7}y = -2$

69. $\frac{2}{5}x + 5y = 9$

70. $8x - \frac{3}{4}y = 1$

SECTION 6.3 Objective 4

Draw the graph of each of the following equations using the slope and the y-intercept:

71. $4x + 5y = 20$

72. $2x + 3y = 9$

73. $7x - 3y = 18$

74. $x - 2y = 4$

75. $2x - 3y = 6$

76. $5x - 6y = 30$

77. $3x - 2y = 10$

78. $x + 3y = 9$

79. $5x + 4y = -12$

80. $2x - 6y = 12$

SECTION 6.4 **Objective 1**

Graph each inequality:

81. $x + y \geq -3$

82. $x - y \leq 5$

83. $2x - y < -6$

84. $x + 3y < -6$

85. $2x - 3y \leq -9$

86. $5x + 3y \leq -15$

87. $\frac{1}{3}x - y < -1$

88. $\frac{1}{4}x - \frac{1}{2}y > -2$

Graph both inequalities on the same coordinate system, and double shade the region that is part of both graphs:

89. $3x - y \leq 6$ and $2x - y > -1$

90. $x + 4y > -8$ and $8x + 3y > 9$

SECTION 6.5 Objective 1

Solve by graphing:

91. $\begin{cases} y = \dfrac{1}{2}x - 6 \\ y = \dfrac{3}{2}x - 10 \end{cases}$

92. $\begin{cases} 2x - y = 3 \\ x - y = -3 \end{cases}$

93. $\begin{cases} 3x + 2y = 1 \\ 2x - 5y = 7 \end{cases}$

94. $\begin{cases} 2x + 5y = -1 \\ x + 2y = 0 \end{cases}$

SECTION 6.6 Objective 1

Solve by substitution:

95. $\begin{cases} y = 2x - 1 \\ 2x + y = 11 \end{cases}$

96. $\begin{cases} y = 3x \\ 2x - 3y = 7 \end{cases}$

97. $\begin{cases} x - y = 8 \\ 2x + 3y = -9 \end{cases}$

98. $\begin{cases} x + y = -1 \\ 5x + 4y = 2 \end{cases}$

99. $\begin{cases} 2x - y = 5 \\ 3x - 2y = 6 \end{cases}$

100. $\begin{cases} x - 4y = 0 \\ 3x - 2y = 20 \end{cases}$

101. $\begin{cases} x - 7y = -5 \\ 2x - 5y = 8 \end{cases}$

102. $\begin{cases} 3x + y = -5 \\ 2x + 5y = 14 \end{cases}$

103. $\begin{cases} 4x + 5y = 9 \\ x + 3y = -3 \end{cases}$

104. $\begin{cases} 2x + 7y = -4 \\ 3x - y = 17 \end{cases}$

SECTION 6.7 Objective 1

Solve by linear combinations:

105. $\begin{cases} 2x - 3y = 11 \\ 5x + y = 19 \end{cases}$

106. $\begin{cases} 5x + 4y = 2 \\ 3x - 2y = -12 \end{cases}$

107. $\begin{cases} 5x + 2y = 3 \\ 3x - 4y = 33 \end{cases}$

108. $\begin{cases} 2x - 7y = 17 \\ 5x + y = 24 \end{cases}$

109. $\begin{cases} 2x + 3y = 23 \\ 3x - 2y = 15 \end{cases}$

110. $\begin{cases} 4x + 3y = 4 \\ 2x - 5y = 54 \end{cases}$

111. $\begin{cases} 9x - 2y = 2 \\ 3x + 8y = 70 \end{cases}$

112. $\begin{cases} 8x - 5y = -4 \\ 2x + 15y = -66 \end{cases}$

113. $\begin{cases} 5x + 11y = 19 \\ 2x - 3y = -22 \end{cases}$

114. $\begin{cases} 7x - 6y = -7 \\ 3x - 5y = -20 \end{cases}$

CHAPTER 6
TRUE–FALSE CONCEPT REVIEW

Check your understanding of the language of algebra. Tell whether or not each of the following is true (always true) or false (not always true).

1. Every linear equation in two variables has exactly two solutions.
2. A single solution for a linear equation in two variables is called an ordered pair.
3. To find a solution for the linear equation $ax + by = c$, $a \neq 0$, $b \neq 0$, we may choose any value we like for x and solve for the corresponding value of y.
4. On a rectangular coordinate system, the point where the x-axis and y-axis intersect is called the ordinate.
5. The point with abscissa, 2, and ordinate, -4, lies in the second quadrant.
6. The distance of a point from the x-axis is the number of units in the y-coordinate.
7. The ordered pair $\left(\frac{3}{2}, -1\right)$ is one of the solutions of $2y + 3x = 0$.
8. The point with coordinates $(-1, -5)$ is located one unit below the x-axis.
9. The graph of every equation of the form $ax + by = c$ (where a and b are not both zero) is a straight line.
10. A line that slopes downward to the left has a positive slope.
11. Every line has exactly one x-intercept and one y-intercept.
12. The rise of a line is the change in its vertical direction.
13. The slope of a line is a ratio.
14. The coordinates of the y-intercept of a line is the point whose ordinate is 0.
15. A vertical line has slope 0.
16. A horizontal line has no slope.
17. The line with equation $x = \frac{6}{7}y - 4$ has slope $\frac{7}{6}$.

18. The line with equation $x = \frac{6}{7}y - 4$ has x-intercept $(-4, 0)$.

19. The ordered pair $(-3, -6)$ is one of the solutions of the linear inequality $3x - 6y > 0$.

20. The graph of $ax + by < c$, a and b not both zero, is always a half-plane.

21. Two equations each containing the same two variables is an example of a system of equations.

22. An independent and consistent system has an infinite number of solutions.

23. The graph of a dependent system of two linear equations is one straight line.

24. The solution set of an inconsistent system is $\emptyset$.

25. A system of linear equations must contain two or more variables.

26. The graph of an inconsistent system of linear equations with two variables is a pair of parallel lines.

27. If a contradiction (false equation) results from solving a system of equations by substitution, the solution set of the system is empty.

28. If an identity results from solving a system of equations by additon, the solution set contains all pairs of real numbers.

29. Graphically, the solution of a system of linear equations is the set of coordinates of the intersection of the graphs of the equations.

30. The ordered pair $(0, 0)$ cannot be the solution of a linear system of equations.

CHAPTER 6
TEST

1. Identify the coordinates of the points on the graph to the right:

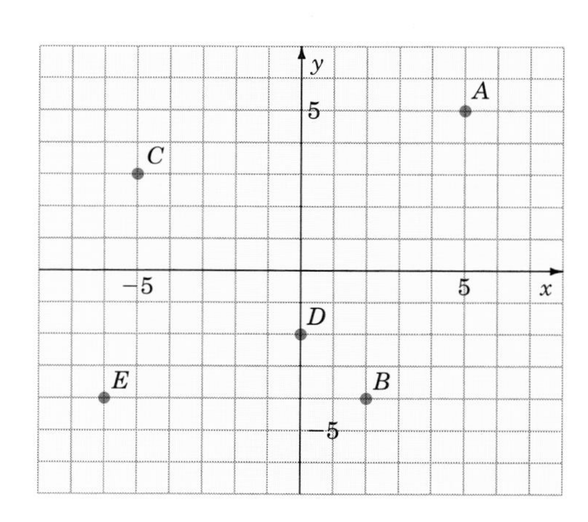

2. Draw the graph of the line through $(0, -3)$ that has slope $m = \frac{2}{3}$.

3. Solve the system by graphing: $\begin{cases} y = -x - 10 \\ y = 2x + 2 \end{cases}$

4. Given the equation $4x - 5y + 20 = 0$, solve for x if $y = -8$.

5. Given the equation $4x - 5y + 20 = 0$, solve for y if $x = -10$.

6. Solve the system by linear combinations:

$$\begin{cases} 2x + 3y = 5 \\ x - y = 7 \end{cases}$$

7. Find the x-intercept and the y-intercept of the graph of the equation $2x - 5y = 15$.

8. Solve the system by substitutions: $\begin{cases} x = 2y \\ x - y = 3 \end{cases}$

9. Draw the graph of the inequality $2x - y > -2$.

10. Solve the system by linear combinations:

$$\begin{cases} \frac{2}{3}x - y = 0 \\ x + \frac{1}{2}y = -5 \end{cases}$$

11. Draw the graph of $x + 2y = 6$.

12. Draw the graph of $y = -\frac{3}{4}x + 2$.

13. Solve the system by substitution: $\begin{cases} y = 2x + 3 \\ 3x - 2y = 7 \end{cases}$

14. Find the slope of the line that passes through the points $(3, -8)$ and $(-2, -5)$.

15. Write the slope of the line whose equation is $3x = 7y - 6$.

16. Locate the following points on a coordinate system: $A(-3, 3)$, $B(4, -3)$, $C(0, 3.5)$, $D(-1.5, -4)$, $E(2, 4.5)$

17. Solve the system: $\begin{cases} x - 2y = -2 \\ 3x - y = 20 \end{cases}$

18. Draw the graph of the inequality $3x - 6y \geq -4$.

19. When the second of two numbers is added to three times the first, the sum is 7. When the first number is subtracted from three times the second, the difference is 1. What are the numbers?

20. The Green Thumb seed store has two mixtures of lawn seed containing bluegrass. One contains 45% bluegrass, and the other contains 60% bluegrass. How many pounds of each are needed to make 15 pounds of a new mixture that is 50% bluegrass?

CHAPTER

7

ROOTS, RADICALS, AND RELATED EQUATIONS

SECTIONS

The lens setting of a camera can be computed by using the formula $\ell\sqrt{s} = k$, where ℓ represents the lens setting, s represents the shutter speed, and k is a constant of variation. In Exercise 65, Section 7.1, the formula is used to find the lens setting for a shutter speed of $\frac{1}{16}$ second. The first set of given data, a shutter speed of $\frac{1}{64}$ second and a lens setting of 16, is used to find the constant of variation. *(Gary Crallé/The Image Bank)*

PREVIEW

The inverse of squaring a number is the operation of finding the square root of a number. In Section 7.1, we show the use of a radical sign to designate the positive square root of a positive number. This leads to a consideration of numbers that are not rational and to the mention of the existence of numbers that are not real numbers. Section 7.2 illustrates the procedure for simplifying radical expressions, and this leads to the operations on radical expressions in Sections 7.3, 7.4, and 7.5. Section 7.6 concludes the chapter with an explanation of the use of the squaring property of equality to solve radical equations and the necessity of checking for extraneous roots.

7.1 ROOTS AND RADICALS

OBJECTIVES

1. Find the square roots of positive numbers.
2. Find the *n*th roots of positive numbers.

To find the square of 15, we write 15^2 and multiply $(15)(15)$.

$15^2 = (15)(15) = 225$

The opposite procedure is to find a square root of a number.

DEFINITION

Square Root

A *square root* of m is a, where $a^2 = m$.

The square roots of 25 are $+5$ and -5 since the square of both is 25.

$(5)^2 = 25 \quad \text{and} \quad (-5)^2 = 25$

A radical sign is used to symbolize square roots.

DEFINITION

Radical

A *radical,* $\sqrt{\ }$, is the symbol used to denote the *positive square root* or *principal square root* of a number. The symbol for the negative square root is $-\sqrt{\ }$.

Thus,

$\sqrt{25} = 5, \qquad -\sqrt{25} = -5, \quad \text{and} \quad \sqrt{225} = 15.$

The following table illustrates some numbers that have rational square roots.

Number	Two Equal Factors			Square Roots
64	$(8)(8)$	or	$(-8)(-8)$	8 and -8
100	$(-10)(-10)$	or	$(10)(10)$	-10 and 10
$\frac{1}{4}$	$\left(\frac{1}{2}\right)\left(\frac{1}{2}\right)$	or	$\left(-\frac{1}{2}\right)\left(-\frac{1}{2}\right)$	$\frac{1}{2}$ and $-\frac{1}{2}$
0.25	$(0.5)(0.5)$	or	$(-0.5)(-0.5)$	0.5 and -0.5

The expression $\sqrt{64}$ is read "the positive square root of 64" or "radical 64."

The expression $-\sqrt{0.25}$ is read "the opposite of the positive square root of 0.25" or "the negative square root of 0.25."

DEFINITION

Radicand

A *radicand* is a number or expression that occurs under a radical.

The radicands of $\sqrt{7}$, $\sqrt{3x}$, and $\sqrt{9x^2y}$ are 7, $3x$, and $9x^2y$ respectively.

The following equations are identities in which a radical sign appears:

$\sqrt{64} = 8$ $\qquad$ $-\sqrt{64} = -8$

$\sqrt{100} = 10$ $\qquad$ $-\sqrt{100} = -10$

$\sqrt{\frac{1}{4}} = \frac{1}{2}$ $\qquad$ $-\sqrt{\frac{1}{4}} = -\frac{1}{2}$

$\sqrt{121} = 11$ $\qquad$ $-\sqrt{121} = -11$

$\sqrt{9} + \sqrt{16} = 3 + 4 = 7$ $\qquad$ $\sqrt{9 + 16} = \sqrt{25} = 5$

Note, that in $\sqrt{9 + 16}$, the radicand must be simplified before determining the square root.

Whole numbers that have integers for square roots are called *perfect squares*. The numbers 64 and 100 are perfect squares. Rational numbers that are ratios of perfect squares have rational square roots, for example,

$$\sqrt{\frac{4}{9}} = \frac{2}{3} \quad \text{and} \quad \sqrt{\frac{25}{36}} = \frac{5}{6}.$$

Not all rational numbers have rational square roots. For example,

$$\sqrt{2}, \quad \sqrt{10}, \quad \sqrt{\frac{1}{3}}, \quad \text{and} \quad \sqrt{\frac{7}{10}}$$

do not represent rational numbers.

EXAMPLE 1

Find the root: $\sqrt{144}$

$\sqrt{144} = 12$

The answer is 12 because $12 \cdot 12 = 144$. The value, 12, can be found by trial and error or by use of a calculator. ■

EXAMPLE 2

Find the root: $\sqrt{289}$

$\sqrt{289} = 17$

The answer is 17 because $17 \cdot 17 = 289$. The value, 17, can be found by trial and error or by use of a calculator. ■

EXAMPLE 3

Find the root: $-\sqrt{36}$

$-\sqrt{36} = -6$

■ **CAUTION**

Remember that $-\sqrt{\ }$ indicates the negative square root.

■

EXAMPLE 4

Find the root: $-\sqrt{\frac{9}{25}}$

$$-\sqrt{\frac{9}{25}} = -\frac{3}{5}$$

■

Square roots of positive numbers, including the square roots of numbers that are not perfect squares, can all be located on the number line. All numbers that can be pictured on the number line are called *real numbers*. For this reason, our usual number line is called the *real number line* (see Figure 7.1).

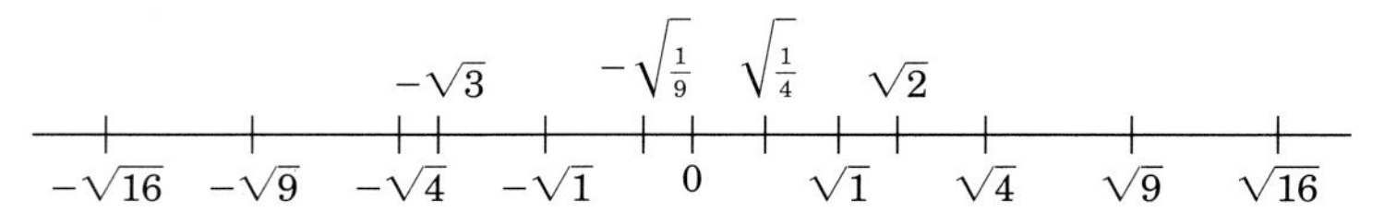

FIGURE 7.1

Square roots of positive numbers that are not perfect squares or ratios of perfect squares are part of a set of numbers called *irrational numbers*. With these new numbers we can expand our classification diagram from Chapter 1, Section 1.4.

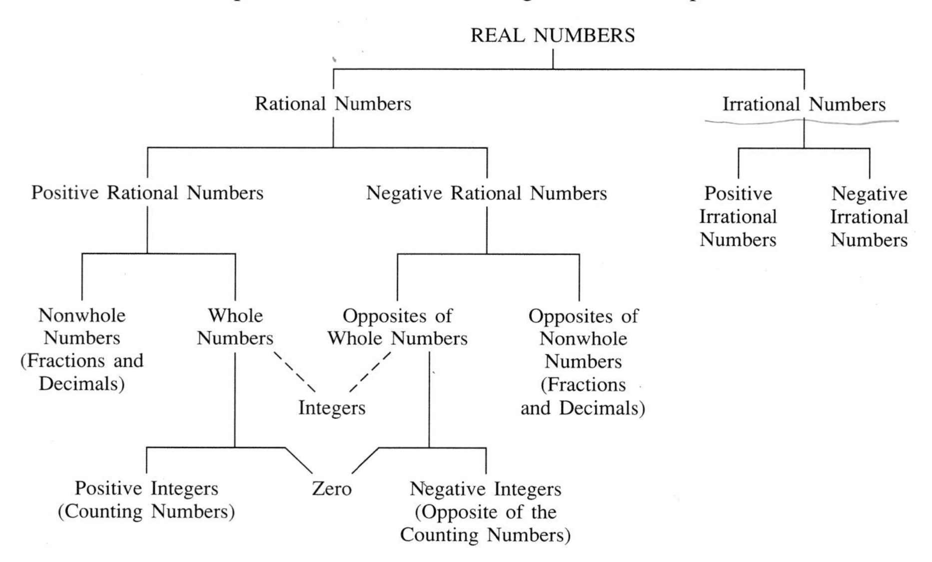

The square roots of negative rational numbers cannot be pictured on the real number line because they are not negative, zero, or positive. For example, the square root of -49 is not positive because the product of two positive numbers cannot be -49.

$\sqrt{-49} \neq 7$ because $(7)(7) \neq -49$

Nor can it be negative or zero.

$\sqrt{-49} \neq -7$ or 0 because $(-7)(-7) \neq -49$ and $(0)(0) \neq -49$

The square roots of negative numbers are not real numbers.

PROPERTY

> If x is positive, $\sqrt{-x}$ is not a real number.

More information about these numbers can be found in Chapter 8.

Since a radical, such as $\sqrt{a}$, $\sqrt{100}$, and $\sqrt{x^2}$, is a symbol for a *positive* square root, it is sometimes necessary to restrict variables when they are included in radical expressions. For instance, because x^2 is never negative, we must say

$\sqrt{x^2} = x$ is true *only if* x is positive or zero, and

$\sqrt{x^2} = -x$ is true *only if* x is negative or zero.

The following table, made with the help of a calculator, shows a way to approximate irrational square roots:

$\sqrt{35}$		
$\sqrt{35}$ is between 5 and 6	$5^2 = 25$	$6^2 = 36$
$\sqrt{35}$ is between 5.9 and 6	$5.9^2 = 34.81$	$6^2 = 36$
$\sqrt{35}$ is between 5.91 and 5.92	$5.91^2 = 34.9281$	$5.92^2 = 35.0464$
$\sqrt{35}$ is between 5.916 and 5.917	$5.916^2 = 34.999056$	$5.917^2 = 35.010889$

In a higher mathematics course, we learn that a table like this has no end. When dealing with such a number, we use either the exact value ($\sqrt{35}$) or an approximation.

Exact Value	**Some Approximations**
$\sqrt{35}$	$\sqrt{35} \approx 6$
	$\sqrt{35} \approx 5.92$
	$\sqrt{35} \approx 5.91608$

Roots and approximations can be found by using a calculator with a square root key or by using a square root table.

EXAMPLE 5

Approximate the root to three decimal places: $\sqrt{88}$

$\sqrt{88} \approx 9.381$

Approximation can be found from a table or with a calculator. Recall that "$\approx$" means approximately equal. ■

EXAMPLE 6

Find the root: $\sqrt{-81}$

$\sqrt{-81}$ is not a real number.

■ **CAUTION**

This is not a real number because the radicand is negative.

■

EXAMPLE 7

Find the root: $\sqrt{3025}$

$\sqrt{3025} = 55$ Positive square root found by calculator. ■

EXAMPLE 8

Approximate the root to four decimal places: $\sqrt{412}$

$\sqrt{412} \approx 20.2978$ Approximation found by calculator and rounded. ■

EXAMPLE 9

The lens setting of a camera can be computed by using the formula $\ell\sqrt{s} = k$, where ℓ represents the lens setting, s represents the shutter speed, and k is a constant of variation. If the speed (s) is 0.01 second for a setting of 16, what is the lens setting for a speed of $\frac{1}{25}$ second?

Formula:

$\ell\sqrt{s} = k$

Substitute:

$16(\sqrt{0.01}) = k$ First, we find the value of k by using $\ell = 16$, $s = 0.01$.

Solve:

$16(0.1) = k$ Since $\sqrt{0.01} = 0.1$.

$1.6 = k$

Formula:

$\ell\sqrt{s} = k$ Using $k = 1.6$, we can now solve for ℓ when $s = 0.04$.

Substitute:

$\ell(\sqrt{0.04}) = 1.6$ $s = 0.04$ and $k = 1.6$.

Solve:

$\ell(0.2) = 1.6$ Since $\sqrt{0.04} = 0.2$.

$\ell = 8$

Answer:

The lens setting is 8. ■

EXAMPLE 10

Linda has entered a Grand Prix race in which the radius of one of the curves is 90 feet. She knows that the formula $V = \sqrt{2.5r}$, where V represents speed in miles per hour and r represents the radius of the curve in feet, can be used to find the approximate maximum speed at which a car can go through the curve without skidding. How fast can Linda drive through the curve without skidding?

Formula:

$V = \sqrt{2.5r}$

Substitute:

$V = \sqrt{2.5(90)}$ $\quad r = 90$

Solve:

$V = \sqrt{225}$

$= 15$

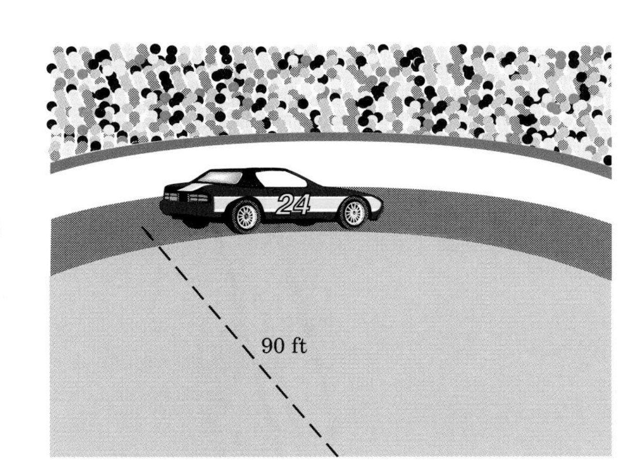

Answer:

Linda can take the curve at 15 miles per hour without skidding. ■

We use nth root radicals to represent roots greater than square roots.

■ DEFINITION

nth Root Index

An *nth root* of m is a, where $a^n = m$. The radical $\sqrt[n]{}$ is used to express such roots. n is called the *index* of the radical.

The table illustrates some nth roots.

Root	Value	Reasoning
Cube root	$\sqrt[3]{8} = 2$	$2^3 = 2 \cdot 2 \cdot 2 = 8$
Cube root	$\sqrt[3]{-27} = -3$	$(-3)^3 = (-3)(-3)(-3) = -27$
Fourth root	$\sqrt[4]{2401} = 7$	$7^4 = 7 \cdot 7 \cdot 7 \cdot 7 = 2401$
Fifth root	$\sqrt[5]{-32} = -2$	$(-2)^5 = -32$
Fifth root	$\sqrt[5]{0} = 0$	$0^5 = 0$
Fourth root	$\sqrt[4]{-16}$, Not a real number	

In general,

PROPERTY

$(\sqrt[n]{x})^n = x$, where $x \geq 0$ and n is an integer greater than 1.

As illustrated in the table, we can find odd roots for both positive and negative numbers. However, even roots of negative numbers are not real numbers (see Chapter 8).

EXAMPLE 11

Find the root: $\sqrt[3]{125}$

$\sqrt[3]{125} = 5$ Because $5^3 = 125$.

Calculators use a variety of labels for the nth root key. You may find one of these on your scientific calculator: $\boxed{x^{1/y}}$, $\boxed{y^{1/x}}$, $\boxed{\sqrt[x]{\ }}$, or $\boxed{\sqrt[x]{y}}$.

EXAMPLE 12

Find the root: $\sqrt[4]{1296}$

$\sqrt[4]{1296} = 6$ Because $6^4 = 1296$. The number, 6, can be found by trial and error or by use of a calculator.

The symbols, $\sqrt{x}$ and $\sqrt[2]{x}$, have the same meaning. The index number 2 is not written when the radical indicates square root.

EXERCISE 7.1

A

Find the roots:

1. $\sqrt{4}$ **2.** $-\sqrt{4}$ **3.** $\sqrt{25}$ **4.** $\sqrt{100}$

5. $-\sqrt{81}$ **6.** $-(-\sqrt{49})$ **7.** $-\sqrt{196}$ **8.** $\sqrt{256}$

9. $\sqrt{\frac{9}{25}}$ **10.** $-\sqrt{\frac{49}{100}}$ **11.** $\sqrt{\frac{4}{169}}$ **12.** $-\left(-\sqrt{\frac{25}{289}}\right)$

13. $\sqrt{0.64}$ **14.** $-\sqrt{0.36}$ **15.** $\sqrt{2.25}$ **16.** $-(-\sqrt{5.29})$

17. $\sqrt[3]{27}$ **18.** $\sqrt[3]{1}$ **19.** $\sqrt[4]{0}$ **20.** $\sqrt[4]{16}$

B

Find the roots:

21. $\sqrt{676}$ **22.** $\sqrt{1024}$ **23.** $-(-\sqrt{400})$ **24.** $\sqrt{900}$

25. $-\sqrt{961}$ **26.** $-\sqrt{1089}$ **27.** $\sqrt{\frac{169}{225}}$ **28.** $\sqrt{\frac{196}{361}}$

29. $\sqrt{\frac{121}{225}}$ **30.** $\sqrt{\frac{324}{441}}$ **31.** $-\sqrt{\frac{484}{289}}$ **32.** $\sqrt{\frac{676}{576}}$

33. $-\sqrt{6.76}$ **34.** $-\sqrt{11.56}$ **35.** $\sqrt{0.0144}$ **36.** $\sqrt{0.0169}$

37. $\sqrt[7]{-1}$ **38.** $\sqrt[3]{-64}$ **39.** $\sqrt[4]{-4}$ **40.** $\sqrt[4]{-1}$

C

Find the approximate root rounded to two decimal places:

41. $\sqrt{7}$ **42.** $\sqrt{11}$ **43.** $-\sqrt{35}$ **44.** $-\sqrt{37}$

45. $\sqrt{200}$ **46.** $\sqrt{201}$ **47.** $\sqrt{340}$ **48.** $\sqrt{345}$

49. $-\sqrt{892}$ **50.** $-\sqrt{596}$ **51.** $\sqrt{0.4}$ **52.** $\sqrt{1.6}$

53. $\sqrt{14.4}$ **54.** $-\sqrt{2.26}$ **55.** $\sqrt{10.9}$ **56.** $-\sqrt{16.4}$

Find the roots:

57. $\sqrt[3]{1331}$ **58.** $\sqrt[3]{-1000}$ **59.** $\sqrt[5]{-3125}$ **60.** $\sqrt[5]{7776}$

D

61. If the radius (r) of the curve of a road is 250 ft, what is the maximum speed (v), in miles per hour, at which an auto can travel through the curve before starting to skid? ($v = \sqrt{2.5r}$)

62. If the radius (r) of a curve of a road is 1000 ft, what is the maximum speed (v), in miles per hour, at which an auto can travel through the curve before starting to skid? ($v = \sqrt{2.5r}$)

63. The radius of a circle can be expressed as $r = \sqrt{\frac{A}{\pi}}$. Find r if $A = 18$ square centimeters. Use $\pi \approx 3.14$. Round your answer to two decimal places.

64. Using the formula in Exercise 63, find r if $A = 35$ square inches. Use $\pi \approx 3.14$, and round your answer to two decimal places.

65. Use the formula $\ell\sqrt{s} = k$ to find the lens setting on a camera. If the speed is $\frac{1}{64}$ sec for a setting of 16, what should be the lens setting for a speed of $\frac{1}{16}$ sec?

66. If the speed of another camera is $\frac{1}{64}$ sec for a setting of 4, what should be the lens setting for a speed of $\frac{1}{49}$ sec?

67. To find the radius of a cylinder, the formula is $r = \sqrt{\frac{V}{\pi h}}$. Find r if $V = 95$ cubic inches and $h = 7$ inches. Use $\pi \approx 3.14$, and round your answer to two decimal places.

68. Using the formula in Exercise 67, find r if $V = 120$ cubic inches and $h = 7$ inches. Use $\pi \approx 3.14$, and round your answer to two decimal places.

STATE YOUR UNDERSTANDING

69. In Example 8 the calculator was used to approximate $\sqrt{412}$. How can we check the answer? Check it.

70. In the property $(\sqrt[n]{x})^n = x$, $x \geq 0$ and n is an integer greater than 1, why is it necessary that $x \geq 0$?

CHALLENGE EXERCISES

Find the roots:

71. $\sqrt{x^4}$

72. $\sqrt[3]{y^{15}}$

73. $\sqrt{\frac{x^6}{y^6}}$

74. $\sqrt[3]{-\frac{y^9}{x^{12}}}$

MAINTAIN YOUR SKILLS (SECTION 6.1)

Given the equation $y = \frac{3}{5}x + 5$, solve for y given the following values of x:

75. $x = 15$

76. $x = -20$

77. $x = -3$

Given the equation $6x - 11y = 22$, solve for y, given the following values of x:

78. $x = 22$

79. $x = 10$

80. $x = -5$

81. The perimeter of a rectangle is 84 ft. If its length is 32 ft, how wide is it?

82. A certain number is five more than six times a second number. If the second number is -4, what is the first number?

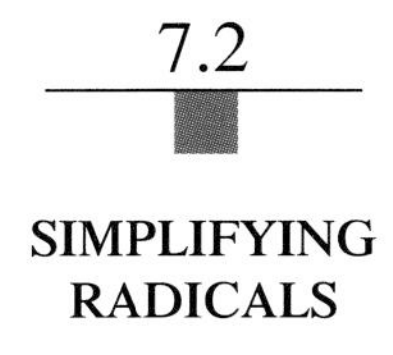

SIMPLIFYING RADICALS

OBJECTIVE

1. Simplify a radical expression.

Irrational numbers such as $\sqrt{15}$ are usually written in one of two ways:

1. We can find an approximation, which is the usual procedure in a practical problem.

$\sqrt{15} \approx 3.87$

2. We can leave the number in radical form, which represents its exact value.

DEFINITION

Radical Expressions

Expressions such as $\sqrt{14xy}$ are called *radical expressions*.

Examples of radical expressions include

$$\sqrt{x}, \quad \sqrt{x^2+8}, \quad \sqrt{\frac{x^2-16}{2}}, \quad \sqrt{14}, \quad \sqrt{64}.$$

DEFINITION

Simplest Radical Form

A radical expression (positive square root) is said to be in *simplest radical form* when:

1. Whole numbers in the radicand have no factors that are perfect squares.
2. Variables in the radicand have no perfect square factors.

The radical expression $\sqrt{18}$ is not in simplest radical form, whereas the expression $3\sqrt{2}$ is in simplest radical form. To simplify a radical, we use the basic principle of radicals.

RULE

Basic Principle of Radicals

The square root of a product is the product of the square roots if neither factor is negative.

$\sqrt{ab} = \sqrt{a}\,\sqrt{b}$, if a and b are not negative

$\sqrt{4100} = \sqrt{100} \cdot \sqrt{41} = 10\sqrt{41}$

A radical is in simplified form when the radicand is an expression that has no perfect-square factors except 1. Radicals with fractional radicands are covered in Section 7.4.

PROCEDURE

To simplify a radical (positive square root) with a whole-number radicand:

1. If possible, factor the radicand so that at least one factor is a perfect square. If the radicand has no perfect-square factors other than 1, it is already in simplest form.
2. Find the square root of the perfect square.
3. Repeat steps 1 and 2 until the radicand is simplified.

When a radical contains a variable factor, we must be careful that it represents a positive number (or zero) so that we can use the previously stated principle for radicals.

$\sqrt{x^3} = \sqrt{x^2} \cdot \sqrt{x} = x\sqrt{x}$, if x represents a nonnegative number

Throughout the remainder of this text, we will assume that variables in the radicand are restricted so that the radicand does not represent a negative number.

EXAMPLE 1

Simplify: $\sqrt{8}$

$\sqrt{8} = \sqrt{4} \cdot \sqrt{2}$
$= 2\sqrt{2}$

Factor 8 by looking for the largest perfect square factor of 8. The factor is 4, so $8 = 4 \cdot 2$. The expression $2\sqrt{2}$ is called the simplest radical form.

or

$\sqrt{8} = \sqrt{2^2 \cdot 2}$
$= \sqrt{2^2} \cdot \sqrt{2}$
$= 2\sqrt{2}$

We factor the radicand using exponents. The square root of 2^2 is 2.

When the radicand contains larger numbers, we can use prime factors to help simplify more quickly.

EXAMPLE 2

Simplify: $\sqrt{392}$

$\sqrt{392} = \sqrt{2^3 \cdot 7^2}$ — The prime factors of 392 are $2 \cdot 2 \cdot 2 \cdot 7 \cdot 7$. These are found by repeated division.

$= \sqrt{2^2}\sqrt{7^2}\sqrt{2}$

$= 2 \cdot 7\sqrt{2}$ — Simplify.

$= 14\sqrt{2}$ ■

Radicands with variables can be simplified by segregating those with even exponents.

EXAMPLE 3

Simplify: $\sqrt{x^{16}}$

$\sqrt{x^{16}} = x^8$ — The radicand has an even exponent, 16, so we write the expression whose square is x^{16}: $x^8 \cdot x^8 = x^{16}$.

■ **CAUTION**

Do not take the square root of the exponent, $\sqrt{x^{16}} \neq x^4$.

■

When a radicand contains both numerals and variables, separate them.

EXAMPLE 4

Simplify: $\sqrt{50y^2}$

$\sqrt{50y^2} = \sqrt{25} \cdot \sqrt{y^2} \cdot \sqrt{2}$ — We write the factors in the following order: the largest perfect square, $\sqrt{25}$, the variable with the largest even exponent, $\sqrt{y^2}$, the rest of the factors, $\sqrt{2}$.

$= 5y\sqrt{2}$ ■

EXAMPLE 5

Simplify: $\sqrt{288y^7}$ Assume that y represents a positive number.

$\sqrt{288y^7} = \sqrt{144} \cdot \sqrt{y^6} \cdot \sqrt{2y} = 12y^3\sqrt{2y}$ ■

EXAMPLE 6

Simplify: $\sqrt{980x^3y^9}$ Assume that x and y represent positive numbers.

$\sqrt{980x^3y^9} = \sqrt{196} \cdot \sqrt{x^2} \cdot \sqrt{y^8} \cdot \sqrt{5xy}$
$= 14xy^4\sqrt{5xy}$

If you did not notice at first that the perfect square, 196, is a factor of 980, the radical can also be simplified as follows:

$\sqrt{4} \cdot \sqrt{x^2} \cdot \sqrt{y^8} \cdot \sqrt{245xy}$
$= 2xy^4\sqrt{245xy}$
$= 2xy^4 \cdot \sqrt{49} \cdot \sqrt{5xy}$
$= 14xy^4\sqrt{5xy}$ ■

Not every radical can be simplified.

EXAMPLE 7

Simplify: $\sqrt{78}$

$\sqrt{78} = \sqrt{2} \cdot \sqrt{3} \cdot \sqrt{13}$ None of the factors are perfect squares.

$\sqrt{78}$ is already in simplest form. ■

In an application involving radicals, it is customary to use an approximation. In Example 8 we show both the exact and the approximate values.

EXAMPLE 8

A city lot has an area of 5600 square feet. If the lot is a square, what is the length of one side? The formula is

$s = \sqrt{A}$,

where s represents the length of one side, and A represents the area.

Formula:

$s = \sqrt{A}$

Substitute:

$s = \sqrt{5600}$ $A = 5600$

Solve:

$s = \sqrt{100} \cdot \sqrt{56}$
$= 10\sqrt{56}$
$= 10\sqrt{4} \cdot \sqrt{14}$
$= 10 \cdot 2\sqrt{14}$
$= 20\sqrt{14}$

Or $s = \sqrt{400} \cdot \sqrt{14}$
$s = 20\sqrt{14}$.

Answer:

The square is $20\sqrt{14}$ feet, or approximately 74.83 feet on each side. $\sqrt{5600} \approx 74.83$, to the nearest hundredth. ■

EXERCISE 7.2

A

Simplify; assume that all variables represent positive numbers:

1. $\sqrt{8}$ **2.** $\sqrt{27}$ **3.** $\sqrt{24}$ **4.** $\sqrt{32}$

5. $\sqrt{40}$ **6.** $\sqrt{50}$ **7.** $\sqrt{26}$ **8.** $\sqrt{34}$

9. $\sqrt{16c^2}$ **10.** $\sqrt{100x^2}$ **11.** $\sqrt{a^2b}$ **12.** $\sqrt{ab^2}$

1 . $\sqrt{32x^2y^2}$ **14.** $\sqrt{45w^4z^4}$ **15.** $\sqrt{121a^3b^6}$ **16.** $\sqrt{144r^6s^3}$

17. $\sqrt{5x}$ **18.** $\sqrt{13y}$ **19.** $\sqrt{5w^2}$ **20.** $\sqrt{25w}$

B

Simplify; assume that all variables represent positive numbers:

21. $\sqrt{80}$ **22.** $\sqrt{125}$ **23.** $\sqrt{45}$ **24.** $\sqrt{128}$

25. $\sqrt{44}$ **26.** $\sqrt{98}$ **27.** $\sqrt{200}$ **28.** $\sqrt{300}$

29. $\sqrt{72x^2}$ **30.** $\sqrt{99y^2}$ **31.** $\sqrt{27t^3}$ **32.** $\sqrt{125s^2}$

33. $\sqrt{288x^2y^4}$ **34.** $\sqrt{242w^6}$ **35.** $\sqrt{300x^4}$ **36.** $\sqrt{405a^8}$

37. $\sqrt{200x^4y^3}$ **38.** $\sqrt{144a^3b^8}$ **39.** $\sqrt{75r^4s^5}$ **40.** $\sqrt{90p^5r^8}$

C

Simplify; assume that all variables represent positive numbers:

41. $\sqrt{192}$ **42.** $\sqrt{1200}$ **43.** $\sqrt{245}$ **44.** $\sqrt{294}$

45. $\sqrt{198w^6}$ **46.** $\sqrt{1125}$ **47.** $\sqrt{150w^2}$ **48.** $\sqrt{156rs}$

49. $\sqrt{156a^4}$ **50.** $\sqrt{148b^4}$ **51.** $\sqrt{245x^8y^5}$ **52.** $\sqrt{108b}$

53. $\sqrt{32p^8}$ **54.** $\sqrt{48q^8}$ **55.** $\sqrt{405a^{12}b^9}$ **56.** $\sqrt{486x^{11}y^7}$

57. $\sqrt{1125r^9s^5}$ **58.** $\sqrt{396a^7b^6}$ **59.** $\sqrt{1764x^{11}y^7}$ **60.** $\sqrt{4050a^3b^5c^8}$

D

61. A city lot has an area of 4500 square feet. If the lot is a square, represent the length of one side as a radical in simplest form. Find the approximate value to the nearest tenth.

62. A sandlot ball diamond (square shape) has an area of 7500 square feet. Express the distance between home plate and first base as a radical in simplest form. Use the formula in Example 8.

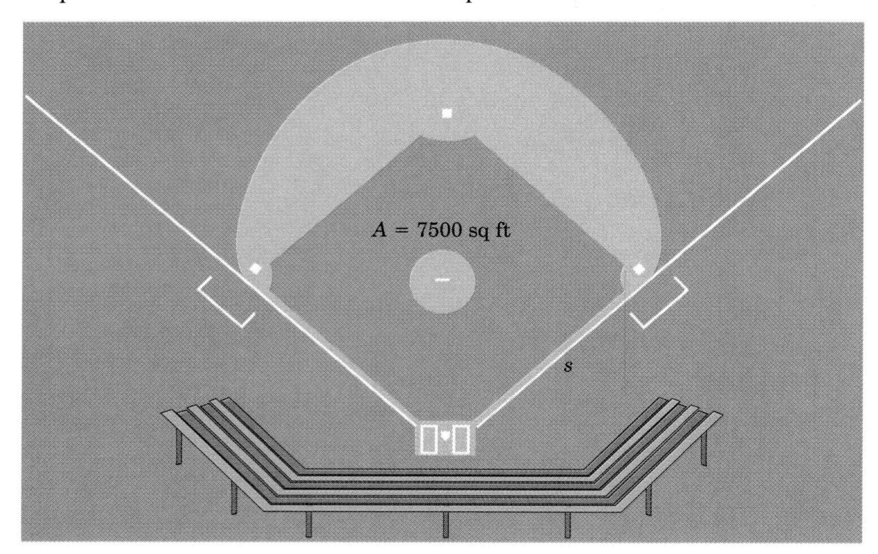

Figure for Exercise 62

63. The formula for the diagonal of a rectangle is given by $d = \sqrt{\ell^2 + w^2}$, where ℓ and w represent the length and the width of the rectangle. A divider is placed diagonally across a rectangular flower bed with dimensions of 6 feet and 10 feet. Find the length of the divider as a radical in simplest form. Also find the approximate length to the nearest tenth of a foot.

64. The formula for the length of a rectangle is given by $\ell = \sqrt{d^2 - w^2}$, where d and w represent the diagonal and the width of the rectangle. A divider of 18 feet is placed diagonally across a rectangular flower bed of width 6 feet. Find the length of the rectangle as a radical in simplest form. Also find the approximate length to the nearest tenth of a foot.

65. An electric space heater requires 600 watts of power. If the resistance is 18 ohms, what is the voltage requirement expressed as a simplified radical? The formula is

$$V = \sqrt{(\text{watts})(\text{ohms})} = \sqrt{PR},$$

where V represents the number of volts, P represents the number of watts of power, and R represents the number of ohms of resistance.

66. Express the voltage needed for a circuit with 30 watts of power and a resistance of 40 ohms in simplest radical form. Use the formula in Exercise 65.

67. The radius of a circle can be expressed as $r = \sqrt{\dfrac{A}{\pi}}$, where r is the radius, and A is the area. If A is 157 square centimeters, find the radius in simplified radical form and to the nearest tenth of a centimeter. Let $\pi \approx 3.14$.

68. Given the formula $V = \sqrt{\dfrac{2k}{m}}$, where k is kinetic energy, m is mass, and V is velocity. Determine the velocity of a car if its mass is 2000 and the kinetic energy is 900,000.

STATE YOUR UNDERSTANDING

69. Why is it not permissible to use the Basic Principle of Radicals to simplify $\sqrt{-18}$?

70. Identify the error in the following example. What is the correct answer.
Simplify $\sqrt{675}$:

$$\begin{aligned}\sqrt{675} &= \sqrt{25} \cdot \sqrt{27} \\ &= 5\sqrt{27}\end{aligned}$$

CHALLENGE EXERCISES

Simplify the following radicals, assume no radicand is negative:

71. $\sqrt{(x-y)^7}$

72. $\sqrt{x^2 + 4x + 4}$ Hint: Factor

73. $\sqrt{(6x-2)(9x-3)}$

74. $\sqrt{(7x-7)(14x-14)}$

75. $\sqrt{(3x^2 - 3y^2)(x+y)}$

MAINTAIN YOUR SKILLS (SECTION 6.2)

Name each of the following points using ordered pairs:

76. A

77. B

78. C

79. D

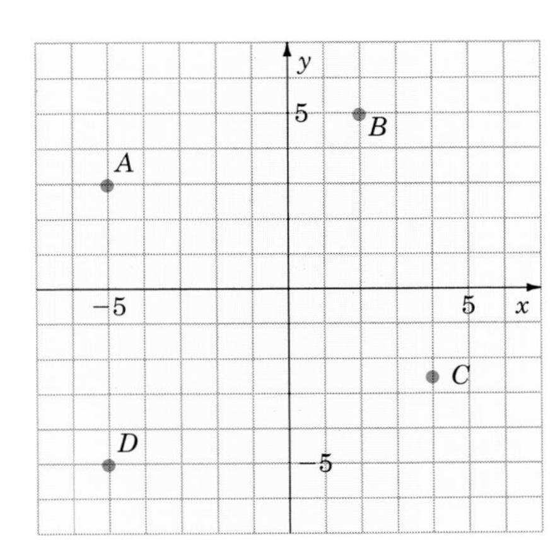

Graph each of the following:

80. $2y - 5x = 10$

81. $y = -x + 2$

82. $3x + 5y = 15$

83. $-y = x$

7.3

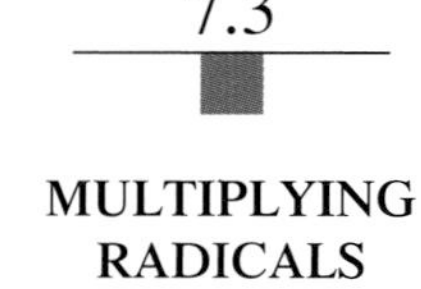

MULTIPLYING RADICALS

OBJECTIVE

1. Multiply two or more radical expressions and simplify.

When we simplify radicals, we use the basic principle of radicals: $\sqrt{ab} = \sqrt{a} \cdot \sqrt{b}$, where a and b are nonnegative. Using the symmetric property of equality, the basic principle reads

$$\sqrt{a} \cdot \sqrt{b} = \sqrt{ab}.$$

Using the basic principle in this form, we can multiply radical expressions.

$$\sqrt{3} \cdot \sqrt{2} = \sqrt{6}$$

$$\sqrt{5} \cdot \sqrt{7} = \sqrt{35}$$

$$\sqrt{6} \cdot \sqrt{3} = \sqrt{18}.$$

Recall that $\sqrt{a}$ is one of two equal factors of a, so

$$(\sqrt{a})^2 = a, \qquad a \geq 0.$$

This is consistent with the multiplication of radicals since

$$\sqrt{a} \cdot \sqrt{a} = \sqrt{a^2} = a, \qquad a \geq 0.$$

When multiplying radicals, it is usually easier to simplify the radicals first (if possible), then multiply. To see why it is usually easier, look at the following table. Note the larger numbers in the column on the left.

Multiply Then Simplify	Simplify Then Multiply
$\sqrt{12} \cdot \sqrt{32} = \sqrt{384} = \sqrt{64} \cdot \sqrt{6} = 8\sqrt{6}$	$\sqrt{12} \cdot \sqrt{32} = 2\sqrt{3} \cdot 4\sqrt{2} = 8\sqrt{6}$
$\sqrt{27} \cdot \sqrt{18} = \sqrt{486} = \sqrt{81} \cdot \sqrt{6} = 9\sqrt{6}$	$\sqrt{27} \cdot \sqrt{18} = 3\sqrt{3} \cdot 3\sqrt{2} = 9\sqrt{6}$
$\sqrt{45} \cdot \sqrt{48} = \sqrt{2160} = \sqrt{144 \cdot 15} = 12\sqrt{15}$	$\sqrt{45} \cdot \sqrt{48} = 3\sqrt{5} \cdot 4\sqrt{3} = 12\sqrt{15}$

Because of this, we use the following procedure:

■ PROCEDURE

To multiply radicals:

1. Simplify each radical if possible.
2. Multiply the coefficients of the radicals.
3. Multiply the radicands.
4. Simplify the radical answer if possible.

EXAMPLE 1

Multiply: $\sqrt{5}\,\sqrt{6}$

$\sqrt{5}\,\sqrt{6} = \sqrt{30}$ — Since both radicands are positive, we can multiply under the radical sign. ■

EXAMPLE 2

Multiply: $\sqrt{6}\,\sqrt{15}$

$\sqrt{6}\,\sqrt{15} = \sqrt{90} = \sqrt{9}\,\sqrt{10} = 3\sqrt{10}$ — Multiply, then simplify. ■

In Example 3 we simplify first, then multiply. The order of steps 1 and 2 in the procedure is not critical.

EXAMPLE 3

Multiply: $\sqrt{18}\,\sqrt{90}$

$\sqrt{18}\,\sqrt{90} = \sqrt{9 \cdot 2} \cdot \sqrt{9 \cdot 10}$

$= 3\sqrt{2} \cdot 3\sqrt{10}$ — Simplify each radical.

$= 9\sqrt{20}$ — Multiply.

$= 9\sqrt{4}\,\sqrt{5}$ — Simplify again.

$= 9(2)\sqrt{5}$

$= 18\sqrt{5}$ — Simplest radical form. ■

If the radicands contain variables, multiply as before assuming the variables represent positive numbers.

EXAMPLE 4

Multiply: $\sqrt{20x}\sqrt{7y}$

$\sqrt{20x}\sqrt{7y} = 2\sqrt{5x}\sqrt{7y}$ — Simplify, then multiply. Check to see if you must simplify again.

$= 2\sqrt{35xy}$ ■

EXAMPLE 5

Multiply: $(\sqrt{3} + 2)\sqrt{3}$

$(\sqrt{3} + 2)\sqrt{3} = \sqrt{3}(\sqrt{3} + 2)$ — Commutative property of multiplication.

$= (\sqrt{3})^2 + \sqrt{3} \cdot 2$ — Use the distributive property to multiply each term in the parentheses by $\sqrt{3}$.

$= 3 + 2\sqrt{3}$ — Simplify. ■

Example 6 represents an important feature of radicals.

EXAMPLE 6

Multiply: $\sqrt{b}\sqrt{b}$

$\sqrt{b}\sqrt{b} = b$ — This example represents the fundamental idea of radicals. $\sqrt{b}$ is the number whose square is b. ■

EXAMPLE 7

Multiply: $\sqrt{x^5} \cdot \sqrt{x^7}$

$\sqrt{x^5} \cdot \sqrt{x^7} = \sqrt{x^{12}}$ — Multiply.

$= x^6$ ■

The same multiplication shortcuts used on polynomials can be used to multiply radical expressions.

EXAMPLE 8

Multiply: $(\sqrt{a} + \sqrt{b})(\sqrt{a} - \sqrt{b})$

$= (\sqrt{a})^2 - (\sqrt{b})^2$ — Note that this is the product of conjugate binomials whose product is the difference of two squares.

$= a - b$ — a is the square of $\sqrt{a}$, and b is the square of $\sqrt{b}$. ■

EXAMPLE 9

Multiply: $(\sqrt{3} - \sqrt{x})^2$

$$(\sqrt{3} + \sqrt{x})^2 = (\sqrt{3})^2 - 2\sqrt{3}\sqrt{x} + (\sqrt{x})^2$$
$$= 3 - 2\sqrt{3x} + x$$

■

The formula for the height of an equilateral triangle contains a radical.

EXAMPLE 10

The height of an equilateral triangle is given by the formula

$$h = \frac{s}{2}\sqrt{3},$$

where h represents the height of the triangle, and s represents one of the three equal sides of the triangle.

Find the height of a triangular medallion, in simplified radical form and to the nearest hundredth, if each side is $\sqrt{6}$ cm long.

Formula:

$$h = \frac{s}{2}\sqrt{3}$$

Substitute:

$$h = \frac{\sqrt{6}}{2}\sqrt{3} \qquad s = \sqrt{6}$$

Solve:

$$h = \frac{\sqrt{6}}{2} \cdot \frac{\sqrt{3}}{1}$$
$$= \frac{\sqrt{18}}{2}$$
$$= \frac{\sqrt{9} \cdot \sqrt{2}}{2}$$
$$= \frac{3\sqrt{2}}{2} \text{ cm} \qquad \text{Simplified radical form.}$$
$$\approx \frac{3(1.41)}{2} \approx 2.12 \text{ cm} \qquad \text{Nearest hundredth.}$$

Answer:

The medallion is $\frac{3\sqrt{2}}{2}$ cm, or approximately 2.12 cm in height.

■

EXERCISE 7.3

A

Multiply and simplify; assume that all variables represent positive numbers:

1. $\sqrt{7}\sqrt{6}$
2. $\sqrt{5}\sqrt{7}$
3. $\sqrt{3}\sqrt{14}$
4. $\sqrt{3}\sqrt{19}$
5. $\sqrt{7}\sqrt{7}$
6. $\sqrt{13}\sqrt{13}$
7. $\sqrt{6}\sqrt{6}$
8. $\sqrt{3}\sqrt{12}$
9. $\sqrt{2}\sqrt{8}$
10. $\sqrt{2}\sqrt{32}$
11. $\sqrt{2}\sqrt{50}$
12. $\sqrt{5}\sqrt{30}$
13. $\sqrt{x}\sqrt{yz}$
14. $\sqrt{ab}\sqrt{c}$
15. $\sqrt{3}\sqrt{5}\sqrt{7}$
16. $\sqrt{11}\sqrt{2}\sqrt{11}$
17. $\sqrt{3}\sqrt{6}\sqrt{18}$
18. $\sqrt{4}\sqrt{3}\sqrt{12}$
19. $\sqrt{2}\sqrt{98}$
20. $\sqrt{3}\sqrt{75}$

B

Multiply and simplify; assume that all variables represent positive numbers:

21. $\sqrt{5x}\sqrt{10y}$
22. $\sqrt{2p}\sqrt{6q}$
23. $\sqrt{3}\sqrt{5}\sqrt{3}$
24. $\sqrt{5}\sqrt{3}\sqrt{5}$
25. $\sqrt{11}\sqrt{10}$
26. $\sqrt{10}\sqrt{11}$
27. $\sqrt{15}\sqrt{21}$
28. $\sqrt{20}\sqrt{30}$
29. $(\sqrt{7} + \sqrt{3})(\sqrt{3})$
30. $(\sqrt{3} - \sqrt{11})\sqrt{7}$
31. $(\sqrt{3} - \sqrt{11})\sqrt{6}$
32. $(\sqrt{5} + \sqrt{6})\sqrt{3}$
33. $\sqrt{10}(\sqrt{5} + \sqrt{2})$
34. $\sqrt{21}(\sqrt{7} - \sqrt{3})$
35. $\sqrt{xy}(\sqrt{wx} + \sqrt{yz})$
36. $\sqrt{ab}(\sqrt{ac} + \sqrt{bc})$
37. $\sqrt{2x}(\sqrt{2x} + 2\sqrt{x} + x\sqrt{2})$
38. $2\sqrt{y}(2\sqrt{y} + \sqrt{2y} - y\sqrt{2})$
39. $\sqrt{2w}(\sqrt{6w} - \sqrt{8w} + 2\sqrt{10w})$
40. $\sqrt{3t}(\sqrt{6t} + 2\sqrt{12t} + 3\sqrt{15t})$

C

Multiply and simplify; assume that all variables represent positive numbers:

41. $\sqrt{x^3}\sqrt{x^4}$
42. $\sqrt{x^8}\sqrt{x^3}$
43. $\sqrt{x^2}\sqrt{x^5}\sqrt{x^4}$
44. $\sqrt{8w}\sqrt{27w^2}$
45. $\sqrt{8r}\sqrt{125r^3}$
46. $(\sqrt{5} + \sqrt{3})(\sqrt{5} - \sqrt{3})$
47. $(\sqrt{10} + \sqrt{3})(\sqrt{10} - \sqrt{3})$
48. $(\sqrt{13} + \sqrt{11})(\sqrt{13} - \sqrt{11})$
49. $\sqrt{2}(\sqrt{6} + \sqrt{8} + \sqrt{10})$
50. $\sqrt{5}(\sqrt{10} + \sqrt{15} + \sqrt{20})$
51. $\sqrt{3x}(\sqrt{x} + \sqrt{3})$
52. $\sqrt{2x}(\sqrt{x^2} + \sqrt{8})$
53. $3\sqrt{x}(\sqrt{x} + 3)$
54. $2\sqrt{y}(\sqrt{2} + \sqrt{y})$
55. $(\sqrt{2} + \sqrt{3})^2$
56. $(\sqrt{7} + \sqrt{8})^2$
57. $(\sqrt{5} - \sqrt{3})^2$
58. $(\sqrt{y} - \sqrt{x})^2$
59. $(\sqrt{7} + \sqrt{3})(\sqrt{14} - \sqrt{3})$
60. $(\sqrt{5} + \sqrt{2})(\sqrt{6} + \sqrt{2})$

D

61. Each side of an equilateral triangle is $\sqrt{8}$ feet. Use the formula $h = \frac{s}{2}\sqrt{3}$, where h is the height and s is the length of each side, to find the height of the triangle (a) in simplified radical form and (b) to the nearest tenth of a foot.

62. Each side of an equilateral triangle is $\sqrt{15}$ meters. Use the formula $h = \frac{s}{2}\sqrt{3}$, where h is the height and s is the length of each side, to find the height of the triangle (a) in simplified radical form and (b) to the nearest tenth of a meter.

63. The surface area of the side of a pressurized tank is given by the formula $S = 2\pi rh$. Find the area in simplified radical form and to the nearest tenth if $r = \sqrt{10}$ meters and $h = \sqrt{15}$ meters. Let $\pi \approx 3.14$.

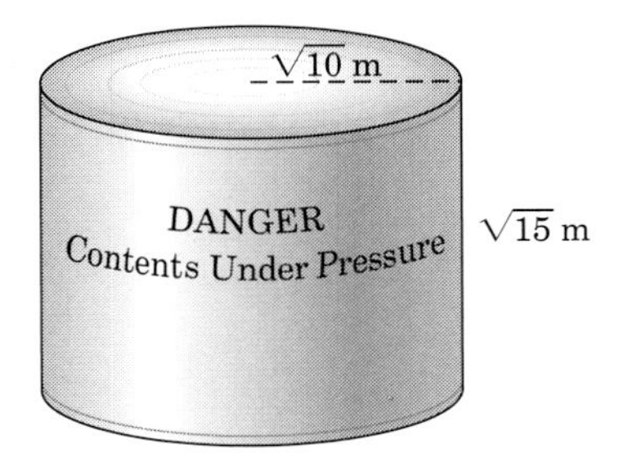

Figure for Exercise 63

64. Using the formula from Exercise 63, find the area in simplified radical form and to the nearest tenth if $r = \sqrt{12}$ inches and $h = \sqrt{18}$ inches.

65. Find the area of a rectangle with base $\sqrt{12}$ inches and height $\sqrt{15}$ inches (a) in simplest radical form and (b) to the nearest tenth.

66. Find the area of a rectangle with base $\sqrt{18}$ decimeters and height $\sqrt{24}$ decimeters (a) in simplest radical form and (b) to the nearest tenth.

67. Find the area of a triangle with base $\sqrt{10}$ cm and height $\sqrt{15}$ cm (a) in simplest radical form and (b) to the nearest tenth.

68. Find the area of a triangle with base $\sqrt{40}$ km and height $\sqrt{50}$ km (a) in simplest radical form and (b) to the nearest tenth.

69. The volume of a rectangular solid is given by the formula $V = \ell wh$. Find the volume if $\ell = \sqrt{8}$ centimeters, $w = \sqrt{2}$ centimeters, and $h = \sqrt{5}$ centimeters. Find your answer in simplest radical form and to the nearest tenth.

70. Given the formula $K = \frac{1}{2}mv^2$, where K is kinetic energy, m is mass, and v is velocity. Find K in simplest form and to the nearest tenth if $m = \sqrt{10}$ and $v = \sqrt{8}$.

STATE YOUR UNDERSTANDING

71. Explain the error in the following example.
Simplify: $\sqrt{-6} \cdot \sqrt{-21}$

$$\begin{aligned}\sqrt{-6} \cdot \sqrt{-21} &= \sqrt{126} \\ &= \sqrt{9} \cdot \sqrt{14} \\ &= 3\sqrt{14}\end{aligned}$$

72. Why is it usually better to simplify radicals before multiplying them?

CHALLENGE EXERCISES

Simplify the following:

73. $(\sqrt{60} - \sqrt{42})(\sqrt{42} + \sqrt{60})$

74. $(\sqrt{12} + \sqrt{48})^2$

75. $(\sqrt{x^4} - 2\sqrt{y^5})^2$

76. $(\sqrt[3]{5} - \sqrt[3]{2})(\sqrt[3]{25} + \sqrt[3]{10} + \sqrt[3]{4})$

MAINTAIN YOUR SKILLS (SECTION 6.3)

For the graph of each of the following equations, find the slope and the y-intercept.

77. $6x - 7y = 14$

78. $3x + 2y = 8$

79. $x + 9y = 12$

80. $5y - 8x = 11$

Are the graphs of the following pairs of equations parallel?

81. $2x - 3y = 8$ and $4x - 6y = -2$

82. $21x - 6y = 20$ and $2x + 7y = -12$

83. $\frac{1}{2}x - 2y = \frac{1}{3}$ and $28x - 7y = 1$

84. $2x - 9y = 4$ and $\frac{2}{3}x - 3y = 1$

7.4 DIVIDING RADICALS

OBJECTIVE

1. Divide two radical expressions.

The indicated quotient of two radical expressions can be written as a single radical.

PROPERTY

$$\sqrt{a} \div \sqrt{b} = \frac{\sqrt{a}}{\sqrt{b}} = \sqrt{\frac{a}{b}}, \; a \geq 0 \text{ and } b > 0.$$

So,

$$\sqrt{5} \div \sqrt{7} = \frac{\sqrt{5}}{\sqrt{7}} = \sqrt{\frac{5}{7}}.$$

In the form $\sqrt{\frac{a}{b}}$, common factors of a and b can be eliminated by reducing.

$$\sqrt{30} \div \sqrt{5} = \sqrt{\frac{30}{5}} = \sqrt{6}$$

A common objective in the division of radicals is the elimination of the radical in the denominator (or numerator). One reason for this is to help us recognize when two radical expressions are equivalent. For example, we can show that

$$\frac{3}{\sqrt{6}} = \sqrt{\frac{3}{2}} = \frac{\sqrt{6}}{2}$$

are three equivalent radical expressions.

We now have three conditions for simplifying radical expressions.

1. A radical containing a positive integer radicand cannot have a perfect square factor other than 1.
2. No radicand can contain a fraction.
3. No denominator of a radical expression may contain a radical.

In the quotient

$$\sqrt{3} \div \sqrt{2} = \frac{\sqrt{3}}{\sqrt{2}} = \sqrt{\frac{3}{2}}$$

the radicand cannot be reduced. One way to eliminate the radical in the denominator is to approximate its value.

$$\frac{\sqrt{3}}{\sqrt{2}} \approx \frac{1.732}{1.414} \approx 1.225$$

The approximation is useful as a measure or to locate an approximate point on the number line when we do not need the exact value. To represent the exact value in simplified form, we rationalize the denominator.

DEFINITION

Rationalize the Denominator

To *rationalize the denominator* of a radical expression means to write an equivalent rational expression in which the denominator is a natural number.

To rationalize the denominator, we multiply the radical expression using the basic principle of fractions. Multiply the numerator and denominator by a factor that produces a radicand in the denominator that is a perfect square. So, $\frac{3}{\sqrt{6}}$, $\sqrt{\frac{3}{2}}$, and $\frac{\sqrt{6}}{2}$ are equivalent because

$$\frac{3}{\sqrt{6}} = \frac{3}{\sqrt{6}} \cdot 1 = \frac{3}{\sqrt{6}} \cdot \frac{\sqrt{6}}{\sqrt{6}} = \frac{3\sqrt{6}}{6} = \frac{\sqrt{6}}{2}$$

$$\sqrt{\frac{3}{2}} = \frac{\sqrt{3}}{\sqrt{2}} = \frac{\sqrt{3}}{\sqrt{2}} \cdot 1 = \frac{\sqrt{3}}{\sqrt{2}} \cdot \frac{\sqrt{2}}{\sqrt{2}} = \frac{\sqrt{6}}{2}.$$

PROCEDURE

To rationalize the denominator of a rational expression containing radicals:

1. Multiply both the numerator and the denominator by a radical expression so that the denominator is a rational number.
2. Simplify.

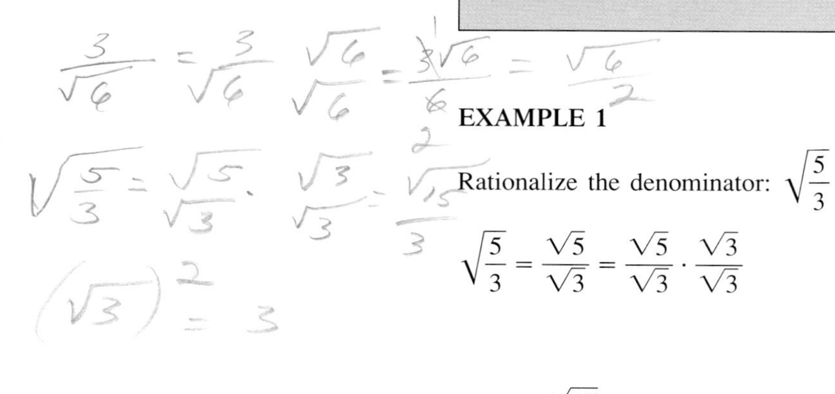

EXAMPLE 1

Rationalize the denominator: $\sqrt{\frac{5}{3}}$

$$\sqrt{\frac{5}{3}} = \frac{\sqrt{5}}{\sqrt{3}} = \frac{\sqrt{5}}{\sqrt{3}} \cdot \frac{\sqrt{3}}{\sqrt{3}}$$

Since $\sqrt{3} \cdot \sqrt{3} = 3$ is a rational number, we multiply both the numerator and denominator by $\sqrt{3}$.

$$= \frac{\sqrt{15}}{3}$$

Simplest radical form. ■

Example 2 illustrates four ways to rationalize the quotient $\sqrt{15} \div \sqrt{18}$.

EXAMPLE 2

Rationalize the denominator: $\frac{\sqrt{15}}{\sqrt{18}}$

$$\frac{\sqrt{15}}{\sqrt{18}} = \sqrt{\frac{15}{18}} = \sqrt{\frac{5}{6}} = \frac{\sqrt{5}}{\sqrt{6}} \cdot \frac{\sqrt{6}}{\sqrt{6}} = \frac{\sqrt{30}}{6}$$

This example can be done four different ways depending on the first step. Here, we first write as a fraction under the radical, reduce, and then simplify as we did in Example 1.

or

$$\frac{\sqrt{15}}{\sqrt{18}} = \frac{\sqrt{15}}{3\sqrt{2}} = \frac{\sqrt{15}}{3\sqrt{2}} \cdot \frac{\sqrt{2}}{\sqrt{2}} = \frac{\sqrt{30}}{6}$$

Here, we first simplify the denominator $\sqrt{18}$ and then multiply the numerator and denominator by $\sqrt{2}$.

or

$$\frac{\sqrt{15}}{\sqrt{18}} = \frac{\sqrt{15}}{\sqrt{18}} \cdot \frac{\sqrt{18}}{\sqrt{18}} = \frac{\sqrt{270}}{18} = \frac{\sqrt{9} \cdot \sqrt{30}}{18}$$

Here, we first multiply the numerator and denominator by $\sqrt{18}$ and then simplify the numerator.

$$= \frac{3\sqrt{30}}{18} = \frac{\sqrt{30}}{6}$$

Reduce the fraction.

or

$$\frac{\sqrt{15}}{\sqrt{18}} = \frac{\sqrt{15}}{\sqrt{18}} \cdot \frac{\sqrt{2}}{\sqrt{2}} = \frac{\sqrt{30}}{\sqrt{36}} = \frac{\sqrt{30}}{6}$$

If we first recognize that twice 18 is a perfect square, multiply the numerator and denominator by $\sqrt{2}$. All four of these procedures yield the same result. ■

If the quotient contains factors outside the radicals, we concentrate our attention on the radicand.

EXAMPLE 3

Rationalize the denominator: $4\sqrt{3} \div \sqrt{8}$

$$4\sqrt{3} \div \sqrt{8} = \frac{4\sqrt{3}}{\sqrt{8}} = \frac{4\sqrt{3}}{\sqrt{8}} \cdot \frac{\sqrt{2}}{\sqrt{2}} = \frac{4\sqrt{6}}{\sqrt{16}}$$

$$= \frac{4\sqrt{6}}{4} = \sqrt{6}$$

We multiply the numerator and denominator by $\sqrt{2}$ (instead of by $\sqrt{8}$) since 2 is smaller than 8 and 16 is a perfect square. The student can verify that the same result is obtained if we multiply by $\sqrt{8}$.

or

$$4\sqrt{3} \div \sqrt{8} = \frac{4\sqrt{3}}{\sqrt{8}} = \frac{4\sqrt{3}}{\sqrt{4 \cdot 2}} = \frac{4\sqrt{3}}{2\sqrt{2}}$$

Simplify the radical in the denominator.

$$= \frac{2\sqrt{3}}{\sqrt{2}} = \frac{2\sqrt{3}}{\sqrt{2}} \cdot \frac{\sqrt{2}}{\sqrt{2}}$$

Multiply the numerator and denominator by $\sqrt{2}$.

$$= \frac{2\sqrt{6}}{2} = \sqrt{6}$$

Simplify. ■

As seen in Section 7.3, example 7, the radicals $\sqrt{a} + \sqrt{b}$ and $\sqrt{a} - \sqrt{b}$ are *conjugates*. Conjugates are useful in simplifying some quotients as their product does not contain a radical.

EXAMPLE 4

Simplify: $\dfrac{\sqrt{6}}{\sqrt{11} - \sqrt{3}}$

$$\frac{\sqrt{6}}{\sqrt{11} - \sqrt{3}} \cdot \frac{\sqrt{11} + \sqrt{3}}{\sqrt{11} + \sqrt{3}} = \frac{\sqrt{66} + \sqrt{18}}{11 - 3}$$

$$= \frac{\sqrt{66} + 3\sqrt{2}}{8}$$

To remove the radicals in the denominator, we multiply both the numerator and denominator by the conjugate of $\sqrt{11} - \sqrt{3}$, which is $\sqrt{11} + \sqrt{3}$. ■

Because of the force of gravity, the formula for the cycle of a pendulum contains a radical expression.

EXAMPLE 5

The clock at the Amtrak station has a pendulum with a length of 30 feet. The time it takes the pendulum to make one cycle is given by

$$T = 2\pi\frac{\sqrt{L}}{\sqrt{32}},$$

where L is the length of the pendulum in feet, and T is the time in seconds. Find T as a product of π and an expression without a radical in the denominator.

Formula:

$$T = 2\pi\frac{\sqrt{L}}{\sqrt{32}}$$

Substitute:

$$T = 2\pi\frac{\sqrt{30}}{\sqrt{32}}$$

$L = 30$

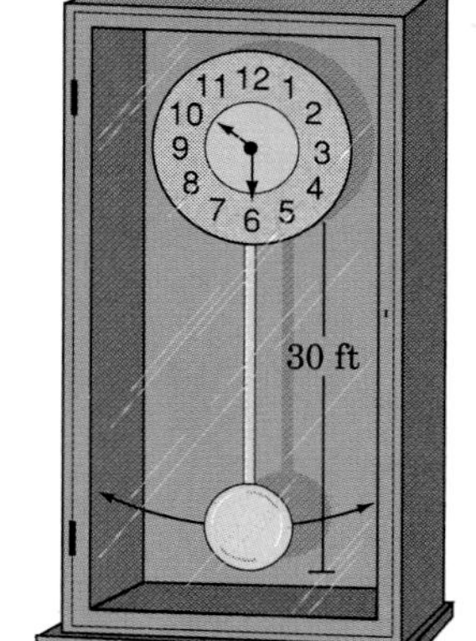

Solve:

$$T = 2\pi\frac{\sqrt{30}}{\sqrt{32}} \cdot \frac{\sqrt{2}}{\sqrt{2}}$$

$$= 2\pi\frac{\sqrt{60}}{\sqrt{64}}$$

$$= 2\pi\frac{\sqrt{4}\sqrt{15}}{8}$$

$$= \frac{4\pi\sqrt{15}}{8}$$

$$= \frac{\pi\sqrt{15}}{2}$$

Answer:

The pendulum takes $\frac{\pi\sqrt{15}}{2}$ sec to complete a cycle, or approximately 6.08 sec. ■

EXERCISE 7.4

A

Divide; rationalize the denominator:

1. $\frac{1}{\sqrt{3}}$
2. $\frac{1}{\sqrt{11}}$
3. $\frac{3}{\sqrt{2}}$
4. $\frac{2}{\sqrt{5}}$
5. $\frac{3}{\sqrt{11}}$
6. $\frac{4}{\sqrt{13}}$
7. $\sqrt{3} \div \sqrt{5}$
8. $\sqrt{2} \div \sqrt{7}$
9. $\frac{\sqrt{6}}{\sqrt{9}}$
10. $\frac{\sqrt{10}}{\sqrt{25}}$
11. $\frac{\sqrt{10}}{\sqrt{5}}$
12. $\frac{\sqrt{30}}{\sqrt{10}}$
13. $\frac{3}{\sqrt{6}}$
14. $\frac{4}{\sqrt{12}}$
15. $\frac{5}{\sqrt{50}}$
16. $\frac{3}{\sqrt{27}}$
17. $\frac{3}{2\sqrt{3}}$
18. $\frac{5}{3\sqrt{5}}$
19. $\frac{2}{3\sqrt{8}}$
20. $\frac{2}{3\sqrt{20}}$

B

Divide; rationalize the denominator:

21. $\frac{\sqrt{6}}{\sqrt{8}}$
22. $\frac{\sqrt{8}}{\sqrt{10}}$
23. $\frac{\sqrt{21}}{\sqrt{7}}$
24. $\frac{\sqrt{12}}{\sqrt{6}}$
25. $\frac{\sqrt{12}}{\sqrt{18}}$
26. $\frac{\sqrt{20}}{\sqrt{15}}$
27. $\sqrt{30} \div \sqrt{54}$
28. $\frac{6}{\sqrt{81}}$
29. $\sqrt{24} \div \sqrt{30}$
30. $\sqrt{15} \div \sqrt{40}$
31. $\sqrt{30} \div \sqrt{24}$
32. $\sqrt{40} \div \sqrt{15}$

33. $\dfrac{\sqrt{5}}{\sqrt{32}}$ **34.** $\dfrac{\sqrt{7}}{\sqrt{60}}$ **35.** $\dfrac{\sqrt{6}}{\sqrt{54}}$ **36.** $\dfrac{\sqrt{15}}{\sqrt{60}}$

37. $\sqrt{18} \div \sqrt{81}$ **38.** $\sqrt{60} \div \sqrt{15}$ **39.** $\dfrac{4}{\sqrt{6} - \sqrt{5}}$ **40.** $\dfrac{5}{\sqrt{13} + \sqrt{7}}$

C

Divide; rationalize the denominator:

41. $\dfrac{3\sqrt{6}}{7\sqrt{8}}$ **42.** $\dfrac{2\sqrt{7}}{\sqrt{21}}$ **43.** $\dfrac{2\sqrt{3}}{3\sqrt{2}}$ **44.** $\dfrac{2\sqrt{3}}{\sqrt{6}}$

45. $\dfrac{\sqrt{5}}{\sqrt{196}}$ **46.** $\dfrac{\sqrt{18}}{\sqrt{225}}$ **47.** $\dfrac{12\sqrt{35}}{5\sqrt{15}}$ **48.** $\dfrac{15\sqrt{6}}{24\sqrt{12}}$

49. $\dfrac{2\sqrt{3}}{3\sqrt{40}}$ **50.** $\dfrac{2\sqrt{5}}{5\sqrt{48}}$ **51.** $\dfrac{3\sqrt{18}}{\sqrt{120}}$ **52.** $\dfrac{9\sqrt{10}}{\sqrt{18}}$

53. $\dfrac{5\sqrt{40}}{\sqrt{80}}$ **54.** $\dfrac{24\sqrt{18}}{40\sqrt{27}}$ **55.** $\dfrac{11\sqrt{54}}{8\sqrt{50}}$ **56.** $\dfrac{20\sqrt{15}}{7\sqrt{75}}$

57. $\dfrac{2}{\sqrt{3} + \sqrt{5}}$ **58.** $\dfrac{2}{\sqrt{3} + 5}$ **59.** $\dfrac{5}{\sqrt{10} - \sqrt{5}}$ **60.** $\dfrac{7}{\sqrt{26} - \sqrt{19}}$

D

61. Using the formula in Example 5, find the time it takes a pendulum to complete a cycle if the pendulum is 2 ft long. (Write the answer in simplest radical form and rounded to the nearest hundredth.)

62. Using the formula in Example 5, find the time it takes a pendulum 12 inches long to complete a cycle. (Write the answer in simplest radical form and rounded to the nearest hundredth.)

63. From physics, there is a formula for the radius of gyration of a pendulum, which is $K = \dfrac{\sqrt{J}}{\sqrt{W}}$. K is the radius of gyration, W is the weight, and J is the moment of inertia. Find K if $J = 18$ and $W = 2$.

64. Using the formula in Exercise 63, find K if $W = 3$ and $J = 75$.

65. Given another formula from physics $s = \dfrac{\sqrt{M_1 M_2}}{\sqrt{F}}$, find s if $M_1 = 12$, $M_2 = 4$, and $F = 20$. Find the answer to the nearest tenth.

66. Using the formula given in Exercise 65, find s if $M_1 = 8$, $M_2 = 25$, and $F = 10$.

67. When the surface area of a cube, S, is known, the length of the edge, e, can be determined by the formula $e = \sqrt{\frac{S}{6}}$. Determine e when $S = 105 \text{ in}^2$. Express in simplest radical form and rounded to the nearest tenth.

68. In an alternating-current circuit, the current, I, in amperes, is given by the formula $I = \sqrt{\frac{P}{R}}$, where P is the power dissipated, in watts, and R is the resistance in ohms. Determine I when the power dissipated is 162 watts and the resistance is 6.0 ohms. Express in simplest radical form and rounded to the nearest tenth.

STATE YOUR UNDERSTANDING

69. In the property $\frac{\sqrt{a}}{\sqrt{b}} = \sqrt{\frac{a}{b}}$, a not negative and b positive, why does it state a not negative and b positive instead of a and b not negative as in the basic principle of radicals?

70. Explain the concept of rationalizing the denominator.

CHALLENGE EXERCISES

Simplify, write each answer in simplest radical form:

71. $\frac{\sqrt{5} - \sqrt{2}}{\sqrt{5} + \sqrt{2}}$

72. $\frac{2\sqrt{3} - 3\sqrt{2}}{2\sqrt{3} + 3\sqrt{2}}$

73. $\frac{\sqrt{x+2} + \sqrt{x-2}}{\sqrt{x+2} - \sqrt{x-2}}$

74. $\frac{1}{\sqrt[3]{2}}$

MAINTAIN YOUR SKILLS (SECTIONS 5.4, 6.3)

Combine:

75. $\frac{3x}{2} + \frac{5x}{2}$

76. $\frac{7abc}{5} - \frac{2abc}{5}$

77. $\frac{3x}{2} + \frac{5x}{3}$

78. $\frac{abc}{5} - \frac{3abc}{7}$

Find the slope of the straight line that contains the given points.

79. $(2, -8), (-1, 0)$

80. $(5, 0), (-1, -1)$

81. $(-3, -4), (5, -1)$

82. $(8, -3), (6, -3)$

7.5 ADDING AND SUBTRACTING RADICALS

OBJECTIVE

1. Combine radical expressions.

As with like terms in polynomials, like radicals can be added.

DEFINITION

Like nth Root Radicals

Like radicals are radicals with the same radicand and the same index.

All the radicals in the following list are like radicals:

$$\sqrt{6} \qquad 7\sqrt{6} \qquad -3\sqrt{6} \qquad \frac{2\sqrt{6}}{7} \qquad \frac{3}{5}\sqrt{6}.$$

The following radicals are not like radicals:

$$\sqrt{6} \qquad \sqrt{5} \qquad \sqrt[3]{2} \qquad \sqrt{7}.$$

Like radicals are combined using the distributive property.

$$4x^3 + 5x^3 - 3x^3 = (4 + 5 - 3)x^3 = 6x^3 \qquad \text{Like terms.}$$

$$4\sqrt{x} + 5\sqrt{x} - 3\sqrt{x} = (4 + 5 - 3)\sqrt{x} = 6\sqrt{x} \qquad \text{Like radicals.}$$

PROCEDURE

To combine like radicals, add or subtract the coefficients, and write the indicated product of the sum and the common radical.

EXAMPLE 1

Combine: $\sqrt{2} + \sqrt{2}$

$\sqrt{2} + \sqrt{2} = (1 + 1)\sqrt{2}$ — Factor using the distributive property. This step is often done mentally.

$= 2\sqrt{2}$ ■

EXAMPLE 2

Combine: $8\sqrt{11} - 3\sqrt{11} + 2\sqrt{11}$

$= (8 - 3 + 2)\sqrt{11}$

$= 7\sqrt{11}$ ■

Radicals should be simplified before deciding whether they are like radicals.

EXAMPLE 3

Combine: $5\sqrt{3} + \sqrt{12}$

$= 5\sqrt{3} + 2\sqrt{3}$ — In their original form, these are not like radicals, but after simplifying, we have like radicals.

$= 7\sqrt{3}$ ■

The sum of unlike radicals is left as an indicated sum.

EXAMPLE 4

Combine: $\sqrt{75} + \sqrt{20} + \sqrt{12}$

$= 5\sqrt{3} + 2\sqrt{5} + 2\sqrt{3}$ — After simplifying, we see that the radicals are not all alike. The like radicals are combined.

$= 7\sqrt{3} + 2\sqrt{5}$ ■

EXAMPLE 5

Combine: $\sqrt{24} - 6\sqrt{6} - \sqrt{150}$

$= 2\sqrt{6} - 6\sqrt{6} - 5\sqrt{6}$ — Simplify each radical first.

$= -9\sqrt{6}$ ■

EXAMPLE 6

Combine: $\sqrt{\frac{2}{3}} + 5\sqrt{\frac{3}{2}}$

$= \frac{\sqrt{2}}{\sqrt{3}} + \frac{5\sqrt{3}}{\sqrt{2}}$

$= \frac{\sqrt{2}}{\sqrt{3}} \cdot \frac{\sqrt{3}}{\sqrt{3}} + \frac{5\sqrt{3}}{\sqrt{2}} \cdot \frac{\sqrt{2}}{\sqrt{2}}$ — Rationalize the denominators.

$= \frac{\sqrt{6}}{3} + \frac{5\sqrt{6}}{2}$

$= \left(\frac{1}{3} + \frac{5}{2}\right)\sqrt{6}$ — Combine terms.

$= \frac{17}{6}\sqrt{6}$ ■

EXAMPLE 7

Combine: $\sqrt{7} + \sqrt{8}$

$= \sqrt{7} + 2\sqrt{2}$

■ **CAUTION**

Unlike radicals may not be combined.

$\sqrt{a} + \sqrt{b} \neq \sqrt{a + b}$

■

A formula for voltage embodies a radical.

EXAMPLE 8

Ethyl plans to heat her living room with two portable heaters. The first heater uses 605 watts and has a resistance of 25 ohms. The second heater uses 845 watts and has a resistance of 36 ohms. If the voltage (V) required to operate each heater is given by

$V = \sqrt{\text{watts} \cdot \text{ohms}}$,

how many volts are needed to operate the heaters?

Simpler word form:

Total voltage = Sum of voltages — The voltage needed to operate the two heaters is the sum of the voltages of both heaters.

Translate to algebra:

$V = V_1 + V_2$ — Let V_1 be the voltage of the first heater and V_2 be the voltage of the second heater.

Solve:

$$V_1 = \sqrt{605 \cdot 25}$$ Voltage of first heater.

$$= \sqrt{121 \cdot 5 \cdot 25}$$

$$= 55\sqrt{5}$$

$$V_2 = \sqrt{845 \cdot 36}$$ Voltage of second heater.

$$= \sqrt{169 \cdot 5 \cdot 36}$$

$$= 78\sqrt{5}$$

$$V = 55\sqrt{5} + 78\sqrt{5}$$ Total voltage.

$$= 133\sqrt{5}$$

Answer:

The total voltage needed to operate the heaters is $133\sqrt{5}$ volts, or approximately 297.4 volts.

EXERCISE 7.5

A

Combine:

1. $3\sqrt{5} + 6\sqrt{5}$
2. $8\sqrt{3} + 6\sqrt{3}$
3. $14\sqrt{10} + 12\sqrt{10}$
4. $9\sqrt{11} - 13\sqrt{11}$
5. $9\sqrt{2} - 7\sqrt{2}$
6. $10\sqrt{3} - 5\sqrt{3}$
7. $7\sqrt{5} - 12\sqrt{5}$
8. $11\sqrt{11} + 18\sqrt{11}$
9. $6\sqrt{3} + \sqrt{3}$
10. $4\sqrt{5} + \sqrt{5}$
11. $\sqrt{7} - 3\sqrt{7}$
12. $5\sqrt{6} + \sqrt{6}$
13. $\sqrt{2} + \sqrt{2}$
14. $\sqrt{11} - \sqrt{11}$
15. $\sqrt{5} + 3\sqrt{5} + 2\sqrt{5}$
16. $8\sqrt{6} + 11\sqrt{6} - \sqrt{6}$
17. $\frac{3\sqrt{7}}{2} + \frac{5\sqrt{7}}{2}$
18. $\frac{6\sqrt{10}}{5} + \frac{\sqrt{10}}{5}$
19. $\frac{4\sqrt{5}}{3} + \frac{4\sqrt{5}}{2}$
20. $\frac{7\sqrt{2}}{4} + \frac{\sqrt{2}}{6}$

B

Combine:

21. $3\sqrt{2} + \sqrt{8}$
22. $5\sqrt{3} + \sqrt{12}$
23. $2\sqrt{6} - \sqrt{24}$
24. $3\sqrt{5} - \sqrt{20}$
25. $\sqrt{45} + \sqrt{20}$
26. $\sqrt{48} - \sqrt{27}$
27. $\sqrt{24x} - \sqrt{6x}$
28. $\sqrt{32x} - \sqrt{18x}$
29. $\sqrt{20} + \sqrt{45} - \sqrt{5}$
30. $3\sqrt{3} - \sqrt{48} + \sqrt{12}$
31. $-\sqrt{50} - \sqrt{8}$
32. $\sqrt{18} - \sqrt{72}$
33. $\sqrt{18} + \sqrt{32}$
34. $-\sqrt{108} - \sqrt{48}$
35. $2\sqrt{80} + 2\sqrt{420}$
36. $9\sqrt{90} + 4\sqrt{40}$
37. $-\sqrt{75} + \sqrt{243}$
38. $-6\sqrt{96} - 8\sqrt{150}$
39. $5\sqrt{18} + 7\sqrt{50} - 2\sqrt{162}$
40. $9\sqrt{40} - 4\sqrt{490} + 3\sqrt{360}$

C

Combine:

41. $5\sqrt{56} + 8\sqrt{224}$
42. $13\sqrt{135} - 7\sqrt{375}$
43. $\sqrt{72} + \sqrt{50} - \sqrt{128}$
44. $\sqrt{27} - \sqrt{75} + \sqrt{192}$
45. $\sqrt{72} - \sqrt{75} + \sqrt{98} - \sqrt{27}$
46. $\sqrt{24} + \sqrt{32} - \sqrt{128} - \sqrt{150}$
47. $\sqrt{50} - \sqrt{27} + \sqrt{72} - \sqrt{108}$
48. $-\sqrt{125} - \sqrt{20} + \sqrt{28} - \sqrt{252}$
49. $\sqrt{28} - \sqrt{700} + \sqrt{63}$
50. $-\sqrt{98} - 2\sqrt{32} + \sqrt{18}$
51. $3\sqrt{75} - 14\sqrt{12} - 2\sqrt{48}$
52. $\sqrt{500} - 8\sqrt{45} - \sqrt{180}$

53. $3\sqrt{45} - 8\sqrt{48} + 11\sqrt{320}$

54. $7\sqrt{162} + 4\sqrt{363} - 6\sqrt{72}$

55. $8\sqrt{\frac{2}{7}} + 3\sqrt{\frac{7}{2}}$

56. $11\sqrt{\frac{3}{5}} - 7\sqrt{\frac{5}{3}}$

57. $3\sqrt{\frac{3}{2}} + 4\sqrt{\frac{1}{6}} - 5\sqrt{\frac{2}{3}}$

58. $12\sqrt{\frac{1}{12}} - 30\sqrt{\frac{1}{75}} + 3\sqrt{\frac{1}{3}}$

59. $6\sqrt{\frac{4}{3}} + 2\sqrt{\frac{3}{4}} - 3\sqrt{\frac{1}{12}}$

60. $\sqrt{108} - \frac{2}{3}\sqrt{243} + 6\sqrt{\frac{1}{2}}$

D

61. Given the formula $\frac{1}{R} = \frac{1}{R_1} + \frac{1}{R_2}$, find R in simplest radical form if $R_1 = \sqrt{\frac{3}{5}}$ and $R_2 = \sqrt{\frac{5}{3}}$.

62. Using the formula from Exercise 61, find R in simplest radical form if $R_1 = \sqrt{\frac{2}{3}}$ and $R_2 = \sqrt{\frac{3}{2}}$.

63. Using the formula in Example 8, find the voltage needed to operate three heaters. The first uses 200 watts and has a resistance of 48 ohms. The second uses 300 watts and has a resistance of 50 ohms. The third uses 625 watts and has a resistance of 24 ohms. Write the answer in simplest radical form.

64. Using the formula in Example 8, find the voltage needed to operate two heaters if the first uses 360 watts and has a resistance of 36 ohms and the second uses 405 watts and has a resistance of 50 ohms. Write the answer in simplest radical form.

65. A brick patio contains two flower gardens that are equilateral triangles. One has sides 25 ft long, and the other has sides 20 ft long. What is the area of the entire plot of ground? (*Hint:* The area of an equilateral triangle is given

Figure for Exercise 65

by the formula

$$A = \frac{1}{4}s^2\sqrt{3},$$

where s is the length of a side.) Write the answer both in simplest radical form and rounded to the nearest tenth.

66. Using the formula in Exercise 65, what is the area of an outdoor stage that is in the shape of two equilateral triangles, one with sides 16 feet long and the other with sides 20 feet long? Write the answer both in simplest radical form and rounded to the nearest tenth.

67. The surface area of a rectangular solid is determined by the formula $S = 2(\ell w + hw + \ell h)$. Determine the surface area of an engagement ring box with $\ell = 5\sqrt{6}$ cm, $w = 3\sqrt{6}$ cm, and $h = 2\sqrt{3}$ cm. Express the answer in simplest radical form and approximated to the nearest tenth.

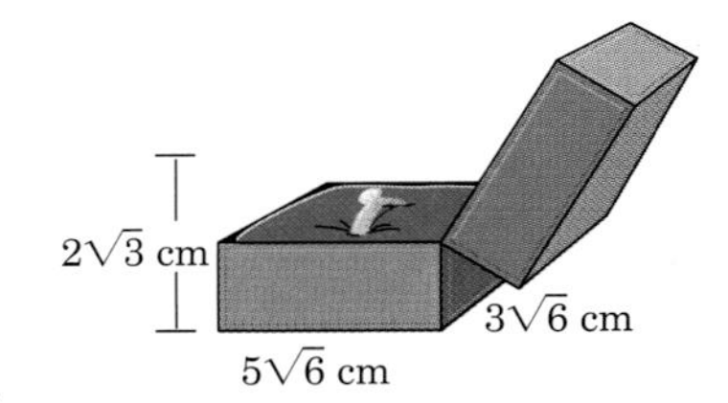

Figure for Exercise 67

68. Use the formula from the previous problem to determine the surface area when $\ell = 5\sqrt{6}$ inches, $w = 2\sqrt{14}$ inches, and $h = 3\sqrt{14}$ inches. Express the answer in simplest radical form and approximated to the nearest tenth.

STATE YOUR UNDERSTANDING

69. Explain what is meant by like radicals.

70. Describe, in detail, how to perform the addition/subtraction of several radicals.

CHALLENGE EXERCISES

Combine:

71. $\sqrt{56x^2y} - x\sqrt{126y}$

72. $\sqrt{72a^4b^5} + 3a\sqrt{18a^2b^5} - 2a^2b\sqrt{8b^3}$

73. $\sqrt{\frac{3y}{2x}} - \sqrt{\frac{2y}{3x}}$

74. $2\sqrt[3]{54x^4y^5} - 4xy\sqrt[3]{16xy^2}$

MAINTAIN YOUR SKILLS (SECTION 5.6)

Solve the following proportions:

75. $\dfrac{1\frac{1}{3}}{2\frac{1}{2}} = \dfrac{3\frac{3}{4}}{x}$

76. $\dfrac{25}{12} = \dfrac{90}{x}$

77. $\dfrac{5}{x} = \dfrac{7.5}{9}$

78. $\dfrac{48}{35} = \dfrac{r}{49}$

79. The tax rate for a certain city is \$29.75 per \$1000 of assessed valuation. What will be the tax on property assessed at \$110,000?

80. The scale on a city map is 1 in. = 500 ft. If a city block is 650 ft long, how many inches on the map are needed to show one city block?

81. Two partners in a business share profits in the ratio of 3 to 5. If the profits for one month were $5000, how much did each partner receive?

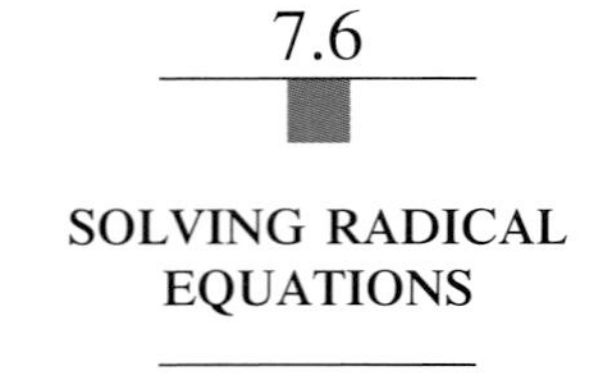

7.6 SOLVING RADICAL EQUATIONS

OBJECTIVE

1. Solve an equation that contains a variable in a radical expression.

Radicals in an equation can be eliminated by squaring each side.

PROPERTY

Squaring Property of Equality

If $a = b$, then $a^2 = b^2$.

Although a very useful property, the squaring property is not reversible. This means that the resulting equation may or may not be equivalent to the original equation. For example, the equations $x = 4$ and $x^2 = 16$ are not equivalent since their solution sets are $\{4\}$ and $\{-4, 4\}$, respectively. Furthermore, the equation $\sqrt{x} = -3$ has no solution because $\sqrt{x}$ never represents a negative number. However, when both sides of this equation are squared, we get

$$(\sqrt{x})^2 = (-3)^2$$
$$x = 9.$$

The solution set is $\emptyset$, and the apparent solution, 9, is called an extraneous solution.

DEFINITION

Extraneous Solution

An *extraneous solution* is a number that is a solution to an equation obtained by squaring each side of an equation that does *not* satisfy the original equation.

The possibility of extraneous solutions makes it necessary to check the solutions after solving by squaring. The solutions that do not check are rejected as extraneous solutions. Extraneous solutions can also be introduced by multiplying both sides of an equation by zero.

EXAMPLE 1

Solve: $\sqrt{x - 1} = 6$

$(\sqrt{x - 1})^2 = 6^2$ — Square each side to eliminate the radical. Every time this step is performed, make a mental note that it will be necessary to check the solutions.

$x - 1 = 36$

$x = 37$

Check:

$\sqrt{37 - 1} = 6$ — Substitute into the original equation.

$\sqrt{36} = 6$ — True.

The solution set is $\{37\}$. ■

■ PROCEDURE

To solve an equation that has a variable in a radical expression:

1. Isolate the radical.
2. Square each side of the equation.
3. Solve.
4. Check for extraneous roots.

If the radical in an equation is not already isolated, the first step is to isolate it.

EXAMPLE 2

Solve: $\sqrt{1 + x} + 8 = 4$

$\sqrt{1 + x} = -4$ — Subtract 8 from each side to isolate the radical and prepare for squaring each side.

$(\sqrt{1 + x})^2 = (-4)^2$

$1 + x = 16$

$x = 15$

Check:

$\sqrt{1 + 15} + 8 = 4$

$\sqrt{16} + 8 = 4$

$4 + 8 = 4$

■ CAUTION

This check is not merely for finding errors. We *must* check for extraneous roots. The number, 15, does not check in the equation. Therefore, 15 is an extraneous root.

The solution set is $\emptyset$.

Since a principal square root is never negative, we could have observed earlier that $\sqrt{1 + x} = -4$ does not have a solution. ■

EXAMPLE 3

Solve: $2\sqrt{2x - 1} + 7 = 10$

$2\sqrt{2x - 1} = 3$ Subtract 7 from both sides to isolate the radical (to prepare for squaring both sides).

$$(2\sqrt{2x - 1})^2 = (3)^2$$

$$4(2x - 1) = 9$$

$$8x - 4 = 9$$

$$8x = 13$$

$$x = \frac{13}{8}$$

Check:

$2\sqrt{2\left(\frac{13}{8}\right) - 1} + 7 = 10$ Substitute $x = \frac{13}{8}$ in the original equation.

$$2\sqrt{\frac{13}{4} - 1} + 7 = 10$$

$$2\sqrt{\frac{9}{4}} + 7 = 10$$

$2\left(\frac{3}{2}\right) + 7 = 10$ True.

The solution set is $\left\{\frac{13}{8}\right\}$. ■

A calculator is useful for quickly checking for extraneous solutions.

EXAMPLE 4

Check the solution in Example 3 with a calculator.

$2\sqrt{2x - 1} + 7 = 10$; $x = \frac{13}{8}$

$2\sqrt{2(13 \div 8) - 1} + 7$ Substitute $x = \frac{13}{8} = 13 \div 8$.

[13] [÷] [8] [×] [2] [−] [1] [=]

[√‾] [×] [2] [+] [7] [=] 10. Each side is equal to 10. ■

It is possible for the solution to be a negative number, even when a variable occurs under a radical.

EXAMPLE 5

Solve: $5\sqrt{10 - y} + 8 = 33$

$5\sqrt{10 - y} = 25$	Subtract 8 from both sides.
$\sqrt{10 - y} = 5$	Divide both sides by 5.
$10 - y = 25$	Square both sides.
$-y = 15$	
$y = -15$	

Check:

$5\sqrt{10 - (-15)} + 8 = 33$	Substitute $y = -15$ in the original equation.
$5\sqrt{25} + 8 = 33$	
$25 + 8 = 33$	
$33 = 33$	

The solution set is $\{-15\}$. ■

If each side of an equation is a radical expression, squaring each side will still eliminate the radicals.

EXAMPLE 6

Solve: $2\sqrt{x - 6} = \sqrt{x + 3}$

$(2\sqrt{x - 6})^2 = (\sqrt{x + 3})^2$	Square both sides to eliminate radicals.
$4(x - 6) = x + 3$	
$4x - 24 = x + 3$	
$3x = 27$	
$x = 9$	

Check:

$2\sqrt{9 - 6} = \sqrt{9 + 3}$	Substitute $x = 9$ in the original equation.
$2\sqrt{3} = \sqrt{12}$	
$2\sqrt{3} = \sqrt{4} \cdot \sqrt{3}$	
$2\sqrt{3} = 2\sqrt{3}$	

The solution set is $\{9\}$. ■

It may be necessary to square each side two times in order to eliminate all radicals.

EXAMPLE 7

Solve: $\sqrt{x} + 2 = \sqrt{x + 8}$

$(\sqrt{x} + 2)^2 = (\sqrt{x + 8})^2$ — Square each side to eliminate the radical on the right.

CAUTION

$(\sqrt{x} + 2)^2 \neq x + 4$

$x + 4\sqrt{x} + 4 = x + 8$ — A radical still remains on the left.

$4\sqrt{x} = 4$ — Simplify then isolate the radical.

$\sqrt{x} = 1$

$(\sqrt{x})^2 = 1^2$ — Square each side a second time.

$x = 1$

The solution set is {1}. — The check is left for the student. ■

EXAMPLE 8

What is the wattage in an electrical system that requires 4 amperes and has 220 ohms of resistance? The formula for the system is

$$\text{amperes} = \sqrt{\frac{\text{watts}}{\text{ohms}}}.$$

Formula:

$$\text{amperes} = \sqrt{\frac{\text{watts}}{\text{ohms}}}$$

Substitute:

$$4 = \sqrt{\frac{w}{220}}$$

Amperes = 4, ohms = 220.
Let w represent the wattage.

Solve:

$$4^2 = \frac{w}{220}$$

Square both sides.

$$16 = \frac{w}{220}$$

$$3520 = w$$

Check:

$$4 = \sqrt{\frac{3520}{220}}$$

$$4 = \sqrt{16}$$

Answer:

The system uses 3520 watts of power. ■

EXERCISE 7.6

A

Solve:

1. $\sqrt{x} = 3$ **2.** $\sqrt{x} = 7$ **3.** $\sqrt{y} = 11$ **4.** $\sqrt{y} = 13$

5. $\sqrt{2w} = 5$ **6.** $\sqrt{3w} = 6$ **7.** $\sqrt{5t} = 15$ **8.** $\sqrt{6t} = 9$

9. $\sqrt{x + 3} = 5$ **10.** $\sqrt{x + 4} = 1$ **11.** $\sqrt{2x + 3} = 5$ **12.** $\sqrt{3x - 1} = 1$

13. $\sqrt{3w} = -9$ **14.** $\sqrt{5w} = -15$ **15.** $\sqrt{x} + 6 = 8$ **16.** $\sqrt{x} - 7 = 3$

17. $2\sqrt{x} + \sqrt{x} = 4$ **18.** $3\sqrt{y} + \sqrt{y} = 2$ **19.** $3\sqrt{x} + 7 = 5$ **20.** $2\sqrt{w} + 2 = -1$

B

Solve:

21. $\sqrt{3x + 1} = 5$ **22.** $\sqrt{2x - 3} = 8$ **23.** $\sqrt{2r + 5} - 3 = 7$

24. $\sqrt{3r - 3} + 8 = 10$ **25.** $\sqrt{m - 7} + 8 = 5$ **26.** $\sqrt{2m + 7} + 5 = 5$

27. $\sqrt{y - 10} - 6 = -2$ **28.** $\sqrt{y + 10} - 6 = -2$ **29.** $5 - \sqrt{x + 1} = -9$

30. $-11 = 6 - \sqrt{2x + 1}$ **31.** $-8 = 7 + \sqrt{3x - 1}$ **32.** $\sqrt{5y + 9} - 9 = -6$

33. $\sqrt{t} + 3\sqrt{2} = 5\sqrt{2}$ **34.** $\sqrt{t} - 8\sqrt{3} = -4\sqrt{3}$ **35.** $2\sqrt{x - 4} = \sqrt{20}$

36. $3\sqrt{x + 1} = \sqrt{12}$ **37.** $7 - \sqrt{5x} = -1$ **38.** $11 - \sqrt{3x} = -2$

39. $3\sqrt{x + 2} = 2\sqrt{x + 3}$ **40.** $4\sqrt{5 - x} = 5\sqrt{4 - x}$

C

Solve:

41. $2\sqrt{x} + 3\sqrt{x} + 6 = 8$

42. $4\sqrt{x} - 7\sqrt{x} + 2 = -13$

43. $8\sqrt{a} + 6\sqrt{a} - 3 = 11$

44. $2\sqrt{a} - 5\sqrt{a} + \sqrt{a} + 7 = -7$

45. $4\sqrt{y} - \sqrt{y} + 8 = 15$

46. $6\sqrt{y} - \sqrt{y} - 3 = 7$

47. $8 - 3\sqrt{w} - 4 = 16$

48. $0 = 18 - 3\sqrt{w} + 2$

49. $2\sqrt{t} - 7\sqrt{t} + 8 = 3\sqrt{t} - 8$

50. $2\sqrt{t+5} = 3$

51. $\sqrt{5x-3} = \sqrt{7x-5}$

52. $3\sqrt{x+5} = 2\sqrt{x-1}$

53. $\sqrt{x+3} = \sqrt{2x-1}$

54. $\sqrt{2x-1} = \sqrt{x+7}$

55. $\sqrt{y+18} = -\sqrt{y-1}$

56. $-\sqrt{y-3} = \sqrt{y+2}$

57. $\sqrt{x-5} = 5 - \sqrt{x}$

58. $\sqrt{y-4} = \sqrt{y} - 1$

59. $\sqrt{32+x} = 16 - \sqrt{x}$

60. $\sqrt{4y-11} = 2\sqrt{y} - 1$

D

61. The length of a rectangle can be expressed as $\sqrt{y+3}$ and the width as $\sqrt{y-2}$. If the area of the rectangle is $\sqrt{14}$ square inches, determine the length and width of the rectangle.

62. The length of a rectangle can be expressed as $\sqrt{x-2}$ and the width as $\sqrt{x-6}$. If the area of the rectangle is $\sqrt{5}$ square inches, determine the length and width of the rectangle.

63. Find the resistance in ohms of an electrical system that uses 360 watts and requires 3 amperes. Use the formula given in the application.

64. Find the wattage in an electrical system that requires 3.5 amperes and has 80 ohms of resistance.

65. Find the stroke (measured in inches) of a 4-cylinder, 180-in^3 displacement engine if the bore of each cylinder is 3.5 in. (to the nearest hundredth).

$$\text{bore} = \sqrt{\frac{1}{0.7854} \cdot \frac{\text{displacement}}{\text{number of cylinders}} \cdot \frac{1}{\text{stroke}}}$$

66. Using the formula in Exercise 65, find the stroke (measured in inches) of a 6-cylinder, 240-in^3 displacement engine if the bore of each cylinder is 3.5 in.

67. The velocity, v, in meters per second, of an object can be determined by the formula $v = \sqrt{\dfrac{3.2 \times 10^{-19} K}{m}}$ where K is the kinetic energy in electron volts (ev) and m is the mass in kilograms (kg). Determine the kinetic energy of a neutron having a mass of 1.7×10^{-27} kg and a velocity of 3.3×10^4 m/s. Express the answer rounded to the nearest tenth.

68. The period of motion, T, in seconds, of an object on the end of an oscillating spring is given by the formula $T = 2\pi\sqrt{\frac{w}{gk}}$ where w is the weight of the object in pounds (lb), g is the gravitational constant in feet per second per second (ft/sec^2), and k is the spring constant in pounds per foot (lb/ft). Approximate the gravitational constant when $T = 0.79$ sec, $w = 1.5$ lb and $k = 3.0$ lb/ft. Use $\pi = 3.14$. Express the answer to the nearest whole number.

STATE YOUR UNDERSTANDING

69. Explain why we must always check the solution to a radical equation.

70. When have we previously encountered extraneous roots? What caused these roots to occur?

CHALLENGE EXERCISES

Solve:

71. $\sqrt{12 - \sqrt{x}} = 3$

72. $\sqrt[3]{x + 1} = 2$

73. $\sqrt[3]{2x - 1} = -5$

74. $\sqrt[4]{\frac{3x + 5}{2}} + 3 = 5$

MAINTAIN YOUR SKILLS (SECTION 2.6, 5.7)

75. 6 is what percent of 30?

76. At a special red-tag sale at the We Got It department store, all merchandise marked with a red tag will be sold at 15% off the last price. What will a lamp cost that was last marked at $23.20?

77. Mr. Riches lost 46% of his investment in the Lost Bull Mine. If he recovered $7020 of his investment, what did he originally invest in the Lost Bull Mine?

78. If m varies directly as n and $m = 20$ when $n = 45$, find m when $n = 27$.

79. If s varies directly as the square of t and $s = 10$ when $t = 6$, find s when $t = 12$.

80. If a varies jointly as b and c and $a = 14$ when $b = 5$ and $c = 6$, find a when $b = 10$ and $c = 12$.

81. The weight of pipe varies directly as its length. If 250 ft of pipe weighs 410 pounds, what will 175 ft of pipe weigh?

82. A real estate salesperson's salary varies directly as his or her total sales. If the salary is $512 for sales of $6800, what will be the salary earned on sales of $21,500 (to the nearest dollar)?

CHAPTER 7
SUMMARY

Square Root	A square root of m is a, where $a^2 = m$.	(p. 471)
Radical Sign	The radical sign, $\sqrt{\ }$, is used to designate the principal square root of a number.	(p. 471)
Radicand	A radicand is the expression that appears under a radical sign. The radicand of $3\sqrt{7pq}$ is $7pq$.	(p. 472)
Radical Expression	A radical whose radicand is an algebraic expression. The expression $\sqrt{5x^2y}$ is a radical expression.	(p. 481)
Perfect Square	A rational number that has rational square roots. The numbers 196 and $\frac{144}{25}$ are perfect squares.	(p. 472)
Irrational Numbers	Numbers that cannot be written in the form $\frac{a}{b}$, where a is an integer, and b is a natural number. Irrational numbers include radicals that are not perfect squares. The radical $\sqrt{6}$ represents an irrational number.	(p. 474)
Real Numbers	The union of the sets of rational and irrational numbers. Real numbers can be graphed on a number line.	(p. 474)
Nonreal Numbers	These are called complex numbers and arise in this chapter from even roots of negative numbers. The symbol $\sqrt{-49}$ does not represent a real number.	(p. 474)
nth Root	An nth root of m is a, where $a^n = m$. The sixth root ($n = 6$) of 729 is 3, $\sqrt[6]{729} = 3$.	(p. 477)
Simplest Radical Form	A radical expression is written in simplest form when: **1.** No integer factor has a perfect square factor other than 1. **2.** No radicand contains a fraction. **3.** No denominator of a fraction contains a radical.	(pp. 481, 482, 495)
Basic Principle of Radicals	$\sqrt{ab} = \sqrt{a}\sqrt{b}$	(p. 481)
Multiplying Radicals	Radical expressions are multiplied and simplified using the basic principle of radicals.	(pp. 488, 489)
Dividing Radicals	Radical expressions are divided by writing in fraction form and rationalizing the denominator.	(pp. 494, 495)
Combining Radicals	Radical expressions are added and subtracted by using the distributive property. Only like radicals may be combined.	(p. 502)

Like Radicals — Like nth root radicals are radicals with the same radicand. The radicals $15\sqrt{17}$ and $-2\sqrt{17}$ are like radicals, whereas the radicals $12\sqrt{13}$ and $13\sqrt{12}$ are unlike radicals. (p. 502)

Radical Equations — Radical equations are solved by isolating the radical expression, then squaring both sides using the squaring property of equality. (p. 509)

Squaring Property of Equality — If $A = B$, then $A^2 = B^2$. Since the use of this property does not guarantee an equivalent equation, it is necessary to check solutions for extraneous roots. (p. 508)

Extraneous Roots — An apparent solution to an equation obtained by squaring each side that is not a root of the original equation. (p. 508)

CHAPTER 7
REVIEW EXERCISES

SECTION 7.1 Objective 1

Find the roots.

1. $\sqrt{121}$ **2.** $\sqrt{\frac{4}{25}}$ **3.** $-\sqrt{324}$ **4.** $-\sqrt{\frac{16}{81}}$

5. $\sqrt{-49}$ **6.** $-\sqrt{-100}$ **7.** $\sqrt{3969}$ **8.** $-\sqrt{2116}$

9. $-\sqrt{1849}$

Find the approximate root rounded to two decimal places.

10. $\sqrt{143}$ **11.** $\sqrt{210}$ **12.** $\sqrt{960}$

SECTION 7.1 Objective 2

Find the roots.

13. $\sqrt[3]{8}$ **14.** $\sqrt[3]{216}$ **15.** $\sqrt[3]{343}$ **16.** $\sqrt[3]{3375}$

17. $\sqrt[3]{-512}$ **18.** $\sqrt[3]{-216}$ **19.** $\sqrt[4]{256}$ **20.** $\sqrt[4]{2401}$

21. $\sqrt[4]{-10{,}000}$ **22.** $\sqrt[5]{1024}$ **23.** $\sqrt[5]{100{,}000}$ **24.** $\sqrt[5]{-243}$

SECTION 7.2 Objective 1

Simplify.

25. $\sqrt{48}$ **26.** $\sqrt{180}$ **27.** $\sqrt{55}$ **28.** $\sqrt{39}$

29. $\sqrt{a^8}$ **30.** $\sqrt{w^{36}}$ **31.** $\sqrt{72z^6}$ **32.** $\sqrt{81z^7}$

33. $\sqrt{175}$ **34.** $\sqrt{250b^5}$ **35.** $\sqrt{504x^5y^4}$

36. A square city lot has an area of 28,800 square feet. Express the length of one side in simplest radical form.

SECTION 7.3 Objective 1

Multiply and simplify.

37. $\sqrt{7}\sqrt{5}$ **38.** $\sqrt{19}\sqrt{6}$ **39.** $\sqrt{3}\sqrt{12}$ **40.** $\sqrt{5}\sqrt{45}$

41. $\sqrt{12}\sqrt{20}$ **42.** $\sqrt{18}\sqrt{27}$ **43.** $\sqrt{5}(7 - \sqrt{5})$ **44.** $\sqrt{7}(7 - \sqrt{5})$

45. $(\sqrt{3} - \sqrt{2})(\sqrt{3} + \sqrt{2})$ **46.** $(\sqrt{5} - \sqrt{7})(\sqrt{5} + \sqrt{7})$

47. $(2\sqrt{5} - \sqrt{11})(2\sqrt{5} + \sqrt{11})$

48. Use the formula $h = \frac{s}{2}\sqrt{3}$ to find the height of a triangular garden plot that is $4\sqrt{15}$ feet on each side. Write the answer in simplest radical form and to the nearest hundredth.

SECTION 7.4 Objective 1

Divide; rationalize the denominator.

49. $\sqrt{\frac{7}{2}}$ **50.** $\sqrt{\frac{2}{7}}$ **51.** $\frac{\sqrt{21}}{\sqrt{15}}$ **52.** $\frac{\sqrt{15}}{\sqrt{21}}$

53. $10\sqrt{22} \div 5\sqrt{6}$ **54.** $6\sqrt{5} \div 22\sqrt{10}$ **55.** $\frac{6\sqrt{15}}{\sqrt{75}}$ **56.** $\frac{7\sqrt{50}}{\sqrt{27}}$

57. $\frac{5}{6\sqrt{11}}$ **58.** $\frac{5}{6 + \sqrt{11}}$ **59.** $\frac{6}{\sqrt{11} - 5}$

60. A clock at city hall has a pendulum that is 28 feet long. Use the formula $T = 2\pi\frac{\sqrt{L}}{\sqrt{32}}$ to find the time it takes to make one cycle. Write the answer in simplest radical form and to the nearest hundredth.

SECTION 7.5 Objective 1

Combine.

61. $4\sqrt{7} + 3\sqrt{7} + 2\sqrt{7}$ **62.** $8\sqrt{17} - \sqrt{17} - 3\sqrt{17}$ **63.** $6\sqrt{50} - 3\sqrt{8}$

64. $10\sqrt{75} - 7\sqrt{12}$ **65.** $\sqrt{80} + \sqrt{27} - \sqrt{20}$ **66.** $\sqrt{24} - \sqrt{45} + \sqrt{96}$

67. $4\sqrt{x} + \sqrt{25x} - 17\sqrt{x}$ **68.** $18\sqrt{a} - \sqrt{25a} + \sqrt{9a} - 14\sqrt{a}$ **69.** $\sqrt{63} - \sqrt{112} + \sqrt{1183}$

70. $\sqrt{20} - 3\sqrt{125}$ **71.** $\sqrt{245} - \sqrt{500}$

72. Ethyl heats her bedroom with two portable heaters. One uses 525 watts and has a resistance of 27 ohms. The other uses 252 watts and has a resistance of 25 ohms. Use the formula $V = \sqrt{\text{watts} \cdot \text{ohms}}$ to find the total voltage needed to operate the two heaters. Write the answer in simplest radical form.

SECTION 7.6 **Objective 1**

Solve.

73. $\sqrt{x} = 9$

74. $\sqrt{y} = 15$

75. $\sqrt{w + 4} = 12$

76. $\sqrt{6 - x} = 5$

77. $\sqrt{2 + y} + 7 = 5$

78. $3\sqrt{x - 5} + 4 = 6$

79. $5\sqrt{2x - 1} = \sqrt{15}$

80. $14 - \sqrt{5y} = 24$

81. $8\sqrt{p - 3} = 4$

82. $3\sqrt{r + 5} = 1$

83. $-6\sqrt{5a - 1} = -18$

84. Find the wattage in an electrical circuit that requires 7 amperes and has 75 ohms of resistance.

CHAPTER 7
TRUE–FALSE CONCEPT REVIEW

Check your understanding of the language of algebra. Tell whether each of the following statements is true (always true) or false (not always true).

1. $\sqrt{25} = -5$.

2. In the expression $\sqrt{50}$, 50 is the radicand.

3. $\sqrt{50}$ is an irrational number.

4. $\sqrt{64}$ is a rational number.

5. $\sqrt{48} = \sqrt{40} + \sqrt{8}$.

6. $\sqrt{x^2} = x$ if $x \geq 0$.

7. $2\sqrt{3}$ and $\sqrt{3}$ are examples of like radical expressions.

8. To combine like radical expressions, combine the coefficients and square the common radical.

9. If $x > 0$ and $y > 0$, then $\sqrt{x} \cdot \sqrt{y} = \sqrt{xy}$.

10. The procedure used to simplify an expression such as $\frac{1}{\sqrt{5}}$ is called rationalizing the denominator.

11. The first step in solving an equation containing a square root radical is to square both sides of the equation.

12. An extraneous root of an equation does not check in the original equation.

13. $\sqrt{125}$ cannot be simplified.

14. $\sqrt{5} \cdot \sqrt{4} = 2\sqrt{5}$.

15. If $a^2 = b^2$, then $a = b$.

CHAPTER 7
TEST

1. Rationalize the denominator: $\sqrt{\frac{7}{18}}$
2. Multiply and simplify: assume the variables represent positive numbers: $\sqrt{3mn} \cdot \sqrt{21m^3n}$
3. Solve: $\sqrt{y - 7} = 12$
4. Find the square root: $-\sqrt{196}$
5. Simplify: $\sqrt{117}$
6. Rationalize the denominator: $\frac{5}{\sqrt{12}}$
7. Solve: $5 + 3\sqrt{2x + 26} = 23$
8. Multiply and simplify: $\sqrt{10}(\sqrt{15} - \sqrt{18})$
9. Combine: $3\sqrt{27} - 2\sqrt{12} - \sqrt{3}$
10. Multiply and simplify: $\sqrt{20} \cdot \sqrt{40}$
11. Combine: $13\sqrt{7} - 18\sqrt{7}$
12. Find the approximate square root rounded to two decimal places (use a square root table or calculator): $\sqrt{179}$
13. Combine: $\sqrt{72} - \sqrt{98} + \sqrt{242}$
14. Simplify; assume the variables represent positive numbers: $\sqrt{98pq^2}$
15. Multiply and simplify: $2\sqrt{8} \cdot 5\sqrt{162}$
16. Find the area of a triangle with base $\sqrt{30}$ ft and height $\sqrt{10}$ ft in simplest radical form.
17. A city lot is in the shape of a square and has an area of 8400 square feet. Find the length of one side of the lot in simplest radical form.
18. Fred has just bought two used portable electric heaters to heat his workshop. One heater uses 500 watts and has a resistance of 40 ohms. The second one uses 200 watts and has a resistance of 25 ohms. The voltage needed to operate one heater is given by the formula

    $V = \sqrt{(\text{watts})(\text{ohms})}$.

    How many volts are needed to operate both heaters? (Write in simplest radical form.)

CHAPTER

8

QUADRATIC EQUATIONS

SECTIONS

The radius of a curve that a train can negotiate safely is dependent upon the type of coupler used on the railroad cars. For cars equipped with Type E or F couplers, the minimum-radius curve is given by $R = \dfrac{B^2 - D^2 - E^2}{2E}$ where R is the minimum-radius curve in feet, B is one-half of the distance over coupling lines in feet, D is one-half of the distance over truck centers in feet, and E is the offset in the coupling line in feet. In Exercise 73, Section 8.1, you are asked to find the minimum radius for a specific type of coupling. *(Nick Nicholson/The Image Bank)*

PREVIEW

Chapter 8 continues the solution of quadratic equations. We progress from taking the square root of both sides of an equation to completing the square to the quadratic formula. In the section on special cases, we include the Pythagorean theorem for solving right triangles. We introduce the complex numbers to make it possible to solve every quadratic equation. We also introduce graphs of quadratic equations, parabolas, to prepare for the study of functions in later courses.

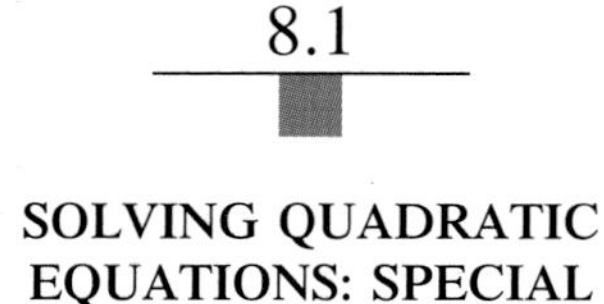

8.1 SOLVING QUADRATIC EQUATIONS: SPECIAL CASES

OBJECTIVES

1. Solve a quadratic equation of the form $ax^2 + bx = 0$.
2. Solve a quadratic equation of the form $ax^2 + c = 0$.

Recall that quadratic equations were discussed in Section 4.1.

DEFINITION

Standard Form of a Quadratic Equation

The *standard form of a quadratic equation* is

$ax^2 + bx + c = 0$, where $a > 0$.

The equation $2x^2 - 14x - 21 = 0$ is a quadratic equation in standard form, where $a = 2$, $b = -14$, and $c = -21$.

It is always possible to write a given quadratic equation in standard form.

In this section, we study special cases of quadratic equations in which $b = 0$ or $c = 0$. These equations are sometimes called incomplete quadratic equations.

DEFINITION

Incomplete Quadratic Equation

An *incomplete quadratic equation* has the form

$ax^2 + c = 0$ or $ax^2 + bx = 0$.

The equation $ax^2 + bx = 0$ can be solved by factoring.

EXAMPLE 1

Solve: $4x^2 + 8x = 0$

$4x(x + 2) = 0$ — Factor; the GCF is $4x$.

$4x = 0$ or $x + 2 = 0$ — Zero-product property.

$x = 0$ or $x = -2$ — The check is left for the student.

The solution set is $\{-2, 0\}$. ■

■ PROCEDURE

> To solve an incomplete quadratic equation of the form
> $ax^2 + bx = 0$:
>
> 1. Factor the left side of the equation.
> 2. Use the zero-product property.
> 3. Solve each equation.

Always remember to write the equation in standard form before trying to solve it.

EXAMPLE 2

Solve: $3x(x - 4) + 8x = (x - 3)^2 - 9$

$3x^2 - 12x + 8x = x^2 - 6x + 9 - 9$ — Multiply to clear parentheses.

$2x^2 + 2x = 0$ — Write in standard form.

$2x(x + 1) = 0$ — Factor.

$2x = 0$ or $x + 1 = 0$ — Zero-product property.

$x = 0$ or $x = -1$ — The check is left for the student.

The solution set is $\{-1, 0\}$. ■

We now consider the second type of incomplete quadratic.

$ax^2 + c = 0$

$x^2 = -\dfrac{c}{a}$ — Isolate x^2 on the left side.

$x^2 = K$ — Let $-\dfrac{c}{a} = K$. This equation may be read, "What number squared is equal to K?"

$x = \sqrt{K}$ or $x = -\sqrt{K}$ — Every positive number has two square roots, a positive root and a negative root.

EXAMPLE 3

Solve: $x^2 = 36$

$x^2 = 36$

$x = 6$ or $x = -6$ The two square roots of 36 are 6 and -6.

The solution set is $\{6, -6\}$.

DEFINITION

$\pm a$

The symbol "$\pm a$" means a or $-a$.

Using this symbol, the solution set in Example 3 is $\{\pm 6\}$.

EXAMPLE 4

Solve: $x^2 = -9$

No real solution There is no real number solution since the square of a real number is positive or 0.

Solutions of equations of the form $x^2 = K$, where $K < 0$, are introduced in Section 8.5.

PROCEDURE

To solve an incomplete quadratic equation of the form $x^2 = K$:

1. If $x^2 = K$, $K \geq 0$, the solutions are $x = \pm\sqrt{K}$.
2. If $x^2 = K$, $K < 0$, there are no real solutions.

EXAMPLE 5

Solve: $4x^2 = 25$

$x^2 = \frac{25}{4}$ Write the equation in the form $x^2 = K$.

$x = \pm\frac{5}{2}$ $x = \pm\sqrt{K}$

The solution set is $\left\{\pm\frac{5}{2}\right\}$.

If the equation is not in standard form, first write it in standard form.

EXAMPLE 6

Solve: $6 + 2x^2 = x^2 + 9$

$x^2 = 3$ — Add $-x^2$ and -6 to each side.

$x = \pm\sqrt{3}$ — $x = \pm\sqrt{K}$

The solution set is $\{\pm\sqrt{3}\}$. ■

A calculator may be used to find approximations of roots that are in radical form.

EXAMPLE 7

Solve, give answer in simplest radical form and as an approximation to the nearest hundredth:

$5x^2 = 324$

$x^2 = \dfrac{324}{5}$ — Write in the form $x^2 = K$.

$x = \pm\sqrt{\dfrac{324}{5}} = \pm\dfrac{18}{\sqrt{5}}$ — If $x^2 = K$ then $x = \pm\sqrt{K}$.

$= \pm\dfrac{18\sqrt{5}}{5}$ — Rationalize the denominator to get the simplest radical form.

$x \approx \pm 8.0498447$ — Use a calculator to approximate.

$\approx \pm 8.05$ — Round to the nearest hundredth.

The solution set is $\left\{\pm\dfrac{18\sqrt{5}}{5}\right\}$ or approximately $\{\pm 8.05\}$. ■

In this text the solutions are given in simplest radical form unless an approximation is specified.

Using a quadratic equation, an engineer can determine the inside diameter of a pipe. For a pipe carrying a fluid, the new diameter can be used to determine the wear on the original diameter. The formula for the inner diameter of a pipe is

$$D^2 = \frac{4F}{\pi V},$$

where D is the inside diameter, F is the flow rate (in cubic feet per second), and V is the velocity of the fluid (in feet per second).

EXAMPLE 8

A pipe that is 0.5 inches thick has an inside diameter of 0.3 ft when installed. After 25 years the flow rate, F, is 3.5 ft^3/sec, and the velocity of the fluid (V) is 40 ft/sec. Find the current inside diameter of the pipe, to the nearest hundredth of a foot, and the current thickness of the pipe to the nearest hundredth of an inch.

Formula:

$$D^2 = \frac{4F}{\pi V}$$

First, find the current diameter of the inside of the pipe.

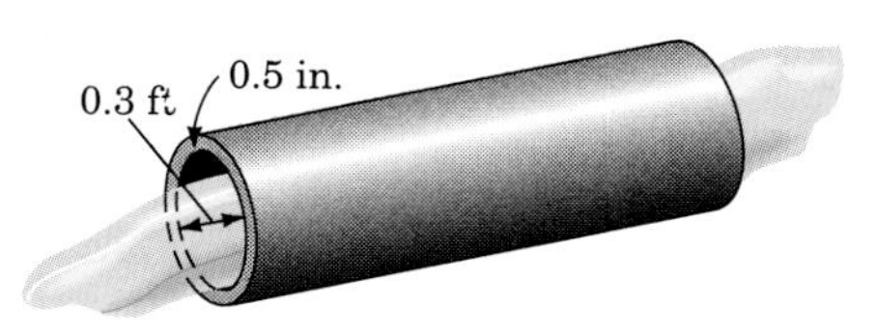

Substitute:

$$D^2 = \frac{4(3.5)}{(3.14)(40)}$$

$F = 3.5\ ft^3/sec$, $V = 40$ ft/sec
$\pi \approx 3.14$

$$D^2 \approx 0.111464968$$

Using a calculator.

$$D \approx \pm 0.333863697$$

Since the diameter of a pipe cannot be negative, we use the positive value.

$$\approx 0.33$$

Round to nearest hundredth.

The inside diameter is approximately 0.33 ft.

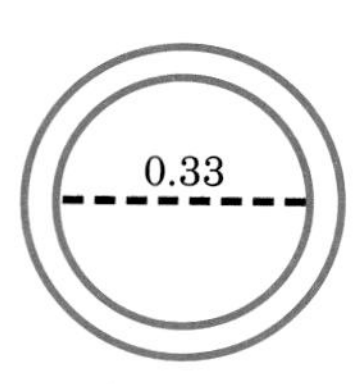

$$0.33 \text{ ft} - 0.3 \text{ ft} = 0.03 \text{ ft}$$

The inside diameter of the pipe has increased by 0.03 ft.

$$\frac{0.03 \text{ ft}}{2} = 0.015 \text{ ft}$$

Assuming even wear, 0.015 ft has been worn off each side.

$$(0.015 \text{ ft})(12 \text{ in./ft}) = 0.18 \text{ in.}$$

Convert to inches.

$$0.5 \text{ in.} - 0.18 \text{ in.} = 0.32 \text{ in.}$$

To find the current thickness, subtract the wear (0.18) from the original thickness (0.5).

Answer:

The current inside diameter is 0.33 ft, and the thickness of the pipe is now 0.32 in. ■

Another useful application of incomplete quadratic equations is the solution of right triangles.

DEFINITION

Right Triangle

A *right triangle* is a triangle that contains a 90° angle. The side opposite the right angle (90° angle) is called the hypotenuse, and the other two sides are called legs.

The Pythagorean theorem describes the relationship between the sides of a right triangle.

FORMULA

Pythagorean Theorem

Given a right triangle with legs a and b and hypotenuse c, we have the relationship shown in Figure 8.1:

$$a^2 + b^2 = c^2$$

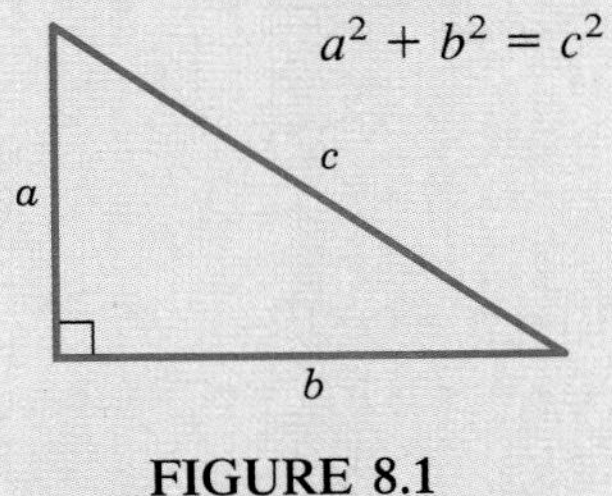

FIGURE 8.1

If any two sides of a right triangle are given, we can use the Pythagorean theorem to find the third side.

EXAMPLE 9

Find the missing side of the following triangle:

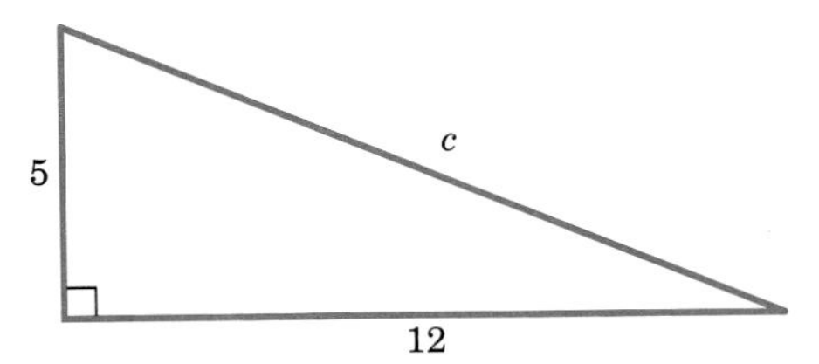

$c^2 = a^2 + b^2$	Pythagorean formula.
$c^2 = (5)^2 + (12)^2$	$a = 5$ and $b = 12$
$c^2 = 25 + 144$	
$c^2 = 169$	
$c = \pm 13$	The negative root is rejected since it does not apply to measuring the side of a triangle.

The missing side (hypotenuse) is 13. ■

If the length of the missing side is not a whole number, it can be approximated or left in radical form.

EXAMPLE 10

Find the length of the leg of a right triangle (to the nearest hundredth) that has one leg of 12 cm and a hypotenuse of 17 cm.

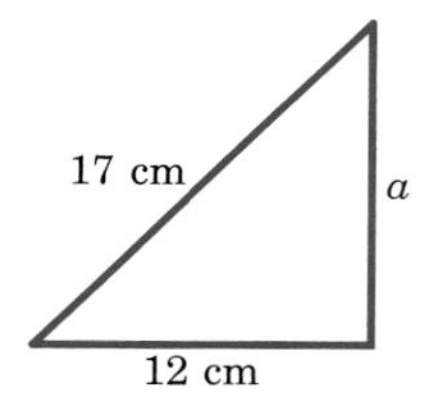

First, sketch the triangle.

Here we call the missing side a.

$c^2 = a^2 + b^2$	
$(17)^2 = a^2 + (12)^2$	$c = 17$ and $b = 12$
$289 = a^2 + 144$	
$a^2 = 145$	
$a = \pm 12.04$	$\sqrt{145} \approx 12.04$

The length of the missing leg is approximately 12.04 cm. ■

The Pythagorean theorem can also be used to show that a triangle is a right triangle.

EXAMPLE 11

Is a triangle with sides of 12 ft, 18 ft, and 24 ft a right triangle?

If the sides form a right triangle, then they must satisfy $a^2 + b^2 = c^2$. The longest measurement is side c.

$a^2 + b^2 = c^2$	
Does $12^2 + 18^2 = 24^2$?	Substitute in the formula.

$144 + 324 \stackrel{?}{=} 576$ The "?" over the equal sign asks the question, "Are the two quantities equal?" We see that they are not.

$468 \neq 576$

The triangle is not a right triangle. ■

Exercise 8.1

A

Solve:

1. $x^2 = 4$ **2.** $b^2 = 64$ **3.** $a^2 = 121$ **4.** $y^2 = 169$

5. $4a^2 = 49$ **6.** $16b^2 = 25$ **7.** $x^2 = -16$ **8.** $a^2 = -9$

9. $-y^2 = 16$ **10.** $-3x^2 = -27$ **11.** $x^2 = 400$ **12.** $y^2 = 225$

13. $x^2 - 11x = 0$ **14.** $x^2 - 8x = 0$ **15.** $6x^2 + 18x = 0$ **16.** $2x^2 + 14x = 0$

Find the missing side of each of the following right triangles:

17. $a = ?, b = 12, c = 13$ **18.** $a = 8, b = 15, c = ?$ **19.** $a = 15, b = ?, c = 39$

20. $a = 24, b = ?, c = 51$ **21.** $a = 20, b = 15, c = ?$ **22.** $a = 16, b = ?, c = 34$

B

Solve:

23. $3a^2 = 21a$ **24.** $12x^2 = -15x$

25. $2x^2 - 35 = x^2 - 9$ **26.** $5x^2 - 21 = 4x^2 + 6$

27. $8x^2 + 62 = 3x^2 + 187$ **28.** $11x^2 - 23 = 4x^2 + 33$

29. $3 - 2x - x^2 = 5 - 2x - 2x^2$ **30.** $11 - 5x + 3x^2 = 21 - 5x + 2x^2$

31. $1 - \frac{36}{25}x^2 = 0$ **32.** $\frac{25}{49}x^2 - 1 = 0$

33. $3(x^2 + 8) = 24 - 15x$ **34.** $5x(2x - 4) = (3x + 1)(3x - 2) + 2$

35. $2x(5x + 1) = (2x + 3)(3x - 4) + 12$ **36.** $(x + 3)^2 = 9 + 11x + 2x^2$

37. $2x(x - 7) = (x - 5)^2 - 25$ **38.** $(x - 2)^2 + (x + 3)^2 = 13$

Find the missing side of each of the following right triangles:

39. $a = ?, b = 5.1, c = 8.5$ **40.** $a = 4, b = ?, c = 8.5$ **41.** $a = 13.5, b = 18, c = ?$

42. $a = 9, b = ?, c = 9.75$ **43.** $a = ?, b = 2, c = 4.25$ **44.** $a = 1.2, b = 1.6, c = ?$

C

Solve; approximate answers to the nearest hundredth (use a calculator or use the square root table in Appendix VI):

45. $a^2 = 23$

46. $b^2 = 41$

47. $3x^2 - 15 = 22$

48. $4y^2 + 17 = 40$

49. $x(3x - 5) = 2x^2 - 5x + 19$

50. $4x^2 - 8x + 11 = x(3x - 8) + 45$

Solve:

51. $4x^2 + 8x - 10 = x(x + 2) + 2(x - 5)$

52. $(x + 3)(x + 4) = -4x(x - 5) + 12$

53. $(x + 2)(x - 2) + (x + 2)^2 = 0$

54. $(3x - 5)(7x + 6) + 15 = 3(x - 5)$

55. $\dfrac{21}{y - 3} = y - 7$

56. $\dfrac{36}{a - 4} = a - 9$

57. $(y + 1)^2 = 2y + 1$

58. $(a - 2)^2 = 3a + 4$

Solve for x:

59. $4x^2 = 100a^2, a \geq 0$

60. $12x^2 = 192b^2, b \geq 0$

Find the missing side of each of the following right triangles:

61. $a = 1, b = 1.25, c = ?$

62. $a = 1.25, b = ?, c = 5$

63. $a = ?, b = 8, c = 11$

64. $a = 12, b = ?, c = 20$

65. $a = 5.7, b = 7.3, c = ?$

66. $a = 9.1, b = ?, c = 12.5$

D

67. Using the formula in the Example 8, find the diameter of the pipe when $F = 12$ ft^3/sec and $V = 6$ ft/sec. (Round the answer to the nearest tenth.)

68. Using the formula in Example 8, find the diameter of the pipe when $F = 10$ ft^3/sec and $V = 5$ ft/sec. (Round the answer to the nearest tenth.)

69. A square plot of land has an area of 2116 ft^2. What is the length of a side?

70. The length of a rectangle can be expressed as $y - 3$ and the width as $y - 5$. Find the length and width if the area is 15 square inches.

71. $K = \dfrac{1}{2}mv^2$ is the formula for kinetic energy, where K is kinetic energy, m is the mass, and v is the velocity. Find v ft/sec if $K = 4{,}000{,}000$ and $m = 2$.

72. A lawn sprinkler covers a circular area of 1809 ft^2. How far must it be placed from a sidewalk so that the walk will not be sprinkled? (Find to the nearest tenth; let $\pi = 3.14$.)

73. For railroad cars equipped with Type E or F couplers, the minimum-radius curve that can be negotiated is given by

$$R = \frac{B^2 - D^2 - E^2}{2E},$$

where R = minimum radius of curve (ft), B = one-half of distance over coupling lines (ft), D = one-half of distance over truck centers (ft), and E = offset at coupling line (ft).* Find B when $E = 1.3$ ft, $D = 23.1$ ft, and $R = 188$ ft (to the nearest tenth).

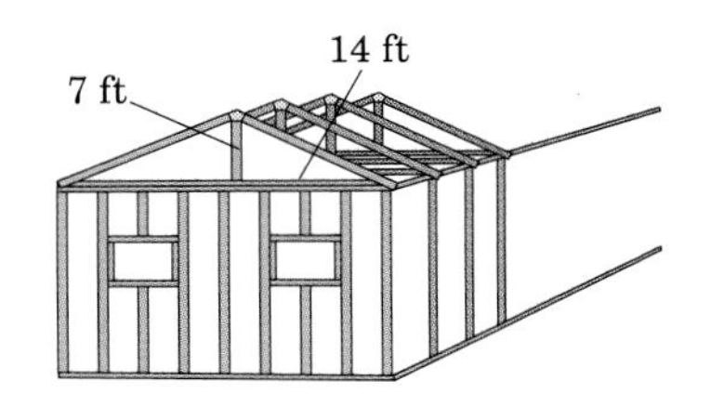

Figure for Exercise 75

74. What is the length of a rafter with a rise of 5′ and a run of 13′ (to the nearest hundredth)?

75. What is the length of a rafter with a rise of 7′ and a run of 14′ (to the nearest hundredth)?

76. What is the length of a rafter with a rise of 8′ and a run of 12′ (in simplified radical form)?

77. What is the rise of a rafter that is 20 feet long and has a run of 16 feet?

78. A guy wire 10 feet from the base of a house is attached to the house 22 feet above the ground. Find the length of the wire to the nearest tenth of a foot.

79. A 10-foot ladder leans against a house with the bottom of the ladder 3.5 feet from the base of the house. How high up the house does the top of the ladder touch? (Find to the nearest tenth of a foot.)

80. A guy wire attached to the top of a 20-foot tree is fastened to the ground 15 feet from the base of a tree. (Find the length of wire to the nearest tenth of a foot.)

81. What is the length of cable needed to replace a brace on a 50′ power pole that is attached to a ground-level anchor that is 35′ from the base of the pole (to the nearest tenth of a foot)?

82. A baseball "diamond" is actually a square that is 90 feet on each side (between the bases). To the nearest tenth of a foot, what is the distance the catcher must throw when attempting to put out a runner who is stealing second base? (The following figure illustrates the problem.)

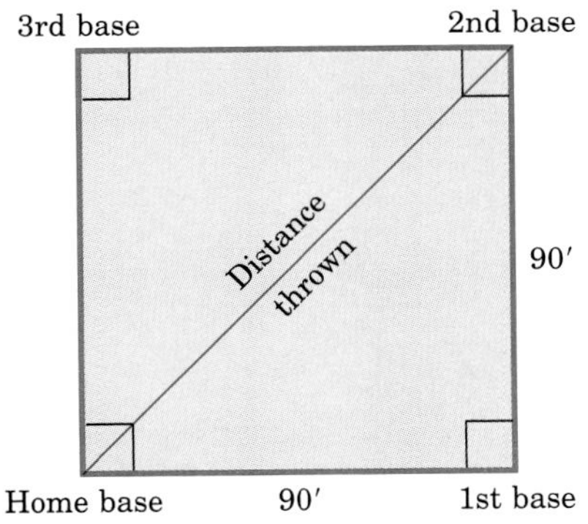

83. If runners are on first base and second base and each takes a 10 ft "lead" from his respective base, how far are they from each other (to the nearest hundredth)? (Use the figure shown in Exercise 82.)

*Formula courtesy of FMC Corp.

STATE YOUR UNDERSTANDING

84. Describe the procedure for solving an incomplete quadratic equation where $c = 0$, i.e., of the form $ax^2 + bx = 0$.

85. Describe the procedure for solving an incomplete quadratic equation where $b = 0$, i.e., of the form $ax^2 + c = 0$.

CHALLENGE EXERCISES

Solve for real solutions:

86. $(x + 5)^2 = 6$ Hint: Let $a = (x + 5)$

87. $(3x - 7)^2 - 20 = 0$

88. $3x^4 = 75$ Hint: Let $a = x^2$

89. $8x^4 - 40 = 0$

MAINTAIN YOUR SKILLS (SECTIONS 7.2, 7.3)

Simplify; assume that all variables represent positive numbers:

90. $\sqrt{252x^4y^5}$

91. $\sqrt{294a^5b^6c^3}$

92. $\sqrt{6} \cdot \sqrt{12} \cdot \sqrt{3} \cdot \sqrt{6}$

93. $15\sqrt{40} \cdot 3\sqrt{15}$

94. $\sqrt{18}(\sqrt{12} - \sqrt{10})$

95. $(\sqrt{8} - \sqrt{3})^2$

96. Find the area of a rectangle with base $\sqrt{15}$ inches and height $\sqrt{35}$ inches (a) in simplest radical form and (b) to the nearest tenth.

97. Find the area of a triangle with a base $\sqrt{30}$ ft and height $\sqrt{70}$ ft (a) in simplest radical form and (b) to the nearest tenth.

8.2

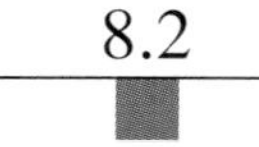

SOLVING QUADRATIC EQUATIONS: COMPLETING THE SQUARE

OBJECTIVE

1. Solve a quadratic equation of the form $ax^2 + bx + c = 0$ by writing it in the form $(x + d)^2 = e$.

An equation that is written in the form $(x + d)^2 = e$ can be solved by using the methods of Section 8.1.

EXAMPLE 1

Solve: $(y + 5)^2 = 16$

$y + 5$ is the number whose square is 16.

$y + 5 = \pm 4$ — Set $y + 5$ equal to each of the square roots of 16.

$y + 5 = -4$ or $y + 5 = 4$

$y = -9$ or $y = -1$ — Solve for y. The check is left for the student.

The solution set is $\{-9, -1\}$. ■

EXAMPLE 2

Solve: $(x - 6)^2 = 25$

$x - 6 = \pm 5$ — If $x^2 = K$, $x = \pm\sqrt{K}$.

$x - 6 = 5$ or $x - 6 = -5$ — Solve for x.

$x = 11$ or $x = 1$ — The check is left for the student.

The solution set is $\{11, 1\}$. ■

Any quadratic equation can be written in the form $(x + d)^2 = e$. The process of writing a quadratic in this form is called completing the square.

■ DEFINITION

Completing the Square

Changing a quadratic equation in standard form to the form $(x + d)^2 = e$ is called *completing the square*.

The goal in completing the square for a quadratic equation is to have a perfect square trinomial on the left side. A perfect square trinomial, with leading coefficient one, has the form

$x^2 + 2dx + d^2$.

The next two examples show the process of completing the square.

EXAMPLE 3

Solve: $x^2 - 14x + 33 = 0$

Step 1

$x^2 - 14x = -33$ — Add -33 to both sides so that we can complete the square on the left side.

Step 2

$x^2 - 14x + (-7)^2 = -33 + (-7)^2$ — In a perfect square trinomial, with leading coefficient of one, the constant term is the square of half the coefficient of the first-degree term.

$\frac{1}{2}(-14) = -7$, so add $(-7)^2 = 49$ to each side.

$(x - 7)^2 = -33 + 49$ Factor the left side, and simplify the right side.

$(x - 7)^2 = 16$

Step 3

$x - 7 = \pm 4$ Solve.

$x - 7 = 4$ or $x - 7 = -4$

$x = 11$ or $x = 3$ The check is left for the student.

The solution set is $\{3, 11\}$. ■

If the leading coefficient is not already one we make it one by division.

EXAMPLE 4

Solve: $4x^2 - 8x - 5 = 0$

Step 1

$4x^2 - 8x = 5$ Add 5 to each side.

Step 2

$x^2 - 2x = \frac{5}{4}$ Divide each side by 4 so the leading coefficient is 1.

Step 3

$x^2 - 2x + 1 = \frac{5}{4} + 1$ Add $(-1)^2$ to both sides. -1 is half the coefficient of the first degree term. $\frac{1}{2}(-2) = -1$

Step 4

$(x - 1)^2 = \frac{9}{4}$ Factor the left side and simplify the right side.

Step 5

$x - 1 = \pm\frac{3}{2}$ Solve.

$x - 1 = \frac{3}{2}$ or $x - 1 = -\frac{3}{2}$

$x = \frac{5}{2}$ or $x = -\frac{1}{2}$

The solution set is $\left\{-\frac{1}{2}, \frac{5}{2}\right\}$. ■

PROCEDURE

To solve the quadratic equation $ax^2 + bx + c = 0$, $a > 0$, $b \neq 0$, $c \neq 0$, by completing the square:

1. Add $-c$ to each side of the equation.
2. If $a \neq 1$, divide both sides of the equation by a.
3. Add to each side of the equation the number that will make the left side a perfect square trinomial. This number is the square of one-half of $\frac{b}{a}$ (the coefficient of x).

   $$\frac{1}{2}\left(\frac{b}{a}\right) = \frac{b}{2a} \quad \text{and} \quad \left(\frac{b}{2a}\right)^2 = \frac{b^2}{4a^2}$$

4. Factor the perfect square on the left side.
5. Solve the resulting equation that is in the form

   $$(x + d)^2 = e$$

   by setting $x + d$ equal to each of the square roots of e. If e is negative, the equation has no real number solutions.

Not all quadratic equations have real roots. When solving by completing the square, this is apparent when $e < 0$.

EXAMPLE 5

Solve: $2x^2 + 2x + 13 = 0$

$$2x^2 + 2x = -13 \qquad \text{Add } -13 \text{ to both sides.}$$

$$x^2 + x = -\frac{13}{2} \qquad \text{Divide both sides by 2.}$$

$$x^2 + x + \left(\frac{1}{2}\right)^2 = -\frac{13}{2} + \left(\frac{1}{2}\right)^2 \qquad \text{Add } \left(\frac{1}{2}\right)^2 \text{ to both sides.}$$

$$\left(x + \frac{1}{2}\right)^2 = -\frac{13}{2} + \frac{1}{4} \qquad \text{Factor and simplify.}$$

$$\left(x + \frac{1}{2}\right)^2 = -\frac{25}{4} \qquad \text{Since } e < 0 \text{, there are no real roots.}$$

There are no real solutions. ■

Examples 1–4 can also be solved by factoring and using the zero-product property. Quadratic equations that cannot be factored over the integers can be solved by completing the square.

EXAMPLE 6

Solve: $x^2 + 10x + 7 = 0$

$x^2 + 10x = -7$	Add -7 to both sides.
$x^2 + 10x + (5)^2 = -7 + (5)^2$	Add $(5)^2$ to both sides.
$(x + 5)^2 = 18$	Factor and simplify.
$x + 5 = \pm\sqrt{18}$	Solve.
$x = -5 \pm 3\sqrt{2}$	Simplify: $\sqrt{18} = 3\sqrt{2}$.
	Since we cannot combine $-5 + 3\sqrt{2}$ or $-5 - 3\sqrt{2}$, we are done.

The solution set is $\{-5 \pm 3\sqrt{2}\}$. ■

EXAMPLE 7

Solve: $-6x^2 - 21x + 3 = 0$

$-6x^2 - 21x = -3$	Add -3 to both sides.
$x^2 + \frac{7}{2}x = \frac{1}{2}$	Divide both sides by -6.
$x^2 + \frac{7}{2}x + \left(\frac{7}{4}\right)^2 = \frac{1}{2} + \left(\frac{7}{4}\right)^2$	Add the square of half $\frac{7}{2}$ to each member.
$\left(x + \frac{7}{4}\right)^2 = \frac{1}{2} + \frac{49}{16}$	Factor and simplify.
$\left(x + \frac{7}{4}\right)^2 = \frac{57}{16}$	
$x + \frac{7}{4} = \pm\frac{\sqrt{57}}{4}$	Solve.
$x = -\frac{7}{4} \pm \frac{\sqrt{57}}{4}$	
$x = \frac{-7 \pm \sqrt{57}}{4}$	

The solution set is $\left\{\frac{-7 \pm \sqrt{57}}{4}\right\}$. ■

EXAMPLE 8

It takes two hours longer to drain a swimming pool than to fill it. If the drain is left open, it takes 12 hours to fill the pool. How long does it take to drain the pool?

Simpler word form:

$$\begin{pmatrix}\text{Fraction of pool}\\ \text{that is filled}\\ \text{in 1 hour with}\\ \text{drain closed}\end{pmatrix} - \begin{pmatrix}\text{Fraction of pool}\\ \text{that is drained}\\ \text{in 1 hour with}\\ \text{water-supply pipe}\\ \text{shut off}\end{pmatrix} = \begin{pmatrix}\text{Fraction of pool}\\ \text{that can be filled}\\ \text{in 1 hour with}\\ \text{supply pipe on and}\\ \text{drain open}\end{pmatrix}$$

Select variable:

If t represents the number of hours to drain the pool with the supply pipe shut off, then $t - 2$ represents the time to fill the pool.

	Time	Fraction of the Pool Completed in One Hour
Fill the pool (drain shut)	$t - 2$	$\frac{1}{t-2}$
Drain the pool (water supply off)	t	$\frac{1}{t}$
Water supply on and drain open	12	$\frac{1}{12}$

Translate to algebra:

$$\frac{1}{t-2} - \frac{1}{t} = \frac{1}{12}$$

Solve:

$$12t - 12(t - 2) = t(t - 2)$$

Multiply both sides by the LCM.

$$12t - 12t + 24 = t^2 - 2t$$

$$0 = t^2 - 2t - 24$$

Simplify.

$$t^2 - 2t - 24 = 0$$

Use the symmetric law of equality to write the equation in standard form.

$$t^2 - 2t = 24$$ Solve by completing the square.

$$t^2 - 2t + 1 = 24 + 1$$

$$(t - 1)^2 = 25$$

$$t - 1 = \pm 5$$

$$t - 1 = 5 \quad \text{or} \quad t - 1 = -5$$

$$t = 6 \qquad\qquad t = -4$$

Since we are solving for a measure of time, we discard the negative root.

Answer:

It will take 6 hours to drain the pool. ■

EXERCISE 8.2

A

Solve:

1. $(x + 3)^2 = 25$

2. $(y - 5)^2 = 100$

3. $(x - 2)^2 = 12$

4. $(y + 6)^2 = 20$

5. $(x + 5)^2 = 20$

6. $(x - 4)^2 = 32$

Solve by completing the square:

7. $y^2 + 6y - 16 = 0$

8. $x^2 - 8x + 12 = 0$

9. $x^2 + 10x - 24 = 0$

10. $y^2 + 16y + 48 = 0$

11. $y^2 - 3y - 40 = 0$

12. $z^2 + 10z + 21 = 0$

13. $x^2 + 6x - 7 = 0$

14. $x^2 + 7x - 8 = 0$

15. $x^2 + 12x + 32 = 0$

16. $x^2 - 15x + 36 = 0$

17. $x^2 - 4x + 13 = 0$

18. $x^2 - 6x + 25 = 0$

19. $x^2 + 6x - 135 = 0$

20. $x^2 + 2x - 143 = 0$

B

Solve:

21. $\left(x - \frac{4}{3}\right)^2 = \frac{25}{9}$

22. $\left(y + \frac{7}{2}\right)^2 = \frac{25}{4}$

23. $\left(a + \frac{2}{3}\right)^2 = \frac{49}{9}$

24. $\left(b - \frac{3}{5}\right)^2 = \frac{16}{25}$

25. $\left(y + \frac{11}{6}\right)^2 = \frac{11}{36}$

26. $\left(x - \frac{5}{7}\right)^2 = \frac{3}{49}$

27. $2x^2 + 12x - 14 = 0$

28. $5x^2 + 20x - 25 = 0$

Solve by completing the square:

29. $x^2 + 5x + 5 = 0$

30. $x^2 + 7x - 1 = 0$

31. $2x^2 - 8x + 10 = 0$

32. $3x^2 + 5x + 11 = 0$

33. $2x^2 - 6x - 5 = 0$

34. $3x^2 + 5x + 1 = 0$

35. $2x^2 + 9x - 35 = 0$

36. $3x^2 + 8x - 16 = 0$

37. $2x^2 + 15x - 27 = 0$

38. $2x^2 + 3x - 27 = 0$

39. $5x^2 - 26x + 5 = 0$

40. $2x^2 - 9x - 18 = 0$

C

Solve by completing the square:

41. $x^2 - 7x - 2 = 0$

42. $2x^2 + 3x - 4 = 0$

43. $2x^2 - 9x + 8 = 0$

44. $7y^2 - 3y - 2 = 0$

45. $2x^2 - 19x + 35 = 0$

46. $4x^2 + 11x - 3 = 0$

47. $3y^2 + 8y + 6 = 0$

48. $4y^2 - 3y + 6 = 0$

49. $5x^2 - 11x - 2 = 0$

50. $3x^2 + 5x - 2 = 0$

51. $3y^2 + 16y + 16 = 0$

52. $4y^2 + 8y + 3 = 0$

53. $9x^2 - 4x + 1 = 0$

54. $16x^2 + 10x + 3 = 0$

55. $7x^2 - 5x + 12 = 0$

56. $8x^2 + 7x + 9 = 0$

57. $3y^2 + 8y - 2 = 0$

58. $4a^2 - 3a - 9 = 0$

59. $5b^2 + 4b - 6 = 0$

60. $6c^2 - 8c - 5 = 0$

D

61. In Example 8, how long will it take to fill the pool with the drain plugged if it takes 8 hours to fill it with the drain open (to the nearest tenth of an hour)?

62. It takes two hours to fill a tank with two faucets. The second faucet takes three more hours to fill the tank alone than the first faucet. How long would it take the first faucet alone?

63. It takes five hours to fill a tank with two faucets. The second faucet takes four more hours to fill the tank alone than the first faucet. To the nearest tenth of an hour, how long will it take the first faucet alone?

64. Two trucks leave Lansing, Michigan, and follow the same 450-mile route to Nashville, Tennessee. One truck averages 10 mph faster than the other one. What was the average speed of each truck if the slower truck took an additional two hours to complete the trip (to the nearest mph)?

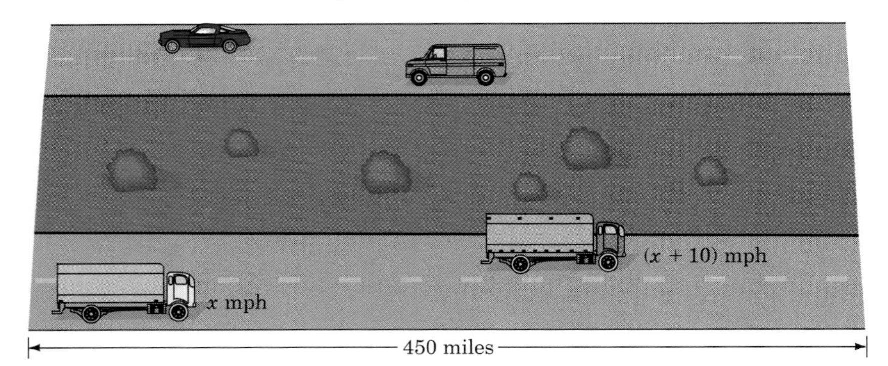

Figure for Exercise 64

65. During the Minnesota Junior Olympics, Rana ran the first 7 miles with the burning torch, and Ben ran the remaining 8 miles. If Ben averaged 5 mph faster than Rana, and if the total time for both runners was $1\frac{1}{2}$ hours, what was the average speed of each (to the nearest tenth of a mph)?

66. The cost per item on the manufacturing of n items at the RJ Specialty Company is given by

$C = 3n^2 - 243n - 620.$

For how many items is the cost \$130 per item (to the nearest whole item)?

67. The revenue per item on the manufacturing of n items at the DM Spool Company is given by

$R = 3n^2 - 300n - 500.$

For how many items is the revenue \$160 per item (to the nearest whole item)?

68. The profit per item on the manufacturing of n items at the "Preppy" Company is given by

$P = -4n^2 + 1200n + 300.$

For how many items is the profit \$36 per item (to the nearest whole item)?

STATE YOUR UNDERSTANDING

69. Describe the general strategy for solving a quadratic equation by completing the square.

70. Locate the error in the following example. What is the correct answer?

Solve: $4x^2 + 6x + 2 = 0$

$4x^2 + 6x = -2$	Add -2 to both sides
$4x^2 + 6x + (3)^2 = -2 + (3)^2$	Add $(3)^2$ to both sides
$(2x + 3)^2 = 7$	Factor and simplify
$2x + 3 = \pm\sqrt{7}$	
$2x = -3 \pm \sqrt{7}$	
$x = \dfrac{-3 \pm \sqrt{7}}{2}$	

The solution set is $\left\{\dfrac{-3 \pm \sqrt{7}}{2}\right\}$

CHALLENGE EXERCISES

71. A holding tank has two inlet pipes that can be used to fill it. It takes the first inlet two hours longer to fill an empty tank than it does the second pipe. The tank also has a drain that can empty a full tank in twice the time that it takes for the second pipe to fill it. When all three valves are open, it takes seven and one-half hours to fill the empty tank. How long does it take for each inlet pipe to fill the tank when operating alone, and how much time is required for the drain to empty the tank when both inlets are closed?

Solve for x by completing the square:

72. $x^2 - 4bx + 4 = 0$

73. $x^2 + x + c = 0$

74. $ax^2 + bx + c = 0$

MAINTAIN YOUR SKILLS (SECTIONS 7.3, 7.4)

Multiply and simplify:

75. $(\sqrt{7} - \sqrt{10})(\sqrt{7} + \sqrt{10})$

76. $3\sqrt{15}(\sqrt{5} + \sqrt{21})$

77. $(\sqrt{5} - \sqrt{2})^2$

78. $14\sqrt{24} \cdot 3\sqrt{8} \cdot 5\sqrt{3}$

Rationalize the denominator:

79. $\dfrac{2\sqrt{24}}{\sqrt{6}}$

80. $\dfrac{\sqrt{3}}{2 - \sqrt{3}}$

81. $\dfrac{\sqrt{2} + 3}{\sqrt{2}}$

82. $\dfrac{1 + \sqrt{11}}{1 - \sqrt{11}}$

8.3

SOLVING QUADRATIC EQUATIONS: THE QUADRATIC FORMULA

OBJECTIVE

1. Solve a quadratic equation using the quadratic formula.

The *quadratic formula* can be used to solve all quadratic equations.

FORMULA

Quadratic Formula

The quadratic formula is

$$x = \frac{-b \pm \sqrt{b^2 - 4ac}}{2a},$$

where a, b, and c are the coefficients in the standard form of a quadratic equation, $ax^2 + bx + c = 0$.

To use the formula, first write the quadratic equation in standard form to identify a, b, and c.

EXAMPLE 1

Solve: $x^2 + 3x = 7x + 60$

$x^2 - 4x - 60 = 0$ — Write the equation in standard form.

$a = 1$, $b = -4$, $c = -60$ — Identify a, b, and c.

$$x = \frac{-b \pm \sqrt{b^2 - 4ac}}{2a}$$ Quadratic formula.

$$= \frac{-(-4) \pm \sqrt{(-4)^2 - 4(1)(-60)}}{2(1)}$$ Substitute in the formula.

$$= \frac{4 \pm \sqrt{16 + 240}}{2}$$ Simplify.

$$= \frac{4 \pm \sqrt{256}}{2}$$

$$= \frac{4 \pm 16}{2}$$ $\sqrt{256} = 16$

$x = 10$ or $x = -6$

Check: $x = 10$ — Check in the original equation.

$$x^2 + 3x = 7x + 60$$

$$(10)^2 + 3(10) = 7(10) + 60$$

$$100 + 30 = 70 + 60 \qquad \text{True.}$$

Check: $x = -6$

$$(-6)^2 + 3(-6) = 7(-6) + 60$$

$$36 - 18 = -42 + 60 \qquad \text{True.}$$

The solution set is $\{-6, 10\}$. ■

When using the quadratic formula it is not necessary to have the leading coefficient equal to 1.

EXAMPLE 2

Solve: $4x^2 + 7x - 1 = 0$

$a = 4,\ b = 7,\ c = -1$ — Identify a, b, and c.

$$x = \frac{-b \pm \sqrt{b^2 - 4ac}}{2a} \qquad \text{Quadratic formula.}$$

$$= \frac{-(7) \pm \sqrt{(7)^2 - 4(4)(-1)}}{2(4)} \qquad \text{Substitute.}$$

$$= \frac{-7 \pm \sqrt{49 + 16}}{8}$$

$$= \frac{-7 \pm \sqrt{65}}{8} \qquad \text{Simplify.}$$

The solution set is $\left\{\dfrac{-7 \pm \sqrt{65}}{8}\right\}$. The check is left for the student. ■

The quadratic formula is derived by completing the square on the general quadratic equation in standard form.

$$ax^2 + bx + c = 0$$

1. Add $-c$ to both sides.

$$ax^2 + bx = -c$$

2. Divide both sides by a to make the leading coefficient equal to 1.

$$x^2 + \frac{b}{a}x = -\frac{c}{a}$$

3. Add the square of one-half of $\frac{b}{a}$ to each side.

$$x^2 + \frac{b}{a}x + \left(\frac{b}{2a}\right)^2 = -\frac{c}{a} + \left(\frac{b}{2a}\right)^2$$

4. Factor the left side, and simplify the right side.

$$\left(x + \frac{b}{2a}\right)^2 = \frac{b^2 - 4ac}{4a^2}$$

5. Set $x + \frac{b}{2a}$ equal to the square roots of the right side.

$$x + \frac{b}{2a} = \pm\sqrt{\frac{b^2 - 4ac}{4a^2}}$$

$$x = \frac{-b}{2a} \pm \frac{\sqrt{b^2 - 4ac}}{2a}$$

$$= \frac{-b \pm \sqrt{b^2 - 4ac}}{2a}$$

6. The solution set is $\left\{\frac{-b \pm \sqrt{b^2 - 4ac}}{2a}\right\}$.

If $b^2 - 4ac < 0$ the solutions are not real numbers.

EXAMPLE 3

Solve: $x^2 - 3x + 5 = 0$

$$x = \frac{-b \pm \sqrt{b^2 - 4ac}}{2a}$$ Quadratic formula.

$$= \frac{-(-3) \pm \sqrt{(-3)^2 - 4(1)(5)}}{2(1)}$$ Substitute, $a = 1$, $b = -3$, and $c = 5$.

$$= \frac{3 \pm \sqrt{9 - 20}}{2}$$

$$= \frac{3 \pm \sqrt{-11}}{2}$$ $\sqrt{-11}$ is not a real number.

The equation has no real solutions. ■

The quadratic formula also holds when $a < 0$. We will not show this since it is very easy to write the equation so that $a > 0$. When using the quadratic equation, always simplify the radical expression and reduce if possible.

EXAMPLE 4

Solve: $-x^2 = -2x - 11$

$$-x^2 + 2x + 11 = 0$$
$$x^2 - 2x - 11 = 0$$
$$x = \frac{-b \pm \sqrt{b^2 - 4ac}}{2a}$$

CAUTION

Write the equation in standard form before identifying a, b, and c.

$$= \frac{-(-2) \pm \sqrt{(-2)^2 - 4(1)(-11)}}{2(1)}$$ Substitute, $a = 1$, $b = -2$, and $c = -11$.

$$= \frac{2 \pm \sqrt{4 + 44}}{2}$$
$$= \frac{2 \pm \sqrt{48}}{2}$$
$$= \frac{2 \pm 4\sqrt{3}}{2}$$ $\sqrt{48} = \sqrt{16 \cdot 3} = 4\sqrt{3}$

$$= \frac{2(1 \pm 2\sqrt{3})}{2}$$ Factor.

$$= 1 \pm 2\sqrt{3}$$ Reduce.

The solution set is $\{1 \pm 2\sqrt{3}\}$. ■

A calculator may be used to solve quadratic equations with large numbers and to approximate irrational solutions.

EXAMPLE 5

An arrow is shot upward from a height of 6 feet. If its distance from the ground is given by the formula

$$s = 6 + 150t - 16t^2,$$

how long will it take the arrow to reach a height of 300 feet (with s in feet, t in seconds)? Find to the nearest hundredth of a second.

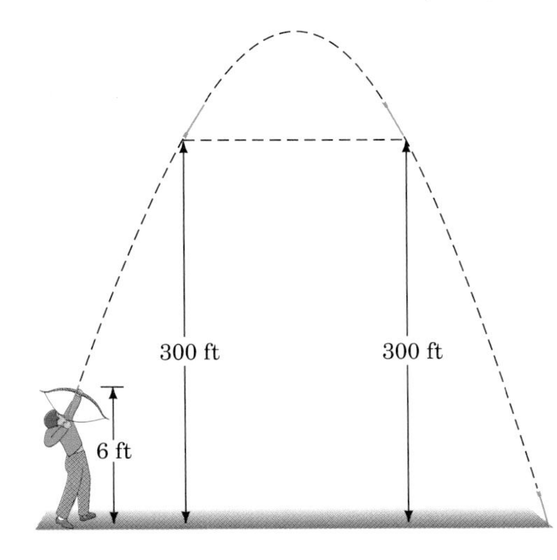

Formula:

$s = 6 + 150t - 16t^2$

Substitute:

$300 = 6 + 150t - 16t^2$ — Substitute, $s = 300$.

Solve:

$16t^2 - 150t + 294 = 0$ — To solve, we first rewrite the equation in standard form.

$8t^2 - 75t + 147 = 0$ — Since 2 is a common factor of the left side, we divide each side by 2 before identifying a, b, and c.

$$t = \frac{-(-75) \pm \sqrt{(-75)^2 - 4(8)(147)}}{2(8)}$$

Substitute, $a = 8$, $b = -75$, and $c = 147$.

$$= \frac{75 \pm \sqrt{921}}{16}$$

$$\approx \frac{75 \pm 30.348}{16}$$

Approximate the radical using a calculator.

$t \approx 6.58$ or $t \approx 2.79$

Answer:

Both answers are appropriate to this application. The arrow is at a height of 300 feet twice, once on the way up after 2.79 seconds and again on the way down after 6.58 seconds. ■

EXERCISE 8.3

A

Solve:

1. $x^2 - 3x - 18 = 0$
2. $x^2 + 6x - 40 = 0$
3. $y^2 + 12y + 35 = 0$
4. $y^2 + 11y + 18 = 0$
5. $b^2 + 14b + 24 = 0$
6. $a^2 - a - 12 = 0$
7. $c^2 + 12c + 20 = 0$
8. $z^2 - 14z + 33 = 0$
9. $x^2 - 10x - 24 = 0$
10. $y^2 - 6y - 27 = 0$
11. $x^2 - 7x - 18 = 0$
12. $x^2 + 15x + 50 = 0$

13. $x^2 + 9x - 70 = 0$
14. $y^2 + 20y + 91 = 0$
15. $6x^2 + x - 15 = 0$
16. $12x^2 - 13x - 4 = 0$
17. $x^2 + 3x + 4 = 0$
18. $x^2 - 5x + 7 = 0$
19. $2x^2 + x - 3 = 0$
20. $2x^2 - 3x - 5 = 0$

B

Solve:

21. $6w^2 - 5w - 6 = 0$
22. $4w^2 - 27w - 7 = 0$
23. $9x^2 - 12x - 5 = 0$
24. $7x^2 - 27x - 4 = 0$
25. $20x^2 - 23x = 21$
26. $2x^2 + 11x = 6$
27. $12x^2 + 1 = 6x$
28. $6x^2 + 6 = 10x$
29. $3x^2 - 4x = 7$
30. $5x^2 - 24x = 5$
31. $3z^2 = -8z - 5$
32. $3b^2 = 14b + 5$
33. $-3t^2 - 8t + 3 = 0$
34. $-7c^2 + 10c - 3 = 0$
35. $-6x^2 + 7x = -3$
36. $-10x^2 - 11x = 3$
37. $6x^2 + 22x = 21x + 77$
38. $25x + 14x^2 = 6x + 3$
39. $3x(2x - 1) + 2x = 8(2 - x) - 13$
40. $(2x - 3)^2 + 2(27 - 10x) = 0$

C

Solve:

41. $x^2 + 10 = 5x$
42. $x^2 - 2x - 16 = 0$
43. $x^2 = 5x + 10$
44. $x^2 = x + 14$
45. $4x^2 - 4x - 5 = 0$
46. $4x^2 + 5 = 4x$
47. $x^2 = -6x + 11$
48. $2y^2 - 3y = 8$
49. $(x + 3)^2 + 2x(x + 3) = 18$
50. $(2x - 5)(2x + 5) + 13x = -16$
51. $(2x + 3)(x - 1) + 3 = (2x + 5)^2 + 2x$
52. $(3x - 5)(x - 1) + 2(x - 2) = x(2x + 3)$

Solve; round the answer to the nearest tenth:

53. $2x^2 - 5x + 1 = 0$
54. $6y^2 - 10y + 3 = 0$
55. $3x^2 - 2x - 7 = 0$
56. $2x^2 - 5x + 5 = 0$
57. $9a^2 + a - 2 = 0$
58. $5z^2 - 10z + 4 = 0$
59. $3x^2 = 5x - 1$
60. $5y^2 - 3y = 1$

D

61. In Example 5, how long will it take the arrow to reach a height of 220 feet (to the nearest hundredth of a second)?

62. How long will it take the arrow in Example 5 to reach 350 feet (to the nearest hundredth of a second)?

63. A rectangular piece of cardboard is 3 inches longer than it is wide. A 2-inch square is cut out of each corner, and the edges are turned up to form a container. If the volume of the container is 496 in^3, what are the dimensions of the piece of cardboard (to the nearest hundredth)?

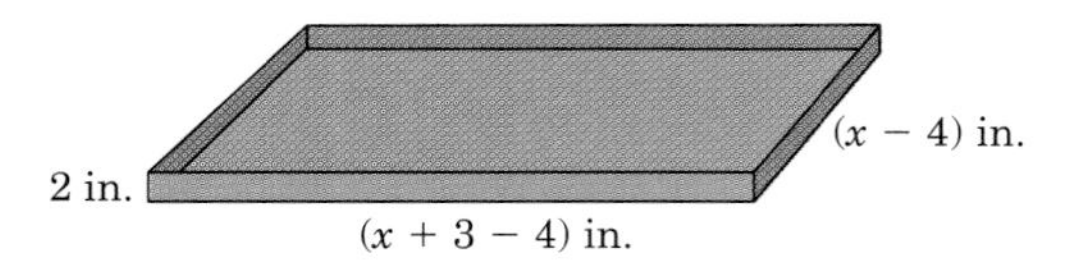

Figure for Exercise 63

64. A rectangular piece of cardboard is 1 inch longer than twice the width. A 1-inch square is cut out of each corner, and the edges are turned up to form a container. If the volume of the container is 250 in^3, what are the dimensions of the cardboard (to the nearest hundredth)?

65. A variable electrical current is given by $i = t^2 - 10t + 20$. If t is in seconds, at what time is the current (i) equal to 12 amperes (to the nearest hundredth)?

66. A variable electrical current is given by $i = t^2 - 8t + 15$. If t is in seconds, at what time is the current (i) equal to 16 amperes (to the nearest hundredth)?

67. A rectangular table has a length that is 1.0 feet more than its width. The diagonal of the table is 4.3 feet. Find the length and width of the table. (to the nearest tenth of a foot)

68. The length of a rectangular sign is 10 feet more than its width. There are the same number of feet in its perimeter as there are square feet in its area. Find its length and width. (to the nearest hundredth of a foot)

STATE YOUR UNDERSTANDING

69. What is the reason for placing a restriction, $b^2 - 4ac \geq 0$, on the quadratic formula?

70. The quadratic formula will work for any values of a, b, and c. However, certain values for a, b, and c will make the mathematical calculations involved in simplifying the formula less complicated. Discuss this statement.

CHALLENGE EXERCISES

Solve for x:

71. $3.1x^2 + 5.2x - 1.25 = 0$ (to the nearest hundredth)

72. $(2x - 1)(x + 1) = x + \sqrt{5}\,x$

73. $ax^2 - \sqrt{a}x + (1 - a) = 0$

74. The electric company has determined that if it increases the distance between each of its poles by 40 feet, it would require 11 fewer poles per mile. How many poles are presently placed each mile?

MAINTAIN YOUR SKILLS (SECTIONS 4.7, 7.5)

Factor:

75. $6x^3 - 9x^2 - 105x$

76. $80xy^3 - 125xy$

77. $14x^4 - 112x^3 + 210x^2$

78. $6xy + 24y + 5ax + 20a$

Combine:

79. $5\sqrt{8} - 2\sqrt{18}$

80. $\sqrt{27} - \sqrt{75} + \sqrt{300}$

81. $\sqrt{567} + \sqrt{112} - \sqrt{175}$

82. $\sqrt{150} - \sqrt{18} + \sqrt{384} + \sqrt{98}$

8.4

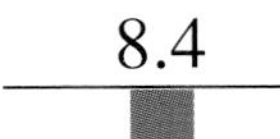

COMPLEX NUMBERS

OBJECTIVES

1. Write complex numbers in standard form.
2. Add and subtract complex numbers.
3. Multiply complex numbers.
4. Divide complex numbers.

We were unable to find the solutions of the equation $x^2 + 1 = 0$ because there is no real number whose square is -1. In order to solve such an equation, we require a new system of numbers. We use the letter i (from the word *imaginary*) as a symbol for one of the solutions of $x^2 = -1$. So, if $x = i$, we have $x^2 = i^2 = -1$ and $i = \sqrt{-1}$.

DEFINITION

The number i is a number whose square is -1. Thus,

$i^2 = -1$ and $i = \sqrt{-1}$.

Using this definition and the rules for simplifying radical expressions, we have symbols for square roots of any negative number.

$$\sqrt{-4} = \sqrt{4(-1)} = \sqrt{4} \cdot \sqrt{-1} = 2i$$

$$-\sqrt{-9} = -\sqrt{9(-1)} = -(\sqrt{9} \cdot \sqrt{-1}) = -3i$$

$$\sqrt{-7} = \sqrt{7(1)} = \sqrt{7} \cdot \sqrt{-1} = \sqrt{7}\, i$$

$$\sqrt{-b} = \sqrt{b(-1)} = \sqrt{b} \cdot \sqrt{-1} = \sqrt{b}\, i, b > 0$$

DEFINITION

Imaginary Numbers

Numbers that are written in the form bi, where b is a real number and $i^2 = -1$, are called *imaginary numbers.*

The algebra of imaginary numbers is the same as the algebra of monomials, with the exception that i^n (n is an integer) can be written as $1, -1, i$, or $-i$.

$$i^3 = i^2(i) = -1(i) = -i$$

$$i^4 = (i^2)(i^2) = (-1)(-1) = 1$$

$$i^5 = (i^4)(i) = (1)(i) = i$$

$$3i + 4i = 7i$$

$$6i - 9i = -3i$$

$$(5i)(-3i) = -15i^2 = -15(-1) = 15$$

$$\frac{12i^5}{4i^6} = \frac{3}{i} = \frac{3}{i} \cdot \frac{i}{i}$$

Multiply by $\frac{i}{i}$ to eliminate the imaginary number i in the denominator since $i^2 = -1$.

$$= \frac{3i}{i^2}$$

$$= \frac{3i}{-1}$$

$$= -3i$$

DEFINITION

Complex Numbers $a + bi$

Numbers that can be written in the form $a + bi$, where a and b are real numbers, are called *complex numbers.* $a + bi$ is called the *standard form* of a complex number.

The real numbers can be written as $a + 0i$ and thus are a subset of the complex numbers. The numbers of the form $0 + bi$ are the imaginary numbers (sometimes called the pure imaginary numbers).

EXAMPLE 1

Write in standard form: $\sqrt{-25}$

$\sqrt{-25} = \sqrt{25} \cdot \sqrt{-1} = 5i$ First, simplify the radical and replace $\sqrt{-1}$ by i.

$= 0 + 5i$ Standard form, $a + bi$. ■

EXAMPLE 2

Write in standard form: $\sqrt{8} - \sqrt{-8}$

$\sqrt{8} - \sqrt{-8} = 2\sqrt{2} - i(2\sqrt{2})$ $\sqrt{-8} = \sqrt{-1}\sqrt{8} = i\sqrt{8} = i(2\sqrt{2})$

$= 2\sqrt{2} - 2\sqrt{2}\,i$ Standard form, $a + bi$. Since $a - bi = a + (-bi)$, $a - bi$ may be considered in standard form. ■

The number system we now have is shown in the following chart.

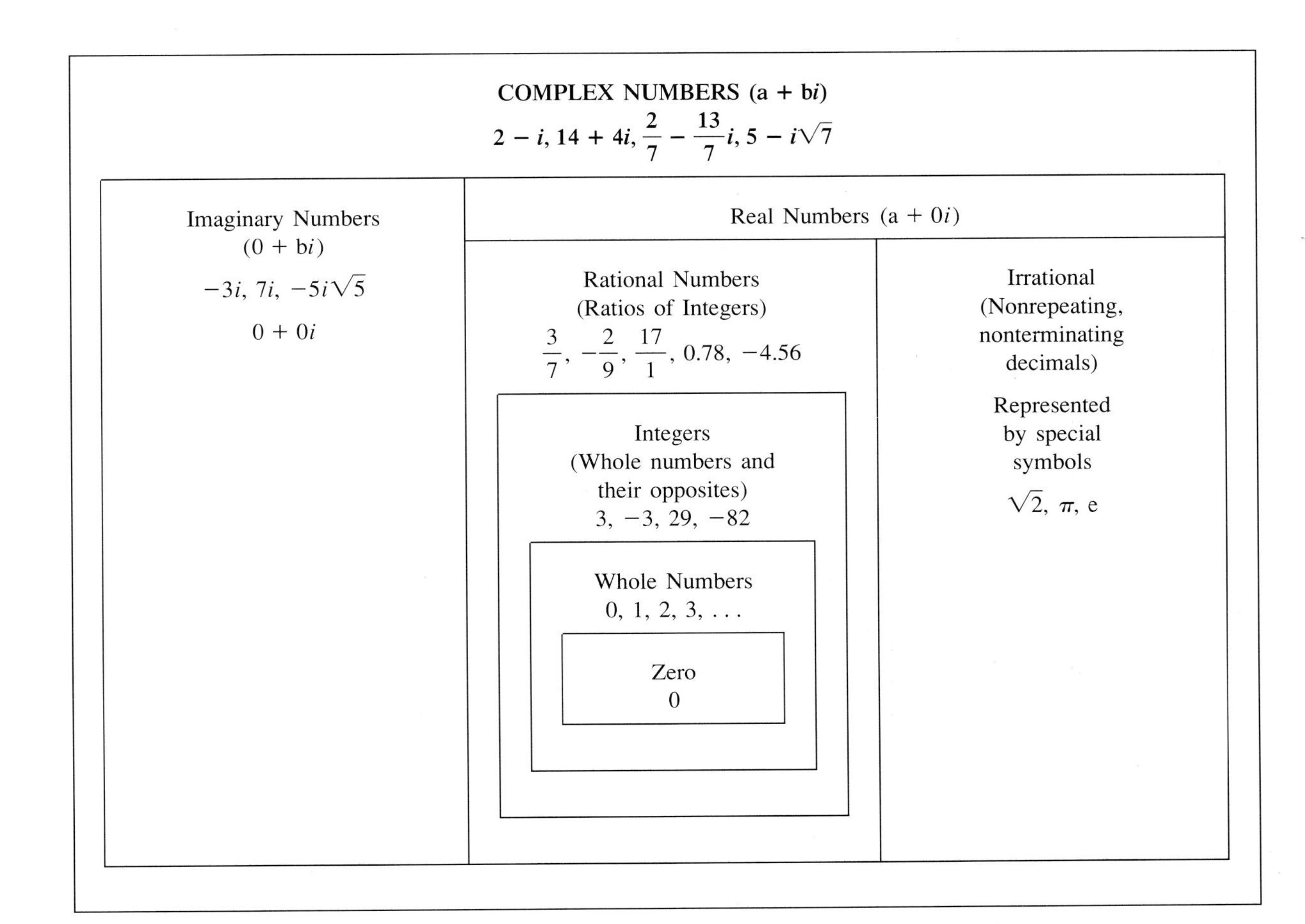

The algebra of complex numbers is the same as the algebra of polynomials.

EXAMPLE 3

Add: $(2 + 3i) + (4 + 5i)$

$$(2 + 3i) + (4 + 5i) = (2 + 4) + (3 + 5)i$$
$$= 6 + 8i$$

■

EXAMPLE 4

Subtract: $(3 + 2i) - (4 - 3i)$

$$(3 + 2i) - (4 - 3i) = (3 - 4) + [2 - (-3)]i$$
$$= -1 + 5i$$

■

If the complex numbers are not in standard form, write them in standard form, then add or subtract.

EXAMPLE 5

Add: $(4 + \sqrt{-20}) + (-2 - 3\sqrt{-20})$

$= (4 + \sqrt{20}\sqrt{-1}) + (-2 - 3\sqrt{20}\sqrt{-1})$ Write each in standard form.

$= (4 + 2\sqrt{5}i) + (-2 - 6\sqrt{5}i)$

$= 2 - 4\sqrt{5}i$ Add by combining terms. ■

■ PROCEDURE

> To multiply complex numbers, apply the same steps used to multiply polynomials, and replace i^2 with -1.

EXAMPLE 6

Multiply: $3(2 + 4i)$

$3(2 + 4i) = 6 + 12i$ Multiply using the distributive property. ■

EXAMPLE 7

Multiply: $2i(3 - 2i)$

$2i(3 - 2i) = 6i - 4i^2$ Multiply.

$= 6i - 4(-1)$ $i^2 = -1$

$= 6i + 4$

$= 4 + 6i$ Standard form. ■

EXAMPLE 8

Multiply: $(2 + 3i)(4 - i)$

$(2 + 3i)(4 - i) = 8 + 10i - 3i^2$	Multiply using FOIL.
$= 8 + 10i - 3(-1)$	$i^2 = -1$
$= 8 + 10i + 3$	
$= 11 + 10i$	Standard form.

■

■ CAUTION

The product property of radicals, $\sqrt{a}\sqrt{b} = \sqrt{ab}$, requires a and b to be nonnegative. This property does not apply if either a or b is negative.

$$\begin{aligned}\sqrt{-4} \cdot \sqrt{-16} &= i\sqrt{4} \cdot i\sqrt{16} \\ &= i^2\sqrt{64} \\ &= (-1)8 \\ &= -8\end{aligned} \qquad \sqrt{-4} \cdot \sqrt{-16} \neq \sqrt{64} = 8$$

To divide two complex numbers, we use a process that is similar to the one used to rationalize the denominator of a fraction. Recall that we multiplied by the conjugate of the denominator. The conjugate of $a - b$ is $a + b$, and the conjugate of $\sqrt{a} + \sqrt{b}$ is $\sqrt{a} - \sqrt{b}$. The conjugate of a complex number follows the same pattern.

Complex Number	Complex Conjugate
$2 + 3i$	$2 - 3i$
$2i$ or $0 + 2i$	$-2i$ or $0 - 2i$
$3 - 4i$	$3 + 4i$
7 or $7 + 0i$	7 or $7 - 0i$

■ PROCEDURE

To divide two complex numbers:

1. Write the division in fraction form.
2. Multiply both the numerator and denominator by the complex conjugate of the denominator.
3. Simplify and write in standard form.

EXAMPLE 9

Divide: $(3 + i) \div (2 - i)$

$$\frac{3+i}{2-i} = \frac{(3+i)(2+i)}{(2-i)(2+i)}$$ Write in fraction form, and multiply by $\frac{2+i}{2+i}$ since $2 + i$ is the conjugate of the denominator, $2 - i$.

$$= \frac{6 + 5i + i^2}{4 - i^2}$$ Multiply.

$$= \frac{6 + 5i + (-1)}{4 - (-1)}$$ $i^2 = -1$

$$= \frac{5 + 5i}{5}$$

$$= 1 + i$$ Standard form. ■

EXAMPLE 10

Divide: $(3 + i) \div (5 - 2i)$

$$\frac{3+i}{5-2i} = \frac{(3+i)(5+2i)}{(5-2i)(5+2i)}$$

$$= \frac{15 + 11i + 2i^2}{25 - 4i^2}$$

$$= \frac{15 + 11i + 2(-1)}{25 - 4(-1)}$$

$$= \frac{13 + 11i}{29}$$

$$= \frac{13}{29} + \frac{11}{29}i$$ Standard form. ■

EXERCISE 8.4

A

Write each of the following complex numbers in standard form:

1. $\sqrt{-16}$
2. $\sqrt{-36}$
3. $\sqrt{-12}$
4. $\sqrt{-48}$
5. $3 + \sqrt{-4}$
6. $8 - \sqrt{-25}$
7. $\sqrt{8} + \sqrt{-12}$
8. $\sqrt{27} - \sqrt{-32}$

Add or subtract:

9. $(4 + 2i) + (3 + 6i)$
10. $(2 - 5i) + (3 + 7i)$
11. $(8 - 2i) + (1 - 6i)$
12. $(10 + 3i) + (4 - 8i)$
13. $(11 - 6i) - (4 - 2i)$
14. $(5 + 2i) - (11 - 7i)$
15. $(3 - i) + (3 + i) + (3 - 3i)$
16. $(2 - 2i) + (4 + 3i) + (5 - 4i)$

Multiply:

17. $2(4 + 3i)$
18. $-1(5 - 2i)$
19. $7(-3 - 4i)$
20. $6(3 + 5i)$

B

Perform the indicated operations; write the answers in standard form:

21. $(5 - 2i) - 6$
22. $(10 + 3i) + 12$
23. $11 - (6 - 4i)$
24. $2 - (6 - 2i)$
25. $(2 - \sqrt{-64}) - (3 + \sqrt{-100})$
26. $(5 + \sqrt{-4}) + (7 - \sqrt{-1})$
27. $(8 - 6i) - (3 + 4i)$
28. $(2 - 7i) - (7 + 3i)$
29. $(17 + 5i) + (3 - 10i)$
30. $(3 - 11i) + (-7 + 2i)$
31. $2(6 - 3i) + 4(2 + i)$
32. $6(4 - 5i) + 2(3 + 2i)$
33. $i(7 - 4i)$
34. $i(2 + 3i)$
35. $2i(2 - 5i)$
36. $3i(2 + 7i)$
37. $-4i(1 + i)$
38. $-5i(3 - 2i)$
39. $\sqrt{-4} \cdot \sqrt{-169}$
40. $\sqrt{-64} \cdot \sqrt{-121}$

C

Perform the indicated operations; write the answers in standard form:

41. $(2 - i)(2 + i)$
42. $(3 - 2i)(3 + 2i)$
43. $(6 - 4i)(6 + 4i)$
44. $(5 - i)(5 + i)$
45. $(2 + i)^2$
46. $(3 - i)^2$
47. $(3 - 5i)(2 + 7i)$
48. $(2 + 5i)(3 - 4i)$
49. $(2 + 7i)(5 - 3i)$
50. $(6 - 5i)(3 - 4i)$
51. $\frac{2}{i}$
52. $\frac{-3}{2i}$
53. $\frac{-5}{3i}$
54. $\frac{7}{5i}$
55. $\frac{2}{1 + i}$
56. $\frac{-3}{2 - i}$
57. $\frac{2i}{5 - 3i}$
58. $\frac{-5i}{6 + 5i}$
59. $\frac{1 - i}{2 + i}$
60. $\frac{1 + 2i}{3 - i}$

D

Two complex numbers $a + bi$ and $c + di$ are equal if and only if $a = c$ and $b = d$. Determine the values of x and y:

61. $2x + 3i = 6 - yi$
62. $6x + 8yi = -12\sqrt{5} + 56\sqrt{3}\,i.$
63. $5 - 7i - x = i + yi$

One of the major applications of complex numbers is in electronics where I represents the electric current.

In an alternating-current circuit the voltage, E, is given by the formula $E = IZ$, where I is the current in amperes and Z is the impedance in ohms.

64. Find E, in volts, when $I = 0.8 - 0.5i$ amperes and $Z = 240 + 160i$ ohms.

65. Find I, in amperes, when $E = 280 + 40i$ volts and $Z = 250 + 170i$ ohms. (numbers to the nearest tenth)

66. Find Z, in ohms, when $I = 0.9 + 0.2i$ amperes and $E = 300 - 20i$ volts. (to the nearest whole number)

In an alternating-current circuit the total impedance, Z_T, produced by two impedances, Z_1 and Z_2, is given by the formula $Z_T = \dfrac{Z_1 Z_2}{Z_1 + Z_2}$.

67. Determine the total impedance when $Z_1 = 2 - 3i$ ohms and $Z_2 = 3 + 3i$. (to the nearest tenth)

68. Determine the total impedance when $Z_1 = 3 + 4i$ ohms and $Z_2 = 2 - i$. (to the nearest tenth)

STATE YOUR UNDERSTANDING

69. Can a complex number be equal to its complex conjugate? Explain your answer.

70. The product of a number and its complex conjugate will always be a real number. Why?

CHALLENGE EXERCISES

Simplify:

71. i^6

72. i^9

73. $\dfrac{1}{i^3}$

74. $(1 - i)^3$

75. $\left(\dfrac{1}{2} + \dfrac{\sqrt{3}}{2}i\right)^3$

76. Is $2 - \sqrt{5}\,i$ a solution of $x^2 = 4x - 9$?

MAINTAIN YOUR SKILLS (SECTIONS 1.9, 6.3)

Evaluate each of the following if $x = -6$, $y = 11$, and $z = 0.75$.

77. $x^2y - xyz + 10z$

78. $yz + xz - y^2 - 23$

79. $(x + y)^2 - (x - z)^2$

80. $x^2z^2 - (y - 4)^2 + 12$

Identify the slope and the y-intercept of the graphs of each of the following.

81. $6x - 7y = 21$

82. $\frac{3x}{5} + \frac{2y}{3} = 1$

Find the slope of the line passing through the following pair of points.

83. $(9, 5), (6, -3)$

84. $(-4, -6), (2, -5)$

8.5

SOLVING QUADRATIC EQUATIONS: COMPLEX ROOTS

OBJECTIVE

1. Solve a quadratic equation with complex roots.

Equations of the form $x^2 = K$, $K < 0$, have complex solutions.

EXAMPLE 1

Solve: $x^2 = -36$

$x = \pm\sqrt{-36}$ — If $x^2 = K$, $x = \pm\sqrt{K}$.

$= \pm 6i$ — $\sqrt{-36} = \sqrt{36} \cdot \sqrt{-1} = 6i$

The solution set is $\{\pm 6i\}$. ■

EXAMPLE 2

Solve: $x^2 + 23 = 0$

$x^2 = -23$

$x = \pm\sqrt{-23}$

$= \pm\sqrt{23}\, i$

The solution set is $\{\pm\sqrt{23}\, i\}$. ■

We can use the quadratic formula to solve quadratic equations with complex roots.

EXAMPLE 3

Solve: $4x^2 - 8x + 13 = 0$

$$x = \frac{-b \pm \sqrt{b^2 - 4ac}}{2a}$$ Quadratic formula.

$$= \frac{-(-8) \pm \sqrt{(-8)^2 - 4(4)(13)}}{2(4)}$$ Substitute, $a = 4$, $b = -8$, and $c = 13$.

$$= \frac{8 \pm \sqrt{-144}}{8}$$

$$= \frac{8 \pm 12i}{8}$$ $\sqrt{-144} = 12i$

$$= 1 \pm \frac{3}{2}i$$ Standard form.

The solution set is $\left\{1 \pm \frac{3}{2}i\right\}$.

■

EXAMPLE 4

Solve: $x^2 - 3x + 10 = 0$

$$x = \frac{-b \pm \sqrt{b^2 - 4ac}}{2a}$$ Quadratic formula.

$$= \frac{-(-3) \pm \sqrt{(-3)^2 - 4(1)(10)}}{2(1)}$$ Substitute, $a = 1$, $b = -3$, and $c = 10$.

$$= \frac{3 \pm \sqrt{-31}}{2}$$

$$= \frac{3 \pm \sqrt{31}i}{2}$$

$$= \frac{3}{2} \pm \frac{\sqrt{31}}{2}i$$ Standard form.

The solution set is $\left\{\frac{3}{2} \pm \frac{\sqrt{31}}{2}i\right\}$.

■

EXAMPLE 5

Solve: $2x^2 - 3x + 5 = 10x - 21$

$2x^2 - 13x + 26 = 0$ First, write in standard form.

$$x = \frac{-b \pm \sqrt{b^2 - 4ac}}{2a}$$ Quadratic formula.

$$= \frac{-(-13) \pm \sqrt{(-13)^2 - 4(2)(26)}}{2(2)}$$ Substitute.

$$= \frac{13 \pm \sqrt{169 - 208}}{4}$$

$$= \frac{13 \pm \sqrt{-39}}{4}$$

$$= \frac{13}{4} \pm \frac{\sqrt{39}}{4} i$$

The solution set is $\left\{\frac{13}{4} \pm \frac{\sqrt{39}}{4} i\right\}$.

EXERCISE 8.5

A

Solve:

1. $x^2 = -16$
2. $x^2 = -36$
3. $x^2 + 3 = 0$
4. $x^2 + 11 = 0$
5. $-5x^2 = 30$
6. $-7x^2 = 42$
7. $x^2 + 112 = 0$
8. $x^2 + 56 = 0$
9. $-7x^2 - 224 = 0$
10. $-9x^2 - 180 = 0$
11. $x^2 - 6x + 10 = 0$
12. $x^2 - 8x + 25 = 0$
13. $x^2 - 4x + 5 = 0$
14. $x^2 - 4x + 29 = 0$
15. $x^2 - 10x + 26 = 0$
16. $x^2 - 14x + 50 = 0$
17. $x^2 - 4x + 68 = 0$
18. $x^2 - 12x + 40 = 0$
19. $x^2 - 2x + 2 = 0$
20. $x^2 - 2x + 10 = 0$

B

Solve:

21. $x^2 + 2x + 2 = 0$
22. $x^2 + 4x + 5 = 0$
23. $x^2 + 2x + 17 = 0$
24. $4x^2 - 24x + 37 = 0$
25. $4x^2 - 16x + 25 = 0$
26. $16x^2 - 8x + 65 = 0$
27. $4x^2 - 4x + 145 = 0$
28. $2x^2 + 2x + 5 = 0$
29. $2x^2 + 2x + 61 = 0$
30. $25x^2 - 150x + 229 = 0$
31. $2x^2 + 8x + 8 = x^2 + 10x - 8$
32. $3x^2 - 7 = 2x^2 - 6x - 47$
33. $2x^2 + x + 3 = x^2 - 5x - 8$
34. $2x^2 + 5x + 4 = x^2 + 3x - 4$
35. $x^2 + 4x + 35 = 0$
36. $3x^2 + x + 4 = 0$
37. $4x^2 + 5x - 3 = -x^2 - 6x - 10$
38. $5x^2 - 5 = 3x - 9$
39. $6x^2 - 11 = 2x^2 + x - 14$
40. $5x^2 - x + 5 = x^2 + x$

C

Solve:

41. $(x-4)(x+4)=(2x+3)(2x-3)$
42. $(2x+1)^2=(x-5)(x+5)+4x$
43. $(x-5)^2=-10x+1$
44. $(x-2)^2=-4x+2$
45. $5x^2+3x+9=0$
46. $7x^2-2x+5=0$
47. $x^2-8x+18=0$
48. $x^2-10x+28=0$
49. $x^2-3x-1=(2x-3)^2$
50. $x^2-5x-8=(3x+1)^2$
51. $2x^2-9x+20=0$
52. $2x^2+7x+19=0$
53. $3x^2-5x+4=0$
54. $3x^2-2x+6=0$
55. $(4x^2-3x-8)-(2x^2+x-12)=0$
56. $(2x^2+4x+6)-(x^2-x-3)=0$
57. $3x(x+3)+10=(x+2)^2$
58. $(x-2)^2=(2x-3)^2+5$
59. $(x-1)(x+1)=(2x+3)^2+4$
60. $(4x-3)(4x+3)=(x+7)^2-62$

STATE YOUR UNDERSTANDING

61. The radicand of the quadratic formula, b^2-4ac, is called the discriminant. If a quadratic equation has real numbers as coefficients, the discriminant can be used to determine whether or not the solutions are real numbers or complex numbers, without solving the equation. How does this work?
62. Whenever a quadratic equation has complex numbers as roots, what is the relationship between the roots?

CHALLENGE EXERCISES

Solve:

63. $x^2-7ix+8=0$
64. $8x^2-26ix=15$
65. $2x^2=10ix+17$
66. $3x^2-14=14ix$
67. $x^4-5x^2=36$ Hint: Let $a=x^2$
68. $4x^4+29x^2+25=0$
69. $15+\frac{11}{x^2}+\frac{2}{x^4}=0$

MAINTAIN YOUR SKILLS (SECTIONS 5.1, 5.2)

Add or subtract:

70. $\frac{2x}{x-6}+\frac{5}{x+7}$
71. $\frac{6}{x-2}-\frac{5x}{x+3}$

72. $\frac{1}{x} - \frac{x-4}{x-3} + \frac{x}{x+3}$

73. $\frac{1}{x-4} - \frac{x}{x+4} + \frac{x+2}{x^2-16}$

Multiply or divide:

74. $\frac{x-1}{x^2-25} \cdot \frac{x^2-3x-10}{x^2+x-2} \cdot \frac{x^2-1}{x^2+4x+4}$

75. $\frac{x}{x+5} \cdot \frac{x^2-8x-65}{x^3-9x} \cdot \frac{x^2+9x+18}{x^2-7x-78}$

76. $\frac{3x^2-27}{x^2+8x+12} \div \frac{2x+6}{x^2+5x-6}$

77. $\frac{x^2+2x-35}{x^2+4x-21} \div \frac{x^2-9x+20}{x^2+6x-27}$

8.6

QUADRATIC EQUATIONS: A REVIEW

OBJECTIVE

1. Solve quadratic equations by factoring, by square roots, or by using the quadratic formula.

You have studied four methods for solving quadratic equations. Selection of the method depends on the particular equation.

PROCEDURE

1. If $b = 0$, write it in the form $x^2 = K$. The solution is $x = \pm\sqrt{K}$.
2. If $c = 0$, factor the quadratic $ax^2 + bx = 0$, and solve by using the zero-product property.
3. If none of a, b, and c are zero, then solve using one of the following methods:
   a. If the factors are obvious, factor and solve by using the zero-product property.
   b. If $a = 1$ and b is an even number, solve by completing the square.
   c. Solve by using the quadratic formula.

$$x = \frac{-b \pm \sqrt{b^2 - 4ac}}{2a}$$

EXAMPLE 1

Solve: $3x^2 - 4x + 17 = 25 - 4x$

$3x^2 - 8 = 0$ — Write in standard form. Note that $b = 0$.

$x^2 = \frac{8}{3}$ — The equation has the form $x^2 = K$.

$x = \pm\sqrt{\frac{8}{3}}$

$= \pm\frac{2}{3}\sqrt{6}$

The solution set is $\left\{\pm\frac{2}{3}\sqrt{6}\right\}$. ■

Some equations containing rational expressions are quadratic. Clear the fractions by multiplying both sides by the LCM of the denominators.

EXAMPLE 2

Solve: $\frac{x}{x+1} + \frac{10x}{(x+1)(x+3)} - \frac{15}{x+3} = 0$

■ **CAUTION**

Note that $x \neq -1$ or -3 since division by zero is not defined.

$x(x + 3) + 10x - 15(x + 1) = 0$ — Multiply both sides by the LCM of the denominators.

$x^2 + 3x + 10x - 15x - 15 = 0$

$x^2 - 2x - 15 = 0$ — Simplify.

$(x - 5)(x + 3) = 0$ — Factor. Always try to factor if $b \neq 0$.

$x - 5 = 0$ or $x + 3 = 0$ — Zero-product property.

$x = 5$ or $x = -3$ — Since $x \neq -3$, it is not a solution.

The solution set is $\{5\}$. ■

Write an equation in standard form before deciding which method of solution to use.

EXAMPLE 3

Solve: $(x - 5)(x + 4) = x(3x - 5) + 3$

$x^2 - x - 20 = 3x^2 - 5x + 3$ — Perform the indicated operations.

$-2x^2 + 4x - 23 = 0$

$2x^2 - 4x + 23 = 0$ — Multiply both sides by -1.

$x = \dfrac{-b \pm \sqrt{b^2 - 4ac}}{2a}$ — The left side cannot be factored over the integers, so we use the quadratic formula.

$= \dfrac{-(-4) \pm \sqrt{(-4)^2 - 4(2)(23)}}{2(2)}$ — Substitute.

$= \dfrac{4 \pm \sqrt{-168}}{4}$ — Simplify.

$= \dfrac{4 \pm 2\sqrt{42}i}{4}$ — $\sqrt{-168} = \sqrt{4} \cdot \sqrt{42} \cdot \sqrt{-1}$

$= 1 \pm \dfrac{\sqrt{42}}{2}i$

The solution set is $\left\{1 \pm \dfrac{\sqrt{42}}{2}i\right\}$. ■

Area problems often involve quadratic equations.

EXAMPLE 4

A 4-in.-by-6-in. photo is mounted in a frame of uniform width. What are the outside dimensions of the frame if the area of the frame and the area of the photo are equal?

Simpler word form:

$$\begin{pmatrix}\text{Area of frame}\\ \text{and photo}\end{pmatrix} = 2\begin{pmatrix}\text{Area of}\\ \text{photo}\end{pmatrix}$$

If the area of the frame and the area of the photo are the same, we can conclude that the combined area is twice the area of the photo.

Select variable:

Let x represent the width of the frame.

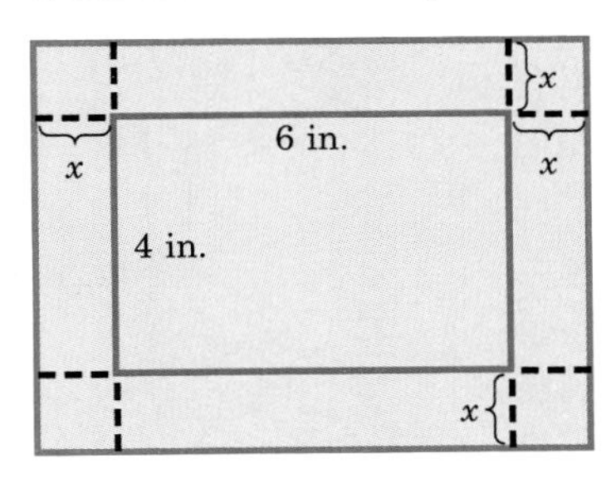

Translate to algebra:

$(2x + 6)(2x + 4) = 2(4 \cdot 6)$	Use the dimensions shown in the figure.
$4x^2 + 20x + 24 = 48$	Multiply.
$4x^2 + 20x - 24 = 0$	Write in standard form.
$x^2 + 5x - 6 = 0$	Divide both sides by 4.
$(x + 6)(x - 1) = 0$	Factor.
$x + 6 = 0$ or $x - 1 = 0$	Zero-product property.
$x = -6$ or $x = 1$	Solve each equation.
	The width of the frame cannot be a negative number.

Answer:

The width of the frame is 1 in. The outside dimensions of the frame are 6 in. by 8 in. ■

EXERCISE 8.6

A

Solve:

1. $x^2 - 4x - 5 = 0$

2. $3x^2 - 8x - 60 = 0$

3. $5x^2 + 5x = -11$

4. $16 = 5x + 2x^2$

5. $3x^2 - x - 9 = 0$

6. $3x^2 + 8 = -12x$

7. $x^2 + 400 = 0$

8. $2x^2 + 100 = 0$

9. $x^2 + 4x = 2(2x + 8)$

10. $5(3x + 10) = x(x + 15)$

11. $x(x - 5) = 5(4 - x)$

12. $x^2 - 6x = 6(6 - x)$

13. $x(x - 5) = x(7 - x)$

14. $x(x + 2) = x(8 - x)$

15. $x(x + 5) = 3(5 - x)$

16. $x(x - 3) = 6(6 - x)$

17. $(x + 5)(x + 4) = 9(x + 7)$

18. $(x - 6)(x - 2) = 8(9 - x)$

19. $x(2x - 1) + 3x(1 - 4x) + 8 = 0$

20. $(y - 6)(2y + 1) = y^2 - 2y + 3$

B

Solve:

21. $x(5x + 8) = x(1 - x) + 5$

22. $(x - 4)(2x + 3) = x^2 - 4x + 8$

23. $3x(x - 6) + 2 = (2x - 4)(x - 7) + 3$

24. $(x + 5)(x + 6) = (2x - 3)(x + 7)$

25. $(x - 3)(x + 4) = (2x + 1)(x - 3) + 6x$

26. $(2x + 5)^2 = (x - 6)(x + 6) + 20x$

27. $(x - 3)(2x + 5) + 10 = (3x + 1)(2x - 1)$

28. $5x(2x + 1) + 2x(3 - 2x) + 5 = 0$

29. $3 + \frac{2}{x} - \frac{5}{x^2} = 0$

30. $1 - \frac{3}{x} - \frac{40}{x^2} = 0$

31. $\frac{4}{a} = \frac{6}{a^2} + 2$

32. $\frac{8}{b} = 2 - \frac{16}{b^2}$

33. $2 + \frac{11}{x} + \frac{5}{x^2} = 0$

34. $15 + \frac{7}{x} - \frac{2}{x^2} = 0$

35. $\frac{15}{x + 1} = 5 + \frac{3}{x}$

36. $\frac{10}{x + 2} = 2 + \frac{4}{x}$

37. $\frac{12}{x - 1} = 10 - \frac{12}{x}$

38. $\frac{7}{x + 2} - \frac{5}{x + 7} = \frac{7}{12}$

39. $\frac{6}{x - 5} + \frac{4}{x - 2} = \frac{8}{3}$

40. $\frac{12}{a + 10} + \frac{12}{a - 10} = -\frac{1}{2}$

C

Solve:

41. $(x + 3)^2 = 21 + 10x$

42. $(x + 5)^2 = 2x + 45$

43. $(x - 1)^2 = 13 - x^2$

44. $10 - x^2 = (4 + x)^2$

45. $(2x + 3)^2 = 53 - x^2$

46. $(3x - 2)^2 = x^2 + 40$

47. $(3x + 1)^2 = 5x^2 - 7$

48. $(2x + 1)^2 + 2x = 3x^2 - 10$

49. $(x - 5)(x + 2) = 139 - (2x - 3)(x + 4)$

50. $8 - \frac{13}{x + 1} = 2x$

51. $(2x + 5)^2 - (x - 4)^2 = 18 - (x + 7)^2$

52. $(4x + 3)^2 + (2x + 1)^2 = 81 + (x - 2)^2$

53. $\frac{7}{x + 3} - 1 = \frac{5}{x - 2}$

54. $8 + \frac{2}{x + 1} = 2x$

55. $\frac{3}{x - 2} + 2 = \frac{2}{x + 5}$

56. $\frac{x^2}{x - 4} = 9 + \frac{56}{x - 4}$

57. $x - \frac{9x}{x - 2} = \frac{-10}{x - 2}$

58. $\frac{10}{x + 3} = x + 6$

59. $\frac{x}{x - 9} + \frac{3}{x + 4} = \frac{-2}{(x - 9)(x + 4)}$

60. $\frac{x}{x - 6} - \frac{2x}{(x - 6)(x - 5)} + \frac{6}{x - 5} = 0$

D

61. A landscaper has a rectangular plot of ground that measures 18 ft by 14 ft. He wishes to put in a petunia bed as a border around the entire plot. What width should the bed be if the landscaper wants 140 ft^2 of the plot remaining for grass?

62. A metalworker must cut a rectangle with an area of 180 in.^{2} from a flat piece of iron. If the rectangle is to be 3 in. longer than it is wide, what are the desired dimensions?

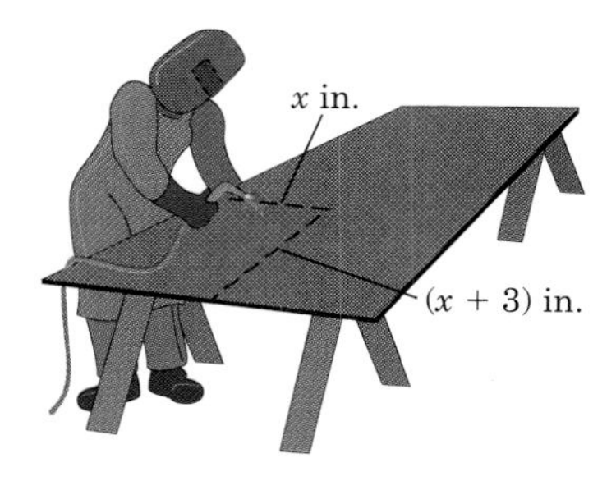

Figure for Exercise 62

63. A right triangle has a base and height as indicated in the drawing. Find the base and height if the area of the triangle is 8 square inches.

$$\left(A = \frac{1}{2}bh\right)$$

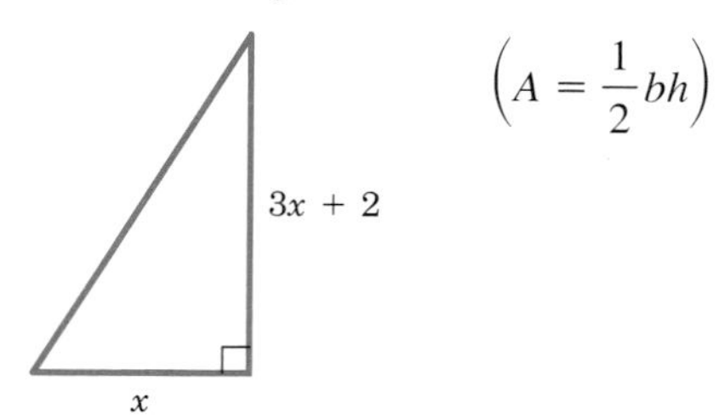

64. A triangle has a base and height as indicated in the drawing. Find the base and height if the area of the triangle is 16 square inches. Round the answer to the nearest tenth.

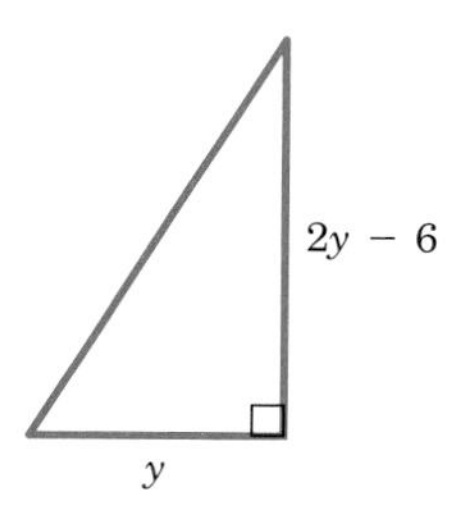

65. The radius of a circular arch of a certain height (h) and span (b) is given by

$$r = \frac{b^2 + 4h^2}{8h}.$$

Find h for a footbridge when $r = 30$ ft and $b = 24$ ft. (Choose the value of h that is less than the radius, and round to the nearest tenth.)

Figure for Exercise 65

66. Find the height of the span in Exercise 65 when $r = 40$ ft and $b = 20$ ft. (Choose the value of h that is less than the radius, and round to the nearest tenth.)

67. A variable electrical current is given by

$i = t^2 - 4t + 16.$

If t is in seconds, at what time is the current (i) equal to 61 amperes?

68. At what time will the current in Exercise 67 equal 100 amperes? (Round to the nearest tenth of a second.)

69. A bus and a car are 180 miles apart, traveling on roads at right angles to each other. The car is traveling 10 miles per hour faster than the bus. They will meet in 3 hours. To the nearest tenth of a mile per hour, find the rate of each vehicle.

70. A truck and a car are 250 miles apart, traveling on roads at right angles to each other. The car is traveling 15 miles per hour faster than the truck. They will meet in 3 hours. To the nearest tenth of a mile per hour, find the rate of each vehicle.

STATE YOUR UNDERSTANDING

71. When is it best to use the "square-root method" to solve a quadratic equation?

72. When is it best to solve a quadratic by factoring? (2 cases)

73. When is it best *not* to use completing the square to solve a quadratic equation?

74. When can the quadratic formula be used to solve a quadratic equation? When is it best to use this technique?

CHALLENGE EXERCISES

Write a quadratic equation in standard form having the given solution set:

75. $\{-2, 3\}$

76. $\left\{\frac{2}{3}, \frac{5}{8}\right\}$

77. $\{5 \pm \sqrt{6}\}$

78. $\{3 \pm 2i\}$

MAINTAIN YOUR SKILLS (SECTIONS 3.7, 4.7, 7.6)

Solve:

79. $(x + 4)(x - 3) - (x - 3)(x + 7) = 67$

80. $24x^2 + 47x - 21 = 0$

81. $3(x + 4) - 3(5 - 2x) - 4(3x + 9) = 2(x + 6)$

82. $3[x - 5(x - 1) + 2] = 4(5 - 2x)$

83. $\sqrt{x} - 3 = 5$

84. $\sqrt{2x + 5} - 9 = -12$

85. $6\sqrt{x} - 4\sqrt{x} + 3 = 5$

86. $\sqrt{y + 7} - 6 = 12$

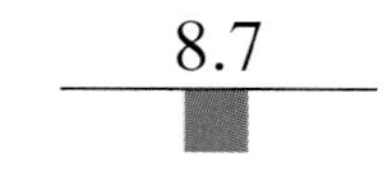

8.7 GRAPHING PARABOLAS

OBJECTIVES

1. Draw the graph of a parabola.
2. Determine the coordinates of the vertex of a parabola.
3. Find the x- and y-intercepts of a parabola.

In this section, we use the properties of quadratic equations to study the graph of a second-degree equation in two variables.

DEFINITION

Parabola

A *parabola* is the graph of an equation that can be written in the form $y = ax^2 + bx + c$, $a \neq 0$.

In the equation defining a parabola, y is equal to the standard form of a quadratic equation. In this course, we study parabolas where $a = 1$.

When we graphed a straight line, we started by plotting some solutions of the equation and then connected these points. Here we do the same thing except we join the points with a smooth curve since the graph is not a straight line.

EXAMPLE 1

Draw the graph of $y = x^2$.

x	x^2	y
-3	$(-3)^2$	9
-2	$(-2)^2$	4
-1	$(-1)^2$	1
0	$(0)^2$	0
1	$(1)^2$	1
2	$(2)^2$	4
3	$(3)^2$	9

Make a table of values by assigning values to x and solving for y.

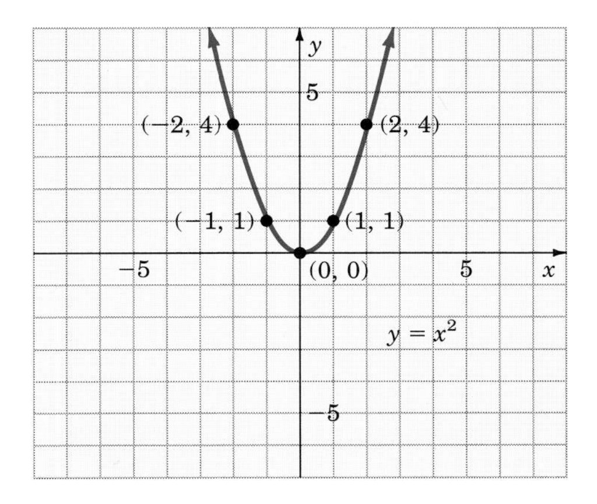

Plot the points on a coordinate system, and connect with a smooth curve.

■

In Example 2, the values of b and c are not both zero.

EXAMPLE 2

Draw the graph of $y = x^2 + 4x + 3$.

Make a table of values.

x	$x^2 + 4x + 3$	y
−5	$(-5)^2 + 4(-5) + 3$	8
−4	$(-4)^2 + 4(-4) + 3$	3
−3	$(-3)^2 + 4(-3) + 3$	0
−2	$(-2)^2 + 4(-2) + 3$	−1
−1	$(-1)^2 + 4(-1) + 3$	0
0	$(0)^2 + 4(0) + 3$	3
1	$(1)^2 + 4(1) + 3$	8

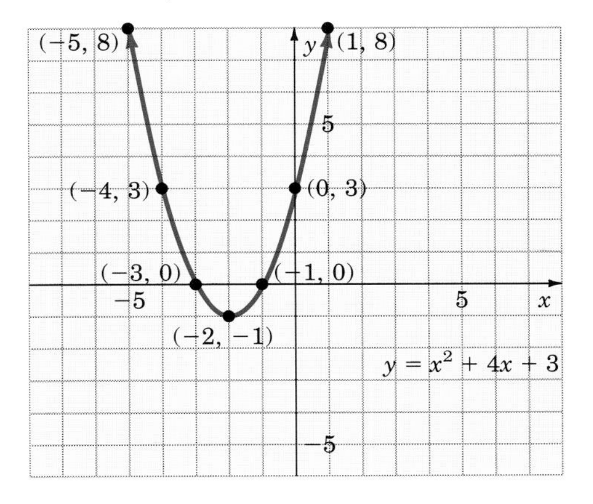

Plot the points, and connect with a smooth curve.

■

In Examples 1 and 2, the graph continues to rise rapidly as x decreases or increases. For example, if we try $x = 5$ in Example 2,

$$y = x^2 + 4x + 3 = (5)^2 + 4(5) + 3 = 48.$$

So, the parabola contains the point (5, 48). Both parabolas will continue to rise indefinitely and are defined for all real values of x.

Notice that each of the graphs has a turning point.

Vertex

DEFINITION

The *vertex* of a parabola is the turning point of the graph of the parabola.

We see from the graph that the vertex of $y = x^2$ is (0, 0), and the vertex of $y = x^2 + 4x + 3$ is $(-2, -1)$.

When given the equation in the form $y = ax^2 + bx + c$ the x-coordinate of the vertex can be found using the following formula:

Vertex

FORMULA

$$x = -\frac{b}{2a} \quad x\text{-coordinate of the vertex}$$

Let's try it using the equation in Example 2:

$$y = x^2 + 4x + 3$$

$a = 1,\ b = 4$ Identify a and b.

$$x = -\frac{b}{2a}$$ Formula for x-coordinate.

$$= -\frac{4}{2(1)}$$ Substitute: $a = 1,\ b = 4$

$$= -2$$

To find the y-coordinate, substitute $x = -2$ in the equation:

$$y = (-2)^2 + 4(-2) + 3$$
$$= 4 - 8 + 3$$
$$= -1$$

The coordinates of the vertex are $(-2, -1)$ as we saw on the graph.

The formula is useful in determining the vertex before we draw a graph. See Example 7.

EXAMPLE 3

Find the coordinates of the vertex of the parabola defined by $y = x^2 - 8x + 2$.

$a = 1, b = -8$ Identify a and b.

$$x = -\frac{b}{2a}$$

$$= -\frac{-8}{2(1)}$$

$$= 4$$ x-coordinate

$$y = (4)^2 - 8(4) + 2$$ Substitute $x = 4$ in the equation to find the y-coordinate.

$$= 16 - 32 + 2$$

$$= -14$$

The coordinates of the vertex are $(4, -14)$. ■

EXAMPLE 4

Find the coordinates of the vertex of the parabola defined by $y = x^2 + 5x - 3$.

$a = 1, b = 5$ Identify a and b.

$$x = -\frac{b}{2a}$$

$$= -\frac{5}{2(1)}$$

$$= -\frac{5}{2}$$ x-coordinate

$$y = \left(-\frac{5}{2}\right)^2 + 5\left(-\frac{5}{2}\right) - 3$$

$$= \frac{25}{4} - \frac{25}{2} - 3$$

$$= -\frac{37}{4}$$ y-coordinate

The coordinates of the vertex are $\left(-\frac{5}{2}, -\frac{37}{4}\right)$. ■

The intercepts of the graph of a parabola can be found the same way the intercepts are found for a straight line, that is by replacing x with zero (y-intercept) and solving for y, or by replacing y with zero (x-intercepts) and solving for x.

EXAMPLE 5

Find the x- and y-intercepts of the graph of $y = x^2 - 7x - 18$.

***y*-intercept**

$y = (0)^2 - 7(0) - 18 = -18$

$(0, -18)$

To find the y-intercept, substitute 0 for x, and solve for y.

***x*-intercept**

$$0 = x^2 - 7x - 18$$

$$0 = (x - 9)(x + 2)$$

$$x - 9 = 0 \quad \text{or} \quad x + 2 = 0$$

$$x = 9 \quad \text{or} \quad x = -2$$

To find the x-intercepts, substitute 0 for y and for x. Since it is a quadratic equation, there will be two intercepts.

The y-intercept is $(0, -18)$, and the x-intercepts are $(9, 0)$ and $(-2, 0)$. ■

It is possible for a parabola to have no x-intercepts. This situation is identified when the entire parabola is above or below the x-axis or when solving for the x-intercepts the solutions are not real numbers.

EXAMPLE 6

Find the x- and y-intercepts of the graph of $y = x^2 - 8x + 18$.

***y*-intercept**

$y = (0)^2 - 8(0) + 18 = 18$ — Let $x = 0$.

$(0, 18)$

***x*-intercept**

$0 = x^2 - 8x + 18$ — Let $y = 0$.

$$x = \frac{-(-8) \pm \sqrt{(-8)^2 - 4(1)(18)}}{2(1)}$$

Solve using the quadratic formula.

$$= \frac{8 \pm \sqrt{-8}}{2}$$

Complex roots, so there are no x-intercepts.

The y-intercept is at $(0, 18)$; there are no x-intercepts. ■

In general,

PROCEDURE

To graph the parabola, $y = ax^2 + bx + c$,

1. Find the vertex.
   a. Use the formula $x = -\frac{b}{2a}$ to find the x-coordinate.
   b. Substitute the value of the x-coordinate in the equation to find the y-coordinate.
2. Find the intercepts.
   a. Find the x-intercepts by substituting $y = 0$ in the equation and solving for s.
   b. The y-intercept is $(0, c)$.
3. Make a table of values to find additional points.
4. Plot the points and connect them with a smooth curve.

EXAMPLE 7

Draw the graph of $y = x^2 - 6x + 5$.

Step 1.

$x = -\frac{b}{2a} = -\frac{-6}{2(1)}$ — Find the vertex.

$= 3$

$y = (3)^2 - 6(3) + 5$

$= 9 - 18 + 5$

$= -4$

$(3, -4)$ — Coordinates of the vertex.

Step 2.

y-intercept: — Find the intercepts.

$(0, 5)$ — y-intercept: $(0, c)$

x-intercepts:

$0 = x^2 - 6x + 5$

$0 = (x - 5)(x - 1)$

$x - 5 = 0$ or $x - 1 = 0$

$x = 5$ or $x = 1$

$(5, 0)$, $(1, 0)$ — x-intercepts

Step 3.

x	$x^2 - 6x + 5$	y
2	$(2)^2 - 6(2) + 5$	-3
4	$(4)^2 - 6(4) + 5$	-3
6	$(6)^2 - 6(6) + 5$	5
7	$(7)^2 - 6(7) + 5$	12
-1	$(-1)^2 - 6(-1) + 5$	12

Find some additional points.

Step 4.

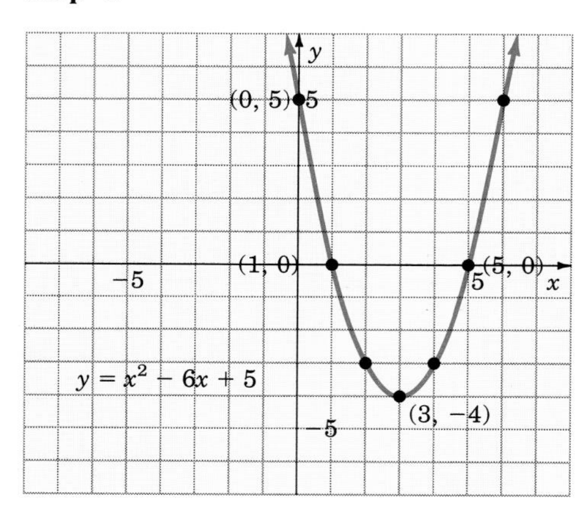

Plot the points and connect them with a smooth curve.

If the coefficient of x^2 in $y = ax^2 + bx + c$ is negative the graph of the parabola opens down.

EXAMPLE 8

Draw the graph of $y = -x^2 - x + 6$

Step 1.

$$x = -\frac{b}{2a} = -\frac{-1}{2(-1)}$$

$$= -\frac{1}{2}$$

Find the vertex. $a = -1$, $b = -1$

$$y = -\left(-\frac{1}{2}\right)^2 - \left(-\frac{1}{2}\right) + 6$$

$$= -\frac{1}{4} + \frac{1}{2} + 6$$

$$= \frac{25}{4}$$

$\left(-\frac{1}{2}, \frac{25}{4}\right)$ Coordinates of the vertex.

Step 2.

$(0, 6)$ y-intercept: $(0, c)$

$$0 = -x^2 - x + 6$$

$$0 = x^2 + x - 6 \quad \text{Multiply both sides by } -1.$$

$$0 = (x + 3)(x - 2)$$

$$x + 3 = 0 \quad \text{or} \quad x - 2 = 0$$

$$x = -3 \qquad\qquad x = 2$$

$(-3, 0), (2, 0)$ x-intercepts.

Step 3.

Find some additional points.

x	$-x^2 - x + 6$	y
-4	$-(-4)^2 - (-4) + 6$	-6
-2	$-(-2)^2 - (-2) + 6$	4
-1	$-(-1)^2 - (-1) + 6$	6
1	$-(1)^2 - (1) + 6$	4
3	$-(3)^2 - (3) + 6$	-6

Step 4.

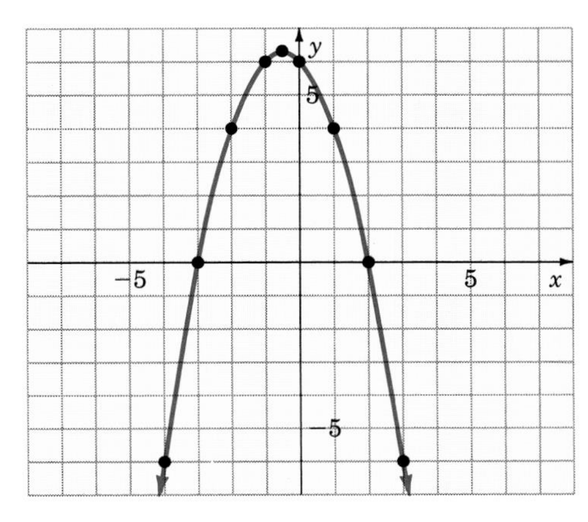

Plot the points and draw the graph. ■

Parabolas are used to describe certain physical phenomena. Example 9 uses a parabola to find the maximum height a ball reaches when thrown from the ground.

EXAMPLE 9

The distance above the ground, s, in feet of a ball launched vertically is given by the equation $s = 120t - 16t^2$, where t is the time in seconds.

a. Determine the maximum height attained by the ball.

b. How many seconds after launching does the ball return to the ground?

Find the vertex.

$$t = -\frac{b}{2a} = -\frac{120}{2(-16)}$$

$$= 3.75$$

Since the curve opens down the vertex is the highest point. The s coordinate is the maximum height obtained by the ball.

$$s = 120(3.75) - 16(3.75)^2$$

$$= 225$$

s coordinate at the vertex, and the maximum height.

Find the x-intercepts.

When $s = 0$, no height, the ball is on the ground.

$$0 = 120t - 16t^2$$

$$2t^2 - 15t = 0$$

$$t(2t - 15) = 0$$

$$t = 0 \quad \text{or} \quad 2t - 15 = 0$$

$$t = 0 \quad \text{or} \quad t = 7.5$$

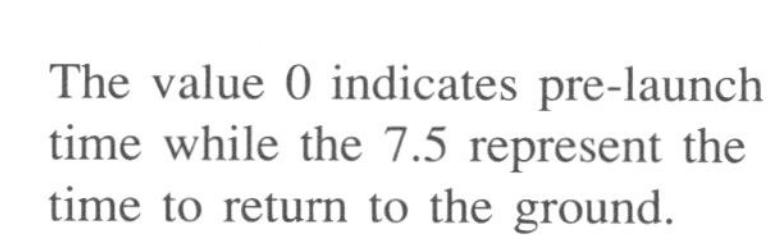
The value 0 indicates pre-launch time while the 7.5 represent the time to return to the ground.

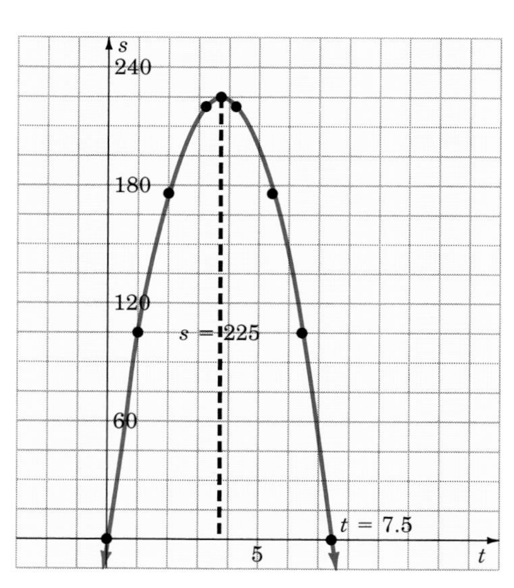

The graph of the parabola illustrate the solutions.

The ball reaches a height of 225 feet and returns to the ground in 7.5 seconds. ■

EXERCISE 8.7

A

Draw the graph of each of the following:

1. $y = x^2 - 2$

2. $y = x^2 + 1$

3. $y = x^2 - 5$

4. $y = x^2 - 7$

5. $y = x^2 + 3$

6. $y = x^2 - 4$

Determine the x- and y-intercepts of each of the following:

7. $y = x^2 + 6$

8. $y = x^2 - 6$

9. $y = x^2 + x - 20$

10. $y = x^2 + x - 2$

11. $y = x^2 - x - 30$

12. $y = x^2 - x - 42$

Determine the coordinates of the vertex of the graph of each of the following:

13. $y = x^2 - 4x + 10$

14. $y = x^2 + 4x + 11$

15. $y = x^2 - 10x + 24$

16. $y = x^2 + 6x + 7$

17. $y = x^2 + 14x + 46$

18. $y = x^2 + 8x + 21$

B

Draw the graph of each of the following:

19. $y = x^2 - 4x + 10$

20. $y = x^2 + 4x + 11$

21. $y = x^2 - 10x + 24$

22. $y = x^2 + 6x + 7$

23. $y = x^2 + 14x + 46$

24. $y = x^2 + 8x + 21$

Determine the x- and y-intercepts of each of the following:

25. $y = x^2 + 3x - 18$

26. $y = x^2 + 2x - 15$

27. $y = x^2 + 11x + 28$

28. $y = x^2 - 9x + 18$

29. $y = x^2 + 4x - 12$

30. $y = x^2 + 3x - 40$

Determine the coordinates of the vertex of the graph of each of the following:

31. $y = x^2 + 6x - 1$

32. $y = x^2 - 6x + 8$

33. $y = x^2 - 4x + 9$

34. $y = x^2 + 4x + 1$

35. $y = x^2 + 10x - 7$

36. $y = x^2 - 8x + 2$

C

If the coefficient of x^2 is negative, the graph of the parabola opens down. Draw the graph of each of the following:

37. $y = -x^2$

38. $y = -x^2 + 5$

39. $y = -x^2 - 3$

40. $y = -x^2 + 2$

41. $y = -x^2 - 4x - 1$

42. $y = -x^2 + 6x - 10$

Determine the x- and y-intercepts of each of the following:

43. $y = x^2 + 2x - 3$

44. $y = x^2 - 10x + 16$

45. $y = x^2 + 10x + 16$

46. $y = x^2 + 6x - 16$

47. $y = x^2 - 2x - 15$

48. $y = x^2 + 8x + 15$

Determine the coordinates of the vertex of the graph of each of the following:

49. $y = x^2 + 3x - 1$

50. $y = x^2 - x + 3$

51. $y = x^2 + x + 2$

52. $y = x^2 + 3x - 2$

53. $y = x^2 + 5x - 3$

54. $y = x^2 - 5x + 4$

D

55. The distance above the ground in centimeters, s, at any time, t, in seconds, of an object projected vertically upward is given by the equation $s = 2450t - 490t^2$.
a. What is the maximum height attained by the object.
b. How long does it take the object to return to the ground?

56. The velocity, v, in meters per second of natural gas in a pipeline is given by the equation $v = 5.2x - x^2$, where x is the distance in centimeters from the wall of the pipe.
a. How far from the wall of the pipe does the maximum velocity occur?
b. What is the maximum velocity of the gas?

57. The distance above the ground in feet, s, at any time, t, of a missile fired vertically upward is given by the equation $s = 150 + 240t - 16t^2$ where t is the time in seconds.
a. How high does it go?
b. When will it hit the ground? (to the nearest tenth)

58. The power, P, in watts that is lost in an electric circuit is given by $P = 50i - 3i^2$ where i is the current in amperes.
a. What current produces the maximum power loss?
b. What is the maximum power loss?

59. The vertical distance, d, in centimeters, of the end of a robot arm above a conveyor belt is given by $d = 4.5t^2 - 36t + 80$. What is the minimum distance between the arm and the conveyor? (NOTE: The distance is a minimum because the graph opens upward.)

60. The area, A, in square meters, of a rectangular lot is given by $A = w(90 - 2w)$. What is the maximum area of the lot?

61. What is the minimum voltage, V, in an electric circuit if $V = 2.4t^2 - 9.6t + 10.2$, $t \leq 5$ min.

62. A parabolic shaped satellite dish has a 10 foot diameter. The equation of the dish is $y = 0.04x^2$.
a. Graph the equation. Remember, the dish is 10 feet wide.
b. What is the depth of the dish?

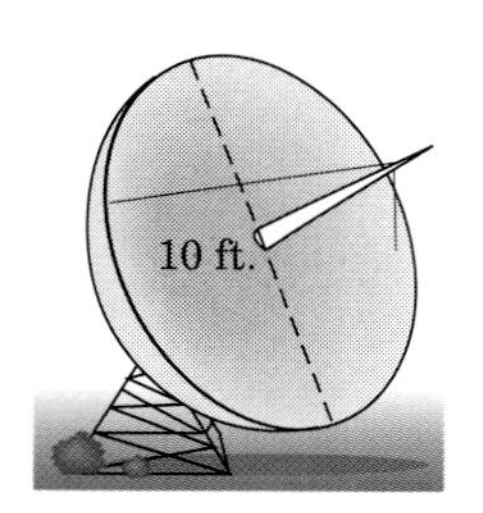

Figure for Exercise 62

STATE YOUR UNDERSTANDING

63. How can you easily determine if a parabola will open upward or downward?

64. A parabola that opens upward has its vertex at (2, 15). What information does this provide?

CHALLENGE EXERCISES

65. Determine b and c so that the graph of $x^2 + bx + c = y$ will contain the points (1, 10) and (−2, −2). Write the equation.

66. Graph $y < x^2 + 4x - 7$:

67. Graph on the same coordinate system. Use double shading to indicate the solution.

$y \geq x^2 - 2x - 8$ and $y < -x^2 + 2x + 4$

68. Graph on the same coordinate system. Use double shading to indicate the solution.

$y \geq -x^2 - 5x + 6$ and $y \geq x^2 + 5x - 3$

MAINTAIN YOUR SKILLS (SECTION 8.4)

Perform the indicated operations; write the answers in standard form:

69. $(2 + i) + (3 - 5i)$

70. $(4 - 5i) - (7 - 8i)$

71. $(6 - i) - (3 + 2i) + (4 - 7i)$

72. $(4 - 12i) + (6 - 8i) - (19 - 14i)$

73. $2i(3 - 7i)$

74. $(4 - 9i)(3 + 2i)$

75. $\dfrac{1 - 3i}{2i}$

76. $\dfrac{11 + 7i}{3 + i}$

CHAPTER 8
SUMMARY

Standard Form of a Quadratic Equation	The standard form of a quadratic equation is $ax^2 + bx + c = 0$, where $a > 0$.	(p. 522)
Incomplete Quadratic Equation	An incomplete quadratic equation has the form $ax^2 + c = 0$ or $ax^2 + bx = 0$.	(p. 522)
Solve by Factoring	To solve a quadratic equation by factoring: **1.** Factor the left side of the equation. **2.** Use the zero-product property. **3.** Solve each equation.	(p. 523)
Solve by Square Roots	To solve a quadratic equation of the form $x^2 = K$, set x equal to the two square roots of K.	(p. 524)
Completing the Square	Changing a quadratic equation in standard form to the form $(x + d)^2 = e$ is called completing the square.	(p. 533)
Solve by Completing the Square	To solve a quadratic equation by completing the square, write in the form $(x + d)^2 = e$, and solve by square roots.	(p. 535)
Quadratic Formula	$x = \dfrac{-b \pm \sqrt{b^2 - 4ac}}{2a}$	(p. 542)
Solve by the Quadratic Formula	To solve a quadratic equation using the quadratic formula, write the equation in standard form, and identify a, b, and c. Now substitute in the formula and simplify.	(p. 542)

Imaginary Numbers	Numbers that can be written in the form bi, where b is a real number, and $i^2 = -1$.	(p. 550)
Complex Numbers	Numbers that can be written in the form $a + bi$, where a and b are real numbers.	(p. 550)
Operations on Complex Numbers	The algebra of complex numbers is the same as the algebra of polynomials.	(pp. 550, 552)
Parabola	The graph of an equation that can be written in the form $y = ax^2 + bx + c$, $a \neq 0$.	(p. 568)
Vertex	The vertex of a parabola is the turning point of the graph of the parabola.	(p. 570)
	When the equation of a parabola is written in the form $y = ax^2 + bx + c$, $-\dfrac{b}{2a}$ is the x-coordinate of the vertex.	(p. 570)
Intercepts of a Parabola	The x-intercepts, where the parabola crosses the x-axis, are found by substituting 0 for y and solving for x. The y-intercept, where the parabola crosses the y-axis, is found by substituting 0 for x and solving for y.	(pp. 571, 572, 573)

CHAPTER 8

REVIEW EXERCISES

SECTION 8.1 Objective 1

Solve:

1. $x^2 - 37x = 0$

2. $x^2 - 35x = 0$

3. $x^2 - 18x = 0$

4. $x^2 - 39x = 0$

5. $3x^2 - 66x = 0$

6. $9x^2 + 45x = 0$

7. $10x^2 - 70x = 0$

8. $11x^2 - 77x = 0$

9. $12x^2 - 60x = 0$

10. $13x^2 + 98x = 0$

SECTION 8.1 Objective 2

Solve:

11. $x^2 = 98$

12. $x^2 = 192$

13. $x^2 = 441$

14. $x^2 = 900$

15. $x^2 - 99 = 0$

16. $x^2 - 176 = 0$

17. $4x^2 - 396 = 0$

18. $5x^2 - 260 = 0$

Find the missing side of each of the following right triangles:

19. $a = 7$, $b = 9$ (Round to the nearest tenth.)

20. $c = 51$, $b = 35$ (Round to the nearest hundredth.)

SECTION 8.2 Objective 1

Solve:

21. $(x-3)^2 = 169$

22. $(x+5)^2 = 100$

23. $\left(x-\frac{1}{2}\right)^2 = \frac{9}{4}$

24. $\left(x-\frac{3}{5}\right)^2 = \frac{16}{25}$

Solve by completing the square:

25. $2x^2 + 13x - 7 = 0$

26. $10x^2 + 9x - 9 = 0$

27. $x^2 + 4x + 1 = 0$

28. $x^2 + 6x + 2 = 0$

29. $2x^2 + 3x - 4 = 0$

30. $3x^2 + x - 5 = 0$

SECTION 8.3 Objective 1

Solve using the quadratic formula:

31. $6x^2 - x - 35 = 0$

32. $4x^2 + 21x + 5 = 0$

33. $2x^2 - 27x - 45 = 0$

34. $2x^2 + 21x - 65 = 0$

35. $x^2 - 4x - 7 = 0$

36. $x^2 + 5x - 8 = 0$

37. $(3x-1)^2 + 3(x-2) = 0$

38. $(3x+1)(2x-1) + 6x - 8 = 0$

39. $\frac{2}{x+1} + \frac{1}{x+2} = \frac{1}{2}$

40. $\frac{1}{x-3} + \frac{3}{x+1} = \frac{1}{3}$

SECTION 8.4 Objective 1

Write each of the following complex numbers in standard form:

41. $\sqrt{-121}$

42. $\sqrt{-144}$

43. $\sqrt{-96}$

44. $\sqrt{-52}$

45. $3 - \sqrt{-36}$

46. $8 + \sqrt{-104}$

47. $\sqrt{20} - \sqrt{-20}$

48. $\sqrt{24} + \sqrt{-12}$

49. i^{13}

50. i^{22}

SECTION 8.4 Objective 2

Add or subtract; write the answers in standard form:

51. $(4-7i) + (3+i)$

52. $(8-5i) + (6-2i)$

53. $(5-6i) + (2-i) + (3+4i)$

54. $(2-11i) + (11-2i) + (6+5i)$

55. $(8-9i) - (2-7i)$

56. $(11-2i) - (9-17i)$

57. $(12-32i) - (-13+40i)$

58. $(15+16i) - (12-3i)$

59. $(8+6i) - (7-4i) - (3-2i)$

60. $(6+7i) - (2-10i) + (5+9i)$

SECTION 8.4 Objective 3

Multiply; write the answers in standard form:

61. $2(8 + 6i)$
62. $-4(9 - 4i)$
63. $4i(2 + 3i)$
64. $-5i(2 - 5i)$
65. $(3 - 7i)(2 + 5i)$
66. $(8 - 2i)(-7 + 5i)$
67. $(15 - i)(15 + i)$
68. $(9 - 5i)(9 + 5i)$
69. $(3 - i)^2(8 + i)$
70. $(2 - 3i)^2(4 + i)^2$

SECTION 8.4 Objective 4

Divide; write the answers in standard form:

71. $-\frac{5}{i}$
72. $\frac{3}{5i}$
73. $\frac{4}{2 + i}$
74. $\frac{3}{3 - i}$
75. $\frac{3i}{4 + i}$
76. $\frac{-2i}{6 - 5i}$
77. $\frac{2 + 3i}{4 - 3i}$
78. $\frac{5 + i}{2 - 3i}$
79. $\frac{4 - 7i}{1 + 2i}$
80. $\frac{1 + i}{12 - 7i}$

SECTION 8.5 Objective 1

Solve:

81. $x^2 = -100$
82. $x^2 = -1600$
83. $x^2 = -\frac{16}{9}$
84. $x^2 = -\frac{7}{16}$
85. $x^2 - 3x + 4 = 0$
86. $x^2 + x + 3 = 0$
87. $x^2 - x + 3 = 0$
88. $x^2 + 6x + 15 = 0$
89. $3x^2 - x + 2 = 0$
90. $2x^2 - 3x + 5 = 0$

SECTION 8.6 Objective 1

Solve:

91. $(x - 3)(x + 5) + 2x - 3 = 7x + 9$
92. $(x + 4)(2x + 3) = 3(4x + 9) - x$
93. $\frac{1}{x^2} + \frac{1}{2x} = 5$
94. $\frac{1}{3x} + \frac{1}{x - 1} = \frac{1}{3}$
95. $(x + 3)(x + 4) - (2x - 1)(x + 4) = 6x^2 + 7$
96. $(x + 1)^2 - (2x + 5)^2 = 8$
97. $3x^2 - 8x + 11 = 0$
98. $5x^2 - 4x - 5 = 0$
99. $\frac{2}{x + 1} + \frac{x}{2x + 3} = 1$
100. $\frac{3}{x - 3} + \frac{x}{3x + 1} = 1$

SECTION 8.7 Objective 1

Draw the graph of each of the following:

101. $y = x^2 - 6$

102. $y = -x^2 + 3$

103. $y = x^2 - 3x + 1$

104. $y = x^2 + 5x + 6$

SECTION 8.7 Objective 2

Determine the x- and y-intercepts of each of the following:

105. $y = x^2 + 6x - 7$

106. $y = x^2 - 8x + 12$

107. $y = 6x^2 - 7x - 5$

108. $y = 12x^2 + x - 1$

109. $y = x^2 - 5x - 14$

110. $y = x^2 + 13x + 42$

111. $y = 10x^2 + 33x - 7$

112. $y = 7x^2 + 73x + 30$

113. $y = x^2 + 9x - 22$

114. $y = x^2 - 5x - 66$

SECTION 8.7 Objective 3

Determine the coordinates of the vertex of the graph of each of the following:

115. $y = x^2 - 22x + 129$

116. $y = x^2 - 10x + 22$

117. $y = x^2 + 6x$

118. $y = x^2 + 2x + 6$

119. $y = x^2 + 4x + 3$

120. $y = x^2 - 12x + 47$

121. $y = x^2 - 10x + 30$

122. $y = x^2 + 2x - 8$

123. $y = x^2 + 14x + 59$

124. $y = x^2 + 18x + 78$

CHAPTER 8
TRUE–FALSE CONCEPT REVIEW

Check your understanding of the language of algebra. Tell whether or not each of the following statements is true (always true) or false (not always true).

1. $5x^2 + 7x + 9 = 0$ is an example of a quadratic equation written in standard form.
2. If $x^2 = 29$, then $x = \pm\sqrt{29}$.
3. $i = \sqrt{-1}$
4. $x^2 = 5$ and $5x^2 = 12x$ are examples of incomplete quadratic equations.
5. The vertex of a parabola is always an x-intercept.
6. The zero-product property states that if $A \cdot B = 0$, then $A = 0$.
7. The hypotenuse of a right triangle is the side opposite the right angle.
8. The Pythagorean theorem states that in a right triangle with hypotenuse c, $a^2 + c^2 = b^2$.
9. The square of an imaginary number is a real number.
10. Completing the square is one of the methods used to solve quadratic equations.
11. To complete the square in the equation $2x^2 + 8x = 5$, divide the coefficient of x (8) by 2, and square that number. The square is then added to both sides.
12. In the quadratic formula $x = \dfrac{-b \pm \sqrt{b^2 - 4ac}}{2a}$, a is the coefficient of x^2.
13. The standard form of a complex number is $a + bi$, $a \in R$, $b \in R$.
14. When using the quadratic formula to solve a quadratic equation, if $b^2 - 4ac < 0$, there is no real solution.
15. $\sqrt{-a}$ is an imaginary number and can be written in the form $\sqrt{a}\, i$.

CHAPTER 8
TEST

1. Write the complex number $3 - \sqrt{-64}$ in standard form.
2. Solve by any method: $\dfrac{3}{y - 2} - 2 = \dfrac{2}{y + 5}$
3. Add: $(3 - 4i) + (6 + 2i) + (4 - 10i)$
4. Solve by completing the square: $x^2 + 16x - 8 = 0$
5. Draw the graph of $y = x^2 + 6x + 5$.

6. Solve: $12x^2 = 768$

7. Find the missing side of a right triangle if $a = 6$ and $b = 7$. Round the answer to the nearest tenth.

8. Use the quadratic formula to solve: $2x^2 + x + 7 = 0$

9. Subtract: $(16 - 8i) - (3 + 2i)$

10. Solve by completing the square: $x^2 - 6x - 14 = 0$

11. Multiply; write the answer in standard form: $(2 - 3i)(4 + 7i)$

12. Determine the x- and y-intercepts of the graph of $y = x^2 - 5x - 24$.

13. Use the quadratic formula to solve: $3x^2 - 3x - 2 = 0$

14. Find the missing side of a right triangle if $b = 24$ and $c = 26$.

15. Divide; write the answer in standard form: $\dfrac{5 - 5i}{2 + i}$

16. Solve: $3x^2 - 17 = 2x^2 + 15$

17. Determine the coordinates of the vertex of the graph of $y = x^2 - 6x + 13$.

18. Use the quadratic formula to solve: $x^2 + 4x - 17 = 0$

19. Solve by any method: $(x - 4)(2x + 3) = x(x - 4) + 8$

20. To get from one corner of a rectangular lot to the opposite corner, Marta must walk 175 yards if she walks along the two sides. She can save a distance of 50 yards if she walks diagonally across the lot. Find the length and width of the lot.

CHAPTERS 5–8
CUMULATIVE REVIEW

CHAPTER 5

Simplify:

1. $\dfrac{3x^2 + 15x + 18}{6x^2 - 24}$

2. $\dfrac{2x^2 + x - 28}{2x^2 - 7x - 30} \cdot \dfrac{x^2 - 8x + 12}{2x^2 - 11x + 14}$

3. $\dfrac{x^3 - y^3}{x^2 - 2xy + y^2} \cdot \dfrac{14y - 14x}{4x^2 + 4xy + 4y^2}$

4. $\dfrac{5ax + 3ay - 5bx - 3by}{9y^2 - x^2} \cdot \dfrac{9y^2 - 3xy - 2x^2}{10x^2 - 9xy - 9y^2}$

5. $\dfrac{x^2 - y^2}{x^2y} \div \dfrac{3x - 3y}{xy^2}$

6. $\dfrac{x^2 + x - 6}{x^2 + 3x - 10} \div \dfrac{x^2 + 2x - 3}{x^2 + 4x - 5}$

7. $\dfrac{a^2 - 3a - 10}{a^2 + 3a - 10} \div \dfrac{12a^2 + 12a - 24}{14a^2 + 56a - 70}$

8. $\dfrac{x^4 - 81}{x^3 - 3x^2 + 9x - 27} \div \dfrac{18 - 2x^2}{6x}$

Determine the LCM (leave in factored form):

9. $2x^2 - 13x - 7;\ 2x^2 + 11x + 5$

10. $4x^3;\ 2x^3 - 8x;\ 4x^4 - 16x^3 + 16x^2$

Solve:

11. $\dfrac{4}{x - 5} - \dfrac{2}{x + 3} = 0$

12. $\dfrac{1}{4x - 12} + \dfrac{1}{2x} = \dfrac{2}{x - 3}$

Add or subtract as indicated:

13. $\dfrac{x - 2}{4x^2 + 4x - 3} + \dfrac{x + 3}{4x^2 + 8x + 3}$

14. $\dfrac{2}{x - 1} - \dfrac{3 - x}{x^2 - x} + \dfrac{1}{x}$

Solve:

15. $\dfrac{4x - 7}{3x} - \dfrac{2x + 3}{-3x} = 4$

16. $\dfrac{5}{x - 4} = \dfrac{4}{4 - x}$

Simplify:

17. $\dfrac{\dfrac{4}{x} - \dfrac{3}{2}}{\dfrac{4}{x} + \dfrac{3}{2}}$

18. $\dfrac{\dfrac{x}{x - 2} + \dfrac{x}{x + 3}}{\dfrac{3x}{x - 2} - \dfrac{2x}{x + 3}}$

19. $\dfrac{5x + 3 - \dfrac{2}{x - 2}}{4x - 3 + \dfrac{3}{x - 2}}$

20. $\dfrac{\dfrac{3x^2 - 14x + 8}{2x^2 - x - 10}}{\dfrac{3x^2 - 5x + 2}{2x^2 - 11x + 15}}$

21. An automobile can travel 370.5 miles on 13 gallons of gasoline. How far can it travel on 7 gallons of gasoline?

22. Yearly taxes on property valued at \$13,000 are \$780. What are the annual taxes on property worth \$20,000?

23. On a map of Ohio $\frac{1}{2}$ inch represents 5.5 miles. How many miles is it from Canton to Akron if the distance on the map is $1\frac{5}{8}$ inches? Express the answer to the nearest tenth.

24. A patient is to receive 15 mg of estradiol. It is available in a solution that contains 30 mg of estradiol per mL of solution. How many mL of solution must be administered?

25. For a certain investment, the interest varies directly with the time. If $27.52 interest is earned in six months, what amount will be earned in 10 months?

26. The time required to build a deck varies inversely with the number of carpenters working on the project. If three carpenters can complete the deck in 10 days, how long would it take 5 carpenters to do the entire job?

27. The amount of water, at a constant pressure, flowing through a pipe varies directly with the square of the diameter of the pipe. If a 3-inch pipe delivers 25 gallons per minute, how many gallons will a 4-inch pipe deliver in one minute? Express the answer to the nearest tenth.

28. The illumination produced by a light source varies inversely with the square of the distance from the light source. If the illumination from a light source is 12 candles at a distance of 3 feet, what is the illumination at a distance of 9 feet?

CHAPTER 6

Given the equation, $4x - 5y - 12 = 0$, solve for y given:

29. $x = 0$

30. $x = 3.2$

Given the equation, $3x - 2y + 8 = 0$, solve for x given:

31. $y = 7$

32. $y = -\frac{2}{3}$

Graph:

33. $4x - y = 8$

34. $0.3x - 0.5y = 1.5$

35. $\frac{3}{4}x - \frac{y}{3} = 2$

36. $2x + 3 = 0$

37. Determine the x-intercept and the y-intercept of $\frac{1}{6}x + \frac{1}{3}y = -2$

38. Graph using the slope and the y-intercept:
$5x - 3y = 6$

39. Determine the slope of the line containing the points $(3, -2)$ and $(-5, -4)$.

40. Determine the slope of the line containing the points $(-4, 2)$ and $(-4, -7)$.

Graph:

41. $2x - 5y \leq 10$

42. $6x + y > 2$

43. $x - y > 5$ and $x + y < 7$

44. $2x - 3y \leq 6$ and $3x - y < 3$

Solve by graphing:

45. $2x + y = 8$
$2x - y = 0$

46. $y = x + \frac{7}{2}$
$4x - 2y = 1$

Solve by substitution:

47. $3x - y = 2$
$x + y = 2$

48. $y = 2x - 4$
$3x + y = 1$

49. $2x - 0.5y = 3$
$y = 4x - 7$

50. $3x - 4y = 7$
$x + 3y = 5$

Solve by linear combinations:

51. $3x - 2y = 7$
$5x + 4y = 8$

52. $5x - 3y = 4$
$y = \frac{5}{3}x - \frac{4}{3}$

53. $\frac{2}{3}x - \frac{1}{6}y = \frac{1}{2}$
$3x - 2y = 4$

54. $7x - 3y = 11$
$4x + 5y = 6$

CHAPTER 7

Find the roots. When necessary, approximate to the nearest hundredth:

55. $\sqrt{572}$

56. $\sqrt{0.93}$

57. $\sqrt[3]{1728}$

58. $\sqrt[5]{-16{,}807}$

Write in simplest radical form:

59. $-\sqrt{162}$

60. $\sqrt{9xy^2}$

61. $\sqrt{450x^3y^6z^8}$

62. $\sqrt{2352x^2y^7z^{11}}$

Multiply and simplify:

63. $\sqrt{5x^2y}\,\sqrt{10xy}$

64. $\sqrt{3x}\left(3\sqrt{x} - \sqrt{3x} + x\sqrt{3}\right)$

65. $\sqrt{8}(\sqrt{2} + \sqrt{6} - \sqrt{10})$

66. $(\sqrt{5} + \sqrt{3})(\sqrt{10} - \sqrt{3})$

67. $(\sqrt{5} + \sqrt{7})^2$

68. $(2\sqrt{3} - \sqrt{7})(2\sqrt{3} + \sqrt{7})$

Write in simplest radical form:

69. $\dfrac{-\sqrt{28}}{4\sqrt{8}}$

70. $\dfrac{7\sqrt{50}}{\sqrt{27}}$

71. $\dfrac{4}{\sqrt{3} - 2}$

72. $\dfrac{7}{\sqrt{21} + \sqrt{7}}$

73. $\sqrt{45} + \sqrt{80} + \sqrt{12}$

74. $7\sqrt{18} - 2\sqrt{8} - 3\sqrt{72}$

75. $3\sqrt{\frac{5}{7}} - 4\sqrt{\frac{7}{5}}$

76. $30\sqrt{\frac{1}{2}} - \frac{9}{2}\sqrt{8} + 9\sqrt{\frac{169}{2}}$

Solve:

77. $7\sqrt{x} - 3 = 4\sqrt{x} + 2$

78. $2\sqrt{3x - 1} - \sqrt{8} = 0$

79. $\sqrt{4x + 7} = \sqrt{x - 5}$

80. $\sqrt{9x - 2} = 3\sqrt{x} - 1$

CHAPTER 8

Solve:

81. $(x - 5)(x + 2) = 3x(x - 5) - 10$

82. $(4x - 1)(2x + 1) = 2x(2x + 1)$

83. $\dfrac{6}{2x + 3} = x + 2$

84. A right triangle has a hypotenuse of 12 cm. and one leg of 8 cm. Determine the length of the other leg to the nearest hundredth.

Solve by completing the square:

85. $x^2 + 4x - 7 = 0$

86. $5x^2 = 10x - 12$

87. $x^2 + 7x + 2 = 0$

88. $3x^2 - 5x - 1 = 0$

Solve by use of the quadratic formula:

89. $28x^2 = 13x + 60$

90. $(2x + 3)(x - 4) = 5(x + 2)$

91. $(x - 4)^2 + 2(x - 2)(x + 2) = 7$

92. $\frac{2}{3}x^2 - \frac{1}{4}x + \frac{1}{2} = 0$

Simplify:

93. $\sqrt{-72}$

94. $(8 - 3i) + (7 + 2i)$

95. $(3 - 4i) - (6 + 7i)$

96. $(5 - 2i)(5 + 2i)$

97. $(1 + 4i)(2 - 3i)$

98. $(3 - 2i)^2$

99. $\frac{8}{5i}$

100. $\frac{3}{2 - i}$

101. $\frac{3i}{9 - 3i}$

102. $\frac{2 + i}{1 - i}$

Solve:

103. $x^2 + 25 = 0$

104. $-3x^2 = 132$

105. $3x^2 - 5x + 8 = 0$

106. $(2x - 5)(3x + 7) = (4x - 1)(2x + 3) - 4x$

Solve by any method:

107. $6x^2 = 13x + 5$

108. $4x^2 - 24x + 7 = 0$

109. $3x^2 - 4x = 8$

110. $6x^2 = 36x$

111. $5x(x + 1) = 4x - 7$

112. $\frac{x - 4}{x - 1} = \frac{2}{x}$

Determine the vertex, the y-intercept, the x-intercepts and graph:

113. $y = x^2 - 4x + 3$

114. $y = x^2 + 4x + 7$

115. $y = -x^2 + x + 6$

116. $y = x^2 + 3$

APPENDIX I

The Properties of Zero

Historically, the number zero is a recent development in mathematics. There is evidence that notched sticks or bones were used for counting as long as 30 million years ago, but the use of a symbol for zero began from about 1200 to 1800 years ago (the development was gradual and not a sudden decision).

Zero is *not* a counting number.

When counting a group of objects we usually count ''one, two, three, four,'' and so on.

Zero is a whole number.

The classification of zero as a whole number is an arbitrary though useful one.

Zero is an integer.

All whole numbers are also integers.

Zero is a rational number.

Zero can be written as a fraction $\frac{0}{b}$, where b is an integer, $b \neq 0$.

Zero is a real number.

All rational (and irrational) numbers are called real numbers.

The zero power of any nonzero real number is 1.
$x^0 = 1, x \neq 0$

Examples: $1^0 = 1$, $2^0 = 1$, $6^0 = 1$, $10^0 = 1$, and so on. This is an arbitrary definition in most arithmetic and algebra texts. It is a useful definition because it fits mathematical patterns such as these:

$3^3 = 1 \cdot 3 \cdot 3 \cdot 3 = 27$		10^3 has place value ''thousand''
$3^2 = 1 \cdot 3 \cdot 3 \quad = 9$	and	10^2 has place value ''hundred''
$3^1 = 1 \cdot 3 \quad = 3$		10^1 has place value ''ten''
$3^0 = 1 \quad = 1$		10^0 has place value ''one'' or ''units''

Any number plus zero is that number.
$a + 0 = a$

Examples: $0 + 0 = 0$, $1 + 0 = 1$ (also $0 + 1 = 1$), $2 + 0 = 2$ (also $0 + 2 = 2$), and so on. This is called the Addition Property of Zero or the Identity Property of Addition.

Any number times zero is zero.
$a \cdot 0 = 0$

Examples: $0 \cdot 0 = 0$, $0 \cdot 1 = 0$ (also $1 \cdot 0 = 0$), $0 \cdot 6 = 0$ (also $6 \cdot 0 = 0$), and so on. This is called the Multiplication Property of Zero.

Zero divided by any nonzero real number is zero.
$0 \div a = 0, a \neq 0$

Examples: $0 \div 1 = 0$, $0 \div 2 = 0$, $0 \div 5 = 0$, and so on. Division does not work in reverse (division is not commutative). Division is the inverse (a kind of opposite) of multiplication. The statement $0 \div 1 = 0$ is true because $1 \cdot 0 = 0$. The statement $0 \div 5 = 0$ is true because $5 \cdot 0 = 0$. Because of this, zero can be used as a numerator for a fraction if the denominator is a non-zero real number.

Division by zero is not defined.
$a \div 0$ has no value

Examples: $1 \div 0$, $2 \div 0$, and $6 \div 0$ have no value.

$1 \div 0$ is not 0 because $0 \cdot 0$ is not 1.
$1 \div 0$ is not 1 because $0 \cdot 1$ is not 1.
$6 \div 0$ is not 0 because $0 \cdot 0$ is not 6.
$6 \div 0$ is not 6 because $0 \cdot 6$ is not 6.
$0 \div 0$ is not defined because it is ambiguous.

By the definition of division, if zero were not an exception, any answer would check.

$0 \div 0 = 17$ (?) because $0 \cdot 17 = 0$.
$0 \div 0 = 2$ (?) because $0 \cdot 2 = 0$.
$0 \div 0 = 0$ (?) because $0 \cdot 0 = 0$.

Such problems are of no use, so division by zero is not defined for *any* dividend. Because of this, zero cannot be used as the denominator of a fraction.

APPENDIX II

Prime Factors of Numbers 1 Through 100

	PRIME FACTORS		PRIME FACTORS		PRIME FACTORS		PRIME FACTORS
1	none	26	$2 \cdot 13$	51	$3 \cdot 17$	76	$2^2 \cdot 19$
2	2	27	3^3	52	$2^2 \cdot 13$	77	$7 \cdot 11$
3	3	28	$2^2 \cdot 7$	53	53	78	$2 \cdot 3 \cdot 13$
4	2^2	29	29	54	$2 \cdot 3^3$	79	79
5	5	30	$2 \cdot 3 \cdot 5$	55	$5 \cdot 11$	80	$2^4 \cdot 5$
6	$2 \cdot 3$	31	31	56	$2^3 \cdot 7$	81	3^4
7	7	32	2^5	57	$3 \cdot 19$	82	$2 \cdot 41$
8	2^3	33	$3 \cdot 11$	58	$2 \cdot 29$	83	83
9	3^2	34	$2 \cdot 17$	59	59	84	$2^2 \cdot 3 \cdot 7$
10	$2 \cdot 5$	35	$5 \cdot 7$	60	$2^2 \cdot 3 \cdot 5$	85	$5 \cdot 17$
11	11	36	$2^2 \cdot 3^2$	61	61	86	$2 \cdot 43$
12	$2^2 \cdot 3$	37	37	62	$2 \cdot 31$	87	$3 \cdot 29$
13	13	38	$2 \cdot 19$	63	$3^2 \cdot 7$	88	$2^3 \cdot 11$
14	$2 \cdot 7$	39	$3 \cdot 13$	64	2^6	89	89
15	$3 \cdot 5$	40	$2^3 \cdot 5$	65	$5 \cdot 13$	90	$2 \cdot 3^2 \cdot 5$
16	2^4	41	41	66	$2 \cdot 3 \cdot 11$	91	$7 \cdot 13$
17	17	42	$2 \cdot 3 \cdot 7$	67	67	92	$2^2 \cdot 23$
18	$2 \cdot 3^2$	43	43	68	$2^2 \cdot 17$	93	$3 \cdot 31$
19	19	44	$2^2 \cdot 11$	69	$3 \cdot 23$	94	$2 \cdot 47$
20	$2^2 \cdot 5$	45	$3^2 \cdot 5$	70	$2 \cdot 5 \cdot 7$	95	$5 \cdot 19$
21	$3 \cdot 7$	46	$2 \cdot 23$	71	71	96	$2^5 \cdot 3$
22	$2 \cdot 11$	47	47	72	$2^3 \cdot 3^2$	97	97
23	23	48	$2^4 \cdot 3$	73	73	98	$2 \cdot 7^2$
24	$2^3 \cdot 3$	49	7^2	74	$2 \cdot 37$	99	$3^2 \cdot 11$
25	5^2	50	$2 \cdot 5^2$	75	$3 \cdot 5^2$	100	$2^2 \cdot 5^2$

APPENDIX III
Formulas

PERIMETER AND AREA

Square

Perimeter: $P = 4s$

Area: $A = s^2$

Rectangle

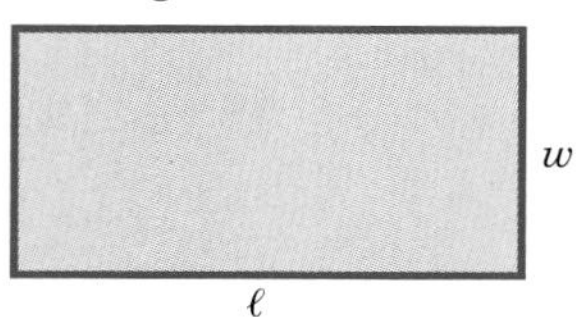

Perimeter: $P = 2\ell + 2w$

Area: $A = \ell w$

Triangle

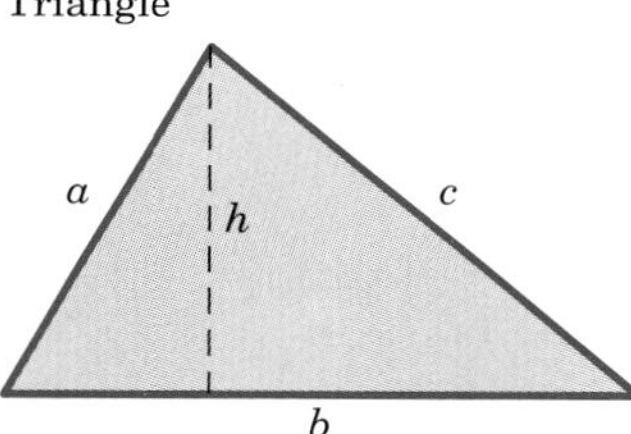

Perimeter: $P = a + b + c$

Area: $A = \dfrac{bh}{2}$

Parallelogram

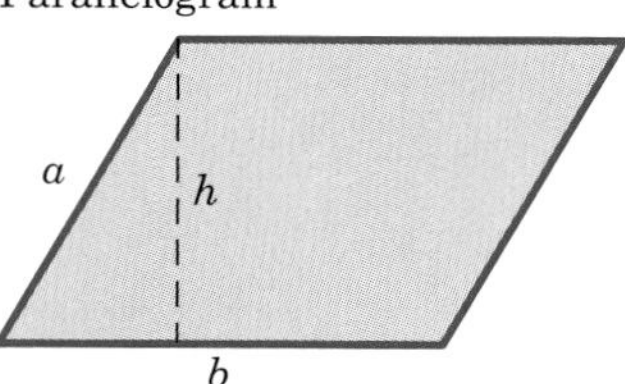

Perimeter: $P = 2a + 2b$

Area: $A = bh$

Trapezoid

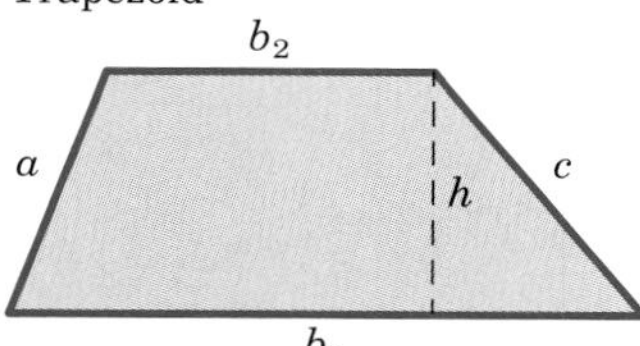

Perimeter: $P = a + b_1 + c + b_2$

Area: $A = \dfrac{1}{2}(b_1 + b_2)h$

Circle

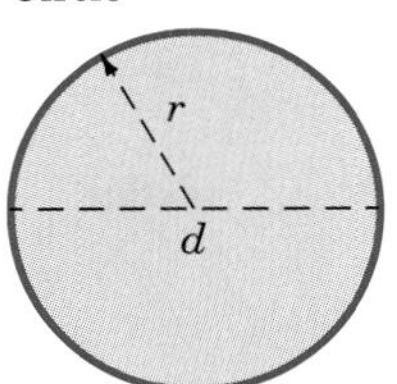

Circumference: $C = 2\pi r$

$C = \pi d$

Area: $A = \pi r^2$

$A = \dfrac{\pi d^2}{4}$

Volume

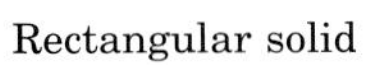

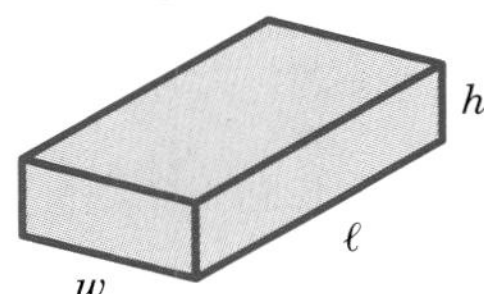

$V = \ell wh$

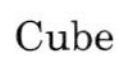

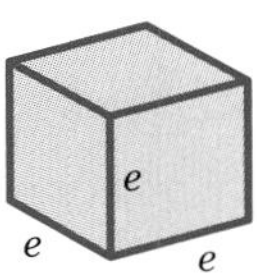

$V = e^3$

Sphere

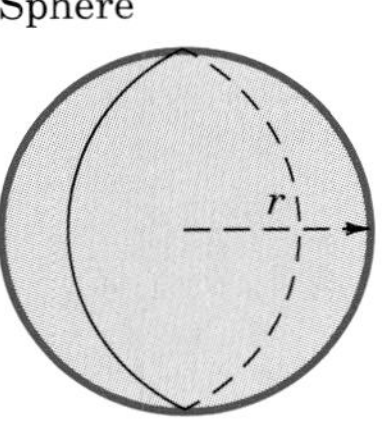

$V = \dfrac{4}{3}\pi r^3$

Cylinder

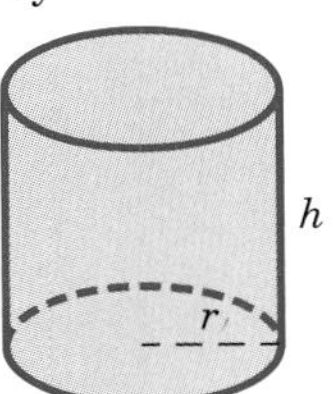

$V = \pi r^2 h$

Cone

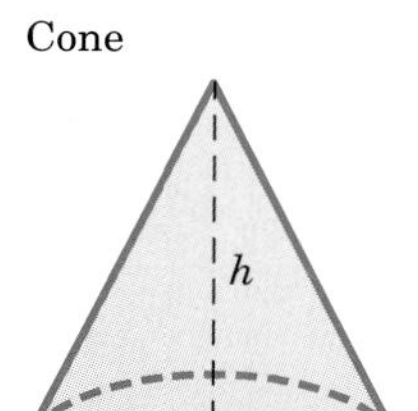

$$V = \frac{1}{3}\pi r^2 h$$

Miscellaneous

FORMULA	DESCRIPTION
$D = rt$	Distance equals rate (speed) times time.
$I = prt$	Interest equals principal times rate (%) times time.
$C = np$	Cost equals number (of items) times price (per item).
$S = c + m$	Selling price equals cost plus markup.
$F = \frac{9}{5}C + 32$	Fahrenheit temperature equals $\frac{9}{5}$ times Celsius temperature plus 32.
$S = v + gt$	Velocity (of object falling) equals initial velocity plus gravitational force times time.
$R \cdot B = A$	Rate (%) times base equals amount (percentage).
$a^2 + b^2 = c^2$	The square of the length of one leg of a right triangle plus the square of the length of the other leg is equal to the square of the length of the hypotenuse.

Many other formulas appear within problems in the text.

APPENDIX IV
Metric Measures and Equivalents

LENGTH
(basic unit is 1 m)

1 millimeter (mm)		=	0.001 m
1 centimeter (cm)	= 10 millimeters	=	0.01 m
1 decimeter (dm)	= 10 centimeters	=	0.1 m
1 METER (m)	= 10 decimeters	=	1 m
1 dekameter (dam)	= 10 meters	=	10 m
1 hectometer (hm)	= 10 dekameters	=	100 m
1 kilometer (km)	= 10 hectometers	=	1000 m

WEIGHT
(basic unit is 1 g)

1 milligram (mg)		=	0.001 g
1 centigram (cg)	= 10 milligrams	=	0.01 g
1 decigram (dg)	= 10 centigrams	=	0.1 g
1 GRAM (g)	= 10 decigrams	=	1 g
1 dekagram (dag)	= 10 grams	=	10 g
1 hectogram (hg)	= 10 dekagrams	=	100 g
1 kilogram (kg)	= 10 hectograms	=	1000 g
1 metric ton	= 1000 kilograms		

LIQUID AND DRY MEASURE
(basic unit is 1 ℓ)

1 milliliter (mℓ)		=	0.001 ℓ
1 centiliter (cℓ)	= 10 milliliters	=	0.01 ℓ
1 deciliter (dℓ)	= 10 centiliters	=	0.1 ℓ
1 LITER (ℓ)	= 10 deciliters	=	1 ℓ
1 dekaliter (daℓ)	= 10 liters	=	10 ℓ
1 hectoliter (hℓ)	= 10 dekaliters	=	100 ℓ
1 kiloliter (kℓ)	= 10 hectoliters	=	1000 ℓ

APPENDIX V

Squares and Square Roots (0 to 199)

n	n^2	$\sqrt{n}$	n	n^2	$\sqrt{n}$	n	n^2	$\sqrt{n}$	n	n^2	$\sqrt{n}$
0	0	0.000	50	2,500	7.071	100	10,000	10.000	150	22,500	12.247
1	1	1.000	51	2,601	7.141	101	10,201	10.050	151	22,801	12.288
2	4	1.414	52	2,704	7.211	102	10,404	10.100	152	23,104	12.329
3	9	1.732	53	2,809	7.280	103	10,609	10.149	153	23,409	12.369
4	16	2.000	54	2,916	7.348	104	10,816	10.198	154	23,716	12.410
5	25	2.236	55	3,025	7.416	105	11,025	10.247	155	24,025	12.450
6	36	2.449	56	3,136	7.483	106	11,236	10.296	156	24,336	12.490
7	49	2.646	57	3,249	7.550	107	11,449	10.344	157	24,649	12.530
8	64	2.828	58	3,346	7.616	108	11,664	10.392	158	24,964	12.570
9	81	3.000	59	3,481	7.681	109	11,881	10.440	159	25,281	12.610
10	100	3.162	60	3,600	7.746	110	12,100	10.488	160	25,600	12.649
11	121	3.317	61	3,721	7.810	111	12,321	10.536	161	25,921	12.689
12	144	3.464	62	3,844	7.874	112	12,544	10.583	162	26,244	12.728
13	169	3.606	63	3,969	7.937	113	12,769	10.630	163	26,569	12.767
14	196	3.742	64	4,096	8.000	114	12,996	10.677	164	26,896	12.806
15	225	3.873	65	4,225	8.062	115	13,225	10.724	165	27,225	12.845
16	256	4.000	66	4,356	8.124	116	13,456	10.770	166	27,556	12.884
17	289	4.123	67	4,489	8.185	117	13,689	10.817	167	27,889	12.923
18	324	4.243	68	4,624	8.246	118	13,924	10.863	168	28,224	12.961
19	361	4.359	69	4,761	8.307	119	14,161	10.909	169	28,561	13.000
20	400	4.472	70	4,900	8.367	120	14,400	10.954	170	28,900	13.038
21	441	4.583	71	5,041	8.426	121	14,641	11.000	171	29,241	13.077
22	484	4.690	72	5,184	8.485	122	14,884	11.045	172	29,584	13.115
23	529	4.796	73	5,329	8.544	123	15,129	11.091	173	29,929	13.153
24	576	4.899	74	5,476	8.602	124	15,376	11.136	174	30,276	13.191
25	625	5.000	75	5,625	8.660	125	15,625	11.180	175	30,625	13.229
26	676	5.099	76	5,776	8.718	126	15,876	11.225	176	30,976	13.266
27	729	5.196	77	5,929	8.775	127	16,129	11.269	177	31,329	13.304
28	784	5.292	78	6,084	8.832	128	16,384	11.314	178	31,684	13.342
29	841	5.385	79	6,241	8.888	129	16,641	11.358	179	32,041	13.379
30	900	5.477	80	6,400	8.944	130	16,900	11.402	180	32,400	13.416
31	961	5.568	81	6,561	9.000	131	17,161	11.446	181	32,761	13.454
32	1,024	5.657	82	6,724	9.055	132	17,424	11.489	182	33,124	13.491
33	1,089	5.745	83	6,889	9.110	133	17,689	11.533	183	33,489	13.528
34	1,156	5.831	84	7,056	9.165	134	17,956	11.576	184	33,856	13.565
35	1,225	5.916	85	7,225	9.220	135	18,225	11.619	185	34,225	13.601
36	1,296	6.000	86	7,396	9.274	136	18,496	11.662	186	34,596	13.638
37	1,369	6.083	87	7,569	9.327	137	18,769	11.705	187	34,969	13.675
38	1,444	6.164	88	7,744	9.381	138	19,044	11.747	188	35,344	13.711
39	1,521	6.245	89	7,921	9.434	139	19,321	11.790	189	35,721	13.748
40	1,600	6.325	90	8,100	9.487	140	19,600	11.832	190	36,100	13.784
41	1,681	6.403	91	8,281	9.539	141	19,881	11.874	191	36,481	13.820
42	1,764	6.481	92	8,464	9.592	142	20,164	11.916	192	36,864	13.856
43	1,849	6.557	93	8,649	9.644	143	20,449	11.958	193	37,249	13.892
44	1,936	6.633	94	8,836	9.659	144	20,736	12.000	194	37,636	13.928
45	2,025	6.708	95	9,025	9.747	145	21,025	12.042	195	38,025	13.964
46	2,116	6.782	96	9,216	9.798	146	21,316	12.083	196	38,416	14.000
47	2,209	6.856	97	9,409	9.849	147	21,609	12.124	197	38,809	14.036
48	2,304	6.928	98	9,604	9.899	148	21,904	12.166	198	39,204	14.071
49	2,401	7.000	99	9,801	9.950	149	22,201	12.207	199	39,601	14.107

n n^2 $\sqrt{n}$ n n^2 $\sqrt{n}$ n n^2 $\sqrt{n}$ n n^2 $\sqrt{n}$

APPENDIX VI
Calculators

The wide availability and economical price of current hand-held calculators make them ideal for doing time-consuming arithmetic operations. You are encouraged to use a calculator as you work through this text. Calculator examples throughout the text show where the use of a calculator is appropriate. Your calculator will be especially useful for

1. doing the fundamental operations of arithmetic (add, subtract, multiply, and divide),

2. checking solutions to equations,

3. checking solutions to problems,

4. finding square roots of numbers, and

5. finding powers of numbers.

To practice solving the problems with your calculator, you will need to know whether the calculator is a basic calculator or a scientific calculator. Here are examples of each.

Basic calculator

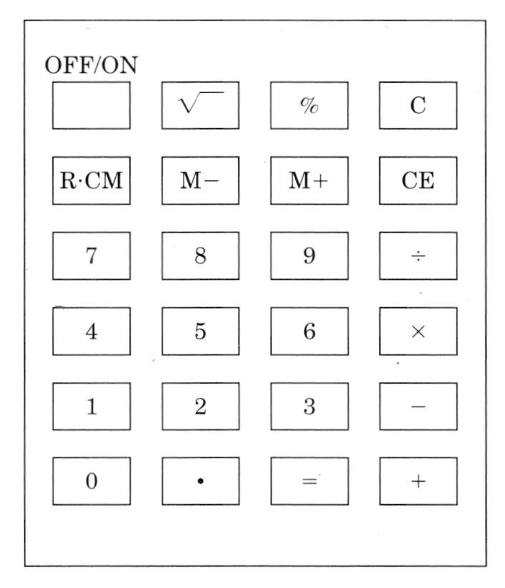

Scientific calculator

$1/x$	x^2	$\sqrt{x}$	OFF	ON/C
INV	SIN	COS	TAN	DRG
K	EE ↓	LOG	ln x	y^x
π	%	(	)	÷
STO	7	8	9	×
RCL	4	5	6	−
SUM	1	2	3	+
EXC	0	•	+/−	=

All computers, many scientific calculators, and some other calculators have the fundamental order of operations built into their circuitry. To tell if your calculator has this feature, do the following exercise:

6 + 4(9)

ENTER	DISPLAY	ENTER	DISPLAY
[6]	6.	[×]	4.
[+]	6.	[9]	9.
[4]	4.	[=]	42.

If the display reads "42," your calculator finds values of combinations according to the agreed-upon order. If the display reads "90," you can get the correct result, 42, by entering the problem in the calculator following the rules for the order of operations.

ENTER	DISPLAY	ENTER	DISPLAY
[4]	4.	[+]	36.
[×]	4.	[6]	6.
[9]	9.	[=]	42.

A calculator that has parentheses keys can also override the order of operations.

ENTER	DISPLAY
[6] [+] [(] [4] [×] [9] [)] [=]	42.

If yours is a scientific calculator, it has other keys that you may find useful in later mathematics courses. Recall that parentheses and fraction bars are used to group operations to show which one is done first.

EXPRESSION	BASIC CALCULATOR Enter		SCIENTIFIC CALCULATOR Enter	DISPLAY
$144 \div 3 - 7$	[144] [÷] [3] [−] [7] [=]		[144] [÷] [3] [−] [7] [=]	41.
$\frac{28 + 42}{10}$	[28] [+] [42] [=] [÷] [10] [=]		[28] [+] [42] [=] [÷] [10] [=]	7.
$\frac{288}{6 + 12}$	[6] [+] [12] [=]	STEP 1		18.
	[288] [÷] [18] [=]	STEP 2		16.
			[6] [+] [12] [=] [$\frac{1}{x}$] [×] [288] [=]	16.
$13^2 + 4(17)$	[13] [×] [13] [=]	STEP 1		169.
	[4] [×] [17] [=]	STEP 2		68.
	[169] [+] [68] [=]	STEP 3		237.
			[13] [×] [13] [+] [4] [×] [17] [=]	237.
			or	
			[13] [x^2] [+] [4] [×] [17] [=]	237.
$\sqrt{5184}$			[5184] [$\sqrt{\ }$]	72.

APPENDIX VII
Sets

Sets.1

SETS, SUBSETS, AND THEIR SYMBOLS

OBJECTIVES

1. Describe a set in words.
2. Describe a set using a roster method.
3. Describe a set using set builder notation.
4. Identify the members of a set.
5. Identify a subset of a given set.

The idea of a *set* is so basic that it is difficult to define in simpler terms. The following group of synonyms will help give meaning to the word set: clump, family, grouping, collection, flock, cluster, and assembly. In mathematics, sets typically contain numbers, geometric figures, and other mathematical objects.

There are several ways to designate a set. Three ways are explained here. First, a set may be described in words. We may talk about "the set of whole numbers between 0 and 6." The same set may be described in words by writing "the set whose elements are one, two, three, four, and five."

The *elements* of a set are the objects in the set. They are also called *members* of the set.

A second choice is to make a roster of the set by listing all the elements or members between a pair of braces. In such a case we often use a capital letter to name the set. The set described above in words can also be described by writing

$A = \{1, 2, 3, 4, 5\}$.

EXAMPLE 1

Use a roster to describe the set of all whole numbers greater than 4 and less than 10. Call this set *G*.

$G = \{5, 6, 7, 8, 9\}$ The set consists of five members. ■

EXAMPLE 2

Describe the set {2, 4, 6, 8} in words.

The set of even whole numbers greater than one and less than nine.

The elements of the set are whole numbers and are even. ■

The roster method is easy to use when the sets contain only a few elements as in Examples 1 and 2. It would be quite tedious, however, to make a roster of all of the whole numbers between 5,748,919 and 5,901,337. Such rosters appear in telephone directories. In mathematics sets with many elements are described using a third method that is called *set builder notation.*

$A =$	$\{x$	$\mid$	x is a whole number and $0 < x < 6\}$
A is the set	of all x	such that	x is a whole number and $0 < x < 6\}$

EXAMPLE 3

Use set builder notation to indicate B, the set of whole numbers larger than 15 and less than 40.

$B = \{x \mid 15 < x < 40$ and x is a whole number$\}$ ■

EXAMPLE 4

Use set builder notation to write all natural numbers less than 57 that are multiples of 4.

$F = \{x \mid x \in N,\ x < 57$ and $x = 4n\}$

We name the set F. We first indicate that x is a natural number, then x is smaller than 57, and finally that x is four times some natural number, n. ■

The symbol $\in$ is used to abbreviate the phrase "is an element of" or "is a member of" so from $G = \{5, 6, 7, 8, 9\}$, we can make the following statements:

The number 7 is a member of set G, or $7 \in G$

The number 9 is an element of set G, or $9 \in G$

The number 5 is an element of set G, or $5 \in G$

The symbol $\notin$ is used to abbreviate the phrase "is not an element of" or "is not a member of," so we make the following statements:

The number 4 is not a member of set G, or $4 \notin G$

The number 10 is not a member of set G, or $10 \notin G$

EXAMPLE 5

True or false. If $A = \{x \mid x \text{ is a whole number and } 15 < x < 68\}$, then $42 \in A$.

The statement is true.

The symbol $\in$ means "is a member of," and 42 is a number that satisfies the conditions. ■

EXAMPLE 6

True or false. If $A = \{x \mid x \text{ is a whole number and } 15 < x < 68\}$, then $36 \notin A$.

The statement is false.

The symbol $\notin$ means "is not an element of," and 36 is larger than 15 and less than 68. ■

Some of the sets of symbols and numbers, together with the letter that identifies each set, that are used in mathematics are

Digits $= \{0, 1, 2, 3, 4, 5, 6, 7, 8, 9\}$

The set of digits is a group of symbols with which we write the place value notation for the different sets of numbers.

Natural or counting numbers	$N = \{1, 2, 3, 4, 5, 6, \ldots\}$
Whole numbers	$W = \{0, 1, 2, 3, 4, 5, 6, \ldots\}$
Integers	$J = \{\ldots, -5, -4, -3, -2, -1, 0, 1, 2, 3, 4, 5, \ldots\}$

In these descriptions we used the roster method. We did this despite the fact that not all the members can be listed. Enough members of the set are listed to establish a pattern. The three dots (ellipses) mean that the listing continues without end.

Study the following sets, which are described in words.

The set of all whole numbers less than zero.

The set of all even numbers that end in 7.

The set of all multiples of 10 that are not divisible by 10.

A set that contains no elements is called the *empty* set or *null* set. We can write the empty set with an empty pair of braces.

Empty set $= \{\ \}$

Another symbol used to indicate the empty set is $\emptyset$.

Empty set $= \emptyset$

When a symbol is needed to indicate an empty set, either symbol can be used. However, in mathematics $\emptyset$ is normally used to indicate an empty set.

CAUTION

The symbol $\{0\}$ does not name the empty set nor does the symbol $\{\emptyset\}$ name the empty set. In each case, the set contains a member.

Let $A = \{0, 1, 2, 3, 4, 5, 6, 7, 8, 9\}$ and $B = \{0, 2, 4, 6, 8\}$. We note that every element of B is also a member of A. In such a case, we say that "B is a subset of A" and we write

$B \subseteq A$

Let $C = \{1, 3, 5, 7, 9\}$ and $D = \{1, 3, 5, 7, 8\}$. Notice that not every member of D is a member of C ($8 \notin C$). Therefore D is not a subset of C and we write

$D \not\subseteq C$

We also make the observation that A is a subset of A since every member of A is contained in A. This satisfies the definition of subset. Since this is true for every set, we make the general statement that every set is a subset of itself. Also $\emptyset$ will be considered to be a subset of every set. (Since the empty set has no elements the definition of subset is not violated because every element of $\emptyset$ is in every set.)

Two sets are *equal* if they contain exactly the same elements.

A set F will be called a *proper subset* of E if $F \subseteq E$ and $F \neq E$. To indicate that proper subset relationship, we write $F \subset E$. The symbol for subset ($\subseteq$) and the symbol for proper subset ($\subset$) are similar to the symbols $\leq$ and $<$. For a first set to be a subset of a second, the first set (subset) can contain fewer or exactly the same number of members as the second set. For a first member to be a proper subset of a second set, the first set (proper subset) must contain fewer members than the second set.

EXAMPLE 7

If $A = \{1, 2, 3, 4, 5, 6, 7, 8, 9, 10\}$, $B = \{1, 2, 3\}$, $C = \{0, 2, 4, 6\}$, $D = \{3, 1, 2\}$, and $E = \{2, 4\}$, are the following true or false?

a. $B \subseteq A$ True All of the members of B are contained in A.

b. $C \subseteq A$ False $0 \notin A$

c. $C \subseteq D$ False Not all of the members of C are contained in D, i.e., $0 \notin D$ and $6 \notin D$.

d. $B \subset A$ True $B \subseteq A$ and $B \neq A$

e. $B = D$ True They each contain the same members. D is stated in a different order, but that makes no difference. ■

EXERCISES

A

Write each of the following sets using a roster.

1. The set of even whole numbers less than 17.

2. The set of even whole numbers less than 22.

3. The set of all whole numbers from 9 to 15 inclusive.

4. The set of all whole numbers from 11 to 21 inclusive.

Describe each set in words.

5. $\{11, 12, 13, 14, 15\}$ **6.** $\{18, 20, 22, 24, 26, 28\}$ **7.** $\{1\}$ **8.** $\{2\}$

Write each set using set builder notation.

9. The set of whole numbers less than 84.

10. The set of whole numbers greater than 48.

11. The set of even numbers between 51 and 133.

12. The set of odd numbers between 42 and 92.

Tell whether each statement is true or false.

13. If $B = \{x | x \text{ is a whole number and } 6 < x < 106\}$, then $3 \in B$.

14. If $B = \{x | x \text{ is a whole number and } 6 < x < 106\}$, then $10 \in B$.

15. If $C = \{x | x \text{ is a whole number and } x > 19\}$, then $19.6 \in B$.

16. If $C = \{x | x \text{ is a whole number and } x > 19\}$, then $300 \in B$.

17. If $S = \{x | x \in W\}$, and $T = \{0\}$, then $T \subseteq S$.

18. If $S = \{x | x \in W\}$, and $T = \{100, 203, 305\}$, then $T \subseteq S$.

19. If $A = \{x | x \in J\}$, and $B = \left\{\frac{2}{3}\right\}$, then $B \subseteq A$.

20. If $A = \{x | x \in J\}$, and $B = \{-31, -28, -15\}$, then $B \subseteq A$.

B

Write each of the following sets using a roster.

21. The set of all multiples of 4 between 15 and 38.

22. The set of all multiples of 3 between 16 and 35.

23. The set of all fractions with numerators that are counting numbers less than 4 and with denominators that are counting numbers less than 5.

24. The set of all fractions with numerators that are counting numbers less than 5 and with denominators that are counting numbers less than 4.

Describe each set in words.

25. $\{9, 12, 15, 18, 21\}$

26. $\{18, 12, 16, 20\}$

27. $\{88, 99, 110, 121, 132\}$

28. $\{96, 108, 120, 132\}$

Write each set using set builder notation.

29. The set of the first five consecutive counting numbers larger than 23.

30. The set of the largest five consecutive counting numbers less than 96.

31. The set of counting numbers that are divisors of 18.

32. The set of counting numbers that are divisors of 20.

Tell whether each statement is true or false.

33. If $J = \{x \mid x \text{ is an integer}\}$, then $-2 \in J$.

34. If $J = \{x \mid x \text{ is an integer}\}$, then $\frac{1}{2} \in J$.

35. If $K = \left\{\frac{x}{y} \,\middle|\, x \text{ is an integer and } y \text{ is a counting number}\right\}$, then $\frac{3}{4} \in K$.

36. If $K = \left\{\frac{x}{y} \,\middle|\, x \text{ is an integer and } y \text{ is a counting number}\right\}$, then $\frac{4}{0} \in K$.

37. If $T = \{x \mid x \text{ is a multiple of } 3\}$, $V = \{x \mid x \text{ is a multiple of } 6\}$, then $T \subseteq V$.

38. If $T = \{x \mid x \text{ is a multiple of } 3\}$, $V = \{x \mid x \text{ is a multiple of } 6\}$, then $V \subseteq T$.

39. If $E = \{x \mid x \text{ is an even number}\}$, and $F = \{x \mid x \text{ is a multiple of } 10\}$, then $E \subseteq F$.

40. If $E = \{x \mid x \text{ is an even number}\}$, and $F = \{x \mid x \text{ is a multiple of } 10\}$, then $F \subseteq E$.

Write each of the following sets using a roster.

41. The set of all multiples of 13 between 200 and 240.

42. The set of all multiples of 13 between 300 and 340.

43. The set of all multiples of 14 between 200 and 240.

44. The set of all multiples of 14 between 300 and 340.

Describe each set in words.

45. $\{128, 132, 136, 140\}$

46. $\{132, 140, 148, 156\}$

47. $\{32, 34, 35, 36, 38, 40\}$

48. $\{35, 36, 38, 40, 42, 44, 46, 49\}$

Write each of the following sets using a roster.

49. $\{t \mid t \in W \text{ and } 36 < t \leq 40\}$

50. $\{t \mid t \in W \text{ and } 36 \leq t < 40\}$

51. $\{s \mid s \in J \text{ and } -4 \leq s < 5\}$

52. $\{s \mid s \in J \text{ and } -5 \leq s \leq 4\}$

Given that $A = \{100, 101, 102, 103, 104, 105, 106, 107, 108, 109, 110\}$ and $B = \{x \mid x \text{ is a multiple of 3 and } 100 < x < 110\}$, and $C = \{x \mid x \in W \text{ and } x < 111\}$, tell whether each statement is true or false.

53. $100 \in A$

54. $100 \in B$

55. $33 \in B$

56. $33 \in C$

57. $109 \in B$

58. $109 \in C$

59. $B \subseteq A$

60. $A \subseteq B$

61. $C \subseteq A$

62. $C \subseteq B$

63. $B \subseteq C$

64. $A \subseteq C$

STATE YOUR UNDERSTANDING

65. Describe the empty set.

66. What is the difference between a subset and a proper subset?

CHALLENGE EXERCISES

Given $A = \{0, 1, 2, 3, \ldots, 50\}$ and $B = \{2, 4, 6, 8, \ldots, 60\}$, write the following using the roster method:

67. $C = \{s \mid s \in A \text{ and } s \text{ is both a multiple of 3 and a multiple of 5}\}$.

68. $D = \{s \mid s \in A,\ s \in B, \text{ and } s \text{ is both a multiple of 3 and } 9 < s < 60\}$.

69. $E = \{s \mid s \in A,\ s \in B, \text{ and } s \text{ is a multiple of 7}\}$.

Sets.2

IDENTIFYING TYPES OF NUMBERS, SET OPERATIONS, AND VENN DIAGRAMS

OBJECTIVES

1. Identify specified subsets of a given set.
2. Find the union of two sets.
3. Find the intersection of two sets.
4. Draw a Venn diagram showing the relationship between two sets.
5. Find the complement of a set.

In the study of mathematics it is often necessary to be able to identify different types of numbers. In addition to the sets of numbers previously defined, we define and give examples of three more sets.

Rational numbers $\quad Q = \left\{ \frac{p}{q} \middle| p \text{ and } q \in J,\ q \neq 0 \right\}$

Rational numbers are referred to as fractions, both positive and negative, such as $\frac{3}{4}, \frac{7}{8}, \frac{11}{15}, \frac{9}{4}, -\frac{15}{16}, -\frac{8}{3}, \frac{5}{1}, -\frac{4}{2}$.

An alternate definition of the set of rational numbers is

$Q = \{x \mid x$ is a terminating or a nonterminating repeating decimal$\}$

Numbers such as 0.333. . . , 0.234234. . . , and 3.121212. . . are rational numbers since they are nonterminating repeating decimals. A nonterminating repeating decimal can also be indicated by drawing a line over the repeating group of digits to indicate the repetition. So, 0.234234. . . = $0.\overline{234}$ and 3.121212. . . = $3.1\overline{21}$.

Irrational numbers $\quad I = \{x \mid x$ is a nonterminating and nonrepeating decimal$\}$

Thus an irrational number is a number that cannot be written as a fraction or as a nonterminating repeating decimal. Numbers such as 0.1011011101111011111. . . , π, $\sqrt{2}$, $\sqrt{5}$, and $\sqrt[5]{5}$ are irrational numbers.

Real numbers $\quad R = \{x \mid x \in Q$ or $x \in I\}$

As a result of this definition, we can say that real numbers are either rational or irrational numbers.

We make the following observations about the six sets of numbers we have discussed:

1. The set of whole numbers contains all of the numbers that are in the set of natural numbers.
2. The set of integers contains all of the numbers that are in the set of whole numbers.
3. The set of rational numbers contains all of the numbers that are in the set of integers. (Write the integer as a fraction with a denominator of 1.)
4. The set of real numbers contains all of the numbers that are in the set of rational numbers and the set of irrational numbers.

As a result of those observations, we make the following statements.

1. $N \subset W$, $N \subset J$, $N \subset Q$, and $N \subset R$
   The set of natural numbers is a proper subset of the sets of whole numbers, integers, rational numbers, and real numbers.
2. $W \subset J$, $W \subset Q$, and $W \subset R$
   The set of whole numbers is a proper subset of the sets of integers, rational numbers, and real numbers.
3. $J \subset Q$ and $J \subset R$
   The set of integers is a proper subset of the sets of rational numbers and real numbers.
4. $Q \subset R$
   The set of rational numbers is a proper subset of the set of real numbers.
5. $I \subset R$
   The set of irrational numbers is a proper subset of the set of real numbers.

In Example 1, this information is used to write subsets of given sets.

EXAMPLE 1

Given $A = \left\{-34.5, -21, -\frac{43}{10}, -\pi, -\sqrt{5}, 0, \frac{3}{4}, \pi, 14, 19.2\overline{2}\right\}$, list using the roster method (a) $\{x \mid x \in N \text{ and } x \in A\}$, (b) $\{x \mid x \in W \text{ and } x \in A\}$, (c) $\{x \mid x \in J \text{ and } x \in A\}$, (d) $\{x \mid x \in Q \text{ and } x \in A\}$, (e) $\{x \mid x \in I \text{ and } x \in A\}$, and (f) $\{x \mid x \in R \text{ and } x \in A\}$.

a. $\{14\}$

b. $\{0, 14\}$

c. $\{-21, 0, 14\}$

d. $\left\{-34.5, -21, -\frac{43}{10}, 0, \frac{3}{4}, 14, 19.2\overline{2}\right\}$

e. $\{-\pi, -\sqrt{5}, \pi\}$

f. $\left\{-34.5, -21, -\frac{43}{10}, -\pi, -\sqrt{5}, 0, \frac{3}{4}, \pi, 14, 19.2\overline{2}\right\}$ ■

There are two operations with sets that we will discuss. They are the operations union and intersection of two or more sets.

The *union* of sets A and B, denoted by $A \cup B$, is a third set containing all of the elements of A and all of the elements of B.

EXAMPLE 2

Given $A = \{-5, -1, 3, 7, 9\}$ and $B = \{-2, -1, 4, 8\}$, find $A \cup B$.

$A \cup B = \{-5, -2, -1, 3, 4, 7, 8, 9\}$

Note the elements of the union are in either or both of the sets. The element -1 was contained in both sets but was listed only one time in the union. ■

The *intersection* of sets A and B, denoted by $A \cap B$, is a third set containing all of the elements that are in both set A and set B. The intersection contains those elements that are common to both sets.

EXAMPLE 3

Using sets A and B of Example 2, find $A \cap B$.

$A \cap B = \{-1\}$

There was only one element, -1, contained in both sets A and B. ■

EXAMPLE 4

Given $A = \{2, 5, 9\}$, $B = \{3, 5, 9, 11\}$, $C = \{4, 6, 11, 12\}$, and $D = \emptyset$, find (a) $A \cup B$, (b) $A \cup B \cup C$, (c) $A \cap B$, (d) $A \cap C$, (e) $B \cap C$, (f) $A \cup D$, (g) $A \cap D$.

a. $A \cup B = \{2, 3, 5, 9, 11\}$

List the elements of A and B.

b. $A \cup B \cup C = \{2, 3, 4, 5, 6, 9, 11, 12\}$

List the elements contained in all three sets.

c. $A \cap B = \{5, 9\}$

5 and 9 are elements of both A and B.

d. $A \cap C = \emptyset$

A and C have no elements in common, therefore the intersection is empty. Two sets are said to be disjoint when their intersection is the empty set.

e. $B \cap C = \{11\}$

11 is common to both B and C.

f. $A \cup D = \{2, 5, 9\} = A$

The union of a set and $\emptyset$ is the set itself. The empty set does not add any new members.

g. $A \cap D = \emptyset$

The empty set contains no elements, therefore the two sets have nothing in common. ■

Venn diagrams, named for the English mathematician John Venn, are pictures that can be used to represent the union or intersection of sets as well as showing subset relationships. Figures 1 and 2 each show a Venn diagram. Figure 1 depicts the union of sets A and B since the circle representing set A and the circle representing set B are each totally shaded. Figure 2 depicts the intersection of sets A and B since only that area which they have in common is shaded. The rectangle containing the two circles is referred to as the *universe* or the *universal set* and is labeled U. The universal set is a set that contains all elements of the type being discussed in the problem.

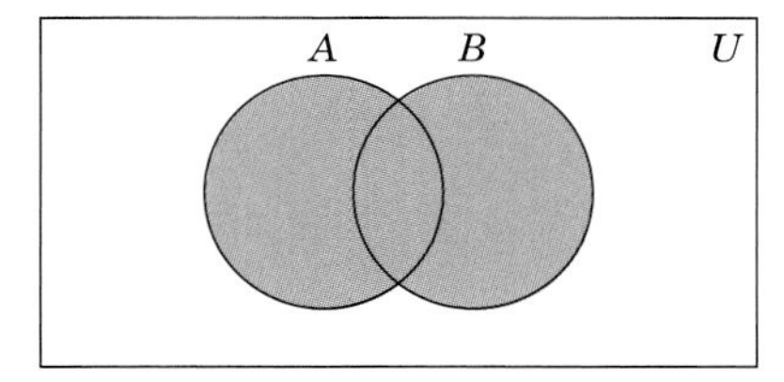

FIGURE 1

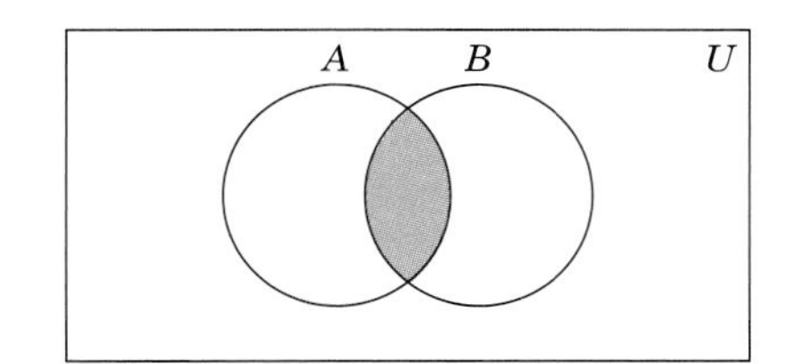

FIGURE 2

Figure 3 is a Venn diagram illustrating two disjoint sets, that is, their intersection is empty. (The two sets, A and B, do not intersect.) Figure 4 illustrates that set A is a proper subset of set B. From the diagram we can see that $A \cap B = A$.

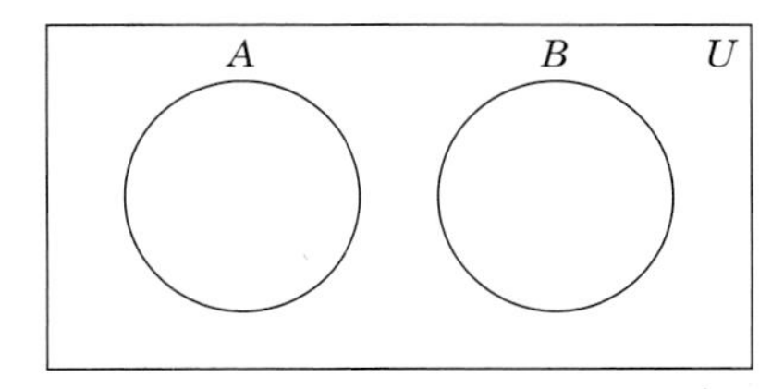

FIGURE 3

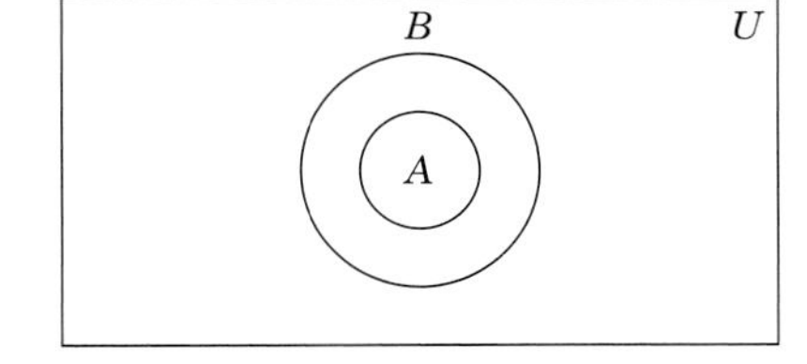

FIGURE 4

EXAMPLE 5

Use a Venn diagram to illustrate the set of all men that are bald.

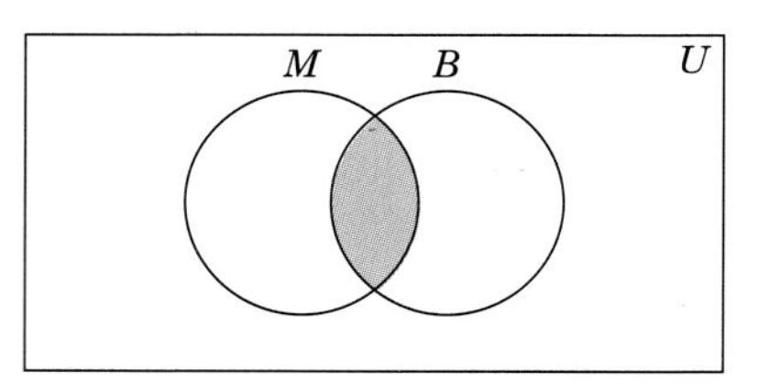

Set M is the set of all men. Set B is the set of all bald people. The universe is the set of all people. Thus all people not in set M are women. The set $M \cap B$ is the set of all men that are bald. ■

COMPLEMENT

Another important set that is often used is called the *complement* of a set. Suppose our universal set is the set of all women, and B is the set of all women with blonde hair. Then the complement of B is the set of all women that do not have blonde hair. The complement of B is written $\sim B$. We can show the meaning of the complement of a set, using a Venn diagram.

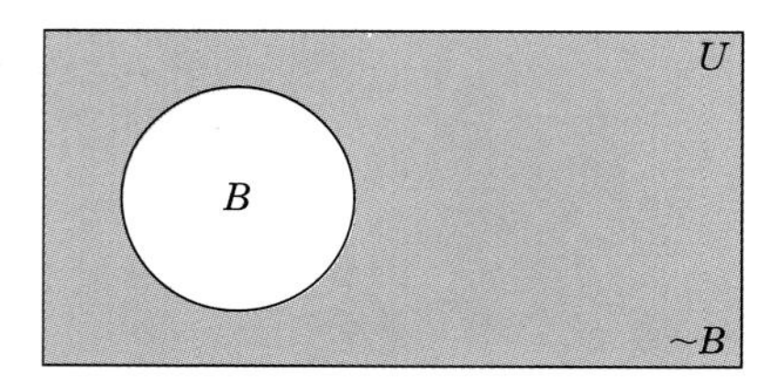

FIGURE 5

EXAMPLE 6

If the universe is the set $U = \{0, 1, 2, 3, 4, 5, 6, 7, 8, 9\}$ and $E = \{2, 4, 6, 8\}$, what are the members of $\sim E$?

$\sim E = \{1, 3, 5, 7, 9\}$

$\sim E$ contains all members of U that are not in E. ■

EXAMPLE 7

If the universe is the set of all automobiles and C is the set of all Chevrolets, describe $\sim C$.

$\sim C$ is the set of all automobiles that are not Chevrolets. ■

EXERCISES

A

Given: $T = \left\{-15.7\overline{7}, -\frac{19}{4}, -4, -\frac{7}{6}, -\pi, 0, 2, 3.121121112\ldots, 10, \frac{42}{3}\right\}$

If $x \in T$, list the following subsets of T.

1. $\{x \mid x \in N\}$ **2.** $\{x \mid x \in J\}$ **3.** $\{x \mid x \in Q\}$ **4.** $\{x \mid x \in I\}$

Given: $A = \{3, 6, 9, 12, 15\}$, $B = \{2, 4, 6, 8, 10, 12\}$, and $C = \{0, 5, 10, 15, 20\}$
Find:

5. $A \cup B$ **6.** $B \cap C$ **7.** $A \cup B \cup C$ **8.** $A \cap B \cap C$

Given: $U = \{0, 1, 2, 3, \ldots, 15\}$, $A = \{0, 5, 10, 15\}$, and $B = \{1, 3, 5, 7, 9, 11, 13, 15\}$
Find:

9. $\sim A$ **10.** $\sim B$

Given: $A = \{0, 1, 2, 3, 4, 5, 6, 7, 8, 9\}$ $B = \{2, 4, 6, 8\}$, and $C = \{3, 6, 9\}$
True or false:

11. $B \subset A$ **12.** $C \subset B$ **13.** $B \subseteq B$ **14.** $A \subset C$

Draw a Venn diagram to show the relationship between the following sets and describe the intersection of the sets.

15. $A = \{x \mid x \text{ is an odd whole number}\}$
$B = \{x \mid x \text{ is an even whole number}\}$

16. $C = \{x \mid x \text{ is a multiple of 3 and } x \in W\}$
$D = \{x \mid x \text{ is a multiple of 2 and } x \in W\}$

B

Given: $M = \left\{-27, -\sqrt{13}, -\frac{9}{4}, -3, -1.01, 0, \frac{5}{11}, 3, 4.57, 9, 16.13\overline{13}\right\}$

If $x \in M$, list the following subsets of M.

17. $\{x \mid x \in W\}$ **18.** $\{x \mid x \in Q\}$ **19.** $\{x \mid x \in I\}$ **20.** $\{x \mid x \in R\}$

Given: $S = \left\{-4, -3, -\frac{1}{2}, 0, \frac{1}{2}, 3, 4\right\}$, $T = \left\{-\frac{1}{2}, -\frac{1}{4}, 0, \frac{1}{4}, \frac{1}{2}\right\}$, and
$V = \{-4, -2, 0, 2, 4\}$
Find:

21. $S \cap V$ **22.** $S \cup T$ **23.** $S \cap T \cap V$ **24.** $S \cup T \cup V$

Given: $U = W = \{\text{all whole numbers}\}$, $A = \{\text{all even whole numbers}\}$, and
$B = \{\text{all whole numbers that are a multiple of 5}\}$
Find:

25. $\sim A$ **26.** $\sim B$ **27.** $\sim(A \cap B)$

28. $\sim(W \cap A)$ **29.** $\sim(W \cap B)$ **30.** $\sim(A \cup B)$

Given: $H = \{x \mid x \in J\}$, $K = \{-5, -4, -3, -2, -1, 0, 1, 2, 3, 4, 5\}$, and
$L = \left\{\frac{1}{4}, \frac{2}{4}, \frac{3}{4}, \frac{4}{4}, \frac{5}{4}, \frac{6}{4}, \frac{7}{4}\right\}$

True or false:

31. $K \subset H$ **32.** $L \subset H$ **33.** $L \subseteq L$ **34.** $K \subseteq H$

Draw a Venn diagram to show the relationship between the given sets and describe the union of the sets.

35. $A = \{\text{all left-handed males}\}$
$B = \{\text{all left-handed females}\}$

36. $C = \{\text{all ducks}\}$
$D = \{\text{all mallard ducks}\}$

Given: $A = \{x \mid x < 20 \text{ and } x \in N\}$, $B = \{x \mid x > 10 \text{ and } x \in W\}$, and $C = \{x \mid x < 18 \text{ and } x \in J\}$
List the following sets:

37. $\{x \mid x \in A \text{ and } x \in Q\}$

38. $\{x \mid x \in C \text{ and } x \in N\}$

39. $\{x \mid x \in B \text{ and } x \in J\}$

40. $\{x \mid x \in C \text{ and } x \in W\}$

41. $A \cap B$ **42.** $B \cap C$ **43.** $A \cap C$ **44.** $A \cap B \cap C$

45. $B \cup C$ **46.** $A \cup C$ **47.** $A \cup B$ **48.** $A \cup B \cup C$

Given: $U = \{\text{all people}\}$, $A = \{\text{all males under 25 years old}\}$, and $B = \{\text{all people with red hair}\}$
Find:

49. $\sim A$ **50.** $\sim B$ **51.** $\sim(A \cap B)$

52. $\sim(A \cup B)$ **53.** $\sim(U \cup B)$ **54.** $\sim[(U \cup A) \cap B]$

Draw a Venn diagram to show the relationship between the following sets and describe the intersection of the sets.

55. $A = \{\text{state capitals}\}$
$B = \{\text{the largest city in each state}\}$

56. $A = \{x \mid x \in N \text{ and } x \text{ is a multiple of } 5\}$
$B = \{x \mid x \in J \text{ and } x < 10\}$

STATE YOUR UNDERSTANDING

57. What is the meaning of $A \cup B$?

58. What can be said about the set A if $A \cap B = B$?

CHALLENGE EXERCISES

Given: $A = \{x \mid x > 25 \text{ and } x \in J\}$, $B = \{x \mid x < 40 \text{ and } x \in W\}$, and $C = \{x \mid x \text{ is a multiple of 5 and } x \in N\}$
Find:

59. $(A \cup B) \cap C$ **60.** $(B \cap C) \cup A$ **61.** $(A \cap C) \cup B$ **62.** $(B \cup C) \cap A$

ANSWERS TO SELECTED EXERCISES

CHAPTER 1

Exercise 1.1

1. $\frac{2}{5}$ **3.** $\frac{1}{2}$ **5.** $\frac{1}{2}$ **7.** $\frac{2}{3}$ **9.** $\frac{1}{6}$ **11.** $\frac{15}{16}$
13. $\frac{49}{25}$ or $1\frac{24}{25}$ **15.** $\frac{23}{12}$ or $1\frac{11}{12}$ **17.** $11\frac{14}{15}$
19. $\frac{1}{4}$ **21.** $2\frac{19}{20}$ **23.** $\frac{6}{7}$ **25.** $\frac{3}{5}$ **27.** $\frac{33}{2}$ or $16\frac{1}{2}$ **29.** $\frac{134}{7}$ or $19\frac{1}{7}$ **31.** $\frac{8}{3}$ or $2\frac{2}{3}$ **33.** $\frac{145}{31}$ or $4\frac{21}{31}$ **35.** 7632 **37.** $\frac{7}{8}$ **39.** 0.53 **41.** 5.808
43. $\frac{3}{4}$ **45.** $\frac{5}{8}$ **47.** 8 **49.** 1.99554 **51.** 2.6
53. 3.35 **55.** 3.819 **57.** 191.738 **59.** 1.656
61. 321.022 **63.** $15\frac{11}{12}$ hours **65.** \$8.25 per hour
67. 5,760 ounces **69.** 6 cartons **71.** In words
73. In words **75.** \$133.75 **77.** The distributive property

Exercise 1.2

1. $4 \cdot 4$; 4^2 or 16 **3.** 3^4; 3^4 or 81 **5.** $\frac{1}{4} \cdot \frac{1}{4} \cdot \frac{1}{4}$; $\left(\frac{1}{4}\right)^3$ or $\frac{1}{64}$ **7.** $\frac{1}{2} \cdot \frac{1}{2} \cdot \frac{1}{2} \cdot \frac{1}{2} \cdot \frac{1}{2}$; $\left(\frac{1}{2}\right)^5$ or $\frac{1}{32}$
9. 7^2; $7 \cdot 7$ **11.** 1^6; 1^6 or 1 **13.** 3^4 **15.** 8^3
17. 7 **19.** 4 **21.** 27 **23.** 2
25. $\frac{2}{3} \cdot \frac{2}{3} \cdot \frac{2}{3} \cdot \frac{2}{3}$; $\left(\frac{2}{3}\right)^4$ or $\frac{16}{81}$ **27.** $x = 144$
29. $z = 10{,}648$ **31.** $x = 1728$ **33.** $z = 83{,}521$
35. 24 **37.** 9 **39.** 7 **41.** 27 **43.** 70
45. 11 **47.** 204 **49.** 13 **51.** 16 **53.** 8
55. $12^3 \cdot 22^3$; $12^3 \cdot 22^3$ or 18,399,744
57. $13 \cdot 13 \cdot 13 \cdot 13 \cdot 5 \cdot 5 \cdot 5$; $13^4 \cdot 5^3$ or 3,570,125
59. $x = 11{,}664$ **61.** $z = 32{,}768$ **63.** $v = 72$
65. 138 **67.** 13 **69.** 19 **71.** 16,037 **73.** 461
75. 7,391,893 **77.** \$400,000 **79.** 700,000,000 years
81. 25,500,000,000,000 miles **83.** 3,620,000 books
85. 1024 investors; \$2,048,000 **87.** 51,727.48865
89. 494.64 **91.** \$447 **93.** \$6834 **95.** \$4200
97. In words **99.** 22 **101.** 425

Exercise 1.3

1. 10 **3.** 11 **5.** 8 **7.** 3 **9.** 441
11. $x - 15$ **13.** 20 less than a given number
15. 24 times a given number
17. Yes **19.** 108 **21.** 4 **23.** 96 **25.** 0
27. $\frac{1}{2}$ **29.** 15 **31.** 2500 **33.** $x - y$ **35.** xy
37. $x + y + 18$ **39.** x^2y **41.** Yes
43. Yes **45.** 12.6075 **47.** 500 **49.** 0.123
51. 232.608375 **53.** 0.00015129 **55.** 29.207375
57. 48.07250186 **59.** Yes **61.** Yes **63.** No
65. 200 **67.** 25.12 **69.** 180 ft^2 **71.** 720 eggs
73. 4 hours **75.** 11.5 amperes **77.** 128
79. In words **81.** 1 **83.** $I = 0.625$ amperes

Exercise 1.4

1. $\{0, +2, +18\}$ **3.** $\left\{+2, \pi, \frac{7}{2}, 9.3, +18\right\}$
5. $\left\{-12, -9.8, -7, -\frac{15}{4}, -1, 0, +2, \frac{7}{2}, 9.3, +18\right\}$
7. $A \leftrightarrow -8$ **9.** $C \leftrightarrow \frac{7}{3}$ **11.** $E \leftrightarrow -\frac{15}{2}$ **13.** $G \leftrightarrow -4$ **15.** $I \leftrightarrow 1$ **17.** True **19.** True **21.** False
23. True **25.** True **27.** True **29.** False
31. True **33.** True **35.** False **37.** True
39. False **41.** True **43.** False **45.** True
47. True **49.** False **51.** False **53.** False
55. True **57.** False **59.** $32 > 25$
61. $8 - 2 < 6 + 2$ **63.** $8 \cdot 4 > 9 + 6$
65. $20 - 2(3 + 1) > 9$ **67.** In words
69. **a.** $6x < 54$ **b.** all numbers less than 9
71. **a.** $x - 12 < 25$ **b.** all numbers less than 37

Exercise 1.5

1. 8 **3.** −37 **5.** −150 **7.** 2 **9.** 9
11. −10 **13.** −8 **15.** $y = 5$ or $y = -5$ **17.** −4
19. −13 **21.** 0 **23.** −13 **25.** −7 **27.** −3
29. 8 **31.** −13 **33.** −25 **35.** −47 **37.** 11
39. $x = 2$ or $x = -2$ **41.** Not possible **43.** −2
45. 10 **47.** $-\frac{1}{6}$ **49.** 7 **51.** −20 **53.** −4.3
55. $-\frac{5}{4}$ **57.** 8.1 **59.** $2\frac{1}{5}$ **61.** 48 **63.** 86
65. $x = 2$ or $x = -2$ **67.** No solution **69.** $-\frac{32}{9}$
71. −53 **73.** −0.375 **75.** $-\frac{2}{5}$ **77.** Yes
79. No **81.** −7 **83.** −18 **85.** $|8|$ **87.** $-|-63|$
89. $-6\frac{3}{4}$ or $-\frac{27}{4}$ **91.** $(-x)y$ **93.** $-x - y$
95. −190 ft **97.** \$43 **99.** 3975 lb
101. 18,041 volumes **103.** $-8 + (-z)$ **105.** $w + z$

107. In words **109.** $(-8 + x) + 4$
111. $4(2 + y^3) + 5[2y + (-7)]$

Exercise 1.6

1. 4 **3.** −19 **5.** −12 **7.** 26 **9.** 4
11. −14 **13.** −9 **15.** 12 **17.** −73 **19.** 23
21. −23 **23.** 73 **25.** −23 **27.** −171
29. −171 **31.** −599 **33.** 171 **35.** −24
37. −2.25 **39.** 22.5 **41.** $-\frac{9}{4}$ **43.** $-\frac{11}{3}$
45. $-\frac{31}{12}$ **47.** 0.8175 **49.** 211 **51.** −511
53. −182 **55.** −5.04 **57.** $\frac{13}{12}$ or $1\frac{1}{12}$
59. \$220.28 **61.** \$189.38 **63.** 24°F **65.** \$619.32
67. 40 miles west **69.** $x - 18$ **71.** $-32 - (-y)$
73. In words **75.** $6(x - y) - 2(4x - 5y)$
77. $\frac{x}{-2} - (2x - x^2)$

Exercise 1.7

1. −6 **3.** 30 **5.** −48 **7.** 36 **9.** 54
11. −9 **13.** −27 **15.** 28 **17.** −110 **19.** 144
21. −258 **23.** 210 **25.** 60 **27.** −44 **29.** 240
31. 24 **33.** 240 **35.** 625 **37.** −625
39. −128 **41.** 1344 **43.** $-\frac{1}{3}$ **45.** $-\frac{2}{7}$
47. −0.6 **49.** 0.0513 **51.** $-\frac{2}{15}$ **53.** −0.04
55. 36 **57.** $-\frac{7}{18}$ **59.** 2.24 **61.** −720
63. −21 lb **65.** −33.96 **67.** 66° **69.** −60°C
71. −10°C **73.** $x(-8)$ **75.** $-(-7)(-y)$
77. $-(xy)$ **79.** In words **81.** 24 **83.** 138

Exercise 1.8

1. −4 **3.** −5 **5.** 7 **7.** −3 **9.** −4
11. −12 **13.** −24 **15.** −7 **17.** −5 **19.** −20
21. 25 **23.** −18 **25.** 12 **27.** −16 **29.** 23
31. −17 **33.** 31 **35.** −24 **37.** 286 **39.** 42
41. −36 **43.** 57 **45.** −23 **47.** −5.5 **49.** 310
51. $\frac{5}{4}$ **53.** −10.1 **55.** $-\frac{16}{9}$ **57.** −125
59. 145 **61.** −220 **63.** 120 **65.** −23
67. −26 **69.** −5.2 **71.** −12° **73.** $-\left(5\frac{27}{40}\right)$
75. −4035 bolts per day **77.** $\frac{-x}{18}$ **79.** $\frac{x}{y}$
81. In words **83.** $(x - 8) - 8y$ **85.** $12x - (y - 8)$

Exercise 1.9

1. 17 **3.** −590 **5.** −75 **7.** $-\frac{1}{2}$ **9.** −108
11. 203 **13.** 154 **15.** 344 **17.** Yes **19.** No
21. −20 **23.** 1 **25.** $-\frac{5}{2}$ **27.** 95.05
29. −2065 **31.** −1594 **33.** −277 **35.** −246
37. $-\frac{8}{3}$ **39.** Yes **41.** No **43.** 53 **45.** 77.46
47. 244.858 **49.** 7185.284 **51.** 51.457 **53.** $\frac{109}{105}$
55. 431.025 **57.** 730.7 **59.** −5263.645 **61.** Yes
63. Yes **65.** 35.5°C **67.** 55°C **69.** −4.9°F
71. \$6.32 profit **73.** Loss of \$30 **75.** In words
77. 159 **79.** 1

Chapter 1 Review Exercises

1. $\frac{3}{5}$ **3.** $\frac{7}{16}$ **5.** $\frac{7}{9}$ **7.** 1.75 **9.** $\frac{35}{36}$
11. 6137 **13.** $\frac{1}{17}$ **15.** 49.4 **17.** 32,768
19. 625 **21.** 7^6 **23.** 8^4 **25.** 2^{10} **27.** 207
29. 248 **31.** 28 **33.** 80 **35.** 48 **37.** $6x + 6$
39. A number raised to the fifth power **41.** Yes
43. No **45.** Yes **47.** All are rational numbers
49. All are real numbers **51.** 4 **53.** −1 **55.** 1.3
57. False **59.** True **61.** −47 **63.** −21
65. −21 **67.** 48 **69.** −72 **71.** −7 **73.** −162
75. Yes **77.** −17 **79.** No **81.** −720
83. −560 **85.** 560 **87.** −19 **89.** $-\frac{3}{5}$
91. −272 **93.** $-\frac{8}{5}$ **95.** −1 **97.** Yes **99.** No

Chapter 1 True–False Concept Review

1. True **2.** False **3.** True **4.** False **5.** False
6. True **7.** False **8.** False **9.** False **10.** True
11. True **12.** False **13.** False **14.** True
15. True **16.** False **17.** False **18.** True
19. True **20.** False

Chapter 1 Test

1. 30.79 **2.** 1.5 **3.** Yes **4.** 3.4 **5.** $\frac{7}{24}$
6. −8.14 **7.** 55 **8.** −7 **9.** 40.71 **10.** −43
11. 2401 **12.** −111 **13.** −86 **14.** 0.85
15. 1134 **16.** 12^6 **17.** 309 **18.** 5.75 **19.** 87
20. Yes **21.** $5x^2$ **22.** 13.5 **23.** 271 **24.** 3
25. $\frac{n}{4p}$ **26.** −49.28 **27.** 65 points **28.** 75°F
29. −10°C

CHAPTER 2

Exercise 2.1

1. {32} **3.** {63} **5.** {−13} **7.** {−47} **9.** {−40} **11.** {−70} **13.** {31} **15.** {10} **17.** $\ell = p - w$ **19.** $x = t + s$ **21.** {102} **23.** {−44} **25.** {−35} **27.** {−72} **29.** {−661} **31.** {604} **33.** {−1139} **35.** {256} **37.** {14} **39.** $a = b + c$ **41.** $\left\{-\frac{3}{4}\right\}$ **43.** $\left\{\frac{13}{20}\right\}$ **45.** $\left\{-\frac{3}{10}\right\}$ **47.** {10.65} **49.** {−12.37} **51.** {−52.7} **53.** {−7.5} **55.** $p = w - r$ **57.** $t = m + 5$ **59.** $p = r - s$ **61.** 44 **63.** 51 **65.** 26 **67.** \$415.66 **69.** \$128.76 **71.** 9.1 meters **73.** In words **75.** {−4, 4} or {±4} **77.** {−18, 18} or {±18} **79.** 12 **81.** 2.22 **83.** −142 **85.** 65

Exercise 2.2

1. {25} **3.** {−2} **5.** {100} **7.** {−9} **9.** {66} **11.** {−48} **13.** {16} **15.** $i = pr$ **17.** $x = 7t$ **19.** $a = 18q$ **21.** {1936} **23.** {−15} **25.** {−180} **27.** {70} **29.** {432} **31.** {112} **33.** {550} **35.** {28} **37.** $x = \frac{m}{5}$ **39.** $x = 8a$ **41.** {−1.29} **43.** {−3.85} **45.** $\left\{-\frac{2}{3}\right\}$ **47.** $\left\{-\frac{9}{8}\right\}$ **49.** $\left\{-\frac{3}{5}\right\}$ **51.** $\left\{\frac{16}{15}\right\}$ **53.** $\left\{-\frac{7}{6}\right\}$ **55.** $n = \frac{C}{p}$ **57.** $A = \ell w$ **59.** $m = 4ab$ **61.** 1961 **63.** 12 **65.** −637 **67.** 0.06 **69.** 21 ft **71.** \$1,200 **73.** 12 feet **75.** \$4.85 **77.** In words **79.** {−4.5, 4.5} or {±4.5} **81.** {−11, 13} **83.** −16 **85.** −51 **87.** −65 **89.** \$2.35 million

Exercise 2.3

1. {5} **3.** {−4} **5.** $\left\{-\frac{25}{2}\right\}$ **7.** {2} **9.** {5} **11.** {30} **13.** {−1} **15.** $x = \frac{d - c}{t}$ **17.** {126} **19.** {120} **21.** {7} **23.** {11} **25.** {5} **27.** $\left\{-\frac{1}{2}\right\}$ **29.** {7} **31.** {54} **33.** {−40} **35.** {52} **37.** $\left\{-\frac{2}{3}\right\}$ **39.** $x = \frac{k + t}{g}$ **41.** $\left\{\frac{5}{12}\right\}$ **43.** $\left\{-\frac{1}{4}\right\}$ **45.** $\left\{-\frac{2}{3}\right\}$ **47.** {5} **49.** {−1.2} **51.** {−3.4} **53.** {−28} **55.** {−0.6} **57.** $w = \frac{P - 2\ell}{2}$ **59.** $p = \frac{c - 20}{18}$ **61.** 5.5 sec **63.** 40 **65.** 28 **67.** 3.2 **69.** 7 payments **71.** 2.5 years **73.** In words **75.** {−2, 2} or {±2} **77.** {−8, 7} **79.** −95 **81.** −34.404 **83.** 6 **85.** −\$775

Exercise 2.4

1. $22x$ **3.** $13t$ **5.** $-6a$ **7.** $34x^2$ **9.** $36ab$ **11.** $13.5p$ **13.** $112y + 6$ **15.** $-1.3t^2 - 3.2$ **17.** $\frac{53}{10}w$ **19.** $\frac{5}{8}c^2d^2$ **21.** $-28x$ **23.** $-52s$ **25.** $17w^2$ **27.** $-7abd^2$ **29.** $17x^2 - 19y + 14$ **31.** $19x + 2y$ **33.** $17x + 4y$ **35.** $19xy - 25xy^2$ **37.** $-3xy - 3xy^2$ **39.** $40a + 27b + 20c$ **41.** $12x - a + 2$ **43.** $8p + q - 9$ **45.** $24a - 3b + c$ **47.** $-0.9x - 2.5p + 8.7$ **49.** $-1.82m + 1.15n - 0.7$ **51.** $7a + b + 12c$ **53.** $\frac{1}{6}p - \frac{1}{6} - \frac{3}{2}q$ **55.** $\frac{7}{6}a + \frac{5}{12}b + \frac{2}{5}c$ **57.** $3x + 8$ **59.** $12x + 12$ **61.** $2x - 7$ **63.** $5x - 9$ **65.** In words **67.** $10y - 4$ **69.** $-4x^2yz - 11xy^2z + 19xyz^2$ **71.** $5(x + 3)$ **73.** $xy - y^2$ **75.** Twice a number and the product decreased by 10. **77.** 0.6

Exercise 2.5

1. {2} **3.** {−7} **5.** {8} **7.** {0} **9.** {4} **11.** {12} **13.** {2} **15.** {−5} **17.** {−1} **19.** {1} **21.** {5} **23.** {−4} **25.** {18} **27.** {0} **29.** {0.5} **31.** $\left\{\frac{20}{3}\right\}$ **33.** {2} **35.** $\left\{-\frac{1}{3}\right\}$ **37.** {5} **39.** {9} **41.** {14} **43.** {−3} **45.** $\left\{\frac{2}{5}\right\}$ **47.** {29} **49.** {−5} **51.** {3} **53.** {9} **55.** {4} **57.** {18} **59.** {−2} **61.** The number is 22 **63.** The number is 15 **65.** Ten years old **67.** \$45,000 **69.** The shortest side is 16.8 cm.; the longest is 33.6 cm., and the third side is 25.2 cm. **71.** In words **73.** R = {all real numbers} **75.** $b = \frac{2A - ah}{h}$ or $\frac{2A}{h} - a$ **77.** {−6, −1} **79.** $\{-\sqrt{2}\}$ **81.** False **83.** True

Exercise 2.6

1. 50% **3.** 2 **5.** 26 **7.** 48 **9.** 120 **11.** 25% **13.** 45 **15.** 200% **17.** 28 **19.** 28 **21.** 23.4 **23.** 50 **25.** $33\frac{1}{3}\%$ **27.** $16\frac{2}{3}\%$ **29.** 9.3% **31.** 0.1% **33.** 12% **35.** 2300 **37.** 2800 **39.** 2050 **41.** 90.7 **43.** 126.2% **45.** 87.4 **47.** 292% **49.** 11.2 **51.** 1281.0 **53.** 88.5 **55.** 17.2% **57.** 334.09 **59.** 82%

61. $66\frac{2}{3}\%$ **63.** \$1191.84 **65.** \$1145.83
67. \$356.96 **69.** \$315 **71.** 69,660 persons
73. \$598 **75.** 50% **77.** In words
79. 40 examples **81.** \$172.58 **83.** $\frac{5}{14}$ **85.** 256
87. 1.07 **89.** −25

Exercise 2.7

1. $\{x \mid x > 20\}$ **3.** $\{y \mid y > -1\}$ **5.** $\{x \mid x \le -9\}$
7. $\{w \mid w < 21\}$ **9.** $\{x \mid x > -9\}$ **11.** $\{x \mid x \ge 15\}$

13. $\{x \mid x \ge 8\}$

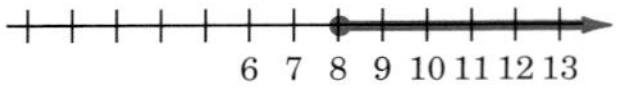

15. $\{x \mid x < -2\}$

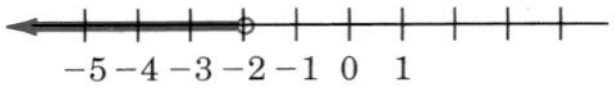

17. $\{x \mid x \ge 1\}$ **19.** $\{y \mid y \ge 3\}$ **21.** $\{x \mid x > -2\}$
23. $\{x \mid x < 4\}$ **25.** $\{x \mid x < -4\}$ **27.** $\{x \mid x < -3\}$

29. $\{x \mid x > -2\}$

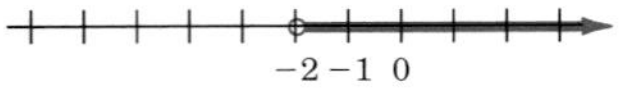

31. $\{x \mid x < 2\}$

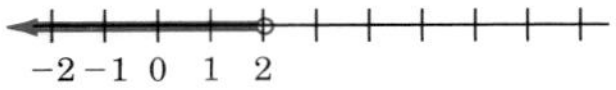

33. $\{x \mid x \le -3\}$

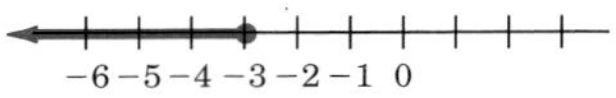

35. $\{x \mid x \le 5\}$

37. $\{x \mid x < 2\}$ **39.** $\{y \mid y \ge 50\}$ **41.** $\{w \mid w \le -1\}$
43. $\{x \mid x \le -16\}$ **45.** $\{x \mid x > -2\}$ **47.** $\{w \mid w > -1\}$
49. $\left\{z \mid z \le \frac{18}{5}\right\}$ **51.** $\{z \mid z \le 2\}$

53. $\{y \mid y < -3\}$

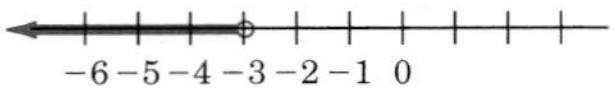

55. $\left\{x \mid x \ge -\frac{4}{5}\right\}$

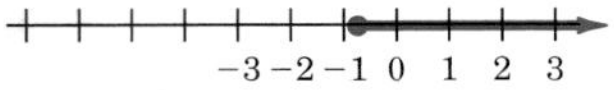

57. All numbers less than or equal to 1
59. All numbers less than 8

61. $C \le 0$

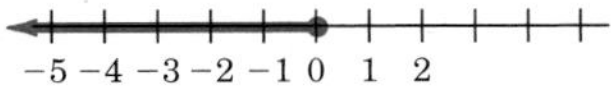

63. $w \le 8$

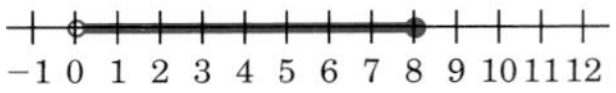

65. $t < 2.25$ sec **67.** In words

69. $\{x \mid x \le -2\}$

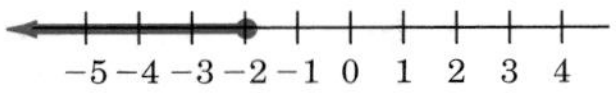

71. It must purchase at least 225 youth tickets.
73. −652 **75.** −6 **77.** $-\frac{5}{12}$ **79.** Yes

Chapter 2 Review Exercises

1. $\{72\}$ **3.** $\{-70\}$ **5.** $\left\{\frac{5}{8}\right\}$ **7.** $\{-1.95\}$
9. $r = g + k$ **11.** $\{13\}$ **13.** $\{9\}$ **15.** $\{-320\}$
17. $\{84\}$ **19.** $b = \frac{v}{a}$ **21.** $\{22\}$ **23.** $\{-14\}$
25. $\{-2.2\}$ **27.** $t = \frac{m - 4s}{5}$ **29.** $17x^2$ **31.** $-28a$
33. $-11a - 17$ **35.** $5y^2 + 11y + 3$
37. $10ab + 3a - 3b$ **39.** $8xy - 17x - 2y - 2$
41. $\{9\}$ **43.** $\{5\}$ **45.** $\{-2\}$ **47.** 87.5%
49. 137 **51.** 1109.1 **53.** $\{x \mid x > 9\}$
55. $\{x \mid x < 7\}$ **57.** $\{x \mid x \ge 24\}$ **59.** $\{x \mid x > -2.5\}$

Chapter 2 True–False Concept Review

1. True **2.** True **3.** False **4.** True **5.** False
6. True **7.** True **8.** True **9.** False **10.** False
11. False **12.** False **13.** True **14.** True
15. True **16.** True **17.** False **18.** False
19. True **20.** False

Chapter 2 Test

1. 3.91 **2.** $\left\{-\frac{7}{2}\right\}$ **3.** $\{-15\}$ **4.** $31x - 18y + 2w$
5. $x = \frac{c - b}{a}$ **6.** $\{-20\}$ **7.** $\{-6.2\}$ **8.** 125%
9. $\left\{\frac{11}{10}\right\}$ **10.** $9x^2 - 23 + 21y^2 + 3xy$ **11.** $\{121\}$
12. $\{x \mid x < 121\}$

13.

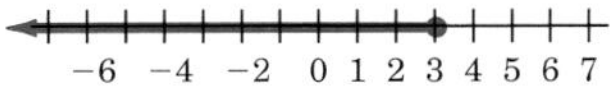

14. 600 **15.** $\left\{m \mid m < \frac{17}{6}\right\}$ **16.** $\{15\}$ **17.** $\{16\}$
18. $P = \frac{R - AM}{M}$ **19.** \$45.40 **20.** \$1.10

CHAPTER 3

Exercise 3.1

1. b^7 **3.** x^8 **5.** x^{15} **7.** a^{17} **9.** x^2y **11.** x^4
13. a^2 **15.** y^4 **17.** $9b^2$ **19.** m^4n^4 **21.** x^{11}
23. y^{12} **25.** b^{22} **27.** a^7x^7 **29.** x^9y^8 **31.** x^{15}
33. a^{30} **35.** z^{24} **37.** 1 **39.** $49a^4b^2$ **41.** y^4z^4
43. a^7 **45.** y^{17} **47.** a^4 **49.** x^{a+b} **51.** a^{3+m}

53. x^{2m} **55.** a^{bc} **57.** y^{n^2} **59.** y^{20}
61. $225x^4y^2z^8$ **63.** $x^6y^4z^6$ **65.** 132.651 in^3
67. $V = w^4$, 81 ft^3 **69.** $V = \pi r^5$, 763.02 in^3
71. 33.5 cm^3 **73.** 162 in^2 **75.** In words **77.** $6x^{5a}$
79. -3^6 **81.** x^{m+1} **83.** $-x^5$ **85.** $72a^{12}b^7c^2$
87. -81 **89.** -116 **91.** $\{-456\}$ **93.** $\left\{-\frac{10}{21}\right\}$

Exercise 3.2

1. $\frac{1}{6}$ **3.** $\frac{1}{6^3}$ **5.** $\frac{1}{z^4}$ **7.** $\frac{1}{2^9}$ **9.** $\frac{1}{x^3}$ **11.** y^2
13. $\frac{1}{x^4}$ **15.** w^6 **17.** $\frac{1}{a^6b^2}$ **19.** $\frac{a^4}{b^4}$
21. $\frac{1}{2^2}$ or $\frac{1}{4}$ **23.** $\frac{1}{10^2}$ or $\frac{1}{100}$ **25.** $\frac{1}{x^6}$ **27.** 1
29. $\frac{1}{x^6}$ **31.** y **33.** a^{x+y} **35.** x^{a-n} **37.** $\frac{a^{12}}{b^8}$
39. $\frac{x^{15}y^6}{w^6z^9}$ **41.** $\frac{25}{a^2}$ **43.** $\frac{1}{4^4a^2}$ or $\frac{1}{256a^2}$
45. $\frac{1}{5^2}$ or $\frac{1}{25}$ **47.** $\frac{1}{3^2}$ or $\frac{1}{9}$ **49.** $\frac{1}{y^7}$ **51.** x^2
53. y^{11} **55.** $\frac{1}{a^{12}}$ **57.** a^m **59.** 1 **61.** $\frac{1}{x^8}$ **63.** x
65. $\frac{a^{30}b^{16}}{c^{14}d^8}$ **67.** $\frac{y^3}{x^9}$ **69.** $\frac{y^9}{10^6}$ or $\frac{y^9}{1{,}000{,}000}$
71. 0.000000005 in. **73.** 0.0015 m **75.** 10
77. 100.48 **79.** In words
81. $\left(\frac{x}{y}\right)^{-a} = \frac{x^{-a}}{y^{-a}} = x^{-a} \cdot \frac{1}{y^{-a}} = \frac{1}{x^a} \cdot y^a = \frac{y^a}{x^a} = \left(\frac{y}{x}\right)^a$
83. 3^{5x+5} **85.** 2 **87.** 165 **89.** $\{-19\}$
91. $\{-48\}$

Exercise 3.3

1. 5.0×10^4 **3.** 4.0×10^{-5} **5.** 9.0×10^6
7. 7×10^3 **9.** 2.1×10^5 **11.** 930 **13.** 89,100
15. 54,000,000 **17.** 46 **19.** 0.000082
21. 3.77×10^5 **23.** 7.01×10^{-5} **25.** 6.11×10^8
27. 4.56×10^{-4} **29.** 9.22×10^{-5} **31.** 0.321
33. 68,900 **35.** 743 **37.** 460,000 **39.** 862,000
41. 3.784×10^3 **43.** 3.448×10^4 **45.** 4.84×10^{-4}
47. 1.11×10^7 **49.** 2.962×10^8 **51.** 38,400,000
53. 0.00000000236 **55.** 0.0000101
57. 0.0000000911 **59.** 57,800,000 **61.** 2×10^{-11}
63. 2.919×10^{10} **65.** 1107 **67.** 400,000
69. 5.87×10^{12} mi **71.** 5.0×10^4 cycles per second
73. 3×10^{-3} in. **75.** 0.00004 cm
77. 25,500,000,000,000 mi **79.** In words
81. 6.5×10^3 **83.** 19 times **85.** 98 **87.** 899
89. \$231 **91.** $\{2\}$

Exercise 3.4

1. Coefficient 5, degree 2 **3.** Coefficient 12, degree 0
5. Coefficient 16, degree 4 **7.** Coefficient $-\frac{7}{9}$, degree 1
9. Coefficient $\frac{14}{3}$, degree 2 **11.** Binomial
13. Monomial **15.** Trinomial **17.** Coefficient -14, degree 10 **19.** Coefficient 1, degree 7
21. Coefficient -3, degree 9
23. Coefficient $-\frac{35}{23}$, degree 3
25. Coefficient $-\frac{15}{2}$, degree 0 **27.** Trinomial
29. Monomial **31.** Binomial **33.** 3 **35.** 8
37. 5 **39.** 9 **41.** 14
43. $-3x^5 - x^4 - 4x^3 + 8x^2 + 2x + 3$
45. $-14x^5 - 3x^2 + 7x - 6$ **47.** $14x^9 - 3x^5 + 1$
49. In words **51.** In words **53.** 2 terms; binomial
55. 2 terms; binomial **57.** 1 term; monomial
59. 38.96 **61.** \$1078 **63.** $x = \frac{e + dy}{c}$
65. $\left\{-\frac{35}{24}\right\}$

Exercise 3.5

1. $9a$ **3.** $2a$ **5.** $-2y^2$ **7.** $-9xyz$
9. $3x^2 - 4y^2 + 6z^2$ **11.** $-2x + 3$ **13.** $x - 5$
15. $3x + 7y + 12$ **17.** $4m + 1$ **19.** $40x - 13y$
21. $\{-2\}$ **23.** $\{2\}$ **25.** $\{3\}$ **27.** $-3x^2 + 7x - 1$
29. $3a + 2b$ **31.** $27y^2 - 19y - 12$
33. $-3r + 7s - 9t$ **35.** $8a^2 - 4ab - 6b^2$
37. $-26a^2 + 147a - 189$ **39.** $29y^2 + 26y - 43$
41. $3a^2 - 5ab - 5b^2$ **43.** $x - 9y + 5$
45. $12x^2 + 70x - 18$ **47.** $\{-4\}$ **49.** $\{2\}$
51. $\left\{-\frac{4}{3}\right\}$ **53.** $16x + 17$ **55.** $x - \frac{9}{4}y + 2z$
57. $-8xy - 4yz + 3xz - 28xyz$
59. $0.46r - 0.46s - 0.45t$ **61.** $-0.1bc - 2ad$
63. $4a^2 - 3ab - 4b^2$ **65.** $13x + 9$ **67.** $11x - 4$
69. $x + 9$ **71.** $0.97a + 0.1b + 0.8$ **73.** $\{1\}$
75. $\left\{-\frac{1}{6}\right\}$ **77.** $\{0\}$ **79.** $\left\{-\frac{11}{3}\right\}$ **81.** 44
83. 11 **85.** $-\frac{5}{2}$ **87.** -16 **89.** 11 ft, 13 ft, 17 ft
91. $P = 5w + 2$, 77 cm **93.** 13, 31 **95.** Pete, 86; Amber, 82 **97.** Cindy, 6124; Minh, 1224 **99.** 5 sec
101. $2\frac{2}{3}$ sec **103.** In words **105.** $x^2 + 9x - 5$
107. $x - 13y$ cm **109.** $\{x \mid x > -6.9\}$ **111.** $\{-3.25\}$
113. $s = \frac{16y - ab}{rt}$ **115.** y^{32}

Exercise 3.6

1. 18 hrs **3.** 22 lb of each type of nut
5. 1.6 hr or 1 hr 36 min **7.** $5\frac{1}{3}$ hr or 5 hr 20 min
9. 135 of each **11.** 22 cases of each
13. \$2560 in each bank **15.** 8 oz **17.** In words
19. The one flies at 300 mph and the other flies at 240 mph.
21. Jill received \$32,000, and Mike received \$24,000.
23. The numbers are 26, 28, and 30.
25. False **27.** $\{x \mid x \leq -2\}$
29. $\left\{x \mid x \leq -\frac{4}{3}\right\}$ **31.** Length at least 62 meters

Exercise 3.7

1. $10x - 15$ **3.** $2x^2 - 6x$ **5.** $28ac - 21bc + 28c^2$
7. $6x^3y - 2x^2y^2 + 2x^2y$
9. $-8x^3y^3z + 24x^3yz^3 - 40xy^3z^3$
11. $-2x^5y^3 + 2.5x^4y^2 - 0.35x^3y$ **13.** $x^2 - 3x + 2$
15. $z^2 + 8z + 15$ **17.** $z^2 + 20z + 96$
19. $a^2 - 18a + 72$ **21.** $y^2 + 7y + 10$ **23.** $\{5\}$
25. $\left\{-\frac{8}{11}\right\}$ **27.** $x^3 + x^2y - xy^2 - y^3$
29. $2x^3 - 7x^2 - 27x - 18$ **31.** $5z^3 - 23z^2 + 47z - 21$
33. $4x^4 - 4x^3y + 3x^2y + 2x - 3xy^2 - 2y$
35. $18x^3 - 45x^2 + 19x + 12$
37. $8a^3 - 26a^2b + 5ab^2 + 3b^3$ **39.** $3y^2 + 11y - 4$
41. $4a^2 - 23a + 15$ **43.** $6b^2 + 7b + 2$
45. $8x^2 - 2x - 15$ **47.** $-21s^2 + 32s + 5$
49. $15c^2 - 59c + 56$ **51.** $\left\{\frac{1}{15}\right\}$ **53.** $\{4\}$
55. $\{-12\}$ **57.** $\{2.2\}$ **59.** $x^3 - 8$ **61.** $8a^3 - b^3$
63. $4x^4 - 4x^3y + 3x^2y - 3xy^2 + 2x - 2y$
65. $0.12x^3 - 0.08x^2 - 0.09x - 0.09$
67. $\frac{1}{6}x^3 + \frac{23}{12}x^2y - \frac{9}{10}xy^2 - \frac{1}{20}y^3$
69. $x^2 + 2xy + y^2 - z^2$ **71.** $20x^2 - 21x - 27$
73. $28x^2 + 33x - 28$ **75.** $20a^2 - 2ab - 6b^2$
77. $0.2x^2 + 0.13x + 0.02$ **79.** $\frac{1}{12}x^2 - \frac{25}{144}x + \frac{1}{12}$
81. $-0.21x^2 + 0.17x + 0.08$ **83.** $\{-8\}$ **85.** $\{-11\}$
87. $\left\{-\frac{43}{11}\right\}$ **89.** $\left\{\frac{2}{3}\right\}$ **91.** $\left\{\frac{2}{3}\right\}$ **93.** $\{2\}$
95. $A = 4w^2 - 7w, A = 330\text{ m}^2$ **97.** 5 **99.** 12 hr
101. -2 **103.** -3 **105.** $-\frac{2}{5}$ **107.** Jean is 10, Pete is 7 **109.** \$5400 at 10%, \$6600 at 12% **111.** In words **113.** $19a^3 - 9b^3$ **115.** $8x^{2n} - 2x^n - 3$
117. $x^{2a} - 2x^{a+1} + 5x^{a-1} - 10$ **119.** $\frac{1}{x^{14}y^2}$
121. $1024x^{40}y^{25}$ **123.** 1.23×10^5 **125.** 59,900,000

Exercise 3.8

1. $x^2 - 16x + 64$ **3.** $a^2 - 81$ **5.** $b^2 + 20b + 100$
7. $d^2 - 36$ **9.** $z^2 - 100$ **11.** $x^2 + 2xy + y^2$
13. $x^2 - y^2$ **15.** $a^2 - 144$ **17.** $4x^2 + 12x + 9$
19. $4c^2 - 25$ **21.** $9a^2 - 24a + 16$ **23.** $36k^2 - 49$
25. $25y^2 - 81$ **27.** $64 - 9c^2$ **29.** $y^2 - 24y + 144$
31. $\{1\}$ **33.** $\{-2\}$ **35.** $\{-4\}$ **37.** $x + 2xy + y^2$
39. $b^2 - c^2$ **41.** $9a^2 - 4b^2$ **43.** $9x^2 - 12xy + 4y^2$
45. $16c^2 + 24cd + 9d^2$ **47.** $\left\{-\frac{5}{2}\right\}$ **49.** $\{5\}$
51. $\{0\}$ **53.** $\{-6\}$ **55.** $\{2\}$ **57.** 1591 **59.** 9999
61. 10 ft **63.** 6 cm by 6 cm **65.** 80 ft by 80 ft
67. The consecutive odd integers are 7 and 9.
69. The pantry is $7' \times 7'$, and the hallway is $3\frac{1}{2}' \times 14'$.
71. In words **73.** $a^4 - 8a^3 + 24a^2 - 32a + 16$
75. $x^{3n} - 9x^{2n} + 27x^n - 27$ **77.** $-4ab - 24bc$
79. $-\frac{77}{36}x^3$ **81.** $9x^2y - 34xy$ **83.** $\left\{\frac{73}{8}\right\}$

Exercise 3.9

1. $6a$ **3.** $-6xy^3z^5$ **5.** $5 + 12y$ **7.** 8
9. $x + 3 + \frac{2}{x + 1}$ **11.** $a - 3$ **13.** $x - 2$
15. $x + 3$ **17.** $-5ab^6c$ **19.** $-10t + 22$
21. $7x + 8$ **23.** $2a + 3b$ **25.** $x + 13 + \frac{55}{x - 5}$
27. $2a - 5 - \frac{12}{a + 3}$ **29.** $2b - 1 - \frac{6}{2b - 1}$
31. $3x + 3 + \frac{2}{2x - 1}$ **33.** $2x + 13 + \frac{15}{x - 5}$
35. $3a - 6 + \frac{11}{a + 2}$ **37.** $3a^2 + 4b - 15a$
39. $2x^2 - 3x + 1$ **41.** $x^3 - x^2 + 2x + 1$
43. $x^3 + 6x^2 + 12x + 8$ **45.** $x^2 + x + 1$
47. $x^2 - x + 1 - \frac{2}{x + 1}$
49. $a^4 + a^3b + a^2b^2 + ab^3 + b^4$ **51.** $x + 4$
53. $x^2 - 4x + 3$ **55.** $4x^2 + 2x + 1$
57. $5y + 2$ inches is the length. **59.** $h = x - 5 + \frac{7}{x}$
61. The polynomial is $x^2 + 5x + 13$. **63.** $n = -35$
65. In words **67.** $4x^2 - 26x + 43$
69. $x^2 - \frac{5}{2}x + \frac{3}{2} - \frac{1}{2x + 1}$ **71.** $-0.6x$
73. $-3.6x^8y^{10}$ **75.** $\left\{-\frac{11}{8}\right\}$ **77.** 800 people

Chapter 3 Review Exercises

1. a^{10} **3.** x^{a+b} **5.** a^{x+y+1} **7.** b^{15} **9.** 1
11. $64a^3$ **13.** $64x^6y^9$ **15.** $8x^{3a}y^{3b}$ **17.** $\frac{1}{x^4y^6}$

19. $\frac{1}{16y^4}$ **21.** x^3 **23.** $16ab$ **25.** $\frac{1}{27x^4}$
27. $\frac{64}{x^3y^6}$ **29.** $\frac{27x^3y^{12}}{64}$ **31.** 1.5×10^8
33. 3.5×10^{-6} **35.** 1.3×10^0 **37.** 0.000000061
39. 71,000 **41.** 5 **43.** 0 **45.** 2 **47.** Binomial
49. Trinomial **51.** $-x + 2$ **53.** $x^3 + x^2 + x + 1$
55. $-3x^5 + 4x^3 - 6x^2 + 5x - 4$ **57.** $63x^3$
59. $15x^2y - 13xy^2$ **61.** $15a^3 + 2a^2 - 5a - 2$
63. $6a^2 - 4a + 3$ **65.** $8a^2bc^2 + 2ab^2c - 4abc$
67. $-x^2 + 7x + 5$ **69.** $x^2 + 3x + 2$ **71.** $a - 5b$
73. -8 **75.** $9x^2 + 12x$ **77.** $\{-2\}$ **79.** $\{-18\}$
81. 25 dimes, 25 nickels **83.** 0.6 hr or 36 min
85. 30 nickels, 30 dimes, 30 quarters **87.** $27x^3 + 1$
89. $6x^3 + 11x^2y - y^3$ **91.** $3x^2 - 11x - 20$
93. $12a^2 + 10ab + 2b^2$ **95.** $5x^2 - 24x - 5$
97. $\{-8\}$ **99.** $\{3\}$ **101.** $9x^2 - 4$ **103.** $25x^2 - 4y^2$
105. $625 - 9a^2$ **107.** $25a^2 - 20a + 4$
109. $16a^2 + 24ab + 9b^2$ **111.** $3x^4$ **113.** $\frac{38a^3c^2}{9b^2}$
115. $\frac{12c}{b^2}$ **117.** $5x - \frac{16y}{5}$
119. $\frac{15yz}{4} + \frac{3yz^2}{x} - \frac{4}{xz}$ **121.** $4x + 2$
123. $2x^2 + 8x + 1 + \frac{2}{2x + 1}$ **125.** $3x^2 + 8x - 12$

Chapter 3 True–False Concept Review

1. False **2.** True **3.** True **4.** False **5.** True
6. False **7.** False **8.** True **9.** False **10.** False
11. True **12.** True **13.** True **14.** False
15. False **16.** True **17.** True **18.** True
19. True **20.** False **21.** False **22.** False
23. True **24.** True **25.** False **26.** False

Chapter 3 Test

1. $9x$ **2.** $\frac{n^3}{m^6}$ **3.** $4x^2 - 81$ **4.** $x - 7$
5. 0.000000036 **6.** $2x^5$ **7.** $20a - 12b - 8$
8. 8.92×10^6 **9.** a^6 **10.** Trinomial **11.** $\left\{-\frac{37}{2}\right\}$
12. x^9y^6 **13.** a^8 **14.** $2x^2 + 5x - 42$
15. $-8x^9 + 5x^5 - 7x^3 - 3x^2 + 4x + 6$ **16.** $\{1\}$
17. w^{21} **18.** $48x^2yz + 28xy^2z - 8xyz^2$
19. $-2x - 7y + 15$ **20.** $x + 5y$
21. $m^2 - 20m + 100$ **22.** $8a^3 - 6a^2b - 13ab^2 + 6b^3$
23. $\{16\}$ **24.** $y^3 - 2y^2 - 13y - 10$ **25.** $\{2\}$
26. $\left\{-\frac{60}{11}\right\}$ **27.** $36x^2 - 65x - 36$
28. $4a^3 - 12a^2 - 8a$ **29.** $14x^2 - 11x + 5$ **30.** $\frac{x^{20}}{y^{16}}$
31. $\{3\}$ **32.** $x^2 - x - 2 + \frac{8}{x + 1}$ **33.** -3
34. $3\frac{1}{2}$ hours

CHAPTER 4

Exercise 4.1

1. $\{1, 7\}$ **3.** $\{-3, 3\}$ **5.** $\{-3, 6\}$ **7.** $\{11, 14\}$
9. $\{-16, -12\}$ **11.** $\{-17, 27\}$ **13.** $\{-24, 24\}$
15. $\{0, 21\}$ **17.** $\{-33, 0\}$ **19.** $\{-17, 25\}$
21. $\{3.4, 6.2\}$ **23.** $\{-3.4, 0.6\}$ **25.** $\{-3.3, 3.3\}$
27. $\{-14.2, 0\}$ **29.** $\left\{-\frac{1}{3}, \frac{2}{5}\right\}$ **31.** $\left\{-\frac{2}{3}, \frac{1}{2}\right\}$
33. $\{0, 25\}$ **35.** $\{0, 2.7\}$ **37.** $\{30, 40\}$
39. $\{-11, 18\}$ **41.** $\left\{\frac{5}{2}, 6\right\}$ **43.** $\left\{-\frac{7}{2}, \frac{9}{2}\right\}$
45. $\left\{\frac{11}{5}\right\}$ **47.** $\left\{-\frac{5}{2}, -\frac{7}{8}\right\}$ **49.** $\left\{-\frac{3}{4}, -\frac{1}{2}\right\}$
51. $\left\{-\frac{13}{3}, -\frac{2}{3}\right\}$ **53.** $\left\{-13, \frac{5}{2}\right\}$ **55.** $\left\{\frac{23}{3}, \frac{46}{7}\right\}$
57. $\left\{-\frac{1}{5}, \frac{1}{10}\right\}$ **59.** $\left\{-\frac{229}{100}, \frac{663}{100}\right\}$ **61.** 0 sec, 5 sec
63. 0 sec, 2 sec **65.** 0 and 13 or 0 and -13
67. 140 ft **69.** In words **71.** $\left\{\frac{5}{2}, -\frac{7}{3}, -\frac{8}{5}\right\}$
73. $\left\{-\frac{1}{2}, \frac{3}{5}, -2\right\}$ **75.** $\left\{-\frac{11}{17}\right\}$ **77.** $\{-63\}$
79. $\{x \mid x \geq 2\}$ **81.** $\{x \mid x \geq -4.9\}$

Exercise 4.2

1. $7(x + w)$ **3.** $a(x + y)$ **5.** $6(x + 2)$
7. $3(3x - 5y)$ **9.** $2b(4a - 7)$ **11.** $6b(c + 2)$
13. $12y(x - 3)$ **15.** $6t(3s + 1)$ **17.** $\{0, 7\}$
19. $\{0, 5\}$ **21.** $xz(15y + 7w)$ **23.** $6a + 5b$
25. $4b(3c - 4d)$ **27.** $2(2ab + 4bc - 3cd)$
29. $5x^2(2y + 5z)$ **31.** $\{0, 1\}$ **33.** $\{0, 5\}$ **35.** $\{0, 4\}$
37. $\{0, 3\}$ **39.** $\{0, 3\}$ **41.** $x^2y^2(1 - xy + x^2y^2)$
43. $\pi h(r^2 + R^2)$ **45.** $8x^2yz(2yz + 3z + 4)$
47. $3mn(mn - 2 - 4m - 5n)$
49. $4rs^2t^2(2rst^3 - 12t - 5r)$ **51.** $\left\{0, \frac{3}{4}\right\}$ **53.** $\left\{0, \frac{1}{3}\right\}$
55. $\left\{0, \frac{7}{3}\right\}$ **57.** $\left\{-\frac{8}{3}, 0\right\}$ **59.** $\{-17, 0\}$ **61.** 4 in.
63. Pete, 3 yr; Thelma, 6 yr **65.** 3 ft by 6 ft **67.** 2 m
69. 9 sec **71.** In words
73. $2(x - 2y)(2x^2 + 4x - 7)$ **75.** $(2a + 3b)(5a + 6b)$
77. $\left\{-7, \frac{3}{2}\right\}$ **79.** $\left\{\frac{8}{3}\right\}$ **81.** $\{-2.73\}$

83. $\left\{x \mid x \geq \frac{19}{11}\right\}$ **85.** $\frac{1}{4}$hr or 15 min

Exercise 4.3

1. $(c + 3)(x + y)$ **3.** $(x - 10)(x + y)$
5. $(x + 1)(x + w)$ **7.** $(y + d)(y - 3)$
9. $(z + 4)(z + 7)$ **11.** $(3x + y)(x - 4)$ **13.** $\{-5, -2\}$
15. $\{-2, 13\}$ **17.** $\{-7, -1\}$ **19.** $\{-8, -3\}$
21. $(x^3 + 1)(x^2 + 1)$ **23.** $(x - 2)(2x + y)$
25. $(a + b)(x + 1)$ **27.** $(4a + c)(b - 2)$
29. $(3ab - 1)(8a + 5c)$ **31.** $\left\{-3, -\frac{2}{3}\right\}$
33. $\left\{-\frac{2}{3}, 8\right\}$ **35.** $\left\{-\frac{1}{2}, 6\right\}$ **37.** $\left\{\frac{1}{6}, 2\right\}$
39. $\{-8, 2\}$ **41.** $(3 - 2a)(3x + 2)$
43. $(4ab + 3c)(2x + 5y)$ **45.** $(4bc + 3)(3a^2bc + 1)$
47. $(5a^2b - 3bd)(2bc + 3a)$ **49.** $(a + 2)(a^2 + b^2)$
51. $\left\{-\frac{3}{2}, -\frac{3}{4}\right\}$ **53.** $\left\{-\frac{3}{2}, -\frac{4}{3}\right\}$ **55.** $\left\{-\frac{4}{3}, \frac{1}{2}\right\}$
57. $\left\{-\frac{7}{5}, \frac{4}{3}\right\}$ **59.** $\left\{-2, \frac{5}{3}\right\}$ **61.** In words
63. $2(x + 2)(2a - 3b)$ **65.** $4c(d - 2e)(2a + b)$
67. $7x^2y(2x - y)(3y + 2z)$ **69.** $16x - 15y + 23$
71. $8x - 23$ **73.** $\{-11\}$ **75.** $\{14\}$

Exercise 4.4

1. $(x + 2)(x + 4)$ **3.** $(y - 1)(y - 14)$
5. $(y + 9)(y - 3)$ **7.** $(w + 7)(w - 3)$
9. $(z - 5)(z - 6)$ **11.** $(x - 11)(x + 6)$
13. $\{-15, -1\}$ **15.** $\{2, 11\}$ **17.** $\{-2, 13\}$
19. $\{4, 6\}$ **21.** $(x + 1)(x + 21)$ **23.** $(w - 13)(w - 3)$
25. $(w - 11)(w + 3)$ **27.** $(t - 7)(t + 5)$
29. $(w - 7)(w - 4)$ **31.** $\{-6, -4\}$ **33.** $\{-27, 1\}$
35. $\{2, 13\}$ **37.** $\{-5, 11\}$ **39.** $\{-9, 10\}$
41. $(x - 10)(x + 4)$ **43.** $(y + 8)(y - 5)$
45. $(w - 40)(w + 1)$ **47.** $(t + 19)(t + 2)$
49. $(b + 17)(b + 2)$ **51.** $\{-21, -2\}$ **53.** $\{-28, -2\}$
55. $\{-7, 8\}$ **57.** $\{-7, 9\}$ **59.** $\{-10, 12\}$
61. The integers are 16 and 17. **63.** The length is 7 ft, and the width is 3 ft. **65.** 4 seconds after launch there are 100 megagrams of fuel remaining.
67. The integers are -3 and -1. **69.** In words
71. $\{-14, 2\}$ **73.** $(6 + x)(3 - x)$ square units
75. $-16xy - 4xz + 13yz$ **77.** $2.64y^2 - 10.2y + 6.74$
79. $x = -20$ **81.** $x = 31$

Exercise 4.5

1. $(x + 2)(2x + 13)$ **3.** $(2x + 5)(x + 7)$
5. $(3x - 7)(x - 4)$ **7.** $(3x - 14)(x - 2)$
9. $(2x + 1)(x + 35)$ **11.** $(3x - 14)(x + 2)$
13. $(11x + 3)(x + 7)$ **15.** $(7x + 64)(x - 1)$
17. $\left\{-7, -\frac{1}{2}\right\}$ **19.** $\left\{-5, \frac{1}{2}\right\}$ **21.** $(5x + 7)(4x - 1)$
23. $(10x + 7)(2x - 1)$ **25.** $(2x + 1)(8x + 1)$
27. $(4x + 7)(x - 2)$ **29.** $(4x + 1)(x - 15)$
31. $(3x + 2)(5x - 11)$ **33.** $(15x + 22)(x - 1)$
35. $(3x - 5)(2x - 11)$ **37.** $\left\{-\frac{1}{5}, \frac{3}{2}\right\}$
39. $\left\{-\frac{2}{3}, \frac{1}{4}\right\}$ **41.** $(4x - 3)(2x - 7)$
43. $(8x - 3)(x - 7)$ **45.** $5(6x + 1)(x + 14)$
47. $2(2x + 1)(3x + 14)$ **49.** $(7x - 5)(5x - 2)$
51. $(3x - 5)(11x + 7)$ **53.** $(9x - 5y)(2x - 3y)$
55. $(4x - 7y)(5x + 4y)$ **57.** $\left\{-1, \frac{35}{2}\right\}$ **59.** $\left\{\frac{15}{18}, 1\right\}$
61. $7\frac{1}{2}$ sec **63.** 4 ft **65.** 10 ft by 35 ft
67. 6 years **69.** 12 **71.** In words
73. $-(2x - 3)(3x + 2)$ **75.** $-3x^2(6x^2 - 1)(5x^2 + 4)$
77. $12y^2 - 31y - 85$ **79.** $25x^2 - 196$ **81.** $\{-2\}$
83. $\left\{\frac{202}{3}\right\}$

Exercise 4.6

1. $(x + 3)(x - 3)$ **3.** $(y + 7)(y - 7)$
5. $(y - 5)(y^2 + 5y + 25)$ **7.** $(s + 3)(s^2 - 3s + 9)$
9. $(w + 2)^2$ **11.** $(t - 8)^2$ **13.** $\{-8, 8\}$
15. $\{-14, 14\}$ **17.** $\{-8\}$ **19.** $\{14\}$
21. $(3x + 2)(3x - 2)$ **23.** $(4y + 5)(4y - 5)$
25. $(2x - 1)(4x^2 + 2x + 1)$ **27.** $(w^2 + 1)(w^4 - w^2 + 1)$
29. $(2w + 7)^2$ **31.** $(3t - 5)^2$ **33.** $\left\{-\frac{6}{5}, \frac{6}{5}\right\}$
35. $\left\{-\frac{14}{3}, \frac{14}{3}\right\}$ **37.** $\left\{-\frac{5}{2}\right\}$ **39.** $\left\{\frac{8}{5}\right\}$
41. Prime **43.** $(7ab + 2c)(7ab - 2c)$
45. $(4x - 3)(16x^2 + 12x + 9)$
47. $(y^2 + 6)(y^4 - 6y^2 + 36)$ **49.** $(13x - 11)^2$
51. $(9y + 7z)^2$ **53.** $\left\{-\frac{4}{3}, \frac{4}{3}\right\}$ **55.** $\{-5, 5\}$
57. $\{-4, 4\}$ **59.** $\{-4\}$ **61.** $\left\{\frac{13}{3}\right\}$ **63.** 12 yr
65. $3\frac{1}{2}$ yr **67.** 5 sec **69.** 7 and 13 **71.** $\frac{9}{4}$ sec
73. In words **75.** $(x^n - 15)(x^n + 15)$
77. $(x - y - 7)(x - y + 7)$ **79.** $(2x - y)(4x - 3y)$
81. $3x^2 - 29x - 44$ **83.** $9a^2 - 42ab + 49b^2$
85. $25y^2 + 70y + 49$ **87.** $45a^2 - 37a - 56$

Exercise 4.7

1. $2a(3a + 4b)$ **3.** $(x + 7)(x - 8)$
5. $(4ab^2 - 5)(4ab^2 + 5)$ **7.** $(a - 23)^2$
9. $(5c + 3d)(c - d)$ **11.** $5(2x - 1)(x - 2)$

13. $7(x-1)(x^2+x+1)$ **15.** $\{-13, 13\}$ **17.** $\{7\}$
19. $(x+17)(x-2)$ **21.** $4x^2y(2x^2-4y^2-1)$
23. $(2s-5t)(3s+7t)$ **25.** $(8m-3t)(3m+t)$
27. $2(10k+1)(10k-1)$
29. $2a(w+11)^2$ **31.** $2w(2y+3z)(4y^2-6yz+9z^2)$
33. $\{3\}$ **35.** $\left\{-\frac{5}{3}, 0\right\}$ **37.** $\left\{-6, -\frac{2}{3}\right\}$
39. $\{-4, 4\}$ **41.** $(5x+2y)(x+y)(x-y)$
43. $9a(a+1)(a-1)$ **45.** $8x(x+2)(x+7)$
47. $7(2t+7)(3t-1)$ **49.** $6ab(3ab+c)(4ab-5c)$
51. $(2x+3)(x+5)(x-5)$
53. $4b^2c(a-2c)(a^2+2ac+4c^2)$ **55.** $\left\{-\frac{7}{3}, \frac{7}{3}\right\}$
57. $\{-1, 9\}$ **59.** $\left\{-\frac{a}{2}, 3a\right\}$ **61.** $y = \pm 2a$ **63.** 5
65. 2 in. **67.** 11 in. by 13 in. **69.** 110 ft by 90 ft
71. 4 in. by 7 in. **73.** In words
75. $(x-y)(x+y)(x^2+xy+y^2)(x^2-xy+y^2)$
77. $(x-2)^2(x+2)^2$ **79.** $(2x^2-53)^2$
81. $8m^2+24m-16mn+6n^2-8n-14$ **83.** $\left\{\frac{14}{19}\right\}$
85. $2x+3$ **87.** x^2-4x+6

Chapter 4 Review Exercises

1. $\{-3, 12\}$ **3.** $\{2, 21\}$ **5.** $\{-2.1, 3.5\}$ **7.** $\left\{0, \frac{9}{4}\right\}$
9. $\left\{-\frac{8}{5}, \frac{7}{3}\right\}$ **11.** $4(y-2z)$ **13.** $7a(3b+d)$
15. $12x(x-3)$ **17.** $2x^2(2-4x+3y^2)$
19. $5ab(ab-2-3a-4b)$ **21.** $\{-9, 0\}$ **23.** $\{0, 3\}$
25. $\{-4, 0\}$ **27.** $(c+d)(t-s)$ **29.** $(w^5+4)(w+1)$
31. $(16x+5z)(2xy-1)$ **33.** $(6ab+7c)(5x+y)$
35. $(s^2+st+t^2)(s+5)$ **37.** $\{-3, -2\}$
39. $\left\{-3, -\frac{5}{4}\right\}$ **41.** $\left\{-\frac{3}{11}, \frac{5}{8}\right\}$ **43.** $(x-7)(x-5)$
45. $(y-16)(y-2)$ **47.** $(a+13)(a-12)$
49. $(x-14)(x+7)$ **51.** $(x+11)(x-10)$
53. $\{-1, 7\}$ **55.** $\{6, 13\}$ **57.** $\{-22, -11\}$
59. $(3x+2)(x-6)$ **61.** $(5x-1)(x+6)$
63. $(2x+7)(5x-9)$ **67.** $(4x-7)(6x+5)$
69. $\left\{-6, \frac{5}{3}\right\}$ **71.** $\left\{-\frac{1}{2}, \frac{5}{3}\right\}$ **73.** $\left\{-\frac{9}{14}, \frac{5}{12}\right\}$
75. $(w-20)(w+20)$ **77.** $(4a-5b)(4a+5b)$
79. $(10y-11z)(10y+11z)$
81. $(7abc-3d)(7abc+3d)$ **83.** $(x^2-3)(x^2+3)$
85. $(s-6)(s^2+6s+36)$
87. $(x+10)(x^2-10x+100)$
89. $(2d-5)(4d^2+10d+25)$
91. $(4abc+5de)(16a^2b^2c^2-20abcde+25d^2e^2)$
93. $(t^2+5)(t^4-5t^2+25)$ **95.** $(x-4)^2$
97. $(x-9)^2$ **99.** $(3x+7)^2$ **101.** $(5x+8y)^2$
103. $(9ab-c)^2$ **105.** $\left\{-\frac{9}{7}, \frac{9}{7}\right\}$ **107.** $\left\{-\frac{5}{6}, \frac{5}{6}\right\}$
109. $\{25\}$ **111.** $\left\{\frac{12}{5}\right\}$ **113.** $3(x-9)(x+5)$
115. $3x(2x-7)(2x+7)$ **117.** $7ab^2(3a+2)(2b-3)$
119. $7(2x+1)(4x^2-2x+1)$
121. $14x(x^2+9)(x+3)(x-3)$ **123.** $\{-7, 3\}$
125. $\left\{-\frac{5}{2}, \frac{11}{3}\right\}$ **127.** $\left\{-\frac{7}{2}, 0, \frac{3}{4}\right\}$

Chapter 4 True–False Concept Review

1. False **2.** False **3.** False **4.** True **5.** False
6. True **7.** True **8.** False **9.** True **10.** False
11. True **12.** True **13.** False **14.** False
15. False **16.** False **17.** True **18.** False
19. True **20.** False

Chapter 4 Test

1. $2abc^2(5ac-6b+8a^2c^2)$ **2.** $3a(5a-9b)^2$
3. $(x+2)(a-b)$ **4.** $(2a-3b)(a+c)$
5. $\{-18, -2\}$ **6.** $(x-9)(x-2)$ **7.** $\{0, 2\}$
8. $(a-9)(a+8)$ **9.** $\{12\}$ **10.** $\{7, -4\}$
11. $22ab(1-2c)$ **12.** $(8xy+7a)(8xy-7a)$
13. $\{-5, 0\}$ **14.** $(r-18)(r-1)$
15. $(3x+2)(2x-3)$ **16.** $7y(2x^2+4x-11w)$
17. $(4x+3y)^2$ **18.** $(y+11)(y-6)$
19. $(4x+y)(2x-1)$ **20.** $(3x+4)(4x-3)$
21. $3(x-5y)(x^2+5xy+25y^2)$ **22.** $(3x-4)(4x-3)$
23. $(6bc-1)(6bc+1)$ **24.** $(3t-4)(2t+5)$
25. $2x(8a-5b)^2$ **26.** $\left\{-\frac{10}{3}, \frac{10}{3}\right\}$ **27.** $\left\{\frac{1}{6}, 6\right\}$
28. $(6x+7y)^2$ **29.** -6 and 15
30. 4 in. wide, 2 in. deep

Chapters 1–4 Cumulative Review

1. $\frac{2}{3}$ **2.** 3.32971 **3.** 2.84 **4.** 10.477 **5.** 7.683
6. $2^4 \cdot 7^2 \cdot 9^3$ **7.** 2 **8.** 50 **9.** $x = -7$ **10.** 176
11. 4.177756 **12.** No **13.** Yes **14.** False
15. True **16.** False **17.** True **18.** 7
19. -1.73 **20.** $-\frac{1}{5}$ **21.** -3 **22.** -7.2
23. $-1\frac{11}{24}$ **24.** -34 **25.** 6 **26.** 70
27. -0.72 **28.** 0 **29.** 144 **30.** -6.3 **31.** $1\frac{1}{8}$
32. -135 **33.** $-4\frac{2}{3}$ **34.** 14 **35.** 19
36. -4.68 **37.** 21.32 **38.** $\left\{\frac{1}{15}\right\}$ **39.** $\{-23.8\}$
40. $c = s - m$ **41.** $x = y + 7$ **42.** $\{-0.5\}$
43. $\left\{-\frac{2}{3}\right\}$ **44.** $\ell = \frac{A}{w}$ **45.** $I = Prt$

46. $\left\{-\frac{1}{120}\right\}$ **47.** $\{2.8\}$ **48.** $C = \frac{5F - 160}{9}$

49. $g = \frac{2v_0t - 2s}{t^2}$ **50.** $11x^2 + 3x$

51. $9x^3y - 4x^3y^2 - 5x^2y^3$ **52.** $-(1/6)a + (1/2)b + (5/12)c$ **53.** $1.73 - 4.74u - 3.6v$ **54.** $\{4\}$

55. $\left\{\frac{5}{8}\right\}$ **56.** $\emptyset$ **57.** $\{-49\}$ **58.** 70.1

59. 53,400 **60.** 72.9% **61.** 133 people are currently employed there.

62. $\{x \mid x > 7\}$

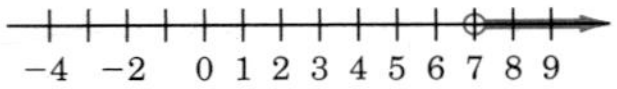

63. $\{x \mid x \leq -6\}$

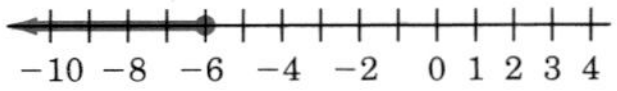

64. $\left\{x \leq -\frac{3}{4}\right\}$

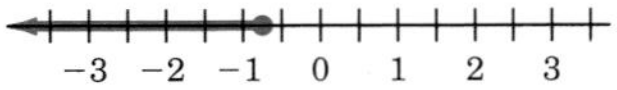

65. $R = \{\text{all real numbers}\}$

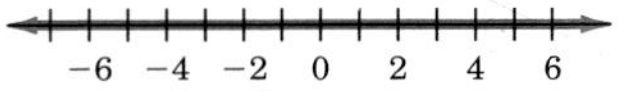

66. x^{11} **67.** a^{m+4} **68.** $9a^4b^2$ **69.** $64x^6z^9$

70. $\frac{1}{4}$ **71.** x^4 **72.** $\frac{1}{x^{21}y^3}$ **73.** $\frac{625b^6}{c^{12}}$

74. 4.715×10^9 **75.** 6.93×10^{-4} **76.** 0.00000874

77. 603,000,000,000,000,000,000,000 **78.** 1.488×10^{-2}

79. 2.5×10^8 **80.** degree 5 **81.** degree 10

82. $-6x^5 - 7x^4 + 8x^3 + 5x^2 - 12x + 11$

83. $13x^3 - 2x^2 - 4x + 6$ **84.** $8x + 11$ **85.** $\{-1\}$

86. $\{17\}$ **87.** It will be 4.25 hours before the drivers meet. **88.** 48 pounds of each cheese were shipped.

89. 40 ml of each solution is required. **90.** There are 89 of each type coin in the bank. **91.** $21y^2 + 29y - 10$

92. $2x^3 - 13x^2 + 34x - 35$ **93.** $\left\{7\frac{1}{4}\right\}$ **94.** $\{-6\}$

95. $4x^2 + 28x + 49$ **96.** $36x^2 - 25$ **97.** $\{0\}$

98. $\{13\}$ **99.** $-3x^2y^2 - 4xy + 1$ **100.** $3x - 7$

101. $10x + 2y + \frac{y^2}{x - 4y}$

102. $x^3 + 3x^2 + 3x + 9 + \frac{35}{x - 3}$ **103.** $\left\{-2, \frac{3}{4}\right\}$

104. $\left\{-\frac{2}{3}, 0\right\}$ **105.** $\left\{-\frac{2}{5}\right\}$ **106.** $\left\{\pm\frac{11}{4}\right\}$

107. $2x(x^2 - 4x + 5)$ **108.** $8x^2y^2z(4yz^3 - 2xz^2 + 3x^2)$

109. $\left\{0, \frac{1}{2}\right\}$ **110.** $\{-12, 0\}$ **111.** $(4b - 3)(2a + 1)$

112. $(y - 2z)(x - 1)$ **113.** $\{-6, 7\}$ **114.** $\left\{-\frac{4}{3}, \frac{1}{2}\right\}$

115. $(x - 4)(x - 8)$ **116.** $(y - 4)(y + 11)$

117. $\{-5, 15\}$ **118.** $\{-24, 3\}$ **119.** $(3x + 2)(6x - 5)$

120. $(5x + 2)(7x + 3)$ **121.** $\left\{-2, \frac{3}{8}\right\}$

122. $\left\{-\frac{4}{5}, \frac{5}{4}\right\}$ **123.** $(9x - 7)(9x + 7)$

124. $(6x - 5)^2$ **125.** $(2x - 5)(4x^2 + 10x + 25)$

126. $\left\{-\frac{3}{2}\right\}$ **127.** $\{\pm 9\}$

128. $(a - 2b)(x - 3)(x + 3)$

129. $4(2y - 1)(2y + 1)(4y^2 + 1)$ **130.** $3(x - 8)^2$

131. $4y^2(4y + 3)(16y^2 - 12y + 9)$

132. $5(12x - 5)(2x - 3)$ **133.** $\left\{\pm\frac{11}{2}\right\}$

134. $\{-4, 8\}$ **135.** $\left\{-4a, \frac{2a}{3}\right\}$

CHAPTER 5

Exercise 5.1

1. $\frac{x}{y}$ **3.** $5rs^2$ **5.** $\frac{5}{9}a^2b^4c^2$ **7.** $\frac{ab}{56}$ **9.** $\frac{20m^2}{7r}$

11. $-\frac{25z}{48w^2}$ **13.** $\frac{x + 3}{x + 2}$ **15.** $\frac{1}{6}$ **17.** 144 in^2

19. 27 ft^3 **21.** $\frac{3}{4b}$ **23.** $\frac{x + y}{z}$ **25.** $\frac{2m}{3}$

27. $\frac{3x^2 - 2x}{15}$ **29.** $\frac{-56m - 28}{2m - 1}$ **31.** $\frac{38x^2 - 57x}{52y^2 - 65y}$

33. $\frac{s^3}{4}$ **35.** $\frac{(2a + 1)(a - 7)}{(a + 3)(a - 5)}$ **37.** 48 in.

39. 10,560 ft **41.** $\frac{a - 2b}{a + 2b}$ **43.** $\frac{p + q}{-3}$ or $\frac{-p - q}{3}$

45. $\frac{3x + 4}{x + 6}$ **47.** 1 **49.** $\frac{3x - 2}{2x - 1}$ **51.** $\frac{3a + 2}{a - 3}$

53. $\frac{1}{a - b}$ **55.** $\frac{xy}{4}$ **57.** $14\frac{2}{3}$ ft **59.** 10

61. $P = \frac{1}{t^6}, P = \frac{1}{46{,}656}$ **63.** 0.12 in. **65.** 14 gal

67. 0.6 oz per in^3 **69.** 28 miles per gallon

71. In words **73.** $-\frac{4y^3 + 30y^2 + 15y + 7}{2y + 1}$

75. $-\frac{a + 2}{x - 2y}$ **77.** $x - 12y - 3$ **79.** $-72x^5y^8$

81. $2x(4x^2 - 6x + 5)$ **83.** $\{0, 6\}$

Exercise 5.2

1. $\frac{3cd}{4xy}$ **3.** $\frac{8y}{27z}$ **5.** $\frac{32y^2z}{3w}$ **7.** $\frac{-28a^2}{15b^3}$ **9.** $\frac{f}{cd^2}$

11. $\frac{p^2q^4}{x^2y}$ **13.** $\frac{7}{5}$ **15.** $\frac{5}{6}$ **17.** $\frac{3}{4}$ **19.** $\frac{10}{9}$

21. $\frac{45a^2b + 27ab^2}{3ax + cx}$ **23.** $\frac{8x^2 + 18x - 56}{15}$

25. $\frac{a^2b - ab^2}{6x + 8y}$ **27.** $\frac{9}{7xy^{12}}$ **29.** $\frac{p}{3}$ **31.** $\frac{5}{11}$

33. $\frac{x - 7}{x + 1}$ **35.** $\frac{x^2 + 3x - 4}{2(x - 2)(x + 3)}$ **37.** $\frac{-7am^2}{6a + 12m}$

39. $\dfrac{8x^2 - 6x - 27}{-12x^2 + 7x + 10}$ **41.** $\dfrac{12a^2 + 16ay - 35y^2}{36z^2 - 44z - 56}$
43. $\dfrac{6az + 9z + 14a + 21}{16az - 10z - 8a + 5}$ **45.** $\dfrac{5a - 1}{2a - 1}$ **47.** 1
49. $\dfrac{(3x + 5)(x - 9)}{(x - 8)(x - 7)}$ **51.** $\dfrac{1}{3}$ **53.** In words **55.** 1
57. $-\dfrac{10x + 10y}{9}$ **59.** $r = \dfrac{13}{2}$ **61.** $r = 0.08$
63. $9t^2(x + 4y - 2z + 8)$ **65.** $(7x + 25)(x - 1)$

Exercise 5.3

1. $15a$ **3.** $12x$ **5.** $24x^2$ **7.** $24(a + b)$
9. $12a^2b^2$ **11.** $\left\{\dfrac{1}{3}\right\}$ **13.** $\left\{\dfrac{5}{4}\right\}$ **15.** $\{2\}$
17. $\left\{\dfrac{3}{4}\right\}$ **19.** $\left\{\dfrac{17}{4}\right\}$ **21.** $8a^2b^2$ **23.** $6(x + y)$
25. $(x + y)(x - y)$ **27.** $7(2a + 3)(2a - 3)$
29. $5b(2b - 3)$ **31.** $\left\{-\dfrac{1}{6}\right\}$ **33.** $\{8\}$ **35.** $\{-7\}$
37. $\{1\}$ **39.** $\{6\}$ **41.** $(b + c)(b - c)$
43. $(y + 3)(y - 3)^2$ **45.** $3(a + b)(a - b)^2$
47. $(x + 2)^2(x + 3)$ **49.** $\left\{\dfrac{13}{3}\right\}$ **51.** $\left\{-\dfrac{23}{3}\right\}$
53. $\left\{\dfrac{32}{7}\right\}$ **55.** $\{-4\}$ **57.** $\emptyset$ **59.** $\{9\}$
61. 5, 6 **63.** $\dfrac{1}{6}$ **65.** 27 mph **67.** 40 mph, 50 mph
69. Pedro, 9 hr; Luis, 18 hr **71.** In words
73. $12(x + y)(x - y)^2(x^2 + xy + y^2)$ **75.** $\dfrac{1}{6x}$
77. $7x + 13$ **79.** $-2x^2 + 4x + 14y - 7y^2$
81. $(x - 8)(x + 7)$ **83.** $(x - 6)(x - 11)$

Exercise 5.4

1. $\dfrac{2}{3 + x}$ **3.** $\dfrac{-m}{2a - 3b}$ **5.** $\dfrac{x - z}{y - 4}$ **7.** $\dfrac{m}{8y}$
9. $\dfrac{3}{2a}$ **11.** $\dfrac{9 + 4p}{3p^2}$ **13.** $\dfrac{a^2 - b^2}{ab}$ **15.** $\dfrac{7x + 4y}{xy}$
17. $\{5\}$ **19.** $\left\{\dfrac{5}{2}\right\}$ **21.** $\dfrac{2a - 3b}{4x + y}$ **23.** $\dfrac{10x - 6}{5x - 3y}$
25. 1 **27.** $\dfrac{12 + x}{10x^2y}$ **29.** $\dfrac{3a + 20b}{12a^2b^2}$
31. $\dfrac{-(3a + 11b)}{a^2 - b^2}$ **33.** $\dfrac{2bc - 18ac - ab}{abc}$
35. $\dfrac{-3a^2 + 16a - 15}{10a^2}$ **37.** $\{19\}$ **39.** $\emptyset$
41. $\dfrac{-5x - 37}{(x + 5)(x - 1)}$ **43.** $\dfrac{3}{x + 7}$ **45.** $\dfrac{3}{(x - 3)(x + 7)}$
47. $\dfrac{x^2 - 3x + 1}{x}$ **49.** $\dfrac{1}{x - y}$ **51.** $\dfrac{7}{(x + 4)(x - 5)}$
53. $\dfrac{x^2 + y^2 + 4}{(x + y)(x - y)}$ **55.** $\dfrac{x^3 + xy^2 + 2x^2 - 2y^2}{x(x^2 - y^2)}$
57. $\{-1\}$ **59.** $\emptyset$ **61.** $\dfrac{1}{R} = \dfrac{3}{r}$ **63.** $\dfrac{1}{t} = \dfrac{3}{2x}$
65. $h = \dfrac{2A}{a + b}$ **67.** $f = \dfrac{ab}{a + b}$ **69.** 5
71. In words **73.** $x = \dfrac{2ab}{b - a}$ **75.** The one travels at 50 mph, and the other at 44 mph. **77.** 2
79. $9(2x - 3)^2$ **81.** $\{2\}$ **83.** $\left\{-\dfrac{3}{4}, \dfrac{5}{2}\right\}$

Exercise 5.5

1. $\dfrac{2b}{5a}$ **3.** $-\dfrac{5b}{4a}$ **5.** $\dfrac{y}{4x}$ **7.** $\dfrac{1}{3}$ **9.** $\dfrac{x}{y}$
11. $\dfrac{16ax^2}{15}$ **13.** $\dfrac{2abx}{3}$ **15.** $-\dfrac{9abr}{8s^3}$ **17.** $\dfrac{x}{x - 1}$
19. $\dfrac{p - 3}{p - 1}$ **21.** $x - 1$ **23.** $\dfrac{(x - 5)(x + 2)}{2}$
25. $\dfrac{2a + 3}{3a - 1}$ **27.** $\dfrac{x}{x - 2}$ **29.** $\dfrac{4x + 5}{6x + 2}$
31. $\dfrac{15x - 1}{10x + 4}$ **33.** $\dfrac{-24x - 15}{34x + 21}$ **35.** $\dfrac{2a - 1}{3a + 1}$
37. $\dfrac{x^2 - y^2}{x^2 + y^2}$ **39.** $\dfrac{60 - 5cd}{4cd + 60}$ **41.** $\dfrac{b^2 - c^2}{b^2 + c^2}$
43. $\dfrac{4y - 6x}{3(16y + 3x)}$ **45.** x **47.** $\dfrac{(x^2 + 4x + 5)(x - 2)}{(x^2 - 4x + 5)(x + 2)}$
49. $\dfrac{12y}{25}$ **51.** $\dfrac{x^2 - 2x - 3}{(x + 3)(x - 2)}$ **53.** 40 mph
55. $I = \dfrac{En}{n + R}$ **57.** $v' = \dfrac{m_1v_1 + m_2v_2}{m_1 + m_2}$
59. 1909 mph **61.** In words **63.** $\dfrac{x}{1 + x}$
65. $1 - x$ **67.** $6x^2 + xy - 35y^2$
69. $21a^2b^2 - 19ab - 12$ **71.** $(2x - 3)(x + 5)$
73. $4(2x + 5)(x - 1)$

Exercise 5.6

1. $x = 16$ **3.** $w = 20$ **5.** $y = 4$ **7.** $x = 18$
9. $x = 4$ **11.** $y = 9$ **13.** $y = 30$ **15.** $x = 5$
17. $x = 45$ **19.** $p = 1$ **21.** $w = 22.5$ **23.** $x = 6$
25. $y = \dfrac{3}{8}$ **27.** $x = \dfrac{2}{5}$ **29.** $x = \dfrac{17}{3}$ **31.** $x = 6$
33. $x = 1$ **35.** $x = 5$ **37.** $x = \dfrac{17}{5}$ **39.** $p = \dfrac{7}{5}$
41. $y = 0.01$ **43.** $w = 20$ **45.** $y = 0.25$
47. $R = 5.5$ **49.** $x = 16$ **51.** $x = 3.36$

53. $x = 20.7$ **55.** $x = \frac{5}{3}$ **57.** $c = 7.5$ **59.** $b = 15$
61. 20 cm **63.** 200 m **65.** 7 additional teachers
67. $17\frac{1}{2}$ cans **69.** 1050 **71.** 4 jobs **73.** No, 96¢
75. 0.3125 cc **77.** 0.0048 in. **79.** \$46,950; \$31,300; \$23,475 **81.** In words **83.** $\frac{2}{3}$ **85.** $\{-1, 7\}$
87. $(x - 16)(x + 9)$ **89.** $(5x - 6y)(5x + 6y)$
91. $(8 - x)(64 + 8x + x^2)$ **93.** $\left\{-\frac{5}{3}, 3\right\}$

Exercise 5.7

1. $d = \frac{k}{r}$ or $dr = k$ **3.** $P = ks$ **5.** $V = kBh$
7. $V = kT$ **9.** $I = kE$ **11.** $I = \frac{k}{D^2}$ **13.** $k = \frac{3}{2}$
15. $k = 72$ **17.** $k = \frac{2}{7}$ **19.** $k = 48$ **21.** $k = 48$
23. $k = 135$ **25.** $k = \frac{1}{4}$ **27.** $k = \frac{1}{3}$ **29.** $a = \frac{112}{3}$
31. $y = 14$ **33.** $m = 307.2$ **35.** $m = 44$
37. $z = 24$ **39.** $y = 22$ **41.** $y = 96$ **43.** 490 lb
45. \$540 **47.** $\frac{15}{22}$ amps **49.** 16 ft **51.** \$60,000
53. In words **55.** It will fall 144 feet in 6 seconds.
57. Tripling the distance between the two objects will divide the gravitational attraction by 9.
59. $25(x + 2)(x - 2)$ **61.** $6(3a + 1)(a - 4)$
63. $\frac{-15a^4b^2}{8c^3}$ **65.** $\frac{175m^3}{48n^4}$

Chapter 5 Review Exercises

1. $\frac{1}{3a + 2}$ **3.** $\frac{1}{x + 4}$ **5.** $\frac{3x - 2}{4x + 1}$ **7.** $\frac{x - 6}{3}$
9. $\frac{1}{x + 1}$ **11.** 66 feet per second **13.** 150 miles per hour **15.** 1,800,000 revolutions per hour **17.** 1
19. $\frac{3}{2}$ **21.** $(x + 2)(x + 3)(x - 3)$
23. $(x + 1)(x + 3)(x + 4)$ **25.** $x(x + 2)(x - 2)^2$
27. $\{3\}$ **29.** $\{2\}$ **31.** 1 **33.** $a + 4$ **35.** $\frac{a + 2}{a - 2}$
37. $\frac{x^2 - 5x + 20}{x(x - 4)}$ **39.** $\frac{3x - 3}{(x + 3)(x + 5)}$ **41.** $\left\{-\frac{8}{3}\right\}$
43. $\{4\}$ **45.** $\{-3, 3\}$ **47.** $\frac{x + 5}{5(x + 3)}$ **49.** $\frac{x - 3}{x - 2}$
51. 1.2 **53.** 1820 **55.** $9\frac{9}{14}$ **57.** 2475 revolutions
59. 1 **61.** ± 12 **63.** 24 **65.** 4 **67.** 75 candlepower

Chapter 5 True–False Concept Review

1. True **2.** False **3.** True **4.** True **5.** False
6. False **7.** False **8.** True **9.** False **10.** True
11. True **12.** False **13.** False **14.** True
15. False **16.** True **17.** True **18.** True
19. False **20.** True **21.** True **22.** True
23. True **24.** False **25.** True

Chapter 5 Test

1. $56xy$ **2.** $\frac{-7w}{w + 10}$ **3.** $y = 7.5$ **4.** $6(a + 7)$
5. $\frac{4wz}{35y}$ **6.** $\frac{42}{wy}$ **7.** $\frac{5r}{r - 5}$ **8.** $\frac{15t}{4p}$
9. $(w + 5)(w - 5)^2$ **10.** $\frac{11}{6y}$ **11.** 132 ft per sec
12. $\{10\}$ **13.** $\frac{5x + 3}{2(x - 1)(x + 1)}$ **14.** $\frac{x + 3}{x - 1}$
15. $\frac{2xy}{3}$ **16.** $x - 3$ **17.** $\{6\}$ **18.** $\frac{1}{x - 1}$
19. 480 mi **20.** 5.5 qt **21.** 28 vibrations per sec

CHAPTER 6

Exercise 6.1

1. $y = 12$ **3.** $y = 5$ **5.** $y = 25$ **7.** $x = 9$
9. $x = 20$ **11.** $x = 0$ **13.** $y = -6$ **15.** $y = 38$
17. $y = -62$ **19.** $x = 3$ **21.** $x = \frac{1}{2}$ **23.** $x = -11$
25. $y = \frac{7}{4}$ **27.** $y = \frac{9}{4}$ **29.** $y = \frac{23}{10}$ **31.** $x = \frac{13}{2}$
33. $x = \frac{17}{3}$ **35.** $x = \frac{25}{3}$ **37.** $y = 11$
39. $y = 31.2$ **41.** $y = 3.5$ **43.** $y = -\frac{15}{2}$
45. $y = -\frac{45}{2}$ **47.** $y = -6$ **49.** $x = -5$
51. $x = -7$ **53.** $x = -1$ **55.** $y = 1.2$
57. $y = \frac{22}{9}$ **59.** $y = -\frac{1}{3}$ **61.** $x = -7.375$
63. $x = -\frac{29}{8}$ **65.** $x = -\frac{14}{5}$ **67.** $y = \frac{91}{5}$
69. $y = 11.9$ **71.** $y = \frac{63}{8}$ **73.** 24.5 **75.** 140
77. 8 five-lb boxes **79.** 6 units **81.** $h = 6$ in.
83. In words **85.** $y = -2/3$ **87.** $x = 5/4$ **89.** $\frac{8}{a}$
91. $\frac{2x + 9}{x^2 - 4}$ **93.** $\frac{x^2 - x - 18}{(x + 3)(x - 1)}$ **95.** $\frac{1}{a + b}$

Exercise 6.2

1.
3.
5.

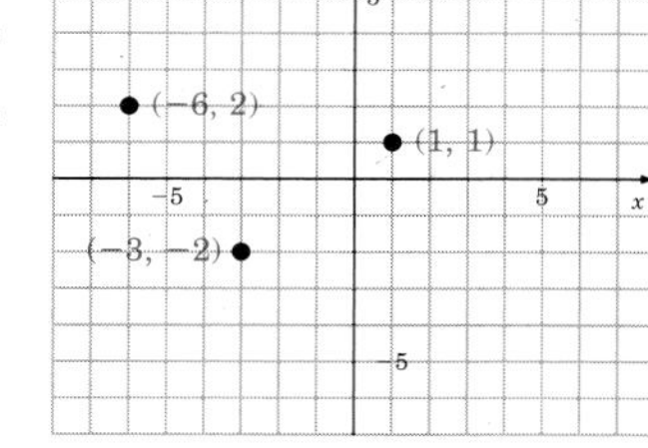

7. (4, 2) **9.** (2, −4) **11.** (−1, −3) **13.** (−3, 1)
15. (2, 0)

17.
19.

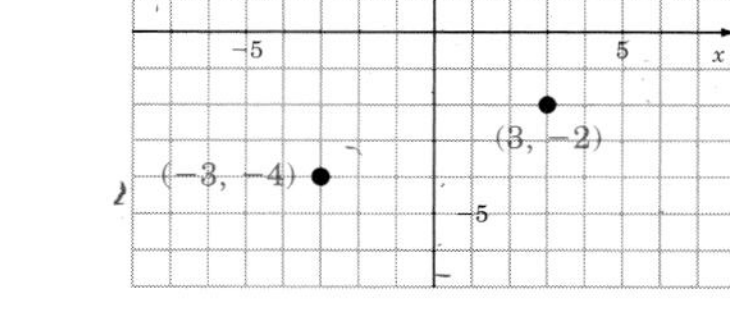

21. (3, 2) **23.** (−4, 3) **25.** (4, 0) **27.** (0, −5)
29. (6, −2)

31.
33.
35.

(1.5, 2)
(4.5, −1)
(−0.5, −3)

37. (2, 2.5) **39.** (0, −2.5) **41.** (−1, −2.5)
43. (−2, 3.5) **45.** (5.5, −4)

47.

x	y
−4	2
0	0
4	−2

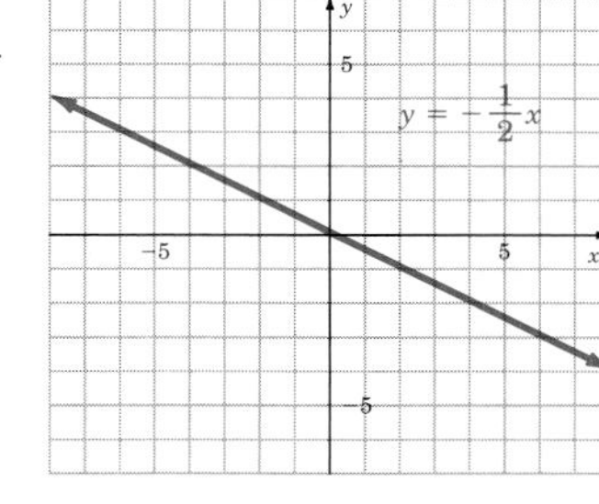

49.

x	y
−1	3
0	0
1	−3

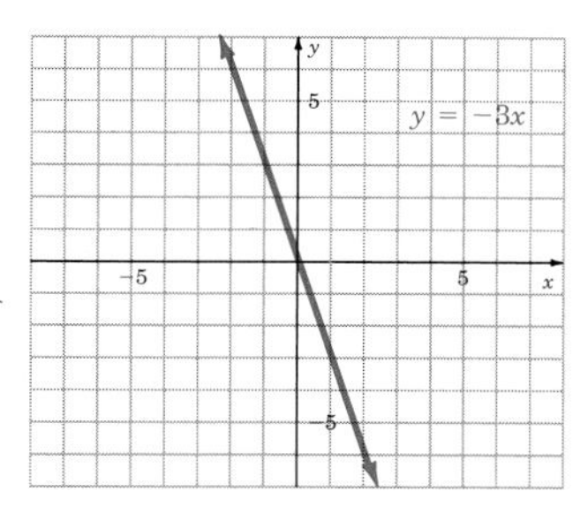

51.

x	y
0	−5
5	0
3	−2

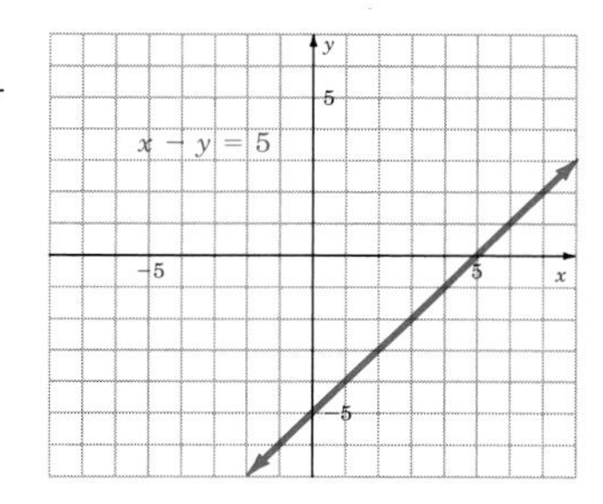

53.

x	y
0	4
4	0
5	−1

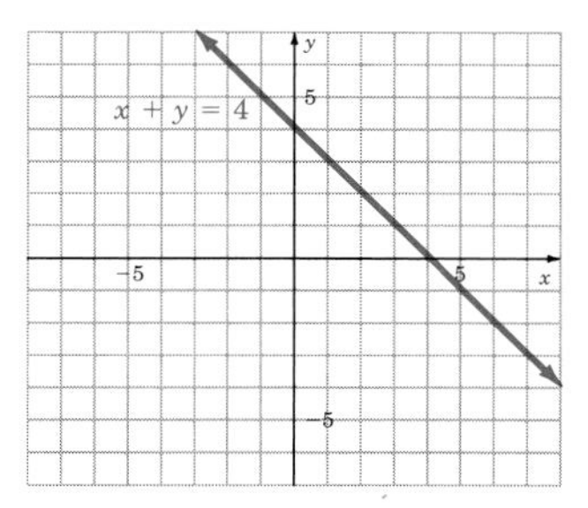

55.

x	y
0	2
−2	0
1	3

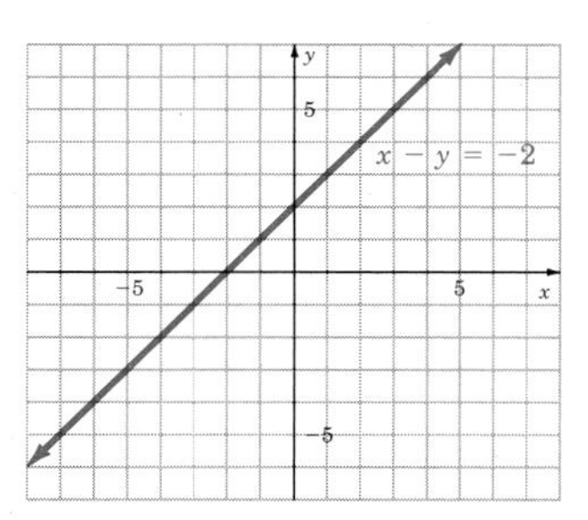

57.

x	y
6	−2
0	−2
−2	−2

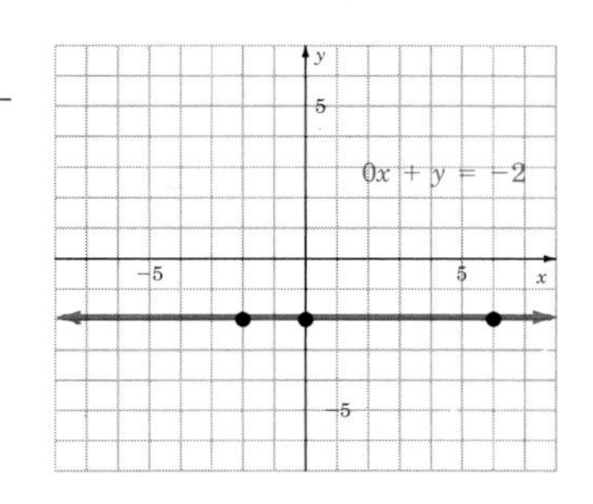

59.

x	y
0	3
6	0
2	2

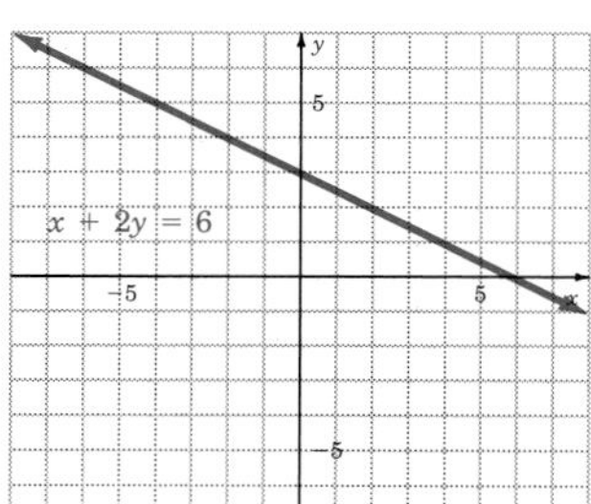

61.

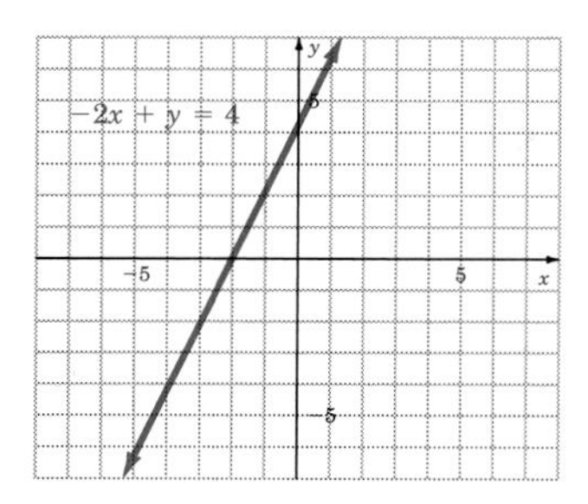

63.

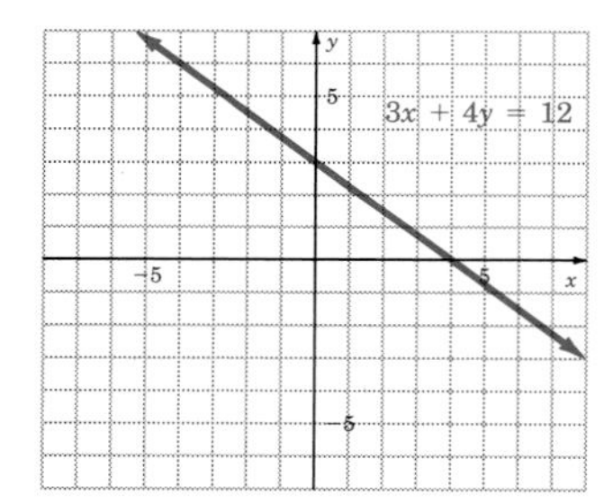

65.

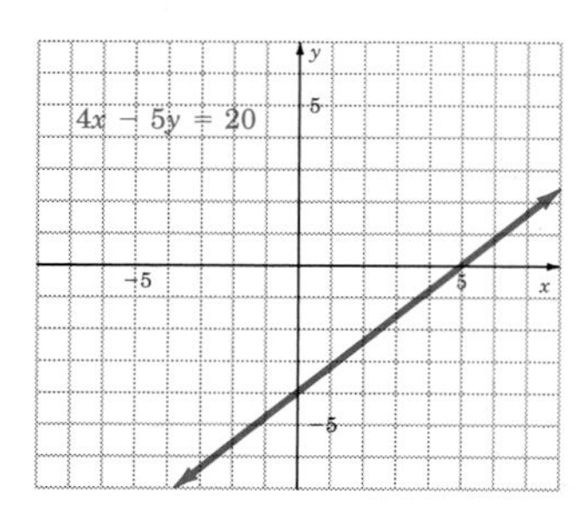

67.

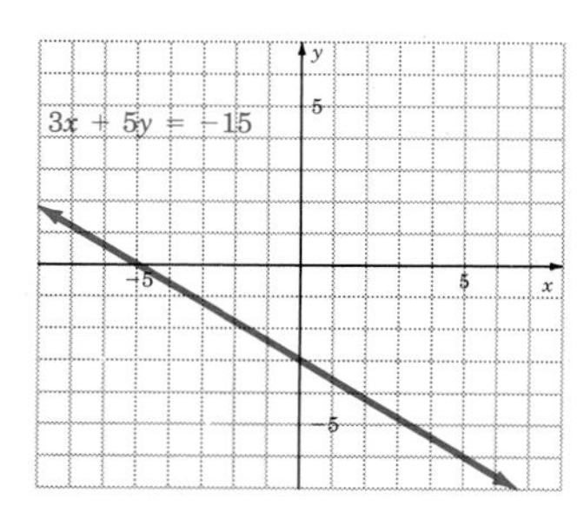

69.

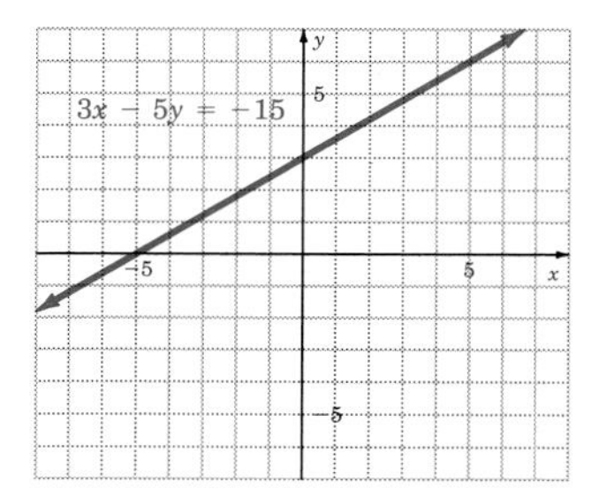

71.

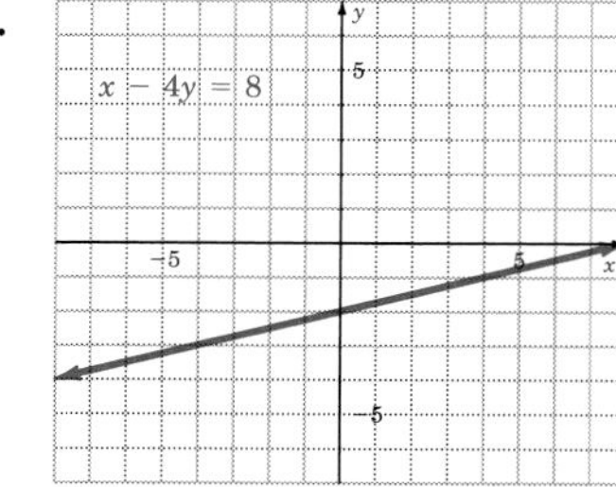

73.

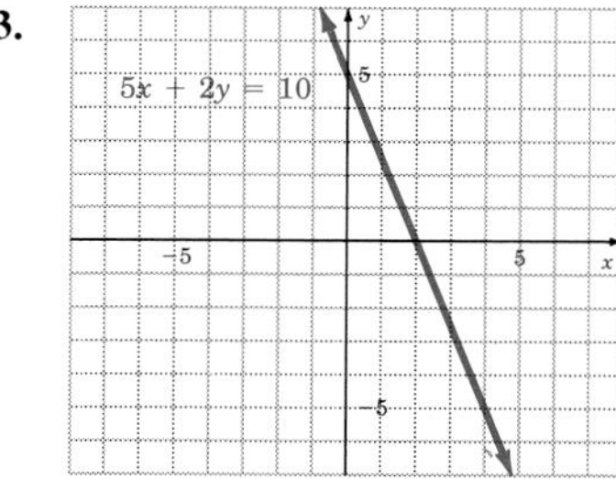

75.

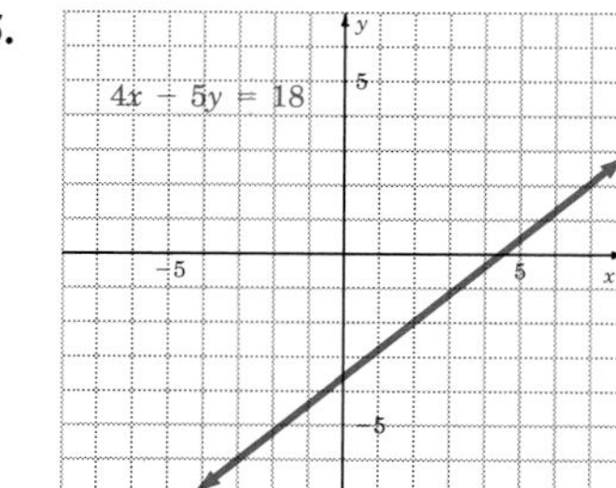

77.

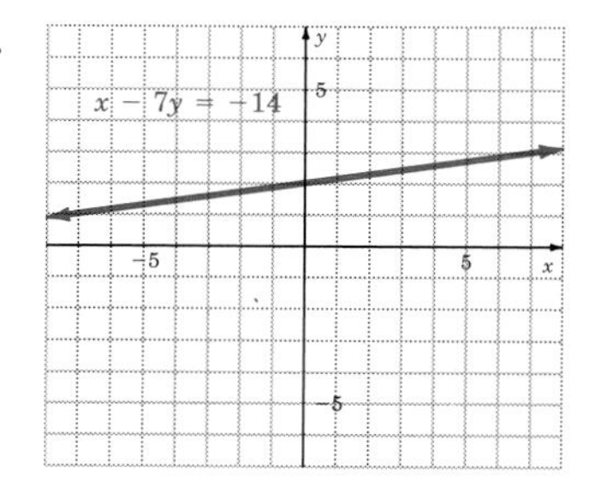

79.

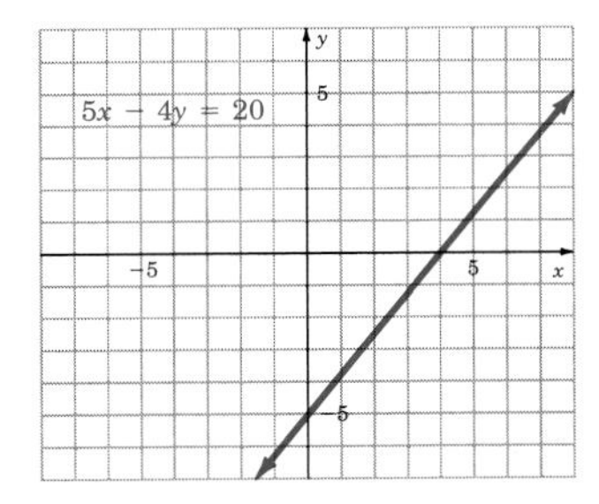

81.

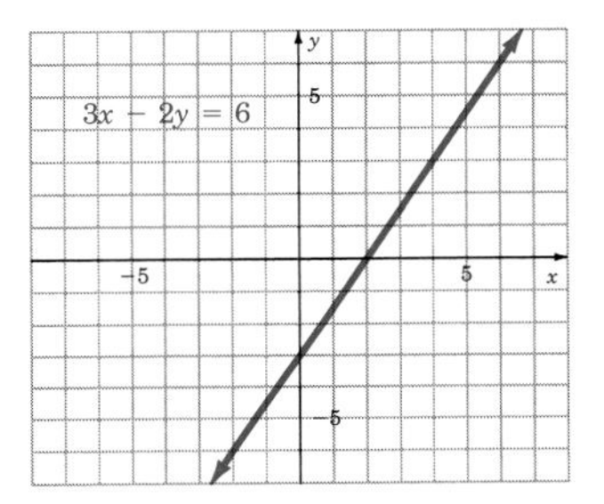

83.

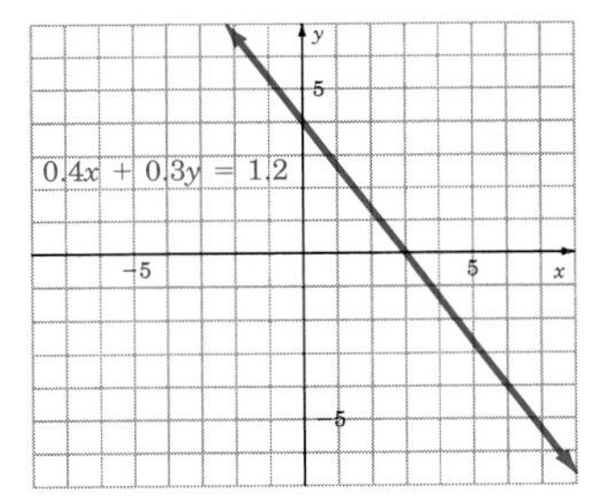

85.

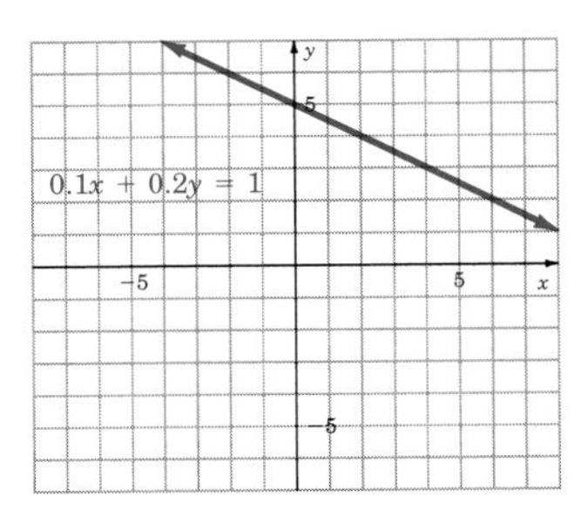

87.

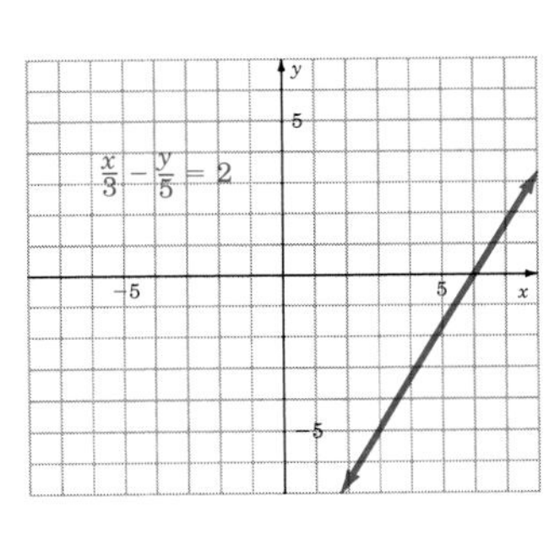

89.

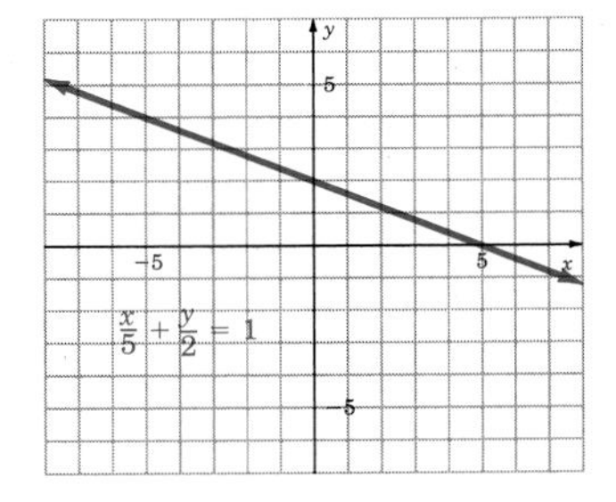

91.

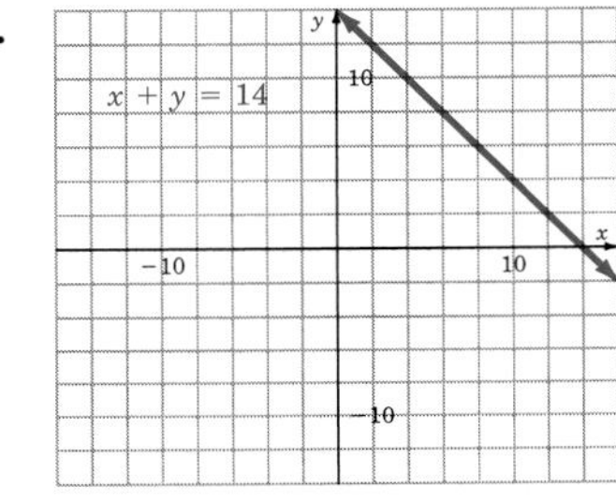

93.

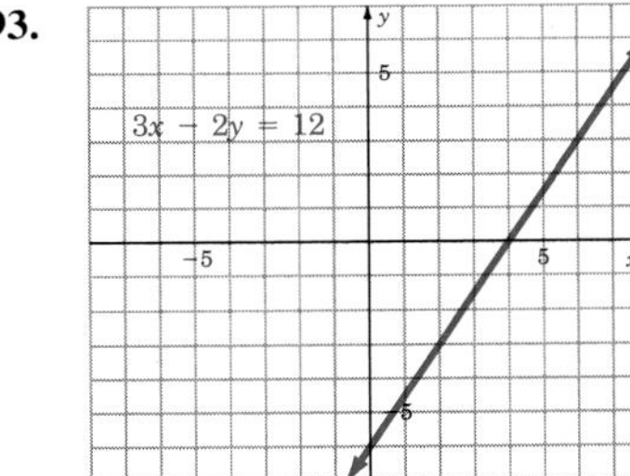

95.

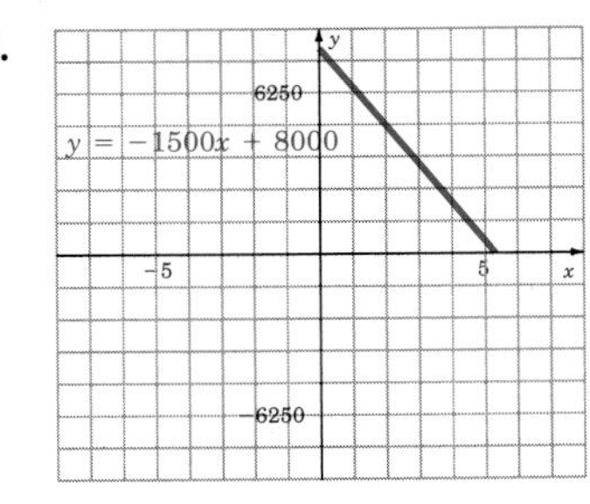

97.

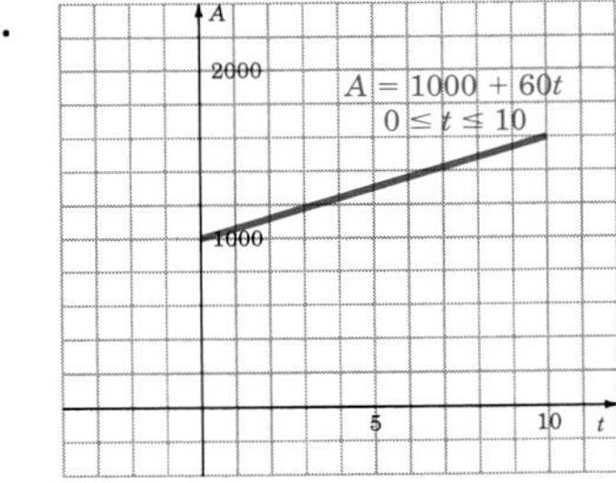

99. In words

101.

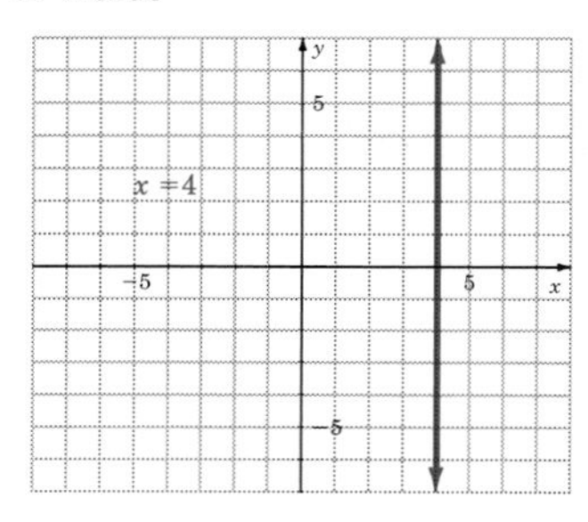

103.

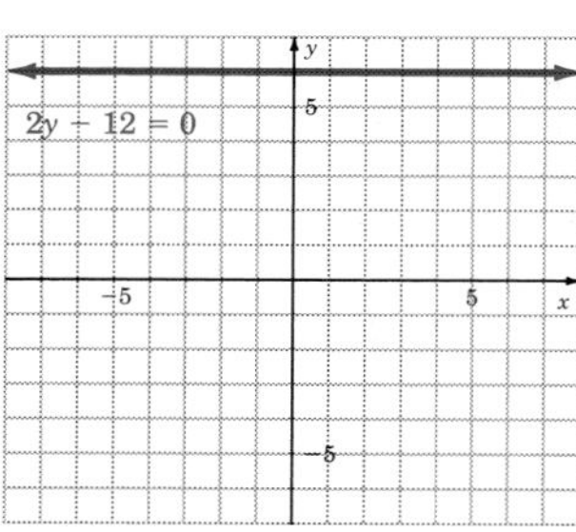

105. $\dfrac{3b}{2a}$ **107.** $\dfrac{3x-1}{5x+7}$ **109.** $75x^2y$ **111.** 35.2 ft

Exercise 6.3

1. (3, 0), (0, 9) **3.** (−12, 0), (0, 2) **5.** (7, 0), (0, 4)

7. $m = -1$ **9.** $m = -\dfrac{1}{6}$ **11.** $m = 1$

13. $m = -\dfrac{1}{2}$, (0, 6) **15.** $m = \dfrac{3}{5}$, (0, −3)

17.

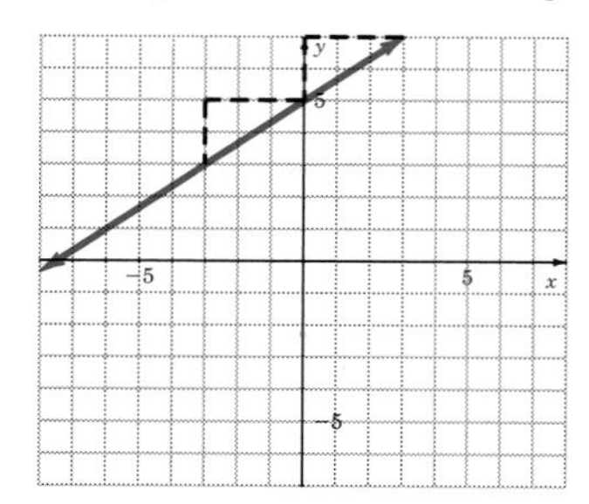

19.

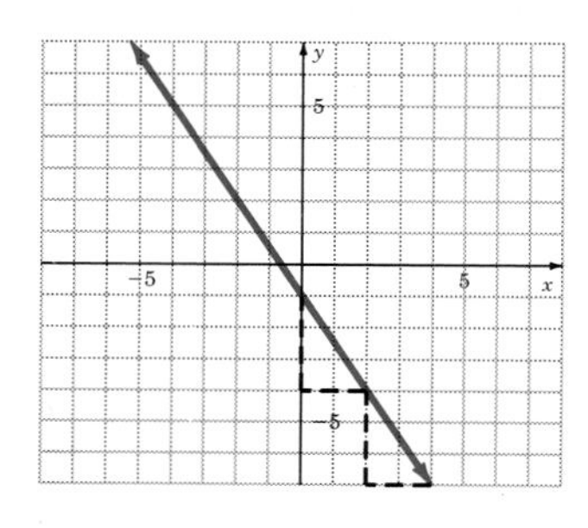

21.

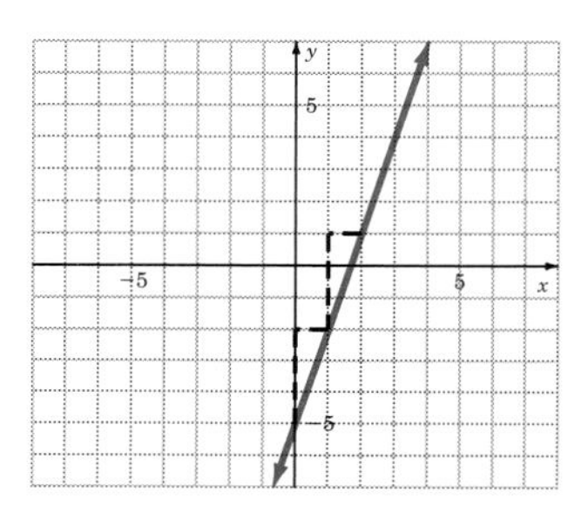

23. $\left(\dfrac{11}{2}, 0\right), \left(0, \dfrac{11}{5}\right)$ **25.** $(2, 0), \left(0, -\dfrac{12}{7}\right)$

27. $(2, 0), \left(0, -\dfrac{20}{7}\right)$ **29.** $m = -\dfrac{1}{3}$ **31.** $m = -\dfrac{4}{7}$

33. $m = \dfrac{7}{12}$ **35.** $m = -\dfrac{3}{7}$, (0, 2)

37. $m = -\dfrac{2}{5}, \left(0, -\dfrac{8}{5}\right)$ **39.** $m = 0$, (0, 3)

41.

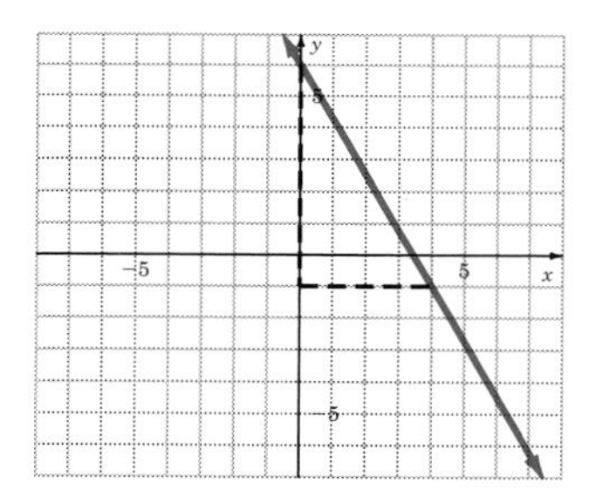

43.

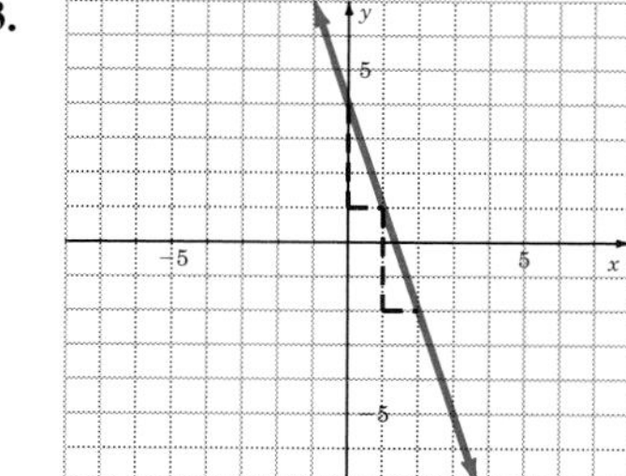

45.

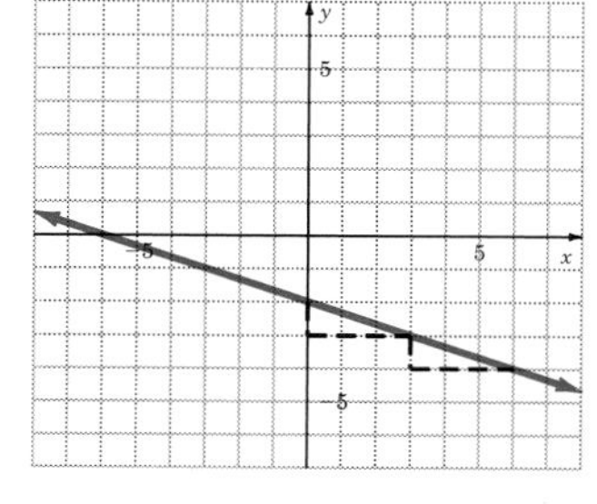

47. $(-18, 0), (0, -12)$ **49.** $\left(\frac{7}{2}, 0\right), \left(0, \frac{3}{2}\right)$

51. $(2, 0), \left(0, -\frac{5}{12}\right)$ **53.** $m = -\frac{23}{5}$ **55.** $m = 0$

57. $m = -\frac{5}{2}$

59.

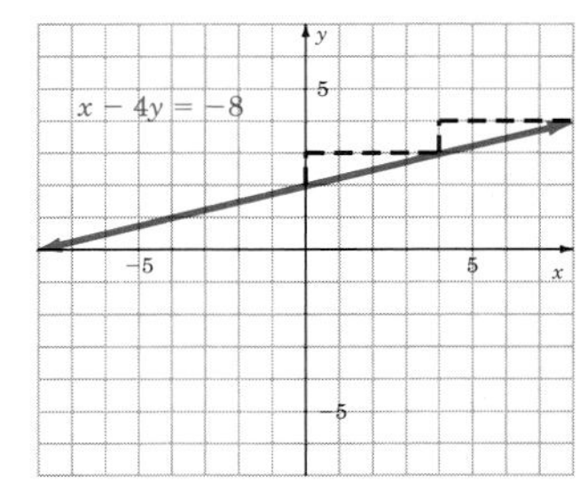

61.

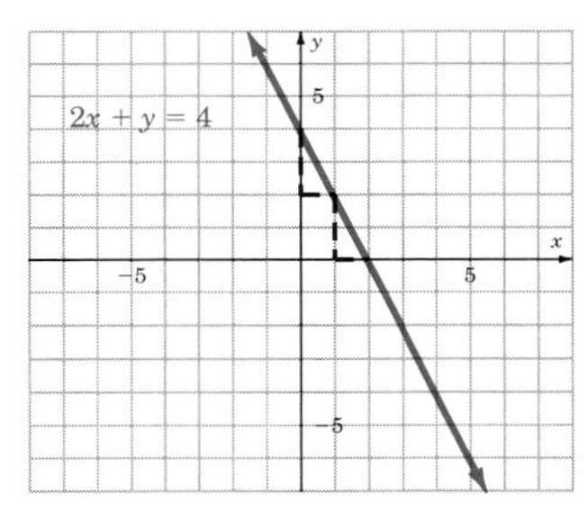

63.

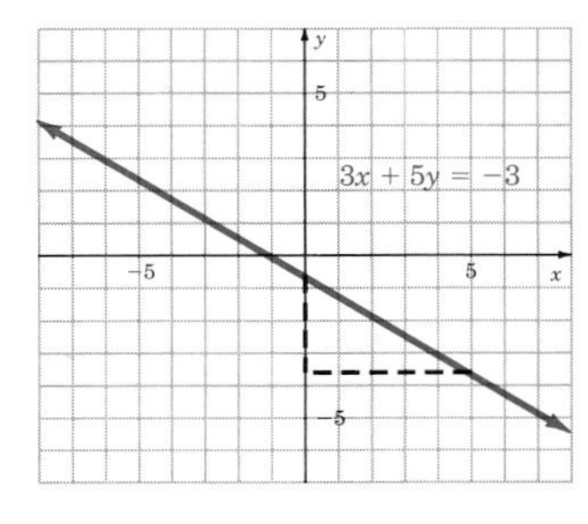

65.

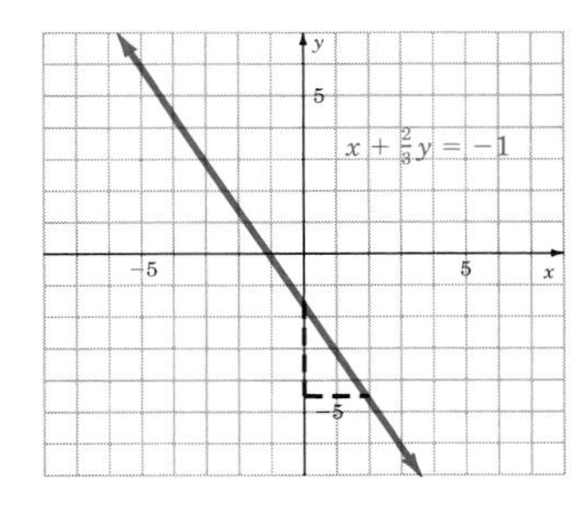

67.

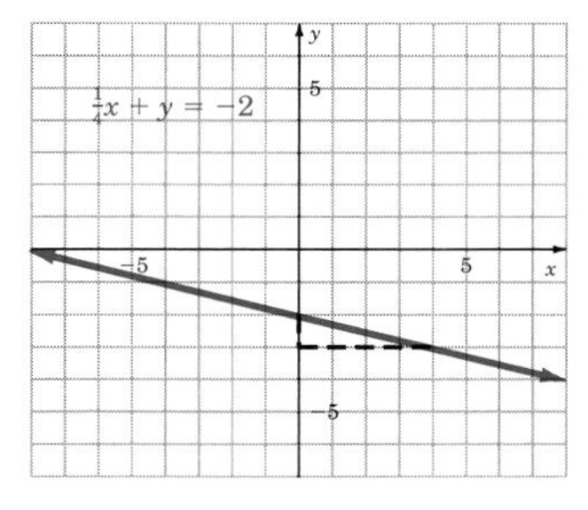

69.

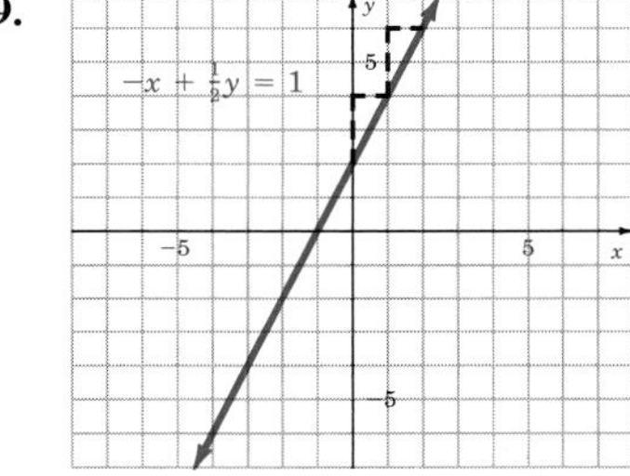

71. $m = \frac{3}{50}$ **73.** $m = \frac{17}{40}$ **75.** $m = \frac{3}{4}$

77. \$0.17, \$10

79.

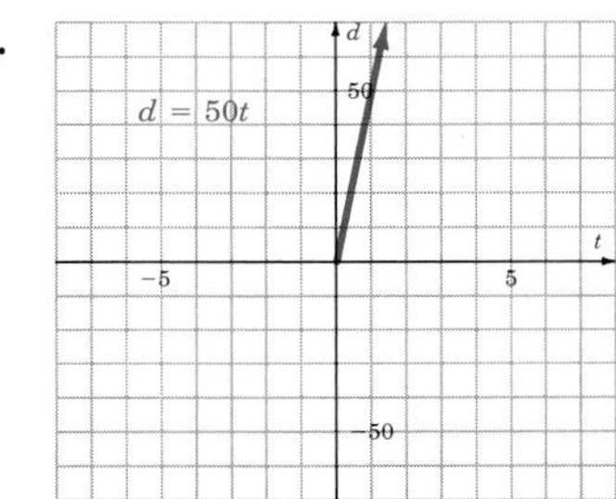

81.

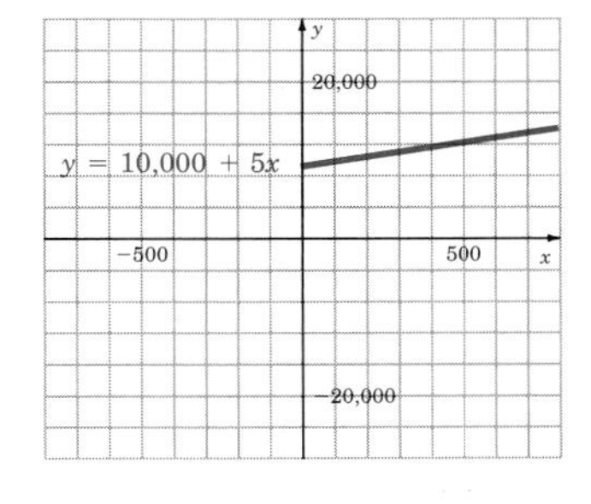

83. In words **85.** $m = 0$ **87.** $m = 0, (0, 3)$

89. $\frac{4}{9xy}$ **91.** $\frac{x+2}{x-2}$ **93.** $\left\{-7, \frac{3}{5}\right\}$ **95.** 3 or -7

Exercise 6.4

1.

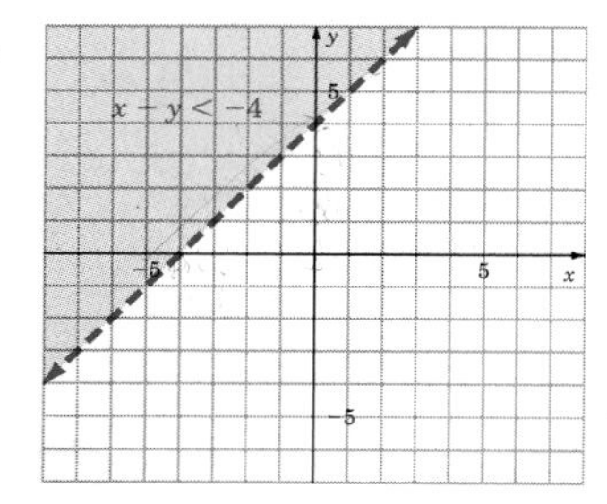

3.

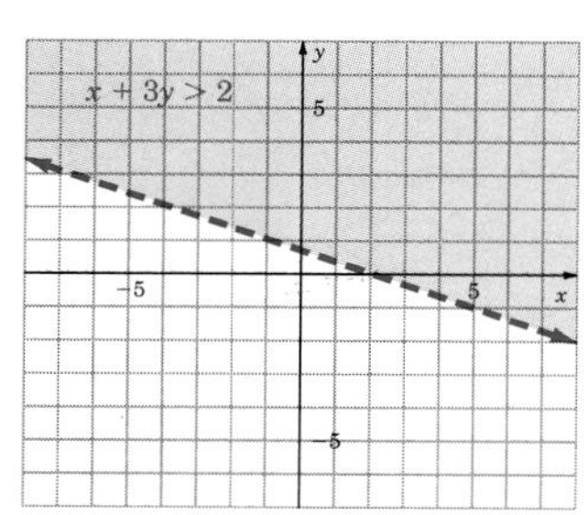

5.

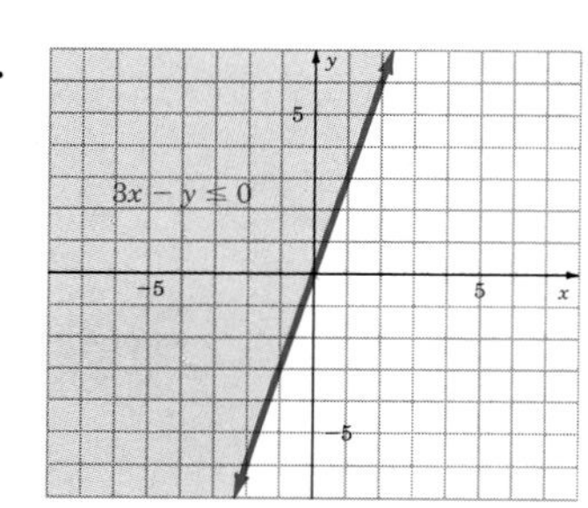

7.

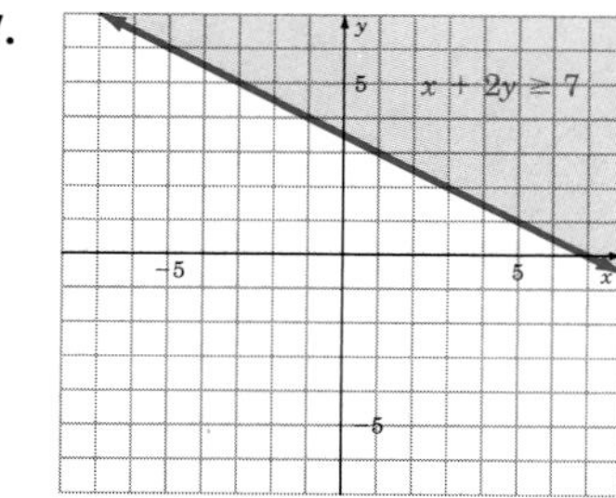

9.

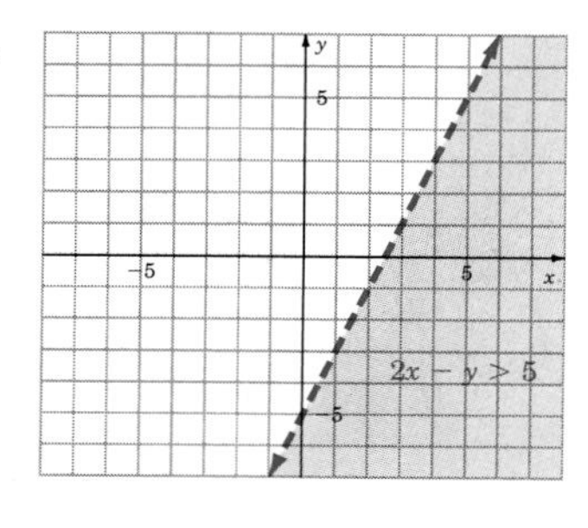

11.

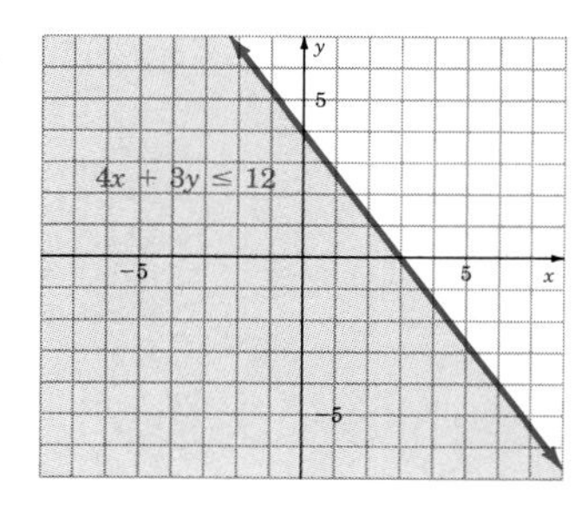

13.

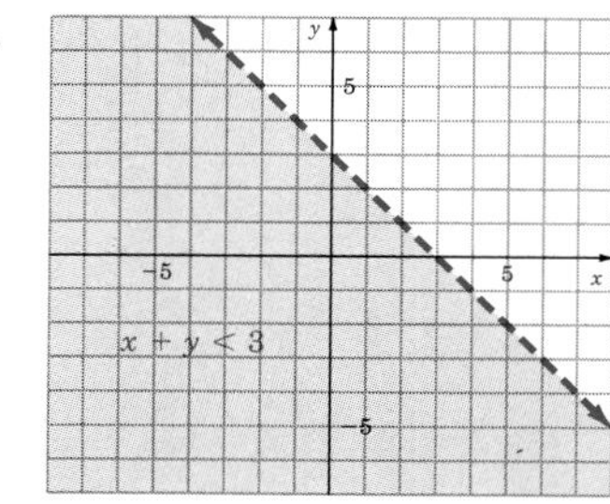

15.

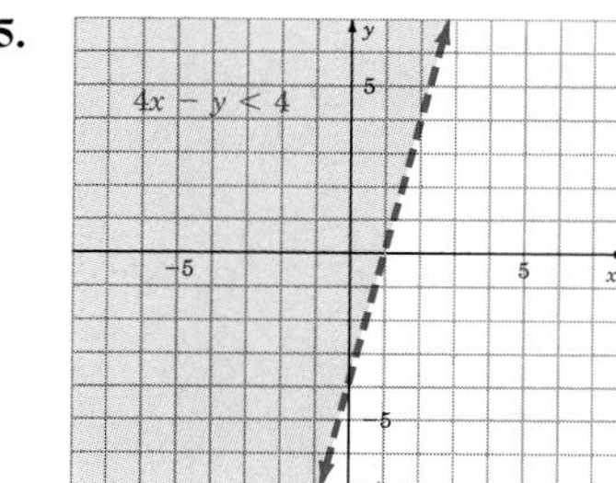

17.

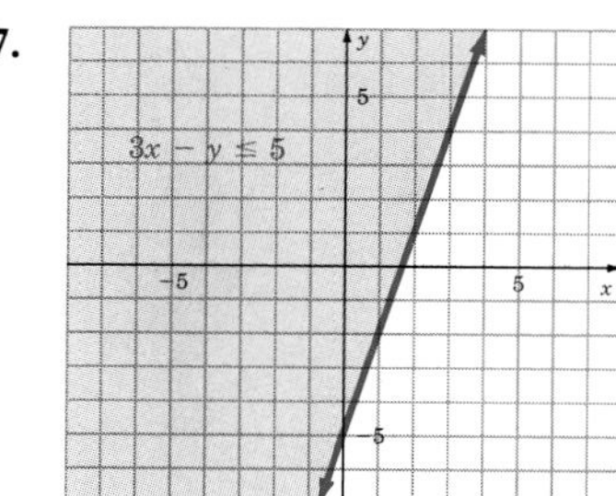

19.

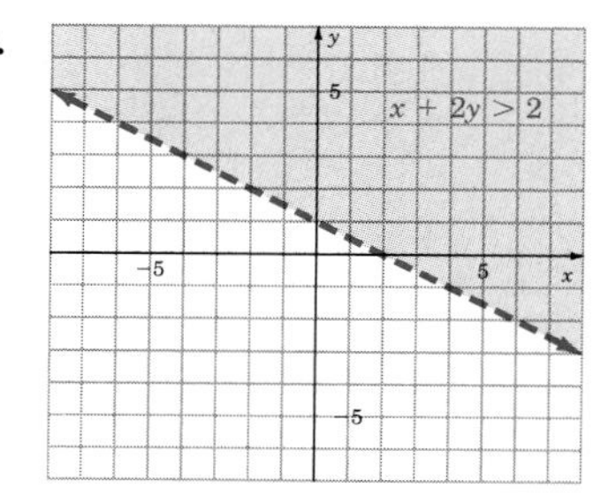

21.

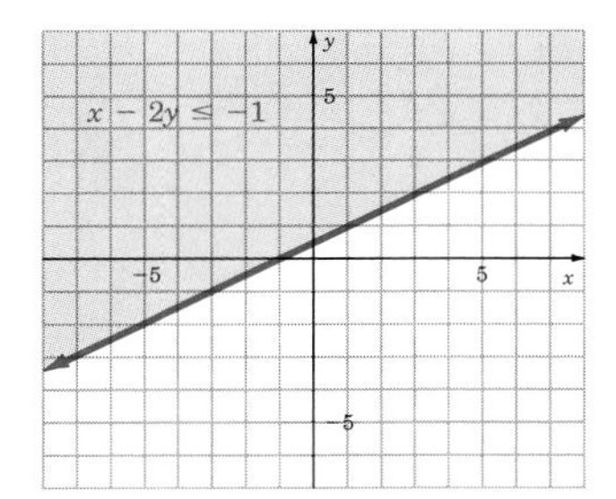

23.

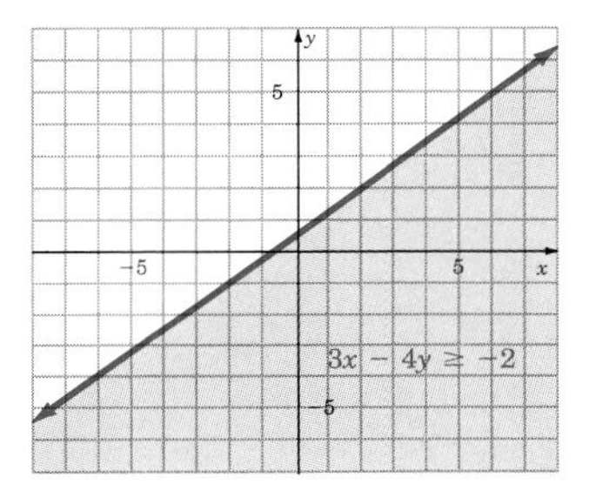

25.

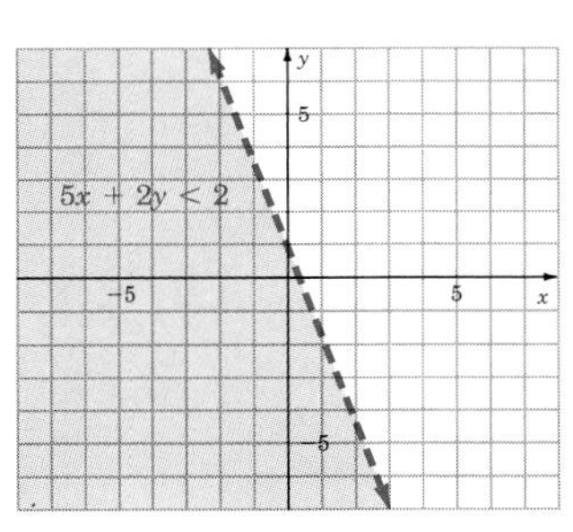

27.

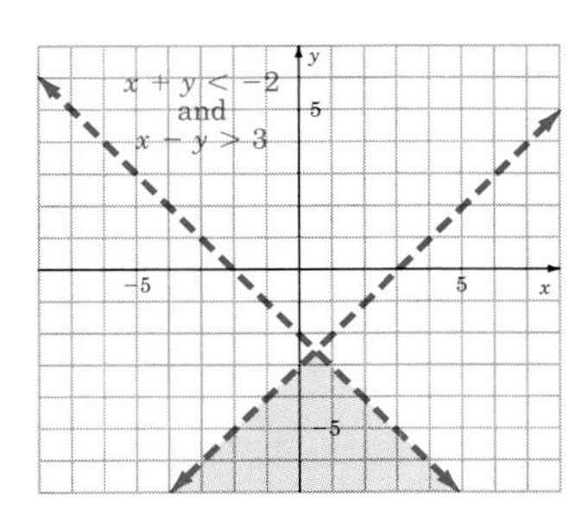

29.

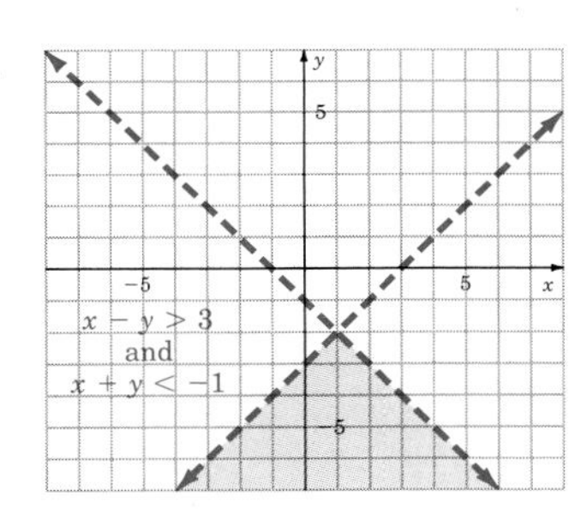

31.

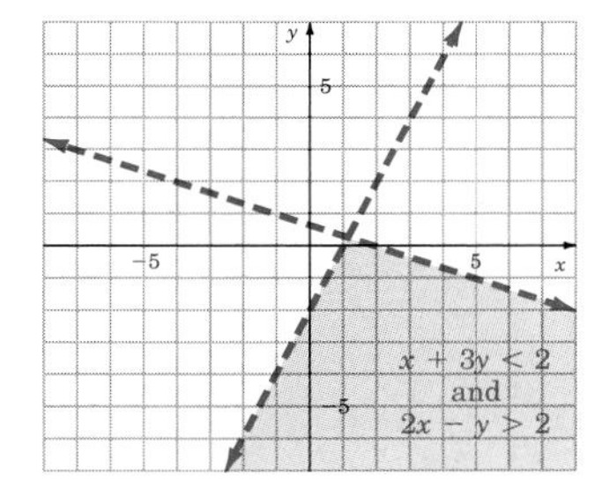

33.

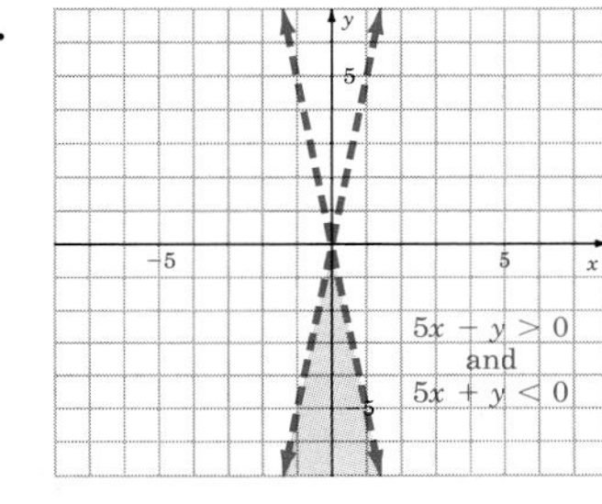

35.

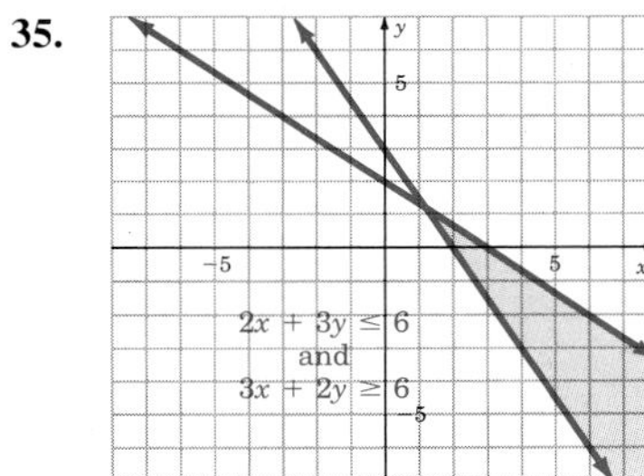

37.

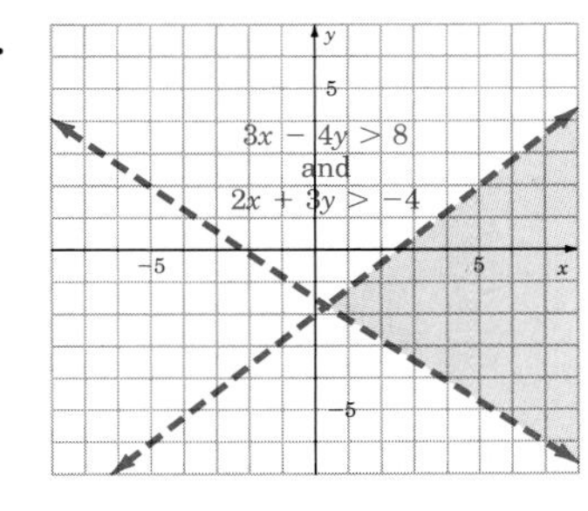

39.

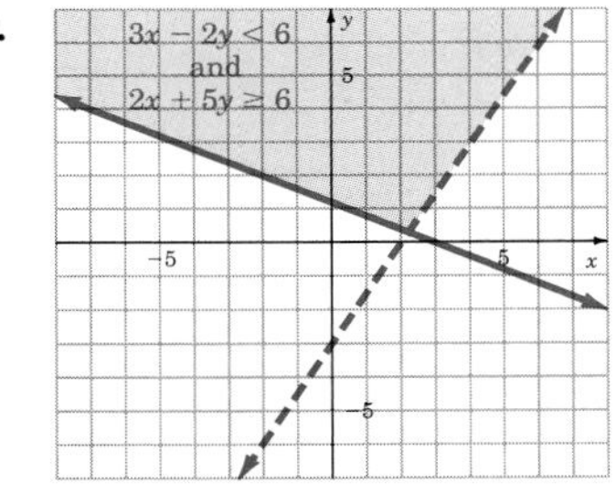

41. $y < x$

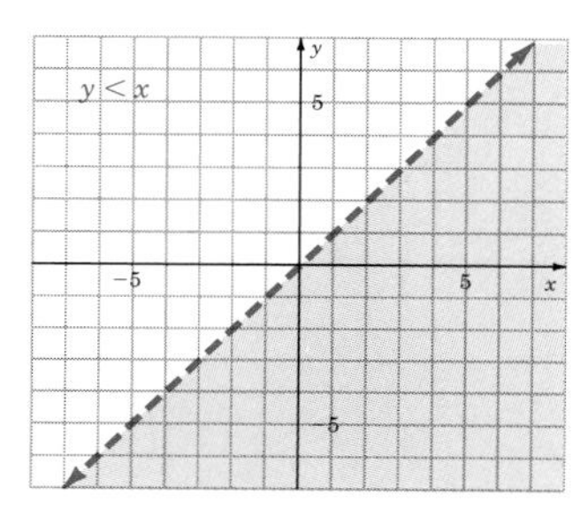

43. $y \le x - 4$

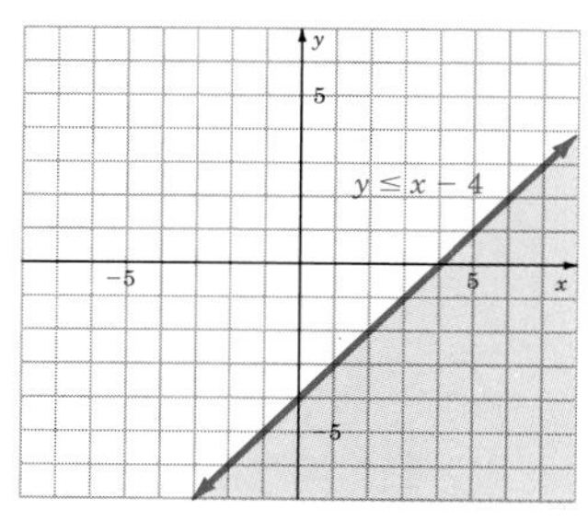

45. $y < \frac{1}{2}x$

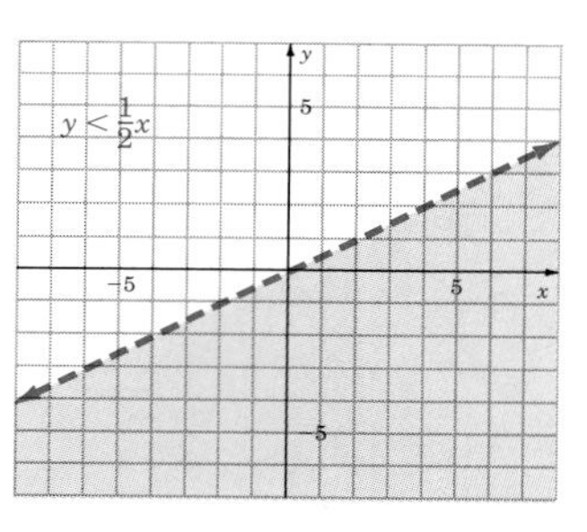

47. $2x + 3y \le 6$

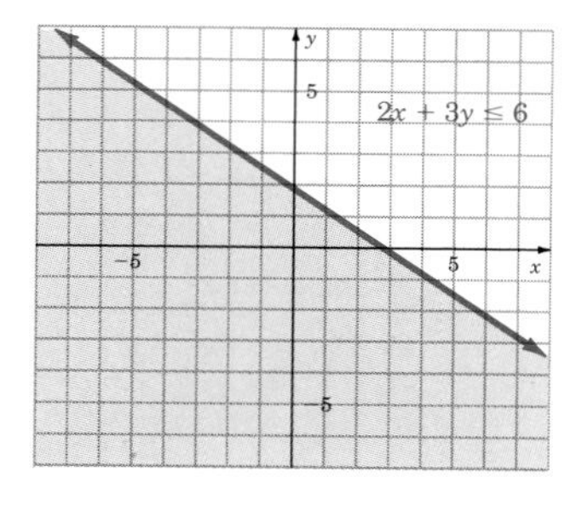

49. In words

51.

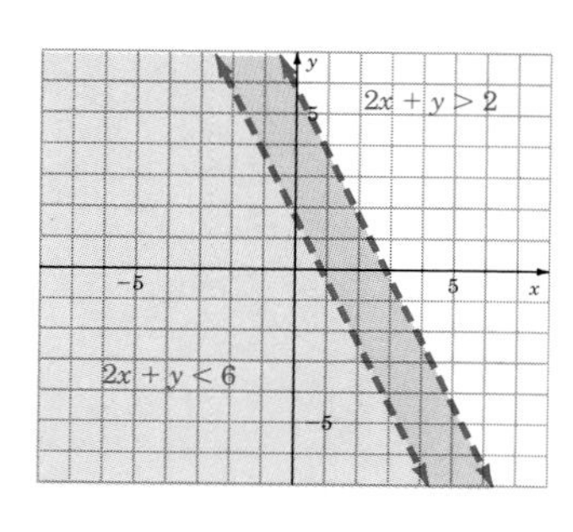

53.

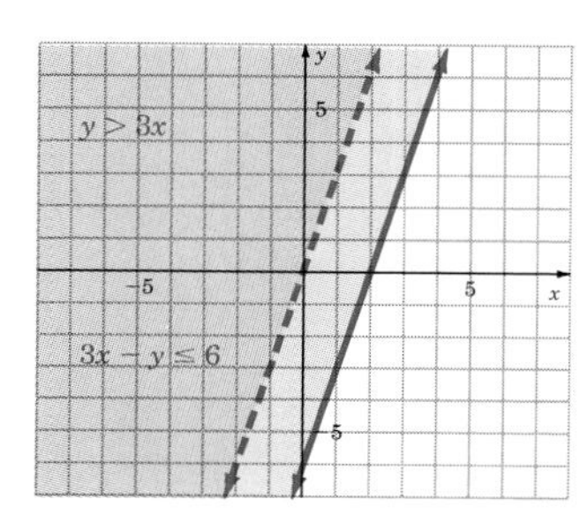

55. $-0.09a$ **57.** $132a + 22$

59. $-2ab + 17ac - 12a + 17bc + 7b - 60c$

61. $-13x^2 + 2xy + 55y^2$

Exercise 6.5

1.

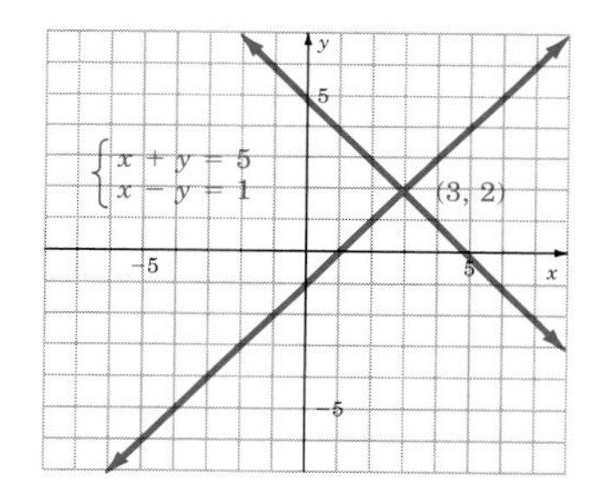

3.

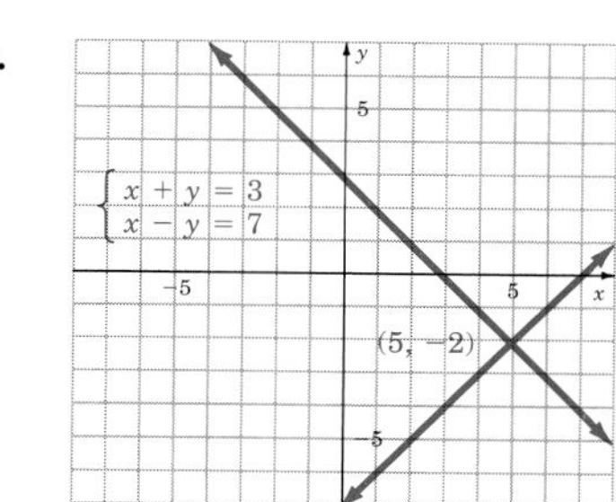

5.

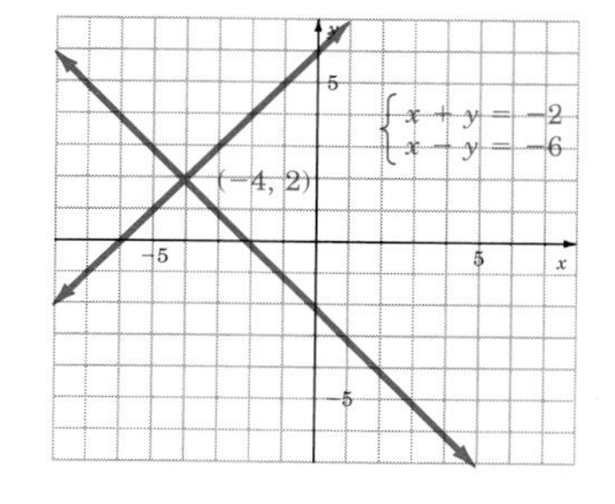

7.

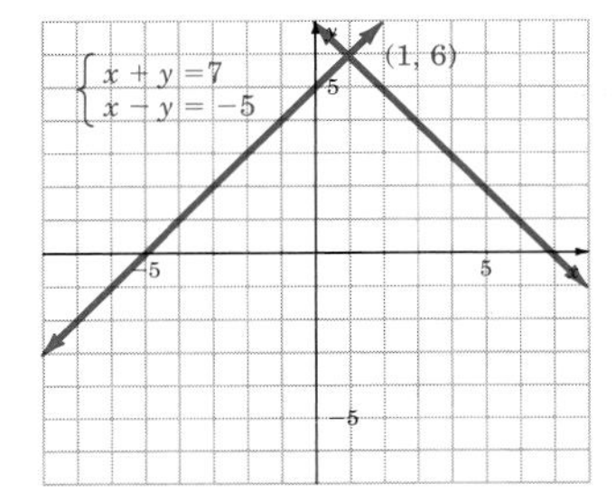

9.

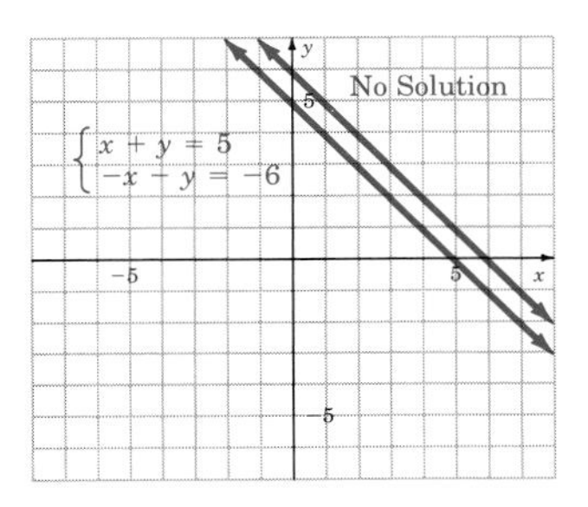

11.

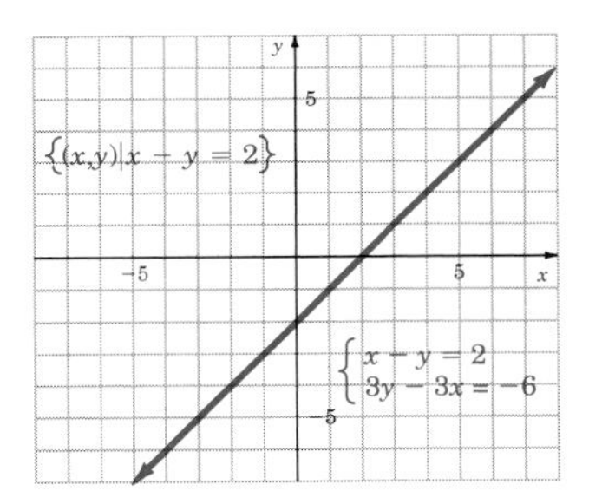

13.

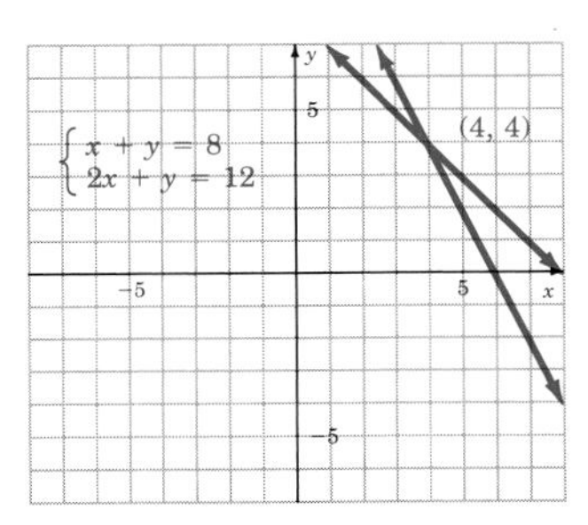

15.

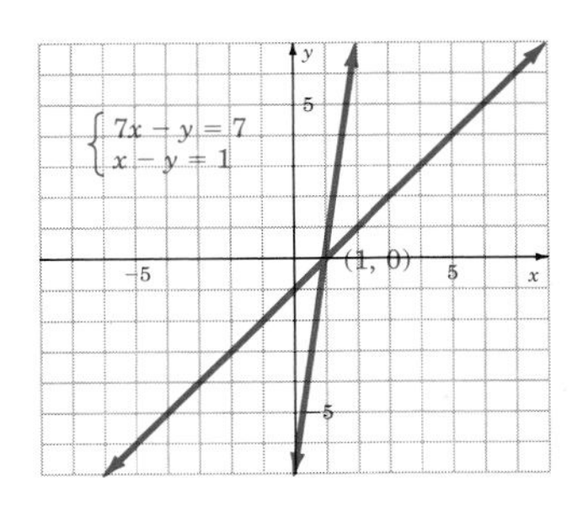

17.

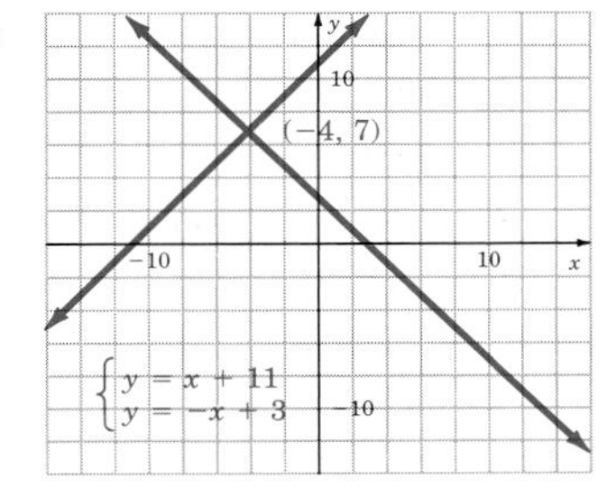

19.

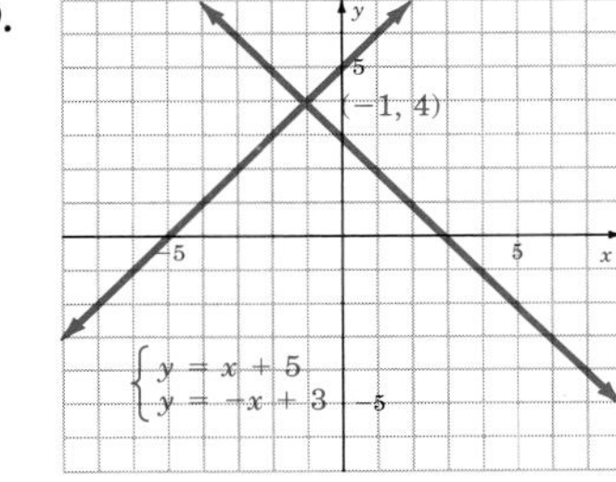

21.

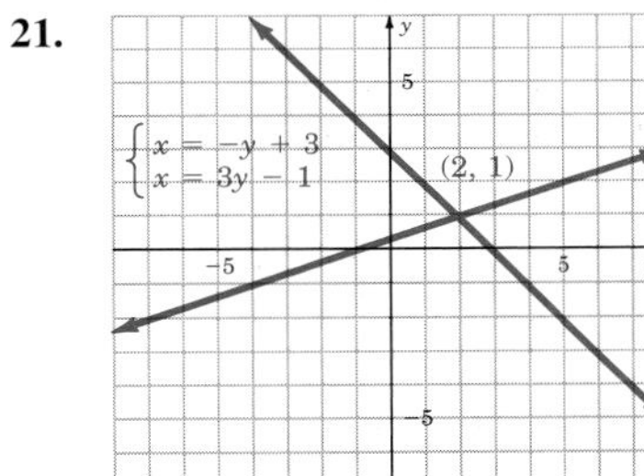

23.

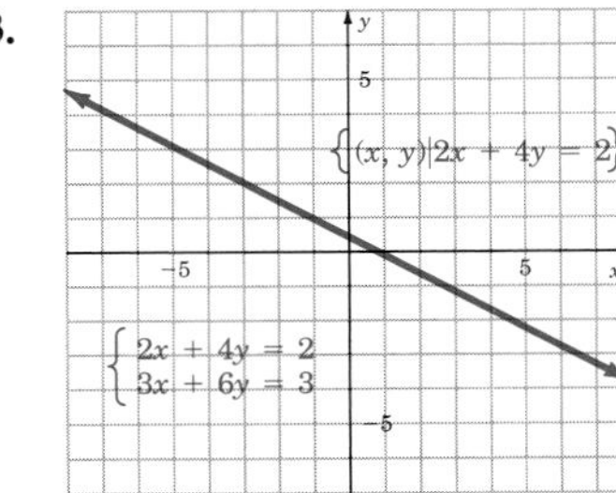

25.

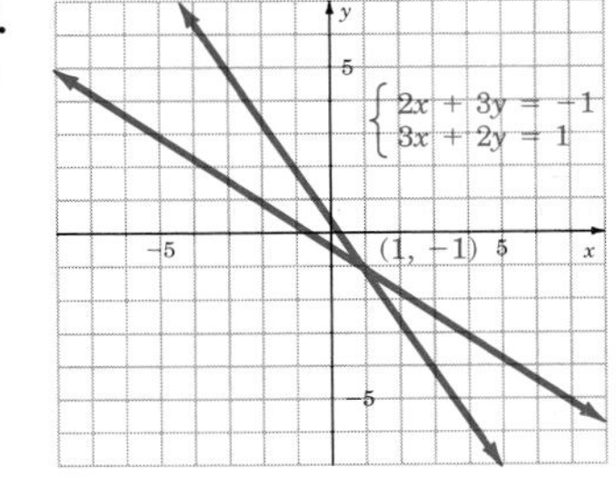

27.

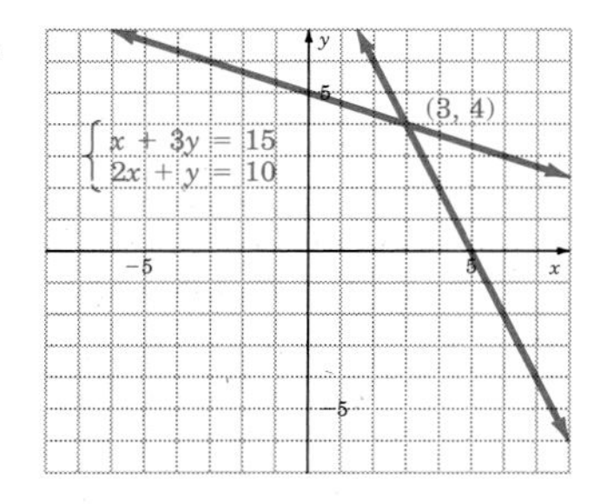

29.

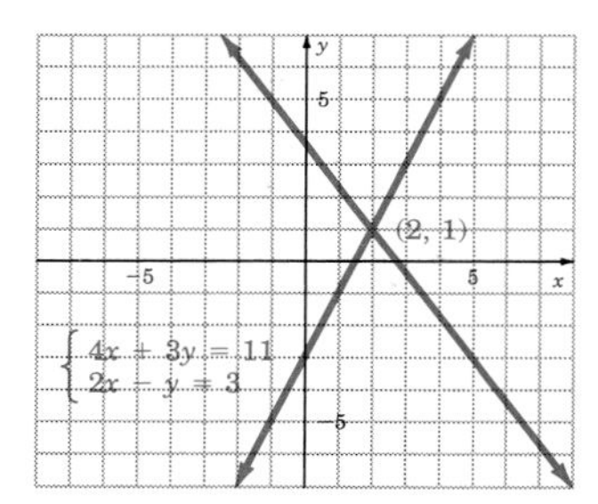

31.

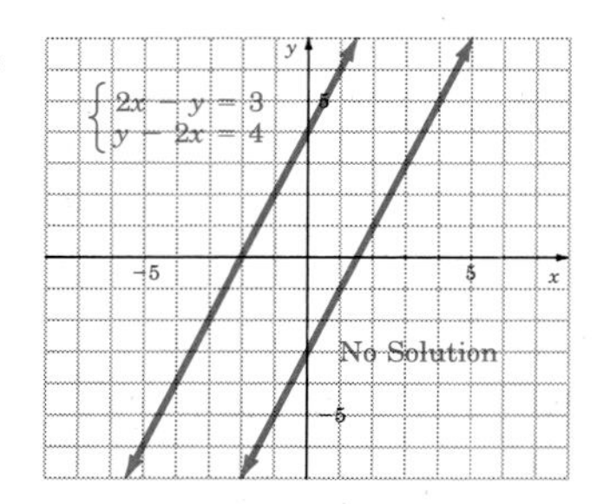

33.

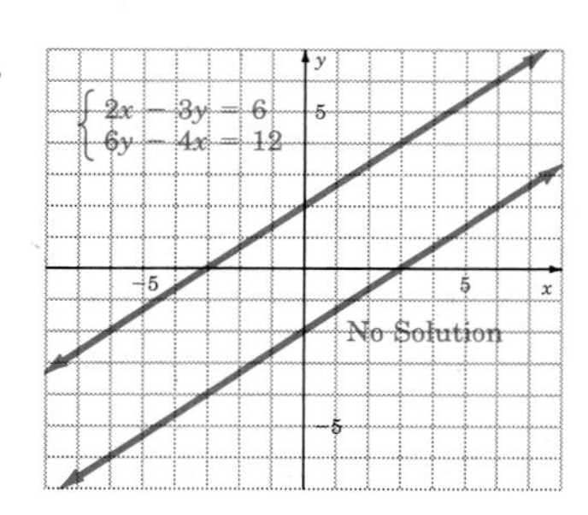

35.

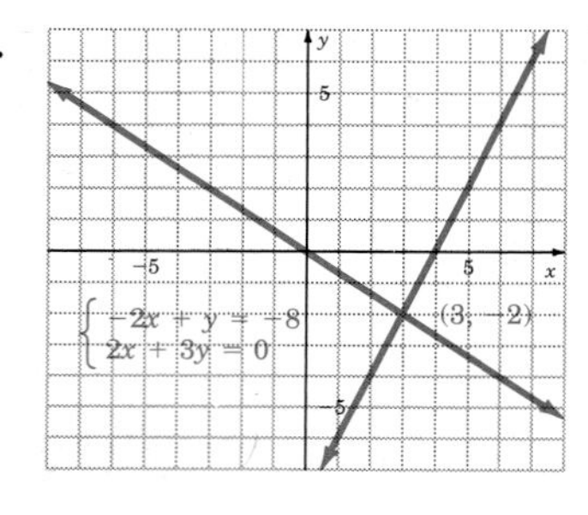

37.

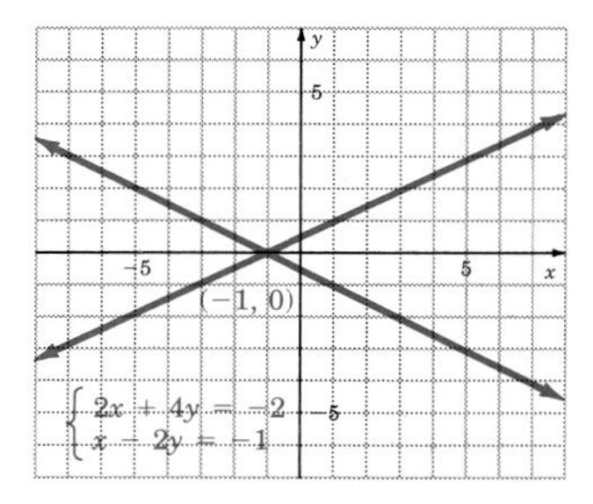

39.

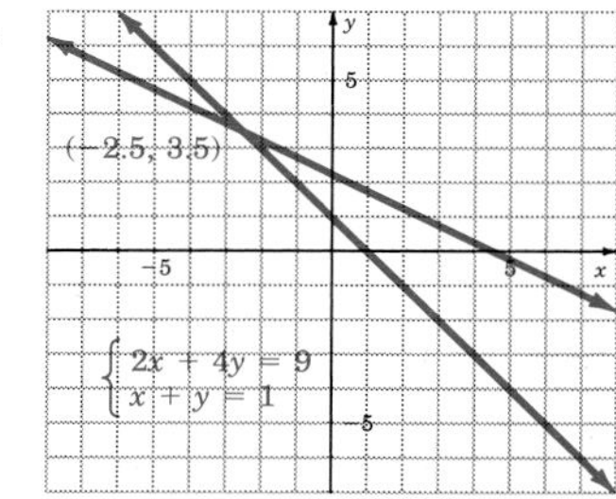

41.

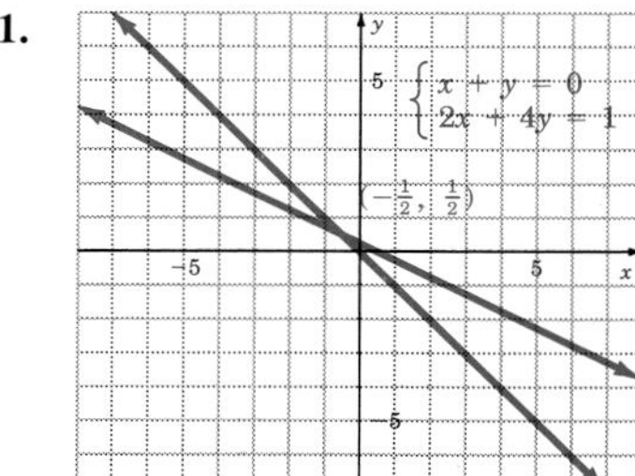

43.

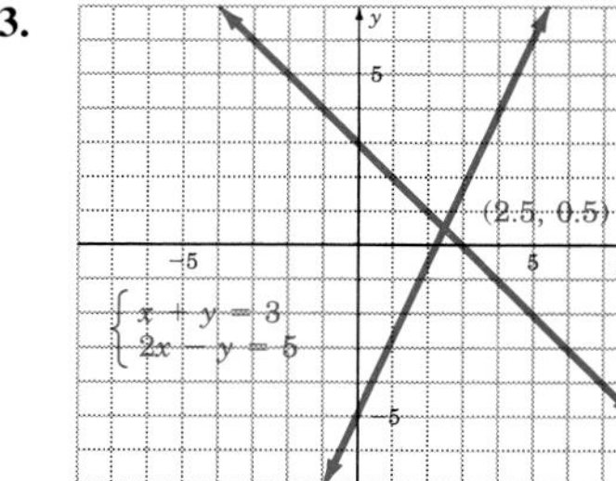

45.

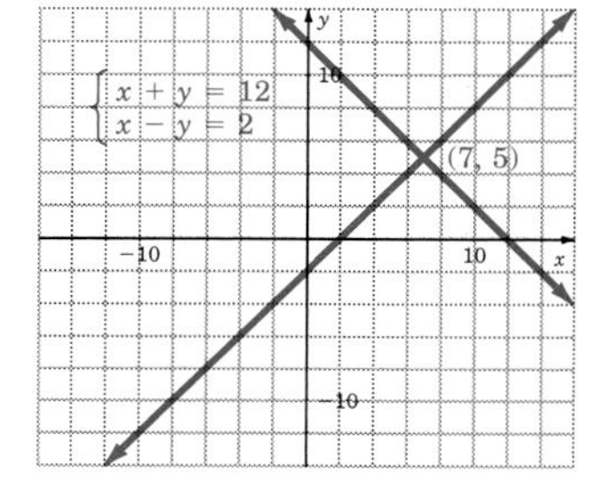

47.

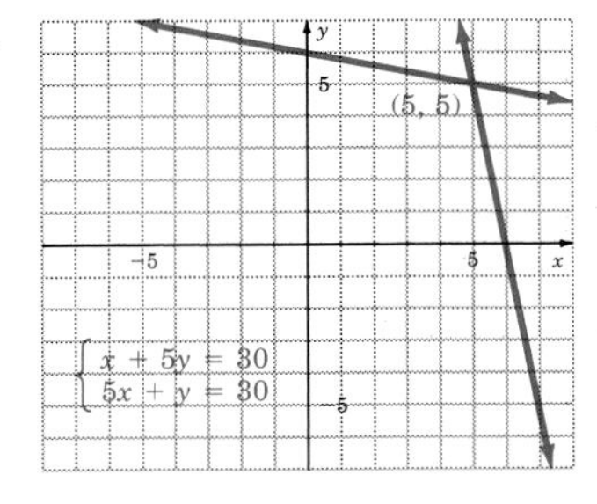

49. The cost for one adult is \$5; the cost for one child is \$3.
51. There are 10 men and 13 women in the class.

53. In words **55.**

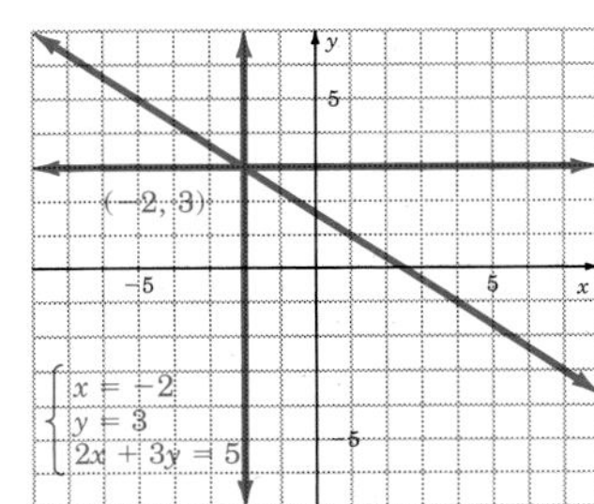

57.

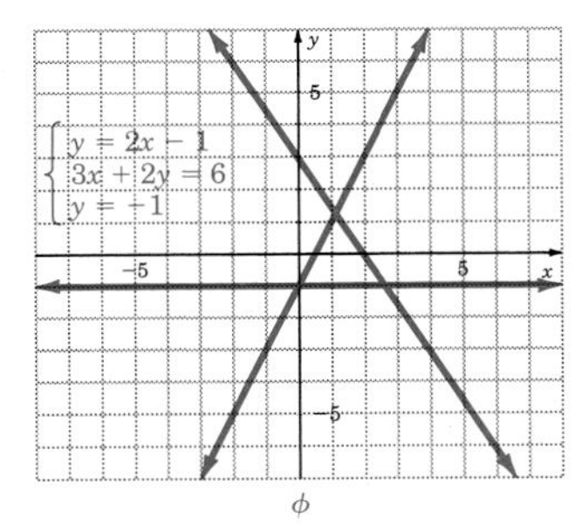

ϕ

59. $\dfrac{8x^5}{5}$ **61.** $\dfrac{x^2+4x+2}{x^2+3x+2}$ **63.** $\dfrac{25}{t-3}$
65. $\dfrac{m^2-9m+4}{m^2-4m+3}$

Exercise 6.6

1. $\{(4, -1)\}$ **3.** $\{(1, 2)\}$ **5.** $\{(9, 2)\}$ **7.** $\{(3, 2)\}$
9. $\{(-5, -3)\}$ **11.** $\{(2, 1)\}$ **13.** $\{(48, 36)\}$
15. $\{(-4, -6)\}$ **17.** $\{(-7, 3)\}$ **19.** $\{(-1, 10)\}$
21. $\{(2, 1)\}$ **23.** $\left\{\left(-4, \frac{5}{4}\right)\right\}$ **25.** $\{(2, -1)\}$
27. $\{(x, y) \mid 4x + y = 12\}$ **29.** $\left\{\left(\frac{14}{5}, \frac{1}{5}\right)\right\}$
31. $\{(-2, 10)\}$ **33.** $\{(7, 3)\}$ **35.** $\{(-2, 3)\}$
37. $\{(9, 2)\}$ **39.** $\{(-2, 3\}$ **41.** $\left\{\left(-\frac{1}{2}, \frac{5}{8}\right)\right\}$
43. $\left\{\left(\frac{13}{10}, \frac{4}{5}\right)\right\}$ **45.** $\emptyset$ **47.** $\left\{\left(-\frac{1}{2}, 4\right)\right\}$
49. $\left\{\left(\frac{1}{2}, \frac{3}{2}\right)\right\}$ **51.** $\left\{\left(\frac{1}{2}, \frac{2}{3}\right)\right\}$ **53.** $\{(8, -1)\}$
55. $\{(3, 2)\}$ **57.** $\{(2, -3)\}$ **59.** $\{(-2, -3)\}$
61. 300 lb of 40%, 300 lb of 70% **63.** 120 adults, 100 children **65.** 10,000 @ 12%, \$8000 @ 15%
67. Cadillac, 410 mi; Datsun, 250 mi **69.** 24, 25¢ stamps; 26, 30¢ stamps **71.** \$1200 @ 18%, \$900 @ 20%
73. In words **75.** $\{(1, 1)\}$ **77.** $a = 2$, $b = 3$, the equation is $2x + 3y = 11$ **79.** $(-2, 0)$, $(0, 7)$
81. $\left(\frac{7}{3}, 0\right)$, $\left(0, -\frac{7}{5}\right)$ **83.** $m = 5$ **85.** $m = 3$

Exercise 6.7

1. $\{(3, 1)\}$ **3.** $\{(4, 1)\}$ **5.** $\emptyset$ **7.** $\{(7, 5)\}$
9. $\{(1, 3)\}$ **11.** $\{(5, 5)\}$ **13.** $\left\{\left(\frac{1}{2}, 4\right)\right\}$
15. $\{(32, 3)\}$ **17.** $\{(-1, 3)\}$ **19.** $\{(-3, 4)\}$
21. $\{(-3, 1)\}$ **23.** $\{(1, -1)\}$ **25.** $\{(4, -1)\}$
27. $\{(4, 3)\}$ **29.** $\emptyset$ **31.** $\left\{\left(\frac{3}{5}, -\frac{2}{3}\right)\right\}$
33. $\left\{\left(-\frac{6}{11}, -\frac{45}{22}\right)\right\}$ **35.** $\{(4, 12)\}$ **37.** $\{(-6, 9)\}$
39. $\{(8, 7)\}$ **41.** $\left\{\left(\frac{20}{13}, -\frac{14}{13}\right)\right\}$ **43.** $\{(3, 2)\}$
45. $\{(3, -8)\}$ **47.** $\left\{\left(\frac{9}{4}, -\frac{3}{16}\right)\right\}$ **49.** $\{(-2, 5)\}$
51. $\{(-4, 3)\}$ **53.** $\{(7, -8)\}$ **55.** $\{(6, 1)\}$
57. $\left\{\left(-\frac{1}{2}, \frac{1}{2}\right)\right\}$ **59.** $\left\{\left(\frac{2035}{427}, -\frac{92}{61}\right)\right\}$ **61.** 5 oz of 85%, 10 oz of 70% **63.** 225 boxseats **65.** 125 lb of rye grass, 375 lb of bluegrass **67.** 150 @ \$27, 60 @ \$48 **69.** 275 g dried fruit, 155 g cereal **71.** In words
73. $\left\{\left(-1, \frac{1}{5}\right)\right\}$ **75.** $a = 8$, $b = -5$, the equation is $8x - 5y = 1$ **77.** $\left\{\frac{5}{2}\right\}$ **79.** $\left\{-\frac{19}{3}\right\}$ **81.** $\{-1\}$

83. $m = -\frac{3}{2}$; $\left(0, \frac{3}{2}\right)$

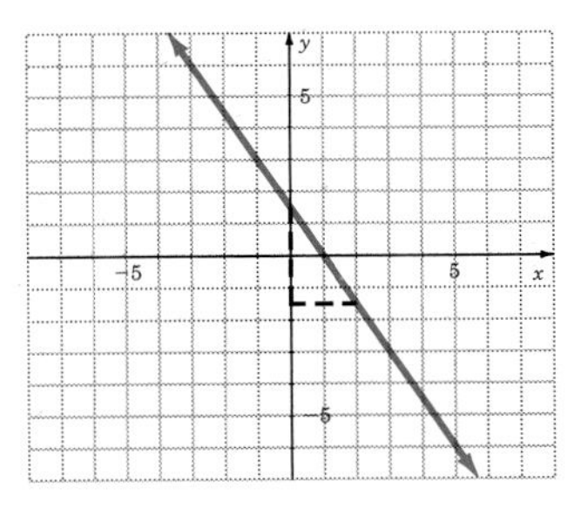

Chapter 6 Review Exercises

1. $y = -3$ **3.** $y = -\frac{9}{5}$ **5.** $y = -5.1$ **7.** $x = 4$

9. $x = \frac{28}{9}$

11.
13.
15.

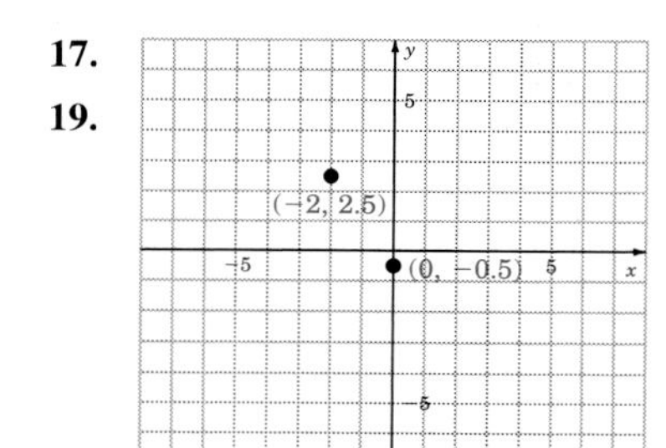

17.
19.

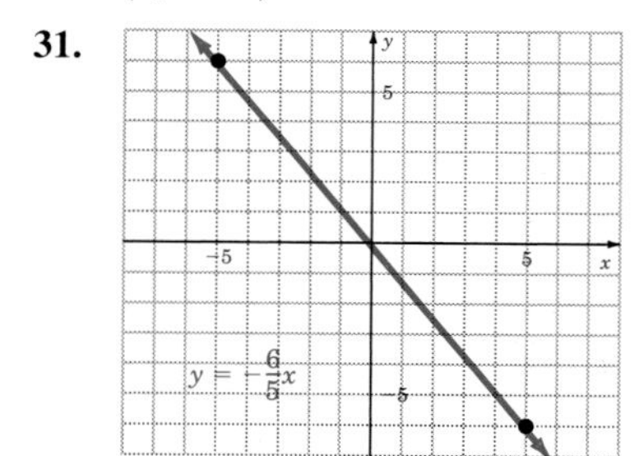

21. (4, 4) **23.** (−1, 2) **25.** (−2, −2) **27.** (2, −3)
29. (6, −4.5)

31.

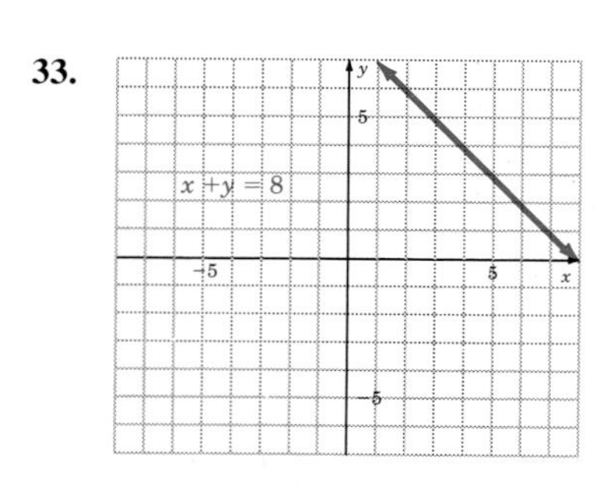

33.

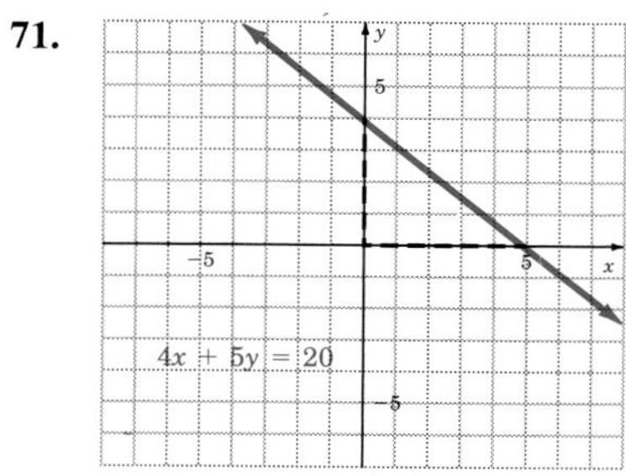

35.

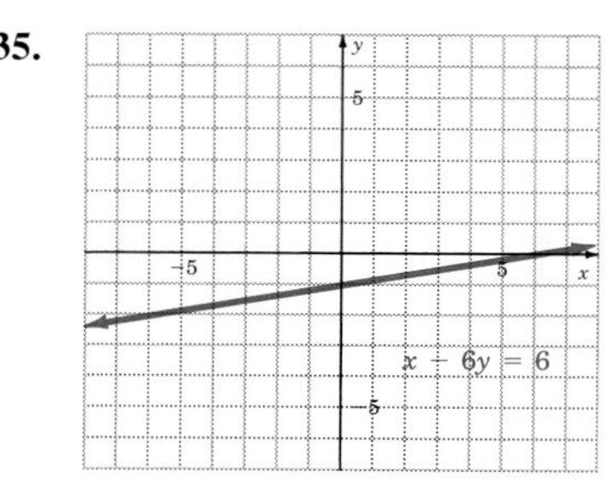

37.

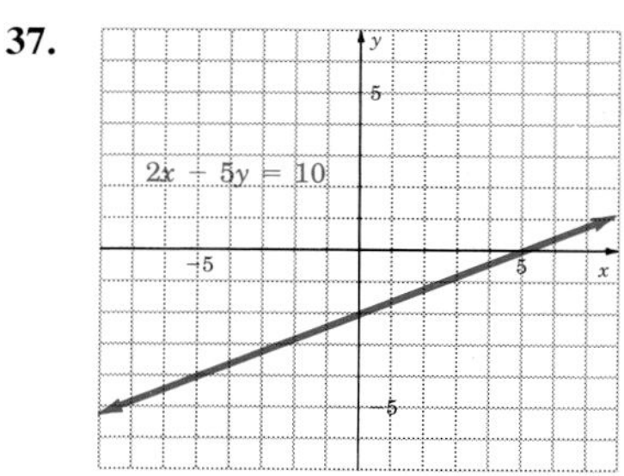

39.

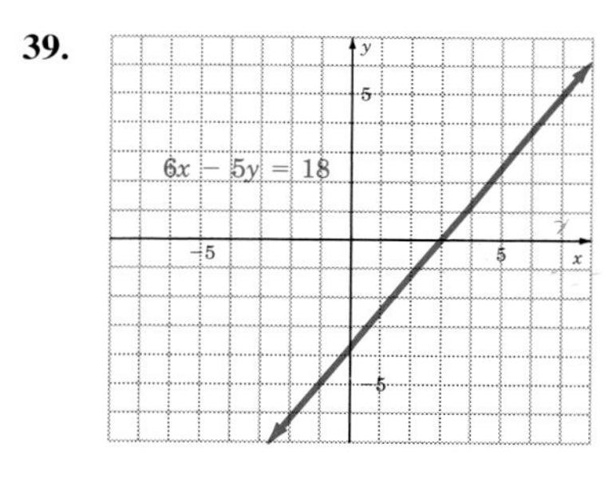

41. (10, 0), (0, 10) **43.** (3, 0), $\left(0, \frac{5}{2}\right)$

45. (−7, 0), (0, −4) **47.** (−5, 0), (0, 9)

49. (12, 0), (0, 2) **51.** $m = \frac{1}{7}$ **53.** $m = -\frac{2}{3}$

55. $m = \frac{3}{7}$ **57.** $m = 6$ **59.** $m = -1.8$

61. $m = \frac{2}{3}$ **63.** $m = 2$ **65.** $m = -\frac{5}{3}$ **67.** $m = \frac{5}{3}$

69. $m = -\frac{2}{25}$

71.

4x + 5y = 20

73.

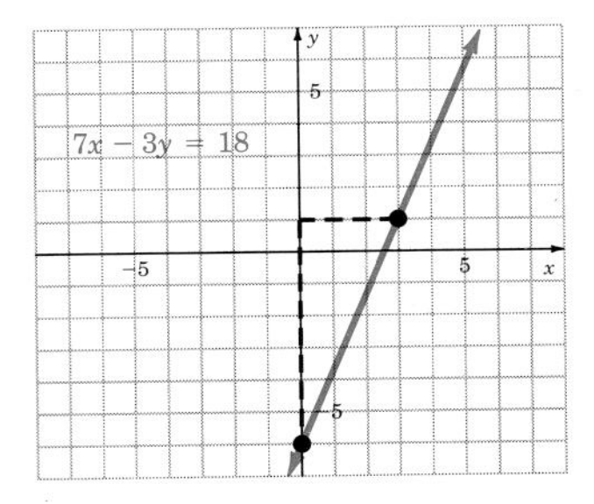

75.

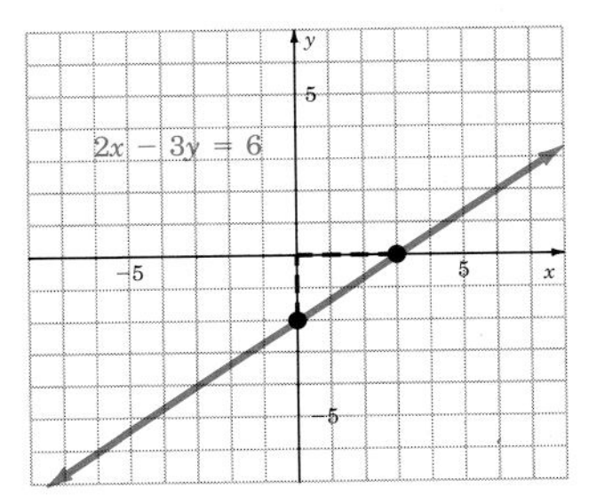

77.

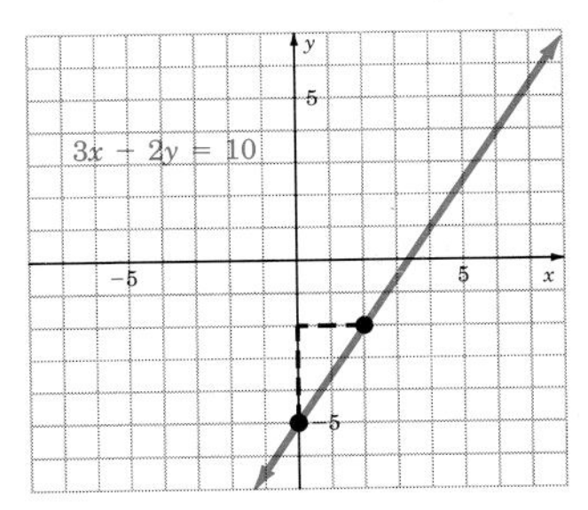

79.

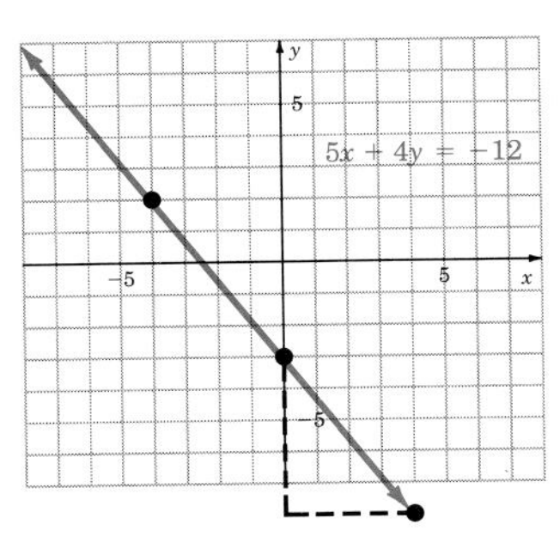

81.

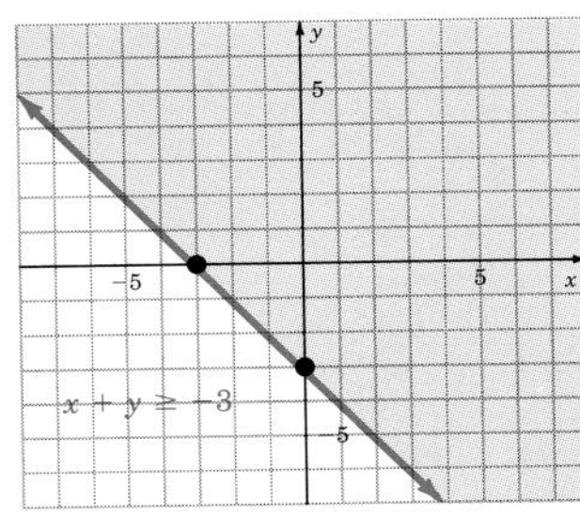

83.

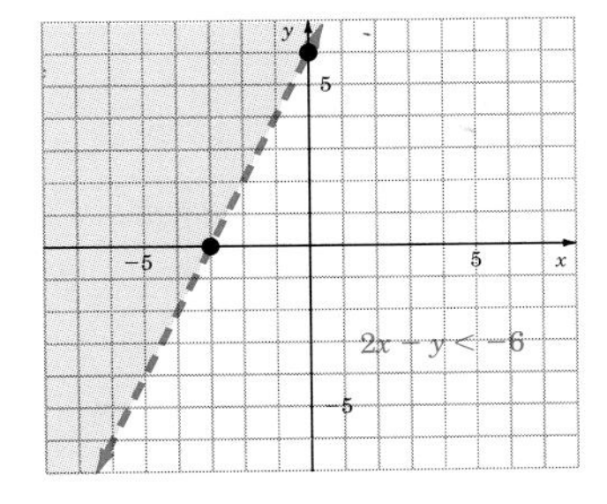

85.

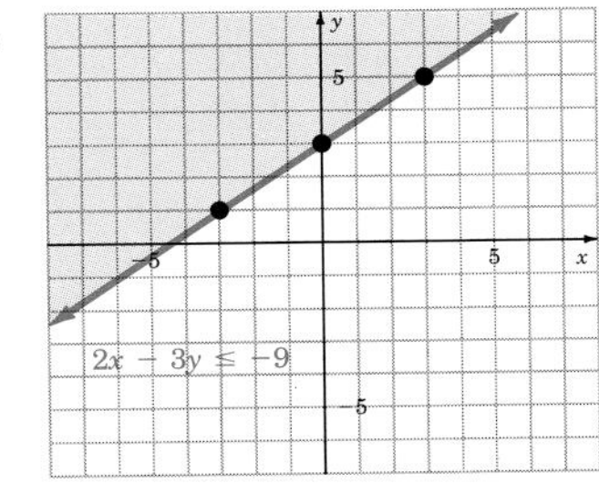

87.

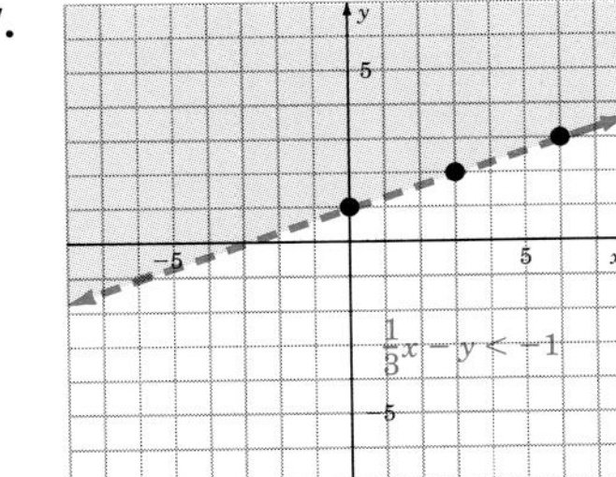

89.

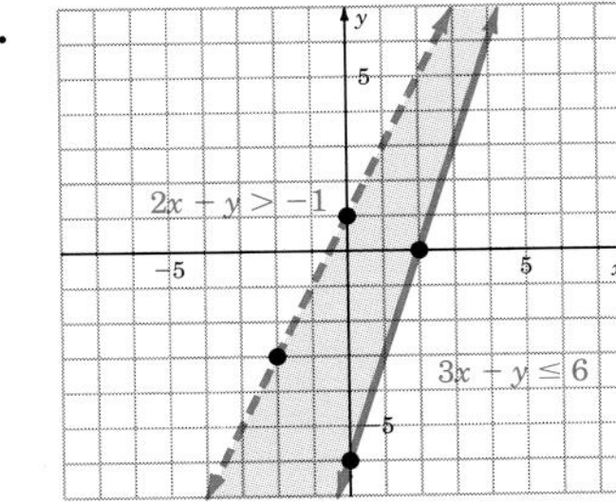

91.

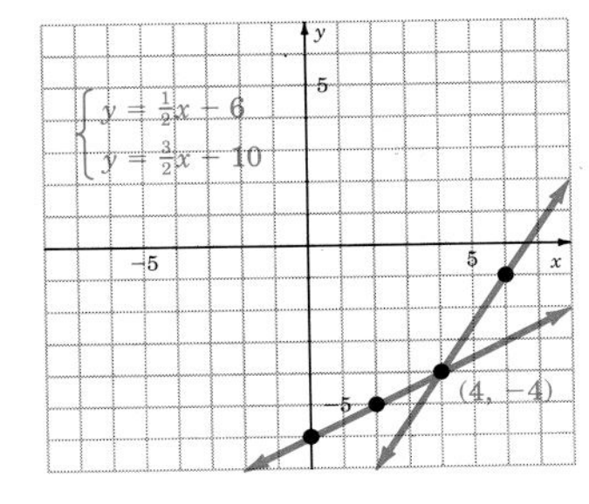

93.

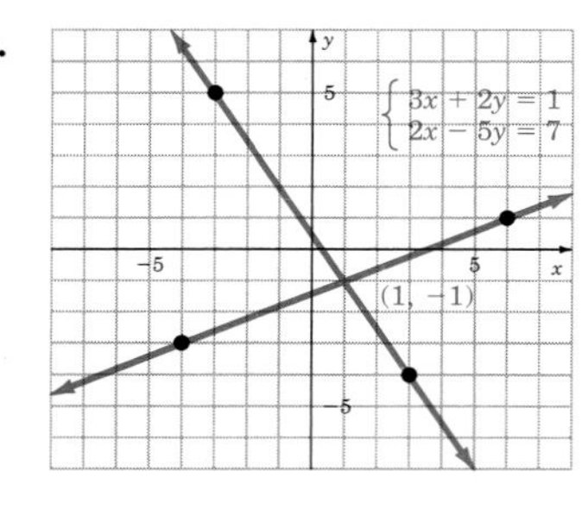

95. $\{(3, 5)\}$ **97.** $\{(3, -5)\}$ **99.** $\{(4, 3)\}$
101. $\{(9, 2)\}$ **103.** $\{(6, -3)\}$ **105.** $\{(4, -1)\}$
107. $\{(3, -6)\}$ **109.** $\{(7, 3)\}$ **111.** $\{(2, 8)\}$
113. $\{(-5, 4)\}$

Chapter 6 True–False Concept Review

1. False **2.** True **3.** True **4.** False **5.** False
6. True **7.** False **8.** False **9.** True **10.** True
11. False **12.** True **13.** True **14.** False
15. False **16.** False **17.** True **18.** True
19. True **20.** True **21.** True **22.** False
23. True **24.** True **25.** False **26.** True
27. True **28.** False **29.** True **30.** False

Chapter 6 Test

1. $A(5, 5)$, $B(2, -4)$, $C(-5, 3)$, $D(0, -2)$, $E(-6, -4)$

2.

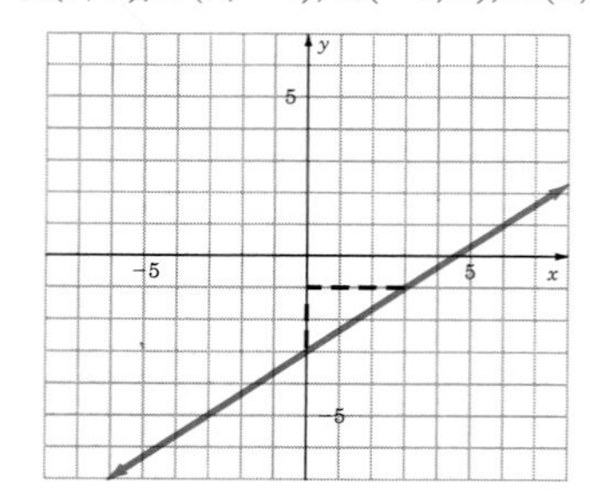

3. $\{(-4, -6)\}$ **4.** $x = -15$ **5.** $y = -4$

6. $\left\{\left(\frac{26}{5}, -\frac{9}{5}\right)\right\}$ **7.** $(7.5, 0)$, $(0, -3)$ **8.** $\{(6, 3)\}$

9.

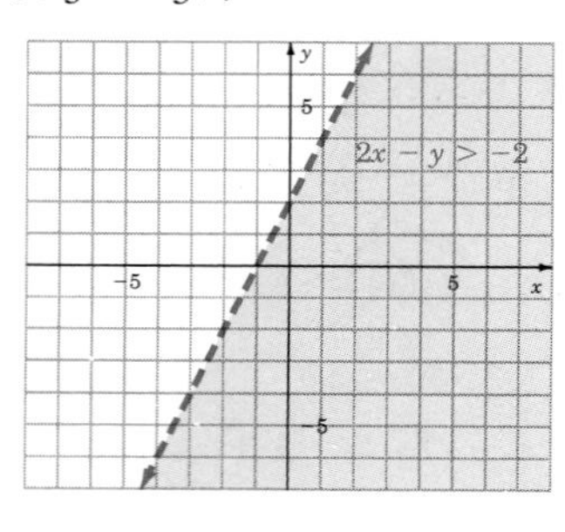

10. $\left\{\left(-\frac{15}{4}, -\frac{5}{2}\right)\right\}$

11.

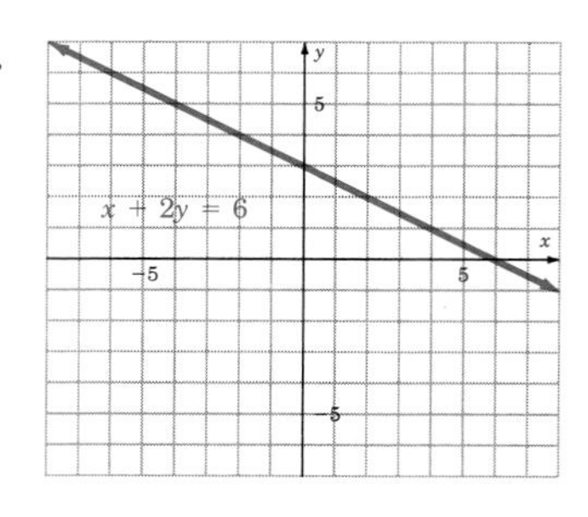

12.

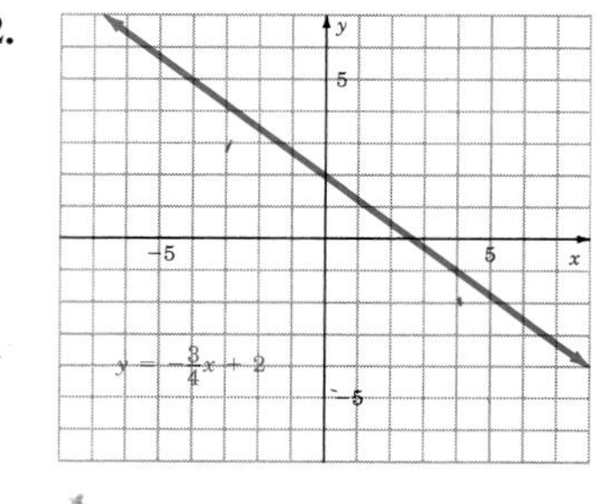

13. $\{(-13, -23)\}$ **14.** $m = -\frac{3}{5}$ **15.** $m = \frac{3}{7}$

16.

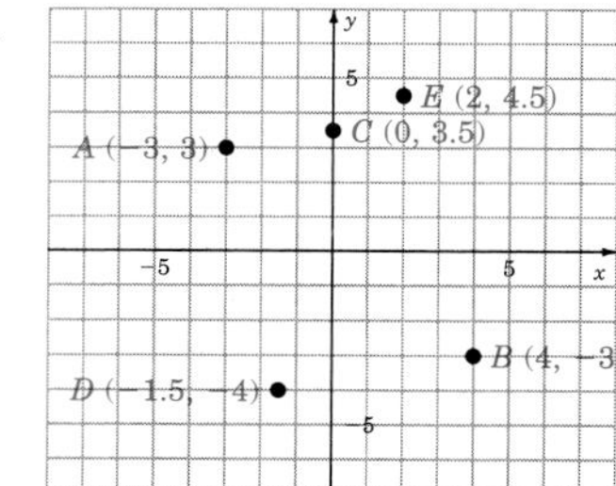

17. $\left\{\left(\frac{42}{5}, \frac{26}{5}\right)\right\}$

18.

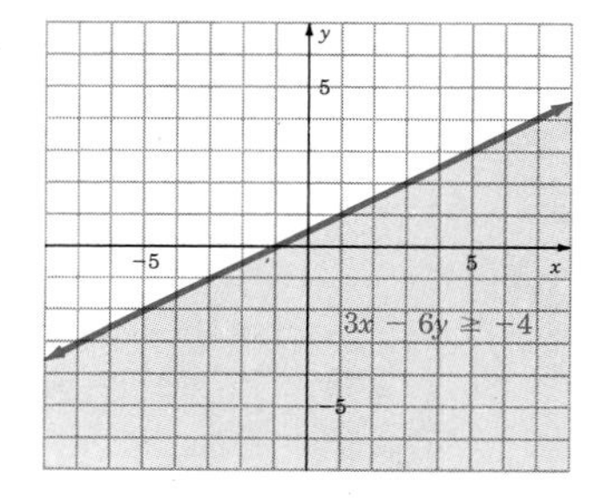

19. 2 and 1 **20.** 5 lb of 60% bluegrass, 10 lb of 45% bluegrass

CHAPTER 7

Exercise 7.1

1. 2 **3.** 5 **5.** −9 **7.** −14 **9.** $\frac{3}{5}$ **11.** $\frac{2}{13}$
13. 0.8 **15.** 1.5 **17.** 3 **19.** 0 **21.** 26
23. 20 **25.** −31 **27.** $\frac{13}{15}$ **29.** $\frac{11}{15}$ **31.** $-\frac{22}{17}$
33. −2.6 **35.** 0.12 **37.** −1 **39.** Not a real number **41.** 2.65 **43.** −5.92 **45.** 14.14
47. 18.44 **49.** −29.87 **51.** 0.63 **53.** 3.79
55. 3.30 **57.** 11 **59.** −5 **61.** 25 mph
63. 2.39 cm **65.** 8 **67.** 2.08 in. **69.** In words
71. x^2 **73.** $x \geq 0, y > 0$ **75.** $y = 14$ **77.** $y = \frac{16}{5}$
79. $y = \frac{38}{11}$ **81.** 10 ft

Exercise 7.2

1. $2\sqrt{2}$ **3.** $2\sqrt{6}$ **5.** $2\sqrt{10}$ **7.** $\sqrt{26}$ **9.** $4c$
11. $a\sqrt{b}$ **13.** $4xy\sqrt{2}$ **15.** $11ab^3\sqrt{a}$ **17.** Already in simplest form **19.** $w\sqrt{5}$ **21.** $4\sqrt{5}$ **23.** $3\sqrt{5}$
25. $2\sqrt{11}$ **27.** $10\sqrt{2}$ **29.** $6x\sqrt{2}$ **31.** $3t\sqrt{3t}$
33. $12xy^2\sqrt{2}$ **35.** $10x^2\sqrt{3}$ **37.** $10x^2y\sqrt{2y}$
39. $5r^2s^2\sqrt{3s}$ **41.** $8\sqrt{3}$ **43.** $7\sqrt{5}$ **45.** $3w^3\sqrt{22}$
47. $5w\sqrt{6}$ **49.** $2a^2\sqrt{39}$ **51.** $7x^4y^2\sqrt{5y}$
53. $4p^4\sqrt{2}$ **55.** $9a^6b^4\sqrt{5b}$ **57.** $15r^4s^2\sqrt{5rs}$
59. $42x^5y^3\sqrt{xy}$ **61.** $30\sqrt{5}$ ft ≈ 67.1 ft
63. $2\sqrt{34}$ ft ≈ 11.7 ft **65.** $60\sqrt{3}$ volts
67. $5\sqrt{2}$ cm ≈ 7.1 cm **69.** In words
71. $(x - y)^3\sqrt{(x - y)}$ **73.** $(3x - 1)\sqrt{6}$
75. $(x + y)\sqrt{(3x - 3y)}$ **77.** (2, 5) **79.** (−5, −5)
81.

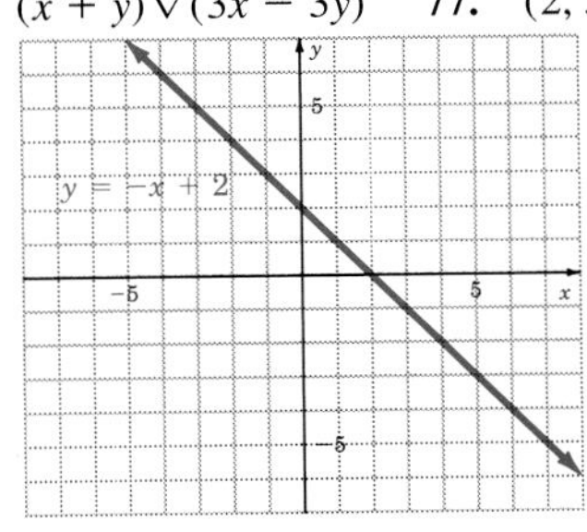

83.

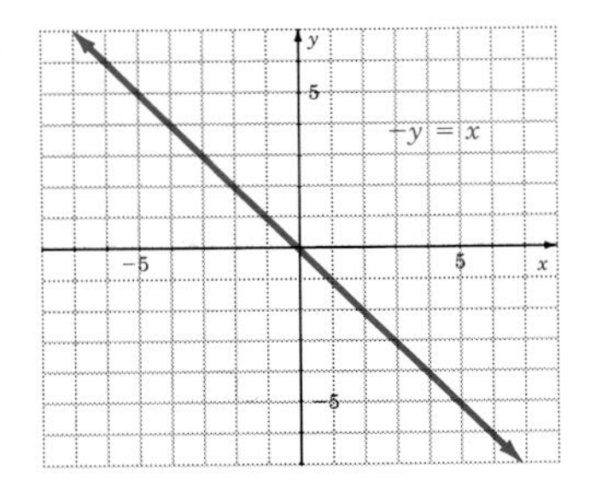

Exercise 7.3

1. $\sqrt{42}$ **3.** $\sqrt{42}$ **5.** 7 **7.** 6 **9.** 4 **11.** 10
13. $\sqrt{xyz}$ **15.** $\sqrt{105}$ **17.** 18 **19.** 14
21. $5\sqrt{2xy}$ **23.** $3\sqrt{5}$ **25.** $\sqrt{110}$ **27.** $3\sqrt{35}$
29. $\sqrt{21} + 3$ **31.** $3\sqrt{2} - \sqrt{66}$
33. $5\sqrt{2} + 2\sqrt{5}$ **35.** $x\sqrt{wy} + y\sqrt{xz}$
37. $2x + 2x\sqrt{2} + 2x\sqrt{x}$ **39.** $2w\sqrt{3} - 4w + 4w\sqrt{5}$
41. $x^3\sqrt{x}$ **43.** $x^5\sqrt{x}$ **45.** $10r^2\sqrt{10}$ **47.** 7
49. $4 + 2\sqrt{3} + 2\sqrt{5}$ **51.** $x\sqrt{3} + 3\sqrt{x}$
53. $3x + 9\sqrt{x}$ **55.** $5 + 2\sqrt{6}$ **57.** $8 - 2\sqrt{15}$
59. $7\sqrt{2} - \sqrt{21} + \sqrt{42} - 3$ **61.** (a) $\sqrt{6}$ ft (b) 2.4 ft
63. $10\pi\sqrt{6}$ m² ≈ 76.9 m² **65.** (a) $6\sqrt{5}$ in² (b) 13.4 in² **67.** (a) $\frac{5\sqrt{6}}{2}$ cm² (b) 6.1 cm²
69. $4\sqrt{5}$ cm³ ≈ 8.9 cm³ **71.** In words **73.** 18
75. $x^4 - 4x^2y^2\sqrt{y} + 4y^5$ **77.** $m = \frac{6}{7}$, (0, −2)
79. $m = -\frac{1}{9}, \left(0, \frac{4}{3}\right)$ **81.** Yes **83.** No

Exercise 7.4

1. $\frac{\sqrt{3}}{3}$ **3.** $\frac{3\sqrt{2}}{2}$ **5.** $\frac{3\sqrt{11}}{11}$ **7.** $\frac{\sqrt{15}}{5}$ **9.** $\frac{\sqrt{6}}{3}$
11. $\sqrt{2}$ **13.** $\frac{\sqrt{6}}{2}$ **15.** $\frac{\sqrt{2}}{2}$ **17.** $\frac{\sqrt{3}}{2}$
19. $\frac{\sqrt{2}}{6}$ **21.** $\frac{\sqrt{3}}{2}$ **23.** $\sqrt{3}$ **25.** $\frac{\sqrt{6}}{3}$
27. $\frac{\sqrt{5}}{3}$ **29.** $\frac{2\sqrt{5}}{5}$ **31.** $\frac{\sqrt{5}}{2}$ **33.** $\frac{\sqrt{10}}{8}$
35. $\frac{1}{3}$ **37.** $\frac{\sqrt{2}}{3}$ **39.** $4\sqrt{6} + 4\sqrt{5}$ **41.** $\frac{3\sqrt{3}}{14}$
43. $\frac{\sqrt{6}}{3}$ **45.** $\frac{\sqrt{5}}{14}$ **47.** $\frac{4\sqrt{21}}{5}$ **49.** $\frac{\sqrt{30}}{30}$
51. $\frac{3\sqrt{15}}{10}$ **53.** $\frac{5\sqrt{2}}{2}$ **55.** $\frac{33\sqrt{3}}{40}$
57. $\sqrt{5} - \sqrt{3}$ **59.** $\sqrt{10} + \sqrt{5}$
61. $\frac{\pi}{2}$ sec ≈ 1.57 sec **63.** $K = 3$ **65.** $s = 1.5$

67. The length of the edge is $\frac{\sqrt{70}}{2}$ in. or approximately 4.18 in. **69.** In words **71.** $\frac{7 - 2\sqrt{10}}{3}$
73. $\frac{x + \sqrt{x^2 - 4}}{2}$ **75.** $4x$ **77.** $\frac{19x}{6}$
79. $m = -\frac{8}{3}$ **81.** $m = \frac{3}{8}$

Exercise 7.5

1. $9\sqrt{5}$ **3.** $26\sqrt{10}$ **5.** $2\sqrt{2}$ **7.** $-5\sqrt{5}$
9. $7\sqrt{3}$ **11.** $-2\sqrt{7}$ **13.** $2\sqrt{2}$ **15.** $6\sqrt{5}$
17. $4\sqrt{2}$ **19.** $\frac{10\sqrt{5}}{3}$ **21.** $5\sqrt{2}$ **23.** 0
25. $5\sqrt{5}$ **27.** $\sqrt{6x}$ **29.** $4\sqrt{5}$ **31.** $-7\sqrt{2}$
33. $7\sqrt{2}$ **35.** $8\sqrt{5} + 4\sqrt{105}$ **37.** $4\sqrt{3}$
39. $32\sqrt{2}$ **41.** $42\sqrt{14}$ **43.** $3\sqrt{2}$
45. $13\sqrt{2} - 8\sqrt{3}$ **47.** $11\sqrt{2} - 9\sqrt{3}$ **49.** $-5\sqrt{7}$
51. $-21\sqrt{3}$ **53.** $97\sqrt{5} - 32\sqrt{3}$ **55.** $\frac{37}{14}\sqrt{14}$
57. $\frac{\sqrt{6}}{2}$ **59.** $\frac{9\sqrt{3}}{2}$ **61.** $\frac{\sqrt{15}}{8}$ **63.** $140\sqrt{6}$ volts
65. $\frac{1025\sqrt{3}}{4}$ ft^2 ≈ 443.8 ft^2 **67.** The surface area of the box is $180 + 96\sqrt{2}$ cm^2 or approximately 315.8 cm^2.
69. In words **71.** $-x\sqrt{14y}$ **73.** $\frac{\sqrt{6xy}}{6x}$
75. $x = \frac{225}{32}$ **77.** $x = 6$ **79.** \$3272.50
81. \$1875, \$3125

Exercise 7.6

1. {9} **3.** {121} **5.** $\left\{\frac{25}{2}\right\}$ **7.** {45} **9.** {22}
11. {11} **13.** ∅ **15.** {4} **17.** $\left\{\frac{16}{9}\right\}$ **19.** ∅
21. {8} **23.** $\left\{\frac{95}{2}\right\}$ **25.** ∅ **27.** {26} **29.** {195}
31. ∅ **33.** {8} **35.** {9} **37.** $\left\{\frac{64}{5}\right\}$ **39.** $\left\{-\frac{6}{5}\right\}$
41. $\left\{\frac{4}{25}\right\}$ **43.** {1} **45.** $\left\{\frac{49}{9}\right\}$ **47.** ∅ **49.** {4}
51. {1} **53.** {4} **55.** ∅ **57.** {9} **59.** {49}
61. $\ell = \sqrt{7}$ in., $w = \sqrt{2}$ in. **63.** 40 ohms
65. 4.68 in. **67.** The kinetic energy is 5.8 electron volts. **69.** In words **71.** {9} **73.** {−62}
75. 20% **77.** \$13,000 **79.** 40 **81.** 287 lb

Chapter 7 Review Exercises

1. 11 **3.** −18 **5.** Not a real number **7.** 63
9. −43 **11.** 14.49 **13.** 2 **15.** 7 **17.** −8
19. 4 **21.** Not a real number **23.** 10 **25.** $4\sqrt{3}$
27. In simplest form **29.** a^4 **31.** $6z^3\sqrt{2}$
33. $5\sqrt{7}$ **35.** $6x^2y^2\sqrt{14x}$ **37.** $\sqrt{35}$ **39.** 6
41. $4\sqrt{15}$ **43.** $7\sqrt{5} - 5$ **45.** 1 **47.** 9
49. $\frac{\sqrt{14}}{2}$ **51.** $\frac{\sqrt{35}}{5}$ **53.** $\frac{2\sqrt{33}}{3}$ **55.** $\frac{6\sqrt{5}}{5}$
57. $\frac{5\sqrt{11}}{66}$ **59.** $-\frac{3\sqrt{11} + 15}{7}$ **61.** $9\sqrt{7}$
63. $24\sqrt{2}$ **65.** $2\sqrt{5} + 3\sqrt{3}$ **67.** $-8\sqrt{x}$
69. $12\sqrt{7}$ **71.** $-3\sqrt{5}$ **73.** {81} **75.** {140}
77. ∅ **79.** $\left\{\frac{4}{5}\right\}$ **81.** $\left\{\frac{13}{4}\right\}$ **83.** {2}

Chapter 7 True–False Concept Review

1. False **2.** True **3.** True **4.** True **5.** False
6. True **7.** True **8.** False **9.** True **10.** True
11. False **12.** True **13.** False **14.** True
15. False

Chapter 7 Test

1. $\frac{\sqrt{14}}{6}$ **2.** $3m^2n\sqrt{7}$ **3.** $y = 151$ **4.** −14
5. $3\sqrt{13}$ **6.** $\frac{5\sqrt{3}}{6}$ **7.** $x = 5$ **8.** $5\sqrt{6} - 6\sqrt{5}$
9. $4\sqrt{3}$ **10.** $20\sqrt{2}$ **11.** $-5\sqrt{7}$ **12.** 13.38
13. $10\sqrt{2}$ **14.** $7q\sqrt{2p}$ **15.** 360 **16.** $5\sqrt{3}$ ft^2
17. $20\sqrt{21}$ ft **18.** $150\sqrt{2}$ volts

CHAPTER 8

Exercise 8.1

1. {±2} **3.** {±11} **5.** $\left\{\pm\frac{7}{2}\right\}$ **7.** No real solutions **9.** No real solutions **11.** {±20}
13. {0, 11} **15.** {−3, 0} **17.** $a = 5$ **19.** $b = 36$
21. $c = 25$ **23.** {0, 7} **25.** $\{\pm\sqrt{26}\}$ **27.** {±5}
29. $\{\pm\sqrt{2}\}$ **31.** $\left\{\pm\frac{5}{6}\right\}$ **33.** {−5, 0}
35. $\left\{-\frac{1}{4}, 0\right\}$ **37.** {0, 4} **39.** $a = 6.8$
41. $c = 22.5$ **43.** $a = 3.75$ **45.** {±4.80}
47. {±3.51} **49.** {±4.36} **51.** $\left\{-\frac{4}{3}, 0\right\}$
53. {−2, 0} **55.** {0, 10} **57.** {0} **59.** $x = \pm 5a$
61. $c \approx 1.60$ **63.** $a \approx 7.55$ **65.** $c \approx 9.26$

67. The dia. is $\frac{2\sqrt{2\pi}}{\pi}$ ft $\approx$ 1.6 ft **69.** 46 ft **71.** $v = 2000$ ft/sec **73.** 32.0 ft **75.** 15.65 ft **77.** 12 ft **79.** 9.4 ft **81.** 61.0 ft **83.** 80.62 ft **85.** In words **87.** $\left\{\frac{7 \pm 2\sqrt{5}}{3}\right\}$ **89.** $\{\pm\sqrt{\sqrt{5}}\}$ **91.** $7a^2b^3c\sqrt{6ac}$ **93.** $450\sqrt{6}$ **95.** $11 - 4\sqrt{6}$ **97.** $5\sqrt{21} \approx 22.9$ ft

Exercise 8.2

1. $\{-8, 2\}$ **3.** $\{2 \pm 2\sqrt{3}\}$ **5.** $\{-5 \pm 2\sqrt{5}\}$ **7.** $\{-8, 2\}$ **9.** $\{-12, 2\}$ **11.** $\{-5, 8\}$ **13.** $\{-7, 1\}$ **15.** $\{-8, -4\}$ **17.** No real solutions **19.** $\{-15, 9\}$ **21.** $\left\{-\frac{1}{3}, 3\right\}$ **23.** $\left\{-3, \frac{5}{3}\right\}$ **25.** $\left\{\frac{-11 \pm \sqrt{11}}{6}\right\}$ **27.** $\{-7, 1\}$ **29.** $\left\{\frac{-5 \pm \sqrt{5}}{2}\right\}$ **31.** No real solution **33.** $\left\{\frac{3 \pm \sqrt{19}}{2}\right\}$ **35.** $\left\{-7, \frac{5}{2}\right\}$ **37.** $\left\{-9, \frac{3}{2}\right\}$ **39.** $\left\{\frac{1}{5}, 5\right\}$ **41.** $\left\{\frac{7 \pm \sqrt{57}}{2}\right\}$ **43.** $\left\{\frac{9 \pm \sqrt{17}}{4}\right\}$ **45.** $\left\{\frac{5}{2}, 7\right\}$ **47.** No real solution **49.** $\left\{\frac{11 \pm \sqrt{161}}{10}\right\}$ **51.** $\left\{-4, -\frac{4}{3}\right\}$ **53.** No real solution **55.** No real solution **57.** $\left\{\frac{-4 \pm \sqrt{22}}{3}\right\}$ **59.** $\left\{\frac{-2 \pm \sqrt{34}}{5}\right\}$ **61.** 3.1 hr **63.** 8.4 hr **65.** 7.9 mph and 12.9 mph **67.** 102 items **69.** In words **71.** It takes the first pipe 12 hours to fill the tank; the second pipe, 10 hours; the drain can empty a full tank in 20 hours. **73.** $\left\{\frac{-1 \pm \sqrt{1 - 4c}}{2}\right\}$ **75.** -3 **77.** $7 - 2\sqrt{10}$ **79.** 4 **81.** $\frac{2 + 3\sqrt{2}}{2}$

Exercise 8.3

1. $\{-3, 6\}$ **3.** $\{-7, -5\}$ **5.** $\{-12, -2\}$ **7.** $\{-10, -2\}$ **9.** $\{-2, 12\}$ **11.** $\{-2, 9\}$ **13.** $\{-14, 5\}$ **15.** $\left\{-\frac{5}{3}, \frac{3}{2}\right\}$ **17.** No real solutions **19.** $\left\{-\frac{3}{2}, 1\right\}$ **21.** $\left\{-\frac{2}{3}, \frac{3}{2}\right\}$ **23.** $\left\{-\frac{1}{3}, \frac{5}{3}\right\}$ **25.** $\left\{-\frac{3}{5}, \frac{7}{4}\right\}$ **27.** No real solution **29.** $\left\{-1, \frac{7}{3}\right\}$ **31.** $\left\{-\frac{5}{3}, -1\right\}$ **33.** $\left\{-3, \frac{1}{3}\right\}$ **35.** $\left\{-\frac{1}{3}, \frac{3}{2}\right\}$ **37.** $\left\{-\frac{11}{3}, \frac{7}{2}\right\}$ **39.** $\left\{-\frac{3}{2}, \frac{1}{3}\right\}$ **41.** No real solution **43.** $\left\{\frac{5 \pm \sqrt{65}}{2}\right\}$ **45.** $\left\{\frac{1 \pm \sqrt{6}}{2}\right\}$ **47.** $\{-3 \pm 2\sqrt{5}\}$ **49.** $\{-2 \pm \sqrt{7}\}$ **51.** $\left\{\frac{-21 \pm \sqrt{241}}{4}\right\}$ **53.** $\{0.2, 2.3\}$ **55.** $\{-1.2, 1.9\}$ **57.** $\{-0.5, 0.4\}$ **59.** $\{0.2, 1.4\}$ **61.** 7.62 sec or 1.76 sec **63.** 18.32 in. by 21.32 in. **65.** 0.88 sec or 9.12 sec **67.** The width of the table is 2.5 feet, and its length is 3.5 feet. **69.** In words **71.** $\{-1.89, 0.21\}$ **73.** $\frac{\sqrt{a} \pm \sqrt{4a^2 - 3a}}{2a}$ **75.** $3x(2x + 7)(x - 5)$ **77.** $14x^2(x - 3)(x - 5)$ **79.** $4\sqrt{2}$ **81.** $8\sqrt{7}$

Exercise 8.4

1. $0 + 4i$ **3.** $0 + 2\sqrt{3}\,i$ **5.** $3 + 2i$ **7.** $2\sqrt{2} + 2\sqrt{3}\,i$ **9.** $7 + 8i$ **11.** $9 - 8i$ **13.** $7 - 4i$ **15.** $9 - 3i$ **17.** $8 + 6i$ **19.** $-21 - 28i$ **21.** $-1 - 2i$ **23.** $5 + 4i$ **25.** $-1 - 18i$ **27.** $5 - 10i$ **29.** $20 - 5i$ **31.** $20 - 2i$ **33.** $4 + 7i$ **35.** $10 + 4i$ **37.** $4 - 4i$ **39.** $-26 + 0i$ **41.** $5 + 0i$ **43.** $52 + 0i$ **45.** $3 + 4i$ **47.** $41 + 11i$ **49.** $31 + 29i$ **51.** $0 - 2i$ **53.** $0 + \frac{5}{3}i$ **55.** $1 - i$ **57.** $-\frac{3}{17} + \frac{5}{17}i$ **59.** $\frac{1}{5} - \frac{3}{5}i$ **61.** $x = 3$ and $y = -3$. **63.** $x = 5$ and $y = -8$. **65.** The current is $0.8 - 0.4i$ amperes. **67.** The total impedance is $3.0 - 0.6i$ ohms. **69.** In words **71.** -1 **73.** i **75.** -1 **77.** 453 **79.** -20.5625 **81.** $m = \frac{6}{7}$, $(0, -3)$ **83.** $m = \frac{8}{3}$

Exercise 8.5

1. $\{\pm 4i\}$ **3.** $\{\pm\sqrt{3}i\}$ **5.** $\{\pm\sqrt{6}i\}$ **7.** $\{\pm 4\sqrt{7}\,i\}$ **9.** $\{\pm 4\sqrt{2}i\}$ **11.** $\{3 \pm i\}$ **13.** $\{2 \pm i\}$ **15.** $\{5 \pm i\}$ **17.** $\{2 \pm 8i\}$ **19.** $\{1 \pm i\}$ **21.** $\{-1 \pm i\}$ **23.** $\{-1 \pm 4i\}$ **25.** $\{2 \pm 1.5i\}$ **27.** $\left\{\frac{1}{2} \pm 6i\right\}$ **29.** $\left\{-\frac{1}{2} \pm \frac{11}{2}i\right\}$ **31.** $\{1 \pm \sqrt{15}i\}$ **33.** $\{-3 \pm \sqrt{2}i\}$ **35.** $\{-2 \pm \sqrt{31}i\}$ **37.** $\left\{-\frac{11}{10} \pm \frac{\sqrt{19}}{10}i\right\}$ **39.** $\left\{\frac{1}{8} \pm \frac{\sqrt{47}}{8}i\right\}$ **41.** $\left\{\pm\frac{\sqrt{21}}{3}i\right\}$ **43.** $\{\pm 2\sqrt{6}i\}$ **45.** $\left\{-\frac{3}{10} \pm \frac{3\sqrt{19}}{10}i\right\}$ **47.** $\{4 \pm \sqrt{2}i\}$ **49.** $\left\{\frac{3}{2} \pm \frac{\sqrt{39}}{6}i\right\}$ **51.** $\left\{\frac{9}{4} \pm \frac{\sqrt{79}}{4}i\right\}$ **53.** $\left\{\frac{5}{6} \pm \frac{\sqrt{23}}{6}i\right\}$ **55.** $\{1 \pm i\}$ **57.** $\left\{-\frac{5}{4} \pm \frac{\sqrt{23}}{4}i\right\}$

59. $\left\{-2 \pm \frac{\sqrt{6}}{3}i\right\}$ **61.** In words **63.** $\{-i, 8i\}$

65. $\left\{\pm\frac{3}{2} + \frac{5}{2}i\right\}$ **67.** $\{\pm 3, \pm 2i\}$

69. $\left\{\pm\frac{\sqrt{10}}{5}i, \pm\frac{\sqrt{3}}{3}i\right\}$ **71.** $\frac{-5x^2 + 16x + 18}{(x-2)(x+3)}$

73. $\frac{-x^2 + 6x + 6}{(x-4)(x+4)}$ **75.** $\frac{1}{x-3}$ **77.** $\frac{x+9}{x-4}$

Exercise 8.6

1. $\{-1, 5\}$ **3.** $\left\{-\frac{1}{2} \pm \frac{\sqrt{195}}{10}i\right\}$ **5.** $\left\{\frac{1 \pm \sqrt{109}}{6}\right\}$
7. $\{\pm 20i\}$ **9.** $\{\pm 4\}$ **11.** $\{\pm 2\sqrt{5}\}$ **13.** $\{0, 6\}$
15. $\{-4 \pm \sqrt{31}\}$ **17.** $\{\pm\sqrt{43}\}$ **19.** $\left\{-\frac{4}{5}, 1\right\}$
21. $\left\{-\frac{5}{3}, \frac{1}{2}\right\}$ **23.** $\{\pm\sqrt{29}\}$ **25.** $\{\pm 3i\}$ **27.** $\{\pm i\}$
29. $\left\{-\frac{5}{3}, 1\right\}$ **31.** $\{1 \pm \sqrt{2}i\}$ **33.** $\left\{-5, -\frac{1}{2}\right\}$
35. $\left\{\frac{7}{10} \pm \frac{\sqrt{11}}{10}i\right\}$ **37.** $\left\{\frac{2}{5}, 3\right\}$ **39.** $\left\{\frac{11}{4}, 8\right\}$
41. $\{-2, 6\}$ **43.** $\{-2, 3\}$ **45.** $\left\{-\frac{22}{5}, 2\right\}$
47. $\left\{-\frac{3}{4} \pm \frac{\sqrt{23}}{4}i\right\}$ **49.** $\left\{-\frac{23}{3}, 7\right\}$
51. $\left\{\frac{-21 \pm \sqrt{281}}{4}\right\}$ **53.** $\left\{\frac{1}{2} \pm \frac{\sqrt{91}}{2}i\right\}$
55. $\left\{\frac{-7 \pm \sqrt{57}}{4}\right\}$ **57.** $\{1, 10\}$ **59.** $\left\{\frac{-7 \pm \sqrt{149}}{2}\right\}$
61. 2 ft **63.** Base = 2 in., height = 8 in. **65.** 2.5 ft
67. 9 sec **69.** Bus = 37.1 mph, car = 47.1 mph
71. In words **73.** In words **75.** $x^2 - x - 6 = 0$
77. $x^2 - 10x + 19 = 0$ **79.** $\left\{-\frac{58}{3}\right\}$ **81.** $\left\{-\frac{51}{5}\right\}$
83. $\{64\}$ **85.** $\{1\}$

Exercise 8.7

1.

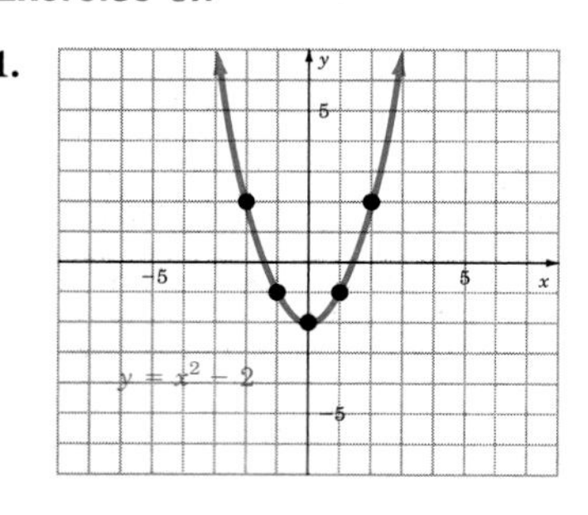

3.

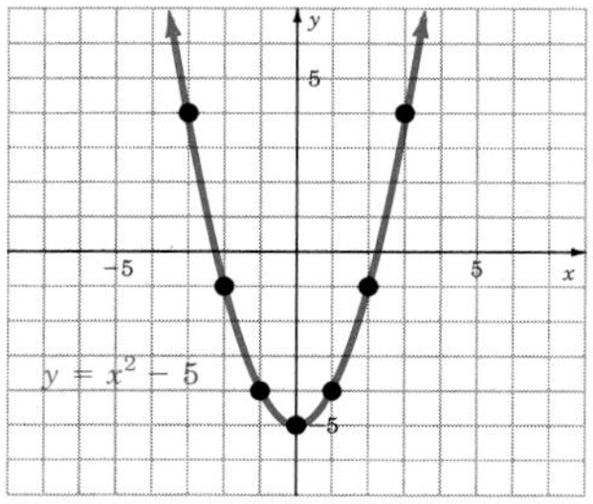

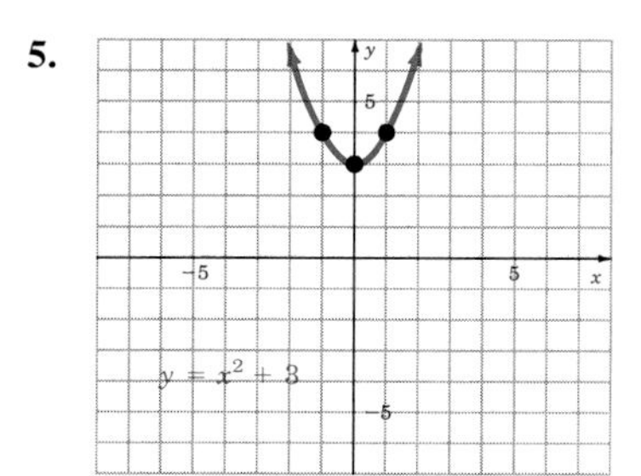

5.

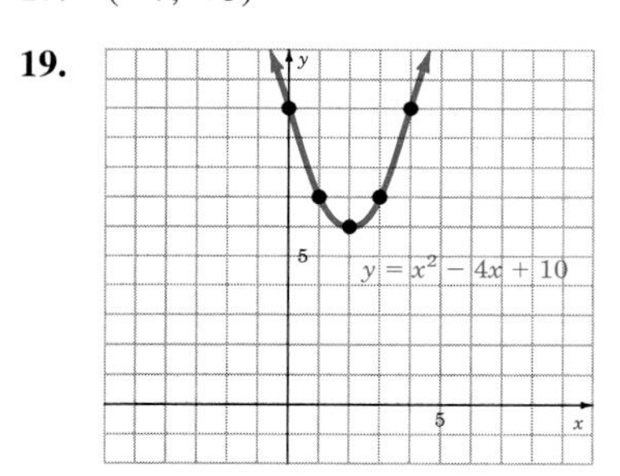

7. No x-intercept, $(0, 6)$ **9.** $(-5, 0)$, $(4, 0)$, $(0, -20)$
11. $(6, 0)$, $(-5, 0)$, $(0, -30)$ **13.** $(2, 6)$ **15.** $(5, -1)$
17. $(-7, -3)$

19.

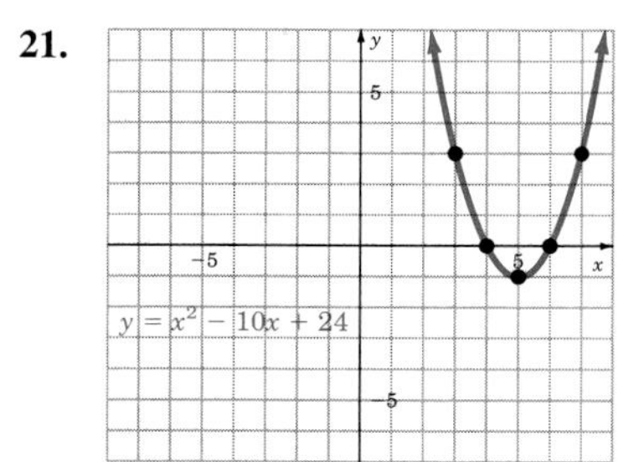

21.

23.

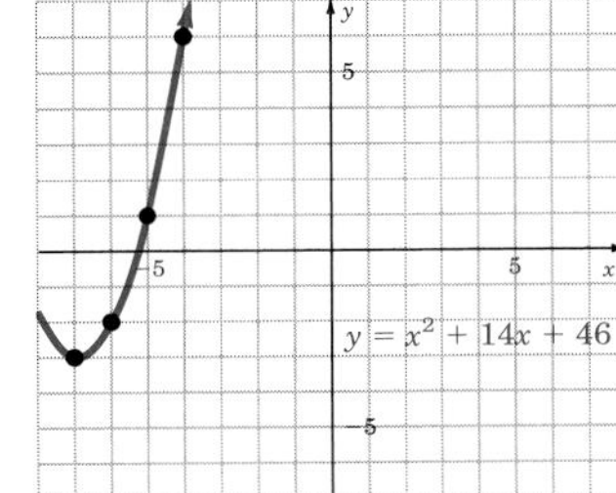

25. $(-6, 0)$, $(3, 0)$, $(0, -18)$
27. $(-7, 0)$, $(-4, 0)$, $(0, 28)$
29. $(-6, 0)$, $(2, 0)$, $(0, -12)$ **31.** $(-3, -10)$
33. $(2, 5)$ **35.** $(-5, -32)$

37.

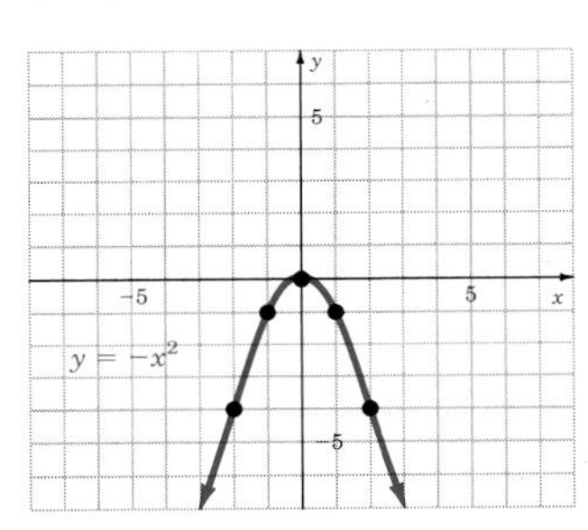

39.

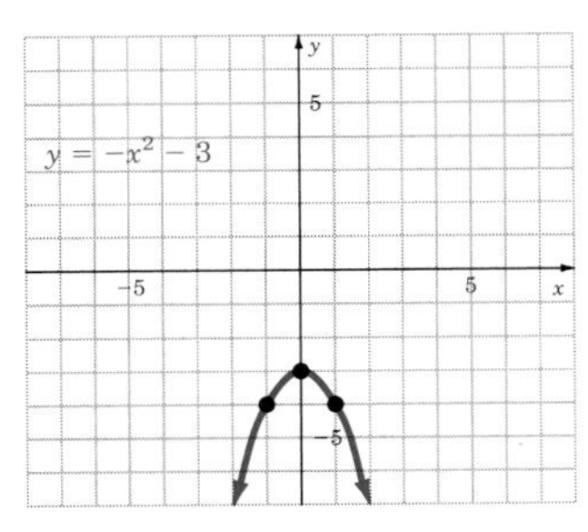

41.

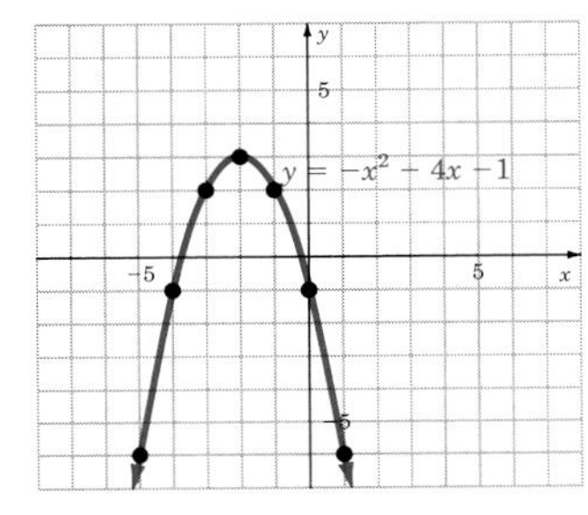

43. $(-3, 0)$, $(1, 0)$, $(0, -3)$ **45.** $(-2, 0)$, $(-8, 0)$, $(0, 16)$
47. $(-3, 0)$, $(5, 0)$, $(0, -15)$ **49.** $\left(-\frac{3}{2}, -\frac{13}{4}\right)$
51. $\left(-\frac{5}{2}, \frac{7}{4}\right)$ **53.** $\left(-\frac{5}{2}, -\frac{37}{4}\right)$ **55.** The maximum height is 3062.5 cm and it returns to earth in 5 seconds.

57. It goes 1050 feet high and it returns to the ground 15.6 sec after launch. **59.** The minimum distance between the arm and the belt is 8 cm. **61.** The minimum voltage in the circuit is 0.6 volts. **63.** In words
65. $b = 5$ and $c = 4$; the equation is $x^2 + 5x + 4 = 0$.

67.

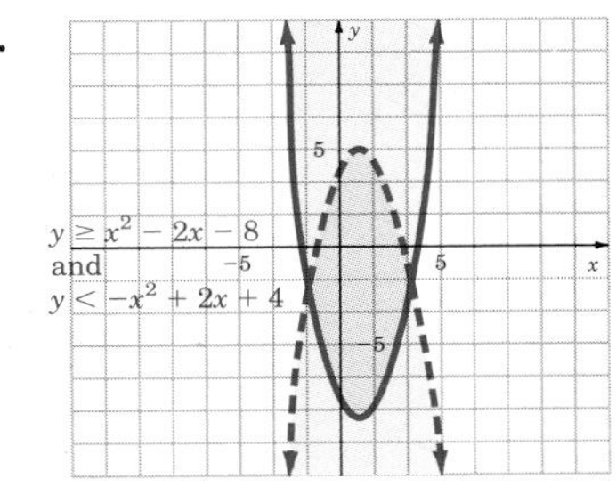

69. $5 - 4i$ **71.** $7 - 10i$ **73.** $14 + 6i$
75. $-\frac{3}{2} - \frac{1}{2}i$

Chapter 8 Review Exercises

1. $\{0, 37\}$ **3.** $\{0, 18\}$ **5.** $\{0, 22\}$ **7.** $\{0, 7\}$
9. $\{0, 5\}$ **11.** $\{\pm 7\sqrt{2}\}$ **13.** $\{\pm 21\}$ **15.** $\{\pm 3\sqrt{11}\}$
17. $\{\pm 3\sqrt{11}\}$ **19.** $c \approx 11.4$ **21.** $\{-10, 16\}$
23. $\{-1, 2\}$ **25.** $\left\{-7, \frac{1}{2}\right\}$ **27.** $\{-2 \pm \sqrt{3}\}$
29. $\left\{\frac{-3 \pm \sqrt{41}}{4}\right\}$ **31.** $\left\{-\frac{7}{3}, \frac{5}{2}\right\}$ **33.** $\left\{-\frac{3}{2}, 15\right\}$
35. $\{2 \pm \sqrt{11}\}$ **37.** $\left\{\frac{1 \pm \sqrt{21}}{6}\right\}$ **39.** $\left\{\frac{3 \pm \sqrt{41}}{2}\right\}$
41. $0 + 11i$ **43.** $0 + 4\sqrt{6}i$ **45.** $3 - 6i$
47. $2\sqrt{5} - 2\sqrt{5}i$ **49.** $0 + i$ **51.** $7 - 6i$
53. $10 - 3i$ **55.** $6 - 2i$ **57.** $25 - 72i$
59. $-2 + 12i$ **61.** $16 + 12i$ **63.** $-12 + 8i$
65. $41 + i$ **67.** $226 + 0i$ **69.** $70 - 40i$
71. $0 + 5i$ **73.** $\frac{8}{5} - \frac{4}{5}i$ **75.** $\frac{3}{17} + \frac{12}{17}i$
77. $-\frac{1}{25} + \frac{18}{25}i$ **79.** $-2 - 3i$ **81.** $\{\pm 10i\}$
83. $\left\{\pm \frac{4}{3}i\right\}$ **85.** $\left\{\frac{3}{2} \pm \frac{\sqrt{7}}{2}i\right\}$ **87.** $\left\{\frac{1}{2} \pm \frac{\sqrt{11}}{2}i\right\}$
89. $\left\{\frac{1}{6} \pm \frac{\sqrt{23}}{6}i\right\}$ **91.** $\left\{\frac{3 \pm 3\sqrt{13}}{2}\right\}$ **93.** $\left\{-\frac{2}{5}, \frac{1}{2}\right\}$
95. $\left\{\pm \frac{3\sqrt{7}}{7}\right\}$ **97.** $\left\{\frac{4}{3} \pm \frac{\sqrt{17}}{3}i\right\}$ **99.** $\{\pm \sqrt{3}\}$

101.

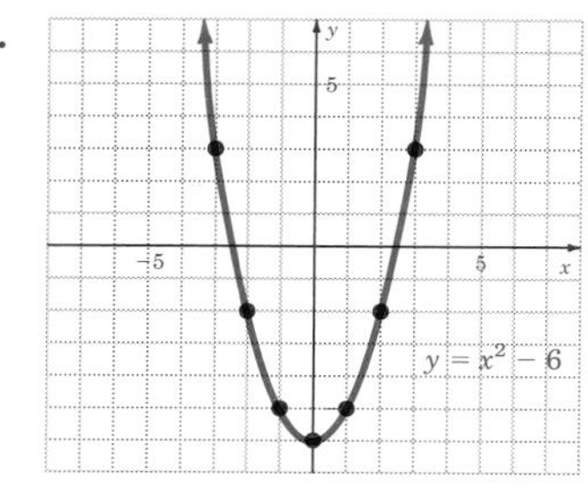

103.

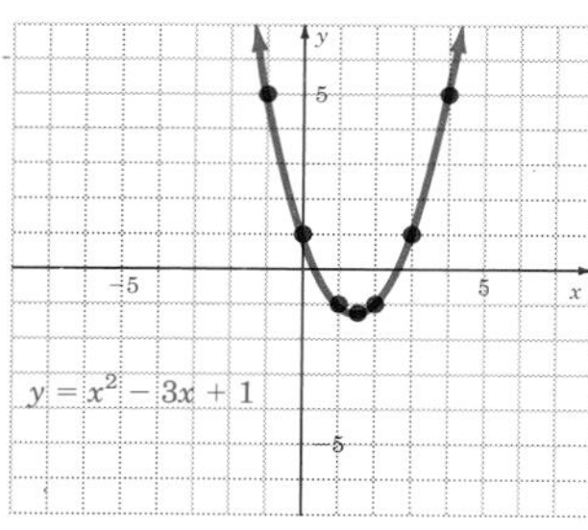

105. $(-7, 0), (1, 0), (0, -7)$ **107.** $\left(-\frac{1}{2}, 0\right), \left(\frac{5}{3}, 0\right)$, $(0, -5)$ **109.** $(-2, 0), (7, 0), (0, -14)$
111. $\left(-\frac{7}{2}, 0\right), \left(\frac{1}{5}, 0\right), (0, -7)$ **113.** $(-11, 0), (2, 0)$, $(0, -22)$ **115.** $(11, 8)$ **117.** $(-3, -9)$
119. $(-2, -1)$ **121.** $(5, 5)$ **123.** $(-7, 10)$

Chapter 8 True–False Concept Review

1. True **2.** True **3.** True **4.** True **5.** False
6. False **7.** True **8.** False **9.** True **10.** True
11. False **12.** True **13.** True **14.** True
15. False

Chapter 8 Test

1. $3 - 8i$ **2.** $\left\{\frac{-5 \pm \sqrt{337}}{4}\right\}$ **3.** $13 - 12i$
4. $\{-8 \pm 6\sqrt{2}\}$
5. $y = x^2 + 6x + 5$

6. $\{\pm 8\}$ **7.** $c = 9.2$ **8.** $-\frac{1}{4} \pm \frac{\sqrt{55}}{4}i$

9. $13 - 10i$ **10.** $\{3 \pm \sqrt{23}\}$ **11.** $29 + 2i$
12. $(8, 0), (-3, 0), (0, -24)$ **13.** $\frac{3 \pm \sqrt{33}}{6}$
14. $a = 10$ **15.** $1 - 3i$ **16.** $\{\pm 4\sqrt{2}\}$ **17.** $(3, 4)$
18. $\{-2 \pm \sqrt{21}\}$ **19.** $\{-4, 5\}$ **20.** 75 yd by 100 yd

Chapters 5–8 Cumulative Review

1. $\frac{x + 3}{2x - 4}$ **2.** $\frac{x + 4}{2x + 5}$ **3.** $-\frac{7}{2}$ **4.** $\frac{b - a}{3y - x}$
5. $\frac{xy + y^2}{3x}$ **6.** 1 **7.** $\frac{7a - 35}{6a - 12}$ **8.** $\frac{3x}{3 - x}$
9. $(2x + 1)(x - 7)(x + 5)$ **10.** $4x^3(x + 2)(x - 2)^2$
11. $\{-11\}$ **12.** $\left\{-\frac{6}{5}\right\}$
13. $\frac{4x^2 + 2x - 5}{(2x - 1)(2x + 1)(2x + 3)}$ **14.** $\frac{4}{x}$ **15.** $\left\{-\frac{2}{3}\right\}$
16. $\emptyset$ **17.** $\frac{8 - 3x}{8 + 3x}$ **18.** $\frac{2x + 1}{x + 13}$
19. $\frac{5x^2 - 7x - 8}{4x^2 - 11x + 9}$ **20.** $\frac{x^2 - 7x + 12}{x^2 + x - 2}$ **21.** It can travel 199.5 miles on 7 gallons of gasoline. **22.** The annual taxes are $1,200. **23.** It is 17.9 miles from Canton to Akron. **24.** The patient must receive 0.5 mL of solution. **25.** $45.87 will be earned in 10 months.
26. It would take 5 carpenters 6 days to build the deck.
27. A 4-inch pipe would deliver 44.4 gallons per minute.
28. The illumination at a distance of 9 feet is $1\frac{1}{3}$ candles.
29. $y = -\frac{12}{5}$ **30.** $y = 0.16$ **31.** $x = 2$
32. $x = -\frac{28}{9}$
33. $4x - y = 8$
34. $0.3x - 0.5y = 1.5$

35.

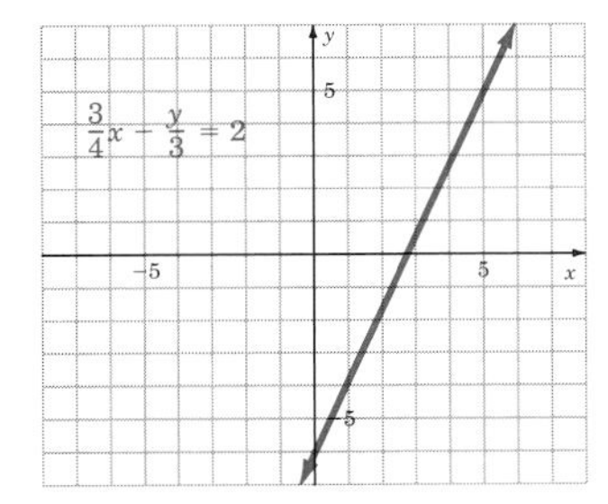

36.

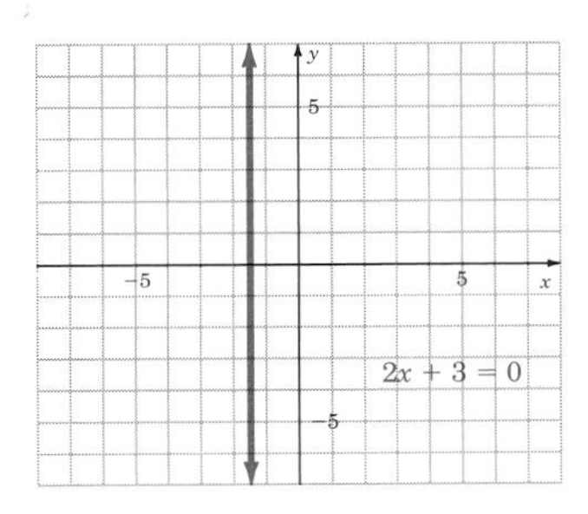

37. $(-12, 0)$, $(0, -6)$

38.

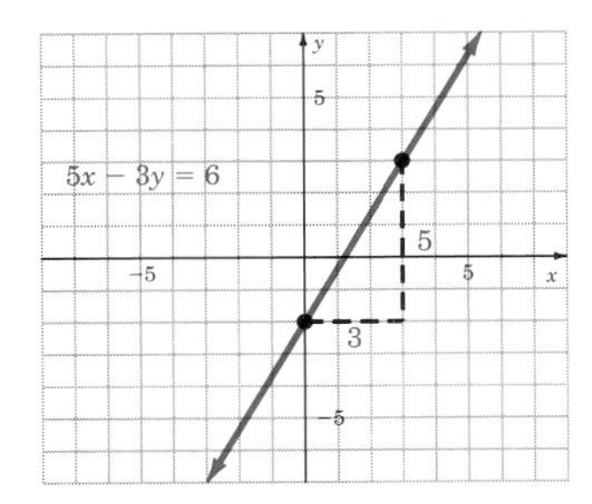

39. The slope is $\frac{1}{4}$

40. There is no slope. The line is vertical.

41.

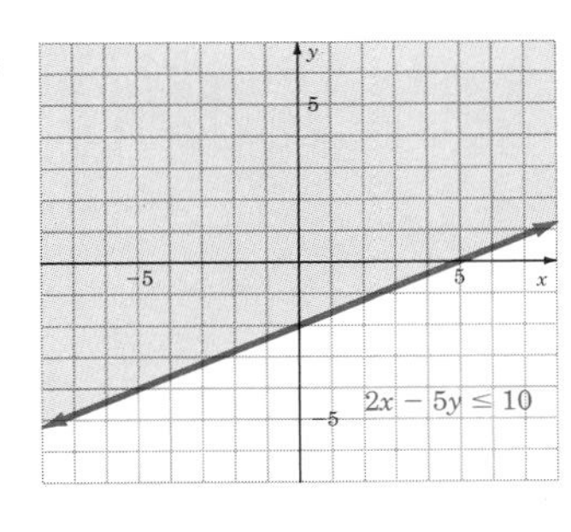

42.

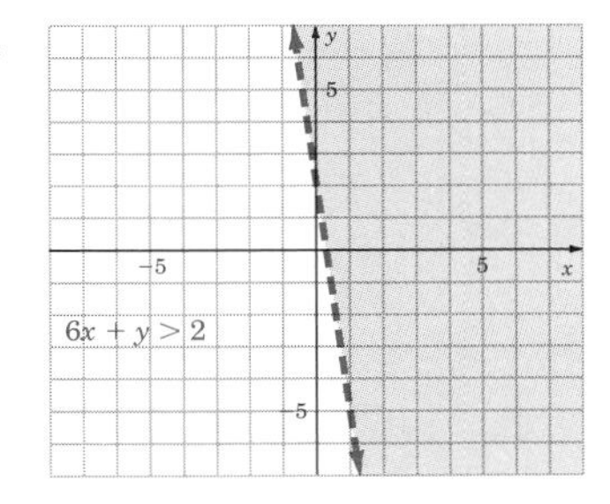

43.

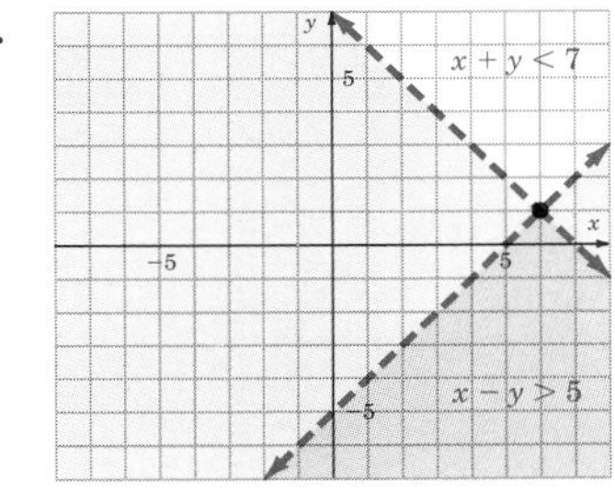

44.

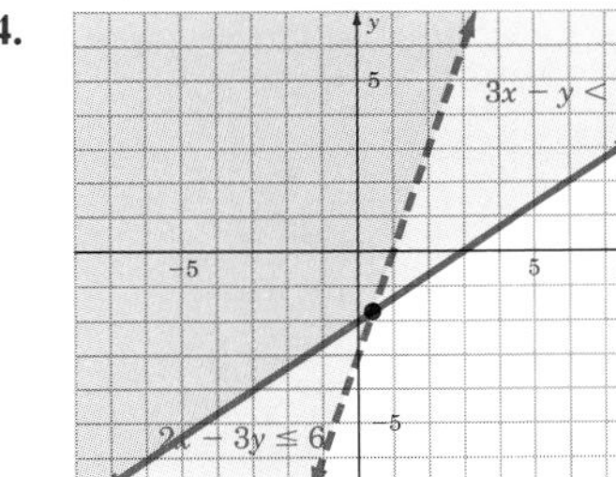

45. $\{(2, 4)\}$

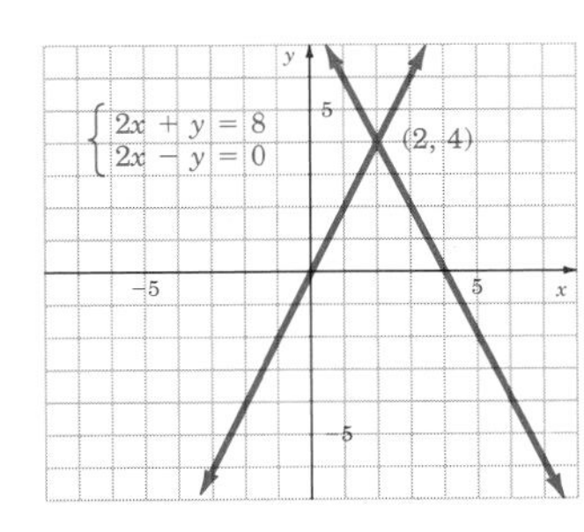

46. $\{(4.0, 7.5)\}$

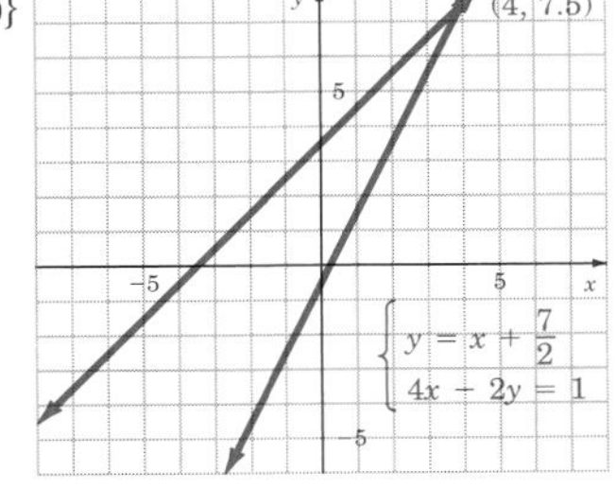

47. $\{(1, 1)\}$ **48.** $\{(1, -2)\}$ **49.** $\emptyset$

50. $\left\{\left(\frac{41}{13}, \frac{8}{13}\right)\right\}$ **51.** $\left\{\left(2, -\frac{1}{2}\right)\right\}$

52. $\{(x, y) \mid 5x - 3y = 4\}$ **53.** $\left\{\left(\frac{2}{5}, -\frac{7}{5}\right)\right\}$

54. $\left\{\left(\frac{73}{47}, -\frac{2}{47}\right)\right\}$ **55.** 23.92 **56.** 0.96 **57.** 12

58. -7 **59.** $-9\sqrt{2}$ **60.** $3y\sqrt{x}$ **61.** $15xy^3z^4\sqrt{2x}$

62. $28xy^3z^5\sqrt{3yz}$ **63.** $5xy\sqrt{2x}$

64. $3x\sqrt{3} - 3x + 3x\sqrt{x}$ **65.** $4 + 4\sqrt{3} - 4\sqrt{5}$

66. $5\sqrt{2} - \sqrt{15} + \sqrt{30} - 3$ **67.** $12 + 2\sqrt{35}$

68. 5 **69.** $-\frac{\sqrt{14}}{8}$ **70.** $\frac{35\sqrt{6}}{9}$ **71.** $-8 - 4\sqrt{3}$

72. $\frac{\sqrt{21} - \sqrt{7}}{2}$ **73.** $7\sqrt{5} + 2\sqrt{3}$ **74.** $-\sqrt{2}$

75. $-\frac{13\sqrt{35}}{35}$ **76.** $\frac{129\sqrt{2}}{2}$ **77.** $\left\{\frac{25}{9}\right\}$ **78.** $\{1\}$

79. $\emptyset$ **80.** $\left\{\frac{1}{4}\right\}$ **81.** $\{0, 6\}$ **82.** $\left\{\pm\frac{1}{2}\right\}$

83. $\left\{-\frac{7}{2}, 0\right\}$ **84.** The remaining leg has a length of 8.94 cm. **85.** $\left\{-2 \pm \sqrt{11}\right\}$ **86.** $\left\{1 \pm \frac{\sqrt{35}}{5}i\right\}$

87. $\left\{\frac{-7 \pm \sqrt{41}}{2}\right\}$ **88.** $\left\{\frac{5 \pm \sqrt{37}}{6}\right\}$

89. $\left\{-\frac{5}{4}, \frac{12}{7}\right\}$ **90.** $\left\{\frac{5 \pm \sqrt{69}}{2}\right\}$ **91.** $\left\{\frac{4 \pm \sqrt{13}}{3}\right\}$

92. $\left\{\frac{3}{16} \pm \frac{\sqrt{183}}{16}i\right\}$ **93.** $6\sqrt{2}i$ **94.** $15 - i$

95. $-3 - 11i$ **96.** 29 **97.** $14 + 5i$ **98.** $5 - 12i$

99. $-\frac{8}{5}i$ **100.** $\frac{6}{5} + \frac{3}{5}i$ **101.** $-\frac{1}{10} + \frac{3}{10}i$

102. $\frac{1}{2} + \frac{3}{2}i$ **103.** $\{\pm 5i\}$ **104.** $\left\{\pm 2\sqrt{11}\, i\right\}$

105. $\left\{\frac{5}{6} \pm \frac{\sqrt{71}}{6}i\right\}$ **106.** $\left\{-\frac{7}{4} \pm \frac{3\sqrt{23}}{4}i\right\}$

107. $\left\{-\frac{1}{3}, \frac{5}{2}\right\}$ **108.** $\left\{\frac{6 \pm \sqrt{29}}{2}\right\}$

109. $\left\{\frac{2 \pm 2\sqrt{7}}{3}\right\}$ **110.** $\{0, 6\}$

111. $\left\{-\frac{1}{10} \pm \frac{\sqrt{139}}{10}i\right\}$ **112.** $\left\{3 \pm \sqrt{7}\right\}$

113. $V = (2, -1)$; y-int. $= (0, 3)$; x-int. $= (1, 0), (3, 0)$

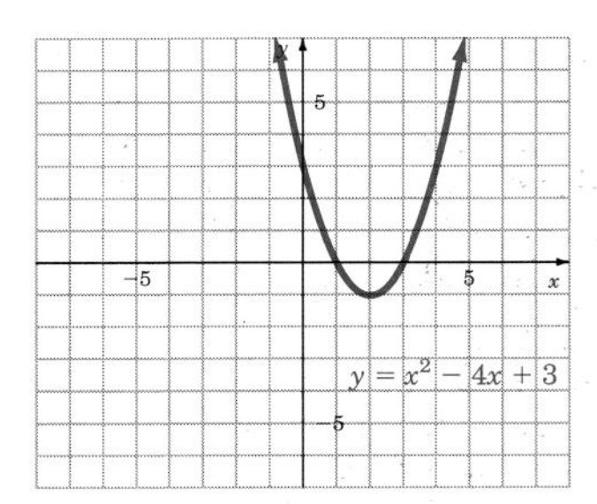

114. $V = (-2, 3)$; y-int. $= (0, 7)$; no x-intercepts

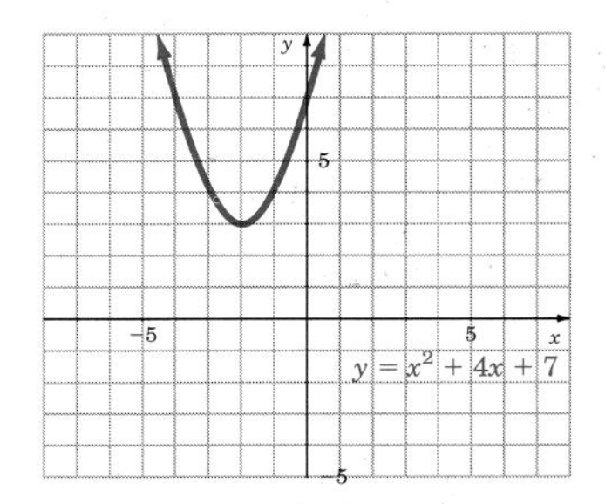

115. $V = (0.5, 6.25)$; y-int. $= (0, 6)$; x-int. $= (-2, 0), (3, 0)$

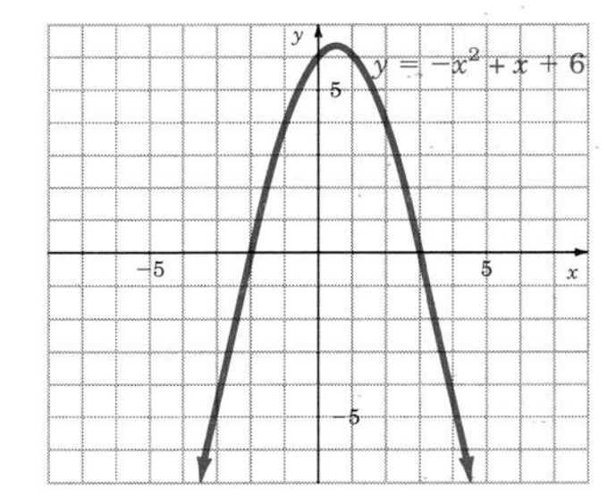

116. $V = (0, 3)$; y-int. $= (0, 3)$; x-int. $-$ none

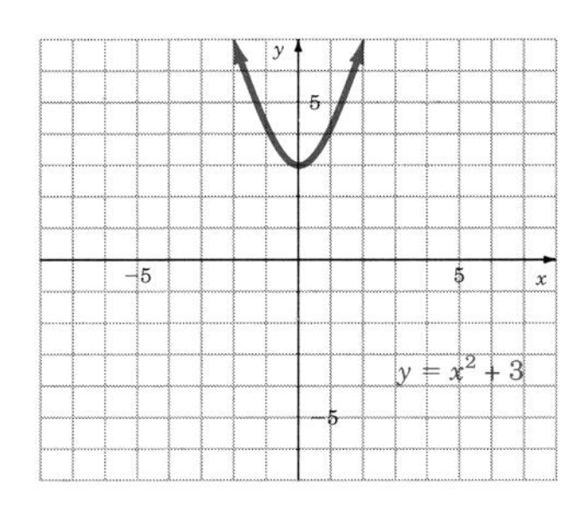

ANSWERS TO SETS.1

1. $\{0, 2, 4, 6, 8, 10, 12, 14, 16\}$ **3.** $\{9, 10, 11, 12, 13, 14, 15\}$ **5.** The set of whole numbers between 10 and 16. **7.** The set containing the number 1. **9.** $\{x \mid x \in W \text{ and } x < 84\}$ **11.** $\{x \mid x \text{ is an even number and } 51 < x < 133\}$ **13.** False **15.** False **17.** True **19.** False **21.** $\{16, 20, 24, 28, 32, 36\}$ **23.** $\left\{\frac{1}{1}, \frac{1}{2}, \frac{1}{3}, \frac{1}{4}, \frac{2}{1}, \frac{2}{2}, \frac{2}{3}, \frac{2}{4}, \frac{3}{1}, \frac{3}{2}, \frac{3}{3}, \frac{3}{4}\right\}$ **25.** The set of multiples of 3 from 9 to 21 inclusive. **27.** The set of multiples of 11 from 88 to 132 inclusive. **29.** $\{24, 25, 26, 27, 28\}$ **31.** $\{1, 2, 3, 6, 9, 18\}$ **33.** True **35.** True **37.** False **39.** False **41.** $\{208, 221, 234\}$ **43.** $\{210, 224, 238\}$ **45.** The set of multiples of 4 between 129 and 141. **47.** The set of multiples of 2 or 5 between 31 and 41. **49.** $\{37, 38, 39, 40\}$ **51.** $\{-4, -3, -2, -1, 0, 1, 2, 3, 4\}$ **53.** True **55.** False **57.** False **59.** True **61.** False **63.** True **65.** In words **67.** $C = \{15, 30, 45\}$ **69.** $E = \{14, 28, 42\}$

ANSWERS TO SETS.2

1. $\{2, 10\}$ **3.** $\left\{-15.7\overline{7}, -\frac{19}{4}, -4, -\frac{7}{6}, 0, 2, 10, \frac{42}{3}\right\}$
5. $\{2, 3, 4, 6, 8, 9, 10, 12, 15\}$
7. $\{0, 2, 3, 4, 5, 6, 8, 9, 10, 12, 15, 20\}$
9. $A = \{1, 2, 3, 4, 6, 7, 8, 9, 11, 12, 13, 14\}$ **11.** True
13. True
15.

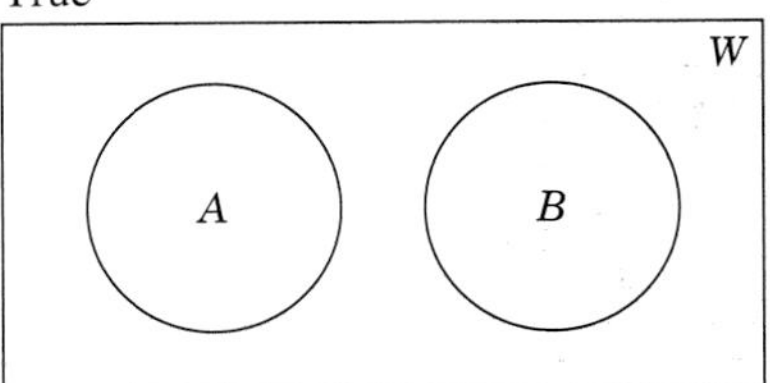

$A \cap B = \emptyset$

17. $\{0, 3, 9\}$ **19.** $\{-\sqrt{13}\}$ **21.** $\{-4, 0, 4\}$ **23.** $\{0\}$ **25.** $A = \{$all odd whole numbers$\}$ **27.** {all whole numbers not a multiple of 5} **29.** {all whole numbers not a multiple of 10} **31.** True **33.** True
35.

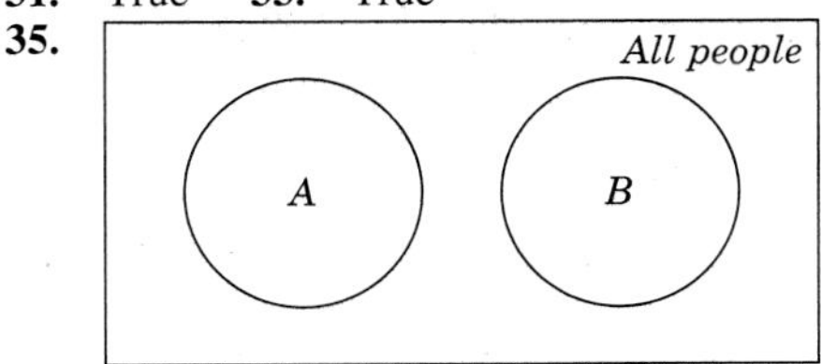

$A \cup B = \{$all left-handed people$\}$

37. A **39.** B **41.** $\{11, 12, 13, 14, 15, 16, 17, 18, 19\}$ **43.** $\{1, 2, 3, \ldots, 17\}$ **45.** J **47.** N **49.** $A = \{$all females and males who are 25 years of age or older$\}$
51. {all males 25 years of age and older whose hair is not red and *all* females whose hair is not red}
53. {all males whose hair is not red, all males 25 years of age and older, and all females whose hair is not red}
55.

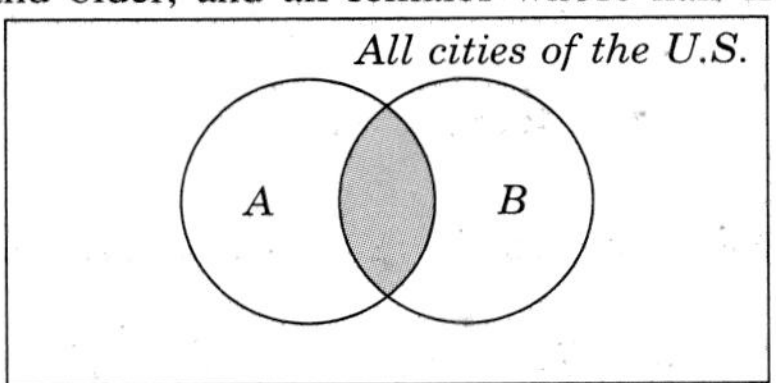

$A \cap B = \{$capitals that are also the largest city in the state$\}$

57. In words **59.** C
61. $\{0, 1, 2, 3, \ldots, 40, 45, 50, 55, \ldots\}$

INDEX